The Galápagos for University Students

The Galápagos for University Students:

A Comprehensive Exploration

By

Guillermo Paz-y-Miño-C

and Avelina Espinosa

Cambridge
Scholars
Publishing

The Galápagos for University Students: A Comprehensive Exploration

By Guillermo Paz-y-Miño-C and Avelina Espinosa

This book first published 2026

Cambridge Scholars Publishing

Lady Stephenson Library, Newcastle upon Tyne, NE6 2PA, UK

British Library Cataloguing in Publication Data
A catalogue record for this book is available from the British Library

ISBN: 978-1-0364-7110-1
ISBN (Ebook): 978-1-0364-7111-8

To the college and university students who visit the Galápagos to explore an unparalleled archipelago; to learn about its natural wonders and rich human history, from its discovery by Tomás de Berlanga in 1535 to Charles Darwin's arrival in 1835, Herman Melville's presence as a whaler in 1841, and World War II during the 1940s, when the United States established a military base in Baltra Island; to witness its breathtaking volcanoes, lava flows, and caves; to reflect on the origins of the enormous prickly-pear *Opuntia* or the *Scalesia* forests in the islands' highlands; to swim with amusing sea lions, fur seals, and penguins; to scuba dive among sharks, dolphins, orcas, giant mantas, and sea turtles; to hike on rocky shores covered in thousands of marine iguanas; to take close-up shots of giant tortoises, land iguanas, lava lizards, and albatrosses; to see up-close flamingos, boobies, frigatebirds, hawks, owls, finches, vermilion flycatchers, or carpenter bees; and to have one of the most thrilling experiences a World Heritage Site—the Galápagos—can offer.

To the university students of Ecuador and Galápagos—*ecuatorianos galapagueños*—who possess the archipelago as their own and have the forever responsibility of protecting.

To the college and university instructors—from all over the world—who motivate their students to travel to Galápagos and see, with their youthful yet inquisitive eyes, the "little world within itself", as Charles Darwin characterized it in *Journal of Researchers* (1839). Darwin himself was quite young—26 years old—when visiting the islands from September 15 to October 20, 1835.

TABLE OF CONTENTS

ACKNOWLEDGEMENTS

We thank the Galápagos National Park (GNP) for leading and facilitating the conservation, research, and outreach programs in the Galápagos Archipelago and its Marine Reserve (GMR). The park officials and rangers, as well as the naturalist guides [licensed by GNP], are essential to the administration, preservation, and management of the fragile terrestrial and aquatic ecosystems. GNP's work is central to seeking harmony between environmental protection and sustainable human development, a constant challenge. The 60-year partnership between the GNP Directorate (GNPD) and the Charles Darwin Foundation (CDF, an international non-profit) has been vital to the scientific work, design, and implementation of management plans and conservation programs in Galápagos.

Several Ecuadorian institutions, organizations, and colleagues have welcomed our participation at local scientific events during the past four decades: Pontifical Catholic University of Ecuador (PUCE), University San Francisco of Quito (USFQ), University of the Americas (UDLA), and Ecuador's Academy of Sciences (ACE). We thank Laura Arcos-Terán, Tjitte de Vries, Carlos Soria, Eugenia del Pino, Giovanni Onore (PUCE); Carlos Montúfar, Diego Quiroga, Andrea Encalada, Stella de la Torre, Gabriel Trueba, Lourdes Torres, Carlos Valle, Gunter Reck, Carlos Mena (USFQ); Carlos Soria, Tjitte de Vries, Eugenia del Pino (ACE).

We thank the institutions and organizations that have sponsored our academic work in multiple science fields and countries (United States, United Kingdom, Denmark, and Ecuador): Roger Williams University (RWU – New England Science Public and the New England Center for the Public Understanding of Science), University of Missouri St. Louis, Washington University, University of Nebraska-Lincoln, University of Kent in Canterbury, St. Jude Children's Research Hospital, University of Aarhus, and Pontifical Catholic University of Ecuador. In addition, the Animal Behavior Society (Conservation Committee), American Society for Microbiology, American Society of Tropical Medicine and Hygiene, International Society of Protistologists, and European Society for Evolutionary Biology have been supportive of our contributions to their scientific and public outreach activities.

Funding for our academic development and research programs can be traced back to the 1980s and up to 2020s, chronologically: Jersey Zoo,

Jersey Wildlife Preservation Trust (=Durrell Wildlife Conservation Trust) and Wildlife Preservation Trust International; Wildlife Conservation Society; International Center for Tropical Ecology [current Harris World Ecology Center] and Compton Foundation (via UM-St. Louis); MERCK-AAAS USRA (American Association for the Advancement of Science); American Society of Microbiology URF; Rhode Island INBRE and National Institutes of Health (NIH-NIGMS #2P20GM103430); Rhode Island National Science Foundation NSF EPSCoR (EPS-1004057); Rhode Island Inclusive Network for Excellence in Science & Technology (NSF RII-NEST #2433276); and the RWU Foundation to Promote Scholarship.

We value the work conducted by the New Bedford Whaling Museum; its exhibits highlight the relevance of world whaling over centuries. Most of the images included in Chapter Four (*What Happened in Galápagos Before and After Darwin?*), in which we discuss the industry of whaling (from pursuing whales to hunting them and processing their parts) and its past environmental impacts on the Galápagos, were taken at the museum. On various occasions, Bob Rocha, former head of Education and Science Programs and current Associate Curator of Science and Research, introduced our students to the museum's displays and shared fascinating stories about whaling, whalers, and whales.

For several years, Niles Eldredge, past Editor in Chief of Evolution, Education & Outreach, welcomed our manuscripts to the journal. In 2013, Niles was particularly supportive of our submission of "Galapagos III World Evolution Summit: Why Evolution Matters" (EE&O 6:28), which summarized the outcomes of an international scientific meeting organized by USFQ in San Cristóbal Island and to which we were invited.

We are grateful to the scientists who have influenced our academic development: Henrik Balslev, Benjamin Øllgaard (d), Tjitte de Vries, Luis Albuja, Stuart Strahl, David Waugh, Peter Raven, George Schaller, Zuleyma Tang-Martínez, Stan Braude (d), Bob Marquis, Carol Kelly, George Taylor, Bob Ricklefs, Susanne Renner, Terry Erwin (d), Larry Salkoff, Jerry Wolff (d), Al Kamil, Alan Bond, Russ Balda (d), Carlos Soria, Sam Stanley, Ellen Li, David Clark, Jim Alfano (d), and Lynn Margulis (d).

We thank Metropolitan Touring, particularly Francisco Dousdebés, Luis Maldonado-Robles (d), the officers, crew, and naturalists at Motto Vessel Santa Cruz, as well as the staff at Finch Bay Hotel (a "National Geographic Unique Lodges of the World"), for facilitating our eco-experiences in Galápagos for so many years. We especially enjoyed leading workshops and training sessions for their naturalists on various occasions.

Over decades, lodging facilities have made special arrangements for us to explore the Galápagos; thanks to Finch Bay Hotel, Casa de Marita, Albemarle Hotel, Casa Opuntia Hotel, Galápagos Sunset Hotel, and Encantadas Guest House.

Non-governmental organizations in Ecuador sponsored—back in the 1980s—our conservation-biology-related work in the country: Fundación Natura, thanks to Yolanda Kakabadse [current President of the Charles Darwin Foundation] and Roque Sevilla; CECIA (Consejo Ecuatoriano para la Conservación e Investigación de las Aves, nowadays Aves y Conservación); and EcoCiencia.

G. Paz-y-Miño-C thanks Ecuador's former President, Rodrigo Borja, and former Secretary of Energy and Mines, Rafael Almeida (d), for appointing Paz-y-Miño-C Undersecretary for the Environment back in 1991. At that time, the appointment was the highest public service position in environmental affairs in the country. As an outcome of that experience, a few years later, César Plaza, Douglas Graham, and George Ledec, members of the Latin America and Caribbean Environment Division (LATEN) of the World Bank, Washington, DC, included G. Paz-y-Miño-C in the initial steps of the project Conservation Assessment of Bioregions in Latin America and the Caribbean.

We thank the hundreds of scientists, conservation biologists, and wildlife managers who authored the peer-reviewed articles cited in this book (496 total references). We relied on their work to synthesize, integrate, and conceptualize (1) the most recent natural history research being conducted in the Galápagos' terrestrial and aquatic environments and (2) the conflicts resulting from human interactions with nature, including local population growth and tourism practices in the context of short- and long-term conservation efforts. We also (3) make predictions about the destiny of the Galápagos' unique biodiversity and landscapes under various scenarios of climate change impacts, urbanization trends, diversification of tourism, and conservation investments.

We thank Francisco Dousdebés for sharing with us the photo of Tomás de Berlanga's statue (Chapter Two, *How Were the y^{a}s de galapagos Discovered?*). The Syndics of Cambridge University Library for allowing us to reproduce two images (the leather cover and page number 36) from Charles Darwin's 1837-8 Notebook B: Transmutation (Chapter Three, *Why Was Charles Darwin's Visit to Galápagos Significant?*). Photographer Ian Wiese for the aerial image of "Blue Whale Releasing a Fecal Plume" (Chapter Four, *What Happened in Galápagos Before and After Darwin?*)! Gustavo Jiménez-Uzcátegui, Ecuadorian wildlife veterinarian and researcher at the Charles Darwin Foundation, helped us identify checklists and studies on the endemic species of the Galápagos (Chapter Eight, *What Is the Conservation Value of the Galápagos?*). Alex Hearn shared with us information about the 2021 proposal for the expansion of the Galápagos Marine Reserve (Chapter Eleven, *What Can You Do to Help Galápagos?*). Arturo Izurieta, former Director of the Galápagos National Park, responded to our queries about the future of conservation efforts in the Galápagos; his perspective helped us shape a more comprehensive view of the challenges of managing the National Park and its Marine Reserve (Chapter Twelve, *Can the "Little World Within Itself" Be Saved Forever?*).

Several of the historic photographs in Chapter Five (*Why Did the United States Establish a Military Base in Galápagos During World War II?*) are broadly available online at multiple depositories, as we provide links in Appendix A. We highlight that book author and historian John Woram compiled a great initial collection of these images on his Galápagos website (www.galapagos.to), unfortunately no longer available.

We acknowledge the significance of having interacted with thousands of students in our science courses (fieldwork, laboratories, and classrooms). We are proud that many of them, graduate and undergraduate, are now researchers in academia at world-renowned institutions.

During the first year of the COVID-19 pandemic, in 2020, caused by coronavirus SARS-CoV-2, we spent much time at our home in Rhode Island, social distancing and working intensely on our "Galápagos books". We followed the daily statistics on global cases, fatalities, recovered patients, and financial sequels of the disease; we shared that information on social media and discussed it with our colleagues and friends. Looking back, SARS-CoV-2 transformed our communities, and Galápagos was no exception. Tourism, the main or only source of income for thousands of galapagueños, ceased [visitors dropped from 271,000 in 2019 to 73,000 in 2020; ≈80% of the local economy depended on tourism at that time, with $356 US million/year generated by tourism alone]; airports and ports closed, except for transporting essentials; islanders became even more isolated; and visiting scientists, students, and tourists were trapped for months in the islands. From February 2020 to September 2021, 1,485 cases of coronavirus were accounted for in the archipelago and 17 deaths (Ecuador's total cases were 503,767 and 32,351 deaths). Galápagos' tourism recovered by 2022 and reached record-high numbers in 2023.

This is our fifth academic volume printed at Cambridge Scholars Publishing, United Kingdom. The books are distributed worldwide.

Thanks to all.

PREFACE

This volume is an abridgement of our academic books, *Arrivals of Life to the Galápagos* and *A Synthesis of the Galápagos*, printed as hardbacks and paperbacks (2024/2025) by Cambridge Scholars Publishing. What is new in the current volume, *Galápagos for University Students*? The abridgment itself, which has been adapted for motivated undergraduates and their instructors, as well as the approach to the book, which relies on questions, answers, discussions, and observation projects. Each chapter begins with a question. The volume offers 280 figures and 733 subfigures, 15 tables, 30 analytical questions, including perspectives from which to answer them, 20 discussion topics (argumentative format), 24 observation projects to be conducted while visiting the islands [always respecting the regulations that protect the Galápagos National Park and its Marine Reserve] or as library-work activities, 9 terminology/content boxes, and 496 references. We have also structured the book for a broad college and university audience, including research undergraduates, graduate students (i.e. master's degrees), their instructors and mentors (professors, lecturers, and postdocs interested in teaching), as well as study-abroad students and international field-trip instructors (i.e. multidisciplinary academic programs) who travel to the Galápagos from all over the world.

The length and chapter organization make the book suitable for seminars and/or laboratory discussions facilitated by a college/university instructor; keep in mind that some chapters are extensive, e.g., Chapter One, *What Are the Galápagos Islands?* (101 pages), or Chapter Eight, *What Is the Conservation Value of the Galápagos?* (149 pages). Because the chapters increase in complexity, we recommend the reader explore them in order. However, we have also organized the material to be read independently by chapter (i.e. by the more experienced reader). For that reason, throughout the book, we deliberately reinforce concepts and summarize content (e.g. meaning of acronyms, synopsis of figures, tables, and data from previous sections) and frequently advise the reader to explore figures in a specific manner. We include dialog boxes to highlight the central message of the figures and/or expedite the interpretation of the statistics (the latter kept to a minimum).

Structure and Content of the Book

This volume has 12 chapters and 2 appendixes (*Appendix A: Figures' Notes and Sources*, and *Appendix B: Galápagos National Park Rules*). All chapters include a conclusion(s) section followed by questions, answers, discussions, and observation projects. The questions are analytical and challenging to the reader. As we state at the beginning of each **Questions** section, "*the following questions merit generous written answers. You might need to do additional reading or library work to enrich your answers (we suggest exploring...* the following references*)*". In the answers section of each chapter (i.e. **How to Answer the Questions Above**), we indicate, "*The questions above have various possible answers. Here we suggest perspectives from which to address the queries*". In other words, the student reader is encouraged to envision alternative answers to the questions. In the **Discussion Topics** section, we indicate, "*The discussion topics below should be addressed with your peers. We encourage college instructors to facilitate the dialogue*". And in the **Suggested Observation Projects**, we emphasize, "*...The suggested observation projects below take into consideration that you will travel to Ecuador and the Galápagos legally; that you will visit the islands as a tourist [holding a Transit Control Card, a mandatory document issued by Ecuador and specifically by the Governing Council of Galápagos]; and that you will respect the regulations that protect the Galápagos National Park and the Marine Reserve (see complete list of rules in Appendix B). Only accredited scientists, academics, and/or naturalists who have obtained official permits can do research in Galápagos. You can take pictures, but flash photography is not permitted when taking photos of wildlife. Professional photography and videos recorded for commercial purposes must be authorized by the Galápagos National Park Directorate. The ideas below are for enthusiastic college explorers interested in learning about the archipelago*".

We begin the book by asking, *What Are the Galápagos Islands?* (**Chapter One**). The chapter is extensive, and in it we cover seven main topics: (i) Plate tectonics as they relate to the archipelago's formation; subtopics include: *The Origin of the Islands. The Galápagos Hotspot. The Galápagos Plateau and Its Proximate Ridges. Age and Formation of the Current Islands.* (ii) Volcanism; subtopics: *Volcanoes Uplift or Subside. Caldera-Volcanoes and Volcanoes without Calderas. Underwater Volcanoes, Seamounts, and Guyots.* (iii) Caves' formation and soils; subtopics: *Lava-Tube Caves or "Pyroducts". Soils: Younger vs. Older Islands.* (iv) An overview of the archipelago as seen from space and how to outline it; subtopics: *Zooming Out: Islands Visible from Space. Sketching Today's Archipelago*. (v) Features of the surrounding ocean; subtopics: *The Ocean and Its Turbulent Waters. The Upwelling*. (vi) Biogeography; subtopics: *Origins of the Galápagos Biota*. And (vii) islands' topography, seasonality, and vegetation; subtopics: *Zooming In: The Islands' Appearance at First Glance. The "Hot" and "Cool" Seasons. Vegetation Cover. A Uniquely Well-Preserved Mangrove Forest.*

In **Chapter Two**, *How Were the* $y^{a}s$ *de galapagos Discovered?*, we travel back in time 490 years, to 1535, when Tomás de Berlanga—Bishop of Panama—and his crew arrived adrift in unknown islands. In a colorful letter to Carlos/Charles I King of Spain, dated April 26, 1535, the Bishop wrote: "*IT seemed right to me to let Your Majesty know about the course of my trip since I departed from Panama, on the twenty-third of February of the current year* [1535]... [Our] *ship sailed with very good breezes for seven days, and the pilot kept course near land and* [then] *we had calm for six days; the currents were so strong, and engulfed us in such a way, that on Wednesday the tenth of March we sighted an island* [likely today's Española/Hood]*; and, because on board there was water for only two more days, it was agreed to take down the lifeboat and go on land for water and grass for the horses. And once there, the men found nothing but sea lions, and turtles and galápagos* [gə'läpə͵ gəs=the giant tortoises] *so big that each could carry a man on top, and many iguanas that look like serpents...*" Berlanga's adventure also helps us introduce readers to the Galápagos early cartography, like the "Portolan Chart of the Pacific Coast from Guatemala to Northern Peru" (1565), in which the "$y^{a}s$ *de galapagos*" or "*isles of galapagos*" [isles of the tortoises] were depicted (*galapagos* = tortoise).

In *Why Was Charles Darwin's Visit to Galápagos Significant?* (**Chapter Three**), we examine Darwin's exploration of the Galápagos Archipelago from September 15 to October 20, 1535. But first, we set the record straight: Robert FitzRoy, Captain of Her Majesty's Ship (HMS) Beagle and head of the vessel's surveying voyage around the world (1831 to 1836), which included the exploration of the Galápagos Islands, had requested and been granted by the British Admiralty permission to bring Mr. Darwin—as naturalist—on board. Four years later, Darwin wrote in *Journal of Researchers* (1839) "*The natural history of this archipelago is very remarkable: it seems to be a little world within itself; the greater number of its inhabitants, both vegetable and animal, being found nowhere else*". The Beagle's voyage, FitzRoy's command, Darwin's brilliance, and the uniqueness of the Galápagos Islands all merged in history. The notion of evolution by means of natural selection, which

Darwin conceptualized later in life (*The Origin* 1859), can now be broadly applied to nature. The "*great monsters*" and "*a little world within itself*", as he referred to the giant tortoises and the isles (1839), are now exemplars of natural history. Today, the theory of evolution by means of natural selection is attributed to both Charles Darwin and Alfred Russell Wallace (co-discoverers of natural selection). Ever since Darwin and Wallace, however, our collective understanding of evolution has changed significantly at the molecular, genetic, organismal, ecosystem, global, and cosmic levels.

In **Chapter Four**, What Happened in Galápagos Before and After Darwin?, we remind readers that "[m]uch happened in [the archipelago] before and after Darwin's short visit—as a naturalist on board HMS Beagle under the command of Captain Robert FitzRoy…" For example, "[T]hree hundred years earlier, the islands were accidentally discovered, in 1535, by Tomás de Berlanga and his crew… And just three years prior to the Beagle's arrival, Ecuador claimed possession of the Archipelago of Ecuador (…) on February 12, 1832... During World War II, the United States established a military base on Baltra [Seymour] Island from 1942 to 1946… In 1959, Ecuador created the Galápagos National Park; by 1978, UNESCO declared Galápagos a World Heritage Site… A few decades later, coinciding with the tourism boom (1980s to 2000s), the local population increased significantly, from 5,545 in 1982 to 28,583 in 2024. Tourism grew from 42,000 visitors in 1989 to 276,000 in 2018. The COVID-19 pandemic reduced tourism to 73,000 visitors in 2020, but trends recovered by 2025…" And finally, we state, "There is one distinctive human activity that took place globally and affected the Galápagos' natural history prior to, during, and after Darwin's visit: whaling!" In Chapter Four, "we discuss industrial whaling as it covered the world's oceans ("Global Whaling: 1760s to 1920s") and its impacts on the Galápagos Islands—also known as "Whaling Ground 38".

In **Chapter Five**, we address:. *Why Did the United States Establish a Military Base in Galápagos During World War II?* By the end of the 1800s, the Galápagos Archipelago had gained geopolitical value due to its strategic location in the Tropical Eastern Pacific and with respect to the soon-to-be-constructed Panama Canal. Ecuador was approached by England (early 1900s), France (1898), and the United States (1899) to negotiate the islands as either payment for Ecuador's debt with England (after England financed Ecuador's war of independence with Spain), purchase an island for France, or lease one island for the United States. In 1911, the US proposed another lease—for 99 years, which was not accepted by Ecuador. After the Japanese attack on Pearl Harbor in 1941,

Ecuador signed an agreement with the United States for the establishment, in 1942, of US military bases on Seymour (Baltra or "The Rock") Island (Galápagos) and Salinas (mainland Ecuador) to protect the Panama Canal and its surroundings. By 1944, the United States had prepared a draft of a "*bilateral agreement for the leasing of* [military] *bases on Ecuadorian territory over a period of 99 years*". The idea was not accepted by Ecuador, and two years later, in 1946, Ecuador requested the US "*withdrawal from Galápagos*". The US officially returned Baltra to Ecuador on July 1, 1946. All American personnel had left Galápagos by 1948. "The Rock" of the 1940s is nowadays a memory of WWII as well as a symbol of sovereignty—for Ecuador—with a complex past. Geopolitically, however, the destiny of the Tropical Eastern Pacific—and its multiple strategic locations, including the Galápagos Archipelago—will remain uncertain. Regional or global interests in this particular geographic space could manifest again in response to circumstantial change.

In **Chapter Six**, Who Lives in Today's Galápagos Islands?, we explain that "The Galápagos human population [has] increased from 17,451 inhabitants in 2001 to 23,046 in 2010 and to 25,244 in 2015. By 2022, it reached 28,583". We also highlight historical trends; for example, "[F]rom 1950 to 2015, the islands' population increased 19-fold. The fastest rate occurred between 1982 (4.0%) and 2001 (6.4%). By 2010, the speed of population growth declined to 3.1% and to 1.8% in 2015. It was 1.6% in 2022. The boost during the 1980s-1990s overlapped with the tourism boom. The decline by 2010/2015 corresponded to enhanced age-cohort-specific emigration (ages 15/17-24), presumably to pursue higher education or jobs elsewhere. Despite the decline in growth rate from 2010 to 2015, the population experienced a net gain of 2,576 individuals during that time. It was driven by higher immigration (3,368 people) than emigration (792 people). Between 2015 and 2022, the population increased by 3,339 individuals". We also state in the chapter that "When humans arrive, transformations of the environment follow", [and that] "[i]n contrast to other island groups discussed in [the chapter], Galápagos remains [relatively] well preserved... But [we alert readers that] just like Galápagos, [other] archipelagos mentioned [in Chapter Six] (Seychelles, Madeira & Selvagens, Azores, Cape Verde, Mascarenes, Canaries, or Hawai'i), once imposing and with matchless wilderness, started their journeys into the Anthropocene with very few human settlers. They were, too, discovered as uninhabited islands, although at different times during history, and gradually became colonized…"

The question, *Why Is Nature Tourism Essential to Galápagos?*, is addressed in **Chapter Seven**. Galápagos' subsistence is powered by tourism. In 2018—a record-high year before the COVID-19 pandemic—275,817 visitors arrived in the archipelago. By 2019, total arrivals dropped to 271,238 visitors. By 2020, arrivals decreased to 72,519 visitors (3.8 times below 2018). By 2021, arrivals had partially recovered to 136,336 visitors (49.4% with respect to 2018). And by 2022, they reached 267,688 visitors (97% with respect to 2018). An expected post-pandemic boost occurred in 2023, with an all-time record high of 329,477 arrivals. During 2024, the numbers decreased to 279,277 visitors, suggesting tourism trends were normalizing after COVID-19. In 2019, the Galápagos National Park alone generated $356 million USD, ≈60% of the profit produced by all protected areas in Ecuador. There was a direct benefit connection between the "entrance fee" [for tourists] to the Galápagos National Park and *in situ* conservation and administration work. In 2024, the fee generated $21.1 million USD; half of it went to the National Park itself (conservation programs, infrastructure, education and outreach activities, and salaries = jobs). In 2018-2019, 63% of Galápagos' tourism-related businesses (*N*=874) were hotels, lodges, and travel agencies; the rest were restaurants and bars (15%), small boats for transportation (12%), boats and ships with on-board stays while exploring/sightseeing the archipelago (9%), and transportation on land (1%). An estimated total of 3,955 jobs were directly created by Galápagos' tourism in 2018-2019, 60% of them at hotels, lodges, and on-board boats and ships. An important component of the tourism workforce was the naturalist guides; about 817 were licensed by the Galápagos National Park by 2019; however, not all were active or employed concurrently. Note that the boats and ships with on-board stays (cruises) employed 394 naturalists (48% of the total licensed ones) during 2019.

Now, the cautionary conservation and sustainability message, in terms of Galápagos' local population growth (Chapter Six) and tourism trends (Chapter Seven), can be summarized as follows: (1) If most of the Galápagos' ecosystems remain legally protected and restoration and reintroduction programs continue, the essence of the archipelago has a chance to make it into the future. (2) If the State of Ecuador continues to own and administer the land under low-impact development ideals, the exceptionally unique Galápagos will be exposed to manageable habitat alteration. (3) If human presence in the islands remains at low numbers and density, Galápagos could persist long-term. (4) If tourism continues to be eco-friendly, small-scale, and respectful of both the fragile terrestrial/marine habitats and the local people, Galápagos could offer for centuries the possibility of being explored by those who love and value nature. But (5) if Galápagos chooses the path of fast unsustainable change, urban crowding leading to occupation of additional land, and eco-unfriendly tourism (amusement, entertainment), or if it is persuaded by local or international investors to mimic or resemble other oceanic, volcanic archipelagos discussed in chapters Six and Seven [e.g. Hawai'i, Cape Verde, Canaries, Madeira & Selvagens, Azores, Seychelles, Mascarenes, including in some cases their mall-shopping, spring-break, or night-life tourism; or their outdoor recreation practices and water sports: windsurfing, kiteboarding, jet-skiing, parasailing, wakeboarding, waterskiing, bumper-tube riding, fly-boarding with hydro-flight aqua-jet propulsion or jet-pack flying], its destiny will be comparable. It will be a matter of time.

The question, *What Is the Conservation Value of the Galápagos?*, is addressed in **Chapter Eight**, the longest of the book. In 1978, the United Nations Educational, Scientific and Cultural Organization (UNESCO) designated Galápagos a World Heritage Site. "*…World Heritage is the designation for places on Earth that are of outstanding universal value to humanity and, as such, have been inscribed on the World Heritage List to be protected for future generations to appreciate and enjoy…*" UNESCO describes Galápagos' "Outstanding Universal Value" as follows: "*…This archipelago and its immense marine reserve is known as the unique 'living museum and showcase of evolution'. Its geographical location at the confluence of... ocean currents makes it one of the richest marine ecosystems in the world. Ongoing seismic and volcanic activity reflects the processes that formed the islands. These processes, together with the... isolation of the islands, led to the development of* [matchless] *plant and animal life—such as marine iguanas, flightless cormorants, giant tortoises, huge cacti, endemic trees and the many different subspecies of mockingbirds and finches…*" The World Heritage Convention (WHC, an international treaty) highlights four criteria (VII, VIII, IX, and X) that warrant Galápagos' designation as a World Heritage Site. Criterion VII is about the spectacular underwater wildlife and seascape features of the Galápagos. Criterion VIII highlights its plate tectonics, volcanism, and ongoing geomorphological processes. Criterion IX focuses on the ecological, evolutionary, and biogeographic [processes and] characteristics of terrestrial and marine environments. And Criterion X underlines the local biodiversity and endemism and the conservation status of the flora and fauna. In the chapter, we integrate all these premises with yet another question: "*By preserving the Galápagos Archipelago and its Marine Reserve, what are we really safeguarding?*" We respond by examining three particular areas: biodiversity, endemism, and ecosystem functions.

What Are the Main Conservation Problems in Galápagos? We address this question in **Chapter Nine**. The archipelago has multiple conservation-related problems, broadly including: The impacts of climate change on the local habitats, the people living on the islands, and their means of subsistence (agriculture, artisanal fishing, and tourism). Large-scale predatory fishing in the vicinity of the Galápagos Marine Reserve (particularly by international fisheries). Macro- and microplastics in the ocean, permeating all trophic levels (globally and in Galápagos). Unmanageable local population growth in the islands of Santa Cruz, San Cristóbal, and Isabela. Increasing eco-unfriendly tourism. Wildlife trade (e.g. baby giant tortoises, which are smuggled for the global market). Illegal drug trafficking in the Tropical Eastern Pacific that has targeted the archipelago as a steppingstone on its way to international consumers. And geopolitical instability in the region (part of it derived from drug trafficking). These problems pose unprecedented threats to the Galápagos and all entail human conflict with nature! In Chapter Nine, we examine, in detail, 12 case studies regarding conservation problems in the islands. Some are local problems; others are regional or global and affect the archipelago directly or indirectly. Their subtitles include: *Introduced Species: 1535 Onwards*. *Invasive Species at World Heritage Sites*. *The Global Decline of Shark Populations*. *The Fu Yuan Yu Leng 999 and the Shark-Fin Market*. *Tuna Fisheries Around the Galápagos Marine Reserve*. *The Decline of Sea Lions and Fur Seals*. *Wildlife Trade: Tortoises and Iguanas for Sale*. *Wildlife Roadkills: Homo the Driver in Galápagos*. *Oceans' Microplastics Reach Galápagos*. *Microplastics in the Guts and Tissues of Marine Life: Coastal Ecuador and Galápagos*. *Macroplastics Present in All Galápagos Islands*. And *Plastics Ingested and Uptaken by All Trophic Levels*.

In **Chapter Ten**, *How Will Climate Change Affect Galápagos?*, we explain that, in Galápagos, climate change will impact almost all aspects of human life and people's interactions with nature. We highlight that the local economy is powered—almost entirely—by nature tourism. Extreme weather (severe hot-rainy and cool-dry seasons) and sea-level rise will bring stress to both natural habitats and human settlements. Frequent rainstorms will affect buildings, harbors, homes, and infrastructure connectivity; sewage pipes will break and lead to further contamination of the drinking water (already a problem in Galápagos). Nature hiking or biking trails in the areas open to visitors will have to be repaired or rebuilt, as will viewing platforms, observation decks, and interpretive signs (additional costs to the Galápagos National Park). Economic losses will impact tour/expedition providers, hotels, restaurants, groceries and fish markets, shops, and cultivated areas. Food supply to the islands, either coming from the continent or being distributed within the archipelago, will be affected by extreme weather (e.g. more turbulent seas will lead to longer journeys, extra use of fuel, and higher market prices for goods and supplies). Depending on the severity of climate change impacts on the human population (e.g. if tourism drops significantly or collapses due to natural disasters, combined with large-scale loss of farmlands to urbanization as well as inflated costs of living), future sizeable emigration back to the continent could result—the eco-exodus or environmental refugees. Climate change will accentuate differential financial stress among dissimilar income groups. Socio-economic inequality and unemployment could grow, as could social tensions and the lack of safety and security for the public and tourists. In Chapter Ten, we also elaborated on two particular topics regarding climate-change impacts on both terrestrial and marine ecosystems in the Galápagos: *Extreme Weather and Its Sequels* and *Impacts of Climate Change on Coral Reefs*. The central point of these topics is that the Galápagos Islands are warming up and—for now—becoming rather dry, but "extreme wet" will prevail long-term. The archipelago appears to be destined to less rain during the cool-dry season (drier archipelago for now), combined with a delay in the onset of the hot-wet season (rains will start later), plus a mix of severe hot-wet seasons during more frequent El Niño years and intense cool-dry seasons during La Niña years. Climate change will bring extreme variation in seasonality to Galápagos and affect both wildlife and humans!

In **Chapter Eleven**, *What Can You Do to Help Galápagos?*, we remind readers that the principle "think globally, act locally" applies to the archipelago and its Marine Reserve. We state *"Now, to be most effective helping the Galápagos, you will need to think and act globally as well as think and act regionally and locally. The global context will be, of course, planet Earth. The regional setting will be the Tropical Eastern Pacific (TEP), where Galápagos is located; TEP will also include the proximate countries, those with jurisdiction over the TEP maritime zones (i.e. territorial waters/sea and exclusive economic zones), like Ecuador, Colombia, Panama, and Costa Rica. And the local scale will correspond to both Galápagos itself, whenever you visit it and explore it as a citizen of the world, and your own community wherever you reside"*. In the chapter, we make specific recommendations under the following subtitles: *Think and Act Globally: Planet Earth. Think and Act Regionally: TEP and Proximate Countries*. And *Think and Act Locally: Galápagos and Your Own Community Wherever You Reside*.

We end the book by addressing: *Can the "Little World Within Itself" Be Saved Forever?* (**Chapter Twelve**). And we ask and answer yet again familiar questions to the reader, such as, Will this World Heritage Site be the one that forever-survives? Will Galápagos be different? Will its caretakers and policymakers not mimic the mistakes that nature custodians have made in other volcanic archipelagos, where pristine ecosystems and endemic wildlife have substantially disappeared while being replaced by large towns and cities, conventional tourism resorts, highways, and introduced species? Will the Galápagos' people, Ecuador, and the world commit to safeguarding and managing the archipelago and its marine reserve so that sound sustainable development practices limit human population expansion on the islands, sponsor solely eco-friendly tourism, and secure revenue fairness for the residents? We close by returning to Darwin and stating, "*Regardless of humans surviving the Anthropocene (e.g. the consequences of climate change, mass extinctions, or warfare) or global catastrophic events (e.g. Earth's collisions with asteroids, which will probabilistically happen long-term), ecological and evolutionary processes will continue in future Galápagos archipelagos, of course on a planet already scarred by human presence. But these realities can inspire us to pledge our existence to the health of the Earth"* (…). *As Charles Darwin reflected in the Journal of Researchers (1839), just four years after exploring the Galápagos in 1835,* [t]*he natural history of this archipelago is very remarkable: it seems to be a little world within itself; the greater number of its inhabitants, both vegetable and animal, being found nowhere else* (…). *Indeed, a little world, the only one we have, is worth saving forever*".

Guillermo Paz-y-Miño-C
Avelina Espinosa

CHAPTER ONE

WHAT ARE THE GALÁPAGOS ISLANDS?

The Galápagos Islands are a volcanic oceanic archipelago located in the Tropical Eastern Pacific, 900 km (560 mi) off coastal Ecuador, near the equator (latitude 0°), and at longitude 91° West (Fig. 1.1; terminology in Box 1.1). The Republic of Ecuador claimed jurisdiction over the islands in 1832. Galápagos belongs to Ecuador (Paz-y-Miño-C and Espinosa 2024a)!

Fig. 1-1 Location of the Galápagos Islands with respect to the Tropical Eastern Pacific and the proximate mainland (South and Central America). The volcanic and oceanic archipelago is situated 900 km (560 mi) off coastal Ecuador (latitude 0° and longitude 91° West). Galápagos belongs to Ecuador. Design: GPC.

In Fig. 1.2, we summarize some descriptive distances characteristic of the archipelago and provide both the most common Spanish and English names of the main islands, as follows: (1) East and Southeastern Islands: San Cristóbal (Chatham), Floreana (Charles), and Española (Hood); (2) Central Islands: Santiago or San Salvador (James), Santa Cruz (Indefatigable), Baltra (Seymour South or South Seymour), Seymour (Seymour North or North Seymour), Rábida (Jervis), Pinzón (Duncan), and Santa Fé (Barrington); (3) Western Islands: Isabela (Albemarle) and Fernandina (Narborough); (4) Northern Islands: Marchena (Bindloe), Pinta (Abingdon), and Genovesa (Tower); and (5) Northwestern Islands: Darwin (Culpepper) and Wolf (Wenman; Paz-y-Miño-C and Espinosa 2024a).

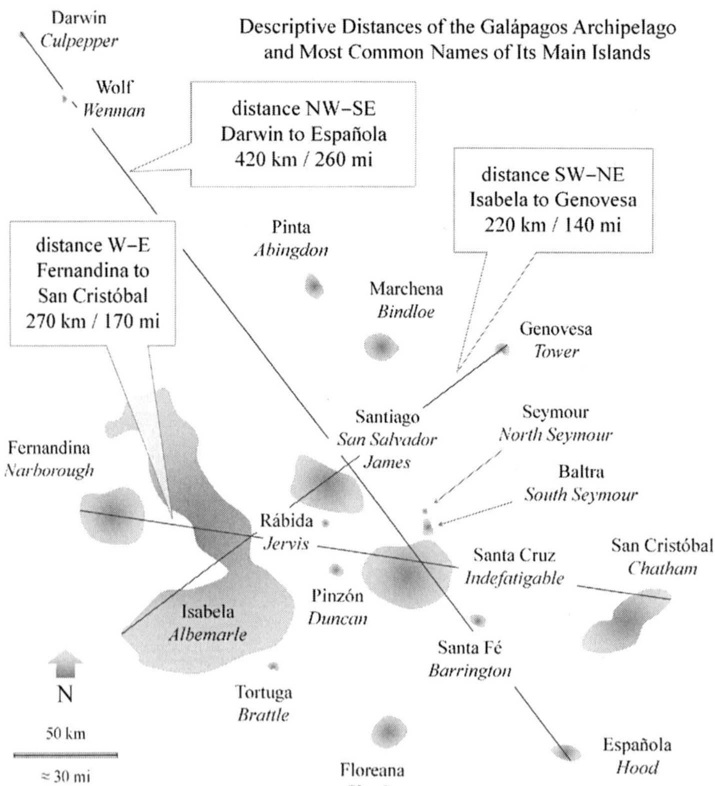

Fig. 1-2 Descriptive distances of the Galápagos Archipelago and most commonly used Spanish and English (*italics*) names of the main islands. Sketch: GPC.

Box 1-1 Essential terminology for this chapter.

Bioregions. Areas with distinctive biodiversity, biogeography, and ecology. Here we discuss bioregions within the Galápagos Marine Reserve (GMR). Biologists have proposed the existence of five bioregions in GMR, i.e. based on the species composition and abundance of fishes and macro-invertebrates (Far North, North, West, Elizabeth Bay, and Southeast; sketch in Box 1.3).

Erosional terraces. Flat or sub flat areas along the slopes of an underwater mountain (seamount) formed after the accumulation of sediments. In Galápagos, underwater mountains (i.e. long-gone islands primarily located to the East of the archipelago) have erosional terraces. Galápagos' seamounts have been exposed to seashore wear [inducer of sedimentation] during millions of years.

Hydrothermal vents. Bottom-of-the-ocean cracks [proximate to volcanic areas] through which hot water discharges. As the vent-water ascends, it reacts with magmatic rocks beneath the seabed and mixes up with chemicals (multiple compounds, particles, and gases result). This "hot buoyant mix" rises, often vigorously. Some hydrothermal vents have "chimneys", which are built up from the sedimentation of particles (sketch in Box 1.2).

Microbes (bacteria, Archaea, and protists) and their viruses. Unicellular organisms and their specialized viruses. Bacteria and Archaea lack nuclei. They are phylogenetically distinct from one another. Protists are unicells with nuclei. Microbes and their viruses are the foundation of ecological communities.

Oceanic vs. continental islands. Oceanic islands like the Galápagos originate from volcanic activity, i.e. a hotspot that pierces the bottom-of-the-ocean crust (Fig. 1.4). These islands have not been part of a continent. By contrast, continental islands form from the fragmentation of a continental landmass (e.g. Tasmania was once part of the Australian continent).

Pyroducts. Or lava tubes. Pyroducts form when lava (molten rock) descends the flanks of a volcano (like a "stream" or "river"). If the surface of the lava-flow cools down and crystallizes, a rock "ceiling" would form, while molten rock continues to flow beneath. After the eruption ceases and the pyroduct empties, a "lava cave" will remain (sketches in Figs. 1.14 to 1.16).

Seamounts and guyots. Below-the-sea-level mountains [or volcanoes] are often called seamounts. Former volcanic islands now submerged are called guyots.

Subduction. Sinking of a tectonic plate [its edge] with respect to another plate [and in the direction of the Earth's mantle]. Plates' edge/edge=subduction zone.

Tectonic plates. Or lithospheric plates. Huge slabs of rock (Earth's crust). Each continent or ocean has its own plate (continental or oceanic lithosphere).

The Origin of the Islands

How did the Galápagos Archipelago originate? As volcanic islands, the Galápagos are product of interactions between tectonic plates (e.g. Nazca, Cocos, South American, Pacific, Antarctic; Fig. 1.3) and a hotspot (i.e. the source of magma that fuels the volcanoes; Fig. 1.4). Examine Fig. 1.3. In it, we offer a close-up look at the location of the Galápagos Archipelago on the Nazca Plate [the islands rest on Nazca] and in relation to the Cocos, South American, Pacific, and Antarctic plates; note the position of the East Pacific Rise (EPR, a submarine volcanic ridge). The Nazca Plate moves East (5.2-7 cm/year or 2-3 in/year); it collides with the South American Plate, which moves in the opposite direction. As Nazca subducts [sinks] under South America, an entire cordillera—nourished by the geological effects of the collision and subduction—rises: the Andes! The interaction with the other plates is less antagonistic. The Cocos Plate travels Northeast, moving away from Nazca, but it does collide with the Caribbean/North American plates in what today is Central America. The Pacific Plate drifts West, away from its boundary with Nazca, the EPR, while the Antarctic Plate slides with Nazca as the former travels East and stumbles upon South America (Paz-y-Miño-C and Espinosa 2024a).

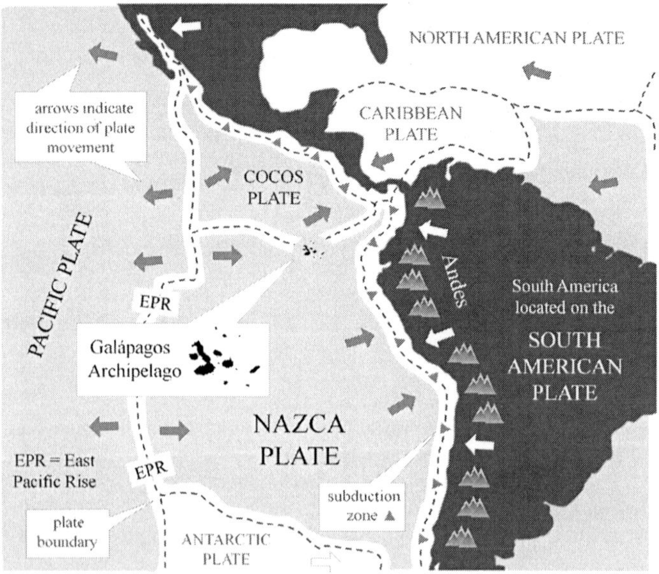

Fig. 1-3 Location of the Galápagos Archipelago on the Nazca Plate and in relation to the Cocos (N), South American (E), Pacific (W), and Antarctic (S) plates. Arrows=direction of plate movement; triangles=subduction zones. Sketch: GPC.

The Galápagos Hotspot

Now, bring your attention to Fig. 1.4. In it, we depict a possible mechanism for the origin of the Galápagos hotspot [a hypothesis that applies to other hotspots on Earth, like Hawai'i or the Canaries]. The current islands are young (0.032-4 Ma; =million years), but the hotspot dates to 139 million years. It manifests itself at the crust, although its source of energy is traceable to the planet's mantle-core boundary. The Nazca Plate drifts above the hotspot. The archipelago and hotspot are connected via a heat column—or plume—which expands from the Earth's [outer] core to the upper mantle (after Toulkeridis and Addison 2014; for comprehensive discussion, see Paz-y-Miño-C and Espinosa 2024a).

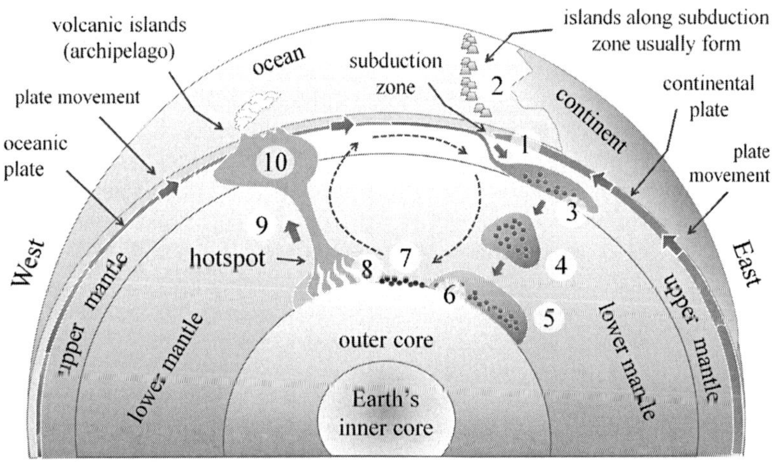

Fig. 1-4 Possible mechanism for the origin of a hotspot, as applied to the Galápagos Archipelago. (1) an oceanic plate [Nazca] subducts under a neighboring continental plate (gravity takes the oceanic plate down); (2) part of the oceanic plate can form small islands along the subduction zone (proximate to the continent), or (3) it re-melts in the upper mantle; (4) remnants of the plate sink toward the outer core; (5) plate remnants linger at the heat boundary between the lower mantle and the outer core (this takes geological time); (6) heavy elements (e.g. iron) separate as the plate remnants continue to melt; (7) separated heavy elements accumulate at the lower-mantle/outer-core boundary; (8) the lighter, less dense remnants of the former plate rise; (9) the up-rising materials form a heat-column (or plume), anchored at the mantle-core boundary; since the now "hotspot" is hotter than the surrounding mantle, its magma continues to move up; (10) upon contact with the above crust (i.e. the existing plate [Nazca]), the magma perforates it, giving origin to volcanic islands, which are later destined to travel Eastward, erode, sink, and die off as underwater mountains or "seamounts" (this phenomenon can take Ma =

millions of years). Layout: GPC; redrawn and modified after Toulkeridis and Addison (2014). For comprehensive discussion about the geological origin of the Galápagos, see Paz-y-Miño-C and Espinosa (2024a).

The Galápagos Plateau and Its Proximate Ridges

What does the bottom of the ocean look like in the Galápagos? In Figs. 1.5 and 1.6, we depict the seafloor topography around the archipelago (after Paz-y-Miño-C and Espinosa 2024a). The islands rest on a 1000-m (≈3300-ft) deep platform [plateau], which becomes shallower toward the East (reaching 400-m/≈1300-ft depth; dialog box in Fig. 1.5, center left). We have highlighted contour lines at 500-m to 3000-m depth (≈1600-ft to ≈10,000-ft depth). Keep in mind that the deepest and steepest ocean floor is located to the West and Southwest of the platform (off the plateau), where it reaches >3000 m/10,000 ft depth. Now, focus your attention on the Galápagos Spreading Center, GSC (spreading = diverging, separation between the Nazca and Cocos plates), a mid-ocean ridge apparently formed 23-25 Ma [keep in mind that both GSC and EPR, the latter mentioned above, are submarine geological formations]. GSC expands ≈1400 km (≈900 mi) from longitude 98° to 86° West (full expansion not shown in Fig. 1.5). To its West (at 98° W), the spreading rate of W-GSC is 45 mm/year (1.8 in); and to the East (at 86° W), 63 mm/year (2.5 in). At its midpoint, which bumps into the W 91° Fracture Zone (center top of Fig. 1.5), the spreading rate of GSC is 57 mm/year (2.3 in; review in Sinton et al. 2003). About 10 Ma, the GSC was located closer to the hotspot (roughly where the islands of Fernandina and Isabela rest nowadays; see "hotspot-plume zone" in Fig. 1.6); back then, most of the plume-derived magma was directed to the GSC (Harpp et al. 2003). At present, the magma primarily fuels the young volcanoes—which have calderas—in Fernandina and Isabela islands. But since older islands lack calderas (East of the archipelago), it appears that their process of formation was different because the hotspot was bound to directing magma mainly to the ridge—in a GSC-caused siphon effect (more on the islands' volcanoes below; Harpp and Geist 2018).

Along the Galápagos Spreading Center (West GSC and East GSC), there are numerous deep-sea hydrothermal vents. These vents form out of seabed cracks [located proximate to volcanic areas] through which hot water (350-°C/660-°F and even higher temperature) discharges. In Box 1.2, we sketch an exemplar hydrothermal-vent field with its microbial habitats and some of its macroorganisms. Chemosynthetic microbes [not photosynthetic] are the foundation of deep-ocean hydrothermal-vent ecosystems (full discussion in Paz-y-Miño-C and Espinosa 2024b).

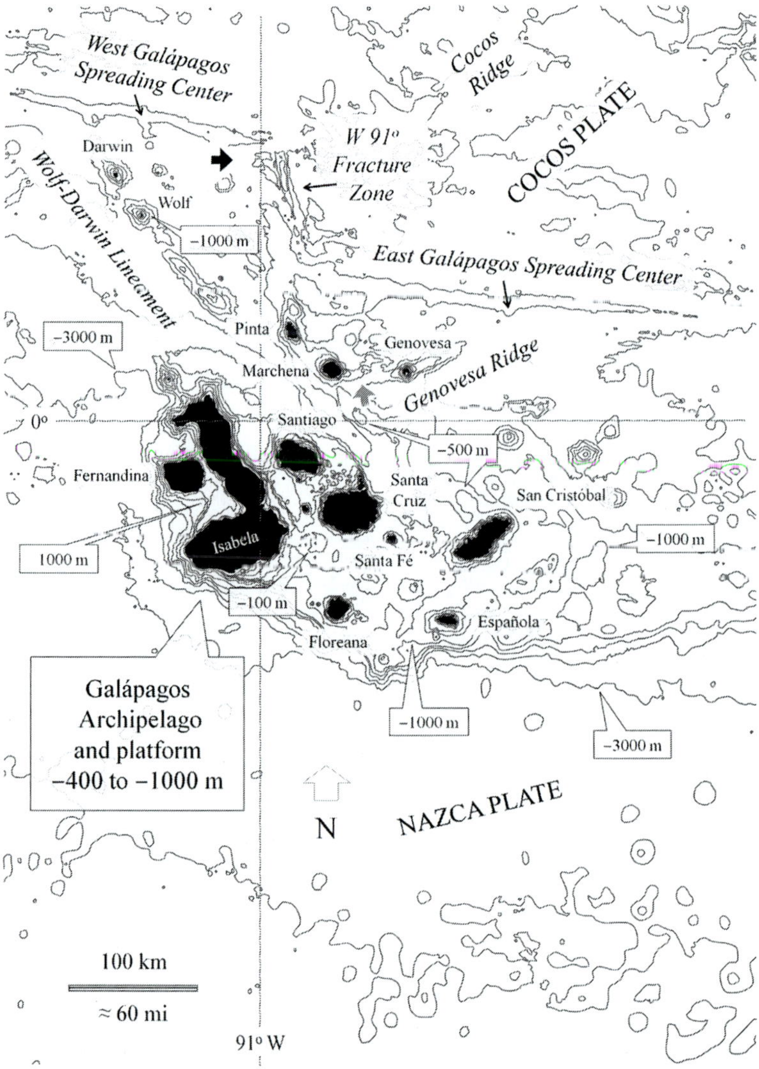

Fig. 1-5 Seafloor topography around the Galápagos Archipelago. Note the West and East Galápagos Spreading Centers (W/E-GSC); W 91° Fracture Zone/ Galápagos Transform Fault (GTF); Wolf-Darwin Lineament (WDL); and Genovesa Ridge (GR). Thick black arrow=GSC-GTF stress region. Thick gray arrow=extension of GR. Sketch: GPC, after General Bathymetric Chart of the Oceans (GEBCO); NOAA, National Centers for Environmental Information (NCEI), and Charles Darwin Foundation (links in Appendix A).

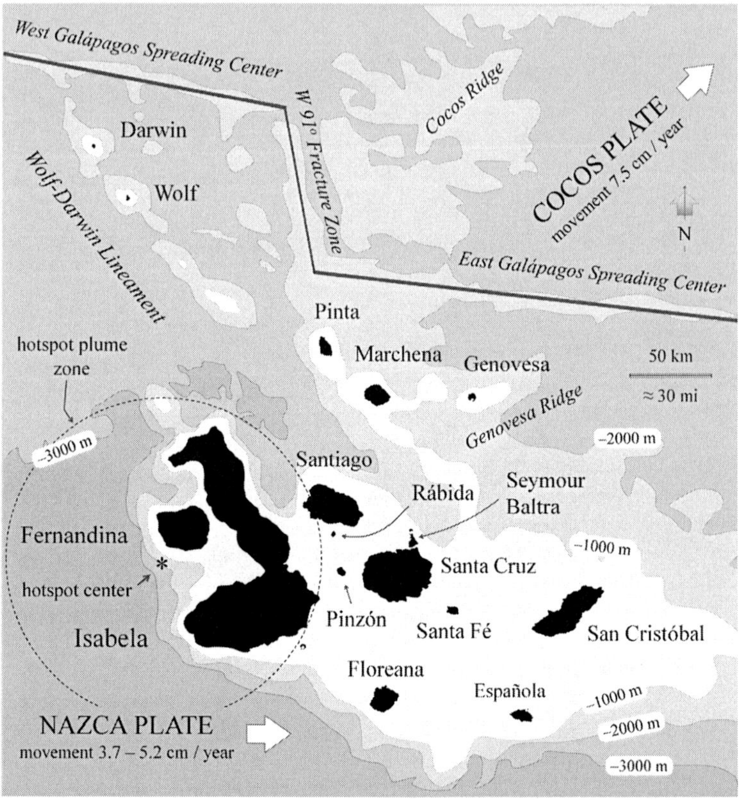

Fig. 1-6 Close-up of the Galápagos Archipelago with respect to its immediate seafloor and hotspot-plume zone (depths 1000-3000 m or 3300-10,000 ft). Layout GPC, after Villagómez et al. (2007); Graham et al. (2014); Sinton et al. (2017).

Box 1-2 What are deep-sea hydrothermal vents?

Hydrothermal vents form along ocean ridges. They were confirmed to exist in 1977 when the submersible *Alvin* reached 2,700-m (9,000-ft) depth and its crew spotted numerous vent fields at the Galápagos Rift. After two months of sea-work and 24 dives to five discernible vent fields, the *Alvin* produced >100,000 color images and multiple collections of unknown-to-science specimens, including giant clams, mussels, snails, and worm-like invertebrates, as well as crabs, octopuses, and even fish (Ballard 1977).

Hydrothermal vent fields are full of life forms. Scientists have documented 592 species, and 332 genera of organisms distributed across 63 world-exemplar vent fields (Bachraty et al. 2009).

How do hydrothermal vent fields form (particularly those with chimneys: "white" or "black" smokers)? Examine the sketch in this Box as follows (here we follow Dick 2019; Paz-y-Miño-C and Espinosa 2024b):

(1) Very cold (2°C/36°F) seawater percolates through cracks in the ocean crust (keep in mind that ocean ridges are distributed globally and conditions for hydrothermal-vent-field formation occur ubiquitously).

(2) The water heats up, reacts with hot magmatic rocks beneath the seabed (surrounding temperature 400°C/750°F), it mixes up with chemicals, and multiple compounds, particles, and gases are formed (=hydrothermal fluid plus gases).

(3) This "hot buoyant mix" (350°C/660°F) ascends, often vigorously; over time, rock-solid tower-chimneys can be formed; they build up from the sedimentation of particles around the vents.

(4) Depending on the chemical (including pH) and particle make-up of the buoyant mix, "white" or "black" flows (plumes) are released through the chimneys (they are exposed to a thermochemical gradient along their way up).

(5) Chemical reactions continue as the plumes rise (e.g. biotic metal oxidation) as well as sedimentation of minerals (e.g. via formation of iron hydroxide $Fe(OH)_3$ or manganese dioxide MnO_2); thus, the seafloor becomes enriched with minerals ("metalliferous sediments") that microorganisms later use in oxidation-reduction cellular physiology.

(6) The very chimneys' surfaces and proximate plumes offer suitable habitats for bacteria and archaea to proliferate, including taxa typical of extreme environments (Delta-, Gammaproteobacteria, Methanococcales, Aquificae, Methanosarcinales, Archaeoglobi, Thermococci, Epsilonproteobacteria).

(7) Some neutrally buoyant plumes (i.e. at much colder waters away from the chimneys' tops, 2°C/36°F) host "suspension feeders" like Alpha-, Delta- and Gammaproteobacteria, as well as Crenarchaeota (archaea).

(8) The seafloor around the vents becomes covered by chemosynthetic mats of Aquificales, Methanogens, Epsilonproteobacteria, and Archaeoglobaceae; if able to colonize these substrates, the giant *Riftia* tubeworms, clams, and mussels can become abundant.

(9) These worms, clams, and mussels depend on symbiotic microbes to harness geochemical energy (Alpha-, Gamma- and Epsilonproteobacteria); in fact, symbiotic interactions with chemosynthetic partners are central to hydrothermal vent communities; the entire ecosystem is powered energetically by both the vents' heat and the chemosynthetic microorganisms.

Thus, (10) these ecological associations among life forms (from unicells to animals) take place in the context of a thermochemical gradient (h=hotter; c=colder in the sketch below), which influences bioprocesses at various habitat scales: the chimneys themselves (their towers and openings), the plumes (white-to-black smokers; close or distant from the chimneys), and the seafloor (details in Dick 2019; Paz-y-Miño-C and Espinosa 2024b).

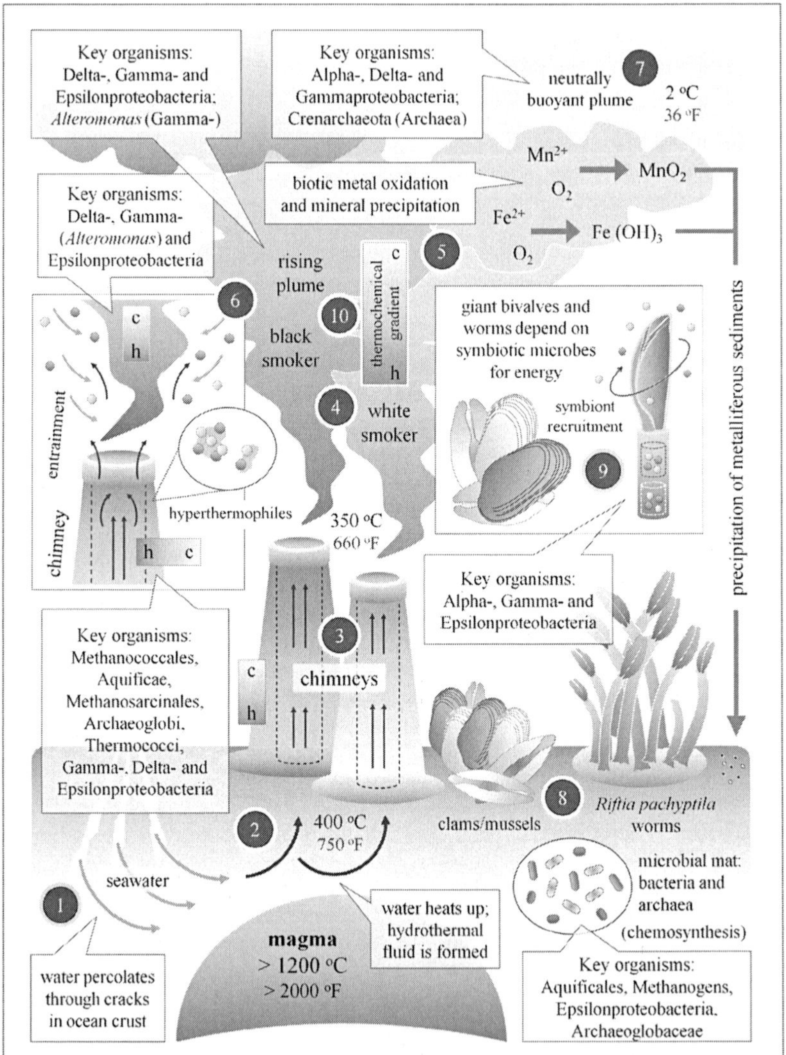

Above: Structure of an exemplar deep-sea hydrothermal vent with its microbial habitats. Not all vent fields have chimneys; when they do, the sub-cylindrical towers are formed bottom-up via sedimentation of minerals and particles previously released and/or set in motion by the vents' activity. Ecological associations among organisms take place in the context of a thermochemical gradient (h= hotter; c=colder), which influences live processes at various habitat scales: chimneys (towers and openings), plumes (close or distant from the chimneys), and seafloor. Layout: GPC; redrawn and modified after Dick (2019).

Age and Formation of the Current Islands

How old are the current islands in the Galápagos Archipelago? Between 0.032 and 4.0 million years, as shown in Fig. 1.7. The youngest is Fernandina (0.032-0.06 Ma) and the oldest are Española (3.0-3.5 Ma) and San Cristóbal (2.4-4.0 Ma). The archipelago gains age from West (younger islands) to East (older islands).

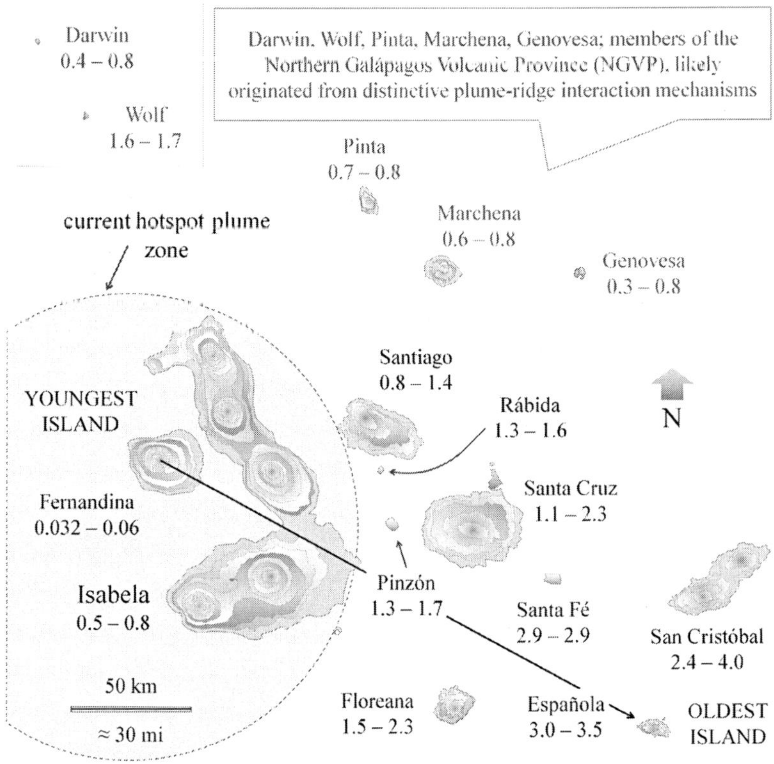

Fig. 1-7 Minimum to maximum emergence ages of the Galápagos Islands (Ma = millions of years) with respect to the current hotspot-plume zone. Layout: GPC; emergence estimates come from Geist et al. (2014); plume-ridge interaction zone comes from Graham et al. (2014).

How did the current islands form? In Fig. 1.8, we summarize a model for the formation of the Galápagos during the past 3 Ma (after Geist et al. 2008; Orellana-Rovirosa and Richards 2016; for full description see Paz-y-Miño-C and Espinosa 2024a). Central to the model is the interaction among: (1) a stationary, medium-strength hotspot plume (hp in Fig. 1.8); (2) the proximity of a tectonic plate (Nazca) and a mid-ocean ridge (Galápagos Spreading Center, GSC) to the hotspot plume; and (3) the movement of the plate (Nazca) and ridge (GSC) away from the hotspot plume (Nazca to the East, GSC to the Northeast).

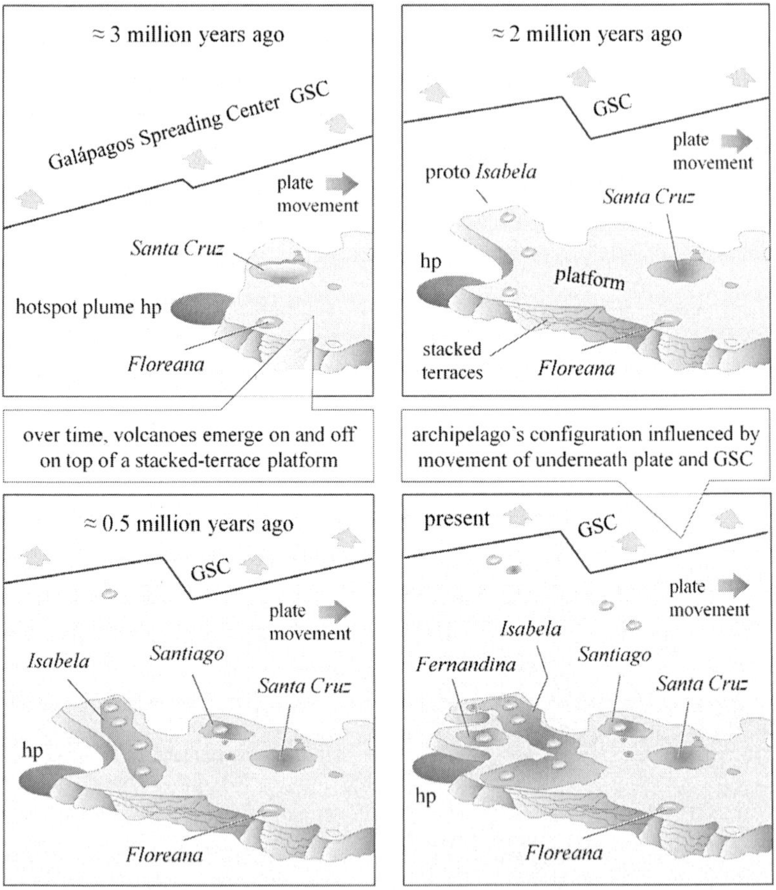

Fig. 1-8 A dynamic archipelago: a model for the formation of the Galápagos Islands during the last 3-million years. Layout GPC; redrawn and modified after Geist et al. (2008) and Orellana-Rovirosa and Richards (2016).

Episodic submarine magma flows of variable volume and duration (toward the East of the hotspot) likely built a plateau (i.e. Galápagos platform) of stacked lava [molten rock] terraces. In Fig. 1.8 (top left), we depict such scenario ≈3 Ma. Notice in the sketch the plateau partially built and the islands of Santa Cruz and Floreana on top. The Galápagos Spreading Center (GSC) appears closer to the hotspot (at ≈3 Ma) than in the other depictions on Fig. 1.8 (at ≈2 Ma, ≈0.5 Ma, or present). The model suggests that, over geologic time, islands or volcanoes gradually emerge—on and off—on the platform.

About 2 Ma (Fig. 1.8, top right), Santa Cruz and Floreana islands drifted East with respect to the hotspot (as Nazca continued to move toward South America), the platform expanded and reached considerable size, and three new volcanoes, corresponding to proto-Isabela Island, emerged. GSC continued to move away from the hotspot.

By ≈0.5 Ma (Fig. 1.8, bottom left), Santiago Island had already been formed and moved East with respect to the hotspot, and Isabela's volcanoes were fully connected by their surrounding slopes. GSC was even farther with respect to both the hotspot and the entire archipelago.

At present (Fig. 1.8, bottom right), Fernandina is the closest island to the hotspot plume. The five volcanoes of Isabela have aged and moved East, as a single island, away from the plume. And the Central, East and Southeastern Islands have traveled even farther toward South America; these islands are now smaller—eroded, less connected to the hotspot—and substantially submerged with respect to their past forms. Yet, the current Galápagos plateau and its archipelago's configuration continue to be influenced by the interplay of factors described earlier: existence of the hotspot, proximity of Nazca and GSC to the hotspot (and to each other), and drift of both in relation to the plume. Interestingly, the Northwestern (Darwin and Wolf) and Northern (Marchena, Pinta, and Genovesa) islands, members of the Northern Galápagos Volcanic Province (NGVP), appear to have originated in the tectonic stress (tension) resulting from the interaction between GSC and the Galápagos Transform Fault (GTF) or along the GTF in the 91° West Fracture Zone (Figs. 1.5 and 1.6).

Caldera-Volcanoes and Volcanoes without Calderas

Do all islands have volcanoes? Do all volcanoes have calderas? No and no! In Fig. 1.9, we sketch the Galápagos Archipelago with its current islands and volcanoes. Only the Western Islands have caldera-volcanoes: Fernandina (La Cumbre) and Isabela (Wolf, Darwin, Alcedo, Sierra Negra, and Cerro Azul). These volcanoes are young (<1-Ma; Fig. 1.7) and located just above the hotspot-plume zone; they receive magma primarily from it (Figs. 1.6 and 1.7). The older islands (Central, East, and Southeastern, 1-4 Ma; Fig. 1.7) lack calderas. As indicated earlier, these islands seem to have formed while the hotspot was still directing magma to the Galápagos Spreading Center (GSC), which was located closer to the hotspot (siphoning magma from the plume). Thus, Central, East, and Southeastern Islands likely developed under low-rate magma supply (i.e. "magma starved") and could not build large calderas (Harpp and Geist 2018). The Northern Islands (Genovesa Ridge) and Northwestern Islands (Wolf-Darwin Lineament) emerged via different geological mechanisms,

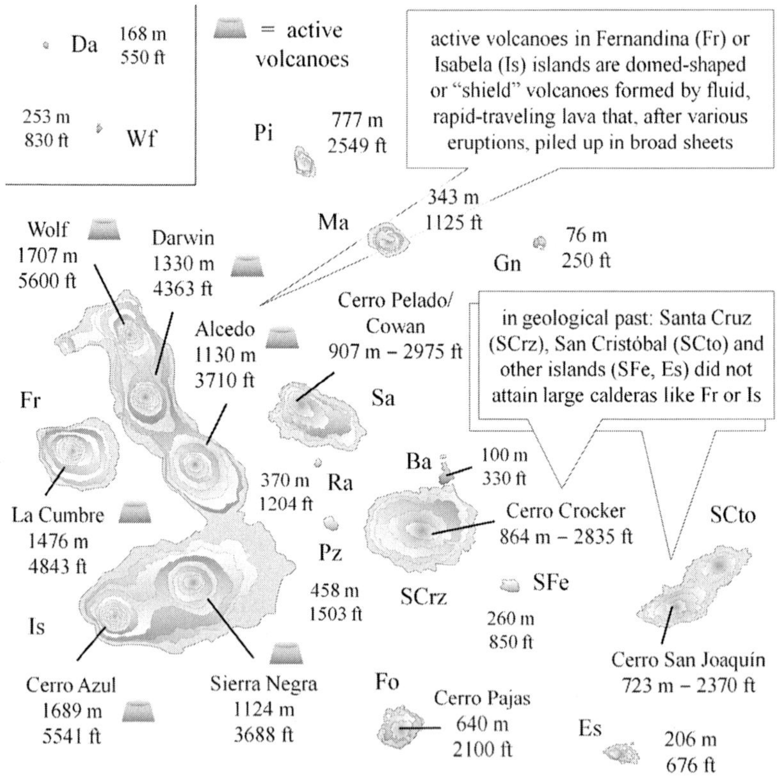

Ba = Baltra (*South Seymour*); Da = Darwin (*Culpepper*); Es = Española (*Hood*); Fo = Floreana (*Charles*); Fr = Fernandina (*Narborough*); Gn = Genovesa (*Tower*); Is = Isabela (*Albemarle*); Ma = Marchena (*Bindloe*); Pi = Pinta (*Abingdon*); Pz = Pinzón (*Duncan*); Ra = Rábida (*Jervis*); Sa = Santiago (*San Salvador, James*); SFe = Santa Fé (*Barrington*); SCrz = Santa Cruz (*Indefatigable*); SCto = San Cristóbal (*Chatham*); Wf = Wolf (*Wenman*)

Fig. 1-9 The Galápagos Archipelago with its major islands and volcanoes. Note that only the Western Islands (Fernandina and Isabela) have volcanoes with calderas. The Central, East, and Southeastern Islands lack calderas (apparently, they never had them). Design: GPC (after Paz-y-Miño-C and Espinosa 2024a).

as discussed above (Fig. 1.5; also, Harpp and Geist 2002). [Ample discussion and references in Paz-y-Miño-C and Espinosa (2024a)].

Volcanoes Uplift or Subside

Do Galápagos volcanoes change shape? Yes, volcanoes uplift or subside over time, as shown in Fig. 1.10. During a seven-year observation period (1992-1999), five volcanoes uplifted, as inferred from satellite-radar scanning: Wolf 9-cm (3.5-in) uplift, Darwin 22 cm (8.7 in), Fernandina 90 cm (35.4 in), Alcedo >90 cm, and Sierra Negra 240 cm (94.5 in). Only one volcano, Cerro Azul, subsided 15 cm (5.9 in). All these volcanoes have

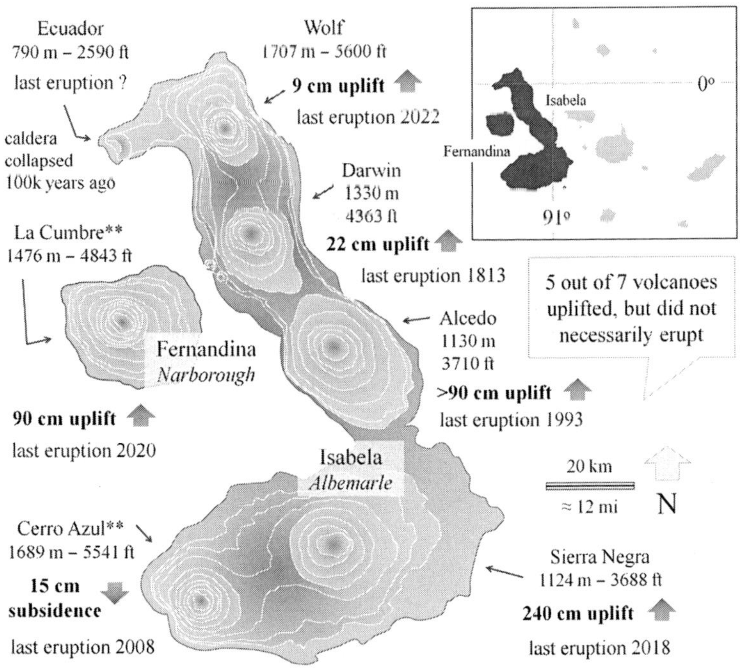

* scientists generate interferograms from satellite scans of the Earth's surface; overlapping images from different times allows researchers detect small changes on the landscape surface or elevation (uplifting or subsidence); detailed imaging in Amelung et al. (2000); ** two volcanoes erupted during the satellite monitoring period (1992-1999), La Cumbre in 1995 (Southwest flank) and Cerro Azul in 1998 (inside the caldera, Southeast flank)

Fig. 1-10 Uplifting or Subsidence of Fernandina and Isabela Volcanoes, Galápagos Archipelago: uplifting or subsidence inferred from satellite-radar-interferometry* (scans obtained from 1992 to 1999); eruption year per volcano corresponds to most recent, up to 2000. Design: GPC, after Amelung et al. (2000).

erupted in the recent past: Wolf in 2022, La Cumbre 2020, Sierra Negra 2018, Cerro Azul 2008, Alcedo 1993, and Darwin—arguably—1813 (Fig. 1.10). Note in Fig. 1.10 what is left of a small volcano, Ecuador, in Northern Isabela. It collapsed 100,000 years ago; its last eruption is uncertain. Volcanoes expand when magma accumulates beneath them (in reservoirs). Based on the uplift events documented in the Galápagos calderas and the absence of subsequent eruptions, geologists have hypothesized that sporadic uplifting might have relaxed stress in the magma chambers and delayed (inhibited) summit eruptions (Amelung et al. 2000). Ample discussion in Paz-y-Miño-C and Espinosa (2024a).

Underwater Volcanoes, Seamounts, and Guyots

Are there underwater volcanoes or seamounts in the Galápagos? Yes, they are ubiquitous in the Galápagos platform and its proximate or distant ridges (e.g. Wolf-Darwin Lineament, Genovesa Ridge, or Cocos Ridge, as in Fig. 1.5). In the geologic past, some seamounts [guyots] were islands with caldera-volcanoes, or sizeable mountains located just above or nearby the hotspot (chronology of studies in Sinton et al. 1996, 2014, 2018). In Fig. 1.11, we sketch the location of seven exemplar seamounts (Pico, Grande, Sunray, Beagle, FitzRoy, Largo, and Iguana) with respect to the current Galápagos Archipelago (North of San Cristóbal Island; off the Northeast of the Galápagos plateau; see "sampled area" in Fig. 1.11). The seamounts range in age from Pico's 5.2 Ma to Iguana's 0.5 Ma. Note that three seamounts (Beagle, Largo, and Iguana) did not reach sea level and remained submerged as underwater mountains (Table 1.1). Others [now guyots] attained considerable altitude: Pico 867 m (≈2800 ft) above sea level, Grande 888 m (≈2900 ft), Sunray 704 m (≈2300 ft), and FitzRoy 403 m (≈1300 ft; Table 1.1). At present, peak-depths of the seven seamounts vary widely, from as deep as 1215 m (≈4000 ft) in the case of Iguana to as shallow as 150 m (≈500 ft) in FitzRoy (Table 1.1). Estimated submergence ages of FitzRoy (0.8-0.3 Ma), Grande (2.2-0.8 Ma), Sunray (2.2-1.0 Ma), and Pico (4.9-2.6 Ma) suggest they remained as above-sea-level caldera-volcanoes, or mountains, until relatively recently (Table 1.1).

[As indicated in Box 1.1, below-the-sea-level mountains (or volcanoes) are often called seamounts (a generic term applied to sizeable formations that rise from the seafloor). Former volcanic islands now submerged (and sometimes flat-topped due to tidal erosion) are often called guyots (a specific term for past above-the-sea-level volcanoes). You can always argue about proper terminology].

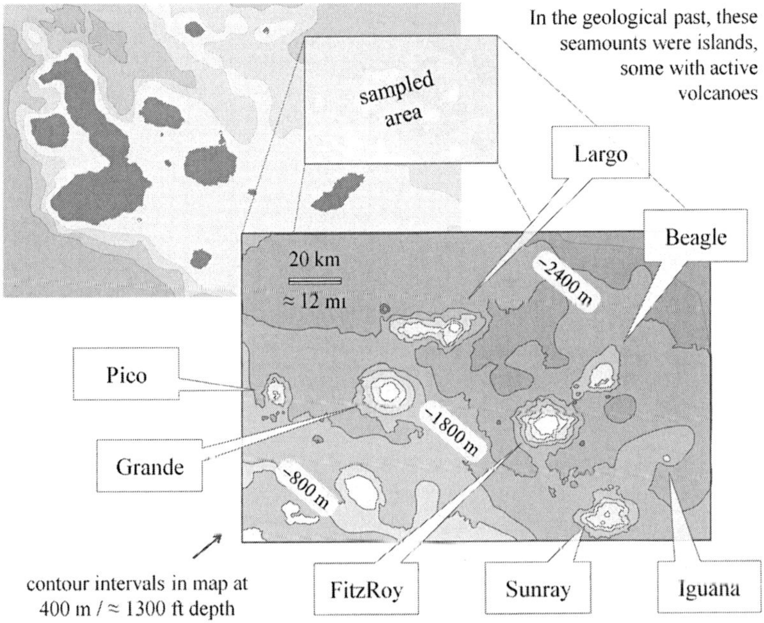

Fig. 1-11 Seamounts located to the East and Northeast of the Galápagos Archipelago. Layout: GPC, redrawn and modified after Sinton et al. (2018).

Note in Table 1.1 that we provide subsidence rates (meters of "sinking" as a function of time) for both the ocean crust (i.e. lithosphere) and the Galápagos platform (last two columns' values: "oceanic"/"platform"). The ocean crust subsides at a slower rate than the platform. This helps us explain the two values under the column "initial elevation above sea level". The lower values (left) correspond to estimates obtained based on the oceanic crust subsidence rate; the higher values (right) correspond to estimates obtained based on the platform subsidence rate. As you should realize, when we referred to the "altitude" of seamounts Pico, Grande, Sunray, and FitzRoy (above), we provided the maximum elevation estimates based on the platform subsidence rate (Paz-y-Miño-C and Espinosa 2024a).

Table 1.1 Characteristics of the East and Northeast Galápagos Seamounts

	minimum age (Ma)	submergence age (Ma)	initial elevation above sea level (m)	current peak depth (m)	subsidence (m) oceanic / platform	
Pico	5.2	[4.9 - 2.6]	32 - 867	800	832	1667
Grande	3.3	[2.2 - 0.8]	413 - 888	250	663	1138
Sunray	3.1	[2.2 - 1.0]	273 - 704	370	643	1074
Beagle	2.2	na	submerged	800	541	758
FitzRoy	1.7	[0.8 - 0.3]	326 - 403	150	476	553
Largo	1.3	na	submerged	650	416	367
Iguana	0.5	na	submerged	1215	258	127

na = "non applicable" corresponds to seamounts that did not reach sea-level; evidence suggests they remained submerged as underwater mountains (data from Sinton et al. 2018)

Are there younger [than those discussed above] seamounts [or guyots] in the Galápagos? Shallow seamounts of comparable age to proximate islands can be found almost everywhere within the plateau of today's archipelago, ranging from near sea level to 1000-m (≈3300-ft) depth (Schwartz et al. 2018). They can also reveal the geologic history of the islands. For example, recent assessments of the seafloor shared among Fernandina, Isabela, Santiago, Santa Cruz, and Floreana suggest that numerous shallow seamounts (20-700-m/65-2300-ft depth) were above-sea-level formations during the Last Glacial Maximum (LGM; 20,000 years ago), when the sea level dropped significantly (Schwartz et al. 2018). The presence of erosional terraces on these seamounts indicates they were islands or islets exposed to seashore wear (paleo-shorelines). Thus, during LGM not only the Galápagos Islands had shorelines 125 m (≈410 ft) below today's sea level (as depicted in Fig. 1.12), but they were also proximate to nearby islands/islets currently under water (Geist et al. 2014; Schwartz et al. 2018).

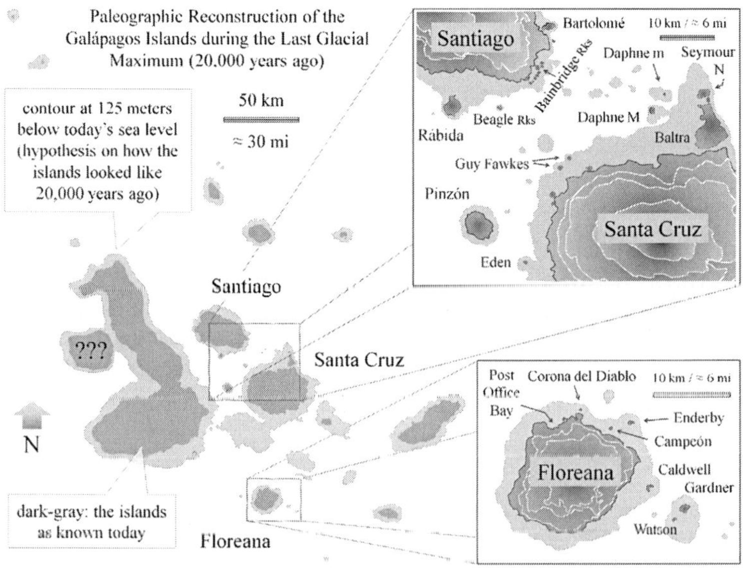

Fig. 1-12 Paleographic reconstruction of the Galápagos Islands during the Last Glacial Maximum (LGM; 20,000 years ago). Today's archipelago is depicted underneath the suspected LGM-islands. ??? = Fernandina uncertain topology considering its recent emergence 32-60 ka. Note the abundance of islets (now seamounts) between the shores of Santiago and Santa Cruz (top right) or the large size of Floreana during LGM (bottom right). Layout GPC (as per Paz-y-Miño-C and Espinosa 2024a), redrawn and modified after Geist et al. (2014).

What does the Galápagos Islands' topography look like below the ocean's surface? Modern computer technology helps us address this question. One of the most impressive and didactic 3D-image-animations of the Galápagos—which compiles multiple-satellite-imagery data—has been put together by NASA (National Aeronautics and Space Administration) Scientific Visualization Studio: "Galapagos Islands Flyby" (NASA-SVS 2009). In it, the archipelago is depicted from aerial and below-the-ocean views in a 2:30-min sequence of computer-generated graphics (link in Appendix A). The clip illustrates the islands rising from the sea floor as set on top of a marine platform (<1000 m/≈3300 ft in depth). The deepest and steepest ocean floor is located to the West and Southwest, off the plateau (it reaches >3000 m/10,000 ft). The platform itself becomes gradually shallower toward the East. In Fig. 1.13, we provide two still images extracted from the animated flyover. For clarity, we have labeled the islands, volcanoes of Isabela, Cocos Ridge (located to the Northeast of the Galápagos plateau), and the bottom of the ocean depths. Note the distance between Darwin (Culpepper) and Española (Hood): 420 km (≈260 mi), as marked by a dashed line (after Paz-y-Miño-C and Espinosa 2024a).

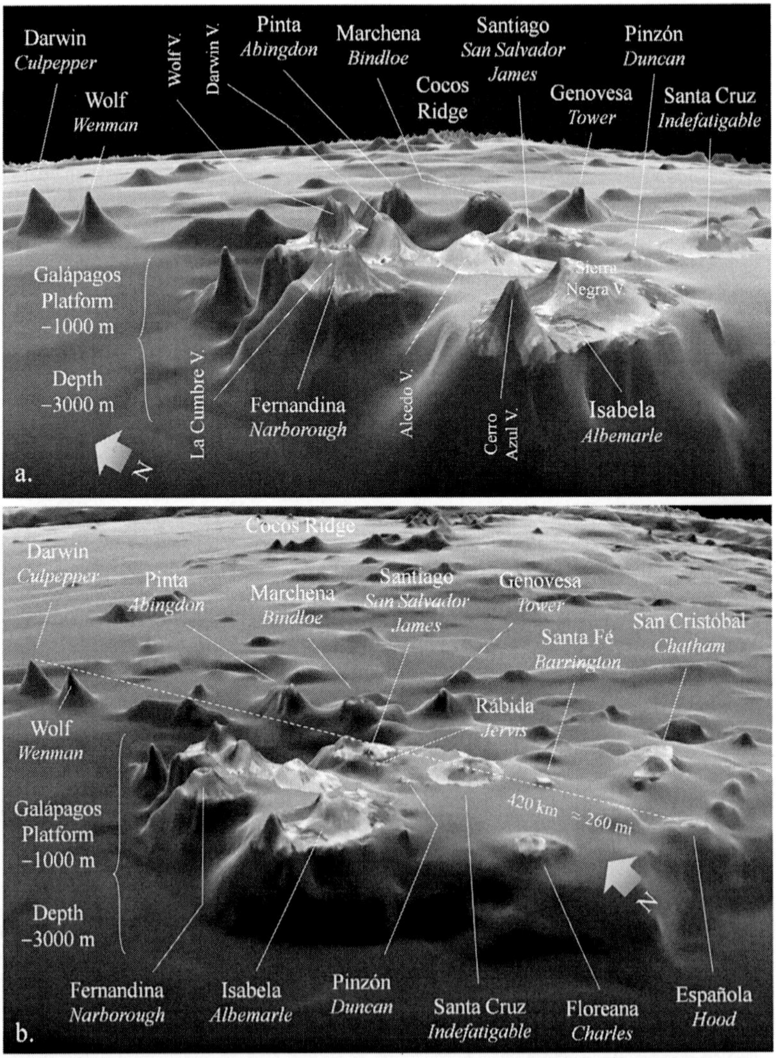

Fig. 1-13 Still images of the animated flyover of the Galápagos Islands produced by NASA/Goddard Space Flight Center, the SeaWiFS Project and GeoEye, Scientific Visualization Studio (NASA-SVS 2009). Names, labels, and dimensions added for clarity (full discussion in Paz-y-Miño-C and Espinosa 2024a).

Lava-Tube Caves or "Pyroducts"

Are there any caves in the Galápagos Islands? There are numerous lava-tube caves, or "pyroducts", in the archipelago. How do pyroducts form? When lava (molten rock) flows, gravity takes it down. Upon contact with air—or water—the ≈1000-ºC (≈1800-ºF) hot lava cools and gradually turns into rock. As lava "streams" or "rivers" descend the flanks of a volcano, they follow the immediate topography and carve their way downhill. A lava-tube, or pyroduct, can form if the surface lava of the "stream" or "river" crystallizes, building an upper crust of rock (the ceiling/s), while the molten rock continues course beneath (encased by the surrounding crust; Jordá-Bordehore et al. 2016; Kempe et al. 2021). Once the eruption ceases and the pyroduct empties, a "lava cave" will remain (Fig. 1.14; see ample discussion in Paz-y-Miño-C and Espinosa 2024a).

In Fig. 1.15, we offer a more detailed depiction of the structure and formation sequence of lava-tube caves in the Galápagos. Two caves, *Cueva del Cascajo* and *Cueva de Gallardo*, both on Santa Cruz Island, are exemplars of the inferred configuration process (after Kempe et al. 2021).

You should first become familiar with three terms: *pāhoehoe* in reference to fluid lava with a smooth to ropy surface; *'a'ā* or flows of rough broken lava; and *puka* the opening or entrance to a lava-tube cave or pyroduct (Fig. 1.15). In Fig. 1.15 (top), we sketch a vertical section of a lava *flow* in which layers of pāhoehoe and 'a'ā are shown. The section depicts the deep-to-surface underground at "time ONE" (as labeled on the right of the sketch). Spot the "molten lava" tag at the center of the drawing; it corresponds to an active pyroduct, which is running beneath and above multiple layers of pāhoehoe and/or 'a'ā from previous eruptions (now solid rocks). Notice how the deep underground is built of pāhoehoe, thicker and older than the above deposits. Toward the surface, massive layers of 'a'ā have piled up.

In the bottom of Fig. 1.15, we show the vertical section of an already-formed lava-*tube* cave and the didactic ten-step sequence that shaped it. The drawing corresponds to "time TWO" (as tagged on the right of the sketch). Notice the multiple passages in the cave and imagine the sequence of formation: (1) an active pyroduct removed 'a'ā rubble from previous, older flows; (2) the pāhoehoe sheet collapsed; (3) 'a'ā rubble on top of the pāhoehoe sheet also collapsed; (4) subsequent downcutting of the older 'a'ā flow narrowed the cave (a canyon formed), while glassy lining (thick line) covered its walls; (5) the top of the 'a'ā rubble and secondary 'a'ā flow (cave's initial roof/ceiling) cool-crack-collapsed; (6) cool air entered

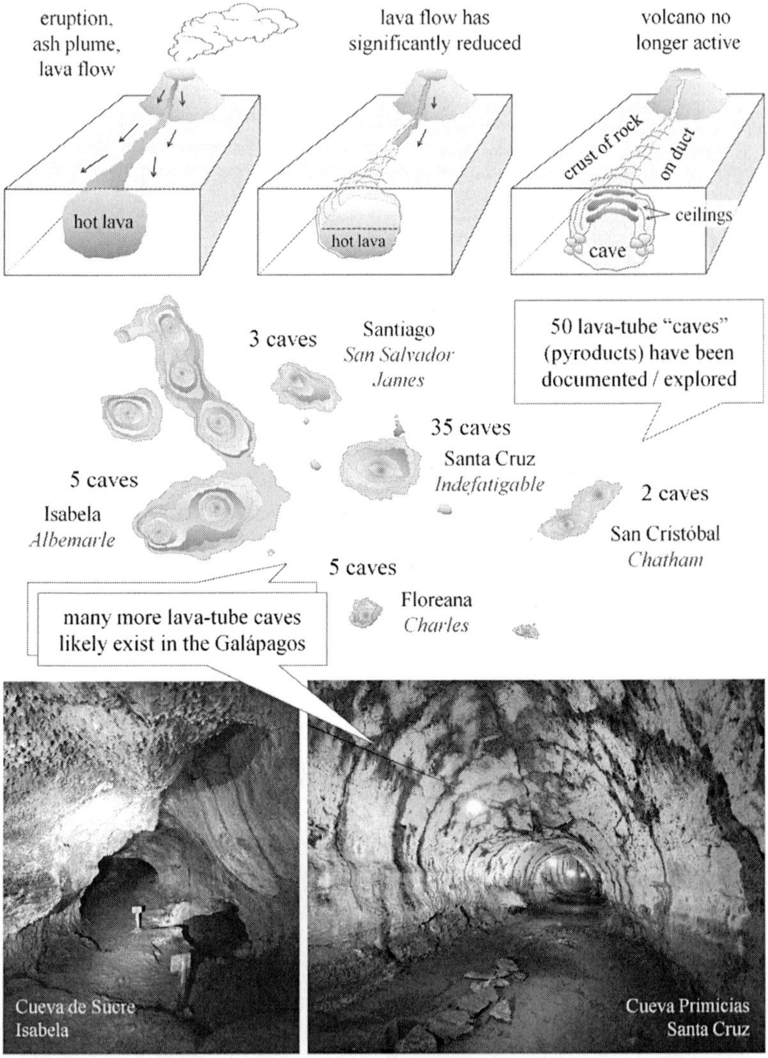

Fig. 1-14 Formation of a lava-tube cave (pyroduct) and number of documented and explored caves in some islands of the Galápagos (many more caves likely exist in the archipelago). Layout/photos: GPC (as per Paz-y-Miño-C and Espinosa 2024a), redrawn and modified after Jordá-Bordehore et al. (2016).

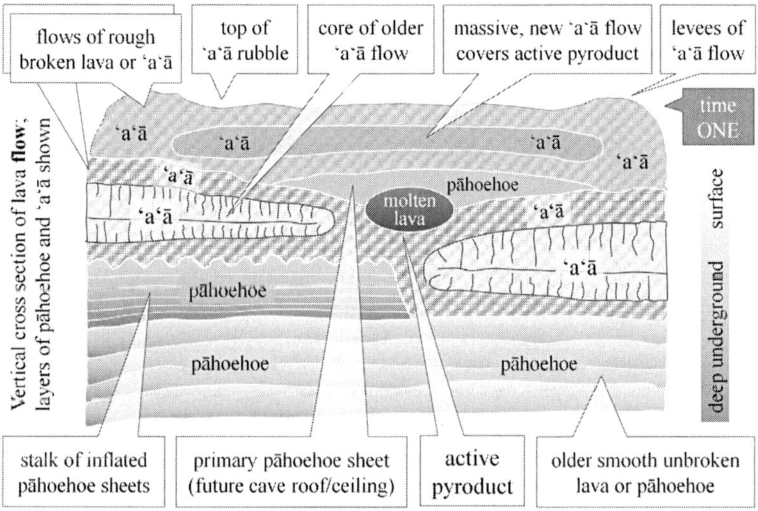

Below, sequence: 1 active pyroduct removed ʻaʻā rubble from older flows; 2 pāhoehoe sheet collapsed; 3 ʻaʻā rubble on top of pāhoehoe sheet also collapsed; 4 downcutting of older ʻaʻā flow narrowed cave (canyon formed), glassy lining (thick line) covered walls; 5 top of ʻaʻā rubble and secondary ʻaʻā flow (cave's initial roof/ceiling) cool-crack-collapsed

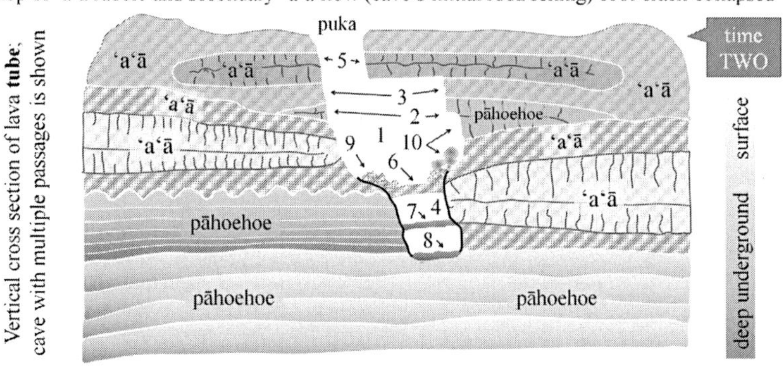

6 cool air entered and solidified a secondary roof/ceiling while molting lava continued to flow at the bottom of canyon; 7 the same process cool-crack-collapse formed yet another roof/ceiling; 8 eruption ended, lowest passage evacuated; 9 more collapse of ʻaʻā rubble walls; 10 further collapse of primary pāhoehoe sheet; thus, multiple-passages cave resulted

Fig. 1-15 Structure and formation sequence of lava-tube "caves" (pyroducts) in the Galápagos: exemplars *Cueva del Cascajo* and *Cueva de Gallardo* (Santa Cruz Island). Top refers to the cross section of a lava flow; bottom of a lava tube. Hawai'ian terms used in volcanology: *pāhoehoe* [paːˈhoweˈhowe] = fluid lava with smooth to ropy surface; *ʻaʻā* [ˈäˌä] = rough broken lava; *puka* [pu ka] = hole or opening/entrance/gate to a pyroduct. Layout: GPC (as per Paz-y-Miño-C and Espinosa 2024a), redrawn and modified after Kempe et al. (2021).

and solidified a secondary roof/ceiling while molting lava continued to flow at the bottom of the canyon; (7) the same process cool-crack-collapse formed yet another roof/ceiling; (8) the eruption ended, and the lowest passage evacuated; (9) additional collapse of the ‘a’ā rubble walls took place, followed by (10) further collapse of the primary pāhoehoe sheet. Thus, a multiple-passage cave resulted (as per Kempe et al. 2021).

About 50 lava caves have been documented and/or explored in the Galápagos (Fig. 1.14); there might be hundreds of them in the archipelago, particularly in uninhabited, remote areas (settlers often discover pukas when clearing the ground, a practice restricted to the agricultural, rural, or urban zones =3% of the Galápagos land surface). In Table 1.2, we list the currently known and/or explored (i.e. biological or geologically surveyed) lava caves in the islands of Santiago (N = 3 caves), Santa Cruz (N = 35), Isabela (N = 5), Floreana (N = 5), and San Cristóbal (N = 2; after Hernández et al. 1991). They range in length from just a few meters (12 m/36 ft) to >3 km (1.9 mi); in most cases, only fragmentary sections of the caves are known (as shown in Fig. 1.16).

Table 1.2 Currently Known and/or Explored (biological / geological surveyed) Lava-tube "Caves" (pyroducts) in the Galápagos Archipelago (after Hernández et al. 1991)

island	most common cave name	length (m)	surveyed biological	surveyed geological
Santiago	Cueva Bucanero I	567	no	yes
n = 3	Cueva Bucanero II	?	no	yes
	Cueva del Cráter de Sal	≈ 350	yes	yes
Santa Cruz	Cueva de Andrés	205	no	yes
n = 35	Cueva de Iguana	100 (-12)	yes	yes
	Cueva de Raúl Aguirre	115	no	yes
	Cueva de Gallardo	2316*	yes	yes
	Cueva de Señora Colombia	47	yes	yes
	Cueva de Gilberto Moncayo	590	yes	yes
	Cueva de Kübler	852	yes	yes
	Grieta del Pozo de Puerto Ayora	50 (-15)	yes	no
	Grietas de Bahía Tortuga	30 (-10)	yes	no
	Grietas en el camino por la Bahía Tortuga	12 (-10)	yes	no
	Cueva de Lentenech	28 (-5)	no	yes
	Cueva de la Finca Kastdalen	1500 + 500	yes	yes
	Cueva de La Curva	80	no	no
	Cueva de Las Cyatheas	18	yes	no
	Cueva de Huesos	750	yes	yes
	Cueva de Fincas Vargas	20	yes	no
	Cueva de la Finca Devine	≈ 100	no	no
	Cueva de La Miconia	276	no	yes
	Cueva de Elena	677	yes	yes
	Sima de La Pirámide	− 44	no	yes
	Cueva 2-Km al Sur de El Chato	500	yes	no
	Cueva de Tres Entradas	400	yes	no
	Cueva de Cerro Banderas	1000	yes	no
	Cuevitas al Sur Oeste del Cerro Banderas	15	yes	no
	Sima del Pozo de Los Gemelos	− 64	yes	yes
	Cueva Sin Nombre	≈ 60	no	no
	Cueva del Cascajo	3010	yes	yes
	Cueva del Monte Cascajo	≈ 100	no	no
	Grieta de la Punta de las Palmas	≈ 12	yes	no
	Cueva 4 cavidades en Zona Naranja	?	no	no
	Cueva al Norte de El Chato	?	no	no
	Cueva del Caballo	?	yes	no
	Cueva de Rovalino	50	no	yes
	Agujeros de Agua	?	yes	no
	Cueva de Miguel Arias	≈ 1000	no	yes

Table 1.2 (continuation)

island	most common cave name	length (m)	surveyed biological	surveyed geological
Isabela	Cueva de Sucre	355	yes	yes
n = 5	Cueva de Macas	92	yes	yes
	Cueva de La Cadena	114	yes	yes
	Simas de Las Torres	≈ 46	no	yes
	Simas de Cerro Grande	≈ 20	no	yes
Floreana	Cueva de Post-Office Superior	38	yes	no
n = 5	Cueva de Post-Office Inferior	202	yes	yes
	Cueva del Pinzón	110	yes	yes
	Cueva de La Lechuza	60	yes	yes
	Cavidades de la Bahía de las Cuevas	16 10 4	no	yes
San Cristóbal	Pozos de Hundimiento	≈ 15	no	no
n = 2	Cueva de Cerro Pelado	?	no	no

* for Cueva de Gallardo = 2316 m length, we follow Katz et al. (2016); these authors list additional and/or synonym names of caves in Santa Cruz (including lengths), all with biological/geological surveying: Cueva Aguirre (≠ Cueva de Raúl Aguirre) = 574 m; Cueva Cañón = 203 m; Cueva Chato 1 = 515 m; Cueva Chato 2 = 457 m; Cueva JoAnn = 80 m; Cueva Primicias = 640 m; Cueva Soyla = 1038 m; and Cueva La Llegada = 2066 m

In Fig. 1.16, we sketch longitudinal and cross sections of Cueva del Cascajo (>3-km long) in three depictions: (a) lateral view of a longitudinal section, its first uphill 150 m (490 ft); (b) top view of the upper passage and cross sections along the pyroduct; and (c) top view of the lower passage and cross sections along the pyroduct. Keep in mind that the drawing scale for the longitudinal vs. cross sections differs to facilitate depiction (scale bar: 20 m/≈ 66 ft for longitudinal sections; 10 m/≈ 33 ft for cross sections). Note in Fig. 1.16a the dialog boxes pointing at the upper and lower passages, secondary ceilings, and the three cave openings (pukas 1-3); there is an old lavafall about one-third distance from puka 1. Fig. 1.16b shows the tortuous path (as seen from above) followed by the lava. Locate the eight cross-section sketches along the lava flow (labelled a-h); sketch f depicts four ceilings and the spillhole connection between ceilings 1 and 2; explore sketch d, which crosscuts puka 2. Fig. 1.16c illustrates the wide-to-narrow irregularity of the lower passage (also as seen from above); the cross sections are labeled i-s; locate cross sections j and k and their walls made of both pāhoehoe and ‘a‘ā lava-rocks.

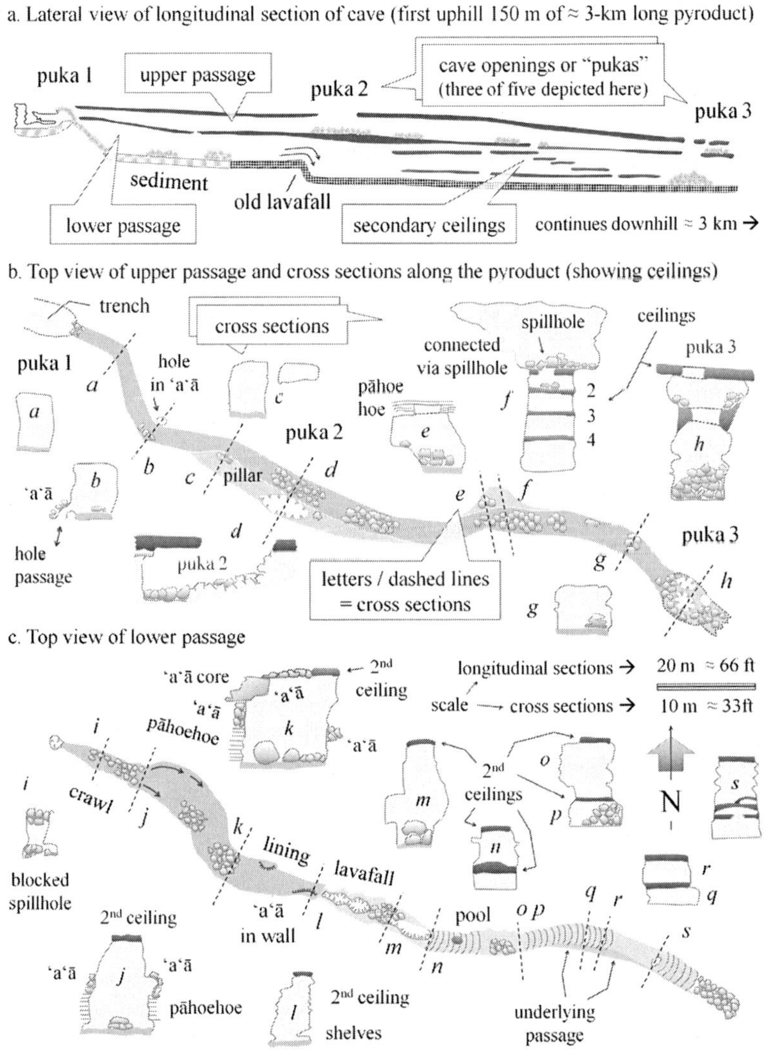

Fig. 1-16 Longitudinal and cross-sections of the lava-tube (pyroduct) *Cueva del Cascajo* in Santa Cruz Island. The first 150 m (500 ft) of a ≈3-km- (1.9-mi-) long cave are sketched. a. Lateral view of longitudinal section (note upper and lower passages, secondary ceilings, and entrances/gates or "pukas"). b. Top view of the upper passage (with 8 cross sections a-h). c. Top view of the lower passage (with 11 cross sections i-s). Layout: GPC (as per Paz-y-Miño-C and Espinosa 2024a), redrawn and modified after Kempe et al. (2021).

Soils: Younger vs. Older Islands

How do researchers extract soil samples in an archipelago like Galápagos? How do they make meaningful comparisons among the islands' soils? Chronosequence studies in which soil samples are collected in a West-to-East direction (away from the hotspot), from younger to older islands (=chronology), and at variable-elevation sampling sites with distinctive climate or vegetation type have helped scientists characterize—at least in part—the Galápagos soils. In Fig. 1.17, we summarize such an approach (after Zehetner et al. 2020). Four islands and eight sampling sites were assessed: Isabela (sites C1, C2), Floreana (C3, C3a), Santa Cruz (C4), and San Cristóbal (C5, C5a, C6; Fig. 1.17, top and center). The elevation of the sampling sites ranged from 170-m (560-ft) above sea level in Floreana (site C3a) to 586 m (1900 ft) in San Cristóbal (site C6; Fig. 1.17 bottom). The climate zones were either very dry (10 to 12 dry months per year), like in Floreana C3a; dry (8 to 10 dry mos/yr), like in San Cristóbal C5a; or humid (4 to 8 dry mos/yr), like in all other sites (Fig. 1.17 bottom). The surrounding vegetation-cover per site varied from shrubs and trees (in the very-dry to dry-climate zones) to grasslands with trees or forest (in the humid-climate zones; Fig. 1.17 bottom).

In Fig. 1.18, we summarize the ages of the soil parent materials (=geological origin from which the soils formed, e.g., weathered bedrock) as well as soil characteristics at the sampling sites. The soils in Isabela were quite recent (C1=1450-years old; C2=4290 yr), followed by Floreana (C3 and C3a=26,000 yr), Santa Cruz (C4=165,500 yr), and San Cristóbal (C5=825,600 yr; C5a=812,900 yr; C6=1,070,000 yr). Thus, younger islands had younger soils, while older islands had older soils (on a logarithmic scale; Zehetner et al. 2020). Floreana (1.5-2.3-Ma-old island) is peculiar since it has undergone late-stage volcanism, which likely explains its younger soils with respect to Santa Cruz (1.1-2.3-Ma-old), also a late-stage volcano, but with no substantial recent activity (Global Volcanism Program 2022).

The thickness of the soil (depth) varied from just 17 cm (7 in) at Isabela C1 to 230 cm (90 in) at San Cristóbal C5 (Fig. 1.18 bottom). Coarse fragments in the soil (particles >2-mm/>0.08-in diameter) were prevalent at Isabela C1 (65.2% of volume) and Floreana C3a (32.8% of volume). Sand content was conspicuous at Isabela C1 (48.0% of weight) and C2 (40.7% of weight), as well as at Floreana C3a (78.7% of weight). Silt content was overall comparable among samples (range 16.2-37.2% of soil weight). And considerable presence of clay (i.e. >50% of soil weight) ranged from 59.4% at San Cristóbal C5 to 71.3% at Floreana C3 (Fig. 1.18 bottom; data extracted from Zehetner et al. 2020).

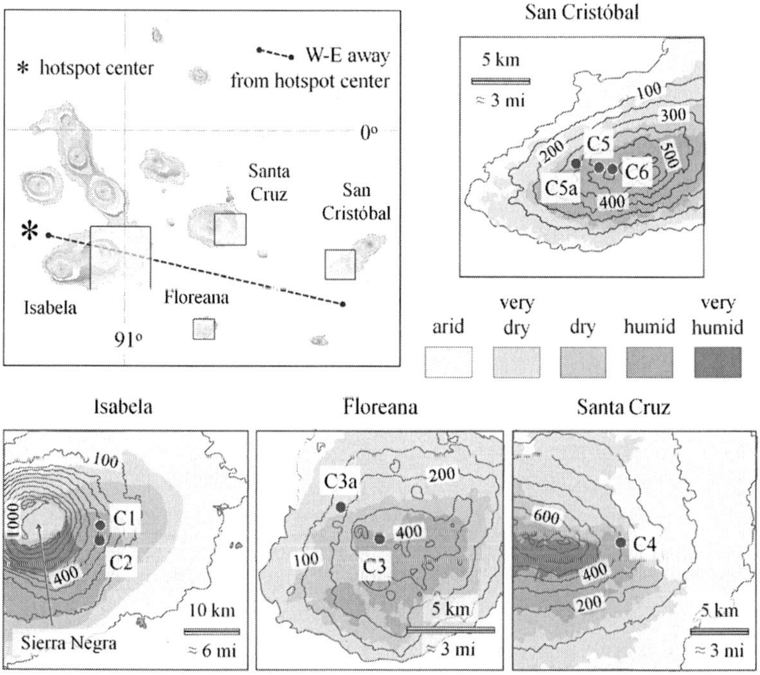

site	island	elevation (m)	slope (%)	aspect* from North	climate zone**	type of vegetation
C1	Isabela	317	5	180 (South)	humid	forest
C2	Isabela	375	6	100 (East)	humid	forest
C3	Floreana	344	8	260 (West)	humid	forest
C3a	Floreana	170	8	290 (West)	very dry	shrubs, trees
C4	Santa Cruz	497	14	180 (South)	humid	grassland, trees
C5	San Cristóbal	548	30	250 (West)	humid	grassland, trees
C5a	San Cristóbal	385	23	180 (South	dry	shrubs, trees
C6	San Cristóbal	586	20	270 (West)	humid	forest

* "aspect" or "exposure" refers to compass direction the terrain faces (degrees from North); ** climate zones: arid = 12 dry months/yr; very dry = 10 to 12 dry months/yr; dry = 8 to 10 dry months/yr; humid = 4 to 8 dry months/yr; very humid < 4 dry months/yr

Fig. 1-17 Soil-sampling sites (dots) on Isabela, Floreana, Santa Cruz, and San Cristóbal islands. Note how arid areas (often covered with shrubby vegetation) tend to be in the lowlands, while humid areas (with grasslands, trees, or tall forests) are in the highlands. Layout: GPC (as per Paz-y-Miño-C and Espinosa 2024a), redrawn and modified after Zehetner et al. (2020).

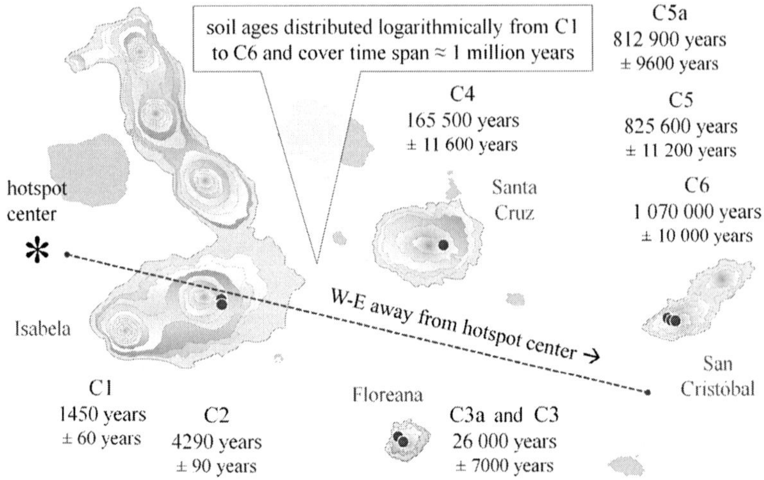

* soil age-determination methods: carbon 14 (^{14}C) for sampling sites C1, C2; helium (He) for sampling sites C3, C3a; argon 40/39 ($^{40}Ar/^{39}Ar$) for sampling sites C4, C5, C5a, C6; ** soil parent materials refer to geological origin from which the soils formed (e.g. bedrock)

Soil characteristics at sampling sites:

site	solum (soil) thickness (cm)	color variables L	a	b	coarse fragments volume %	sand	silt weight %	clay
C1	17	24.0	3.8	6.9	65.2	48.0	35.1	16.9
C2	55	29.4	4.8	11.7	7.9	40.7	36.0	23.4
C3	85	22.5	2.5	5.6	0.9	5.9	22.7	71.3
C3a	15	39.7	6.1	16.4	32.8	78.7	16.2	5.1
C4	120	25.2	6.7	8.8	1.1	2.6	35.5	62.0
C5	230	28.9	10.6	11.1	0.1	3.4	37.2	59.4
C5a	85	24.9	6.1	4.1	2.5	2.9	28.9	68.2
C6	175	30.5	12.9	15.1	0.2	2.4	31.1	66.4

coarse fragments: particles > 2 mm in diameter
b = color, from blue (−) to yellow (+)
a = color, from green (−) to red (+)
L = lightness, from black (0) to white (100)

Fig. 1-18 Ages* of the soil parent materials** (top) as well as soil characteristics at sampling sites (bottom) in Isabela (C1, C2), Floreana (C3, C3a), Santa Cruz (C4), and San Cristóbal (C5, C5a, C6) islands. Geologists rely on various descriptive parameters to characterize soils: from thickness (depth) to color variation; from particle size to relative proportions of sand, silt, or clay. Noteworthy values in gray background as discussed in the text. Layout: GPC (as per Paz-y-Miño-C and Espinosa 2024a), redrawn and modified after Zehetner et al. (2020).

What type of scoria was identified beneath the superficial soil? Basalt. We summarize the scoria's chemical characteristics in Fig. 1.19. Scoria is a dark rock with a spongy appearance resulting from gas bubbles trapped in its matrix as the predecessor molten rock cooled and hardened, thus forming the scoria (of ubiquitous distribution and abundance in the Galápagos). Be aware that geologists determine the type of volcanic rock by assessing its chemical composition, as per plotting the amount (%) of sodium oxide (Na_2O) plus potassium oxide (K_2O) found in a sample (*y* axis in Fig. 1.19 top) as function of the amount (%) of silica (SiO_2) detected in that sample (*x* axis in Fig. 1.19 top). According to this approach, all the scoria samples collected in the Galápagos (C1 to C6) were catalogued as basalts, either alkali or tholeiite basalts, which are typical of volcanic oceanic islands (Zehetner et al. 2020).

The amorphous content (i.e. lack of crystal structure) of the scorias ranged from 85-83% of weight at Isabela C1 and C2, respectively, to 18% of weight at San Cristóbal C5a (Fig. 1.19 bottom). The overall mineralogy was characterized by plagioclase (andesine $NaCa[(AlSi)_4O_8]$ in all samples), pyroxene (close to diopside $CaMg[Si_2O_6]$ in all samples), considerable amounts of olivine (iron bearing forsterite $(MgFe)_2[SiO_4]$ in samples C3 to C6), and noticeable hematite (samples C4 to C6; Fig. 1.19 bottom). Thus, the scorias (C1 to C6), although not identical, were geochemically and mineralogically comparable among all sites regardless of the climate zones and/or type of vegetation typical of the sites (Figs. 1.17 to 1.19; Zehetner et al. 2020).

In Fig. 1.20, we show the soil stratigraphic profiles of sampling sites Isabela C1 and C2, which correspond to locations on the Eastern slopes of Sierra Negra Volcano (as depicted in Fig. 1.17). The profiles cover 150 cm (60 in) of soil depth. Note in site C1 how the soil ages increase with depth, from 503 years before present in the recent topsoil to 845 yr for the indurated piece in the upper lapilli and to 1884 yr in the top of the paleosol (Fig. 1.20 left). By contrast, in site C2, the soil ages increase from 467 yr in the recent topsoil to 1812 yr in the recent subsoil and to 4690 yr in the top of the paleosol. Now, take a close look at the lava-tube Cueva de Sucre (see image in Fig. 1.14, bottom left), which is located 60 m (200 ft) away from sampling site C2; for didactic purposes, this pyroduct is depicted in Fig. 1.20 under the soil profile at C2 (bottom right). Be aware that, at the bottom of the cave, fossils of the extinct Galápagos giant rat (*Megaoryzomys*) have been found (calibrated ages of fossils=4205 yr or 5030 yr before present; after Zehetner et al. 2020).

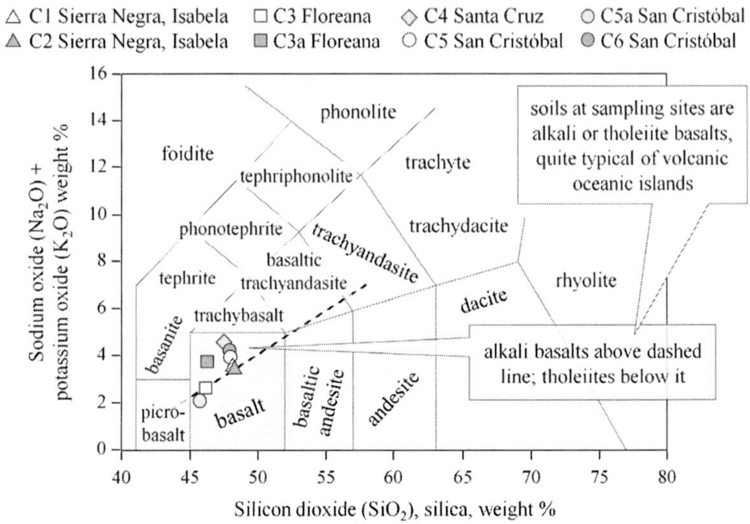

Mineralogy of the scoria underlying the most recent soils at sampling sites:

site	amorphous weight %	plagioclase weight %	pyroxene weight %	olivine weight %	hematite weight %	mica weight %
C1	85	9	6	–	–	–
C2	83	9	8	–	–	–
C3	42	17	24	17	–	–
C3a	59	15	11	15	–	–
C4	36	39	14	10	1	–
C5	43	33	13	7	4	traces
C5a	18	38	24	19	1	traces
C6	32	40	14	3	8	3

↓ iron-bearing forsterite $(MgFe)_2[SiO_4]$

↓ close to diopside $CaMg[Si_2O_6]$

↓ andesine: NaCa $[(AlSi)_4O_8]$

Fig. 1-19 Geochemical characteristics (top) and mineralogy (bottom) of the scoria underlying the soils at sample sites in Isabela (C1, C2), Floreana (C3, C3a), Santa Cruz (C4), and San Cristóbal (C5, C5a, C6) islands. *alkali* basalt = basic, with a relatively higher content of sodium + potassium oxides with respect to the content of silicon dioxide; *tholeiite* basalt = relatively higher content of silicon dioxide with respect to the content of sodium + potassium oxides. Note that the scorias at C1 to C6 were geochemically and mineralogically comparable, regardless of the climate zones and/or type of vegetation typical to the sites in which they were collected (as per Figs. 1.17 and 1.18). Layout: GPC (as per Paz-y-Miño-C and Espinosa 2024a), redrawn and modified after Zehetner et al. (2020).

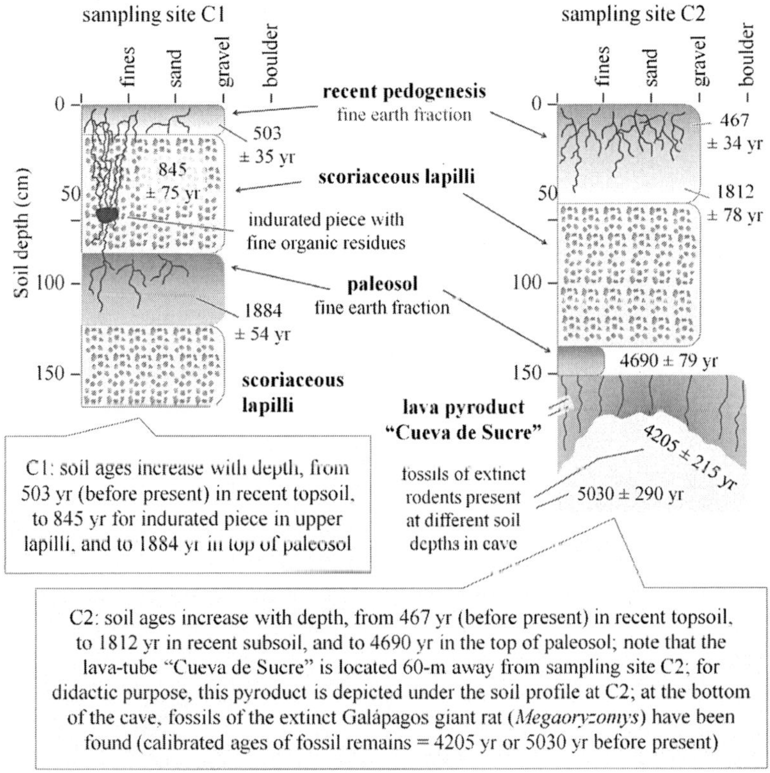

Fig. 1-20 Soil stratigraphic profiles of sampling sites Isabela C1 and C2 (Eastern slopes of the Sierra Negra Volcano, as depicted in Fig. 1.17). Layout: GPC (as per Paz-y-Miño-C and Espinosa 2024a), redrawn/modified after Zehetner et al. (2020).

Zooming Out: Islands Visible from Space

Can satellites spot the Galápagos Islands? Can astronauts see the archipelago from space? Yes, and yes. NASA's Mercury-bound Messenger spacecraft captured on August 2, 2005, an impressive image of Earth in which the archipelago can be spotted at a distance (Fig. 1.21). Close-up photographs of individual islands also exist and have been taken via satellites (Terra, Landsat 7) or directly by astronauts at the International Space Station (ISS; Figs. 1.22-1.23; after Paz-y-Miño-C and Espinosa 2024a). One of the most stunning shots comes from 350 km (220 mi) above our planet and was obtained by the ISS' Expedition 25 crew on October 28, 2010 (Fig. 1.23). In it, the Soyuz TMA-19 spacecraft (COIO3) is docked to the ISS (center of image), and the islands of Isabela (Albemarle), Fernandina (Narborough), Santiago (James or San Salvador), and part of Santa Cruz (Indefatigable) appear "upside down" with respect to the conventional North—see outlined silhouettes of islands in the magnification of the photo, bottom of Fig. 1.23. Keep in mind that all Figs. in this section come from Paz-y-Miño-C and Espinosa (2024a).

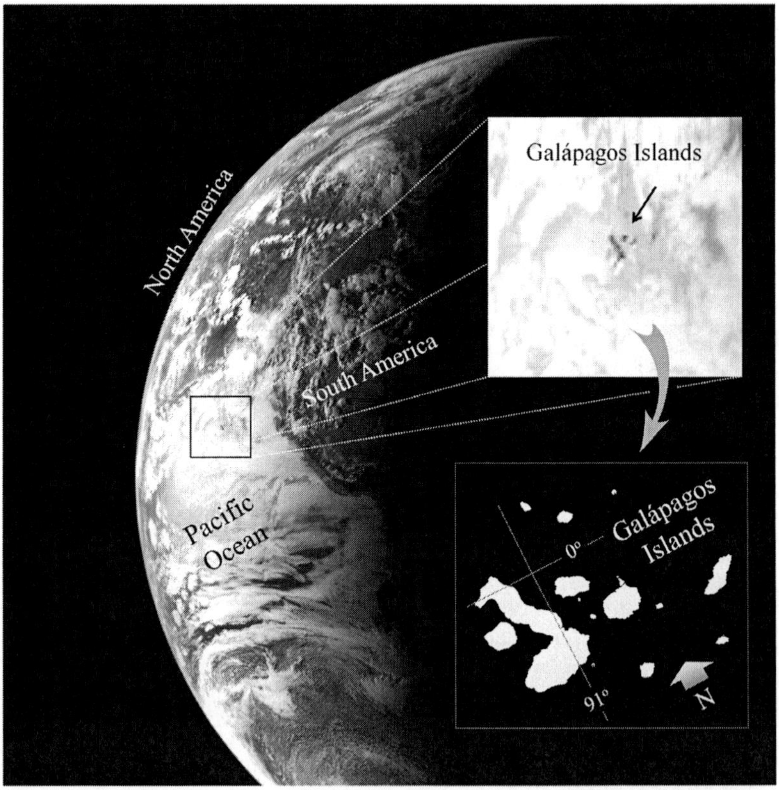

Fig. 1-21 Earth and the Galápagos Islands. Image captured by NASA's (National Aeronautics and Space Administration) Mercury-bound Messenger spacecraft (350 km/220 mi distance from our planet); August 2, 2005 (link in Appendix A). Note the position of North America (top left) and South America (center; its Northwestern coast), the Pacific Ocean, and, therefore, the location of the Galápagos Islands. The zoom-in inset (top right) is factual; in it, at least five islands can be spotted: Fernandina (Narborough), Isabela (Albemarle), Santiago (James or San Salvador), Santa Cruz (Indefatigable), and San Cristóbal (Chatham). Can you see them? Figure layout: GPC.

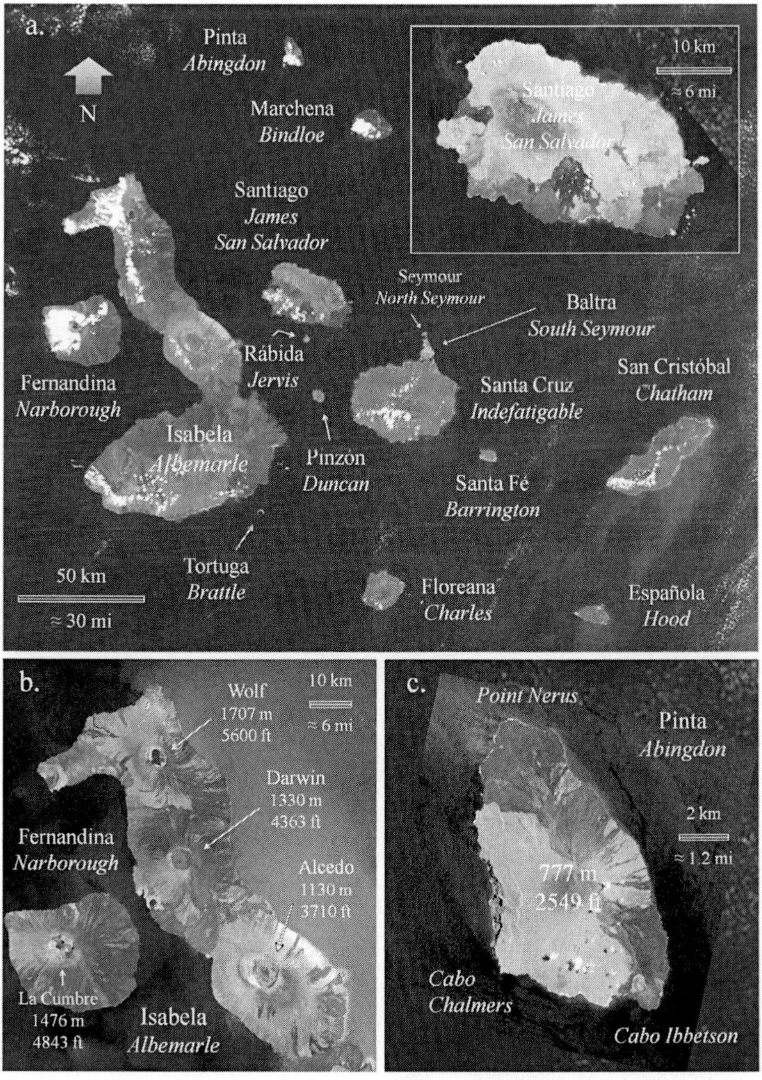

Fig. 1-22 Satellite and International-Space-Station (ISS) images of the Galápagos Islands. a. NASA's Terra satellite, Moderate-resolution Imaging Spectroradiometer (MODIS), March 12, 2002; top-right Santiago (James or San Salvador), photo by ISS' Expedition Crew 33, October 18, 2012. b. Isabela (Albemarle) and Fernandina (Narborough) with volcanoes Wolf, Darwin, Alcedo, and La Cumbre, NASA's Landsat 7 satellite, October 16, 2001. c. Pinta (Abingdon), NASA's Enhanced Thematic Mapper (ETM+), Landsat 7 satellite, April 7, 2009.

Fig. 1-23 International Space Station (ISS) overpassing the Galápagos Islands. Soyuz TMA-19 spacecraft (COIO3) is docked to the ISS (center); the islands of Isabela, Fernandina, Santiago, and part of Santa Cruz appear "upside down" with respect to the conventional North (notice image magnification and islands' silhouettes in bottom). Photo by Expedition Crew 25 of ISS, October 28, 2010 (credit NASA; link in Appendix A). Layout: GPC.

In Fig. 1.24 we show two exemplar satellite images of Wolf Volcano, Northern Isabela (Albemarle) Island (a, b), and a photo of the archipelago taken from the International Space Station (c). In May 2015, Wolf erupted after a 33-year pause, sent ash and gases 15 km (9 mi) into the sky, and lava flowed toward the East and Southeast slopes; it reached the ocean by early June. The image in Fig. 1.24a was taken by the Terra satellite on June 11, 2015 (NASA 2015). To determine if the ground had moved as a consequence of the eruption, two radar images were processed (May 3 and June 23, 2015) by satellite Sentinel 1A—European Space Agency, ESA—and later overlapped (Fig. 1.24b). The resulting dissimilar tone in light-dark bands on the merged image (thick arrow in Fig. 1.24b), known as interferometric fringes, confirmed that the ground had indeed been displaced, a maximum of 28 cm (11 in), by the eruption (EFFIGIS 2016).

The photo in Fig. 1.24c was captured by the Expedition 38 crew of the ISS. The oblique shot was taken with a 180-mm lens on November 11, 2013 (NASA 2013). The islands of Fernandina (Narborough) and Isabela (both with their volcanoes visible) appear on the foreground bottom, while Marchena (Bindloe), Santiago (San Salvador or James), and Santa Cruz (Indefatigable) are in the background. Note the considerable size of Isabela: 130 km (≈80 mi), as marked by a dashed line (Fig. 1.24c).

But the splendor of both the archipelago and what satellites can capture of it is shown in Fig. 1.25: an image procured by Copernicus Sentinel-2 (ESA) on September 3, 2020. As a two-satellite mission, Sentinels carry high-resolution cameras (10-m/30-ft) that scan Earth's surface in 13 spectral channels/bands (visible/near infrared and short-wave infrared). The satellites help monitor land use, vegetation cover, soil, inland waterways, and the coastal areas of our planet (ESA 2021).

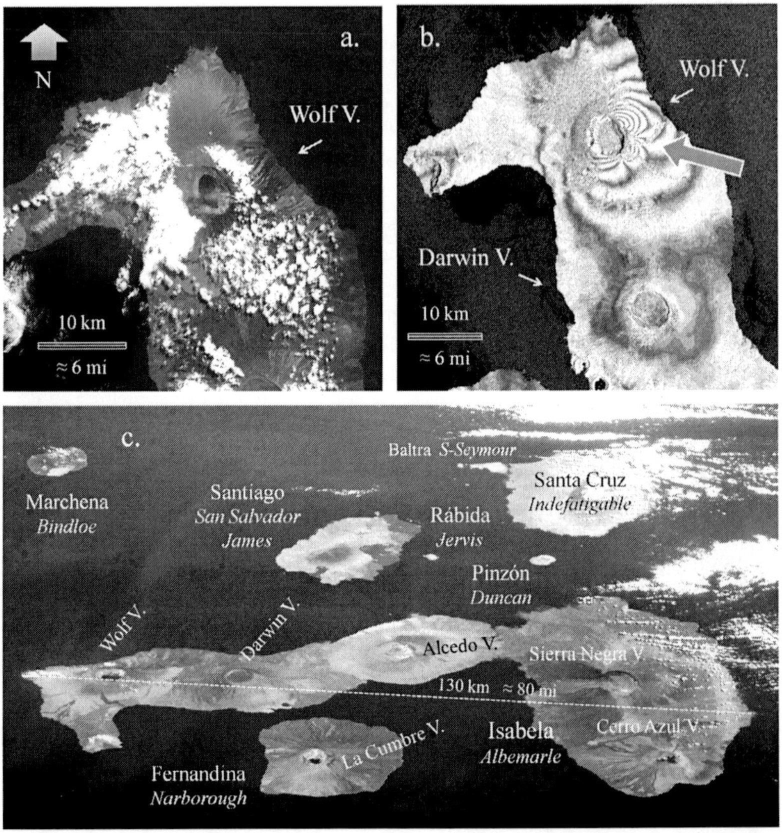

Fig. 1-24 Exemplar images of the Galápagos Islands generated by modern technologies. a. Eruption of Wolf Volcano (Isabela Island); Advanced Spaceborne Thermal Emission and Reflection Radiometer (ASTER) on board of the Terra Satellite; NASA, June 11, 2015. b. Interferogram resulting from overlapping two radar images obtained by Sentinel 1A satellite, May 3 and June 23, 2015 (processed by EFFIGIS GeoSolutions; NASA, US Geological Survey, and European Space Agency; Copernicus program); thick arrow points at different tones in light and dark bands (after the overlap), known as interferometric fringes. c. Oblique view of the Galápagos Islands from the International Space Station (November 11, 2013); Northwest (bottom left), Southeast (top right); note the size of Isabela 130 km (≈80 mi).

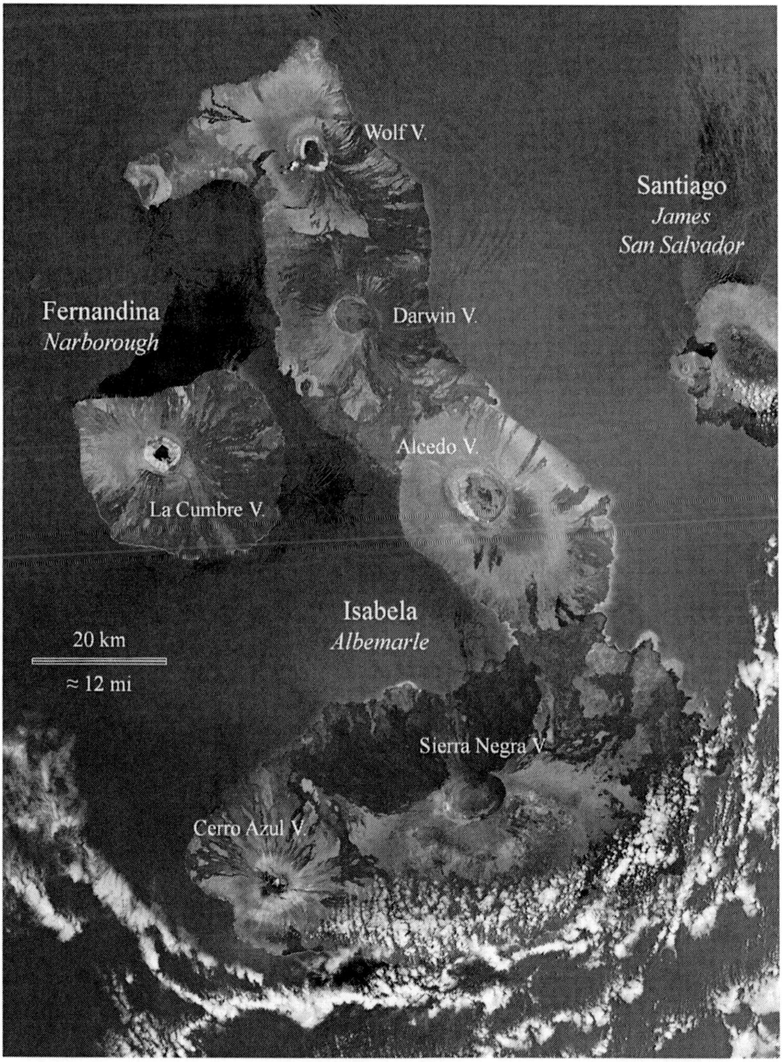

Fig. 1-25 The splendor of Fernandina, Isabela, and Santiago as captured by Copernicus Sentinel-2 of the European Space Agency, September 23, 2020 (link in Appendix A).

An instance of vegetation-cover satellite-scanning of the Galápagos is shown in Fig. 1.26. The scan comes from February 27, 2017, and was obtained by the miniaturized ESA satellite Proba-Vegetation (Proba-V); its main camera can capture images in bands up to 2250-km width (≈1400 mi) in the blue, red, near-infrared, and mid-infrared wavebands (ESA 2019). The spatial resolution in Fig. 1.26 is 100 m (≈330 ft); note that in that image we have marked with dashed lines the contour of the vegetation patches (after Paz-y-Miño-C and Espinosa 2024a).

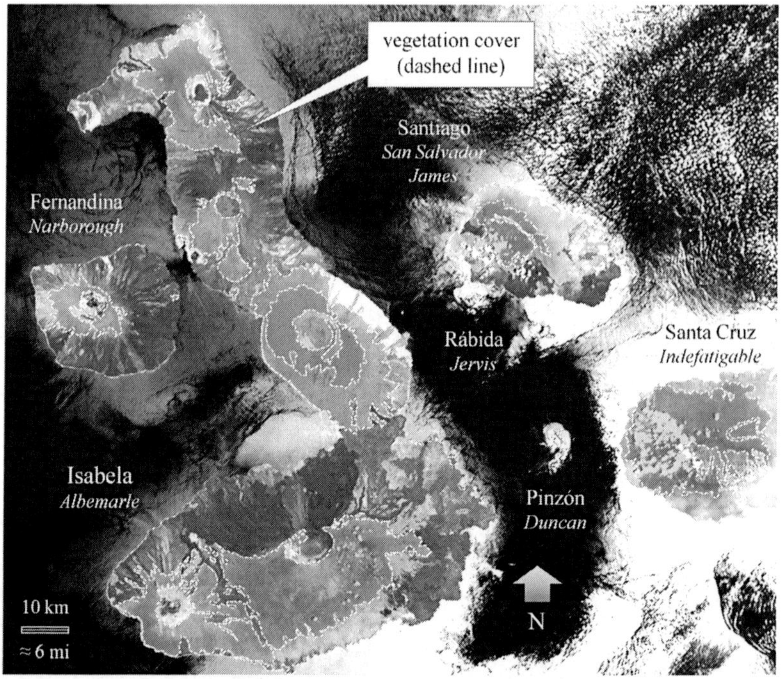

Fig. 1-26 Proba-Vegetation (Proba-V) satellite image highlighting the Galápagos vegetation cover. European Space Agency/Belspo, Produced by VITO Remote Sensing Belgium (2017).

Sketching Today's Archipelago

Can we sketch-simplify essential geographic information about today's archipelago? Of course. You should know by now that, for practical or didactic reasons, we refer to the Galápagos as islands, archipelago, or isles. And that, in this volume, we use Spanish and English names (the most applied nowadays) when referring to individual islands. In Figs. 1.27 to 1.31, we provide our own sketches of the islands (with labels of relevant geographic features), as per their relative location in the archipelago and our favored educational approach. Figs. 1.27 to 1.31 are modifications of Paz-y-Miño-C and Espinosa (2024a); they include the latest census data for the human populated islands (as per INEC 2022).

(1) *East and Southeastern Islands* (Fig. 1.27): San Cristóbal (Chatham), Floreana (Charles) and Española (Hood). These are the oldest islands in the archipelago, ranging from 1.5 to 4.0 million years since formation. The largest in the group, San Cristóbal (558 km^2/215 sq mi), is also the tallest (723 m/2370 ft above sea level). Its main settlement is Puerto Baquerizo Moreno (in the Southwest). By 2022, the total human population on the island reached 8,300 inhabitants (INEC 2022). San Cristóbal Airport (SCY) is proximate to Puerto Baquerizo Moreno. Floreana and Española are much smaller, low-elevation islands. Only Floreana is populated (+100 inhabitants by 2022) and has a small heliport.

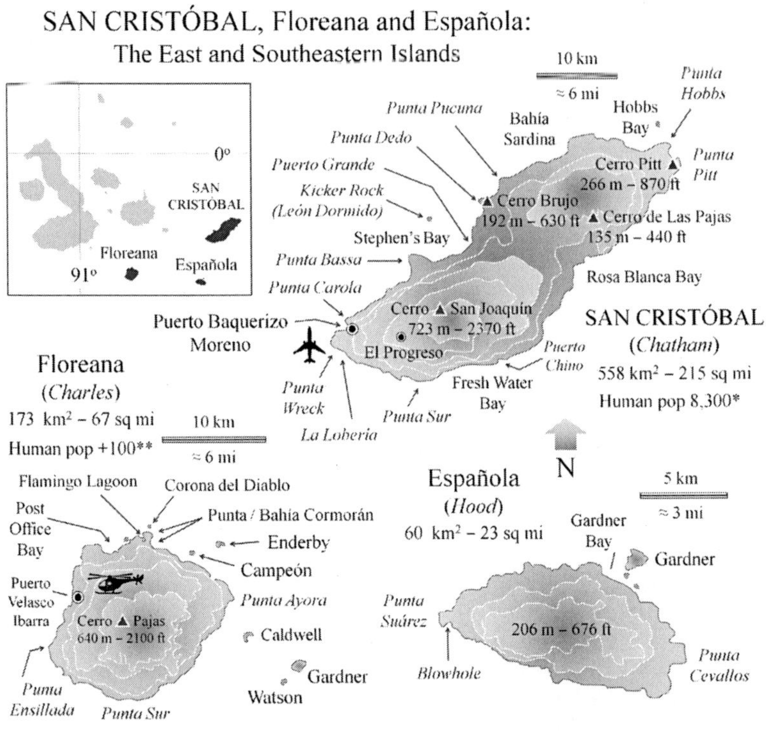

Fig. 1-27 Galápagos' East and Southeastern Islands (San Cristóbal, Floreana, and Española). *Census; **estimated based on INEC (2022). Sketch: GPC.

(2) *Central Islands* (Fig. 1.28): Santiago or San Salvador (James), Santa Cruz (Indefatigable), Baltra (Seymour South or South Seymour), Seymour (Seymour North or North Seymour), Rábida (Jervis), Pinzón (Duncan), and Santa Fé (Barrington). Middle-age islands, ranging from 0.8 to 2.9 million years since formation. Santa Cruz is the largest (986 km^2/381 sq mi) and tallest (864 m/2835 ft) in the group. Its main settlement is Puerto Isidro Ayora (in the South). The total human population on the island, by 2022, reached 17,233 inhabitants. The closest runway is Seymour Galápagos Ecological Airport (GPS), located on Baltra Island (in the North). Santiago is the second largest (585 km^2/226 sq mi) and tallest (907 m/2975 ft) island in the group; the rest are small, low-elevation ones.

SANTIAGO, SANTA CRUZ, Baltra, Seymour, Rábida, Pinzón and Santa Fé: The Central Islands

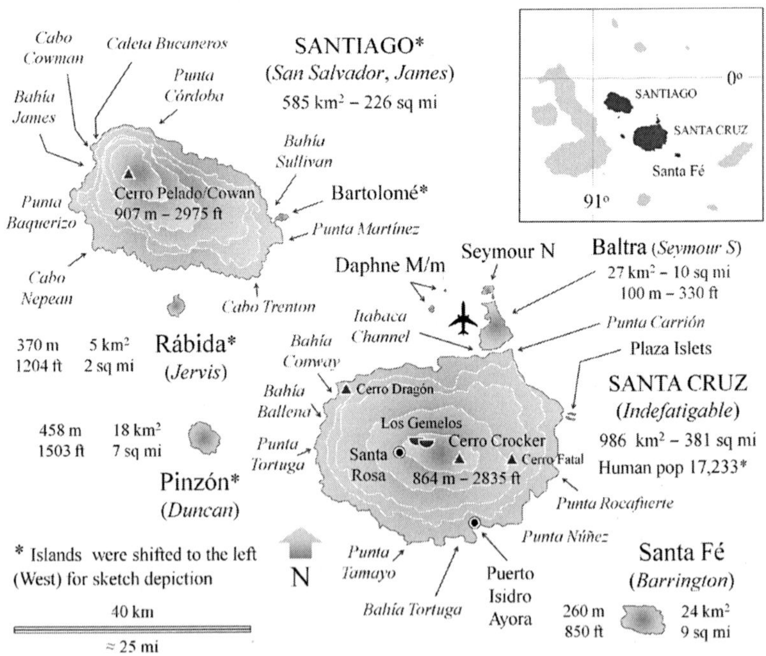

Fig. 1-28 Galápagos' Central Islands (Santiago, Santa Cruz, Baltra, Seymour, Rábida, Pinzón, and Santa Fé). *Census INEC (2022). Sketch: GPC.

(3) *Western Islands* (Fig. 1.29): Isabela (Albemarle) and Fernandina (Narborough). Younger islands, ranging from 0.032 to 0.8 million years since formation. Isabela is the largest (4670 km^2/1806 sq mi) and tallest (Wolf

Volcano = 1707 m/5600 ft) island in the archipelago. It has five volcanoes: Wolf, Darwin (1330 m/4363 ft), Alcedo (1130 m/3710 ft), Sierra Negra (1124 m/3688 ft), and Cerro Azul (1689 m/5541 ft). Its main settlement is Puerto Villamil (in the South). Total human population on the island, by 2022, reached 3,050 inhabitants. A small runway is located near Puerto Villamil [this runway shall eventually turn into a fully functional airport comparable to Baltra's GPS or San Cristóbal's SCY, above]. Fernandina is the youngest island in the Galápagos (0.032 to 0.06 million years since formation). Its volcano, La Cumbre, reaches 1476 m/4843 ft elevation.

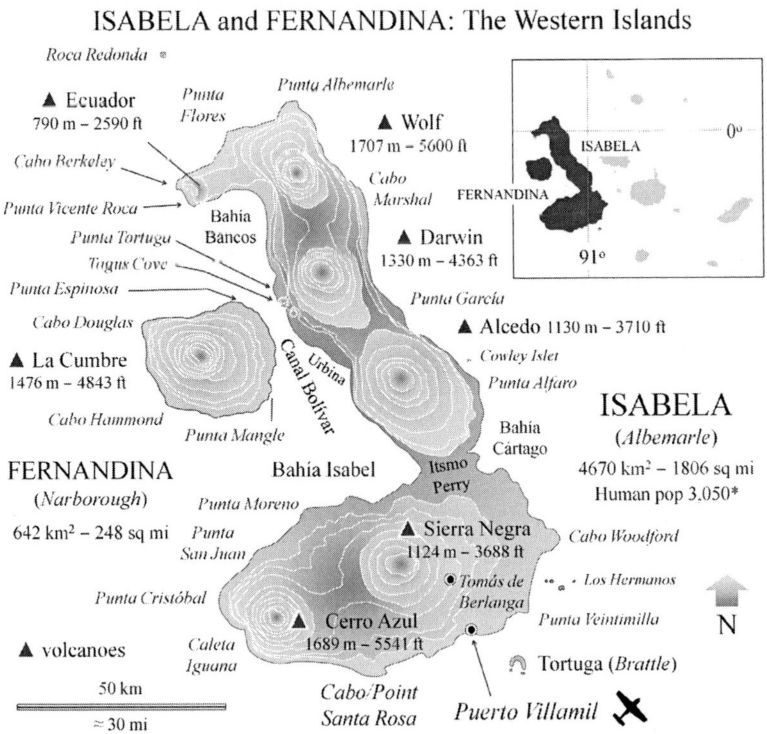

Fig. 1-29 Galápagos' Western Islands (Isabela and Fernandina). *Census INEC (2022). Sketch: GPC.

(4) *Northern Islands* (Fig. 1.30): Marchena (Bindloe), Pinta (Abingdon), and Genovesa (Tower). Fairly young islands, ranging from 0.3 to 0.8 million years since formation. Small, low-elevation islands. The largest in this group is Marchena (130 km^2/50 sq mi), the tallest Pinta (777 m/2549

ft), and the smallest and lowest—yet one of the most beautiful islands in the Galápagos—Genovesa!

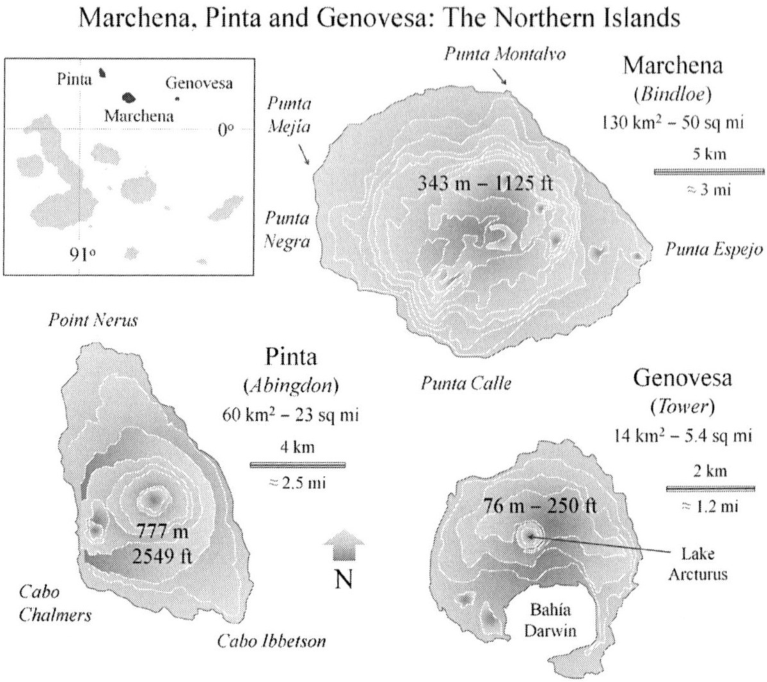

Fig. 1-30 Galápagos' Northern Islands (Marchena, Pinta, and Genovesa). Sketch: GPC.

(5) *Northwestern Islands* (Fig. 1.31): Darwin (Culpepper) and Wolf (Wenman). Younger to middle-age islands, ranging from 0.4 to 1.7 million years since formation. They are the smallest and lowest-elevation islands in the archipelago and the most distant with respect to the central islands. Their reefs and rocky-shallow bottom-of-the-ocean are among the most spectacular in the world (biodiversity, biomass). The iconic "Darwin's Arch" (Fig. 1.31) collapsed in 2021; nowadays, it is a pillar formation and has been renamed "Los Pilares de la Evolución" or "Pillars of Evolution", a label that may endure.

To close this section, can you make your own sketch of the Galápagos Archipelago? Can you label the islands correctly, i.e. with their most used Spanish and English names? Which islands have caldera-volcanoes? Why? Can you label such volcanoes on your sketched map? Can you identify on your sketch the human-inhabited islands, the locations of the main towns, or cities-in-the-making?

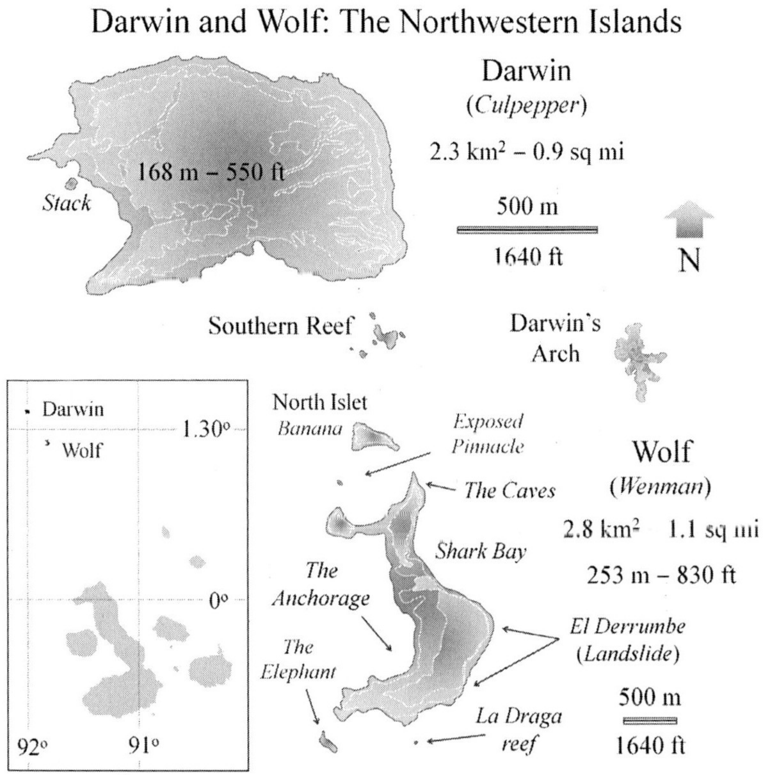

Fig. 1-31 Galápagos' Northwestern Islands (Darwin and Wolf). "Darwin's Arch" (middle right) collapsed in 2021; a pillar formation remains. Sketch: GPC.

The Ocean and Its Turbulent Waters

How does the surrounding ocean influence the Galápagos Archipelago? As we explained above, the Galápagos Islands exist, ultimately, as a cause-and-effect outcome of the interplay between Earth-crust-above and hotspot-below. Thus, one should infer that the islands' geological past and present, combined with their current geographic location and interactions with the ocean, are crucial to understanding their biological processes and patterns, as well as their biogeographic affinities (i.e. the colonization of the archipelago by all organisms; their occupation of marine or terrestrial habitats; their phylogenetic relatives in the continent). Thus, let us take a closer look at the ocean and its turbulent waters; the "arrivals of life" to the Galápagos co-depend on them (Paz-y-Miño-C and Espinosa 2024a).

In Fig. 1.32, we sketch-summarize ocean currents in the Tropical Eastern Pacific (TEP), with emphasis on the Galápagos. Note in Fig. 1.32a (top) that the North Pacific Gyre follows a clockwise direction, while the South Pacific Gyre follows a counterclockwise direction (gyres' water-circulation patterns result from diverse factors, including the direction of

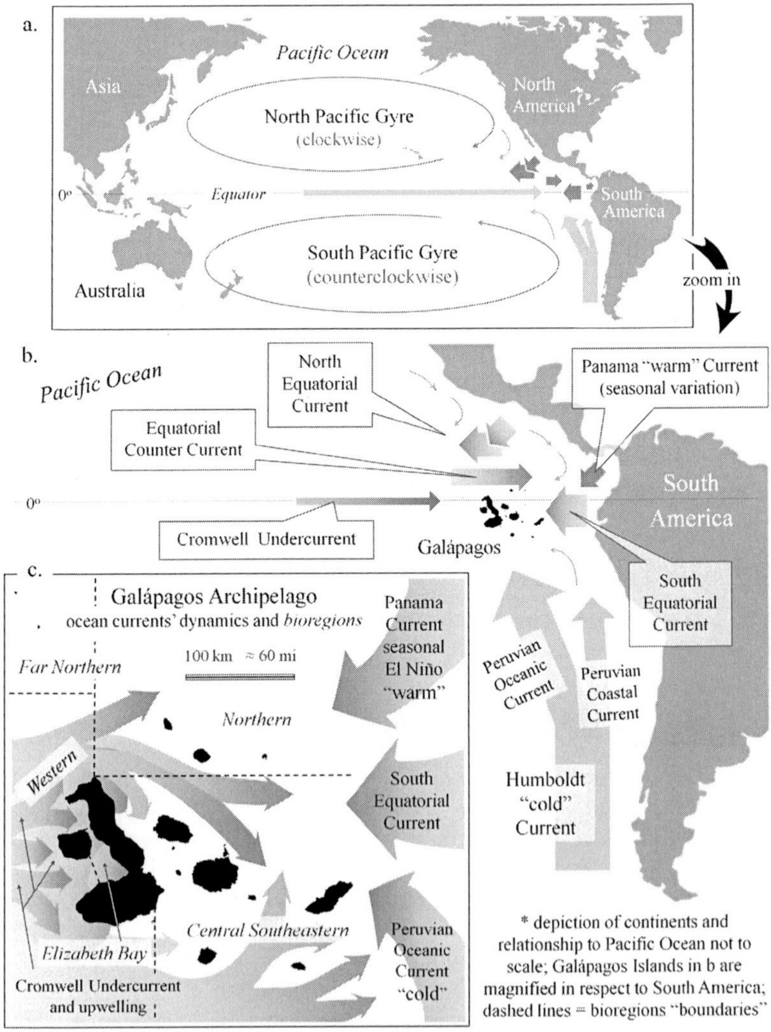

Fig. 1-32 Ocean currents in the Tropical Eastern Pacific and Galápagos. a. North and South Pacific Gyres. b. Tropical Eastern Pacific. c. Galápagos. Sketch: GPC.

Earth's rotation). The Galápagos are located just in the middle, on the equator. When zooming in and examining the TEP, you should spot several ocean currents converging toward the archipelago (Fig. 1.32b,c). Locate the North Equatorial Current and its East-to-West direction with respect to the Galápagos (Fig. 1.32b); do the same with the Panama "warm" Current (which arrives to the islands from the Northeast), South Equatorial Current (from the East), and the Peruvian "cold" Oceanic Current (also known as Humboldt, which comes from the Southeast). Incoming from the opposite direction, locate the Equatorial Counter Current (which runs West-to-East) and the Cromwell Undercurrent (with the same West-to-East movement but as a deep-ocean current; Fig. 1.32b). Upon collision with the archipelago, these currents follow complex patterns (Fig. 1.32c; Paz-y-Miño-C and Espinosa 2024a,b).

Fig. 1.32c (bottom) illustrates the ocean currents' dynamics around the Galápagos Archipelago. The waters move in predictable directions and are subject to seasonal variation, like El Niño or La Niña events (i.e. warm-cool fluctuations in climate patterns in the Pacific Ocean), which can bring unusually warm- or cool-surface waters to the islands, respectively. Both the Panama "warm" Current and the Peruvian "cold" Oceanic Current interact during El Niño or La Niña years. Depending on the intensity of the interaction, the Galápagos marine and terrestrial ecosystems can be affected by extreme weather (as discussed below, "The 'Hot' and 'Cool' Seasons", Figs. 1.43-1.44).

Notice in Fig. 1.32c how the Cromwell Undercurrent (which travels at depths >3000 m/10,000 ft) bumps into the Galápagos platform (a 1000-m-/3300-ft-depth plateau on which the islands rest; as shown in Figs. 1.5, 1.8, and 1.13, above). The collision, and/or wind patterns, cause deep-seawater upwelling [we discuss the upwelling below], which brings nutrients to the surface. In fact, this entire region (around Fernandina and West of Isabela Islands) corresponds to two bioregions (i.e. areas with distinctive biodiversity, biogeography, and ecology), the most nutrient-rich in the Galápagos: Western and Elizabeth Bay! Three additional bioregions are recognized in the Galápagos marine environments, all influenced by ocean currents and seasonality: Central Southeastern, Northern, and Far Northern (as characterized by reef-fish and macro-invertebrate communities; Edgar et al. 2004; comprehensive discussion in Paz-y-Miño-C and Espinosa 2024a,b). In Box 1.3, we provide supplementary information about these bioregions [you should become familiar with them]. Galápagos' bioregions are central to discussing the archipelagos' biodiversity, conserving, and managing natural resources, planning tourism activities, and designing contingency responses to impacts caused by climate change.

The Upwelling

Seasonal upward surges of cold, nutrient-rich deep-sea waters—upwelling—fuel the biological cycles of microorganisms (primary production), which, in turn, sustain all other life forms in the islands (Fig. 1.33; Forryan et al. 2021). What triggers the upwelling? Two hypotheses have been proposed. (*Hi*) The ocean-current-driven upwelling (1960s) suggests that the seawater surge on the West side of the Galápagos is driven by the collision of the sub-surface Equatorial Undercurrent with the archipelago (Fig. 1.34). Intensity in the upwelling depends on remote/stochastic events (currents' interactions; seasonality El Niño/La Niña). By contrast, (*Hii*) the wind-driven upwelling (2020s) suggests that the seawater surge is driven by local Northward winds, which activate turbulence at upper ocean fronts to the West of the archipelago; the turbulence forces deep, cold waters toward the surface while warm waters go down (Fig. 1.34). Intensity in the upwelling depends on local atmosphere-and-ocean interactions (Forryan et al. 2021).

Scientists concur that any sound preservation, restoration, or management program of the Galápagos' habitats must take into consideration the close interaction between the "seasonal upward surges of cold, nutrient-rich deep-sea waters" and the archipelago's marine and terrestrial environments. Ultimately, the upwelling nourishes sea and land organisms, from micro to macro life. The upwelling keeps the essence of the Galápagos alive (Paz-y-Miño-C and Espinosa 2024b).

observation: upward surges of cold, nutrient-rich deep waters fuel the growth of microscopic organisms upon which the entire ecosystem thrives

inference: the long-term survival and reproductive success of both marine organisms and terrestrial life forms are bound to the upwelling

Fig. 1-33 The entire Galápagos' ecosystem is sustained by a seasonal oceanic upwelling. Ultimately, the upwelling nourishes sea and land organisms, from micro to macro life. The upwelling keeps the essence of the Galápagos alive. Digital sketches: GPC (as per Paz-y-Miño-C and Espinosa 2024b); concepts after Forryan et al. (2021).

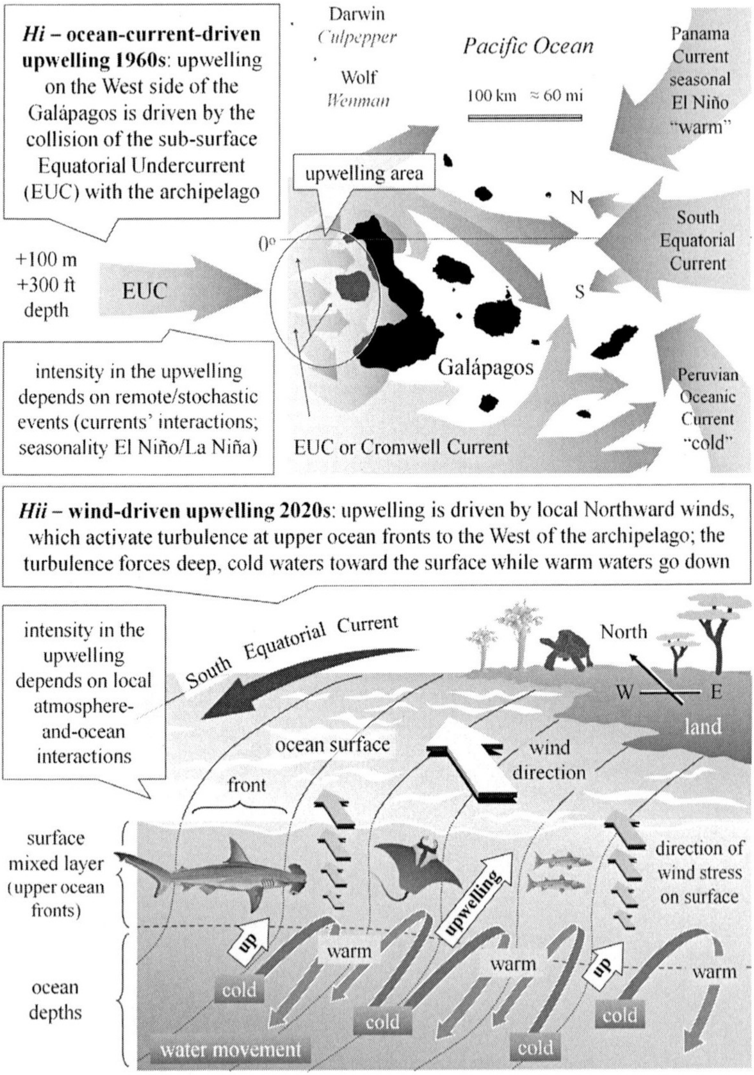

Fig. 1-34 Hypotheses about the drivers of the Galápagos' oceanic upwelling. *Hi* ocean-current-driven upwelling; *Hii* wind-driven upwelling. [For zoom-out depiction of ocean currents in the Tropical Eastern Pacific, see Fig. 1.32. For discussion about "hot" vs. "cool" seasons in Galápagos (El Niño/La Niña), see Figs. 1.43-1.44]. Layout: GPC (as per Paz-y-Miño-C and Espinosa 2024b); redrawn and modified after Forryan et al. (2021).

Box 1-3 Proposed bioregions within the Galápagos Marine Reserve

Based on offshore assessments of fish and macro-invertebrate diversity (i.e. species composition, relative abundance as per geographic occurrence), biologists have proposed the existence of five bioregions within the Galápagos Marine Reserve (sketch below): Far North (around Darwin and Wolf), North (around Pinta, Marchena, and Genovesa), West (on the West side of Fernandina and Isabela), Elizabeth Bay (the channel between Fernandina and Isabela), and Southeast (East side of Isabela, around Santiago, Santa Cruz, San Cristóbal, Floreana, and Española; Edgar et al. 2004; Kislik et al. 2017).

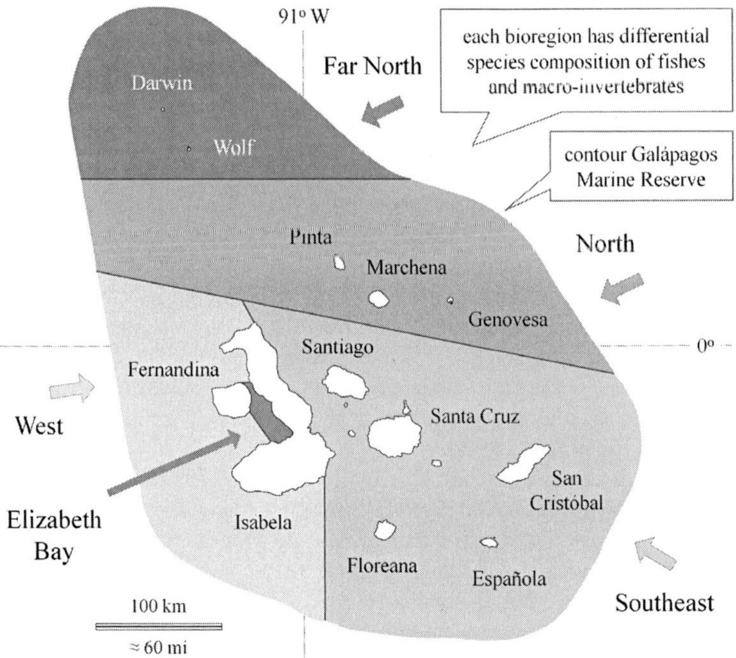

* bioregions have been characterized by Edgar et al. (2004) based on differential species composition (fish and macro-invertebrates) in the subtidal communities on the shallow reefs (fishes: 579 census transects, 50x10 m [≈164x33 ft], along contour lines 2–20 m [≈7–66 ft] depth; macro-invertebrates: 569 census transects, 50x2 m [≈164x7 ft], along the same depth-contour lines); the Far North bioregion is rich in fish species with Indo-Pacific affinity; the West bioregion has high numbers of endemic fish taxa with cool-temperate Peruvian affinity, as well as high macro-invertebrates diversity; the Elizabeth Bay bioregion is the richest in endemics; the Southeast bioregion includes multi-origin species with distinctive affinity with Panama and Central America; the North bioregion shares fish and macro-invertebrates ("overlap region") with both the Far North and Southeast (yet, in idiosyncratic and measurable proportions)

Each of the bioregions has distinctive fishes and/or macro-invertebrates with either unique endemism and/or discernible affinities with nearby biogeographic provinces, as follows: the Far North bioregion is rich in fish species with Indo-Pacific affinity; the West bioregion has high numbers of endemic fish taxa with cool-temperate Peruvian affinity as well as high macro-invertebrates diversity; the Elizabeth Bay bioregion is the richest in endemics; the Southeast bioregion includes multi-origin species with distinctive affinity with Panama and Central America; the North bioregion shares fish and macro-invertebrates with both the Far North and Southeast. Layout: GPC (Paz-y-Miño-C and Espinosa 2024b); redrawn/modified after Kislik et al. (2017) with input from Edgar et al. (2004).

Origins of the Galápagos Biota

Where do the archipelago's organisms come from? Three working hypotheses explain the origins of the Galápagos biota (Fig. 1.35): *Hi* overwater dispersal from mainland—in which organisms flew, swam, or were carried to the islands by ocean currents (Fig. 1.36); *Hii* stepping-stone dispersal from mainland (a variant of *Hi*)—in which organisms dispersed via the now-submerged paleo islands, or seamounts [or guyots; Box 1.1], and made it to the current archipelago after "stepping" from older to younger seamounts [when they were above-sea-level volcanoes]; and *Hiii* metapopulation vicariance—in which groups of organisms originated *in situ* during the breakup of widespread ancestral populations (in the deep geological past; reviews in Grehan 2001; Heads and Grehan 2021; Paz-y-Miño-C and Espinosa 2024a).

The metapopulation vicariance hypothesis (*Hiii*) is the most elaborate (here we follow Heads and Grehan 2021). In Fig. 1.35 (bottom), we explain the process that apparently gave origin to organisms *in situ* (rather than solely via arrivals of dispersing individuals from the mainland, *Hi* or *Hii*) by means of spatial separation (vicariance) of populations (called metapopulations), whose members belonged to a taxonomic group. Assume that at Time 1, an ancestral metapopulation was widely spread (intercontinentally or globally) in a single terrain; such terrain had two areas, A and B, which were relatively close to one another and had continuity between them. The "area phylogeny" for the distribution of the ancestral metapopulation can be represented by the notation "(A+B)".

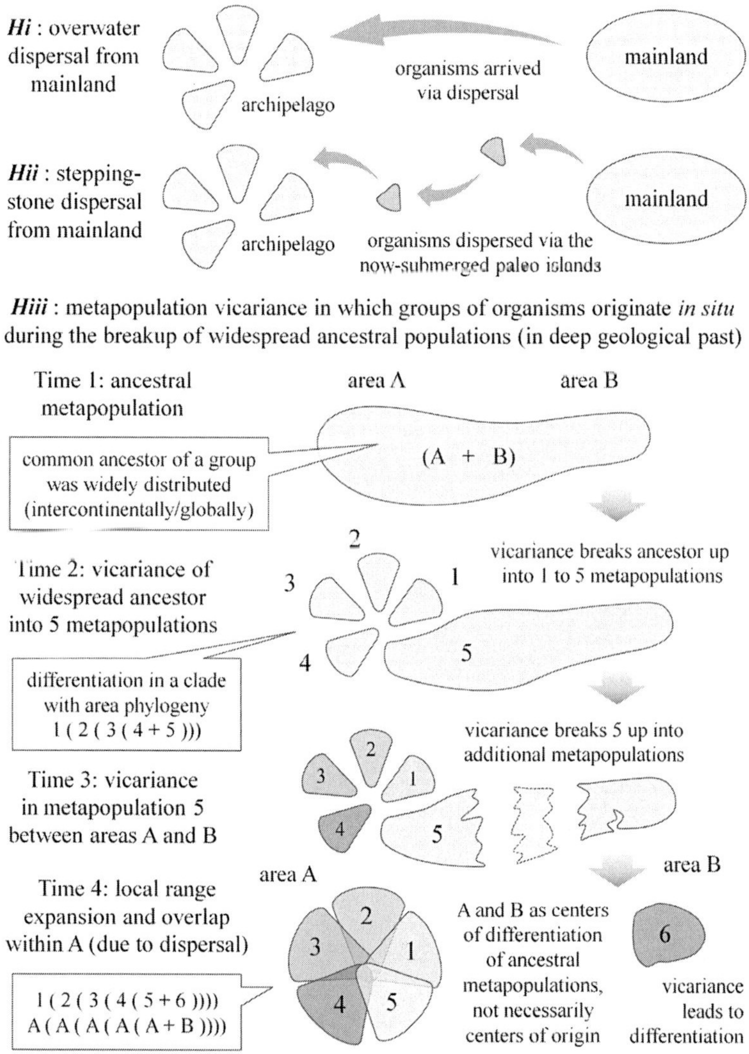

Fig. 1-35 Hypotheses (*Hi*, *Hii*, and *Hiii*) about the origins of the Galápagos biota. The overwater dispersal from mainland (*Hi*) or the stepping-stone dispersal from mainland (*Hii*) hypotheses are straightforward yet simplistic. The metapopulation vicariance hypothesis (*Hiii*) is the most elaborate (rationale in text). Layout: GPC (as per Paz-y-Miño-C and Espinosa 2024a), redrawn and modified after Heads and Grehan (2021).

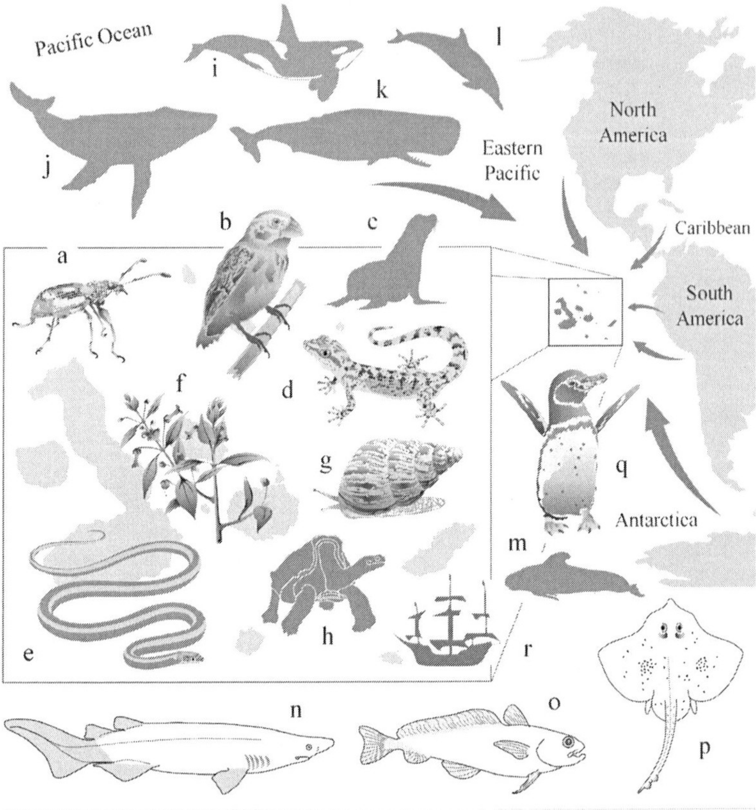

a Galápagos weevils (*Galapaganus*); **b** Galápagos/Darwin finches (*Geospiza*, *Camarhynchus*, *Certhidea*, *Platyspiza*); **c** sea lions (*Zalophus*); **d** leaf-toed geckos (*Phyllodactylus*); **e** racer snakes (*Pseudalsophis*); **f** snapdragon shrubs (*Galvezia*); **g** land-snails (*Naesiotus*); **h** giant tortoises (*Chelonoidis*); cetaceans: **i** killer whales/orcas (*Orcinus*), **j** humpbacks (*Megaptera*), **k** sperm whales (*Physeter*), **l** common dolphins (*Delphinus*), **m** pilot whales (*Globicephala*); deep-sea fishes: **n** Echinorhinidae, **o** Etmopteridae, **p** Rajidae; **q** Galápagos penguins (*Spheniscus*); and, of course, **r** Tomás de Berlanga's and his crew's adrift, yet natural arrival in 1535 (Chapter Two), which triggered human colonization of the islands.

Fig. 1-36 The "overwater dispersal from mainland hypothesis" (*Hi*) is often invoked to explain the origins of the Galápagos biota. *Hi* is simplistic; it suggests that the archipelago's current life forms either flew, swam, or were carried to the islands by ocean currents during the recent geological past. As shown in Fig. 1.35, hypotheses *Hii* "stepping-stone dispersal from mainland" (a variant of *Hi*) and *Hiii* "metapopulation vicariance" offer more elaborate, yet conceivable, explanations for the origins and evolution of the Galápagos' organisms. Sketches: GPC.

At Time 2, fortuitous events fragmented this well-connected ancestral metapopulation into five smaller groups, i.e., metapopulations 1 to 5. The model assumes that, over time, geographic isolation among these groups would lead to differentiation (i.e. genetic, phenotypic). But before addressing that, keep in mind that we can represent the area phylogeny of the five metapopulations at Time 2 as A(A(A(A+B))), or using the number labels as 1(2(3(4+5))).

At Time 3, vicariance further broke metapopulation 5 into additional metapopulations, a total of three, as illustrated in the sketch in Fig. 1.35. One of these metapopulations, marked by a dotted line and placed in the middle of the sketch, became extinct (say, its little island sank).

By Time 4, natural expansion due to dispersal of metapopulations 1-5 led to their overlap within area A, while metapopulation 6 emerged within the left-over fragment of area B. Thus, areas A and B functioned as centers of differentiation for ancestral metapopulations, not necessarily centers of origin. Vicariance led to overall *in situ* differentiation and formation of distinctive taxa. The area phylogeny of the six metapopulations at Time 4 would be A(A(A(A(A+B)))) or 1(2(3(4(5+ 6)))), as shown in Fig. 1.35.

Hypotheses *Hi* and *Hii* (dispersal) assume that the Galápagos Islands emerged just a few million years ago, which is conceivable for the current archipelago (0.032-4 Ma). But, as we discussed earlier, the Galápagos hotspot has been active for 139 Ma, and various igneous plateaus, provinces, and plates have drifted nearby, or passed just above it, before the current Galápagos existed. Thus, the metapopulation vicariance model (*Hiii*) suggests that taxa (species, genera) can persist almost indefinitely in an ocean region ["hovering around"], as long as new islands continue to emerge (Heads and Grehan 2021). In Fig. 1.37, we summarize 10 patterns of biogeographic affinities of the Galápagos biota (219 taxa) inferred from associations between taxa distributions and the geological past. These patterns coincide with the space-time trajectory of various tectonic structures, including: East Pacific Rise; West Pacific and American subduction zones; multiple igneous plateaus in the Pacific; Alisitos terrane (Baja California); Guerrero terrane (Mexico); rifting of North and South America; Caribbean Plateau; Andean uplift; and Romeral fault system (Heads and Grehan 2021). It is plausible that numerous Galápagos taxa originated during the fragmentation of ancestral metapopulations, which were widely distributed in the Pacific, the nearby proto-Americas, the proto-Caribbean platform, or elsewhere. The vicariance model attributes the geographic range of taxa as well as their overlapped distributions to *dispersal*, while phylogenetic differentiation results from *vicariance* (Fig. 1.35). By contrast, *Hi* and *Hii* attribute both the range of taxa distribution and phylogenetic differentiation to dispersal. *Hi* and *Hii* assume dispersal events to be rare and infrequent, a prolonged absence of gene flow, and differentiation primarily due to chance, with no major environmental factors breaking up populations. This differs from vicariance since it affects entire communities, which become fragmented (isolated) due to substantial environmental change (e.g. tectonics; Heads and Grehan 2021).

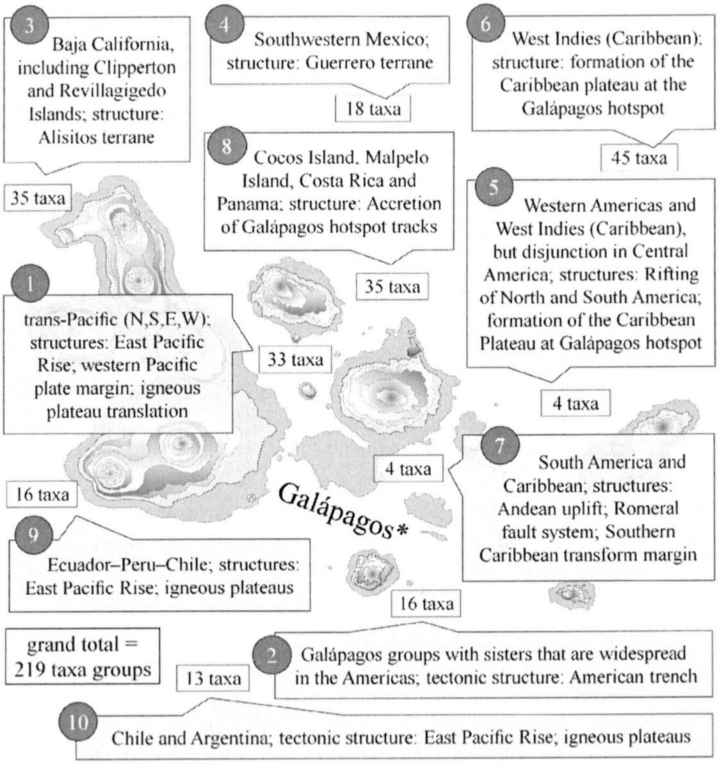

Fig. 1-37 Biogeographic affinities of the Galápagos biota inferred from associations between taxa distributions and the [distant] geological past. *Galápagos archipelago (background) depicts island configuration during the last glacial maximum, just 20,000 years ago, with contour shores at 125 meters below today's sea level (details in Fig. 1.12); however, the 1 to 10 patterns of biogeographic affinities correspond to global, much older connections (considering the activity of the Galápagos hotspot, up to 139 Ma) than the current islands in which the oldest (San Cristóbal, *Chatham*) dates back to 2.4-4 Ma; the present archipelago has been preceded by paleo-islands (now submerged), which themselves were preceded by even older land masses located proximate or distant to one another, or to the mainland, as the hotspot continued to shape the above Earth's crust (Fig. 1.4). Layout: GPC, after Heads and Grehan (2021).

Keep in mind that the vicariance model brings perspective to our understanding of the origins of the Galápagos biota. Because the current islands are recent (youngest Fernandina = 0.032-0.06 Ma; oldest San Cristóbal = 2.4-4 Ma; and Española = 3-3.5 Ma), biogeographic relations tend to be dated according to the age of the present archipelago. Such approach disregards that today's islands were preceded by earlier ones (now seamounts/guyots) and that ancestors of some contemporary taxa likely "hovered around" (above) and diversified *in situ* as new islands continued to emerge. Thus, although many taxa—perhaps the majority—arrived via dispersal from proximate or distant mainland, not all did (Heads and Grehan 2021).

In Table 1.3, we list examples of species colonizations (terrestrial and marine) of the archipelago, including birds, reptiles, mammals, arthropods, and plants. Scientists rely on genetic markers (e.g. mtDNA mitochondrial DNA; rRNA ribosomal RNA; nDNA nuclear DNA; rDNA ribosomal DNA; cpDNA chloroplast DNA) to reconstruct phylogenies, infer relationships among members of diverse taxa and their closest relatives on the mainland (e.g. Central/South/North America, the Caribbean, or elsewhere), and deduce the number of successful colonizations of the islands. Although the majority of taxa seem to have colonized the archipelago only once (and dispersed among islands afterwards), others did it two, three, or multiple times. That is the case of lava lizards or carpenter bees (2 colonizations), leaf-toed geckos (3 colonizations), and common Guava or humans (multiple colonizations; Table 1.3). Note in Table 1.3 that many of such colonizations occurred within the time frame of existence of the current islands (up to 4 Ma). But other taxa seem to be older than the present archipelago: mockingbirds (1.6-5.5 Ma), land and marine iguanas (5.85-11.06 Ma), lava lizards (1.69-4.54 Ma), leaf-toed geckos (3.03-13.8 Ma), land snakes (4.4-6.9 Ma), weevils (8.6-11.5 Ma), and sand spiders (15 Ma; Table 1.3). Thus, these "older-than-the-islands taxa" must have arrived at paleo islands (at present seamounts or guyots) and continued to disperse and colonize newer islands in a "stepping-stone" fashion (*Hi* and *Hii*, above), or, alternatively, their ancient lineages "hovered around" and diversified *in situ* (*Hiii*) as tectonics modified and shaped the landscape above the hotspot. Consequently, the Galápagos biogeographic affinities result from complex interactions among multiple variables: geological/tectonic past; proximate or distant relations with the mainland; ocean currents (seasonality); patterns of organisms' colonizations of the islands; and *in situ* differentiation.

Table 3.1 Exemplar Colonizations of Terrestrial and Marine Species to the Galápagos Archipelago: some taxa have "arrived" more than once

Species [E]	No. Successful Colonizations	How long ago? millions of years	Marker(s) Used for Determination	Study
Birds				
Galápagos/Darwin Finches	1	1.2 to 2.3	mtDNA	A
Mockingbirds	1	1.6 to 5.5	mtDNA	B
Galápagos Penguins	1	1.5	nDNA mtDNA	C
Galápagos Hawks	1	0.051 to 0.254	mtDNA	D
Reptiles				
Giant tortoises	1	2.0 to 3.0	mtDNA rRNA	E
Iguanas *	1	5.85 to 11.06	mtDNA nDNA	F
Lava lizards	2	1.69 to 4.54**	mtDNA nDNA	G
Leaf-toed geckos	3	3.03 to 13.8	mtDNA nDNA	H
Land-snakes	1	4.4 to 6.9	rDNA mt/nDNA	I
Mammals				
Sea lions / Fur seals	1	< 1.0	combined evidence	J
Humans	+1	0.000485	NA	NA
Arthropods				
Moths (*Galagete*)	1	2.9 to 3.7	mtDNA nDNA	K
Weevils (*Galapaganus*)	1	8.6 to 11.5	mtDNA	L
Carpenter bees (*Xylocopa*)	2	<1.5 (?)	mtDNA	M
Sand spiders (*Sicarius*)	1	15 (?) ***	DNA/biogeography	N
Terrestrial Molluscs				
Land-snails (*Naesiotus*)	1	3.0 to 3.5	nDNA	O
Plants				
Snapdragon (shrub)	1	0.09 to 0.66	cpDNA nDNA	P
Prickly Pears (*Opuntia*)	1	1.0 (?)	nDNA r/cpDNA	Q
Giant Daisies (*Scalesia*)	1	0.13 to 0.74	rDNA	R
Common Guava (*Psidium*)	1 - 2 or many	0.00015 to 0.00012	SSR markers	S

Species [E] = all species listed are endemics, except for humans and the common guava (*Psidium*); mtDNA mitochondrial DNA; rRNA ribosomal RNA; nDNA nuclear DNA; rDNA ribosomal DNA; cpDNA chloroplast DNA; A = Sato et al. 2001; B = Arbogast et al. 2006; C = Gavryushkina et al. 2017; D = Bollmer et al. 2006; E = Caccone et al. 1999, 2002; F = MacLeod et al. 2015; G = Benavides et al. 2009; H = Torres-Carvajal et al. 2014, 2016; I = Zaher et al. 2018; J = Berta et al. 2018 and Churchill et al. 2014, combined evidence = morphological, nDNA, mtDNA; K = Schmitz et al. 2007; L = Sequeira et al. 2008; M = Vargas et al. 2015; N = Magalhaes et al. 2021; O = Phillips et al. 2020; P = Guzmán et al. 2017; Q = Helsen et al. 2009; R = Fernández-Mazuecos et al. 2020; S = Urquía et al. 2019, 2021; * both land and marine iguanas share a common ancestor, a terrestrial iguana that, after arriving from Central America, diversified into the Galápagos species; ** includes two overlapping ranges of estimated clade-age: "Western" (1.69 to 3.69 Ma) and "Eastern (2.09 to 2.8 Ma) within the 1.69 to 4.54 range; (?) = high uncertainty about time of arrival; NA = non applicable, it refers to Tomás de Berlanga and his crew's arrival in 1535; *** arrival to paleo-islands (currently submerged in ocean)

Zooming In: The Islands' Appearance at First Glance

How can one best describe the Galápagos Islands? Naturalists and scientists have traditionally sketched Galápagos' "vegetation zones" using Santa Cruz (*Indefatigable*) Island as an exemplar of the archipelago (Fig. 1.38; e.g. Black 1973; Itow 1992; Mueller-Dombois 2002; Rial et al. 2017). The approach continues to be didactic and of historical value (i.e. Santa Cruz is one of the most studied islands; currently the most populated; the Charles Darwin Research Station and offices of the Galápagos National Park were first established on Santa Cruz). Modern research, however, demonstrates that vegetation cover is complex and characteristic of each island (the emphasis of this section).

[Keep in mind that the "life cycle" (ontogeny) of an island (or volcanic archipelago) determines the fate of taxa arrivals, speciation, and extinction. Thus, vegetation cover is, too, expected to correlate with island ontogeny (Whittaker 2017a,b; Paz-y-Miño-C and Espinosa 2024a). For the duration of their life cycles, hotspot islands or archipelagos—like the Galápagos—are expected to go through: emergence (new underwater volcanoes formed as seamounts), development (maturation of younger-to-older subaerial volcanoes), and subsidence (dying off, sinking of volcanic islands as guyots). Islands/archipelagos will go through such life cycles regardless of their unique characteristics or practical classifications, e.g., mantle-plume-hotspot islands or archipelagos, or continental island arcs (Whittaker 2017a,b; Paz-y-Miño-C and Espinosa 2024a)].

Look at Fig. 1.38, which includes a top- and lateral-view (North to the left; South to the right) of Santa Cruz Island. On both views, five vegetation (or climatic) zones are sketched: littoral (coastal); arid lowlands (with conspicuous *Opuntia* tree-like life form); transition (with *Psidium* trees or shrubs); moist uplands (*Scalesia* trees; *Frullania* liverworts; *Miconia* shrubs); and highlands. Note the differential slope on the North (leeward, dryer) and South (windward, rainy) sides of the island. On the North side, the arid lowlands reach 300-m/980-ft elevation; the transition zone continues up to about 500-m/1,640-ft elevation; the moist uplands up to 600-m/1,970-ft; and the highlands with Cerro [hill, mount] Crocker up to 864-m/2,835-ft. By contrast, on the South side of Santa Cruz, the arid lowlands quickly merge into the transition zone, which itself merges into the moist uplands (Fig. 1.38, bottom). Researchers emphasize that all these habitats have been altered by humans, particularly the transition and moist uplands (agriculture, introduced taxa; Trueman and d'Ozouville 2010).

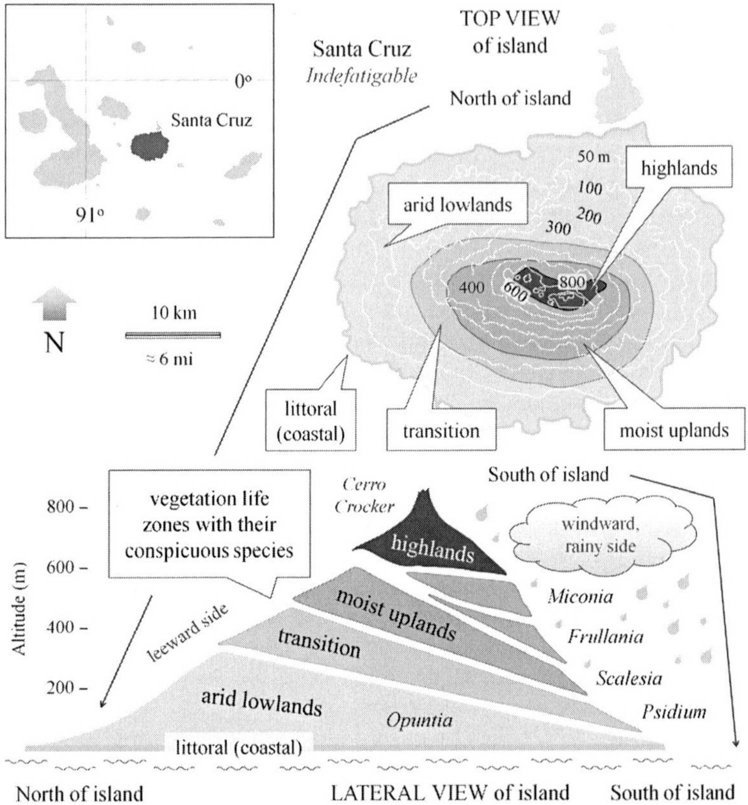

Fig. 1-38 Didactic, historic characterization of vegetation (climatic) zones in Santa Cruz (*Indefatigable*) Island. Design: GPC (as per Paz-y-Miño-C and Espinosa 2024a); redrawn and modified after a sample of sources, including Black (1973); Itow (1992); Mueller-Dombois (2002); Rial et al. (2017). [Note that vegetation cover is complex and characteristic of each island].

In Figs. 1.39 to 1.42, we offer a sample of vegetation images that are typical of some Galápagos habitats, including notes on their natural history. They illustrate contrasting environments (deciduous forests, evergreen forests, dense forests, shrublands, and wetlands) and islands (Santa Cruz, Isabela, Rábida, and Fernandina). Note that the flora of the Galápagos is relatively well known (Wiggins and Porter 1971; McMullen 1999). Later in this section, we will characterize seasonal weather affecting vegetation cycles (i.e. Trueman and d'Ozouville 2010) as well as outline vegetation cover (i.e. Rivas-Torres et al. 2018; Moity et al. 2019).

Fig. 1-39 Examples of terrestrial and wetland habitats on Santa Cruz, Rábida, and Isabela islands, Galápagos. a. *Opuntia* prickly-pear tree, the most iconic life form of the Galápagos flora; distributed in the entire archipelago. b. Prominent incense tree (*palo santo*) *Bursera* on rocky grounds (*Opuntia* shrub on the back right). c. *Scalesia* trees at *Los Gemelos* (twin pit craters). d. Flamingo Lagoon (*Poza de los Flamingos*). Photos: GPC.

Fig. 1-40 Dense lowland forests and wetlands in Isabela (*Albemarle*) Island, Galápagos. Wetlands' Trail (part of *Complejo de Humedales*), South of the island; proximate to *Puerto Villamil*, the main settlement. Bottom right Candelabra Cactus (*Jasminocereus*), endemic to Galápagos; varieties *Jasminocereus thouarsii* var. *sclerocarpus* (Isabela, Fernandina); var. *thouarsii* (Floreana, San Cristóbal, cf. Santa Cruz, cf. Isabela); var. *delicatus* (Santiago, Santa Cruz; Wiggins and Porter 1971; McMullen 1999). Photos: GPC.

Fig. 1-41 Some common and conspicuous plants in the Galápagos. Top left: yellow cordia (*muyuyo*) or *Cordia lutea*, native, of ubiquitous distribution. Top right: bush violet, *Browallia americana*, introduced from tropical America. Middle left: hand fern *Doryopteris pedata* var. *palmata*, highlands. Middle right: bracken fern *Pteridium aquilinum* var. *arachnoideum,* highlands. Bottom left: hart's tongue fern *Polypodium phyllitidis*, highlands. Bottom right: lava cactus *Brachycereus nesioticus*, endemic, coastal Fernandina. Photos: GPC.

Fig. 1-42 Lowlands of Isabela (*Albemarle*) Island: home to prominent cacti and giant tortoises. Top: deciduous forest and shrubland dominated by Candelabra Cactus (*Jasminocereus*). Bottom: a giant tortoise (*Chelonoidis*) crossing trail near *Cerro Orchilla*, Southern Isabela, near *Puerto Villamil*. Note that in Galápagos giant tortoises migrate altitudinally. Not all tortoises do it, nor do they migrate in all islands (Bastille-Rousseau et al. 2016, 2019). Photos: GPC.

The “Hot” and “Cool” Seasons

How is the weather in Galápagos? Local weather patterns result from interactions between ocean currents and trade winds (Trueman and d’Ozouville 2010). [For ocean currents in the Tropical Eastern Pacific and Galápagos, see Figs. 1.32, 1.34]. Galápagos is located in what atmosphere scientists call the Inter Tropical Convergence Zone (ITCZ), an area around Earth’s equator where Northeast and Southeast trade winds meet (direction of such winds is caused by the planet’s rotation; Fig. 1.32a). The position of this “warm-air band” (ITCZ) varies seasonally between latitude 3°N (winter in the Northern Hemisphere) and 10°N (summer). From June to December of each year, when ITCZ is located 10°N, Galápagos is exposed to cold trade winds arriving from the Southeast (“cool” season; Fig. 1.43). By contrast, from January to May, when ITCZ is located 3°N, the cold trade winds diminish and ocean currents—from the near North (Fig. 1.32c)—bring in warm water, humid air, rainfall, and higher temperature (“hot” season; Fig. 1.43; Trueman and d’Ozouville 2010).

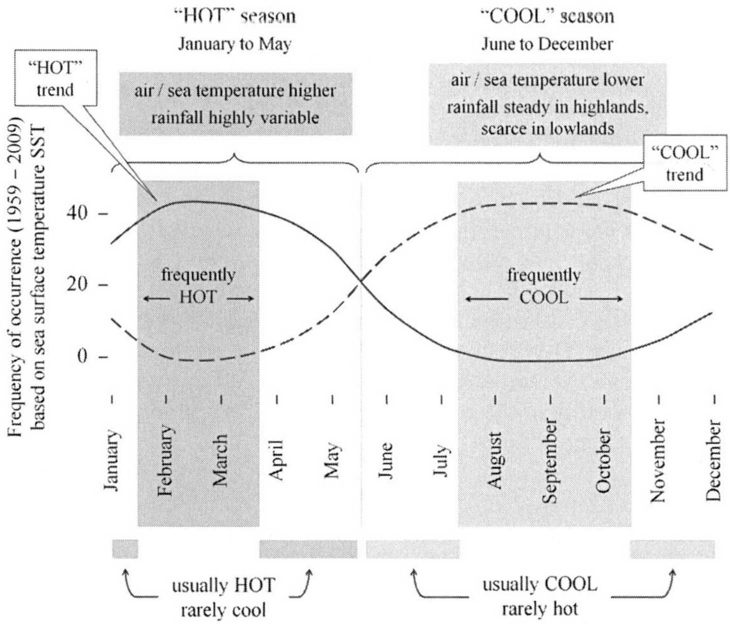

Fig. 1-43 The “hot” and “cool” seasons in the Galápagos Archipelago: data 1959-2009 (seasonality inferred from 50-year sea surface temperature SST variation). Layout: GPC (as per Paz-y-Miño-C and Espinosa 2024b); redrawn and modified after Trueman and d’Ozouville (2010).

In Fig. 1.43, we sketch the frequency of occurrence of the "hot" and "cool" seasons as per monthly variation in sea surface temperature (SST). The data corresponds to a 50-year period (1959-2009; after Trueman and d'Ozouville 2010). Note the peak in the "hot" trend (solid line) during February-March, while the peak in the "cool" trend (dashed line) corresponds to August-September. During the "hot" February-March, air/sea temperature is generally higher than during other months of the year, while rainfall is highly variable. By contrast, during the "cool" August-September, air/sea temperature is usually lower than during other months of the year, while rainfall is steady in the highlands and scarce in the lowlands (details below; see also Trueman and d'Ozouville 2010). Now, island altitude also plays a role in the local weather. As suggested by the sketch of Santa Cruz Island in Fig. 1.38-bottom ("lateral view"), the highlands are more humid than the lowlands, particularly on the windward, rainy side (South of the island). During the "cool" season, cold air arriving from the ocean collides with the island; it rises and gets trapped below the warmer/humid air already accumulated (clouds) in the highlands. The arriving cold breeze triggers water-vapor condensation at about 250-m/820-ft altitude, which is followed by dense fog formation that descends and piles up above ground. Locals call it *garúa* [ga'ru.a]. It is a form of precipitation since it coats all vegetation with minute water drops that later reach the ground (Trueman and d'Ozouville 2010; Domínguez et al. 2017). *Garúa* is typical of Galapagos' forests in the highlands' mist!

[As discussed earlier, ocean currents' dynamics around the Galápagos Archipelago are complex (Figs. 1.32 and 1.34). The waters move in predictable directions and are subject to seasonal variation, like El Niño or La Niña events (i.e. warm-cool fluctuations in climate patterns in the Pacific Ocean), which can bring unusually warm- or cool-surface waters to the islands, respectively. Both the Panama "warm" Current and the Peruvian "cold" Oceanic Current interact during El Niño or La Niña years. Depending on the intensity of the interaction, the Galápagos marine and terrestrial ecosystems can be affected by extreme weather].

Bring your attention to Fig. 1.44. Its plots help us characterize seasonality in the Galápagos as per multiple climate variables. The data corresponds to two localities in Santa Cruz Island (Fig. 1.44 bottom right): seashore CDRS at the Charles Darwin Research Station (littoral/arid lowlands) and Bellavista at 200-m/656-ft elevation (a settlement in the moist uplands; after Trueman and d'Ozouville 2010). Mean, median, or maximum/minimum values per month (50-year period) are plotted for: (a) sea surface temperature (SST); (b) air temperature; (c) wind speed and direction; (d) daily sunshine hours; (e) relative humidity; (f) days of rain;

and (g) rainfall (complete data in Trueman and d'Ozouville 2010). We encourage you to integrate the information provided in Figs. 1.38, 1.43, and 1.44. Be aware that we refer to Galápagos' weather as "hot" and "cool" seasons; other authors use different terms (wet/dry; rainy/dry; below).

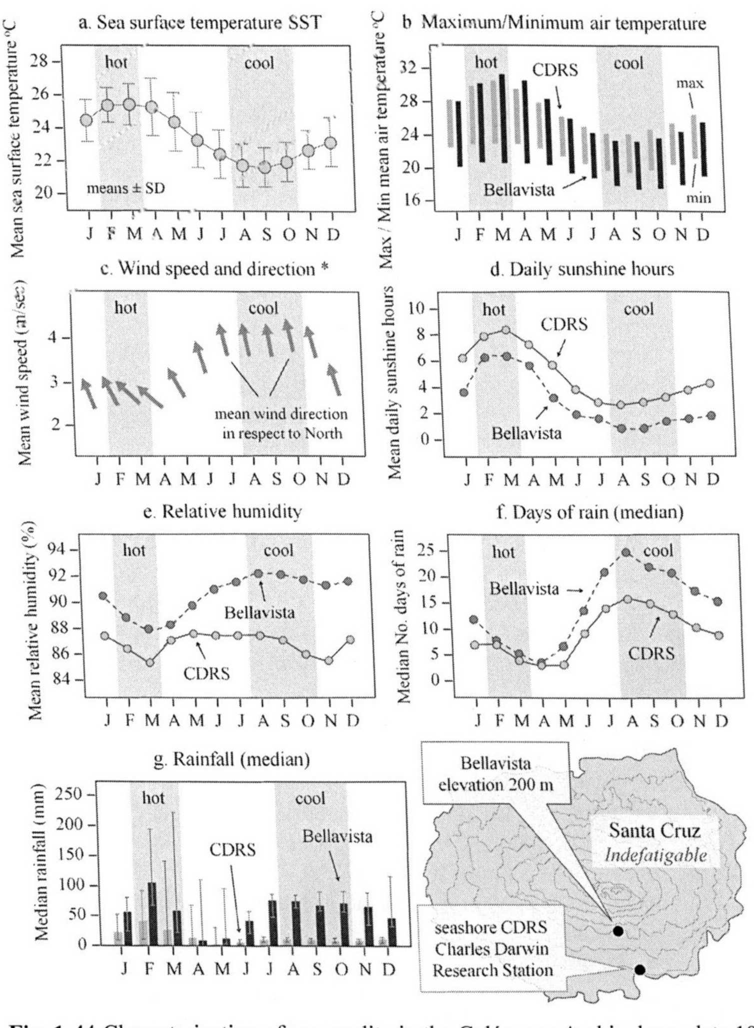

Fig. 1-44 Characterization of seasonality in the Galápagos Archipelago: data 1959-2009. Layout: GPC (as per Paz-y-Miño-C and Espinosa 2024b); redrawn and modified after Trueman and d'Ozouville (2010).

Note in Fig. 1.44a,b,d that mean sea surface temperature (SST), maximum/minimum air temperature, and daily sunshine hours peak during the "hot" season (the last two either at CDRS or Bellavista; Fig. 1.44b,d). Realize that daily sunshine is higher year-round at CDRS than at Bellavista (Fig. 1.44d). Wind speed increases and wind direction enhances with respect to the North during the "cool" season (Fig. 1.44c). Relative humidity is higher year-round at Bellavista than at CDRS; it increases at Bellavista during the "cool" season, but it remains steady year-round at CDRS (Fig. 1.44e). Days of rain increase at both Bellavista and CDRS during the "cool" season, but there are more days of rain at Bellavista than at CDRS (Fig. 1.44f). Year-round, it rains more (rainfall amount) at Bellavista than at CDRS (Fig. 1.44g); there is more variation in rainfall at both localities during the "hot" season than during the rest of the year; however, rainfall remains steadily higher at Bellavista and steadily lower at CDRS during the entire "cool" season (Trueman and d'Ozouville 2010).

Thus, Galápagos' life forms are exposed year-round to hot-and-cool seasonal variation in both terrestrial and aquatic environments. Vegetation cover (the next topic) is directly influenced by regular or severe weather and climate fluctuations (e.g. El Niño or La Niña events).

Vegetation Cover

What does vegetation cover look like on the islands? Plant ecologists and geographers have mapped Galápagos vegetation cover (Rivas-Torres et al. 2018; Moity et al. 2019). By combining satellite imagery (Landsat 8-OLI; Operational Land Imager, a remote sensing device) with high-resolution drone imagery, groundwork, and visual confirmation of the landscape, a series of maps have been produced (www links in Appendix A). We have redrawn and simplified such maps [as per Paz-y-Miño-C and Espinosa 2024b] for didactic purposes (including labels for quick identification of the type of vegetation cover), sketched them in black-and-white (distinctive gray-color tones and textures), and grouped them as per Northwestern and Northern Islands (Fig. 1.45); Western Islands (Fig. 1.46); and Central, East, and Southeastern Islands (Fig. 1.47). We encourage you to explore the original color maps as well as the studies that support them (i.e. Rivas-Torres et al. 2018; Moity et al. 2019).

Look at the key to vegetation cover and other features of the environment in Fig. 1.45, which also applies to Figs. 1.46 and 1.47. The "native ecosystems" (*N*=9) correspond to: evergreen forest and shrubland (EFS); evergreen seasonal forest and shrubland (ESFS); coastal humid forest and shrubland (CHFS); deciduous forest (DF); deciduous shrubland (DS);

humid tallgrass (HT); highland deciduous tallgrass (HDT); deciduous tallgrass (DT); and mangrove forest (MF). One type of vegetation cover corresponds to invasive species (IS). Three "features of the environment" correspond to urban settings (US); agricultural lands (AL); and water bodies (WB; some in MF wetlands). And two types of lava flows: recent lava (RL); and old/er lava (OL).

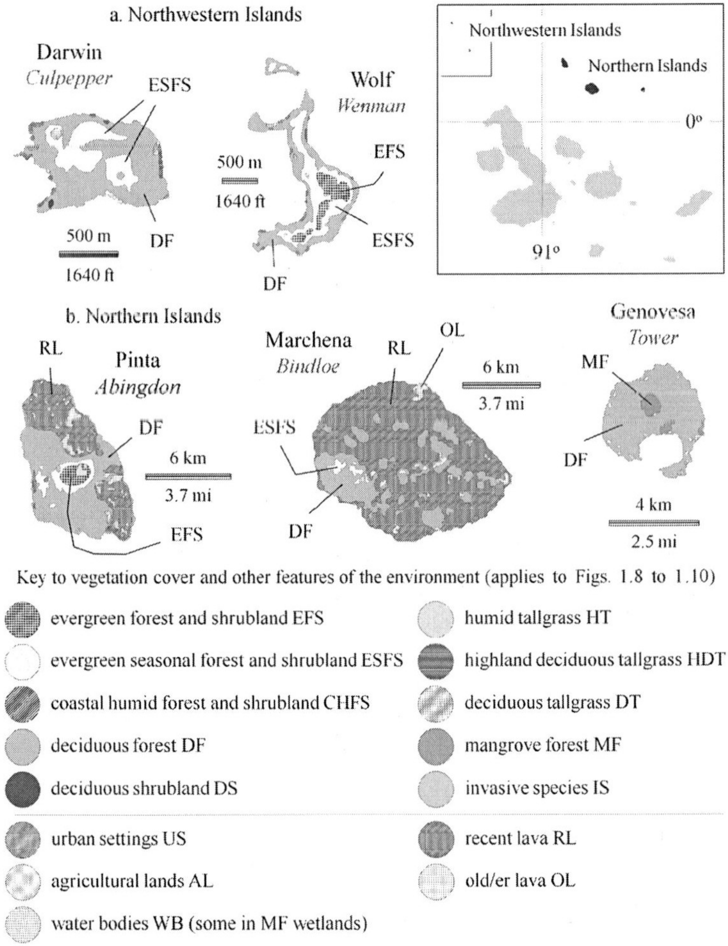

Fig. 1-45 Vegetation cover Galápagos Archipelago: Northwestern (a) and Northern Islands (b). Key of gray-color tones and textures also applies to Figs. 1.46 and 1.47. Layout: GPC (as per Paz-y-Miño-C and Espinosa 2024b); redrawn and simplified after Rivas-Torres et al. (2018); Moity et al. (2019).

Note that 96% of the terrestrial Galápagos environments are protected by law (i.e. Galápagos National Park; total archipelago's area 7,900 km^2/3,050 sq mi), while the remaining 4% corresponds to agricultural and urban areas (Paz-y-Miño-C and Espinosa 2024a,b). The mapping of vegetation cover (Figs. 1.45 to 1.47) reveals that 53% of the Galápagos National Park is covered by native ecosystems. The most dominant correspond to deciduous forest (DF 34%), evergreen seasonal forest and shrubland (ESFS 8%), evergreen forest and shrubland (EFS 4%), deciduous shrubland (DS 4%), deciduous tallgrass (DT 2%), and highland deciduous tallgrass (HDT 1%). Lava flows cover 43% of the archipelago (recent lava RL 32%; old/er lava OL 11%). Agricultural lands (AL) 3%. And urban settings (US) 1% (complete data in Rivas-Torres et al. 2018).

Examine the vegetation cover in the Northwestern Islands of Darwin and Wolf (Fig. 1.45a). Darwin is rich in deciduous forest (DF) and evergreen seasonal forest and shrubland (ESFS), while Wolf contains a shared mix of deciduous forest (DF), evergreen seasonal forest and shrubland (ESFS), and evergreen forest and shrubland (EFS). The Northern Islands are larger and more complex (Fig. 1.45b). Both Pinta and Marchena have noticeable cover of recent lava flows (RL) plus deciduous forest (DF), while Genovesa is almost entirely covered by deciduous forest (DF) with patchy yet conspicuous mangrove forest (MF). [We already discussed the peculiar volcanic origin of the Northwestern and Northern Islands, e.g. Fig. 1.5; Darwin and Wolf are 0.4- to 1.7-Ma old; Pinta, Marchena, and Genovesa 0.3- to 0.8-Ma old; Fig. 1.7].

Fernandina and Isabela (the youngest islands; 0.032- to 0.8-Ma old; Fig. 1.7) are typical for their vast cover of recent (RL) and old/er lava (OL; Fig. 1.46). Fernandina has little deciduous forests (DF) with respect to its recent lava flows (RL), the latter occupying most of the terrain (the entire island is an active volcano; Fig. 1.9). Isabela, by contrast, is the largest, most complex island in Galápagos, rich in almost all native ecosystems distributed across its five volcanoes (i.e. altitudinal and habitat variation; Fig. 1.9). Isabela's terrain is rich in recent (RL) and old/er lava (OL) flows entwined with deciduous forest (DF), evergreen seasonal forest and shrubland (ESFS), evergreen forest and shrubland (EFS), highland deciduous tallgrass (HDT), deciduous tallgrass (DT), humid tallgrass (HT), mangrove forest (MF), and—of course—striking presence of invasive species (IS), particularly in the Southern part of the island, proximate to the agricultural lands (AL; Fig. 1.46). The most abundant (as per vegetation cover) introduced taxa to Isabela—and other islands—include: *Psidium guajava* (common guava shrub/tree), *Pennisetum purpureum* (Napier/elephant grass), *Cedrella odorata* (Spanish/Cuban cedar tree), *Rubus niveus* (blackberry shrub), and *Cinchona pubescens* (red cinchona, quinine, or quina tree; Rivas-Torres et al. 2018; Laso et al. 2020).

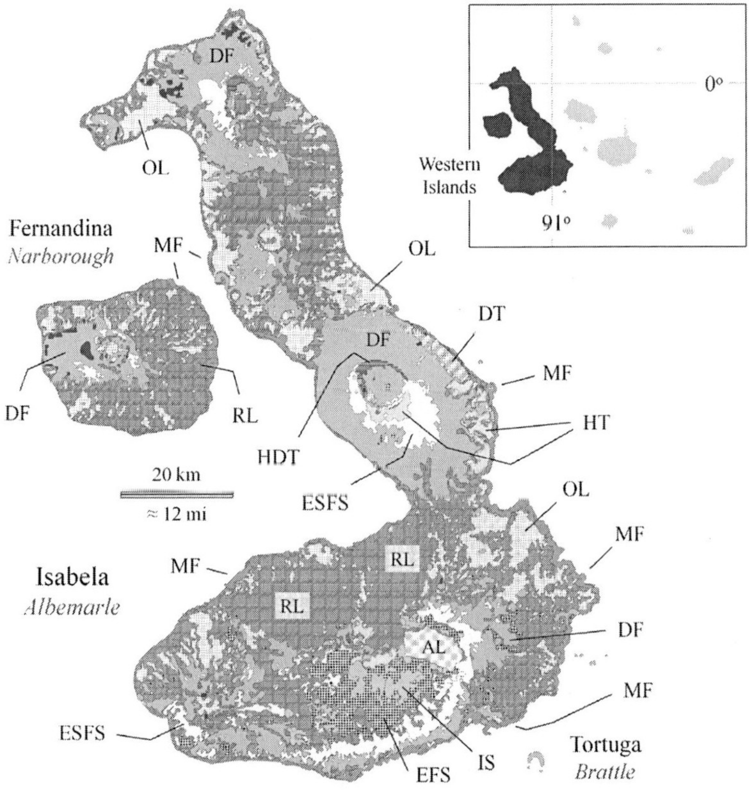

Fig. 1-46 Vegetation cover Galápagos Archipelago: Western Islands. Key to vegetation cover and other features of the environment in Fig. 1.45. Layout: GPC (as per Paz-y-Miño-C and Espinosa 2024b); redrawn and simplified after Rivas-Torres et al. (2018); Moity et al. (2019).

Look at Fig. 1.47. The larger islands of Santiago (0.8-1.4-Ma old), Santa Cruz (1.1-2.3-Ma), San Cristóbal (2.4-4.0-Ma), and Floreana (1.5-2.3-Ma) are more complex—as per vegetation cover—than all the smaller islands, i.e., Rábida (1.3-1.6-Ma), Pinzón (1.3-1.7-Ma), Santa Fé (2.9-2.9-Ma), Española (3.0-3.5-Ma), and Baltra (1.1-2.3-Ma; Fig. 1.7. [As discussed earlier, the Central, East, and Southeastern Islands lack prevalent calderas; apparently, they never had them]. Except for Rábida (rich in deciduous tallgrass DT), Santa Fé (also DT) and Baltra (also DT), the other islands are vastly covered by deciduous forest (DF). Santa Cruz and San Cristóbal have sizeable agricultural lands (AL); both islands are affected by introduced invasive species (IS).

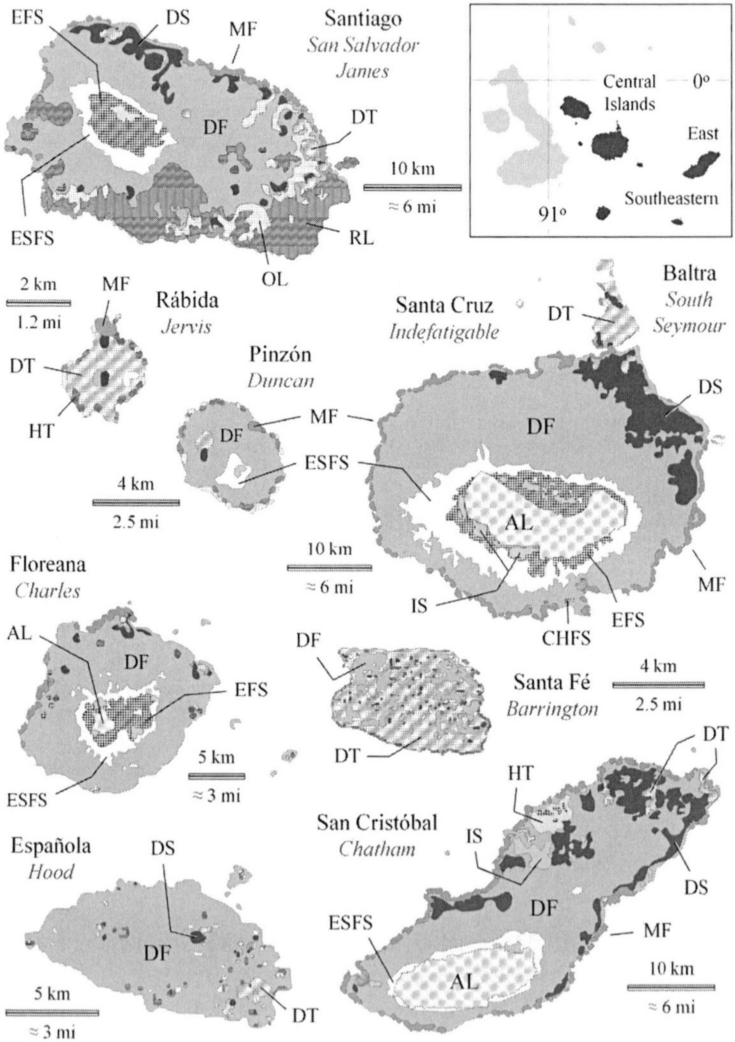

Fig. 1-47 Vegetation cover Galápagos Archipelago: Central, East, and Southeastern Islands. Key to vegetation cover and other features of the environment in Fig. 1.45. Layout: GPC (as per Paz-y-Miño-C and Espinosa 2024b); redrawn and simplified after Rivas-Torres et al. (2018); Moity et al. (2019).

We can generalize that Galápagos' vegetation cover is complex, contingent on island ontogeny (above), size, altitude, and weather/climate patterns. Half of the archipelago (53%) is covered by native ecosystems; 43% corresponds to recent or old/er lava flows; 3% is agricultural land (particularly in Santa Cruz, San Cristóbal, Isabela, and Floreana islands); and 1% is urban settings (Rivas-Torres et al. 2018).

A Uniquely Well-Preserved Mangrove Forest

Why are mangroves so important? One-third (35%) of the Galápagos' coastline is covered by mangrove forests, about 37 km^2/14 sq mi. Most mangroves are located around Isabela Island (79.4%), Santa Cruz (9.3%), Santiago (4.3%), and San Cristóbal and Fernandina (3.2% each; Fig. 1.48; Moity et al. 2019). It may not seem like much, but mangroves are vital ecosystems. They contribute to carbon fixation (above and below ground), habitat and reproduction sites for terrestrial and marine organisms, and natural protection in coastal areas subject to intertidal cycles (Friess et al. 2019; Tanner et al. 2019; Paz-y-Miño-C and Espinosa 2024b).

About 778 thousand tons of carbon are stored in Galápagos' mangroves, with an economic ecosystem service value per hectare ranging from US$2,940 to $22,838 (range total $10.8 to $83.5 million for the entire archipelago). Mangroves' benefits to local fisheries represent $245 hectare/year (total $896 thousand year/entire archipelago). Benefits to local tourism represent $16,958 hectare/year (total $62 million year/entire archipelago; Tanner et al. 2019).

Three species of mangrove dominate in Galápagos: red (*Rhizophora mangle*), black (*Avicennia germinans*), and white (*Laguncularia racemosa*). Other taxa commonly associated with intertidal communities include buttonwood mangrove (*Conocarpus erectus*), manchineel or poison apple (*Hippomane mancinella*), and salt bush (*Cryptocarpus pyriformis*; Moity et al. 2019; comprehensive work about the Galápagos flora in Wiggins and Porter 1971; McMullen 1999). Contrasting most of the world's mangrove forests, which are threatened by anthropogenic impacts, including pollution, overextraction, and conversion to aquaculture or agriculture (Friess et al. 2019), the Galápagos intertidal ecosystems are uniquely well preserved [they have been protected by law since 1959, when the Galápagos National Park was established]. In fact, Galápagos' mangroves are growing! Between 2004 and 2014, mangrove cover on three sampled islands (Fernandina, San Cristóbal, Santa Cruz) increased by 24% (2.4%/year), primarily via expansion of already existing forest patches. The phenomenon was more pronounced in Fernandina (Moity et al.

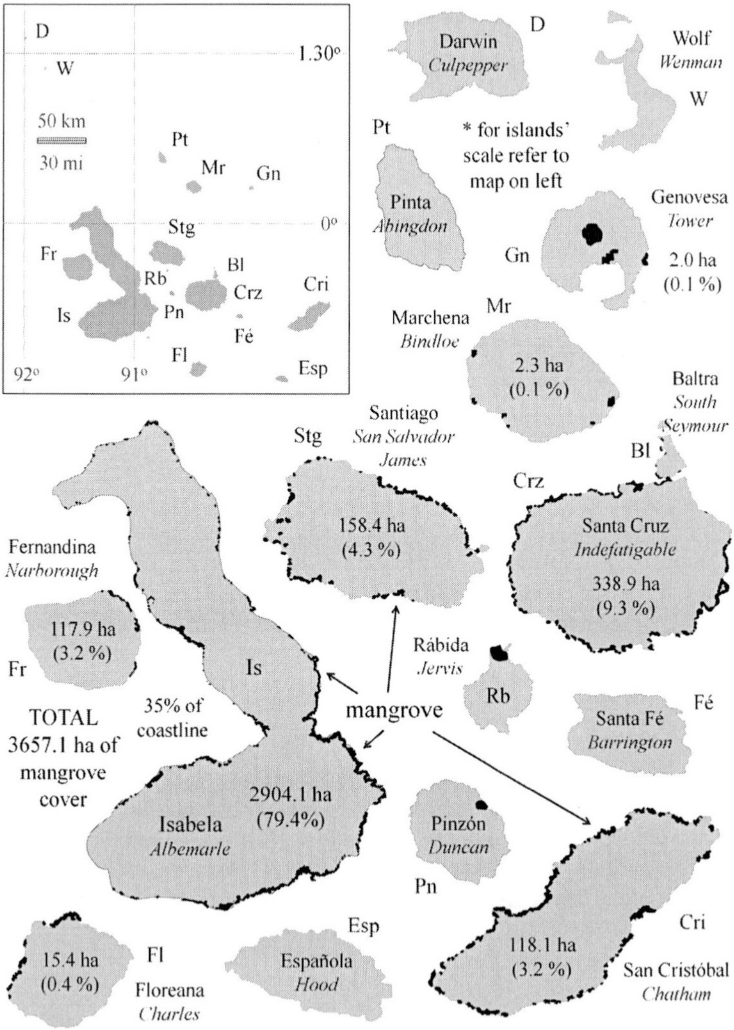

Fig. 1-48 Mangrove cover in the Galápagos Archipelago*: 3657.1 hectares (37 km²/14 sq mi), 35% of coastline. Isabela 79.4% of total cover (29 km²/11 sq mi); Santa Cruz 9.3% (3.4 km²/1.3 sq mi); Santiago 4.3% (1.6 km²/0.6 sq mi); San Cristóbal and Fernandina 3.2% each (1.2 km²/0.5 sq mi each). Rábida and Pinzón do have mangrove forest patches (not accounted for in total percent estimate); by contrast, Española has little if any mangrove; the same applies to Darwin, Wolf, and Pinta. Layout: GPC (as per Paz-y-Miño-C and Espinosa 2024b); redrawn and modified after Moity et al. (2019), with input from Rivas-Torres et al. (2018).

2019). Realize that this increase in area might be temporary and associated with island ontogeny (above) since, as pioneer vegetation, mangroves may still be colonizing the younger Galápagos islands (Moity et al. 2019); Fernandina is the youngest (Fig. 1.7). Additionally, mangroves on the Western islands (Fernandina, Isabela) benefit directly from the proximate, nutrient-rich deep-sea-water upwelling powered by the [Cromwell] Equatorial Undercurrent and/or by winds (Fig. 1.32c and hypotheses *Hi* and *Hii* in Fig. 1.34). But Galápagos mangroves are not completely safe; like all global seashore habitats, climate change is their main threat (i.e. sea level rising; extreme weather; Friess et al. 2019). We discuss the effects of climate change on Galápagos terrestrial and marine ecosystems in Chapters Nine (*What Are the Main Conservation Problems in Galápagos?*), Ten (*How Will Climate Change Affect Galápagos?*), and Eleven (*What Can You Do to Help Galápagos?*).

Conclusions

We minimize citations below to abridge the text [sources for abridgment: this chapter and Paz-y-Miño-C and Espinosa 2024a,b]. Each conclusion point summarizes topics substantiated by citations in this chapter.

(1) The Galápagos Islands are a volcanic oceanic archipelago located in the Tropical Eastern Pacific, 900 km (560 mi) off coastal Ecuador, near the equator (latitude 0°), and at longitude 91° West (Figs. 1.1). The Republic of Ecuador claimed jurisdiction over the islands in 1832. Galápagos belongs to Ecuador!

(2) For practical or didactic reasons, we refer to the Galápagos as islands, archipelago, or isles. When alluding to individual islands, we use both the official names (a mix of Spanish and English callings) and the most common English names, as follows (Fig. 1.2): (i) East and Southeastern Islands: San Cristóbal (Chatham), Floreana (Charles) and Española (Hood); (ii) Central Islands: Santiago or San Salvador (James), Santa Cruz (Indefatigable), Baltra (Seymour South or South Seymour), Seymour (Seymour North or North Seymour), Rábida (Jervis), Pinzón (Duncan), and Santa Fé (Barrington); (iii) Western Islands: Isabela (Albemarle) and Fernandina (Narborough); (iv) Northern Islands: Marchena (Bindloe), Pinta (Abingdon), and Genovesa (Tower); and (v) Northwestern Islands: Darwin (Culpepper) and Wolf (Wenman).

(3) The East and Southeastern Islands (Fig. 1.27) are the oldest in the archipelago, ranging from 1.5 to 4.0 million years since formation. San Cristóbal is the largest (558 km^2/215 sq mi) and tallest (723 m/2370 ft above sea level). Its main settlement is Puerto Baquerizo Moreno. By

2022, the total human population on the island was 8,300 inhabitants. Floreana's population was +100 inhabitants.

(4) The Central Islands (Fig. 1.28) are "middle-age", ranging from 0.8 to 2.9 million years since formation. Santa Cruz is the largest (986 km^2/381 sq mi) and tallest (864 m/2835 ft). Its main settlement is Puerto Isidro Ayora. By 2022, the total human population on the island was 17,233 inhabitants. Santiago is the second largest (585 km^2/226 sq mi) and tallest (907 m/2975 ft).

(5) The Western Islands (Fig. 1.29) are "younger", ranging from 0.032 to 0.8 million years since formation. Isabela is the largest (4670 km^2/1806 sq mi) and tallest (Wolf Volcano = 1707 m/5600 ft) of all islands in the archipelago. It has five volcanoes: Wolf, Darwin, Alcedo, Sierra Negra, and Cerro Azul. Its main settlement is Puerto Villamil. By 2022, the total human population on the island was 3,050 inhabitants. Fernandina is the youngest island in the Galápagos (0.032 to 0.06 million years since formation). Its volcano, La Cumbre, reaches 1476 m/4843 ft elevation.

(6) The Northern Islands (Fig. 1.30) are "fairly young", ranging from 0.3 to 0.8 million years since formation. The largest is Marchena (130 km^2/50 sq mi), the tallest Pinta (777 m/2549 ft), and the smallest and lowest—yet uniquely beautiful—Genovesa!

(7) The Northwestern Islands (Fig. 1.31) are "younger to middle-age", ranging from 0.4 to 1.7 million years since formation. They are the smallest and lowest-elevation islands in the archipelago and the most distant with respect to the central islands.

(8) Geologically speaking [tectonics], the Galápagos Islands move East; they age and sink in the same direction (Figs. 1.7 and 1.9); and some turn into sizable underwater mountains (seamounts or guyots; Figs. 1.11-1.12). A few million years from now, the entire archipelago will be replaced by new islands. Today's iconic features of the landscape, like the instantly recognizable Pinnacle Rock in Bartolomé Island (Fig. 1.49) or Kicker Rock (León Dormido), another spectacular formation off the Western shores of San Cristóbal (*Chatham*) Island (Fig. 1.50), will be displaced by even more monumental geologic formations or by less imposing ones. And, as fresh volcanoes continue to emerge, others will collapse or quietly become eroded and subside.

(9) The Galápagos result from the interactions between the Nazca Plate (on top of which the Galápagos platform travels), the Galápagos Spreading Center (GSC; the Northern border of the Nazca Plate), the Cocos Plate (the Northern neighbor of Nazca), and a stationary hotspot beneath them (Figs. 1.3-1.6). The current islands are young (0.032-4 Ma), but the hotspot (i.e. the source of magma that fuels the volcanoes) dates to 139

Fig. 1-49 The instantly recognizable Pinnacle Rock, Bartolomé Island. This volcanic plug (volcanic neck or lava neck) is destined to erode, collapse, and sink as the entire Galápagos Archipelago continues to drift East on top of the Nazca Plate. The Galápagos hotspot will continue to build new islands on the above-moving crust. Photos: GPC (as per Paz-y-Miño-C and Espinosa 2024a).

million years (Fig. 1.4). Hotspot and today's volcanoes are connected via a "heat column" or plume, which expands from the Earth's outer core to the upper mantle (Fig. 1.4; see also Fig. 1.8).

(10) The Western volcanoes in Fernandina (La Cumbre) and Isabela (Wolf, Darwin, Alcedo, Sierra Negra, and Cerro Azul) islands have calderas, but the Central (Santiago, Santa Cruz, Baltra, Seymour, Rábida, Pinzón, and Santa Fé) and East and Southeastern (San Cristóbal, Floreana, and Española) islands lack calderas (Figs. 1.9-1.10). The summits of Fer-

Fig. 1-50 The spectacular Kicker Rock, or León Dormido (tuff cone 150-m/490-ft tall), is located off the Western shores of San Cristóbal (*Chatham*) Island (2.4-4.4 Ma). For scale, look at the divers' boats on the bottom right. On and off, rocks fall to the ocean from this "tuff cone". Kicker Rock is also destined to erode, collapse, and sink. Photo: GPC (as per Paz-y-Miño-C and Espinosa 2024b).

nandina and Isabela are younger, proximate to the hotspot-plume zone, and have received sufficient magma to build calderas. By contrast, the East and Southeastern islands (older) were formed while the GSC was located close to the hotspot-plume zone (before GSC drifted North) and still "siphoning" substantial amounts of magma from it. Thus, the Central,

East, and Southeastern islands developed under low-rate magma supply (magma starved) and could not build calderas. They never had them.

(11) The Northwestern (Darwin and Wolf) and Northern (Marchena, Pinta, and Genovesa) islands seem to have originated in the tectonic stress (tension) resulting from the interaction between GSC and the Galápagos Transform Fault (GTF) or along the GTF in the 91° West Fracture Zone (Figs. 1.5-1.6). Darwin and Wolf belong to the Wolf-Darwin Lineament, while Marchena, Pinta, and Genovesa (members of the Northern Galápagos Volcanic Province) are part of the Genovesa Ridge. Both distinct ridges.

(12) Seamounts are—and were—ubiquitous in the Galápagos platform and its proximate or distant ridges (i.e. Wolf-Darwin Lineament, Genovesa Ridge, Carnegie, or Cocos ridges). In the geologic past, some seamounts [now guyots] were islands with caldera-volcanoes, or sizeable mountains located just above or nearby the hotspot (Fig. 1.11). During the Last Glacial Maximum (20,000 ya), when the sea level dropped significantly and the Galápagos Islands had shorelines 125 m (≈410 ft) below today's sea level, numerous shallow seamounts (20-700-m / 65-2300-ft depth) were subaerial formations (Fig. 1.12). Consequently, they contributed to enhancing inter-island connectivity in a "stepping-stones" fashion. Biologists highlight the significance of such paleo-connectivity (or, alternatively, paleo-fragmentation) for today's organisms' biogeography.

(13) Deep-sea hydrothermal vents form along ocean ridges. They were confirmed to exist in 1977 when the submersible *Alvin* reached 2,700-m (9,000-ft) depth and its crew spotted numerous vent fields at the Galápagos Rift. Hydrothermal vent fields are full of life forms. Scientists have documented 592 species, and 332 genera of organisms distributed across 63 world-exemplar vent fields (Box 1.2).

(14) There might be hundreds of lava caves (lava tubes or pyroducts) in the Galápagos, particularly in uninhabited, remote areas (Figs. 1.14-1.16). About 50 pyroducts have been scientifically documented and/or explored (Table 1.2). Some are spectacular.

(15) Chronosequence surveys of the Galápagos soils (West-to-East sampling from younger to older islands) indicate that the soils in Isabela are quite recent (1450-4290 yr), followed by older soils in Floreana (26,000 yr), Santa Cruz (165,500 yr), and San Cristóbal (812,900-1,070,000 yr) islands. Floreana (a 1.5-2.3-Ma-old island) is peculiar since it has undergone late-stage volcanism, which probably explains its much younger soils with respect to Santa Cruz (a 1.1-2.3-Ma-old island), also a late-stage volcano, but with no substantial recent activity (Figs. 1.17-1.20).

(16) All the scorias [i.e. dark rocks with a spongy appearance] collected in the Galápagos (at Isabela, Floreana, Santa Cruz, or San Cristóbal

islands) are basalts, either alkali or tholeiite basalts, which are typical of volcanic oceanic islands (Fig. 1.19). And, although not identical to one another, these scorias are geochemically and mineralogically comparable, regardless of the climate zones and/or type of vegetation that characterize the [collection] sites from which they come.

(17) In synthesis (points 1 to 16), plate tectonics and a hotspot were—still are—central to the formation and persistence of the Galápagos Archipelago. The islands' geological past and present, combined with their current geographic location (equator, proximity to Central and South America), and interactions with the ocean (currents, seasonality), are crucial to understanding their biological processes and patterns, as well as the biogeographic affinities of their life forms with respect to recent or deep ancestry (terrestrial and marine environments). The archipelago exists, ultimately, as a cause-and-effect outcome of the interplay between the Earth's crust above and the hotspot below.

(18) The Galápagos Islands are visible from space (Figs. 1.21-1.26). Satellite images (Mercury-bound Messenger spacecraft, Terra, Landsat 7, Sentinel 1A, Sentinel 2, Proba-V) credited to the National Aeronautics and Space Administration (NASA) and/or European Space Agency (ESA), as well as photographs of the islands taken by astronauts from the International Space Station (ISS Expeditions 25, 33, and 38 crews), have documented multiple geographic features of the archipelago: geology, volcanic activity, soils, land use, vegetation cover, inland waterways, and characteristics of the coastal areas. NASA Scientific Visualization Studio (SVS) has also produced high-definition 3D aerial- and below-the-ocean views of the Galápagos in computer-generated animations (e.g. "Galapagos Islands Flyby"; Fig. 1.13).

(19) Three working hypotheses explain the origins of the Galápagos biota (Figs. 1.35-1.37): *Hi* overwater dispersal from mainland—in which organisms flew, swam, or were carried to the islands by ocean currents; *Hii* stepping-stone dispersal from mainland (variant of *Hi*)—in which organisms dispersed via the now-submerged paleo islands, or seamounts, and made it to the current archipelago after "stepping" from older to younger seamounts; and *Hiii* metapopulation vicariance—in which groups of organisms originated *in situ* during the breakup of widespread ancestral populations (in the deep geological past).

(20) Hypotheses *Hi* and *Hii* (dispersal) assume that the Galápagos Islands emerged just a few million years ago, which is conceivable for the current archipelago (0.032-4 Ma; although *Hii* acknowledges that seamounts are older than 4Ma). But the Galápagos hotspot has been active for 139 Ma, and various igneous plateaus, provinces, and plates have drifted

nearby or passed just above it before the current Galápagos existed. The metapopulation vicariance model (*Hiii*) suggests that taxa (species, genera) can persist almost indefinitely in an ocean region ["hovering around"], as long as new islands continue to emerge. *Hiii* attributes the geographic range of taxa as well as their overlapped distributions to *dispersal*, while phylogenetic differentiation results from *vicariance*. By contrast, *Hi* and *Hii* attribute both range of taxa distribution and phylogenetic differentiation to dispersal. *Hi* and *Hii* assume dispersal events to be rare and infrequent, a prolonged absence of gene flow, and differentiation primarily due to chance, with no major environmental factors breaking up populations. This differs from vicariance since it affects entire communities, which become fragmented (isolated) due to substantial environmental change (e.g. tectonics; Fig. 1.35).

(21) The vicariance model brings perspective to our understanding of the origins of the Galápagos biota. Although many taxa—perhaps the majority—arrived via dispersal from proximate or distant mainland (Fig. 1.37), not all did.

(22) The Galápagos biogeographic affinities result from complex interactions among multiple variables: geological/tectonic past; proximate or distant relations with the mainland; ocean currents (seasonality); patterns of organisms' colonizations of the islands; and *in situ* differentiation. Examples of species colonizations (terrestrial and marine) of the archipelago, include birds, reptiles, mammals, arthropods, and plants (Table 1.3).

(23) Scientists have relied on genetic markers (mtDNA; rRNA; nDNA; rDNA; cpDNA) to reconstruct phylogenies, infer relationships among members of diverse taxa and their closest relatives on the mainland (e.g. Central/South/North America, the Caribbean, or elsewhere), and deduce the number of successful colonizations of the islands. Although the majority of taxa seem to have colonized the archipelago only once (and dispersed among islands afterwards), others did it two, three, or multiple times (Table 1.3). That is the case of lava lizards or carpenter bees (2 colonizations), leaf-toed geckos (3 colonizations), and common Guava or humans (multiple colonizations). Many of such colonizations occurred within the time frame of existence of the current islands (up to 4 Ma). But other taxa seem to be older than the present archipelago: mockingbirds (1.6-5.5 Ma), land and marine iguanas (5.85-11.06 Ma), lava lizards (1.69-4.54 Ma), leaf-toed geckos (3.03-13.8 Ma), land snakes (4.4-6.9 Ma), weevils (8.6-11.5 Ma), and sand spiders (15 Ma). Thus, these "older-than-the-islands taxa" must have arrived at paleo islands (at present seamounts or guyots) and continued to disperse and colonize newer islands in a "stepping-stone" fashion (*Hi* and *Hii*, above), or, alternatively, their an-

cient lineages "hovered around" and diversified *in situ* (*Hiii*) as tectonics modified and shaped the landscape above the hotspot (above).

(24) Seasonal upward surges of cold, nutrient-rich deep-sea waters—"upwelling"—fuel the biological cycles of microorganisms (primary production), which, in turn, sustain all other life forms in the islands (Fig. 1.33). Two hypotheses have been proposed to explain the upwelling. (*Hi*) The ocean-current-driven upwelling (1960s) suggests that the seawater surge on the West side of the Galápagos is driven by the collision of the sub-surface Equatorial Undercurrent with the archipelago (Fig. 1.34). Intensity in the upwelling depends on remote/stochastic events (currents' interactions; seasonality El Niño/La Niña). By contrast, (*Hii*) the wind-driven upwelling (2020s) suggests that the seawater surge is driven by local Northward winds, which activate turbulence at upper ocean fronts to the West of the archipelago; the turbulence forces deep, cold waters toward the surface while warm waters go down (Fig. 1.34). Intensity in the upwelling depends on local atmosphere-and-ocean interactions. Ultimately, the upwelling nourishes sea and land organisms, from micro to macro life. The upwelling keeps the essence of the Galápagos alive.

(25) Based on offshore assessments of fish and macro-invertebrate diversity (species composition, relative abundance as per geographic occurrence), biologists have proposed the existence of five bioregions within the Galápagos Marine Reserve (GMR; Box 1.3): Far North (around Darwin and Wolf), North (around Pinta, Marchena, and Genovesa), West (on the West side of Fernandina and Isabela), Elizabeth Bay (the channel between Fernandina and Isabela), and Southeast (East side of Isabela, around Santiago, Santa Cruz, San Cristóbal, Floreana, and Española). The Far North bioregion is rich in fish species with Indo-Pacific affinity; the West bioregion has high numbers of endemic fish taxa with cool-temperate Peruvian affinity, as well as high macro-invertebrates diversity; the Elizabeth Bay bioregion is the richest in endemics; the Southeast bioregion includes multi-origin species with distinctive affinity with Panama and Central America; the North bioregion shares fish and macro-invertebrates ("overlap region") with both the Far North and Southeast (yet, in idiosyncratic and measurable proportions).

(26) Weather patterns in Galápagos result from interactions between ocean currents and trade winds (Fig. 1.32c). Galápagos' life forms are exposed year-round to hot-and-cool seasonal variation in both terrestrial and aquatic environments. Vegetation cover is directly influenced by regular or severe weather and climate fluctuations (e.g. El Niño or La Niña events; warm-cool fluctuations in climate patterns in the Pacific Ocean, which can bring unusually warm- or cool-surface waters to the islands,

respectively). [Keep in mind that both the Panama "warm" Current and the Peruvian "cold" Oceanic Current interact during El Niño or La Niña years. Depending on the intensity of the interaction, the Galápagos marine and terrestrial ecosystems can be affected by extreme weather].

(27) During the peak of the hot season (February-March), air/sea temperature is generally higher than during other months of the year, while rainfall is highly variable. During the peak of the cool season (August-September), air/sea temperature is usually lower than during other months of the year; rainfall is steady in the highlands and scarce in the lowlands (Figs. 1.43-1.44).

(28) Island altitude plays a role in the local weather (Fig. 1.38). The highlands are more humid than the lowlands, particularly on the windward, rainy side (e.g. South of the exemplar Santa Cruz Island). During the cool season, cold air arriving from the ocean collides with the island(s); it rises and gets trapped below the warmer/humid air already accumulated in the highlands. The arriving cold breeze triggers water-vapor condensation at about 250-m/820-ft altitude, which is followed by dense fog formation that descends and piles up above ground. This form of precipitation, *garúa*, coats all vegetation with minute water drops that later reach the ground.

(29) Galápagos' vegetation cover [Figs. 1.45-1.47] is complex and contingent on island ontogeny, size, altitude, and weather/climate patterns. Sample images of some common Galápagos habitats (as per vegetation cover) are shown in Figs. 1.39-1.42. Half of the archipelago (53%) is covered by native ecosystems. [The most dominant ecosystems are deciduous forest (34%), evergreen seasonal forest and shrubland (8%), evergreen forest and shrubland (4%), deciduous shrubland (4%), deciduous tallgrass (2%), and highland deciduous tallgrass (1%)]. Lava flows cover 43% of the archipelago (recent lava 32%; old/er lava 11%). Agricultural lands 3% (in Santa Cruz, San Cristóbal, Isabela, and Floreana islands). And urban settings 1%. [Note that the Galápagos National Park includes 96% of the terrestrial environments, which are protected by law; total archipelago's area 7,900 km^2/3,050 sq mi].

(30) One-third (35%) of Galápagos coastline is covered by mangrove forests, about 37 km^2/14 sq mi (Fig. 1.48). Most mangroves are located around Isabela (79.4%), Santa Cruz (9.3%), Santiago (4.3%), and San Cristóbal and Fernandina (3.2% each). Three species of mangrove dominate in Galápagos: red (*Rhizophora mangle*), black (*Avicennia germinans*), and white (*Laguncularia racemosa*).

(31) 778 thousand tons of carbon are stored in Galápagos' mangroves, with an economic ecosystem service value per hectare ranging from US$2,940 to $22,838 (range total $10.8 to $83.5 million for the entire

archipelago). Mangroves' benefits to local fisheries represent \$245 hectare/year (total \$896 thousand year/entire archipelago). Benefits to local tourism represent \$16,958 hectare/year (total \$62-million year/entire archipelago).

(32) Contrasting most of the world's mangroves, which are threatened by anthropogenic impacts, including pollution, overextraction, conversion to aquaculture, or agriculture, the Galápagos intertidal ecosystems are uniquely well preserved (protected by law since 1959, when the Galápagos National Park was established). Between 2004 and 2014, mangrove cover on three sampled islands (Fernandina, San Cristóbal, and Santa Cruz) increased by 24%, primarily via expansion of already existing forest patches. The phenomenon was more pronounced in Fernandina. This increase in area might be temporary and associated with island ontogeny since, as pioneer vegetation, mangroves may still be colonizing the younger Galápagos islands (Fernandina is the youngest). Additionally, mangroves on the Western islands (Fernandina, Isabela) benefit directly from the proximate, nutrient-rich deep-sea-water upwelling. But Galápagos mangroves are not completely safe; like all global seashore habitats, climate change is their main threat (imminent sea level rising; extreme weather).

Questions

The following questions (Q1–5) merit generous written answers. You might need to do additional reading or library work to enrich your answers (we suggest exploring Paz-y-Miño-C and Espinosa 2024a,b).

Question 1: There are numerous volcanic oceanic archipelagos in the world; why is Galápagos considered *unique*?

Question 2: What predictions can you make about the *geologic future* of the Galápagos Archipelago [bear in mind that the interactions between the archipelago's proximate tectonic plates (Nazca, Cocos, South American, or Pacific) and the Galápagos hotspot will continue for millions of years]?

Question 3: The topic "origins of the Galápagos biota" often includes the accidental or intentional introduction of species into the islands by humans. [The islands were discovered in 1535]. Over time, some of the introduced taxa turned invasive and harmful to the archipelago's habitats (terrestrial and marine). (a) What factors *triggered* and *accelerated* the introduction of *invasive alien species* to the Galápagos? (b) *Why* can some invasive alien taxa cause *extinctions* of Galápagos' endemics?

Question 4: You are aware of the two [alternative] hypotheses that explain the nutrient-rich deep-seawater surge on the West part of the Galápagos Archipelago—the *upwelling* (Fig. 1.34). *Hi*, or the ocean-current-driven upwelling (1960s), and *Hii*, the wind-driven upwelling (2020s). Proponents of *Hii* (Forryan et al. 2021) state that the "Galápagos upwelling is controlled by highly localized atmosphere-ocean interactions". These researchers indicate that the upwelling "is primarily regulated by northward winds blowing on quasi-permanent upper-ocean fronts to the west of the archipelago". They attribute a minor, if any, role to an ocean current (*Hi*) as the cause of the upwelling. If climate change is expected to affect Galápagos and its entire marine reserve, *what difference* will it make [from the viewpoint of assessing and mitigating Galápagos' ecosystems' vulnerability to climate change] if the upwelling is caused by ocean currents or by winds?

Question 5: (a) *How* will climate change affect the *intensity* of the hot and cool seasons in the Galápagos? (b) What will be the *consequences* of such an impact for the *Galápagos' environments* (terrestrial and marine)? (c) How will *human settlements* be affected by climate change (urban, rural, or agricultural areas)? (d) How will *nature tourism* [the essence of Galápagos' economy] be *transformed* by climate change?

How to Answer the Questions Above

The questions above have various possible answers (A1–5). Here we suggest perspectives from which to address the queries.

Answer 1: You might want to compare Galápagos to other volcanic oceanic archipelagos. For example, in the Northern hemisphere (Hawai'i, Cape Verde, Canaries, Madeira & Selvagens, Azores) or Southern hemisphere (Marquesas, Juan Fernández, Seychelles, Mascarenes). These archipelagos vary in distance from/to the mainland; archipelago configuration (e.g. cluster of islands, "chain" of islands); number of main islands; land area occupied by the entire archipelago; range of islands' size; minimum to maximum distances among islands; altitude of the tallest islands; age of the archipelago; differential age of the islands; location of the islands with respect to the archipelago's hotspot (or the source of magma that fuels the volcanoes, e.g. deep-ocean volcanic ridges); ocean depths around the archipelago; main ocean currents; and presence of seamounts or guyots. Another perspective to answer Q1 could be the existence of human settlements on the archipelago's islands; human population size (total and per island) and its projections for the future; human population density; proportion of the archipelago protected by law (nature reserves, national parks, marine reserves, or sanctuaries) versus agricultural or urban areas; and nature-oriented tourism (trends of number of visitors per year). Here is another take on Q1: "Uniqueness" of an archipelago like Galápagos is often discussed—among naturalists—from the perspective of its matchless biodiversity (e.g. "endemic" species = found nowhere else). You can focus your answer to Q1 from the viewpoint of Galápagos' endemism and compare/contrast this feature of the islands to other volcanic oceanic archipelagos. Regardless of the perspective(s) you adopt to answer Q1 (e.g. geographic features of the archipelago; human population presence combined with conservation-related challenges; or endemism in the islands), keep in mind that you ought to address *Galápagos uniqueness* with respect to other volcanic oceanic archipelagos in the world!

Answer 2: Here you need to think in terms of geologic time, i.e., on the scale of thousands or millions of years. To make predictions about the geologic future of the Galápagos Archipelago, you need to first think about the past. Keep in mind that the youngest island, Fernandina, is just 0.032 to 0.06 million years (Ma) old [or 32,000 to 60,000 years] and the oldest island, San Cristóbal, is about 2.4 to 4.0 Ma (Fig. 1.7). Thus, you can cautiously guess that the current archipelago shall be replaced by new islands in about +4.0 Ma (a rough estimate). But remember that although today's Galápagos Islands are young (max 4 Ma), the hotspot that fuels them has been around for 139 Ma (this chapter; Fig. 1.4). Various igneous plateaus, provinces, and plates have drifted nearby or passed just above it before the formation of the existing Galápagos archipelago (reviews in Harpp et al. 2014; Clennett et al. 2020; Heads and Grehan 2021; see also Paz-y-Miño-C and Espinosa 2024a). If our planet's plate tectonics ultimately brought us where we are, i.e., the Galápagos plateau traveling East, resting on the Nazca Plate (Figs. 1.3 and 1.6), which is itself colliding—for the time being—with the South American Plate (as Nazca subducts under South America), while having a less antagonistic interaction with the Cocos Plate or Pacific Plate, you might want to investigate what will be the future of Earth's tectonics ≈250 Ma from now. [By the way, why ≈250 Ma?]. Infer from your research the destiny of the Galápagos. [We suggest exploring Huang et al. (2022); Davies et al. (2018, 2020); Merdith et al. (2019); Martin et al. (2023); and Paz-y-Miño-C and Espinosa (2024b)].

Answer 3: Accidental or intentional introductions of invasive alien species into pristine environments are associated with sudden historical changes, e.g., the arrival of humans. Q3 above is asking you to (a) find out the factors that *triggered* and *accelerated* the introduction of *invasive alien species* to the Galápagos. Take into consideration that human colonization of new areas is often predictable (arrival, settlement, production of goods, growth, and trade); you will need to examine such a process to determine the *factors* leading to the introductions of invasive species, as well as the variables that [over time] *accelerated* the accidental or intentional introductions. Once you tackle that, then you can address (b) the ecological and evolutionary implications of the question: *why* can some [but not all] invasive alien taxa cause *extinctions* of Galápagos' endemics?

Answer 4: The key point to address Q4 is what drives the intensity in the upwelling. According to *Hi*, or the ocean-current-driven upwelling hypothesis (1960s), intensity in the upwelling depends on remote and stochastic events, such as seasonal ocean currents' interactions, e.g., during El Niño or La Niña events [i.e. warm-cool fluctuations in climate patterns in the Pacific Ocean, which can bring unusually warm- or cool-surface waters to the islands, respectively; this chapter]. By contrast, according to *Hii*, or the wind-driven upwelling hypothesis (2020s), intensity in the upwelling depends on local atmosphere-and-ocean interactions [right there in the Galápagos' surroundings]. To answer Q4, you need to acknowledge the spatiotemporal scale at which climate change will manifest.

Answer 5: (a) Climate change [a global phenomenon] is expected to trigger extreme weather. Think of the warm-cool fluctuations in climate patterns in the [Tropical Eastern] Pacific Ocean (El Niño or La Niña events) and their consequences for the Galápagos Archipelago and its Marine Reserve. (b) Think of the local habitats and the organisms (in many cases endemic taxa) that live in the Galápagos' terrestrial and marine environments. (c) Keep in mind that human settlements in Galápagos are located in the lowlands and highlands of Santa Cruz, San Cristóbal, Isabela, and Floreana islands. The ways in which humans respond to the impacts of climate change will depend on both how the entire archipelago and/or specific islands are affected. Since each of the human-inhabited islands is unique (as per location in the archipelago, size, volcanic activity, amount of shoreline relative to island-size, or altitudinal range), you should hypothesize how climate change will differentially affect each island's settlements and how the local populations will respond to such impacts. (d) You should envision scenarios, from short-term [say 10 to 20 to 50 years from now] to long-term [100 to 200 years from now], and how nature tourism will respond.

Discussion Topics

The discussion topics below (D1–3) should be addressed with your peers. We encourage college instructors to facilitate the dialogue.

Discussion 1: During formation, deep-sea hydrothermal-vents (Box 1.2) and their communities (microbes and their viruses, as well as macroscopic organisms) seem to transition through “developmental stages”, i.e. from emergence to persistence, to senescence, and to complete inactivity (when the vents themselves cease to discharge hot water). Scientists think that hydrothermal-vents and their communities are on-and-off dynamic ecosystems, whose beginnings date back to—arguably—the origin of life on Earth (Dick 2019). Propose alternative hypotheses in which life on our planet began in deep-sea hydrothermal-vents [rather than close to the ocean surface, or freshwater hydrothermal field pools proximate to volcanoes, or elsewhere], in the absence of sunlight, and entirely powered by chemosynthesis [not by photosynthesis]. Keep in mind that these biodiversity-rich habitats (i.e. hydrothermal-vent fields) sustain life not only in the dark but also at the bottom of the ocean (150-atmospheres of pressure at 1,500-m/5,000-ft depth), with surrounding seawater temperature near freezing (2°C/36°F).

Discussion 2: Scientists refer to islands as either “Darwinian islands” (under the assumption that Charles Darwin envisioned islands, particularly volcanic ones, to form *de novo*) or “fragment islands” (perhaps because Alfred Russell Wallace, 1880, co-discoverer of evolution by means of natural selection, noticed that many islands formed as fragments of continents; Gillespie and Roderick 2002; Graham et al. 2017). Both Darwinian islands or fragment islands nurture the emergence of endemic taxa (i.e. unique species, genera, or even families), but the mechanisms leading to endemism can be different, and the outcomes distinctive [e.g. neo-endemics or paleo-endemics at the level of species, genera, or families]. Examine Fig. 1.51. Discuss the evolutionary path through which neo-endemic taxa versus paleo-endemic [or relic] taxa can emerge on volcanic islands (like Galápagos or Hawai’i) in comparison to fragment islands (New Zealand or Madagascar). Make broad generalizations about which type of islands, Darwinian or fragment, can give origin to a higher number of endemic taxa; explain why. From the perspective of conserving the world’s biodiversity [which includes high-endemism areas], what level of protection priority should islands have with respect to continents? Explain why. [You might want to explore Kier et al. (2009); Gascuel et al. (2016); Veron et al. (2019); Paz-y-Miño-C and Espinosa (2024b)].

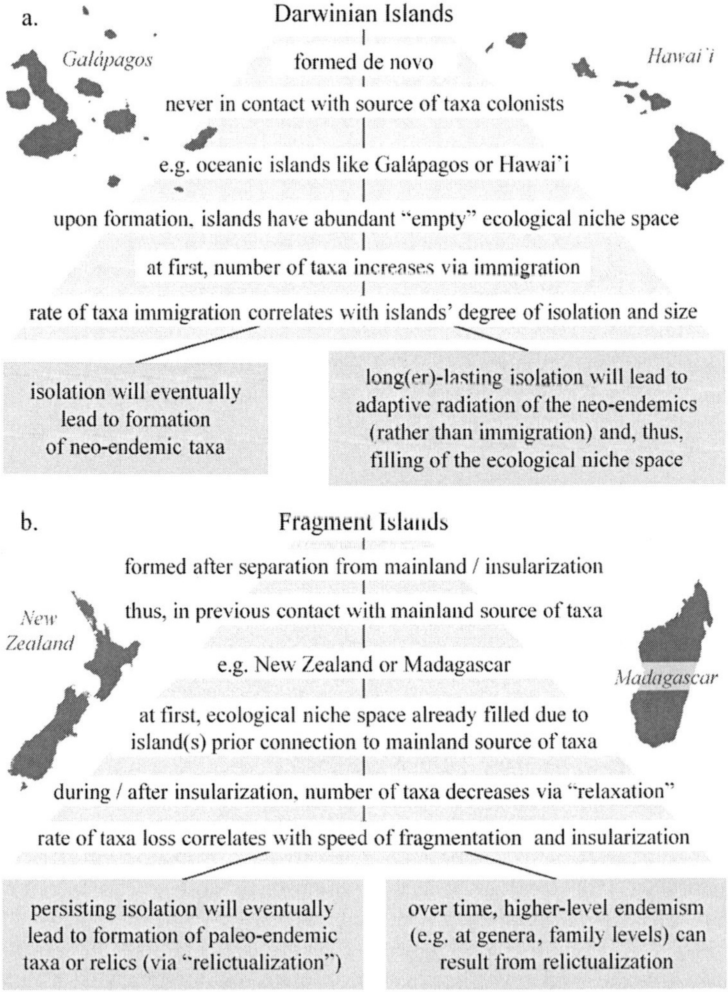

Fig. 1-51 Endemism on "Darwinian" (a) versus "fragment" (b) islands. In the case of fragment islands (FIs), "relaxation" occurs in the context of low carrying capacity, which is intrinsic to small areas (i.e. smaller islands often have lower carrying capacity than larger islands). The rate of taxa loss in FIs correlates with both the speed of fragmentation [the breakup of continents] and insularization [isolation]. Design: GPC (as per Paz-y-Miño-C and Espinosa 2024b); adapted from/modified after Gillespie and Roderick (2002); Graham et al. (2017).

Discussion 3: Examine Fig. 1.5 (Seafloor topography around the Galápagos Archipelago) and Fig. 1.11 (Seamounts located to the East and Northeast of the Galápagos Archipelago). Propose alternative hypotheses explaining the importance of the "underwater topography" [the proper term is bathymetry of the seafloor] for the Galápagos marine life forms, from coral reefs to their invertebrates and fishes (e.g. rays, sharks, or whale sharks); from sea lions and fur seals to other marine mammals like dolphins and whales; and from shallow waters to very deep ones where darkness is the norm and yet life flourishes in hydrothermal-vent fields. Address why the bathymetric complexity of the sea floor around Galápagos (i.e. ridges, spreading centers, seamounts, or guyots) is as important for the local and regional biodiversity as the landscape above the sea level (i.e. the islands themselves with their volcanoes). Reason why marine and terrestrial conservation in Galápagos is intertwined.

Suggested Observation Projects

The suggested observation projects below (P1–5) take into consideration that you will travel to Ecuador and the Galápagos legally; that you will visit the islands as a tourist [holding a Transit Control Card, a mandatory document issued by Ecuador and specifically by the Governing Council of Galápagos]; and that you will respect the regulations that protect the Galápagos National Park and the Marine Reserve (see complete list of rules in Appendix B). Only accredited scientists, academics, and/or naturalists who have obtained official permits can do research in Galápagos. You can take pictures, but flash photography is not permitted when taking photos of wildlife. Professional photography and videos recorded for commercial purposes must be authorized by the Galápagos National Park Directorate. The ideas below are for enthusiastic college explorers interested in learning about the archipelago.

Project 1: From the very questions we asked you before (this chapter). Can you make your own sketch of the Galápagos Archipelago? Can you label the islands correctly, i.e. with their most used Spanish and English names? Which islands have caldera volcanoes? Why? Can you label such volcanoes on your sketched map? Can you identify on your sketch the human-inhabited islands, the locations of the main towns, or cities-in-the-making? If that is the case, great. Here we have an additional question. Can you put together all that information in a formal, artistically original, full-color slide-show presentation (a 10-minute talk) for a general audience? Avoid slides with "text only" [they are often terrible]. Keep it rigorous yet fun and didactic.

Project 2: From Google Maps, and prior to traveling to Galápagos, you can build up a collection of satellite images [screenshots or downloads] of the archipelago (zoom-outs and close-ups of individual islands); keep the collection in your cell phone, tablet, and/or laptop so that you can access it at any time. You can do the same with images generated by NASA or by the European Space Agency (links in Appendix A but do your own www search). Internet connection in the islands (often spotty) will be limited to urban areas or some nature tourism boats/yachts. On-and-off mark on your images (dots) the locations where you have been, including dates and hours. All this information should be organized daily in separate files, together with a detailed travel log and field notes [for the latter, you will need a field notes book]. The files can also have your cell phone photos and video clips. Document everything you see!

Project 3: Keep in mind that volcanism is the essence (as per origin and geographic setting) of the Galápagos Archipelago. Prior to exploring the islands, you should become familiar with their individual [visual] characteristics and geological past (this chapter). In your daily field notes book, you can make sketches of the landscape and label relevant geographic features (this will expedite learning and reinforce your memory). It will be practical to carry binoculars with you. Once on land, try to always identify the type of lava rock on which you are walking, either *pāhoehoe* (former fluid lava with a smooth to ropy surface) or *'a'ā* (rough broken lava). You should visit lava-tube caves, or pyroducts, in the human-inhabited islands of Santa Cruz, Isabela, Floreana, or San Cristóbal (Table 1.2). [Local taxis or licensed naturalist guides can take you there for a reasonable fee. Always travel with a buddy or in a group]. Bring a waterproof flashlight [e.g. a small scuba diving flashlight] to illuminate the caves' ceilings while taking cell-phone photos along your journey and remember to sketch the pyroducts in as much detail as possible (as in Fig. 1.16) and label the location of their *pukas* (openings or entrances). Document everything you see!

Project 4: Here comes again the importance of your field notes book and your observations in Galápagos as a trainee naturalist. Create an Excel file [with various pages] on your laptop. To it, transfer the information from your field notes. You can organize the Excel pages according to the islands you visited and the wildlife you documented on them, e.g., island; sites explored; watched shore birds and reptiles; inland birds and reptiles; shore marine mammals (sea lions, fur seals); iconic arthropods and invertebrates; or iconic vegetation (Prickly Pears, Giant Daisies). Make your Excel database quantitative so that you develop a good sense for the *relative abundance* of Galápagos wildlife, i.e., estimated numbers of individuals in the species you observed; biogeographic affinities and origins of such species (as per Fig. 1.37 and Table 1.3). Do the same for the marine species you observed while snorkeling or scuba diving. At the end of your visit to the archipelago, your database will be comprehensive and your knowledge of Galápagos' biodiversity impressive!

Project 5: Compare and contrast Galápagos versus Hawai'i. Do it entirely from the perspective of both archipelagos being volcanic and oceanic. [Here you will need to do some library work]. Compare and contrast their origins, hotspots, distinctive tectonic-plate dynamics, archipelago configuration, differential volcanism, and destinies as per their geologic future (i.e. millions of years from now). Build up a slideshow presentation (for a 20-minute talk) to be shared with a general audience. Rely on images (e.g. Google Maps, the Smithsonian Institution's Global Volcanism Program, and others) and academic literature to illustrate your story. Avoid slides with "text only" [they are often terrible]. Keep it rigorous yet fun and didactic.

References

Amelung, Falk, Sigurjon Jonsson, Howard Zebker, and Paul Segall. 2000. "Widespread Uplift and 'Trapdoor' Faulting on Gálapagos Volcanoes Observed with Radar Interferometry". *Nature*, No. 407: 993-996.

Arbogast, Brian S., Sergei V. Drovetski, Robert L. Curry, Peter T. Boag, Gilles Seutin, Peter R. Grant, B. Rosemary Grant, and David J. Anderson. 2006. "The Origin and Diversification of Galapagos Mockingbirds". *Evolution*, No. 60: 370-382.

Bachraty, Charleyne, Pierre Legendre, and Daniel Desbruyéres. 2009. "Biogeographic Relationships Among Deep-Sea Hydrothermal Vent Faunas at Global Scale". *Deep Sea Research Part I: Oceanographic Research Papers*, No. 56: 1371-1378.

Ballard, Robert D. 1977. "Notes on a Major Oceanographic Find". *Oceanus*, No. 20: 35-44.

Bastille-Rousseau, Guillaume, Charles B. Yackulic, Jacqueline L. Frair, Freddy Cabrera, and Stephen Blake. 2016. "Allometric and Temporal Scaling of Movement Characteristics in Galapagos Tortoises". *Journal of Animal Ecology*, No. 85: 1171.1181.

—. Charles B. Yackulic, James P. Gibbs, Jacqueline L. Frair, Freddy Cabrera, and Stephen Blake. 2019. "Migration Triggers in a Large Herbivore: Galápagos Giant Tortoises Navigating Resource Gradients on Volcanoes". *Ecology*, No. 100: e02658.

Benavides, Edgar, Rebecca Baum, Heidi M. Snell, Howard L. Snell, and Jack W. Sites. 2009. "Island Biogeography of Galápagos Lava Lizards (Tropiduridae: *Microlophus*): Species Diversity and Colonization of the Archipelago". *Evolution*, No. 63: 1606-1626.

Berta, Annalisa, Morgan Churchill, and Robert W. Boessenecker. 2018. "The Origin and Evolutionary Biology of Pinnipeds: Seals, Sea Lions, and Walruses". *Annual Review of Earth and Planetary Sciences*, No. 46: 203-228.

Black, Juan. 1973. *Galápagos Archipiélago del Ecuador*. Quito: Imprenta Europa.

Bollmer, Jennifer L., Rebecca T. Kimball, Noah Kerness Whiteman, José Hernán Sarasola, and Patricia G. Parker. 2006. "Phylogeography of the Galápagos Hawk (*Buteo galapagoensis*): A Recent Arrival to the Galápagos Islands". *Molecular Phylogenetics and Evolution*, No. 39: 237-247.

Caccone, Adalgisa, James P. Gibbs, Valerio Ketmaier, Elizabeth Suatoni, and Jeffrey R. Powell. 1999. "Origin and Evolutionary Relationships of Giant Galápagos Tortoises". *Proceedings of the National Academy of Sciences of the United States*, No. 96: 13223-13228.

—. Gabriele Gentile, James P. Gibbs, Thomas H. Frirts, Howard L. Snell, Jessica Betts, and Jeffrey R. Powell. 2002. "Phylogeography and History of Giant Galápagos Tortoises". *Evolution*, No. 56: 2052-2066.

Churchill, Morgan, Robert W. Boessenecker, and Mark T. Clementz. 2014. "Colonization of the Southern Hemisphere by Fur Seals and Sea Lions (Carnivora: Otariidae) Revealed by Combined Evidence Phylogenetic and Bayesian Biogeographical Analysis". *Zoological Journal of the Linnean Society*, No. 172: 200-225.

Clennett, Edward J., Karin Sigloch, Mitchell G. Mihalynuk, Maria Seton, Martha A. Henderson, Kasra Hosseini, Afsaneh Mohammadzaheri, Stephen T. Johnston, and R. Dietmar Müller. 2020. "A Quantitative Tomotectonic Plate Reconstruction of Western North America and the Eastern Pacific Basin". *Geochemistry, Geophysics, Geosystems*, No. 20: e2020GC009117.

Davies, Hannah, Mattias Green, and Joao Duarte. 2018. "Back to the Future: Testing Different Scenarios for the Next Supercontinent Gathering". *Global and Planetary Change*, No. 169: 133-144.

—. J. A. Mattias Green, and Joao C. Duarte. 2020. "Back to the Future II: Tidal Evolution of Four Supercontinent Scenarios". *Earth System Dynamics*, No. 11: 291-299.

Dick, Gregory J. 2019. "The Microbiomes of Deep-Sea Hydrothermal Vents: Distributed Globally, Shaped Locally". *Nature Reviews Microbiology*, No. 17: 271-283.

Domínguez, Christian Gonzalo, Miguel Francisco García Vera, Cédric Chaumont, Julien Tournebize, Marcos Villacís, Noemi d'Ozouville, and Sophie Violette. 2017. "Quantification of Cloud Water Interception in the Canopy Vegetation from Fog Gauge Measurements". *Hydrological Processes*, No. 31: 3191-3205.

Edgar, Graham J., Stuart Banks, José M. Farifiat, Manuel Calvopiña, and Consuelo Martínez. 2004. "Regional Biogeography of Shallow Reef Fish and Macro-Invertebrate Communities in the Galapagos Archipelago". *Journal of Biogeography*, No. 31: 1107-1124.

EFFIGIS. 2016. "Volcanic Eruption in the Galapagos Islands: Observed 28-Centimetre Soil Displacement". Accessed March 25, 2022. *https://effigis.com/en/volcanic-eruption-galapagos-islands-observed-28-centimetre-soil-displacement/.*

ESA. 2019. "Proba-V view of Galápagos". Accessed March 25, 2022. *https://www.esa.int/ESA_Multimedia/Images/2019/03/Proba-V_view_of_Galapagos.*

—. 2021. "Galápagos Islands". Accessed March 25, 2022. *https://www.esa.int/ESA_Multimedia/Images/2021/03/Galapagos_Islands.*

Fernández-Mazuecos, Mario, Pablo Vargas, Ross A. McCauley, David Monjas, Ana Otero, Jaime A. Chaves, Juan Ernesto Guevara Andino, and Gonzalo Rivas-Torres. 2020. "The Radiation of Darwin's Giant Daisies in the Galápagos Islands". *Current Biology*, No. 30: 4989-4998.

Forryan, Alexander, Alberto C. Naveira Garabato, Clément Vic, A. J. George Nurser, and Alexander R. Hearn. 2021. "Galápagos Upwelling Driven by Localized Wind-Front Interactions". *Scientific Reports*, No. 11: 1277.

Friess, Daniel A., Kerrylee Rogers, Catherine E. Lovelock, Ken W. Krauss, Stuart E. Hamilton, Shing Yip Lee, Richard Lucas, Jurgenne Primavera, Anusha Rajkaran, and Suhua Shi. 2019. "The State of the World's Mangrove Forests: Past, Present, and Future". *Annual Review of Environment and Resources*, No. 44: 89-115.

Gascuel, Fanny, Fabien Laroche, Anne-Sophie Bonnet-Lebrun, and Ana S. L. Rodrigues. 2016. "The Effects of Archipelago Spatial Structure on Island Diversity and Endemism: Predictions from a Spatially-Structured Neutral Model". *Evolution*, No. 70: 2657-2666.

Gavryushkina, Alexandra, Tracy A. Heath, Daniel T. Ksepka, Tanja Stadler, David Welch, and Alexei J. Drummond. 2017. "Bayesian Total-Evidence Dating Reveals the Recent Crown Radiation of Penguins". *Systematic Biology*, No. 66: 57-73.

Geist, Dennis, Bridget A. Diefenbach, Daniel J. Fornari, Mark D. Kurz, Karen Harpp, and Jerzy Blusztajn. 2008. "Construction of the Galápagos Platform by Large Submarine Volcanic Terraces". *Geochemistry, Geophysics, Geosystems*, No. 9: Q03015.

—. Howard Snell, Heidi Snell, Charlotte Goddard, and Mark D. Kurz. 2014. "A Paleogeographic Model of the Galápagos Islands and Biogeographical and Evolutionary Implications". In *The Galápagos: A Natural Laboratory for the Earth Sciences*, edited by Karen S. Harpp, Eric Mittelstaedt, Noémi d'Ozouville, and David W. Graham, 145-166. Washington DC: American Geophysical Union and John Wiley & Sons Inc.

Gillespie, Rosemary G., and George K. Roderick. 2002. "Arthropods on Islands: Colonization, Speciation, and Conservation". *Annual Review of Entomology*, No. 47: 595-632.

Global Volcanism Program. 2022. "Ecuador". Accessed April 16, 2022. *https://volcano.si.edu/volcanolist_countries.cfm?country=Ecuador*.

Graham, David W., Barry B. Hanan, John E. Lupton, Fay Hoernle, Reinhard Werner, David M. Christie, and John M. Sinton. 2014. "Helium Isotope Variations and Mantle Plume-Spreading Ridge Interactions Along the Galápagos Spreading Center". In *The Galápagos: A Natural Laboratory for the Earth Sciences*, edited by Karen S. Harpp, Eric Mittelstaedt, Noémi d'Ozouville, and David W. Graham, 393-414. Washington DC: American Geophysical Union and John Wiley & Sons Inc.

Graham, Natalie R., Daniel S. Gruner, Jun Y. Lim, and Rosemary G. Gillespie. 2017. "Island Ecology and Evolution: Challenges in the Anthropocene". *Environmental Conservation*, No. 44: 323-335.

Grehan, John. 2001. "Biogeography and Evolution of the Galapagos: Integration of the Biological and Geological Evidence". *Biological Journal of the Linnean Society*, No. 74: 267-287.

Guzmán, Beatriz, Ruben Heleno, Manuel Nogales, Walter Simbaña, Anna Traveset, and Pablo Vargas. 2017. "Evolutionary History of the Endangered Shrub Snapdragon (*Galvezia leucantha*) of the Galápagos Islands. *Diversity and Distributions*, No. 23: 247-260.

Harpp, Karen, and Dennis Geist. 2002. "Wolf-Darwin Lineament and Plume-Ridge Interaction in Northern Galápagos". *Geochemistry, Geophysics, Geosystems*, No. 3: 8504.

—. Daniel J. Fornari, Dennis J. Geist, and Mark D. Kurz. 2003. "Genovesa Submarine Ridge: A Manifestation of Plume-Ridge Interaction in the Northern Galápagos Islands". *Geochemistry, Geophysics, Geosystems*, No. 4: 8511.

—. Eric Mittelstaedt, Noémi d'Ozouville, and David W. Graham, eds. 2014. *The Galápagos: A Natural Laboratory for the Earth Sciences*.

Washington DC: American Geophysical Union and John Wiley & Sons Inc.
—. and Dennis J. Geist. 2018. "The Evolution of Galápagos Volcanoes: An Alternative Perspective". *Frontiers in Earth Science*, No. 6: 50.
Heads, Michael, and John R. Grehan. 2021. "The Galápagos Islands: Biogeographic Patterns and Geology". *Biological Reviews*, No. 96: 1160-1185.
Helsen, Philippe, Robert A. Browne, David J. Anderson, Peter Verdyck, and Stefan van Dongen. 2009. "Galápagos' Opuntia (prickly pear) Cacti: Extensive Morphological Diversity, Low Genetic Variability. *Biological Journal of the Linnean Society*, No. 96: 451-461.
Hernández, Juan José, Isaac Izquierdo, and Pedro Oromí. 1991. "Contribution to the Vulcanospeleology of the Galapagos Islands". In *6th International Symposium on Vulcanospeleology*, edited by G. Tomas Rea, 204-222. Huntsville, Alabama: National Speleological Society Inc.
Huang, Chuan, Zheng-Xiang Li, and Nan Zhang. 2022. "Will Earth's Next Supercontinent Assemble Through the Closure of the Pacific Ocean?" *National Science Review*, No. 9: nwac205.
INEC. 2022. *Censo Ecuador: Resultados Principales Galápagos*. Quito: Instituto Nacional de Estadística y Censos.
Itow, Syuzo. 1992. "Altitudinal Change in Plant Endemism, Species Turnover, and Diversity on Isla Santa Cruz, the Galápagos Islands". *Pacific Science*, No. 46: 251-268.
Jordá-Bordehore, Luis, Theofilos Toulkeridis, Paola Leonor Romero-Crespo, Rafael Jordá-Bordehore, and Iker García-Garizabal. 2016. "Stability Assessment of Volcanic Lava Tubes in the Galápagos Using Engineering Rock Mass Classifications and an Empirical Approach". *International Journal of Rock Mechanics & Mining Sciences*, No. 89: 55-67.
Katz, Aron D., Steven J. Taylor, Felipe N. Soto-Adames, Aaron Addison, Geoffrey B. Hoese, Michael R. Sutton, and Theofilos Toulkeridis. 2016. "New Records and New Species of Springtails (Collembola: Entomobryidae, Paronellidae) from Lava Tubes of the Galápagos Islands (Ecuador)". *Subterranean Biology*, No. 17: 77-120.
Kempe, Stephan F. J., Greg Middleton, Aaron Addison, Theofilos Toulkeridis, and Geoffrey Hoese. 2021. "New Insights into the Genesis of Pyroducts of The Galápagos Islands, Ecuador". *Acta Carsologica*, No. 50: doi.org/10.3986/ac.vi.7587.
Kier, Gerold, Holger Kreft, Tien Ming Lee, Walter Jetz, Pierre L. Ibisch, Christoph Nowicki, Jens Mutke, and Wilhelm Barthlott. 2009. "A Global Assessment of Endemism and Species Richness Across Island

and Mainland Regions". *Proceedings of the National Academy of Sciences of the United States*, No. 106: 9322-9327.

Kislik, Emily, Gabriel Mantilla Saltos, Gladys Torres, and Mercy Borbor-Córdova. 2017. "Biological Hotspots in the Galápagos Islands: Exploring Seasonal Trends of Ocean Climate Drivers to Monitor Algal Blooms". *International Journal of Bioengineering and Life Sciences*, No. 11: 824-834.

Laso, Francisco J., Fátima L. Benítez, Gonzalo Rivas-Torres, Carolina Sampedro, and Javier Arce-Nazario. 2020. "Land Cover Classification of Complex Agroecosystems in the Non-Protected Highlands of the Galapagos Islands". *Remote Sensing*, No. 12: doi:10.3390/rs120 10065.

MacLeod, Amy, Ariel Rodríguez, Miguel Vences, Pablo Orozco-terWengel, Carolina García, Fritz Trillmich, Gabriele Gentile, Adalgisa Caccone, Galo Quezada, and Sebastian Steinfartz. 2015. "Hybridization Masks Speciation in the Evolutionary History of the Galapagos Marine Iguana". *Proceedings of the Royal Society B*, No. 282: 2015.0425.

Magalhaes, Ivan L. F., Adalberto J. Santos, and Martín J. Ramírez. 2021. "Incorporating Topological and Age Uncertainty into Event Based Biogeography of Sand Spiders Supports Paleo-Islands in Galapagos and Ancient Connections among Neotropical Dry Forests". *Diversity*, No. 13: 418.

Martin, Erin L., Peter A. Cawood, J. Brendan Murphy, R. Damian Nance, and Philip J. Heron. 2023. "The Tectonics of Introversion and Extroversion: Redefining Interior and Exterior Oceans in the Supercontinent Cycle". *Geological Society, London, Special Publications*, No. 542: doi.org/10.1144/SP542-2023-5.

McMullen, Conley K. 1999. *Flowering Plants of the Galápagos*. Ithaca: Cornell University Press.

Merdith, Andrew S., Simon E. Williams, Sascha Brune, Alan S. Collins, and R. Dietmar Müller. 2019. "Rift and Plate Boundary Evolution Across Two Supercontinent Cycles". *Global and Planetary Change*, No. 173: 1-14.

Moity, Nicolas, Byron Delgado, and Pelayo Salinas-de-León. 2019. "Mangroves in the Galapagos Islands: Distribution and Dynamics". *PLoS ONE*, No. 14: e0209313.

Mueller-Dombois, Dieter. 2002. "Forest Vegetation Across the Tropical Pacific: A Biogeographically Complex Region with many Analogous Environments". *Plant Ecology*, No. 163: 155-176.

NASA. 2013. "View of Galapagos Islands from Space Station". Accessed March 25, 2022. *https://www.nasa.gov/content/view-of-galapagos-islands-from-space-station*.

—. 2015. "Eruption of Wolf Volcano, Galapagos Islands". Accessed March 25, 2022.
https://www.nasa.gov/image-feature/eruption-of-wolf-volcano-galapagos-islands.

NASA-SVS. 2009. "Galapagos Islands Flyby". Accessed March 25, 2022. *https://svs.gsfc.nasa.gov/3628*.

Orellana-Rovirosa, Felipe, and Mark Richards. 2016. "Evidence and Models for Lower Crustal Flow Beneath the Galápagos Platform". *Geochemistry, Geophysics, Geosystems*, No. 17: 113-142.

Paz-y-Miño-C, Guillermo, and Avelina Espinosa. 2024a. *Arrivals of Life to the Galápagos: From Berlanga to Darwin, Melville, and World War II*. Newcastle upon Tyne: Cambridge Scholars Publishing.

—. 2024b. *A Synthesis of the Galápagos: Natural History, Human Conflict, and the Destiny of an Archipelago*. Newcastle upon Tyne: Cambridge Scholars Publishing.

Phillips, John G., T. Mason Linscott, Andrew M. Rankin, Andrew C. Kraemer, Nathaniel F. Shoobs, and Christine E. Paren. 2020. "Archipelago-Wide Patterns of Colonization and Speciation Among an Endemic Radiation of Galápagos Land Snails". *Journal of Heredity*, No. 111: 92-102.

Rial, Marcela, Antonio Martínez Cortizas, Teresa Taboada, and Luis Rodríguez-Lado. 2017. "Soil Organic Carbon Stocks in Santa Cruz Island, Galapagos, under Different Climate Change Scenarios". *Catena*, No. 156: 74-81.

Rivas-Torres, Gonzalo, Fátima L. Benítez, Danny Rueda, Christian R. Sevilla, and Carlos F. Mena. 2018. "A Methodology for Mapping Native and Invasive Vegetation Coverage in Archipelagos: An Example from the Galápagos Islands". *Progress in Physical Geography*, No. 42: 83-111.

Sato, Akie, Herbert Tichy, Colm O'hUigin, Peter R. Grant, B. Rosemary Grant, and Jan Klein. 2001. "On the Origin of Darwin's Finches". *Molecular Biology and Evolution*, No. 18: 299-311.

Schmitz, Patrick, Alice Cibois, and Bernard Landry. 2007. "Molecular Phylogeny and Dating of an Insular Endemic Moth Radiation Inferred from Mitochondrial and Nuclear Genes: The Genus *Galagete* (Lepidoptera: Autostichidae) of the Galapagos Islands". *Molecular Phylogenetics and Evolution*, No. 45:180-192.

Schwartz, Darin M., S. Adam Soule, V. Dorsey Wanless, and Meghan R. Jones. 2018. "Identification of Erosional Terraces on Seamounts: Implications for Interisland Connectivity and Subsidence in the Galápagos Archipelago". *Frontiers in Earth Science*, No. 6: 88.

Sequeira, Andrea S., Analía A. Lanteri, L. Roque Albelo, Sayantani Bhattacharyam, and Manisha Sijapati. 2008. "Colonization History, Ecological Shifts and Diversification in the Evolution of Endemic Galápagos Weevils". *Molecular Ecology*, No. 17: 1089-1107.

Sinton, Christopher W., David M. Christie, and Robert A. Duncan. 1996. "Geochronology of Galápagos Seamounts". *Journal of Geophysical Research*, No. 101: 13689-13700.

—. Karen S. Harpp, and David M. Christie. 2014. "A Preliminary Survey of the Northeast Seamounts, Galápagos Platform". In *The Galápagos: A Natural Laboratory for the Earth Sciences*, edited by Karen S. Harpp, Eric Mittelstaedt, Noémi d'Ozouville, and David W. Graham, 335-362. Washington DC: American Geophysical Union and John Wiley & Sons Inc.

—. Karen S. Harpp, Dennis J. Geist, Eric Mittelstaedt, Daniel J. Fornari, and S. Adam Soule. 2017. "Banco Tuzo: An Ancient Galapagos Island and Potential Stepping Stone for Species Dispersal". *Galapagos Research*, No. 69: 9-17.

—. Folkmar Hauff, Kaj Hoernle, and Reinhard Werner. 2018. "Age Progressive Volcanism Opposite Nazca Plate Motion: Insights from Seamounts on the Northeastern Margin of the Galapagos Platform". *Lithos*, No. 310-311: 342-354.

Sinton, John, Robert Detrick, J. Pablo Canales, Garrett Ito, and Mark Behn. 2003. "Morphology and Segmentation of the Western Galápagos Spreading Center, 90.5-98 W: Plume-ridge Interaction at an Intermediate Spreading Ridge". *Geochemistry, Geophysics, Geosystems*, No. 4: 8515.

Tanner, Michael K., Nicolas Moity, Matthew T. Costa, Jose R. Marin Jarrin, Octavio Aburto-Oropeza, and Pelayo Salinas-de-León. 2019. "Mangroves in the Galapagos: Ecosystem Services and their Valuation". *Ecological Economics*, No. 160: 12-24.

Torres-Carvajal, Omar, Charles W. Barnes, María J. Pozo-Andrade, Washington Tapia, and Gabriela Nicholls. 2014. "Older than the Islands: Origin and Diversification of Galápagos Leaf-toed Geckos (Phyllodactylidae: *Phyllodactylus*) by Multiple Colonizations. *Journal of Biogeography*, No. 41: 1883-1894.

—. Andrea Rodríguez-Guerra, and Jaime A. Chaves. 2016. "Present Diversity of Galápagos Leaf-toed Geckos (Phyllodactylidae: *Phyl-*

lodactylus) Stems from Three Independent Colonization Events". *Molecular Phylogenetics and Evolution*, No. 103: 1-5.

Toulkeridis, Theofilos, and Aaron Addison. 2014. "Geodynamic and Volcanic Origin of Lava Caves in the Galápagos". In *Abstract Volume & Field Guide, 16th International Symposium of Vulcanology, 2nd Simposio Internacional de Espeleologia, Galápagos, Ecuador*, edited by Theofilos Toulkeridis and Aaron Addison, 54-57. Quito: ESPE Universidad de las Fuerzas Armadas.

Trueman, Mandy, and Noémi d'Ozouville. 2010. "Characterizing the Galápagos Terrestrial climate in the Face of Global Climate Change". *Galapagos Research*, No. 67: 26-37.

Urquía, Diego, Bernardo Gutierrez, Gabriela Pozo, María José Pozo, Analía Espín, and María de Lourdes Torres. 2019. "*Psidium guajava* in the Galapagos Islands: Population Genetics and History of an Invasive Species". *PLoS ONE*, No. 14: e0203737.

—. Bernardo Gutierrez, Gabriela Pozo, Maria Jose Pozo, and Maria de Lourdes Torres. 2021. "Origin and Dispersion Pathways of Guava in the Galapagos Islands Inferred through Genetics and Historical Records". *Ecology and Evolution*, No. 11: 15111-15131.

Vargas, Pablo, Beatriz Rumeu, Ruben H. Heleno, Anna Traveset, and Manuel Nogales. 2015. "Historical Isolation of the Galápagos Carpenter Bee (*Xylocopa darwini*) Despite Strong Flight Capability and Ecological Amplitude". *PLoS ONE*, No. 10: e0120597.

Veron, Simon, Thomas Haevermans, Rafaël Govaerts, Maud Mouchet, and Roseli Pellens. 2019. "Distribution and Relative Age of Endemism Across Islands Worldwide". *Scientific Reports*, No. 9: 11693.

Villagómez, Darwin R., Douglas R. Toomey, Emilie E. E. Hooft, and Sean C. Solomon. 2007. "Upper Mantle Structure Beneath the Galápagos Archipelago from Surface Wave Tomography". *Journal of Geophysical Research*, No. 112: B07303.

Whittaker, Robert J., José María Fernández-Palacios, Thomas J. Matthews, Michael K. Borregaard, and Kostas A. Triantis. 2017a. "Island Biogeography: Taking the Long View of Nature's Laboratories". *Science*, No. 357: eaam8326.

—. 2017b. "Island Biogeography: Taking the Long View of Nature's Laboratories". *Science*, No. 357: 885.

Wiggins, Ira L., and Duncan M. Porter. 1971. *Flora of the Galápagos Islands*. Stanford: Stanford University Press.

Zaher, Hussam, Mario H. Yánez-Muñoz, Miguel T. Rodrigues, Roberta Graboski, Fabio A. Machado, Marco Altamirano-Benavides, Sandro L. Bonatto, and Felipe G. Grazziotin. 2018. "Origin and Hidden Diversity within the Poorly Known Galápagos Snake Radiation (Serpentes: Dipsadidae)". *Systematics and Biodiversity*, No. 16: 614-642.

Zehetner, Franz, Martin H. Gerzabek, J. Gregory Shellnutt, Franz Ottner, Christopher Lüthgens, Daniel P. Miggins, Pei-Hao Chen, I. Nyoman Candra, Greta Schmidt, Maria V. Rechberger, and Tobias Sprafke. 2020. "Linking Rock Age and Soil Cover Across Four Islands on the Galápagos Archipelago". *Journal of South American Earth Sciences*, No. 99: 1025002.

CHAPTER TWO

HOW WERE THE *Y^{A}S DE GALAPAGOS* DISCOVERED?

On March 10, 1535, the Bishop of Castilla de Oro (Panama), Tomás de Berlanga (Fig. 2.1), accidentally arrived in what today is Española (Hood) Island [one of the later-to-be-known as *y^{a}s de galapagos* or isles of the tortoises]—as hypothesized by historian Estrella Figueras Vallés (2010). Berlanga had left Panama on February 23, 1535, aiming for the "lands of Perú", but his ship lost course (Fig. 2.2a). After sailing with "very good breezes for seven days" and the pilot keeping route "near land", as Berlanga later narrated the journey in a colorful letter to Carlos/Charles I, King of Spain (dated April 26, 1535), the winds calmed, and the vessel drifted—due to strong currents—to the Southwest. It reached unknown islands on "Wednesday the tenth of March" of that year (Fig. 2.2).

We have abridged the text of Berlanga's letter in Box 2.1, which is a translation of the old-Spanish original, with a few of our own "[notes to facilitate grasping the content]". Keep in mind that the Spanish Crown had assigned Berlanga to mediate territorial governorship disputes (the purpose of the trip to Peru) between conquistadores Francisco Pizarro and Diego de Almagro, who, after overpowering the Inca Empire, had taken the land for Spain—and themselves (Cieza de Leon 1999). But Berlanga's vicissitudes while traveling frustrated the mission. He and the cargo-ship crew hardly made it alive to Caraques a month and a half later (April 9), and to Puerto Viejo by April 26—when he finally wrote to the King.

To rebuild Berlanga's timeline—together with his adopted crew—while navigating about the islands described in the letter, we relied on five accurate dates extracted from his narrative (Fig. 2.3): departure from Panama on February 23; accidental arrival in a "first island" on March 10; celebration of mass on Passion Sunday at a "second island" on March 14; return to the mainland and arrival in Caraques on April 9; and the writing of the letter once in Puerto Viejo by April 26. All other dates in Fig. 2.3 are estimates inferred from Berlanga's imprecise accounts (Box 2.1; all as per Paz-y-Miño-C and Espinosa 2024a).

Fig. 2-1 Fray Tomás de Berlanga (1487-1551), Bishop of Castilla de Oro, Panama. Statue at Berlanga de Duero, Spain. Photo Francisco Dousdebés (with permission).

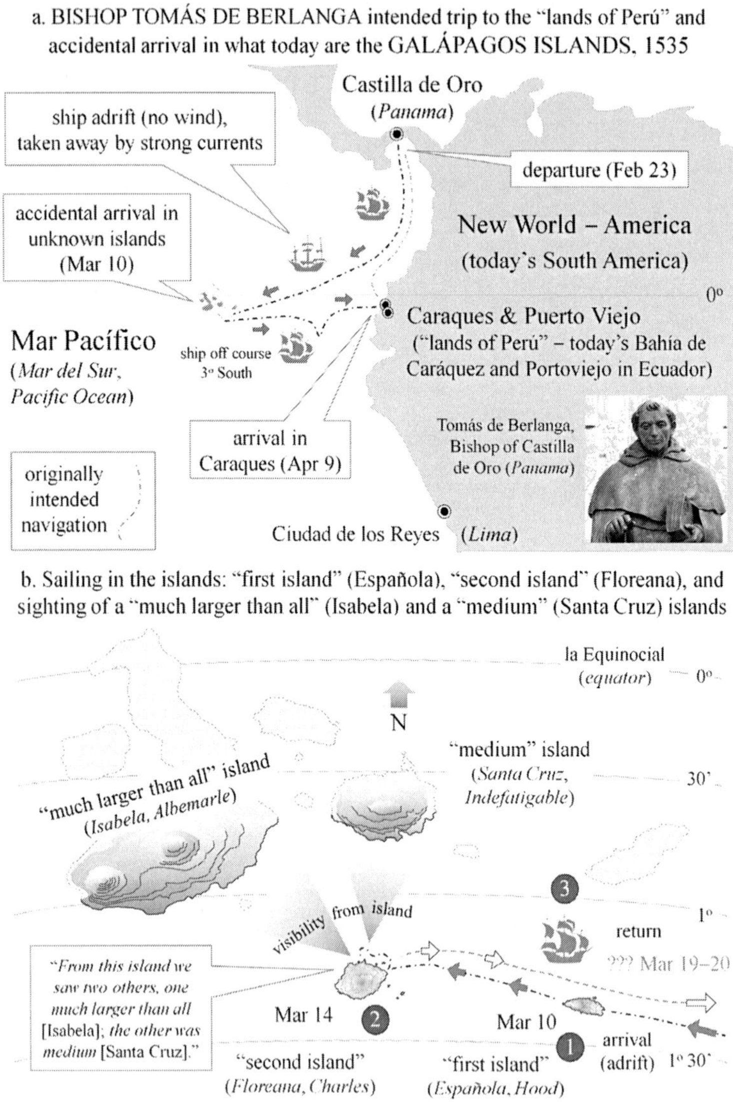

Fig. 2-2 a. Tomás de Berlanga intended navigation to the "lands of Perú" (1535) and accidental arrival on unknown islands. b. Suspected sailing 1-3 among islands (March 10 to 19-20). In hindsight, Berlanga and crew likely disembarked on Española (Hood) and Floreana (Charles or Santa María); from there, the men spotted Isabela (Albemarle) and Santa Cruz (Indefatigable). Design: GPC (as per Paz-y-Miño-C and Espinosa 2024a), after Figueras Vallés' proposal (2010).

Box 2-1 Bishop Tomás de Berlanga's letter* to Carlos/Charles I King of Spain (also ruler of the Holy Roman Empire as Carlos/Charles V).

PUERTO VIEJO. —[Tuesday] *APRIL 26, 1535*
Sacred Imperial [Roman] *Catholic Majesty*
IT seemed right to me to let Your Majesty know about the course of my trip since I departed from Panama, on [Tuesday] *the twenty-third of February* [Feb 23] *of the current year* [1535]*, until I arrived in this new Village of Puerto Viejo.*

The ship sailed with very good breezes for seven days [by Mar 1/2]*, and the pilot kept course near land and* [then] *we had calm for six days* [by Mar 7/8]*; the currents were so strong, and engulfed us in such a way, that on Wednesday the tenth of March* [Mar 10] *we sighted an island* [today's Española/Hood]*; and, because on board there was water for only two more days, it was agreed to take down the lifeboat and go on land for water and grass for the horses. And once there, the men found nothing but sea lions, and turtles and galápagos* [the giant tortoises] *so big that each could carry a man on top, and many iguanas that look like serpents. On another day* [at least starting Mar 11 onwards]*, we saw another island* [today's Floreana/Charles]*, larger than the former* [Española] *and with big sierras; and, thinking that for its size as well as for its mountains it will not fail to have rivers and fruits, we went to it. The contour of the first island* [Española] *was about four or five leagues, and of the other* [Floreana] *ten or twelve leagues. By this time, the water on board gave out and it took us three days to reach the island* [Floreana, by Mar 13]*, owing to the calms, during which all of us, as well as the horses, suffered greatly.*

Once the ship was anchored [in Floreana, Mar 13]*, we all went on land, and some were given charge of digging a well, and others of looking for water around the island: from the well, bitterer water than that of the ocean came out; on land, the men could not find a drop of water for two days; and, due to the need that people had, they grabbed a leaf of some thistles-like-prickly-pears* [likely *Opuntia* stems and fruits]*, and because they were juicy, although not very tasty, we began to eat them, and squeeze them for their water; and once extracted, it looked like slops of lye, and we drank it as if it were rose water.*

On Passion Sunday [Mar 14]*, I had men set on land what was needed to celebrate mass; and once mass was said, I sent the people in twos or threes, over diverse parts* [of the island]*: it was Our celebrated Lord who guided them to find in a ravine, among the rocks, as much as half a pipe of water, and after procuring it they found more and more; in the end, eight pipes were filled,* [plus] *the barrels and jugs that the ship had. But due to the* [lasting] *need of water, a man died; and two days after we left that island, another* [by ???Mar 19/20]*; and ten horses died.*

From this island [Floreana]*, we saw two others, one much larger than all* [today's Isabela/Albemarle]*, which contour was clearly fifteen or twenty leagues; the other was medium* [today's Santa Cruz/Indefatigable]. [In Floreana]*, I took the altitude* [of the sun] *to know where the* [new] *islands were, and they* [were located] *from half-a-degree to a degree-and-a-half of the Equator, in the South. On this second* [island, Floreana]*, the scenery was the same as in the* *first* [Española]*; many sea lions, turtles, iguanas, galápagos* [the giant tortoises], *many birds like those of Spain, but they were so dumb that they did not know to flee, and many were caught by hand. To the other two* [islands] *we did not reach; I do not know what scenery they have. On this one* [Floreana], *on the sand at the beach, there were some small stones* [chinas], *that as we walked we stepped on; they were like diamond-stones, and others of amber color; but on the entire island I do not think there is a place where a bushel of corn could be sown, because most of it* [the island] *is full of very big* [rocks], *that it seems like some time ago God showered* [rocks]*; and the existing soil is like slag, worthless, because it does not have the virtue to nurture some grass, but only some thistles, the leaf of which I said we picked* [the *Opuntia* and its fruits]. *Thinking that we were no more than twenty or thirty leagues away from the land of Perú, we were satisfied with the* [procured] *water, although we could have taken twenty more pipes like those we* [already] *did; but we set sail* [by ???Mar 19-20], *and with mild weather we navigated eleven days without sighting land* [by Mar 30/31]*; and the pilot and master came to me to ask where we were, and to tell me that there was no more than one pipe of water left: I tried to take, that day, the altitude of the sun, and found that we were three degrees to the South; and noticed that for the course we were taking we were becoming more engulfed; that we were not reaching land, because we were heading South: I made them turn* [correct course], *and distribute the butt of water in the following way: half for the animals, and the other half to make a beverage* [to] *be stored in the wine-pipe* [cask]. *Being certain that we could not be far from land and navigating for eight days during which the beverage lasted by rationing it to each of us contentment, and once that pipe was empty and we had no other choice, we sighted land and had calm for two days* [by Apr 6/7], *during which we drank pure wine, but we were joyful for having sighted land. We entered the bay and river of the Caraques on Friday the ninth of April* [Apr 9]. [the letter continues addressing issues unrelated to the accidental arrival in the islands]...

May The Lord fill Your Sacred Majesty with holy love and grace for long times, and with the preservation of your Realms and expansion to new others, as I hope. From the new Village of Puerto Viejo, the twenty-sixth of April [Monday, Apr 26], *of one thousand, five hundred and thirty five years* [1535]. *— Of Your Sacred Imperial* [Roman] *Catholic Majesty. — Most sincere serf and servant and perpetual Chaplain, who kisses your royal feet and hands. — Fray Tomás eps locastelli aurii. — With signature.*

*Adapted from Woram (2005); Spanish transcript *Colección Documentos Inéditos* (1884). See also Paz-y-Miño-C and Espinosa (2024a).

The Bishop described having landed on two islands: "the contour of the first island was about four or five leagues [22-28 km or 14-17 mi], and of the other ten or twelve leagues [55-67 km or 34-41 mi]." He also wrote that, from the second island, "[he] took the altitude [of the sun] to know where the [new] islands were, and they [were located] from half-a-degree to a degree-and-a-half of the Equator, in the South." Where exactly did the men arrive? Relying on the islands' location, contour dimensions, and

Berlanga's depiction of the landscape, historians have suggested that the "first island" corresponded to Española and the "second" to Floreana (Charles or Santa María; Woram 2005; Figueras Vallés 2010). If that was the case, from "Floreana" the crew sighted two more islands: "one much larger than all [possibly today's Isabela or Albemarle], which contour was clearly fifteen or twenty leagues [80-110 km or 50-68 mi]; the other was medium [possibly today's Santa Cruz or Indefatigable]" (Fig. 2.2b).

Thus, unknown islands had been discovered. By 1565, the name "*y*[a]*s de galapagos*" (*isles of the tortoises*) appeared, perhaps for the first time in history, on the "Portolan Chart of the Pacific Coast from Guatemala to Northern Peru" (Fig. 2.4). In it, a cluster of 12 islands of arbitrary shape was roughly painted. It will take 271 years until Captain FitzRoy and the Officers of HMS Beagle [on board of which Darwin also traveled] mapped the Galápagos Islands (1836) with cartographic accuracy (Fig. 2.5).

Much has happened since Berlanga and crew discovered the isles: Ecuador's annexation of the archipelago to its territory (1832); FitzRoy's and Darwin's exploration of the islands (1835); Herman Melville's visit to (1841) and writing of *The Encantadas or, Enchanted Isles* (1856); human colonization (1850s onwards); California Academy of Sciences scientific collection expedition (1905-6); establishment of a military base by the United States Army in Seymour Island (Baltra) during World War II (1942-6); creation of the Charles Darwin Foundation and the Galápagos National Park (1959); tourism boom (1960s onwards); and declaration of World Heritage Site by UNESCO (1979). We refer to these events in Chapter Four (*What Happened in Galápagos Before and After Darwin?*); for comprehensive analyses of Galápagos history, see Luna Tobar (1997), Idrovo (2005), or Woram (2005).

[Note that in Chapter Five, *Why Did the United States Establish a Military Base in Galápagos During World War II?*, we analyze the events leading to the establishment of such a base in Seymour Island (i.e. Baltra, Seymour South, South Seymour, or "The Rock"). In Chapter Six, *Who Lives in Today's Galápagos Islands?*, we discuss human colonization of the islands. And in Chapter Seven, *Why is Nature Tourism Essential to Galápagos?*, we examine tourism beginnings and current trends].

Timeline of Bishop Tomás de Berlanga while in Unknown Islands, 1535

Sun	Mon	Tue	Wed	Thu	Fri	Sat
		23 – Feb	24 – Feb	25 – Feb	26 – Feb	27 – Feb
		Bishop and crew set sail from Panama; destination the "lands of Perú."		Ship sails South with good breezes along the shores of mainland…		
28 – Feb	1 – Mar	2 – Mar	3 – Mar	4 – Mar	5 – Mar	6 – Mar
	…good breezes end.		Ship adrift, taken away from mainland (Southwest direction) by strong ocean currents…			
7 – Mar	8 – Mar	9 – Mar	10 – Mar	11 –	12 – Mar	13 – Mar
…ship still adrift…		…ship arrives in "first island" (today's Española); crew looks for water.			Ship sails toward and arrives in "second island" (today's Floreana).	
14 –	15 – Mar	16 – Mar	17 – Mar	18 – Mar	19 – Mar	20 – Mar
Passion Sunday in "second island" (Floreana).	From "second island," Bishop and crew see two other islands; a "much larger than all" [Isabela], the other "medium" [Santa Cruz].				??? Ship leaves the islands and sails back to mainland…	
21 – Mar	22 – Mar	23 – Mar	24 – Mar	25 – Mar	26 – Mar	27 – Mar
…ship navigates East toward mainland, aiming for the bay of Caraques…						
28 – Mar	29 – Mar	30 – Mar	31 – Mar	1 – Apr	2 – Apr	3 – Apr
		…Bishop and crew realize ship is out of course, sailing South rather than East…		…crew corrects course and continues (East)…		
4 – Apr	5 – Apr	6 – Apr	7 – Apr	8 – Apr	9 – Apr	10 – Apr
		…"land" at sight…			…ship enters the bay and river of Caraques (in today's Ecuador).	
25 – Apr	26 – Apr				"accurate dates" in gray boxes others are estimates	
	In Puerto Viejo (nearby Caraques), Bishop writes "course of the trip" letter to King of Spain, narrating the accidental discovery of the islands.					

Fig. 2-3 Tomás de Berlanga's and crew's vicissitudes while on unknown islands, 1535 (as per Paz-y-Miño-C and Espinosa 2024a).

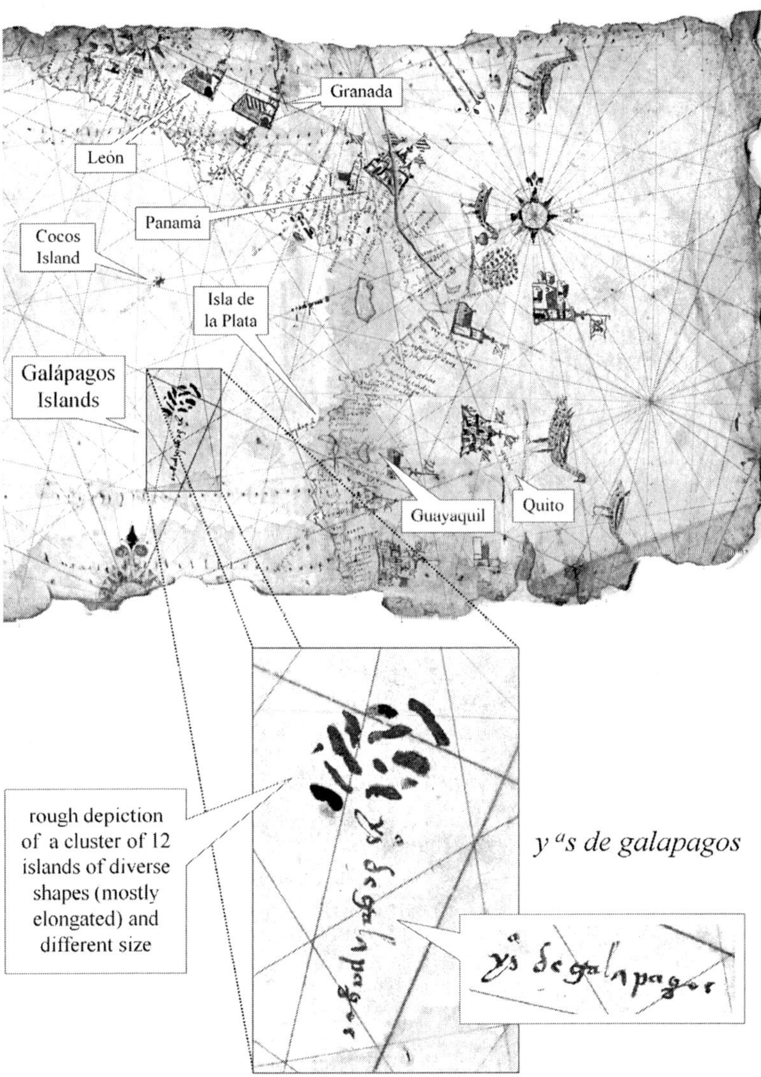

Fig. 2-4 "Portolan Chart of the Pacific Coast from Guatemala to Northern Peru" (1565), in which the "*y^a^s de galapagos*" or "*isles of galapagos*" [isles of the tortoises] were depicted (*galapagos* = tortoise). The nautical chart (portolan = port, harbor) was sketched on vellum (animal skin). Digital image (61.5 x 33.5 cm; 24 x 13 in): Library of Congress of the United States. Figure layout: GPC (as per Paz-y-Miño-C and Espinosa 2024a).

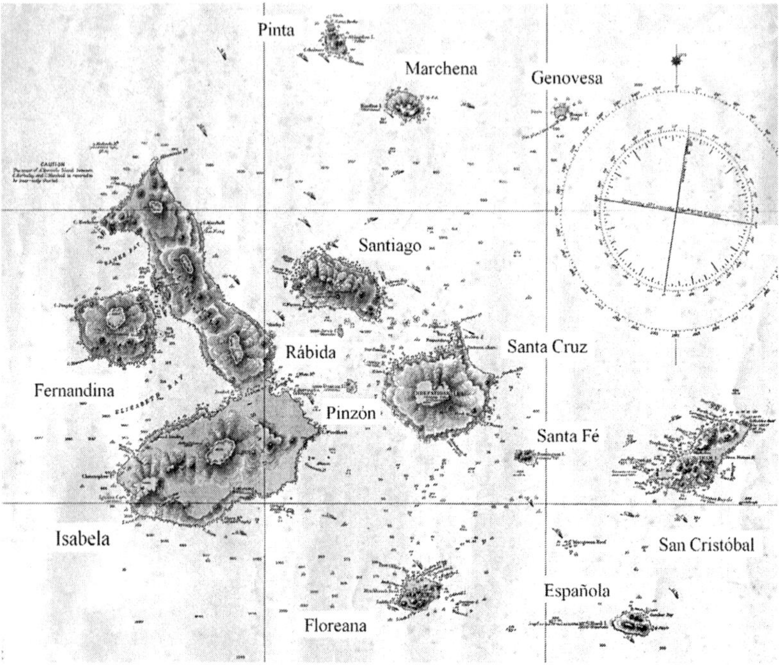

Fig. 2-5 "Galápagos Islands", by Captain Robert FitzRoy and the Officers of HMS Beagle (1836). The original has additions and corrections traceable to 1926 (source: Wikimedia Commons). Spanish names of islands added on map. Layout: GPC (as per Paz-y-Miño-C and Espinosa 2024a).

To close this chapter, we highlight that Tomás de Berlanga and his crew were, according to existing evidence, the first humans to arrive in the Galápagos Islands. Archaeological and/or palaeoecological studies in Floreana (Charles or Santa María), San Cristóbal (Chatham), Santa Cruz (Indefatigable), Isabela (Albemarle), and Santiago (James or San Salvador)—historically the human-populated islands—have demonstrated that all settlements in the archipelago took place during the last 500 years (Anderson et al. 2016; Paz-y-Miño-C and Espinosa 2024a).

Conclusions

Tomás de Berlanga and crew were the first humans to arrive—accidentally—in what today are the Galápagos Islands, on March 10, 1535. The men disembarked on two islands—likely Española (Hood) and

Floreana (Charles or Santa María). From "Floreana", they sighted two more islands: one much larger than all (possibly Isabela or Albemarle); the other was medium (possibly Santa Cruz or Indefatigable).

Archaeological and/or palaeoecological studies in the historically human-populated islands of Floreana, San Cristóbal (Chatham), Santa Cruz, Isabela, and Santiago (James or San Salvador) have demonstrated that all settlements in the archipelago began ≈500 years ago.

Questions

The following questions (Q1–2) merit generous written answers. You might need to do additional reading or library work to enrich your answers (we suggest exploring Paz-y-Miño-C and Espinosa 2024a,b).

Question 1: The Galápagos Islands are located relatively close to mainland South America (927 km or 575 mi), yet they were—according to scientific evidence—not discovered by humans until 1535 (i.e. Tomás de Berlanga's accidental arrival on March 10 of that year; this chapter). There is no evidence that Amerindians ever drifted to [or settled in] Galápagos prior to Berlanga's adventure. By contrast, other volcanic oceanic archipelagos, like Hawai'i, despite being remotely located with respect to a mainland (3,850 km or 2,390 mi off Baja California in North America), were discovered and colonized by Indigenous peoples, the Polynesians, 1,700 to 1,500 years ago (reviews in Loope et al. 1988; Ziegler 2002; see also Paz-y-Miño-C and Espinosa 2024a). [Note that the first Europeans arrived in Hawai'i in 1778]. Can you hypothesize *why* Polynesians managed to discover and settle in Hawai'i, but Amerindians apparently did not discover, nor did they settle in the *yªs de galapagos*?

Question 2: What scientific evidence should be produced to conclude that Amerindians arrived in the *yªs de galapagos* prior to Berlanga?

How to Answer the Questions Above

The questions above have various possible answers (A1–2). Here we suggest perspectives from which to address the queries.

Answer 1: Here you need to consider how long humans have inhabited the proximate "mainland" [or islands] to Galápagos versus Hawai'i. This is because human expansion is often gradual as native peoples colonize new areas. Now, traveling technology is crucial to successful dispersal to a distant location. Thus, think of traveling technologies available to native peoples (Amerindians versus Polynesians) and how that might have correlated with the arrival to and colonization of remote oceanic islands.

Answer 2: You need to investigate what evidence scientists (e.g. archaeologists or paleoecologists) look for to conclude that a particular site has been visited and/or occupied (settlements) by humans. Keep in mind that there are reliable methods to discover past human presence at a site and technologies to date how long humans have inhabited an area.

Discussion Topic

The discussion topic below (D1) should be addressed with your peers. We encourage college instructors to facilitate the dialogue.

Discussion 1: Examine Tomás de Berlanga's letter to Carlos/Charles I King of Spain (Box 2.1). It deserves careful analysis. [To this day, researchers continue to re-read and reexamine the letter]. Each sentence and paragraph should give you clues about the type of landscape Berlanga and crew encounter; the animals with which they interacted (you can actually identify them to the species level); the plants found and from which they tried to extract water; the presence [or absence] of natural creeks with drinking water on the islands [according to Berlanga, only one of the two islands had such a source of water]; the types of small stones [chinas] that were found on the second island; or the estimated size of the two larger islands that Berlanga and crew spotted from the "second island". This discussion will give you enough insight into addressing the questions above and enrich the suggested observation projects below.

Suggested Observation Projects

The suggested observation projects below (P1–2) take into consideration that you will travel to Ecuador and the Galápagos legally; that you will visit the islands as a tourist [holding a Transit Control Card, a mandatory document issued by Ecuador and specifically by the Governing Council of Galápagos]; and that you will respect the regulations that protect the Galápagos National Park and the Marine Reserve (see complete list of rules in Appendix B). Only accredited scientists, academics, and/or naturalists who have obtained official permits can do research in Galápagos. You can take pictures, but flash photography is not permitted when taking photos of wildlife. Professional photography and videos recorded for commercial purposes must be authorized by the Galápagos National Park Directorate. The ideas below are for enthusiastic college explorers interested in learning about the archipelago.

Project 1: In this chapter, you learned that modern historians have hypothesized that the "first island" on which Tomás de Berlanga and crew disembarked—when accidentally drifting toward an unknown archipelago in 1535—was today's Española (Hood), and that the "second" island was today's Floreana (Charles or Santa María; Woram 2005; Figueras Vallés 2010). Realize that this inference, although formalized by qualified scholars, is theoretical and can be challenged. Here is your project: Integrate the knowledge you have acquired about Berlanga's journey in 1535 [this chapter] with the information provided in Fig. 1.32, Chapter One (i.e. Ocean currents in the Tropical Eastern Pacific and Galápagos), and propose *alternative hypotheses* about where Berlanga could have accidentally arrived (i.e. other islands in the archipelago besides Española or Floreana). Rank your hypotheses according to both the probable course Berlanga's adrift vessel could have taken based on the ocean currents in the Tropical Eastern Pacific and the Bishop's descriptions of the islands in his letter to Carlos/Charles I King of Spain (Box 2.1). Sketch [computer diagrams] your hypotheses on a slideshow presentation (for a 20-minute talk) to be shared with a general audience. Avoid slides with "text only" [they are often terrible]. Keep it rigorous yet fun and didactic.

Project 2: This project requires library work (we suggest exploring Idrovo 2005; Woram 2005; Paz-y-Miño-C and Espinosa 2024a). Contrast the first impression that the *y^{a}s de galapagos* caused, upon accidental or deliberate arrival on the archipelago, in the minds of historically famous visitors like Tomás de Berlanga (1535), Charles Darwin (1835), Herman Melville (1841), Eleanor Roosevelt (1944), and others of your choice [not pop celebrities]. What was your own impression when visiting the islands for the first time? How did it differ from Berlanga's, Darwin's, Melville's, or Roosevelt's? Why? Put together a slideshow presentation (for a 20-minute talk) to be shared with a general audience. Avoid slides with "text only" [they are often terrible]. Keep it rigorous yet fun and didactic.

References

Anderson, Atholl, Karen Stothert, Helene Martinsson-Wallin, Paul Wallin, Iona Flett, Simon Haberle, Henk Heijnis, and Edward Rhodes. 2016. "Reconsidering Precolumbian Human Colonization in the Galápagos Islands, Republic of Ecuador". *Latin American Antiquity*, No. 27: 169-183.

Cieza de León, Pedro de. 1999. *The Discovery and Conquest of Peru: Chronicles of the New World Encounter*. Durham: Duke University Press.

Colección de Documentos Inéditos de América y Oceanía. 1884. Tomo XLI: 538-544. Madrid: Imprenta de Manuel G. Hernández. *https:// archive.org/details/coleccindedocum19ultrgoog/page/n7/mode/2up*.

Figueras Vallés, Estrella. 2010. *Fray Tomás de Berlanga. Una Vida Dedicada a la Fe y la Ciencia*. Soria: Ochoa Editores.

Idrovo, Hugo. 2005. *Galapagos: Footsteps in Paradise*. Quito: Libri Mundi.

Loope, Lloyd L., Ole Hamann, and Charles P. Stone. 1988. "Comparative Conservation Biology of Oceanic Archipelagoes: Hawai'i and the Galápagos". *BioScience*, No. 38: 272-282.

Luna Tobar, Alfredo. 1997. *Historia Política Internacional de las Islas Galápagos*. Quito: Ediciones Abya Yala.

Paz-y-Miño-C, Guillermo, and Avelina Espinosa. 2024a. *Arrivals of Life to the Galápagos: From Berlanga to Darwin, Melville, and World War II*. Newcastle upon Tyne: Cambridge Scholars Publishing.

—. 2024b. *A Synthesis of the Galápagos: Natural History, Human Conflict, and the Destiny of an Archipelago*. Newcastle upon Tyne: Cambridge Scholars Publishing.

Woram, John. 2005. *Charles Darwin Slept Here*. Rockville Centre: Rockville Press Inc.

Ziegler, Alan C. 2002. *Hawaiian Natural History, Ecology, and Evolution*. Honolulu: University of Hawai'i Press.

CHAPTER THREE

WHY WAS CHARLES DARWIN'S VISIT TO GALÁPAGOS SIGNIFICANT?

At age 26, in September 1835, Charles Darwin arrived in the Galápagos. [For terminology used in this chapter, see Box 3.1]. Robert FitzRoy, Captain of Her Majesty's Ship (HMS) Beagle (Fig. 3.1) and head of the vessel's surveying voyage around the world (1831 to 1836; Fig. 3.2), which included the exploration of the Galápagos Islands, had requested and been granted by the British Admiralty permission to bring Mr. Darwin—as naturalist—on board (historical analysis in van Wyhe 2013, including why Darwin was a "naturalist" on board of HMS Beagle).

By 1836, FitzRoy and the Beagle's officers had drafted the first reliable map of the Galápagos, which we have sketched in Fig. 3.3. The source (link in Appendix A) includes additions and corrections dating back to 1926 (for chronological mapping of the *y[a]s de galapagos*, see Paz-y-Miño-C and Espinosa 2024a). For a thorough, analytical compilation about *Darwin in Galápagos*, see Grant and Estes (2009), which we use as source for the section below (as per Paz-y-Miño-C and Espinosa 2024a).

Box 3-1 Essential terminology for this chapter.

Adaptive Radiation. An evolutionary process during which a species diversifies, over time, into other distinctive species. E.g., upon arrival to and colonization of new habitats, descendants of the ancestral taxa can occupy available ecological niches and gradually become distinctive as they adapt to new environments and food resources, or compete with conspecifics and other taxa, or cope with predators and pathogens. The Galápagos' or Darwin's finches discussed in this chapter (Figs. 3.16–3.18) are a textbook example of adaptive radiation.

Allopatric Speciation. Also called speciation in isolation or geographic speciation. An evolutionary process in which populations of a species become "physically" isolated [sometimes fragmented] from one another due to physical barriers that separate the populations (e.g. the ocean, rivers, or mountains). Islands of an archipelago [like Galápagos] are often scenarios of allopatric speciation. During isolation, gene flow between/among populations [i.e. cross reproduction] ceases or becomes insignificant. Thus, over time, the populations diverge on their own and become distinctive in genetic structure.

Articulata. An old-fashioned term used by Charles Darwin on his *Notebook B: Transmutation*, on which his celebrated "I think" tree is depicted (Fig. 3.12). In various pages (36–46) of Darwin's 1837 notebook, he refers to the sketch while reasoning about relations [relatedness] among genera, represented by A, B, C or D, although, as he implied in writing, A–D could indicate other groups (i.e. mammals, birds, Articulata [at that time insects and crustaceans], plants).

Divergent Selection. A process in which [two] *populations* of a species are exposed to contrasting selection pressures that lead them to diverge (e.g. large vs. small beak size populations) or even become distinctive species over time.

Introgressive Hybridization. A process in which [two] *species* mate and produce fertile hybrid offspring. Sometimes, the offspring backcross with the parents, and their genomes fuse (via introgression). Occasionally, entire new lineages emerge from the hybrid offspring and turn into new species.

Natural Selection. Differential survival and reproduction of individuals as a function of their fitness [i.e. their vigor to survive environmental hazards, outcompete conspecifics, and leave fertile progeny]. Over time, natural selection can lead to an increase in the frequency of [temporarily] beneficial traits in a population (evolution). Such traits vary from one individual to another [trait variation is often the product of gene variants, called *alleles*, or differential gene expression] and can be inherited by their descendants (i.e. heritable variation).

Phylogenetic Tree. Branching diagram depicting possible [i.e. hypothetical] relatedness of a group of taxa (e.g. species, genera, families, or orders). The branching pattern of a tree, when "rooted", originates in a "common ancestor" [often unknown], which, after adaptive radiation and other evolutionary processes, gives origin to the descendants represented in the diagram by the end-branches [the tips] of the tree. Nowadays, tree topologies have multiple uses.

Species, genera, families, and orders. Taxonomic classification systems rely on hierarchical *categories*, such as species, genera, families, and orders (each is a *taxon*, plural *taxa*), to explain phylogenetic relations [e.g. Fig. 3.17] among organisms. Each category is—in itself—an explanatory hypothesis.

Transmutation (Mutability). Merriam–Webster defines "transmute" as "to change or alter in form, appearance, or nature, and especially to a higher form". And "mutability" as "prone to change". Charles Darwin used the word "transmutation" to suggest that taxa [species] could or did change. In his 1837 *Notebook B: Transmutation* (Fig. 3.12), Darwin reasoned about the possible emergence of new genera from an "ancient [species] type", itself "capable of making [i.e. via transmutation or mutability]... recent [new] forms".

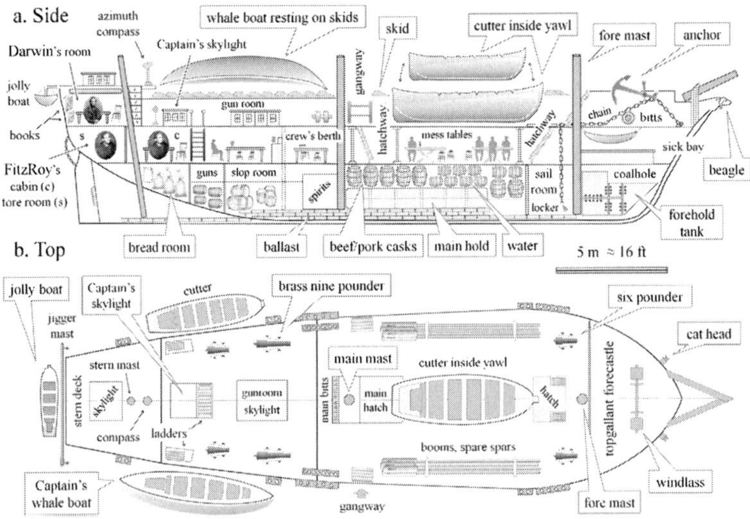

Fig. 3-1 Her Majesty's Ship (HMS) Beagle (28 x 8 m/92 x 26 ft). Captain Fitz-Roy's and Darwin's rooms were in the stern. Redrawn by GPC (as per Paz-y-Miño-C and Espinosa 2024a), after Journal of Researchers (1890).

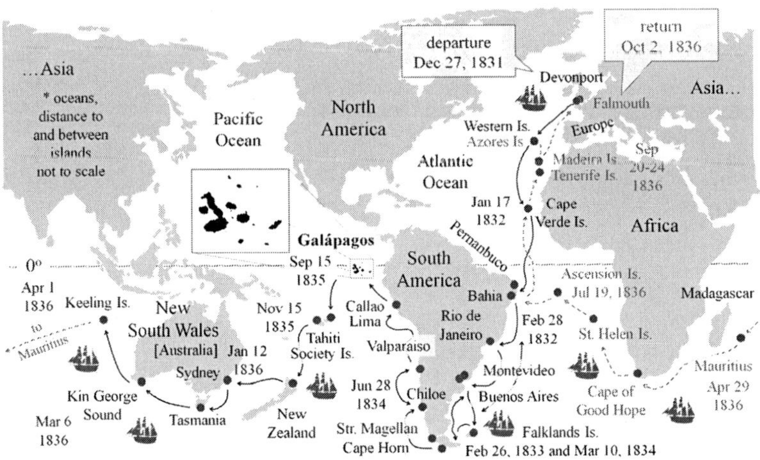

Fig. 3-2 HMS Beagle left Devonport, England, on December 27, 1831, and returned to Falmouth on October 2, 1836. After surveying the East and West coasts of South America for more than three years (1832 to 1835), the ship arrived in the Galápagos on September 15, 1835, and left the archipelago by October 20, the same year. Design: GPC (as per Paz-y-Miño-C and Espinosa 2024a).

To summarize FitzRoy's and his crew's journeying in the Galápagos —including Darwin's—we will rely on the simplified depiction of the islands shown in Fig. 3.3 (as per Paz-y-Miño-C and Espinosa 2024a). More elaborate cartography is shown in other chapters of this volume.

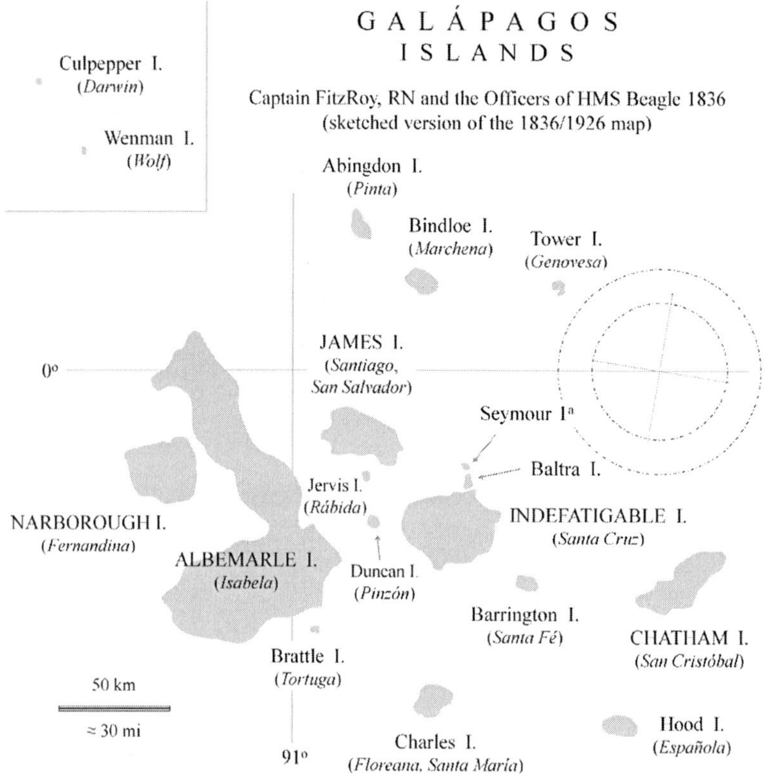

Fig. 3-3 Sketched version of the 1836 map of the Galápagos Islands by Captain Robert FitzRoy and the Officers of HMS Beagle. The 1836 original has additions and corrections traceable to 1926. English and Spanish names of islands included. Design: GPC (as per Paz-y-Miño-C and Espinosa 2024a).

Upon arrival in the archipelago, on September 15, 1835, the Beagle anchored nearby Chatham Island (today's San Cristóbal); its navigation trajectory during the next six weeks is shown in Fig. 3.4, and Darwin's exploration of the islands he visited in Fig. 3.5, which included Chatham, Charles (Floreana or Santa María), Albemarle (Isabela), and James (Santiago or San Salvador). In Fig. 3.6, we have abridged the timeline of HMS

Beagle while on sail or at anchor, with emphasis on Darwin's whereabouts (see Darwin's itinerary in Rookmaaker 2009).

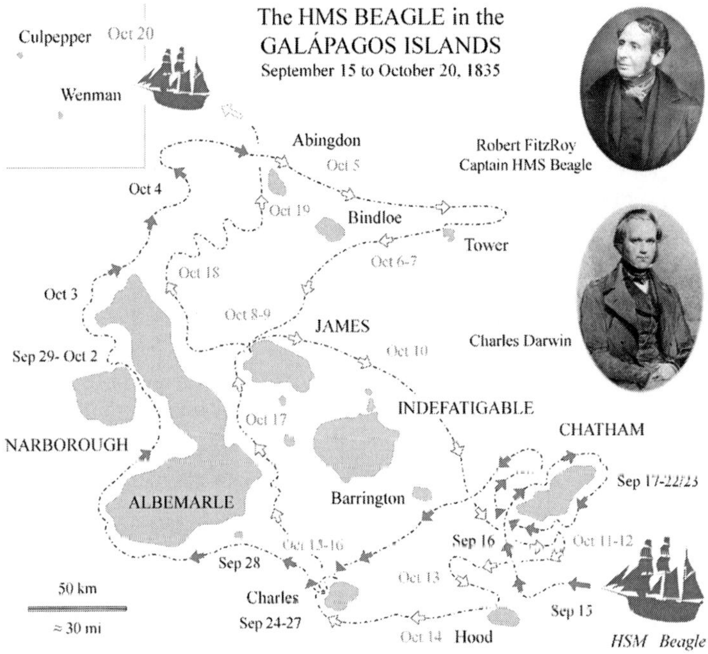

Fig. 3-4 Navigation trajectory of HMS Beagle in the Galápagos. Sketch: GPC (as per Paz-y-Miño-C and Espinosa 2024a).

From September 16 to 20, the Beagle sailed around Chatham (number 1 in Fig. 3.5). Keep in mind that FitzRoy's mission, as commanded by the British Admiralty, was to survey and chart the Galápagos Islands for strategic shipping purposes, while Darwin's was centered on naturalism, particularly geology (Allmon 2016). In the afternoon of September 16, FitzRoy and Darwin climbed what today is known as Frigatebird Hill or Cerro Tijeretas; hard to access back then (Chatham was uninhabited), this spot is currently at relaxed hiking distance from the offices of the National Park, the Gianni Arismendy Environmental Interpretation Center, and the thriving town of Puerto Baquerizo Moreno. On September 17, the Beagle reached St. Stephens Harbor, where it anchored (it returned to this area subsequently, September 21 and 22/23). FitzRoy went ashore searching—without success—for fresh water and climbed Finger Hill (nowadays Cerro Brujo). Darwin collected rocks at this location, likely on September

21 or 22. By September 18, the Beagle anchored off Terrapin Road (Bahía Tortuga de Agua Dulce); Darwin, together with an officer and crew members, went to explore a nearby tuff crater (volcanic cone), now known as Pan de Azúcar. On the surrounding slopes, the men caught 15-18 giant tortoises and brought them back on board as food. After sailing around North and East Chatham for two days (September 19-20), the ship arrived at Fresh Water Bay, where FitzRoy found the source of drinkable water he had been looking for, although not for immediate consumption; the Beagle returned on October 11 to fill its casks prior to leaving Galápagos for good and heading to Tahiti. Darwin's final exploration of Chatham, while the ship reanchored at St. Stephens Harbor (September 21-22/23), included additional geological sampling, plant and animal collecting, and observations at Finger Hill and the iconic Kicker Rock (today's León Dormido). For a geologist, Chatham was significant: the volcanic landscape suggested that tuff craters had formed "while standing immersed in the sea"—he was right (Johnson et al. 2018); for a naturalist, well, a delight of specimens' collections—fishes, birds, insects, and plants.

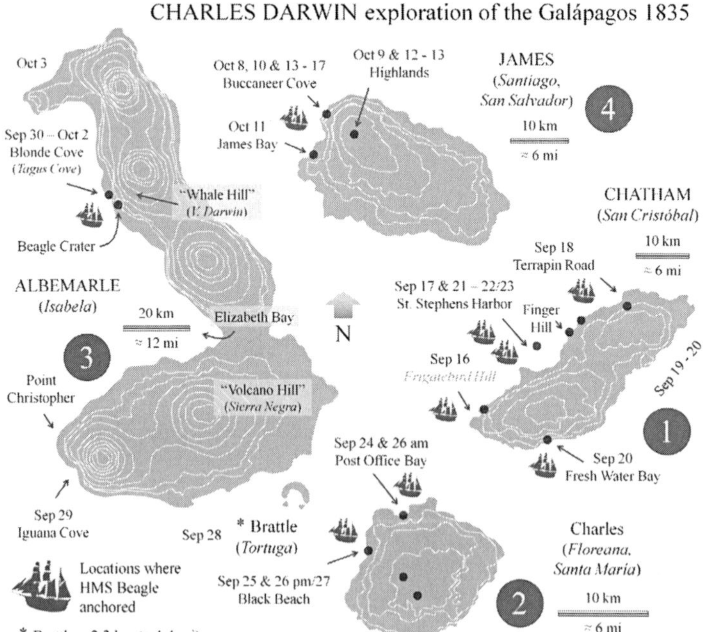

Fig. 3-5 Islands explored by Darwin while visiting the Galápagos. Design: GPC (as per Paz-y-Miño-C and Espinosa 2024a).

Timeline of HMS Beagle while exploring the Galápagos Islands, 1835

Sun	Mon	Tue	Wed	Thu	Fri	Sat
 → HMS Beagle arrives from Lima		15 – Sep HMS Beagle anchors off Chatham	16 – Sep HMS Beagle sails around Chatham, Captain FitzRoy surveys the coast… Darwin (& FitzRoy) land on Chatham…	17 – Sep	18 – Sep	19 – Sep
20 – Sep …exploration of Chatham continues and ends. …Darwin completes 4 landings on Chatham. …Darwin links some birds and plants to counterparts in the continent (South America).	21 – Sep	22 – Sep	23 – Sep	24 – Sep HMS Beagle sails to and arrives in Charles for surveying… In Charles Island, Darwin hears that tortoises on different islands show variations in the form of the shell.	25 – Sep	26 – Sep
27 – Sep …survey-ing of Charles ends.	28 – Sep HMS Beagle sails by Brattle	29 – Sep HMS Beagle sails to and arrives in Albemarle	30 – Sep HMS Beagle anchors in Blonde Cove (Tagus Cove)	1 – Oct (survey of Narborough) Darwin explores Tagus Cove	2 – Oct	3 – Oct …survey-ing of Albemarle and Narborough ends.
4 – Oct HMS Beagle sails North–Northeast toward Abingdon, Bindloe and Tower	5 – Oct	6 – Oct	7 – Oct	8 – Oct HMS Beagle arrives in James and sails Southeast… Darwin and small crew stay in and explore James…	9 – Oct	10 – Oct
11 – Oct HMS Beagle sails off the shores of Indefatigable, toward Chatham and Charles… …HMS Beagle arrives in Chatham… …Darwin and crew (ship's surgeon Benjamin Bynoe, assistants Syms Covington and Harry Fuller) explore James; observations of tortoises.	12 – Oct	13 – Oct …HMS Beagle sails by Hood…	14 – Oct	15 – Oct …HMS Beagle arrives in Charles…	16 – Oct	17 – Oct …HMS Beagle sails to James and picks up Darwin
18 – Oct …HMS Beagle sails North toward Abingdon…	19 – Oct …HMS Beagle sails by Abingdon	20 – Oct …HMS Beagle sails by Wenman and Culpepper	…HMS Beagle leaves the Galápagos Islands…		 …and continues to Tahiti →	

Fig. 3-6 The Beagle's timeline while in the Galápagos. Design: GPC (as per Paz-y-Miño-C and Espinosa 2024a).

Charles Island was explored next, from September 24 to 27 (number 2 in Fig. 3.5). The Beagle sailed into Post Office Bay (evening of the 24th), where Nicholas Lawson, a British resident in the islands, knowledgeable of the archipelago and connected with the authorities, welcomed FitzRoy and Darwin—Lawson was temporarily in charge of the local Governor's affairs. His casual chats will later become important to Darwin and naturalism itself. On September 25, FitzRoy, Darwin, and crewmen boat-sailed to Black Beach, on the Western coast of the island; from there, they hiked to the highlands—Round Hill (today's Cerro Pajas)—to meet Lawson. At the governor's dwelling, in a small settlement, he hosted a feast. During the conversations with Lawson, Darwin learned that the tortoises differed among islands as per body/scales shape and size. But it took Darwin months to realize the significance of Lawson's remarks and link them to the possible instability of species (Grant and Estes 2009). The following day, September 26, Darwin collected specimens around Post Office Bay (fishes, snakes, lizards, birds, insects, land snails, and plants), while the Beagle reached Black Beach to anchor. The last day at Charles, September 27, Darwin and crewmen climbed Round Hill for a final assessment of the landscape. In retrospect, the exchange with Lawson turned historical in that, together with Darwin's own observations of the mockingbirds, which too varied in anatomy from one island to another, led him to suspect that species could indeed change, a novel idea in the 1800s.

The Beagle navigated by Brattle (today's Tortuga) on September 28, aiming for Albemarle, which was surveyed from September 29 to October 3 (number 3 in Fig. 3.5). After a brief stop/pass by Iguana Cove (Caleta Iguana), 29th, the ship continued around Point Christopher (Punta Cristóbal) on the Southwestern part of the island. For any geologist, sightseeing Albemarle's volcanoes from Elizabeth Bay (Bahía Isabel) would have been memorable; Darwin was certainly impressed; he even spotted steam coming out of "Volcano Hill"—likely Sierra Negra. By September 30, the ship anchored off Blond Cove (Tagus Cove). Darwin explored this area on October 1, with particular attention to Beagle Crater (unnamed back then) and "Whale Hill" (later Darwin Volcano). On the morning of October 2, the Beagle left this location, sailed around North Albemarle (3rd), and headed to Abingdon (Pinta), Bindloe (Marchena), and Tower (Genovesa). Darwin wrote at length [i.e. geology] about Blond Cove and Beagle Crater, both while in Galápagos and back in England.

After tortuous routing around the "Northern Islands" (Fig. 3.4), FitzRoy steered the Beagle back South, toward James, where it arrived on October 8 (number 4 in Fig. 3.5). For Darwin, the adventure had just begun. He and three others—the ship's surgeon, Benjamin Bynoe, and assistants Syms Covington and Harry Fuller—disembarked at Buccaneer Cove (Caleta Bucaneros), where the team explored the proximate terrain and hills for nine days! Meanwhile, the Beagle continued Southeast, aiming for Chatham (Fig. 3.4); procuring fresh water was a priority, and FitzRoy knew where to find it at Fresh Water Bay (above). The ship arrived in Chatham on October 11; after refilling its casks, it navigated by Hood (Española), Charles, and the passage between Albemarle and Indefatigable (Santa Cruz; Fig. 3.4). The expectation was to then meet Darwin's hiking team at Buccaneer Cove—which occurred on October 17.

The team's exploration of James reached the Northwest slopes and West lowlands of the island. After setting up camp on rough terrain (October 8), where rocks and lizards' burrows—land iguanas'—were plenty [they are extinct nowadays, but introduction programs do exist to recover their ecological role; Mejía 2019], the men were committed to "observe, collect, and hike" (Grant and Estes 2009). On October 9 and later—12th and 13th, they hiked up to 300 meters (1000 ft) elevation and spotted what today is called Cerro Pelado (Herbert et al. 2009). The nights of October 10 and later—13th to 17th—were spent at Buccaneer Cove, while on October 11, the team was taken by settlers on a fishing boat to a salt-mine crater ("salina"), South of James Bay (Fig. 3.5).

The scenery, craters, rocks, vegetation in the low and highlands, birds, insects, land snails, land and marine lizards [iguanas], and giant tortoises, which Darwin observed again at James and even tested how fast they could walk on the irregular terrain (Fig. 3.7), became relevant and left lasting impressions for future writing (e.g. Chapter XVII of *Journal of Researchers* 1839/45 is entirely dedicated to the Galápagos Archipelago).

FitzRoy sailed the Beagle to Buccaneer Cove on October 17. Darwin's team came on board, and the ship traveled North to Abingdon (18th/19th) to pick up survey-boat crewmen (Fig. 3.4). The remaining islands to explore were Wenman (Wolf) and Culpepper (Darwin), on the far Northwest of the archipelago, which were surveyed by October 20.

To this day, the Beagle's six-week journeying about the Galápagos, meticulous cartographic surveying by crewmen, or copious specimens collected by Darwin and his team continue to amaze scientists. How was that possible in the 1830s? HMS Beagle was equipped with leading-edge instruments for navigation and mapping, handled by smart assistants under the command of a gifted Captain. As the ship sailed the archipelago (Fig. 3.4), the surveying officers fast traveled in yawls to chart the coastlines of near or distant islands (Fig. 3.8); they left for days or weeks and met the Beagle at prearranged locations. As for Darwin, history will later confirm that he was the most brilliant naturalist an expedition vessel could have.

Darwin's observations in the Galápagos as sketched in Holder's 1891 book *"Charles Darwin, His Life and Work"*

Fig. 3-7 Top left: giant tortoises; one is feeding on *Opuntia* cactus, or "prickly pears". Darwin's/Galápagos' finches appear in the bottom. Top right: green sea turtle in the opening of Chapter XI (*Among The Ocean Volcanoes*) in Holder's book. Bottom: Darwin assessing the speed of a giant tortoise (30 yards in 5 minutes!), in reference to Darwin's narrative of having conducted the assessment while on James (Santiago) Island. Source: Holder (1891).

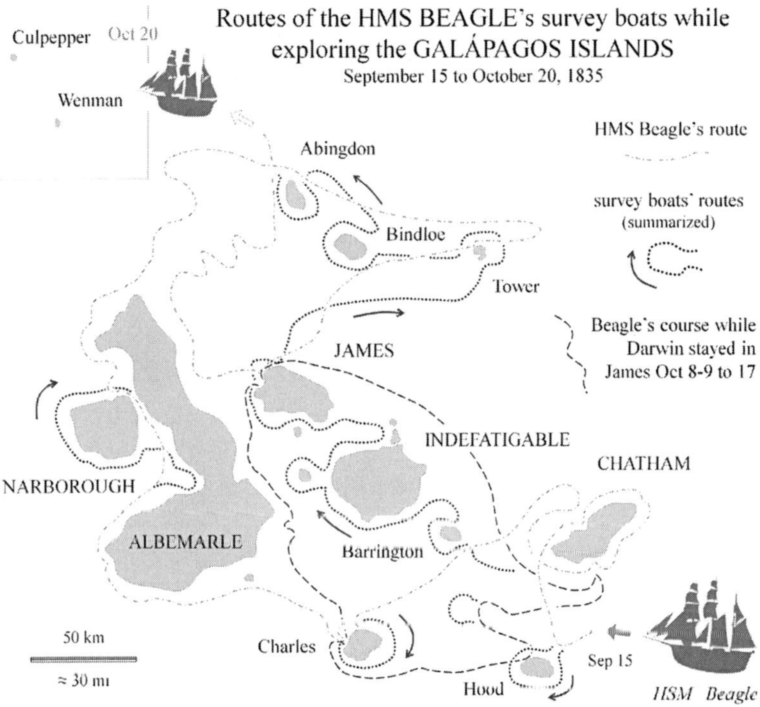

Fig. 3-8 As the Beagle's sailed the Galápagos, its survey boats (yawls and cutters), led by skillful officers, surveyed and mapped individual islands. Design: GPC (as per Paz-y-Miño-C and Espinosa 2024a).

Days at Sea, at Anchor, or On Land

Did Darwin come up with "his theory of evolution by means of natural selection" while in the Galápagos? The short answer is—of course—no! The longer one, admitting generous hindsight, could lead us to find connections between Darwin's thinking once back in England and observations at Chatham, Charles, Albemarle, and James, or even the chat with Lawson about the tortoises' variations among islands (above). Although Darwin did make sharp inferences about geology and naturalism during his entire voyage around the globe (Brinkman 2010), the all-inclusive argument of *On the Origin of Species by Means of Natural Selection, or the Preservation of Favoured Races in the Struggle for Life* matured years later (1859). The visit to the Galápagos was short (Fig. 3.9), and realizations of its significance for evolution took place gradually.

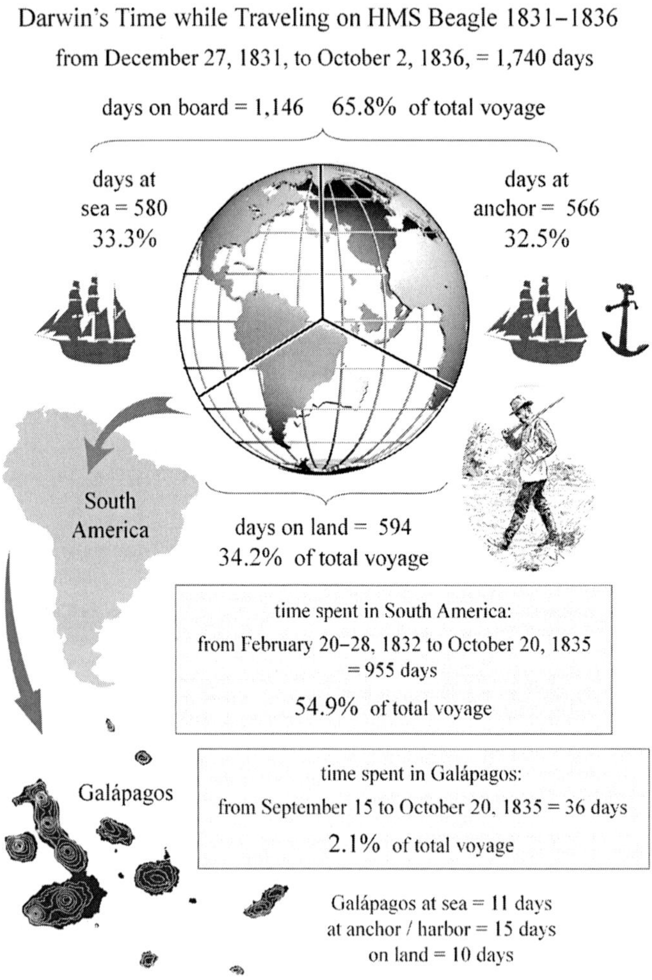

Fig. 3.9 Darwin's time share on board, at anchor, or on land while traveling on HMS Beagle (1831-1836). Note that the time spent in the Galápagos was 36 days, about 2.1% of the total voyage. Sketch: GPC (as per Paz-y-Miño-C and Espinosa 2024a); data extracted from Rookmaaker (2009).

Darwin's time while traveling on HMS Beagle added up to 1,740 days: one third at sea (580 days, 33.3%), another at anchor (566 days, 32.5%), and another on land (594 days, 34.2%). More than half of the voyage (955 days, 54.9%) was spent in South America (Fig. 3.9). The time dedicated to the Galápagos corresponded to 36 days, or 2.1% of the Beagle's complete voyage; Darwin was at sea 11 days, at anchor/harbor 15 days, and on land 10 days (Rookmaaker 2009). While traveling, he kept detailed written records of daily events; 15 notebooks/diaries (1832-1836) exist of these accounts, totaling 2,070 pages of notes, 116,080 words, and 241 sketches (Chancellor and van Wyhe 2009; Fig. 3.10). The notebooks ranged in number of pages from 88 to 242; words from 3,130 to 22,130; and sketches from 4 to 47. The "*Galapagos*" notebook included 100 pages (4.8% of the total in the 15 notebooks), 3,610 words (3.1%), and 4 sketches (3.6%). Notice in Fig. 3.10 that Darwin started to write about specific destinations (Galápagos) months before the Beagle visited them.

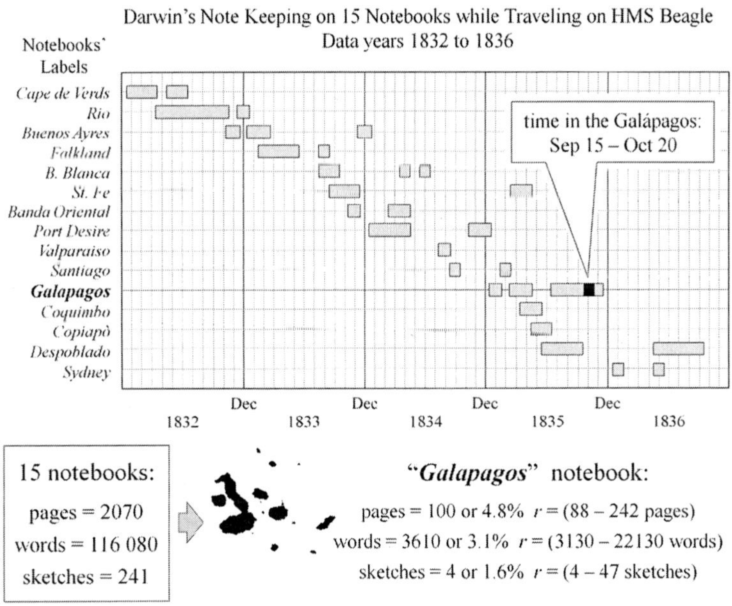

Fig. 3-10 Darwin's note-keeping while traveling the globe on HMS Beagle (data years 1832-1836). Redrawn by GPC (as per Paz-y-Miño-C and Espinosa 2024a), after Chancellor and van Wyhe (2009).

By 1839, *Journal of Researchers* (JoR) had been published (King, FitzRoy, and Darwin 1839): Volume I by P. Parker King (Captain of HMS Beagle prior to FitzRoy), Volume II by FitzRoy, and Volume III by Darwin. In future editions, JoR and particularly Volume III became known as *The Voyage of the Beagle* or *A Naturalist's Voyage* (Fig. 3.11), which was largely based on Darwin's diaries. He subtitled the contents of Chapter

XVII [the book had 21 chapters], *Galapagos Archipelago*, as follows: *The whole Group Volcanic – Number of Craters – Leafless Bushes – Colony at Charles Island – James Island – Salt-lake in Crater – Natural History of the Group – Ornithology, curious Finches – Reptiles – Great Tortoises, habits of – Marine Lizard, feeds on Sea-weed – Terrestrial Lizard, burrowing habits, herbivorous – Importance of Reptiles in the Archipelago – Fish, Shells, Insects – Botany – American [continent] type of organization – Differences in the Species of Races on different Islands – Tameness of the Birds – Fear of Man, an acquired instinct.*

Fig. 3-11 HMS Beagle navigating the Straits of Magellan (Southern South America; see map in Fig. 3.2). Illustration by Robert Taylor Pritchett (Darwin 1890).

Notebook B: Transmutation (1837–8)

As a naturalist, Darwin constantly wrote notebook diaries. Back in England in 1837–8, two years after exploring the Galápagos, he sketched on page 36 of *Notebook B: Transmutation* the now celebrated "I think" tree (Fig. 3.12), which we have redrawn and expanded for clarity in Fig. 3.13. In various pages (36–46) of the notebook, and referring to the sketch, Darwin reasoned about relations [relatedness] among genera, represented by A, B, C or D, although, as he implied in writing, A–D could indicate other groups (i.e. mammals, birds, Articulata [at that time insects and crustaceans], plants). An "ancient type" (circled number 1 at the bottom of Fig. 3.13) represented a species capable of making 13 recent forms [i.e.

giving origin to them], which have a perpendicular line at the end (see dialog box in bottom right Fig. 3.13); the remaining "twelve of the contemporarys [species] must have left no offspring at all" (center left Fig. 3.13). Darwin concluded: "Thus between A & B immense gap of relation. C & B the finest gradation, B & D rather greater distinction. Thus genera would be formed. — bearing relation" (Figs. 3.12 and 3.13).

Charles Darwin's Notebook B: Transmutation (1837–8), page 36

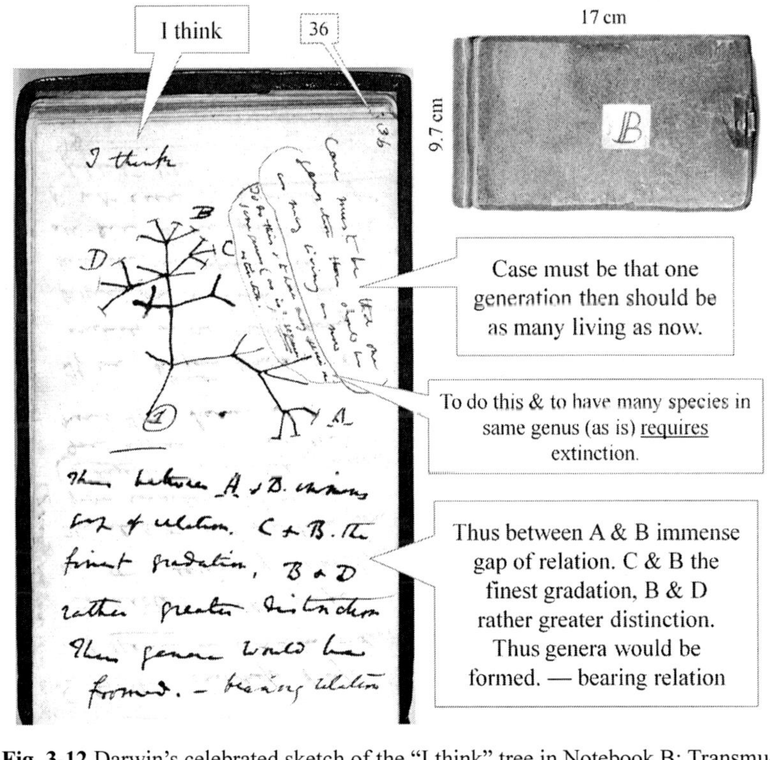

Fig. 3-12 Darwin's celebrated sketch of the "I think" tree in Notebook B: Transmutation (with permission of the Syndics of Cambridge University Library).

Yes, Darwin was talking to himself about how modern genera could emerge in nature, but he did not have—at least not yet—a mechanism to explain the process. That will be provided in *The Origin* of 1859, together with a saturation of examples gathered from the finest literature, correspondence with accomplished colleagues, and his own observations while traveling the world or experimenting at home—Darwin was busy (van Wyhe 2007)! In due course, the Galápagos turned out to be central to the debate about the mutability of species (finches, mockingbirds, flightless cormorants, giant tortoises), and Darwin and the archipelago became inseparable, bound to be understood and discussed together.

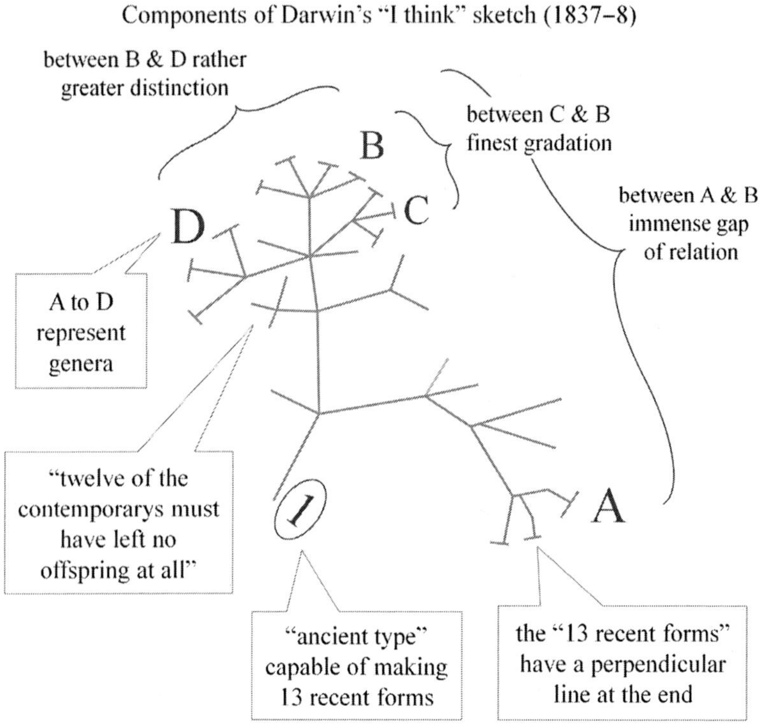

Fig. 3-13 Notes to himself, Darwin's "I think" sketch. Design: GPC (as per Paz-y-Miño-C and Espinosa 2024a).

1859: A More Complex Depiction of Evolution

The first edition of *On the Origin of Species* (1859) had a single diagram, which appeared in Chapter IV *Natural Selection*. We have re-sketched it in Fig. 3.14 and added to it dialog explanatory boxes. It was, evidently, a far more complex drawing than the "I think" tree of 1837–8 (keep in mind that modern evolutionary trees are quite different from Darwin's initial attempts to illustrate species/genera, or any taxa, diversification).

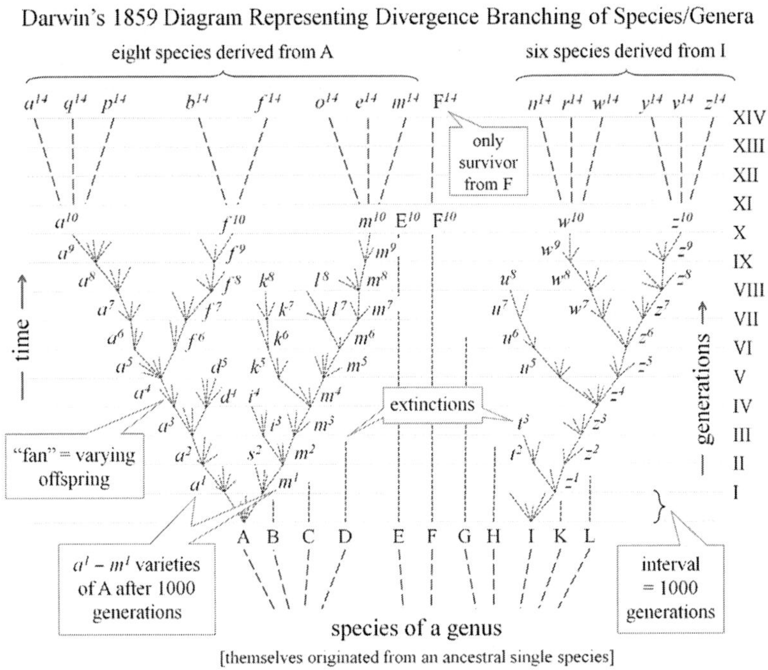

Fig. 3-14 Darwin's (1859) diagram representing divergence branching of species or genera; it was the only illustration in *On the Origin of Species* and a formal attempt to explain patterns of lineages' diversification in nature. Redrawn and modified by GPC (as per Paz-y-Miño-C and Espinosa 2024a).

At the bottom of the illustration (Fig. 3.14), Darwin listed A to L as species of a genus, which themselves originated from an ancestral one. "Time" (left of diagram) ascended from past to present and corresponded to intervals of 1,000 "generations", labeled I to XIV (right of diagram); Darwin alerted that "it would have been better if each had represented ten thousand generations." In *The Origin*, he continued:

> The little fan of diverging dotted lines of unequal lengths proceeding from (A), may represent its varying offspring [i.e. the dialog box "fan = varying offspring," lower left in Fig. 3.14]. The variations are supposed to be extremely slight, but of the most diversified nature; they are not supposed all to appear simultaneously, but often after long intervals of time; nor are they all supposed to endure for equal periods. Only those variations which are in some way profitable will be preserved or naturally selected.

> And here the importance of the principle of benefit being derived from divergence of character comes in; for this will generally lead to the most different or divergent variations (represented by the outer dotted lines) being preserved and accumulated by natural selection.
> When a dotted line reaches one of the horizontal lines, and is there marked by a small numbered letter [e.g. a^1, a^2, a^3...], a sufficient amount of variation is supposed to have been accumulated to have formed a fairly well-marked variety, such as would be thought worthy of record in a systematic [taxonomic] work.

Darwin acknowledged the frequent occurrence of extinctions over time; for example, species D or t^3 (a descendant of t^2, which itself originated in z^1, and both traceable to I) went extinct at generation interval III (dialog box "extinctions", center in Fig. 3.14). Others, like F^{14}, endured the hazards of nature and persisted, although substantially modified by generation interval XIV (dialog box "only survivor from F", top in Fig. 3.14). But some, like species A or I, out of the initial 11 taxa (A to L) diversified greatly and gave origin to 8 (a^{14}, q^{14}, p^{14}, b^{14}, f^{14}, o^{14}, e^{14}, m^{14}) or 6 species (n^{14}, r^{14}, w^{14}, y^{14}, v^{14}, z^{14}), respectively. Darwin clarified:

> But I must here remark that I do not suppose that the process ever goes on so regularly as is represented in the diagram, though in itself made somewhat irregular. I am far from thinking that the most divergent varieties will invariably prevail and multiply: a medium form may often long endure, and may or may not produce more than one modified descendant; for natural selection will always act according to the nature of the places which are either unoccupied or not perfectly occupied by other beings; and this will depend on infinitely complex relations.

Thus, in Chapter IV of *The Origin*, Darwin brought into play the mechanism of natural selection, in concordance with the characteristics of the habitats in which organisms occur, as the force allowing or limiting the prevalence of lineages in nature. He concluded:

> The great Tree of Life – The affinities of all the beings of the same class have sometimes been represented by a great tree. I believe this simile largely speaks the truth. The green and budding twigs may represent existing species; and those produced during each former year may represent the long succession of extinct species. At each period of growth all the growing twigs have tried to branch out on all sides, and to overtop and kill the surrounding twigs and branches, in the same manner as species and groups of species have tried to overmaster other species in the great battle for life.

Ernst Mayr Take on Darwin's Thinking

In 1977, Ernst Mayr, back then a professor of zoology at Harvard University, speculated about "how Darwin may have discovered his highly unconventional theory" (Mayr 1977). In Fig. 3.15, we have summarized Mayr's proposal. It consisted of five observations in nature, or "facts" (1-5), intertwined with three inferences (1st to 3rd).

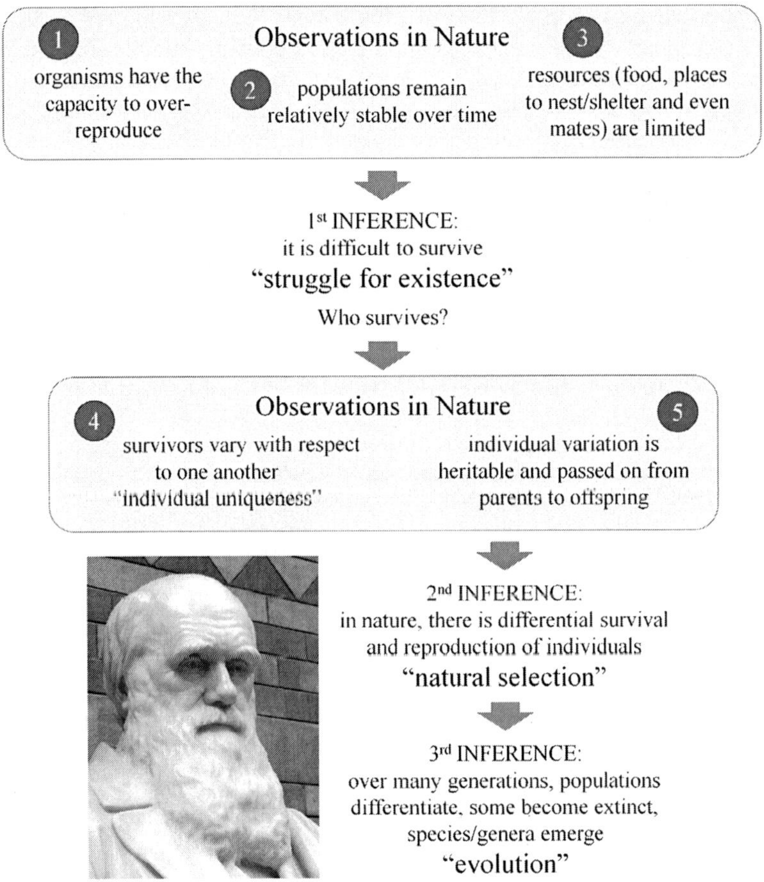

Fig. 3-15 Ernst Mayr's proposal about how Darwin might have conceptualized evolution by means of natural selection. Design and photo: GPC (as per Paz-y-Miño-C and Espinosa 2024a), after Mayr (1977). For a modern and friendly readable version of Darwin's *On the Origin of Species*, see Pechenik (2023).

Observation 1: Darwin knew, through readings and work by Thomas R. Malthus (1798/1836) and others, that organisms had the capacity to over-reproduce (i.e. superfecundity, populations grew exponentially). Observation 2: Despite this intrinsic tendency to breed and generate large progenies, wild populations remained relatively stable. Observation 3: Resources in the wilderness were limited (food, places to nest/shelter, and even mates). Linked to these observations was the 1st inference: it ought to be difficult to survive, or "the struggle for existence". Who survived?

Observation 4: Survivors varied with respect to one another (individual variation), which suggested individual uniqueness. Observation 5: Such variation was evidently heritable since offspring resembled their progenitors. 2nd inference: Darwin postulated that, in nature, there was differential survival and reproduction of individuals who later passed on valuable traits (i.e. advantageous in "the battle for life"); natural selection!

What were the long-term consequences? 3rd inference: Darwin theorized that over many generations, populations would differentiate, some would become extinct, and species, genera, and other taxonomic categories would emerge—evolution!

Today, the theory of evolution by means of natural selection is attributed to both Charles Darwin and Alfred Russel Wallace (co-discoverers of natural selection; Kutschera 2003; but see van Wyhe 2020). Ever since Darwin and Wallace, however, our collective understanding of evolution has changed significantly at the molecular, genetic, organismal, ecosystem, global, and cosmic levels (Vanchurin et al. 2022).

The "Darwin's Finches" Tale

An iconic example of the connection between Darwin and Galápagos is centered around the finches, but the story is intricate. While in the archipelago, Darwin and his assistants collected birds of multiple species (above). By 1837, English ornithologist John Gould catalogued 13 of them as finches, rather than a mix of finches, blackbirds, grossbeaks, or warblers, as Darwin and his team thought they were (Sulloway 1982). In 1935, Percy R. Lowe, bird curator who had examined the collections at the British Museum, read a paper before the British Association (Darwin Centenary): *The Finches of the Galapagos in Relation to Darwin's Conception of Species*; in it, Lowe proposed to name the birds "Darwin's Finches" or the Geospizids (Lowe 1936). He stated:

> I do so because I think it is true to say that there is no group of birds in the whole world which has more right to occupy the attention of zoologists at the present moment; for the problem presented by the very extraordinary

> diversity seen within this group of Finches appears to me to be a problem of first-class biological importance. We know that it was the diversity presented by these Finches, as well as the Mocking-birds, tortoises, and plants, which started Darwin down that brilliant corridor of thought which led to his conception of the origin of species.

Years later, in 1947, British evolutionary biologist David Lack published *Darwin's Finches: An Essay on the General Biological Theory of Evolution*, nowadays a classic. The rest is history; Darwin and the finches became bound forever, but legendarily (Sulloway 1982).

During the past +40 years, Peter and Rosemary Grant, emeritus professors at Princeton University, have sustained the most comprehensive research program on Darwin's finches, particularly on Daphne Major Island. The Grants contributed too to underline the relevance of Lack's 1940s work; they called attention to his attempt to explain the process of differentiation of the finches upon arrival of an ancestor—from nearby mainland—to the Easternmost islands of the archipelago (i.e. the oldest and nearest to Central and South America; Fig. 3.16).

Fig. 3-16 Suspected process of arrival, colonization, dispersal, and speciation of ancestors of the Galápagos' or Darwin's finches. Lack (1947) envisioned speciation in allopatry (i.e. via geographic isolation). Layout: GPC (as per Paz-y-Miño-C and Espinosa 2024a); redrawn after Grant and Grant (2014).

The suspected process included various stages of allopatric speciation (i.e. populations geographically isolated with respect to one another). Stage 1: colonization of ancestors from the mainland and establishment of a breeding population on one of the islands (chosen arbitrarily in Fig. 3.16). Stage 2: establishment of additional populations on nearby islands (the phenomenon could repeat itself and involve multiple islands). Stage 3: secondary contact between divergent populations. Note that since stages 2 and 3 were expected to be redundant over time, numerous species or their varieties could emerge in the archipelago (Grant and Grant 2014).

We know now that speciation might result from multiple routes rather than the allopatric process envisioned for the finches by Lack and others after him (but see disagreement in Bowman 1961). Divergent populations of the same species can or cannot interbreed upon secondary contact and can or cannot generate new species. For example, new species can emerge—not always—upon secondary contact if divergent populations do not interbreed and their differences and incompatibilities (e.g. genetic, behavioral) bring them—over time—even more apart. But if they do interbreed, their descendants can either integrate into a single population (fusion) or depart into distinctive species over time (fission by means of divergent selection; Grant and Grant 2014). The Grants have documented, in recent years, that new lineages have emerged among Darwin's finches via the mixing path of introgressive hybridization. In it, members of distinctive species hybridized, and their hybrid descendants continued to reproduce—in unequal genetic mixing—with the parental species. Thus, multiple phenotypes resulted, in some cases of enriched genetic variability and fully fertile (Lamichhaney et al. 2018).

These days, 15 species of Darwin finches are recognized based on genetic data (14 in Galápagos and 1 in Cocos Island). In Fig. 3.17, we have sketched a phylogenetic reconstruction by Lamichhaney et al. (2015) based on the whole genome sequencing of 120 birds from 13 islands. The researchers suggest that, about 900,000 years ago, the "warbler finches", an unfortunate name—they are not warblers—attributed to birds in the genus *Certhidea* (top Fig. 3.17), split from the other finches (i.e. so called "ground" and "tree" finches: *Platyspiza*, *Pinaroloxias*, *Geospiza*, *Camarhynchus*); the latter subsequently diverged around 100,000 to 300,000 years ago (scale in bottom left of Fig. 3.17). Note that the birds' skulls on the right side of the phylogeny were redrawn after Bowman (1961) and added to the tree—by us—for didactic purposes (as per Paz-y-Miño-C and Espinosa 2024a). Robert J. Bowman authored another classic *Morphological Differentiation and Adaptation in the Galápagos Finches*, which had superb illustrations; we have re-sketched them in Fig. 3.18.

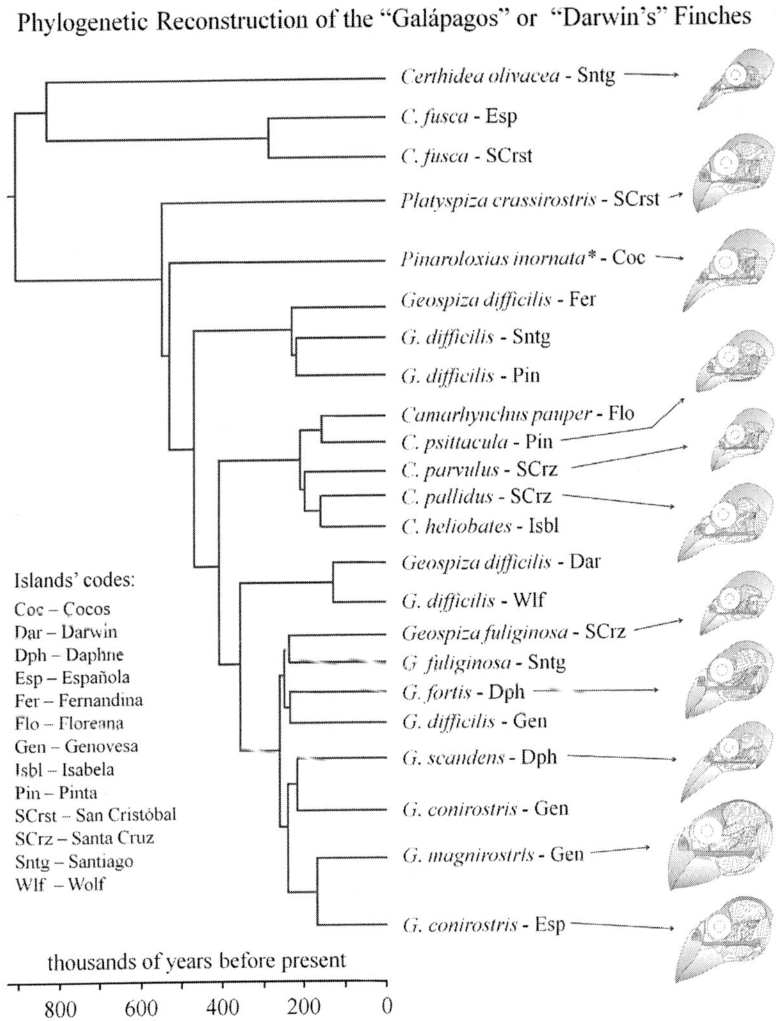

Fig. 3-17 Phylogenetic reconstruction of the Galápagos' or Darwin's finches; maximum likelihood tree based on autosomal genome sequences (16K protein-coding genes, 120 samples); * Cocos finch in Cocos Island (Northwest of the Galápagos archipelago); redrawn after Lamichhaney et al. (2015). Sketches of finches' skulls redrawn after Bowman (1961) and placed next to phylogeny for didactic purposes only. Layout: GPC (as per Paz-y-Miño-C and Espinosa 2024a).

Yet the finches continue to surprise us (Fig. 3.19). Recent taxonomic and biogeographic analyses have placed them within the tanagers' family Thraupidae, subfamily Coerebinae—not finches at all, and with likely origins in the Caribbean or—less likely—South America (Sato et al. 2001; Funk and Burns 2018). [Details in Paz-y-Miño-C and Espinosa 2024a].

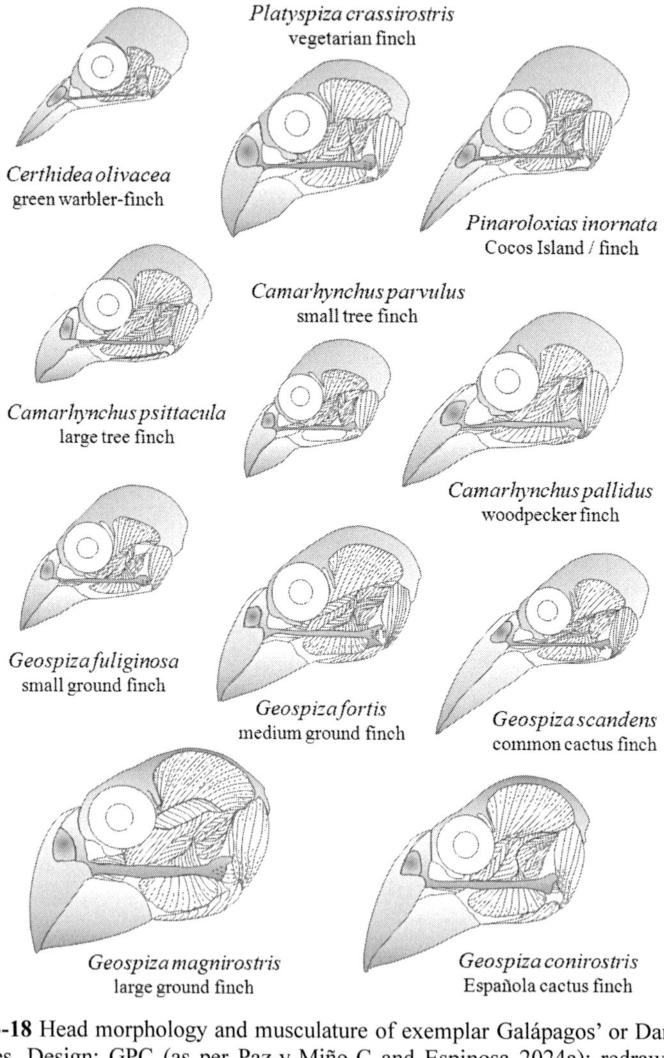

Fig. 3-18 Head morphology and musculature of exemplar Galápagos' or Darwin's finches. Design: GPC (as per Paz-y-Miño-C and Espinosa 2024a); redrawn and modified after Bowman (1961).

Cactus finch (*Geospiza scandens*), one of the Galápagos/Darwin finches

A female medium-ground finch (*G. fortis*) approaching our camera, Santa Cruz

Fig. 3-19 Cactus and medium-ground finch. Recent studies suggest all Darwin's finches belong to the tanagers' family Thraupidae (Sato et al. 2001). Photos: GPC (as per Paz-y-Miño-C and Espinosa 2024a).

To close this chapter, Darwin's life and legacy are celebrated in the Galápagos (Fig. 3.20), perhaps as much as in other latitudes, an evident transformation since we—the authors—first visited the islands in 1980/1, when the human population totaled +5,000. By the 2022 census, it had reached +28,500 and was growing (INEC 2022; see also Chapter Six, *Who Lives in Today's Galápagos Islands?*).

Fig. 3-20 Statues of Charles Darwin in San Cristóbal (Chatham), Santa Cruz (Indefatigable), Isabela (Albemarle); CDRS = Charles Darwin Research Station. Photos: GPC (as per Paz-y-Miño-C and Espinosa 2024a).

Conclusion

As Darwin ends in *The Origin*:

> ...there is grandeur in this view of life, with its several powers, having been originally breathed into a few forms or into one... from so simple a beginning endless forms most beautiful and most wonderful have been, and are being, evolved.

The Beagle's voyage (1831-36), FitzRoy's command, Darwin's brilliance, and the uniqueness of the Galápagos Islands all merged in history. The notion of evolution by means of natural selection, which Darwin conceptualized later in life (1859), can now be applied to nature (but see Laland et al. 2014). The "great monsters" and "a little world within itself", as he referred to the giant tortoises and the [Galápagos] isles (1839; for *The Quotable Darwin*, see Browne 2018), are now exemplars of natural history (but see thorough critical analysis in Sulloway 2009). We will return to them in later chapters of this volume.

Questions

The following questions (Q1–3) merit generous written answers. You might need to do additional reading or library work to enrich your answers (we suggest exploring Paz-y-Miño-C and Espinosa 2024a,b).

Question 1: You have learned from this chapter that "while in the [Galápagos] archipelago, Darwin and his assistants collected birds of multiple species. By 1837, English ornithologist John Gould catalogued 13 of them as finches, rather than a mix of finches, blackbirds, grossbeaks, or warblers, as Darwin and his team thought they were" (as per Sulloway 1982). You have also learned that "recent taxonomic and biogeographic analyses have placed [the Galápagos finches] within the tanagers' family Thraupidae—not finches at all" (as per Sato et al. 2001; Funk and Burns 2018). What does this initial confusion tell you about the *adaptive radiation* (evolution and particularly anatomical convergent evolution) of these birds since their ancestors arrived in Galápagos from the mainland, either the Caribbean or—less likely—South America? [In other words, why did Darwin and his assistants think the birds were "a mix of finches, blackbirds, grossbeaks, or warblers", when they were, according to bird specialist John Gould, 13 related species in a lineage?].

Question 2: As indicated in this chapter, nowadays, the theory of evolution by means of natural selection is attributed to both Charles Darwin and Alfred Russel Wallace (co-discoverers of the mechanism of *natural selection*; Kutschera 2003; although some scholars dispute this view, e.g., van Wyhe 2020). What were the *specific differences* in Darwin's versus Wallace's views about natural selection?

Question 3: Much has been written about Darwin's role during "The Voyage of the Beagle", e.g., the liaison "Darwin" and "evolution" (pop science has publicized it for more than a century). However, little is known—except among academics—about the Beagle's broad mission(s) during the early 1800s as a Royal Navy vessel. What was the *geopolitical purpose* of HMS Beagle's survey voyage around the world (1831 to 1836), under the command of Captain Robert FitzRoy?

How to Answer the Questions Above

The questions above have various possible answers (A1–3). Here we suggest perspectives from which to address the queries.

Answer 1: The key component of this question is convergent evolution, which you should be able to define before addressing why Darwin and his team thought that the birds they collected in Galápagos (1835) were—at first glance—a mix of finches, blackbirds, grossbeaks, or warblers, rather than 13 species within a single lineage ["finches"], as ornithologist John Gould concluded (1837). Now, the fact that in the 2000s, modern scientists determined—via taxonomic and biogeographic analyses—that the Galápagos' or Darwin's finches were tanagers (Sato et al. 2001; Funk and Burns 2018)—not finches at all—should give you a clue about the state of knowledge in the early 1800s versus the 2000s concerning the phylogenetic affinities among the world's birds' taxa. [Note that the latter, although related, is a separate part of the question].

Answer 2: Here you need to do library research and identify specific books and academic articles in which Darwin's versus Wallace's views of evolution are discussed (e.g. Kutschera 2003; Browne 2013). Stay away from websites, although there are some reputable ones that you could explore, including Darwin Online (The Complete Work of Charles Darwin Online), Darwin Correspondence Project, Wallace Online, and The Wallace Correspondence Project [look for them online].

Answer 3: Library work is also required here. You should find out what England's standing in the world was during the early to mid 1800s, as well as the global dynamics among central economies in Europe with respect to the New World, islands in the Pacific, New Zealand, Australia, and coastal Africa (Fig. 3.2). Charles Darwin's responsibilities as naturalist on board HMS Beagle were important [his work later turned historical], but in the context of the ship's broad mission (1831-1836), as a Royal Navy vessel, Darwin's duties were supplementary.

Discussion Topics

The discussion topics below (D1–2) should be addressed with your peers. We encourage college instructors to facilitate the dialogue.

Discussion 1: Examine Fig. 3.17 (Phylogenetic reconstruction of the Galápagos' or Darwin's finches) and Fig. 3.18 (Head morphology and musculature of exemplar Galápagos' or Darwin's finches). Propose alternative hypotheses that explain *why* and *how* an archipelago setting would facilitate the emergence, via adaptive radiation, of distinctive species of finches in Galápagos. Discuss how the "life cycle" (ontogeny) of an island (or volcanic archipelago) would determine the fate of taxa arrivals [i.e. ancestors of today's finches], speciation [i.e. adaptive radiation], and extinction [of finches]. Keep in mind that, as discussed in Chapter One (*What Are the Galápagos Islands?*), for the duration of their life cycles, hotspot islands or archipelagos—like the Galápagos—are expected to go through emergence, development, and subsidence. Thus, your hypotheses should explain how the Galápagos' or Darwin's finches have *diversified* and *persisted* on a constantly changing archipelago as per the ontogeny of its islands.

Discussion 2: There are several myths about Charles Darwin. One of them is that he came up with "his theory of evolution by means of natural selection while in Galápagos or soon after visiting the archipelago". Argue why this is absurd and yet continues to be disseminated as fact in talks, books, articles, or documentaries. Hint: think of why this could not have been the case—for Darwin—in 1835 or soon after.

Suggested Observation Projects

The suggested observation projects below (P1–3) take into consideration that you will travel to Ecuador and the Galápagos legally; that you will visit the islands as a tourist [holding a Transit Control Card, a mandatory document issued by Ecuador and specifically by the Governing Council of Galápagos]; and that you will respect the regulations that protect the Galápagos National Park and the Marine Reserve (see complete list of rules in Appendix B). Only accredited scientists, academics, and/or naturalists who have obtained official permits can do research in Galápagos. You can take pictures, but flash photography is not permitted when taking photos of wildlife. Professional photography and videos recorded for commercial purposes must be authorized by the Galápagos National Park Directorate (GNPD). The ideas below are for enthusiastic college explorers interested in learning about the archipelago.

Project 1: Can you ride a bicycle? Here is a fun project for you and a project partner: "Darwin in Galápagos". For it, you will need bikes [they can be rented for a few dollars a day, together with protective gear, at the local towns, including Puerto Ayora, Puerto Baquerizo Moreno, Puerto Villamil, or Puerto Velasco Ibarra], street maps of the local towns [from Google Maps], a notebook, and your cell phone camera. Plan your bike rides in search of: statues of Charles Darwin (from official to amateur); streets or businesses named after him (e.g. souvenir stores, tour providers, hotels, restaurants, cafes, or bars); schools and education centers; interpretation centers and local museums with references to Darwin; boats and ships on the harbors; and anything you believe honors Charles Darwin on the islands. Ask the locals for feedback and directions. You will be overwhelmed by the amount of information you can gather. Plus, it will be an excellent way to get to know the towns quickly. Take notes and photos, document the location of all identified sites on your maps, and organize the information thoroughly. Upload your observations to an Excel file and turn the data into descriptive statistics [i.e. totals of this or that and the percentage representation of your items; be creative]. Put together a slideshow presentation (for a 30-minute talk) to be shared with a general audience. Avoid slides with "text only" [they are often terrible]. Keep it rigorous yet fun and didactic. You can also upload amateur videos to the web and share them with the world. [But remember, professional photography and videos recorded for commercial purposes must be authorized by the GNPD].

Project 2: While visiting the islands, you will learn to spot the Galápagos' or Darwin's finches. For that, you will need binoculars and a photo-based field guide of the local birds (e.g. Swash and Still 2006; De Roy 2022). You will soon realize that it can be challenging to tell apart the species of finches. The birds vary in size, color (from black to dark brown, light brown, grey, olive, or yellowish), sexual dimorphism (males and females can be different in color and size), beak morphology and dimensions (with many intermediate morphs), occurrence (species vary from island to island or from the lowlands to the highlands), and behavior (ground versus tree finches with distinct feeding habits). Yet, this makes them ideal for an observation project. During your entire visit to the Archipelago, document the finches you observe in the protected areas of the National Park, agricultural lands, side of the road, or urban settings. Take many pictures (cell-phone zoom-ins will be helpful) and save them as per date, location, hour of observation, and type of habitat on which you find finches. [Do not discard blurry photos; most shots will be helpful to identify species]. Keep quantitative records of your observations; this should be transcribed to an Excel file. Relying on your photo-based field guide of the Galápagos birds, identify the finches to the species level (or, at least, ground versus tree finches). Work on your Excel spreadsheet(s) and organize the data according to island, location (i.e. inside the National Park, agricultural lands, side of the road, or urban settings), lowlands or highlands, habitat type (e.g. sandy beach, rocky shore, wetland, old/er or more recent lava flows, or inland), and vegetation cover (e.g. grassland, shrubland, mangrove, dry forest, or humid forest). [For vegetation cover maps of all islands, see Rivas-Torres (2018), Moity et al. (2019), and Paz-y-Miño-C and Espinosa (2024b)]. Put together a slideshow presentation (for a 30-minute talk) to be shared with a general audience. Connect your story of the finches to Charles Darwin's exploration of the islands in 1835 and to the relevance of this group of birds to our current understanding of speciation on oceanic islands. Avoid slides with "text only" [they are often terrible]. Keep it rigorous yet fun and didactic.

Project 3: Prior to the HMS Beagle survey voyage around the world (1831 to 1836), Captain Robert FitzRoy was an accomplished Officer in the Royal Navy. He retired from active service in 1850. He then became a prolific meteorologist until his tragic death in 1865. FitzRoy's relevance in history has been—arguably—outshined by Charles Darwin's monumental legacy to naturalism and modern science. Put together a slideshow presentation (for a 30-minute talk) to be shared with a general audience in which you address "The Relevance and Fate of Robert FitzRoy (1805-1865), Former Captain of HMS Beagle". Avoid slides with "text only" [they are often terrible]. Keep it rigorous yet didactic.

References

Allmon, Warren. 2016. "Darwin and Palaeontology: A Re-evaluation of His Interpretation of the Fossil Record". *Historical Biology*, No. 28: 680-706.

Bowman, Robert I. 1961. *Morphological Differentiation and Adaptation in the Galápagos Finches*. Berkeley and Los Angeles: University of California Press.

Brinkman, Paul D. 2010. "Charles Darwin's Beagle Voyage, Fossil Vertebrate Succession, and 'The Gradual Birth & Death of Species'". *Journal of the History of Biology*, No. 43: 363-399.

Browne, Janet. 2018. *The Quotable Darwin*. Oxford: Princeton University Press.

—. 2013. "Wallace and Darwin"'. *Current Biology*, No. 23: R1071-R1072.

Chancellor, Gordon, and John van Wyhe. 2009. *Charles Darwin's Notebooks from the Voyage of the Beagle*. Cambridge and New York: Cambridge University Press.

Darwin, Charles. 1839. 1845. 1890. *Journal of Researches into the Natural History and Geology of the Countries Visited During the Voyage of H.M.S. Beagle Round the World, under the Command of Capt. Fitz Roy, R.N.* London: John Murray.

—. 1859. *On the Origin of Species by Means of Natural Selection, or the Preservation of Favoured Races in the Struggle for Life.* 1st ed. London: John Murray.

De Roy, Tui. 2022. *A Pocket Guide to Birds of Galápagos*. Princeton and Oxford: Princeton University Press.

Funk, Erik R., and Kevin J. Burns. 2018. "Biogeographic Origins of Darwin's Finches (Thraupidae: Coerebinae)". *The Auk*, No. 135: 561-571.

Grant, Peter R., and Rosemary Grant. 2014. *40 Years of Evolution: Darwin's Finches on Daphne Major Island*. Princeton and Oxford: Princeton University Press.

Grant, Thalia K., and Gregory B. Estes. 2009. *Darwin in Galápagos: Footsteps to a New World*. Princeton and Oxford: Princeton University Press.

Herbert, Sandra, Sally Gibson, David Norman, Dennis Geist, Gregory Estes, Thalia Grant, and Andrew Miles. 2009. "Into the Field Again: Re-examining Charles Darwin's 1835 Geological Work on Isla Santiago (James Island) in the Galápagos Archipelago". *Earth Sciences History*, No. 28: 1-31.

Holder, Charles Frederick. 1891. *Charles Darwin: His Life and Work*. New York & London: G. P. Putnam's Sons.

INEC. 2022. *Censo Ecuador: Resultados Principales Galápagos*. Quito: Instituto Nacional de Estadística y Censos.

Johnson, Markes E., B. Gudveig Baarli, Mário Cachão, Eduardo Mayoral, Ricardo S. Ramalho, Ana Santos, and Carlos M. da Silva. 2018. "On the Rise and Fall of Oceanic Islands: Towards a Global Theory Following the Pioneering Studies of Charles Darwin and James Dwight Dana". *Earth-Science Reviews*, No. 180: 17-36.

King, P. Parker, Robert FitzRoy, and Charles Darwin. 1839. *Narrative of the Surveying Voyages of His Majesty's Ships Adventure and Beagle between the Years 1826 and 1836 Describing their Examination of the*

Southern Shores of South America and the Beagle's Circumnavigation of the Globe. Volumes I–III. London: Henry Colburn.

Kutschera, Ulrich. 2003. "A Comparative Analysis of the Darwin-Wallace Papers and the Development of the Concept of Natural Selection". *Theory in Biosciences*, No. 122: 343-359.

Lack, David. 1947. *Darwin's Finches: An Essay on the General Biological Theory of Evolution*. Cambridge: Cambridge University Press.

Laland, Kevin, Tobias Uller, Marc Feldman, Kim Sterelny, Gerd B. Müller, Armin Moczek, Eva Jablonka, John Odling-Smee, Gregory A. Wray, Hopi E. Hoekstra, Douglas J. Futuyma, Richard E. Lenski, Trudy F. C. Mackay, Dolph Schluter, and Joan E. Strassmann. 2014. "Does Evolutionary Theory Need a Rethink?". *Nature*, No. 514: 161-164.

Lamichhaney, Sangeet, Jonas Berglund, Markus Sällman Almén, Khurram Maqbool, Manfred Grabherr, Alvaro Martinez-Barrio, Marta Promerová, Carl-Johan Rubin, Chao Wang, Neda Zamani, B. Rosemary Grant, Peter R. Grant, Matthew T. Webster, and Leif Andersson. 2015. "Evolution of Darwin's finches and their beaks revealed by genome sequencing". *Nature*, No. 518: 371-375.

—. Fan Han, Matthew T. Webster, Leif Andersson, B. Rosemary Grant, and Peter R. Grant. 2018. "Rapid Hybrid Speciation in Darwin's Finches". *Science*, No. 359: 224-228.

Lowe, Percy R. 1936. "The Finches of the Galapagos in Relation to Darwin's Conception of Species". *Ibis*, No. 78: 310-321.

Malthus, Thomas R. 1798. 1836. *An Essay on the Principle of Population*. 6th ed. London: John Murray.

Mayr, Ernst. 1977. "Darwin and Natural Selection: How Darwin May Have Discovered His Highly Unconventional Theory". *American Scientist*, No. 65: 321-327.

Mejía, Daniela. 2019. "Reintroducción de Iguanas Terrestres en Galápagos se Perfila Como Modelo Mundial de Conservación". *Mongabay*, Mayo 19, 2019.
https://es.mongabay.com/2019/05/ecuador-iguanas-galapagos/.

Moity, Nicolas, Byron Delgado, and Pelayo Salinas-de-León. 2019. "Mangroves in the Galapagos Islands: Distribution and Dynamics". *PLoS ONE*, No. 14: e0209313.

Paz-y-Miño-C, Guillermo, and Avelina Espinosa. 2024a. *Arrivals of Life to the Galápagos: From Berlanga to Darwin, Melville, and World War II*. Newcastle upon Tyne: Cambridge Scholars Publishing.

—. 2024b. *A Synthesis of the Galápagos: Natural History, Human Conflict, and the Destiny of an Archipelago*. Newcastle upon Tyne: Cambridge Scholars Publishing.

Pechenik, Jan A. 2023. *The Readable Darwin: The Origin of Species Edited for Modern Readers*. New York: Oxford University Press.

Rivas-Torres, Gonzalo, Fátima L. Benítez, Danny Rueda, Christian R. Sevilla, and Carlos F. Mena. 2018. "A Methodology for Mapping Native and Invasive Vegetation Coverage in Archipelagos: An Example from the Galápagos Islands". *Progress in Physical Geography*, No. 42: 83-111.

Rookmaaker, Kees. 2009. "Darwin's Itinerary on the Voyage of the Beagle." *Darwin Online*. Accessed March 4, 2022. http://darwin-online.org.uk/content/frameset?viewtype=text&itemID=A575&pageseq=1.

Sato, Akie, Herbert Tichy, Colm O'hUigin, Peter R. Grant, B. Rosemary Grant, and Jan Klein. 2001. "On the Origin of Darwin's Finches". *Molecular Biology and Evolution*, No. 18: 299-311.

Sulloway, Frank J. 1982. "Darwin and His Finches: The Evolution of a Legend". *Journal of the History of Biology*, No. 15: 1-53.

—. 2009. "Tantalizing Tortoises and the Darwin-Galápagos Legend". *Journal of the History of Biology*, No. 42: 3-31.

Swash, Andy, and Rob Still. 2006. *Birds, Mammals, and Reptiles of the Galápagos Islands: An Identification Guide*. New Heaven and London: Yale University Press.

Vanchurin, Vitaly, Yuri I. Wolf, Eugene V. Koonin, and Mikhail I. Katsnelson. 2022. "Thermodynamics of Evolution and the Origin of Life". *Proceedings of the National Academy of Sciences of the United States*, No. 119: e2120042119.

van Wyhe, John. 2007. "Mind the Gap: Did Darwin Avoid Publishing His Theory for Many Years?". *Notes and Records: The Royal Society Journal of the History of Science*, No. 61: 177-205.

—. 2013. "'My Appointment Received the Sanction of the Admiralty': Why Charles Darwin Really Was the Naturalist on HMS Beagle". *Studies in History and Philosophy of Biological and Biomedical Sciences*, No. 44: 316-326.

—. 2020. "A. R. Wallace in the Light of Historical Method". *Journal of the Royal Society of Western Australia*, No. 103: 89-95.

CHAPTER FOUR

WHAT HAPPENED IN GALÁPAGOS BEFORE AND AFTER DARWIN?

Much happened in Galápagos before and after Charles Darwin's short visit to the archipelago—as a naturalist on board HMS Beagle under the command of Captain Robert FitzRoy—from September 15 to October 20, 1835 (Chapter Three). Remember that three hundred years earlier, the islands were accidentally discovered, in 1535, by Tomás de Berlanga and his crew (Chapter Two). And just three years prior to the Beagle's arrival, Ecuador claimed possession of the Archipelago of Ecuador (*Archipiélago del Ecuador*) on February 12, 1832 (Chapter Five). During World War II, the United States established a military base on Baltra [Seymour] Island from 1942 to 1946 (Chapter Five). In 1959, Ecuador created the Galápagos National Park; by 1978, UNESCO declared Galápagos a World Heritage Site (Chapter Five). A few decades later, coinciding with the tourism boom (1980s to 2000s), the local population increased significantly, from 5,545 in 1982 to 28,583 in 2024. Tourism grew from 42,000 visitors in 1989 to 276,000 in 2018. The COVID-19 pandemic reduced tourism to 73,000 visitors in 2020, but trends recovered by 2025 (Chapters Six and Seven). [We discuss these topics in various chapters of this book and in Paz-y-Miño-C and Espinosa (2024a,b)].

Now, we ought to remind readers that prior to Darwin's and FitzRoy's exploration of the Galápagos, or Berlanga's and crew's discovery of the *y*[a]*s de galapagos*, Ecuador's claim of jurisdiction over the archipelago, the events of World War II, the local population's establishment and growth, and the tourism boom, millions of years ago the islands emerged from the bottom of the ocean (Chapter One). They were colonized by life forms that later evolved *in situ*, arguably undisturbed by humans until 1535.

There is one distinctive human activity that took place globally and affected the Galápagos' natural history prior to, during, and after Darwin's visit: whaling! In this chapter, we discuss industrial whaling as it covered the world's oceans ("Global Whaling: 1760s to 1920s") and its impacts on the Galápagos Islands—also known as "Whaling Ground 38".

Global Whaling: 1760s to 1920s

Whaling is an old human practice. Industrial whaling, however, became a profitable economic activity—particularly in North America—relatively recently, from the 1760s to the 1920s (Bannister et al. 2008). Prosperity of the American colonies and the Industrial Revolution of the young republic (United States 1820-1860) were linked to the slaughtering of whales (Moment 1957). By the mid-1800s, whaling was the fifth largest industry in the United States (Dolin 2007), with New England and Yankee whalers dominating the chase, hunt, and processing of whales and their products (Tower 1907; Moment 1957; Davis et al. 1997).

Whales were hunted for their oil (blubber), spermaceti (wax) extracted from "the case" of sperm whales, meat, bones, baleen (keratin "mouth bristles" from baleen whales), teeth (from toothed whales), and ambergris (used as perfume fixative) obtained—if present—from the hindgut of sperm whales (Figs. 4.1-4.7). Sperm whale oil, the most precious good (waxy and thick), burned brightly and was ideal for lanterns; it was also used as lubricant. In fact, before petroleum oil products became popular (late 1800s onwards), almost all industrial machines or home instruments were lubricated with sperm whale oil. The spermaceti wax was shaped into luxurious, long-lasting candles. Whale bones and baleen were crafted into multiple artifacts and utensils, as shown in Figs. 4.7-4.9 (Stackpole 1953; Tower 1907; Moment 1957). In one of the most profitable years in whaling history, 1853, the combined sales of whale oil, spermaceti, and baleen in the United States totaled US$ 11 million (today's money, US$ 413 million; after Moment 1957; Dolin 2007).

At sea, whaling relied on three components: whales to be hunted (toothed or baleen), whale- or whaling-ships (the tools for the hunt), and whalers (skillful seafarers). In Fig. 4.1a, we show the interior and main features of an exemplar 1800s whaleship (after the *Alice Knowles* of New Bedford; Brown-Goode 1887). And in Fig. 4.2, we share images of *Lagoda* at the New Bedford Whaling Museum, a half-size replica of a merchant vessel (1820s), which was converted to a whaling ship by 1841.

Whaleships were efficient (crew 16-30), constructed to sail long distances, chase and hunt whales (using fast whaleboats, 3-5 per ship), process the whales' blubber in try-works (built-in on board), extract, and store the blubber oil in casks (Figs. 4.1-4.6). They played the role of "factory ships" or "whale-processing vessels" during the 18th and early 20th centuries. Once the ships returned to harbor, their cargo was further processed (e.g. oil refinement, crafting of artifacts and utensils from bones/baleen) and commercialized (Figs. 4.7-4.9).

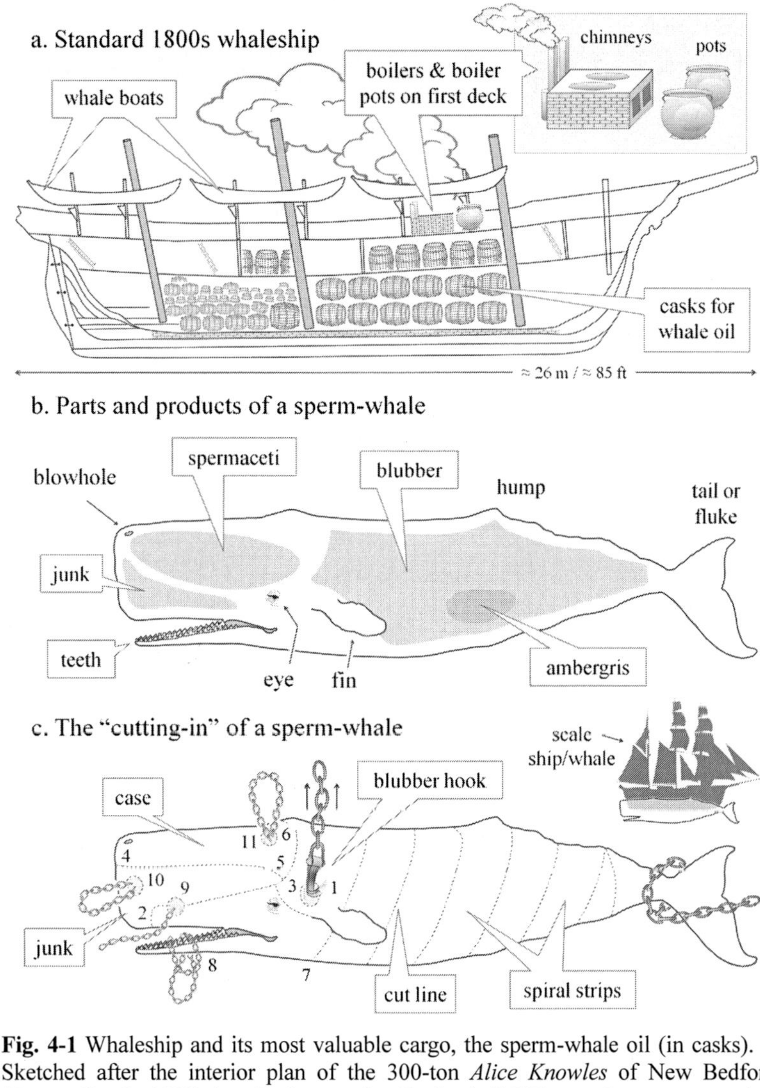

Fig. 4-1 Whaleship and its most valuable cargo, the sperm-whale oil (in casks). a. Sketched after the interior plan of the 300-ton *Alice Knowles* of New Bedford (Brown-Goode 1887). b-c. Redrawn after Starbuck (1878); numbers correspond to the sequence of cuts (1 to 11) once the whale was killed and fastened to the ship with chains (see silhouettes of ship and whale for scale). A "blubber hook" (1; see also Fig. 4.5) allowed whalers to pull the thick skin (blubber) as it was cut around the body (dotted lines), from head to tail/fluke. The blubber was "peeled off" in spiral strips as the whale was rotated; the strips were sliced and boiled in boiler pots (a) to extract the oil (details in Figs. 4.2, 4.5, 4.6). Layout/sketch: G. Paz-y-Miño-C.

Fig. 4-2 Replica (half-scale) of the whale ship *Lagoda* of New Bedford, Massachusetts. *Lagoda* was a merchant ship built in the 1820s. It was converted to a whaling vessel (1841) by adding try-works, whaling gear, and five whaleboats. Whaleships were efficient (crew 16-30), built to sail long distances, chase and hunt whales (using the fast whaleboats), process the whales' blubber in try-works (built-in on board), extract, and store the blubber oil in casks. New Bedford Whaling Museum. Photos: G. Paz-y-Miño-C.

In Fig. 4.3 we show a 1920s whaleboat used during the final years of American whaling. Arguably, the most perfect water craft ever built!

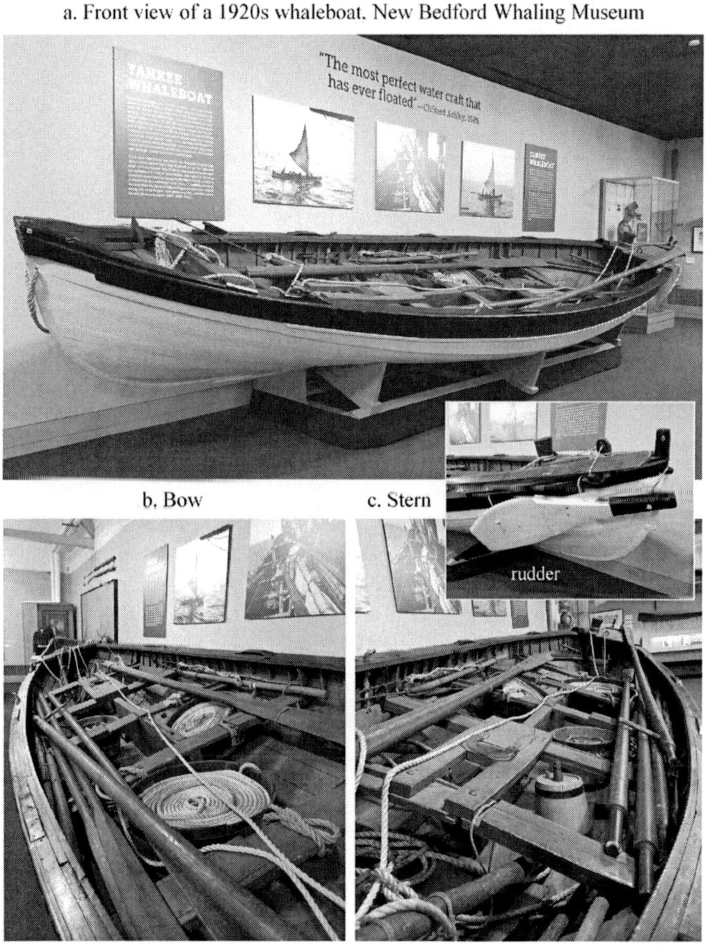

Fig. 4-3 Whaleboat. "The most perfect water craft that has ever floated", as characterized by American writer, sailor and knot expert Clifford W. Ashley (1881-1947). Whaleboats were "double-ended" (for quick maneuvering) and of superb design (crew 6); they could be sailed, paddled or pulled (by the main ship), and were essential for chasing and hunting whales. The 18th Century American "Yankee" whaleboat was light, made of cedar and white oak. Whaleships often carried 3-5 whaleboats, which were equipped with mast and sail, paddles, rudder, drogue, water breaker, grapnel, box compass, lantern, harpoons, lances, spades, ropes, among +30 tools. New Bedford Whaling Museum. Photos: G. Paz-y-Miño-C.

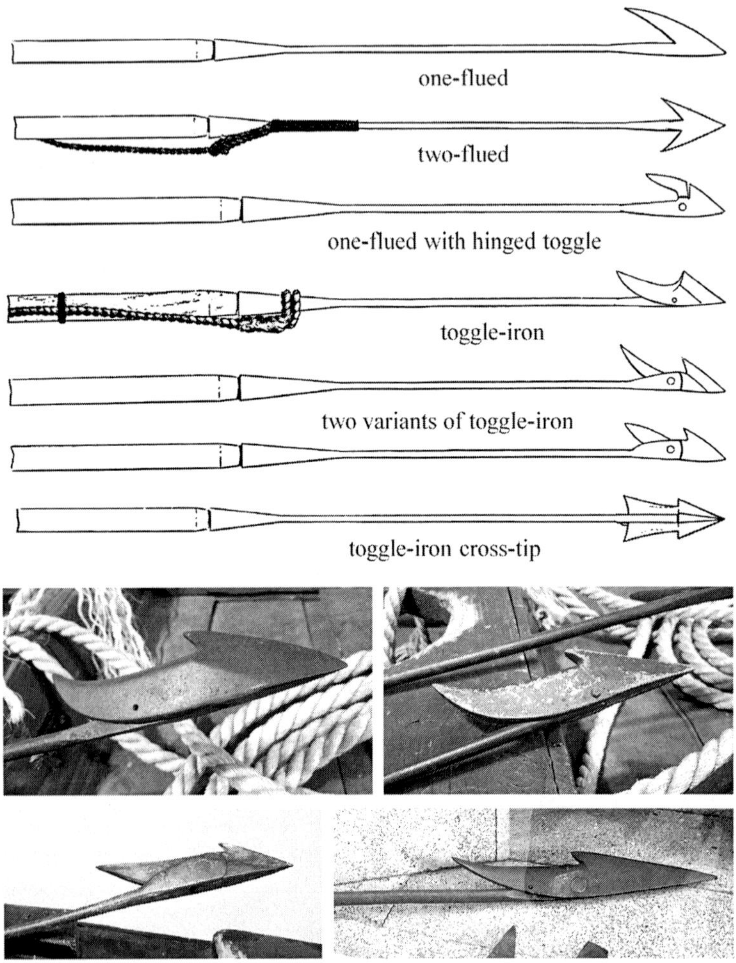

Fig. 4-4 Whaling harpoons 1880s to 1900s (source National Oceanic and Atmospheric Administration, NOAA, Historic Fisheries Collection; see link in Appendix A). Whaleboats carried a variety of harpoons, which were attached to a thick rope (the line); if possible, whales were harpooned more than once to secure the hunt. Yet whales often escaped after breaking the line (taking the harpoon/s inserted in them), or while dragging and sinking the whaleboat, or ramming against it and breaking it into pieces. For any whaler, whaling was a dangerous and modest-income job. For example, in 1851, the net value of the *Benjamin Tucker*'s cargo (whale oil/bones) was US$ 45,000 (today US$ 1.7 million), of which a crew member was entitled to $220-280 (today US$ 8,200-10,500) for the entire voyage without due deductions (Hohman 1926)! Photos: G. Paz-y-Miño-C, New Bedford Whaling Museum.

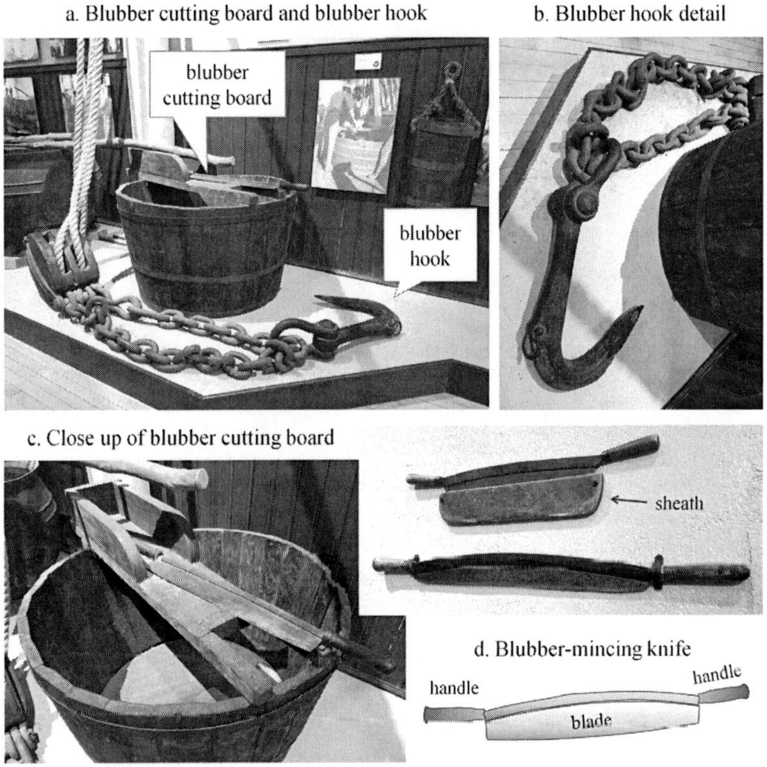

Fig. 4-5 Whaling gear/tools. The "blubber hook" (a-b) was used to pull the whale skin as the blubber was "peeled off" in strips (Fig. 4.1c). The strips were sliced into smaller pieces on the blubber cutting board with a blubber-mincing knife (a-d); the pieces, called "bible leaves" (see Fig. 4.6b), were boiled in large pots on the try-works; the extracted oil was stored in casks (Fig. 4.7d). Photos: G. Paz-y-Miño-C, New Bedford Whaling Museum.

Industrial whaling covered the world's oceans. In Fig. 4.10 we show three of four "charts" (A, C, and D, as per original labeling) put together by Charles H. Townsend (1859 to 1944) in his 1935 work *The Distribution of Certain Whales as Shown by Logbook Records of American Whaleships* (Whaling History 2017). The records ranged from 1761 to 1920. By default, Townsend's compilation turned into a database of the global hunts of Sperm whales (Chart A: 1761 to 1920), Northern- and Southern-Right whales (Chart C: 1785 to 1913), and Bowhead and Humpback whales

(Chart D: mostly 1800s). Each dot on the maps in Fig. 4.10 represents the position of a whaleship on a day when one or more whales were hunted!

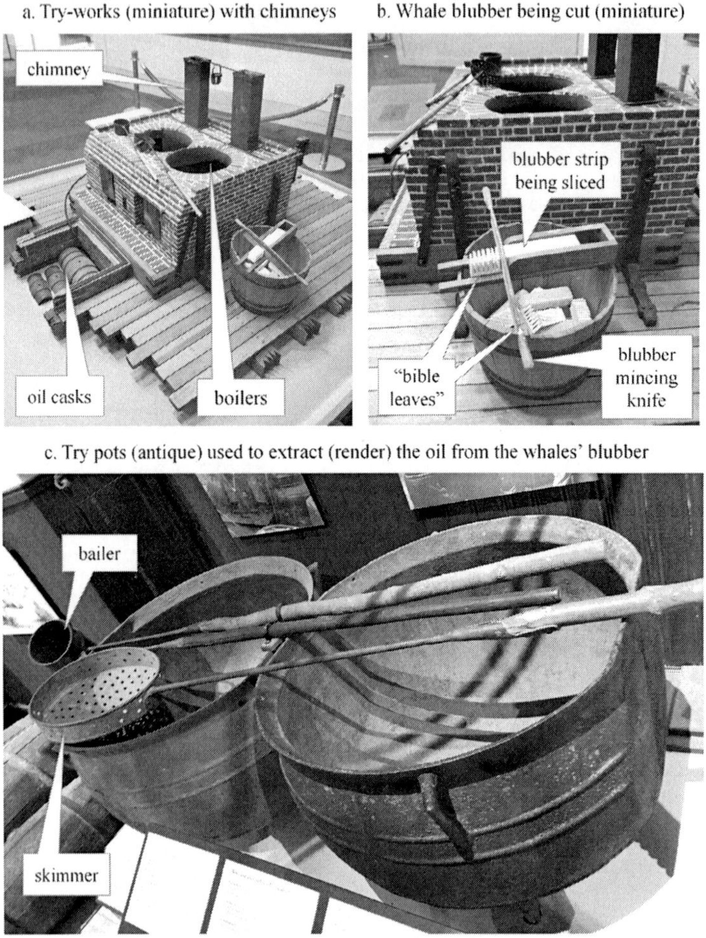

Fig. 4-6 Try-works, try pots, and oil extraction from whales' blubber. The try-works were built with bricks, concrete, and iron on the deck of the ship; they usually had two furnaces (burners), each with their own chimney. The process of oil extraction was laborious; whale blubber strips were cut into smaller pieces ("bible leaves"), which were boiled to separate the oil. Firewood was used as fuel to start the fire; once some blubber was tried, scraps of it were then used for fuel. Whaleships stored whale oil in casks for months/years and return to harbor after filling all possible casks. Large vessels like the *Alice Knowles* of New Bedford (Fig. 4.1) carried +300 tons in casks. Photos: G. Paz-y-Miño-C, New Bedford Whaling Museum.

Fig. 4-7 Sperm whale oil, spermaceti wax, ambergris, and fine oil for machines and instruments. Note that ambergris (grey amber) was extracted by whalers—if present—from the sperm whale gut (see Fig. 4.1b); it is a solid, lipid-rich amorphous byproduct of undigested matter; sperm whales can naturally eliminate ambergris, which can be found occasionally ashore (Macleod et al. 2020). Photos: G. Paz-y-Miño-C, New Bedford Whaling Museum.

Townsend concluded that, for 159 years (1761 to 1920), 744 whaleships completed 1,665 voyages. They captured 53,877 whales; including: 36,908 sperm whales (*Physeter macrocephalus*); 5,114 bowheads (*Balaena mysticetus*); 2,118 Northern Rights (*Eubalaena glacialis*) in the Pacific Ocean; 1,685 Southern Rights (*E. australis*) in the Pacific Ocean; 35 Northern Rights in the Atlantic Ocean; 2,981 Southern Rights in the Atlantic Ocean; 1,596 Southern Rights in the Indian Ocean; 2,883 Humpbacks (*Megaptera novaeangliae*); and 557 Californian Grays (*Eschrichtius robustus*). And Townsend identified the hunting grounds: North Atlantic, South Atlantic, Pacific, and Indian Ocean.

a. Scrimshaw walking canes and walking sticks (whale bone and Sperm Whale teeth)

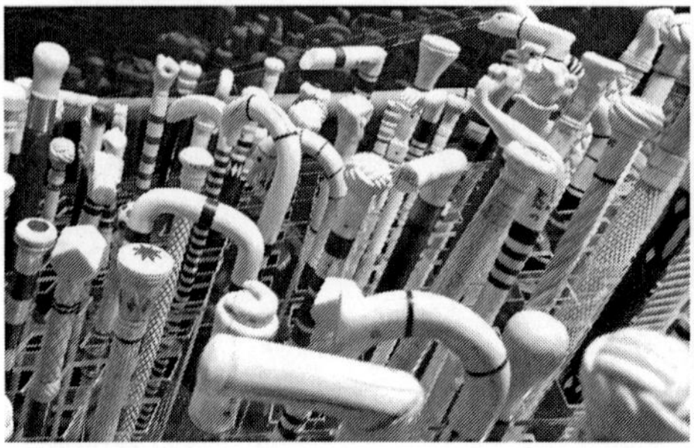

b. Engravings on Sperm Whale teeth

c. Table clock (bone, wood)

d. Pie crimpers or "jagging wheels" (bone)

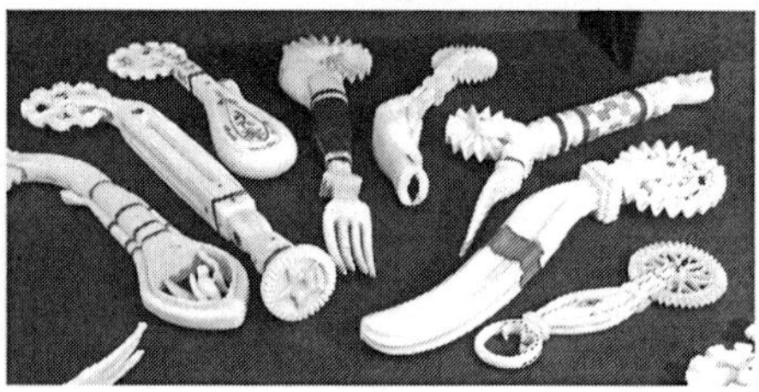

e. Bird cage (bone)

f. Lantern (whale oil)

g. Swift for yarn (bone)

Fig. 4-8 Artifacts and utensils made of whale parts (bones, teeth, oil). Photos: G. Paz-y-Miño-C, New Bedford Whaling Museum.

In the Pacific, Townsend noted the significance of various hunting locations: Coast of Japan, Bonin Islands, Hawaiian Islands, Lower [Baja] California, off the shores of Central Mexico, as well as the grounds known

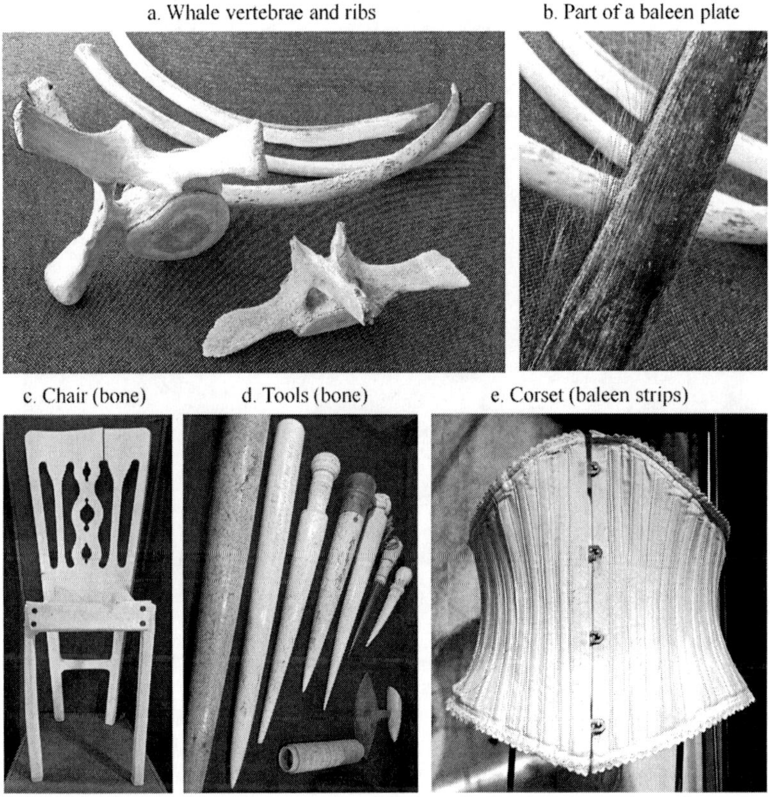

Fig. 4-9 Artifacts and tools made of whale parts (bones and baleen). Photos: G. Paz-y-Miño-C, New Bedford Whaling Museum.

as Panama, Galápagos, Offshore, On the Line, and others located across the equatorial Pacific and from which "sperm whales were taken in great numbers during all seasons of the year" (Chart A, Fig. 4.10)! In fact, whaleships chased and hunted whales nonstop while navigating among grounds; from Japan to Lower California; from Western Mexico to Panama; from equatorial-coastal South America to Galápagos; and along the equator back to Asia, toward the Sulu Sea, Celebes Sea, Molucca Passage, and Coast of New Holland (West/Southwestern Australia). Sperm whaling nearby Australia and New Zealand (Middle Ground) was seasonal, from December to March; the same occurred on the Callao (Peru) or Coast of Chile grounds (Townsend 1935; review in Smith et al. 2012). Yes, Ground 38, the Galápagos, was yet another profitable ground!

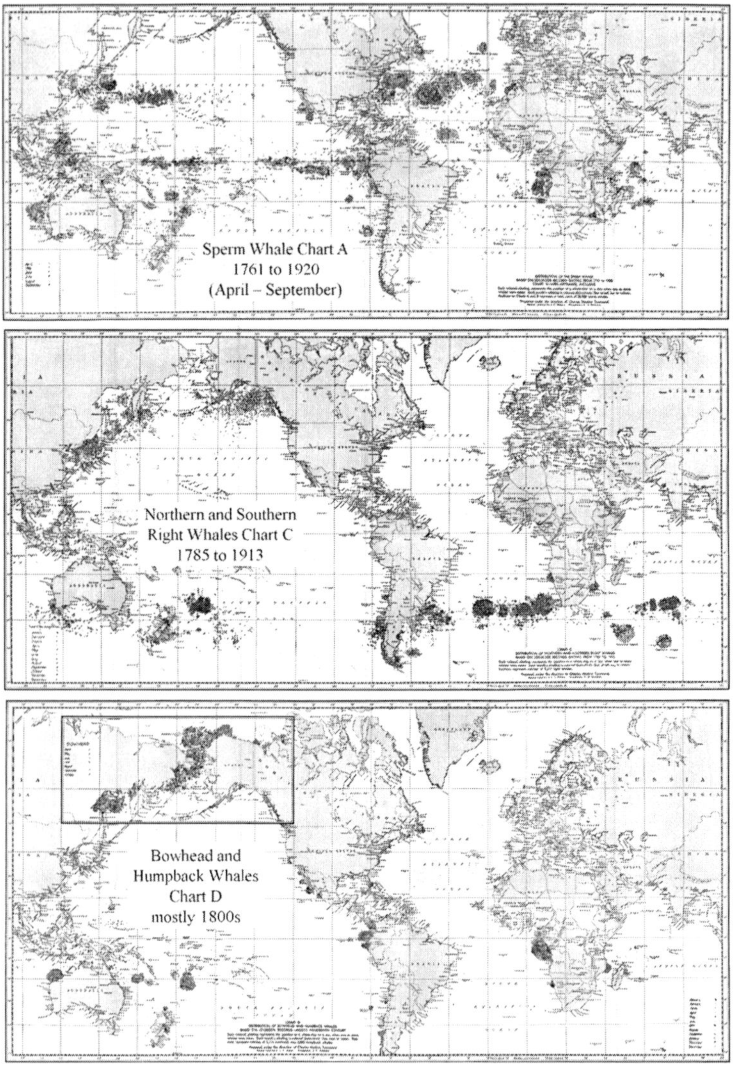

Fig. 4-10 Distribution of Sperm-, Northern-, and Southern-Right-, Bowhead- and Humpback whales based on logbook records dating from 1761 to 1920; data compiled by Townsend (1935). Each dot represents the position of a whaleship on a day when one or more whales were hunted. Digitized maps by Whaling History (2017; Mystic Seaport Museum and New Bedford Whaling Museum). See assessment of Townsend's work in Smith et al. (2012).

Galápagos: Whaling Ground 38

Twenty-seven species of Cetaceans (whales and dolphins) have been confirmed to occur in the Galápagos (31% of the world's 87 species). The majority of taxa in the archipelago (21 spp.) are toothed-Cetaceans or Odontoceti (dolphins, orcas, sperm whales); the rest (6 spp.) are baleen-whales or Mysticeti (humpbacks, rights, blues; Fig. 4.11; after McGowen et al. 2009 with input from Denkinger et al. 2013). The most commonly observed Cetaceans in today's Galápagos Marine Reserve (1993 to 2010) are listed in Fig. 4.12 (sightings coincide with predicted suitable habitats; Eguiguren et al. 2019). Note that 70% of the sightings correspond to: Bottle-nosed Dolphin (*Tursiops truncatus*), Bryde's Whale (*Balaenoptera edeni*), Common Dolphin (*Delphinus delphis*), Killer Whale or Orca (*Orcinus orca*), Humpback Whale (*Megaptera novaeangliae*), unidentified baleen species (*Balaenoptera* sp.), Short-finned Pilot Whale and/or unidentified pilot whale (*Globicephala macrorhynchus*/sp.), and Sperm Whale or Cachalot (*Physeter macrocephalus*; Denkinger et al. 2013).

In the archives of American whaling, the Galápagos were cataloged as Ground 38 (Fig. 4.13; Bannister et al. 2008). To its North was Panama Bay (Ground 35); to the Southeast, Callao (Ground 52); to the South, Archer (Ground 51); and to the West, Offshore (Ground 39). Somewhat more distant and located to the Northwest of the archipelago was Ground 36, also known as "Southwest of Clipperton Island". Note in Fig. 4.13 (top) how whaling grounds 1 to 17 corresponded to the North Atlantic, the closest hunting waters to the New England and Yankee harbors. Historically, the North Atlantic was whaled first (early 1700s onwards), and its whales became considerably reduced (early 1800s); over time, whaleships needed to find new grounds in all other oceans. By 1840-1850 (The Golden Age), whaling became global, with particular intensity in the Pacific: Japan, Hawai'i, and along the equator from West to East (Bannister et al. 2008). Thus, whalers sailed longer distances and remained at sea for months or years. The ecocide of the "leviathans" was on, but it was only acknowledged in the 20th century!

Among whalers, the "whale of whales" was the Sperm Whale (*P. macrocephalus*), the most profitable catch for its quality of oil, spermaceti, and the infrequently found ambergris (above; Dolin 2007). In Fig. 4.14, we show a close-up (emphasizing Central and South America) of Charles H. Townsend's (1935) Chart B "Distribution of the Sperm Whale Based on Logbook Records Dating from 1761 to 1920—October-March, Inclusive". Note how "sperm-whaling" was ubiquitous off coastal Central and South America and the significance of the Galápagos and Panama Bay grounds.

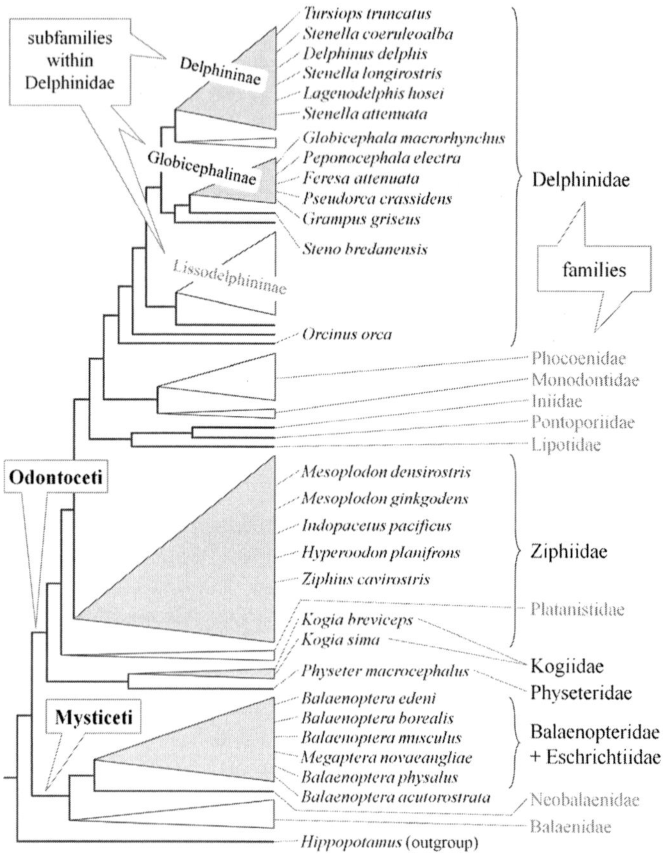

Fig. 4-11 Diversity of whales and dolphins (Cetaceans) in the Galápagos. Figure layout: GPC, redrawn and modified after McGowen et al. (2009) with input from Denkinger et al. (2013). Phylogeny based on nDNA, mtDNA, and transposon data (Bayesian and Maximum Likelihood methods; McGowen et al. 2009). Twenty-seven species of Cetaceans have been observed in the Galápagos (one-third of the world's species).

Focus your attention on Fig. 4.15. In it, we depict the whales hunted 1700s-1900s (i.e. sperms, rights, bowheads, and humpbacks) as per three data sets: (a) Townsend's (1935); (b) Woolmer's (2013) and Halpin et al.'s (2009); and (c) our compilation for Galápagos/Panama Bay (based on a/b).

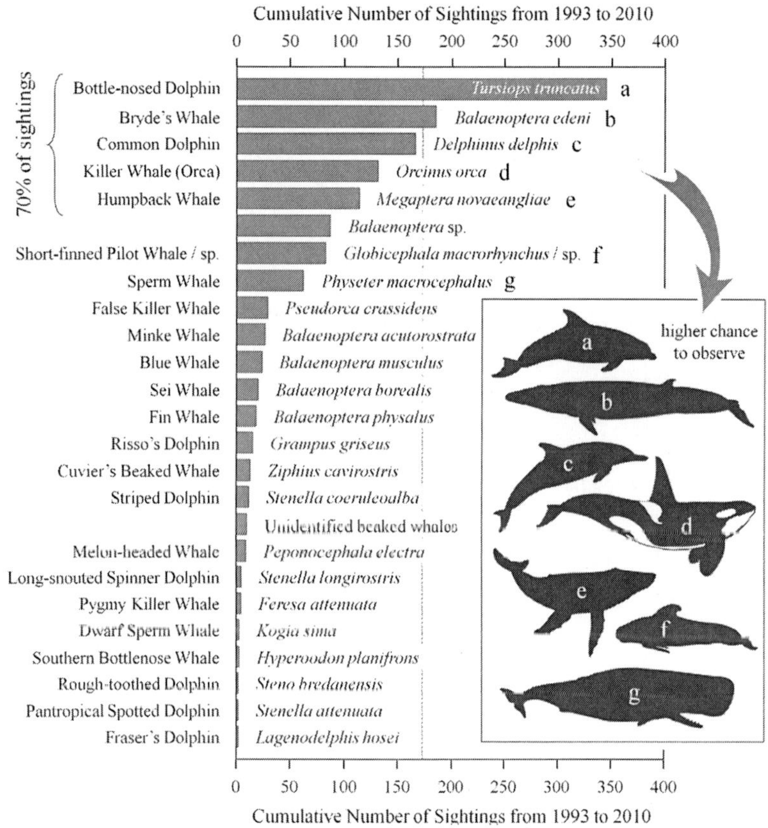

Fig. 4-12 Most commonly observed whales and dolphins in the Galápagos. Figure layout: GPC, redrawn and modified after Denkinger et al. (2013).

The data sets by Townsend (1935), Woolmer (2013), and Halpin et al. (2009) are public (see references and Appendix A for links). Be aware that Woolmer (2013) and Halpin et al. (2009) curated Townsend's data and entered it in various files. By the time we accessed and downloaded the information, it contained 64% of Townsend's original work (nowadays it covers 95%; project sponsored by Wildlife Conservation Society).

Note in Fig. 4.15a (Townsend's data) the number of whales hunted by American whaleships (worldwide) according to logbook records (1761-1920): grand total 53,320 (excluding 557 grays reported above); including sperms 36,908 or 69.2%; Northern and Southern rights 8,415 or 15.8%; bowheads 5,114 or 9.6%; and humpbacks 2,883 or 9.3%.

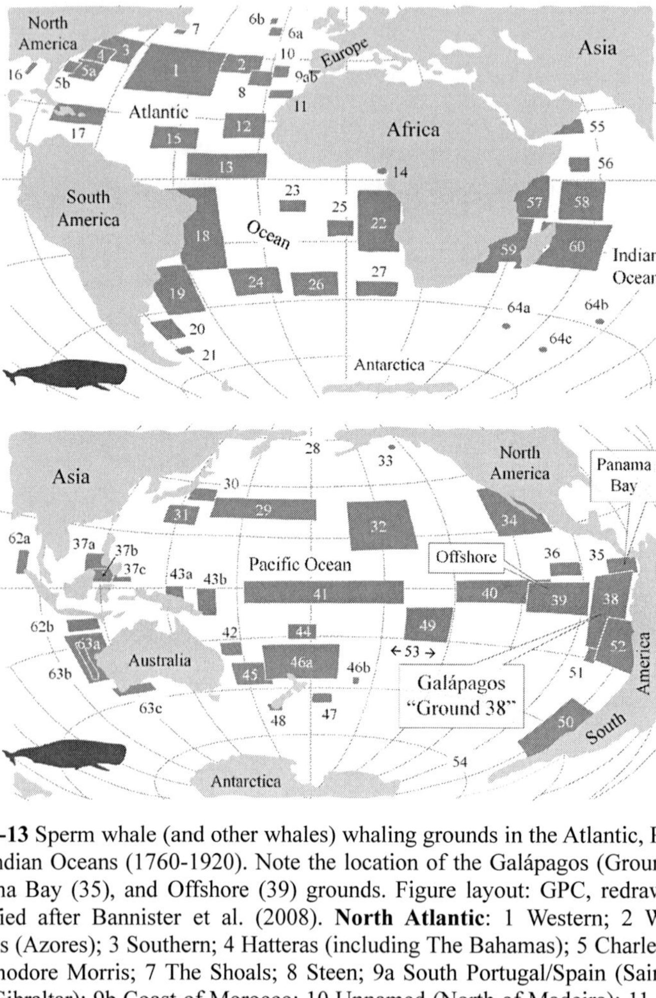

Fig. 4-13 Sperm whale (and other whales) whaling grounds in the Atlantic, Pacific, and Indian Oceans (1760-1920). Note the location of the Galápagos (Ground 38), Panama Bay (35), and Offshore (39) grounds. Figure layout: GPC, redrawn and modified after Bannister et al. (2008). **North Atlantic**: 1 Western; 2 Western Islands (Azores); 3 Southern; 4 Hatteras (including The Bahamas); 5 Charleston; 6 Commodore Morris; 7 The Shoals; 8 Steen; 9a South Portugal/Spain (Saint Vincent-Gibraltar); 9b Coast of Morocco; 10 Unnamed (North of Madeira); 11 Canaries-African Coast; 12 San Antonio (the Twenty Twenties, Cape Verde); 13 Cornell; 14 Gulf of Guinea (near Fernando Po Island); 15 The Twelve-Forty; 16 Gulf of Mexico; 17 Caribbean Sea/West Indies. **South Atlantic**: 18 Coast of Brazil; 19 Platte; 20 False Banks; 21 Walvis (Walfisch) Bay; 22 Coast of Africa; 23 Around Ascension; 24 Pigeon; 25 Carroll (including Saint Helena); 26 Tristan; 27 Tristan-C of Good Hope. **North Pacific**: 28 High Latitudes (50-60° North); 29 Japan; 30 Coast of Japan; 31 Japanese Coast to the Bonin Island (including Bashee Islands/Bataan); 32 [Near] Hawaiian Islands; 33 Northwest Coast; 34 California Coast; 35 Panama Bay; 36 Southwest of Clipperton Island; 37 [East Indies] (in-

cluding China Sea/Sulu Sea/Celebes Sea/Molucca Passage). **South Pacific**: 38 Galapagos; 39 Offshore; 40 On the Line [East]; 41 Equatorial Pacific; 42 [South Coral Sea]; 43 around New Guinea (including New Ireland, New Britain, Solomon Islands, North Coral Sea, New Caledonia); 44 [Fiji-Samoa-Tonga]; 45 Middle; 46 New Zealand, Vasquez, French Rock; 47 Chatham Islands; 48 South of New Zealand (Solander); 49 [Tuamotus, Marquesas, Society Island (Tahiti)]; 50 Coast of Chile; 51 Archer; 52 Callao; 53 across the South Pacific between 21 and 27° South (not plotted); 54 Cape Horn. **Indian Ocean**: 55 Coast of Arabia (including Red Sea, Persian Gulf); 56 unnamed (off Somaliland); 57 Zanzibar; 58 Mahe Banks; 59 Delagoa Bay/Mozambique Channel [South Madagascar]; 60 East Madagascar; 61 South of Ceylon (not plotted); 62 Andaman Islands to Sumatra; 63 Coast of New Holland; 64 Northern Crozet Islands/Saint Paul/Kerguelen.

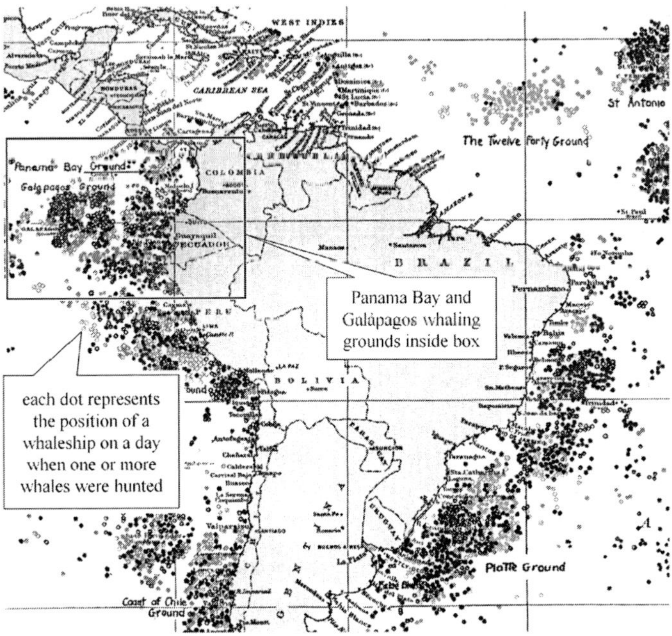

Fig. 4-14 Sperm whale hunts around Central and South America as plotted in Charles H. Townsend (1935) Chart B 1761 to 1920 (October-March). Each dot represents the position of a whaleship on a day when one or more whales were hunted. Close-up of digitized map by Whaling History (2017; Mystic Seaport Museum and New Bedford Whaling Museum). The Panama Bay (Ground 35) and Galápagos (Ground 38) whaling locations are enclosed within the top left rectangle. Both covered a large area off the coasts of Central and Northwestern South America. Dark or lighter dots correspond to different times of the year (as per October-March) during which the hunts occurred. Note how whaling was ubiquitous around South America. For links to maps and www information, see Appendix A.

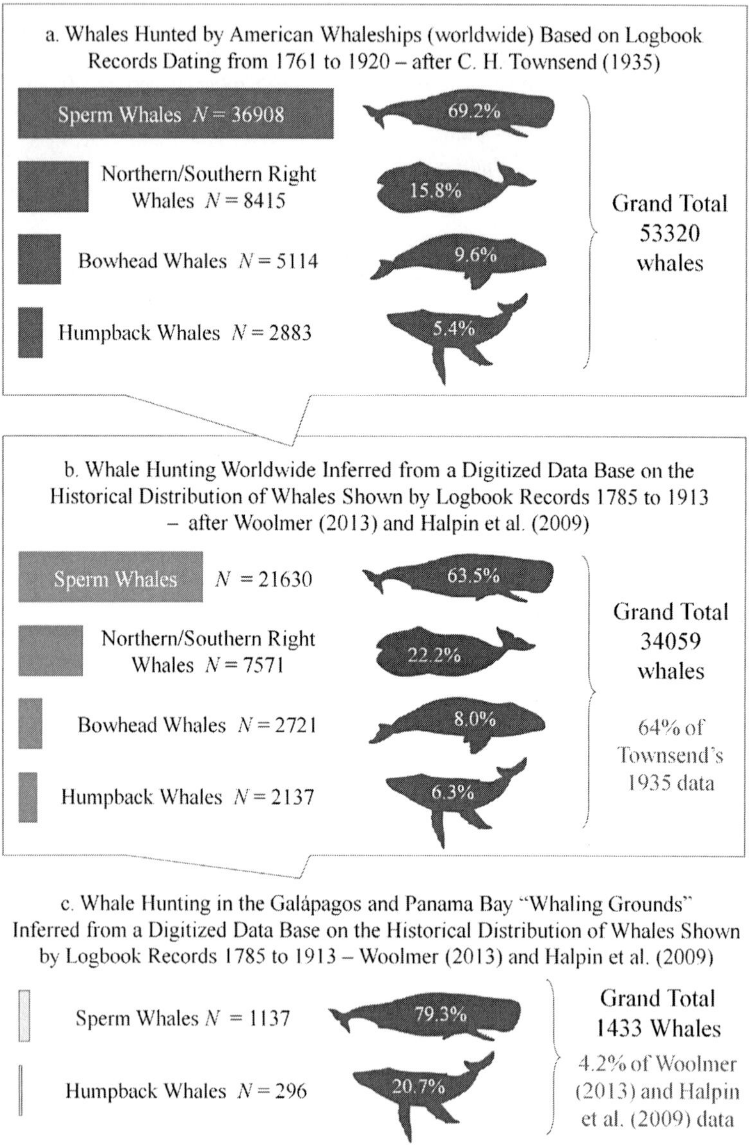

Fig. 4-15 Whales hunted (sperms, rights, bowheads, and humpbacks) as per Townsend (1935), Woolmer (2013), and Halpin et al. (2009) data sets (1700s-1900s), with emphasis on whaling grounds Galápagos (Ground 38) and Panama Bay (Ground 35). Layout: GPC.

In Fig. 4.15b (Woolmer's 2013 data with input from Halpin et al. 2009), the number of whales hunted by American whaleships (worldwide) is based on 64% of curated data from Townsend (limited to 1785 to 1913). Grand total 34,059; including sperms 21,630 or 63.5%; Northern and Southern rights 7,571 or 22.2%; bowheads 2,721 or 8.0%; and humpbacks 2,137 or 4.3%. The ratios are in overall agreement with Townsend's data.

And in Fig. 4.15c (our data extraction from Townsend 1935, Woolmer 2013 with input from Halpin et al. 2009), the number of whales hunted by American whaleships in Galápagos and Panama Bay grounds (limited to 1785 to 1913): grand total 1,433 (4.2% of Woolmer 2013 and Halpin et al. 2009 data); including sperms 1,137 or 79.3%; and humpbacks 296 or 20.7%. Yes, at Grounds 38 and 35, whales were hunted by the thousands!

With the purpose of documenting exemplar whaleships entering the Galápagos in the 1800s, we identified 13 ships (Alexander AL; Balaena BL; Cadmus CD; Edward Quesnel EQ; Frances FR; George Howland GH; Isabella IS; Java II JA; Mercury ME; Ploughboy PL; Roscoe RO; Stanton ST; William and Eliza WE) in the Woolmer (2013) and Halpin et al. (2009) databases. The vessels sailed the archipelago between 1819 and 1862. In Fig. 4.16, we plotted their locations (as per latitude/longitude), plus month/year of navigation on top of a map of today's Galápagos Marine Reserve (133,000-km^2 or 51,352-sq-mi protected area established in 1998; Paz-y-Miño-C and Espinosa 2024b). The figure also includes locations of sperm/pilot whale hunts. Eight out of the 13 ships caught whales: Balaena BL (2 sperms); Isabella IS (2 sperms); Edward Quesnel EQ (3 pilots); George Howland GH (7 sperms); Ploughboy PL (1 sperm); and Roscoe RO (1 pilot; Fig. 4.16). See whale-oil production in Fig. 4.17.

Hunting whales was a challenge. In Fig. 4.18, we show a single exemplar, the Frances FR. The ship navigated the Galápagos during April-June and September 1844. Log records indicate Frances arrived in the archipelago on April 24. It sailed into Chatham (San Cristóbal) on April 25-28, and into Charles (Floreana) on April 30 to May 3. From there, it navigated North, passing by Albemarle (Isabela) and Narborough (Fernandina) on May 8. By May 12, Frances was in Rock Redondo and continued in a tortuous navigation (evidently looking for whales) toward Culpepper (Darwin) and Wenman (Wolf). The ship returned to Rock Redondo on May 30. Between June 10 and 13, Frances was again in the vicinity of Wenman (June 12). It left the Galápagos during the rest of June and came back three months later. By September 22, the ship was in James (Santiago). Frances left the archipelago for good on September 24 (data from Townsend 1935; Woolmer 2013). There are no records that the vessel captured or processed any whales while in the Galápagos!

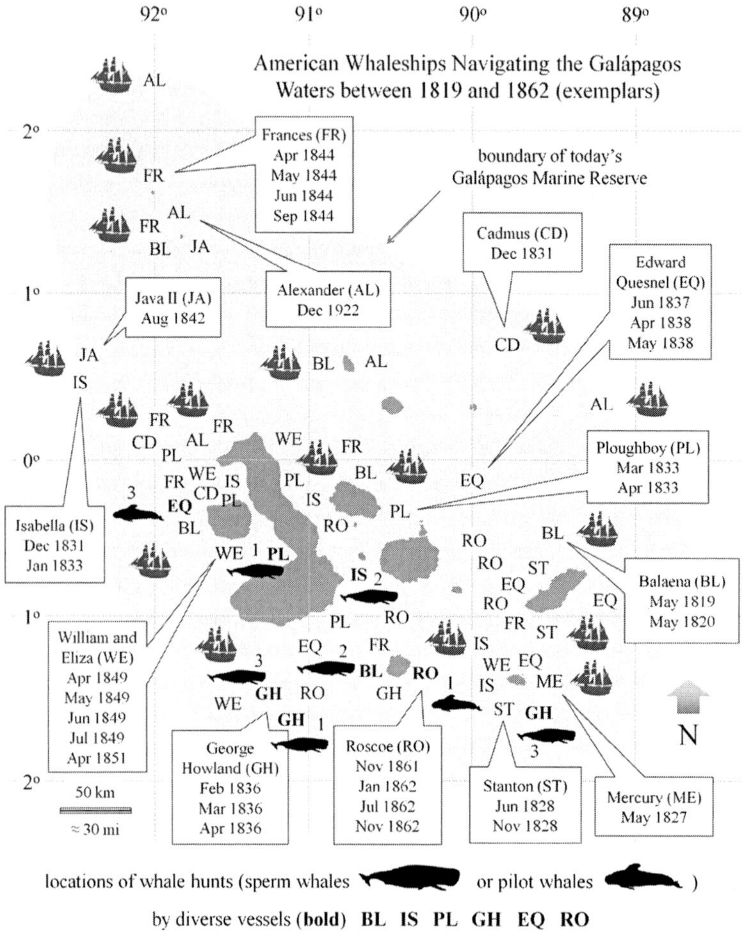

Fig. 4-16 Exemplar American whaleships navigating the Galápagos waters from 1819 to 1862. Data extracted from Townsend (1935), Woolmer (2013), and Halpin et al. (2009). Eight out of 13 ships caught sperm or pilot whales: Balaena BL (2 sperms on May 3, 1819); Isabella IS (2 sperms on December 11, 1831); Edward Quesnel EQ (3 pilots on April 24, 1838); George Howland GH (7 sperms: 3 on February 22, 1836; 3 on March 6, 1836; and 1 on April 18, 1836); Ploughboy PL (1 sperm on March 11, 1833); and Roscoe RO (1 pilot on July 9? or 11, 1862). The presence and location of the vessels indicate that whales were chased and hunted almost everywhere in the archipelago. The ships returned during different months (from January to December) and during different years; sometimes, they remained in the Galápagos for months (e.g. Frances FR from April to June 1844; or William and Eliza WE from April to July 1849). For oil-barrels/capture see Fig. 4.17. Layout: GPC.

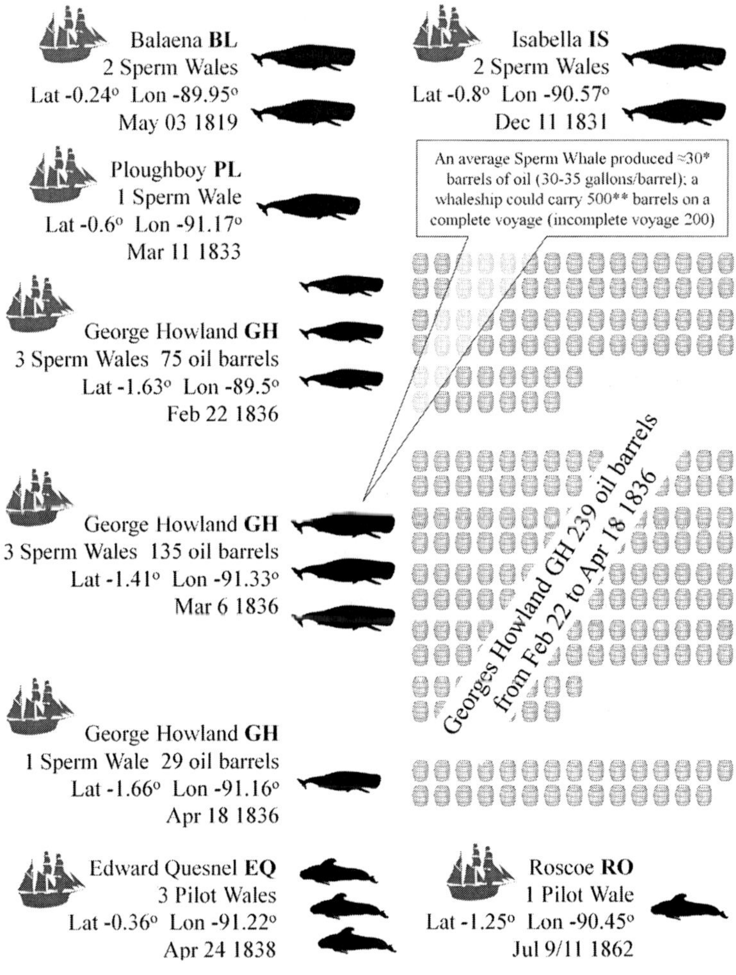

Fig. 4-17 Exemplar whaleships (from Fig. 4.16), the whales they hunted, the amount of whale oil barrels processed after hunts, and the geographic coordinates and dates where/when the hunts occurred in Galápagos (1819-1862). Data extracted from Townsend (1935), Woolmer (2013), and Halpin et al. (2009). Note how the George Howland GH, after hunting 7 large sperm whales (Feb-Apr 1836), secured 239 oil barrels. Number of whale oil barrels comes either from whaleship logbooks or are estimates based on the literature, as follows: *The 30 barrels of oil per whale correspond to the Galápagos, after Whitehead and Hope (1991). The average whale size varied (globally) as per whaling ground, a sequel to hunting intensity over time (i.e. more intensely hunted grounds were eventually left with smaller whales). **Data on a whaleship carrying 500 barrels on a complete voyage comes from Smith et al. (2008).

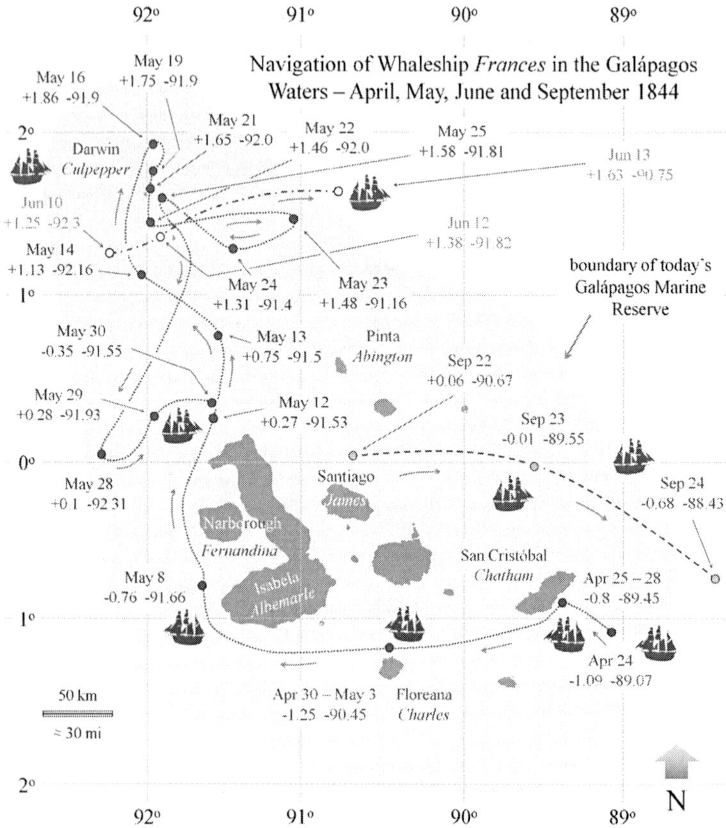

Fig. 4-18 Exemplar navigation of the American whaleship Frances (FR in Fig. 4.15) in the Galápagos, April-June and September 1844. The ship arrived in the archipelago on April 24 (bottom right; numbers -1.09 and -89.07 correspond to latitude and longitude coordinates, respectively); it then sailed to Chatham (San Cristóbal) on April 25-28 (dark-gray dots and dotted line). Frances reached Charles (Floreana) on April 30 to May 3, and from there it navigated North, passing by Albemarle (Isabela) and Narborough (Fernandina) on May 8. By May 12, the ship was in Rock Redondo and continued in tortuous navigation (evidently looking for whales) toward the Northwestern Islands of Culpepper (Darwin) and Wenman (Wolf). It returned to Rock Redondo by May 30. Between June 10 and 13, the ship's records indicate it was in the vicinity of Wenman (June 12; white dots and dashed-dotted line). Frances must have left the Galápagos during the rest of June (no log records) and returned three months later. By September 22, Frances was in James (Santiago); it left the archipelago for good on September 24 (light-gray dots and dashed line). There are no records indicating that the vessel captured or processed whales while in the Galápagos! Data after Townsend (1935), Woolmer (2013), and Halpin et al. (2009). Layout: GPC.

Whales Captured in Galápagos Waters

From the public database Whaling History (2017), which compiles Townsend's (1935) work on American whaling 1761-1920, we identified 313 (2%) out of 16,227 entries (years 1829-1889). We placed such entries (as dots on a map) within the boundaries of today's Galápagos Marine Reserve (Fig. 4.19). The entries corresponded to hunts of 505 whales: 497 sperms (98.4%) and 8 humpbacks. The hunts took place everywhere, particularly off Albemarle (Isabela): 122 sperms in Elizabeth Bay alone!

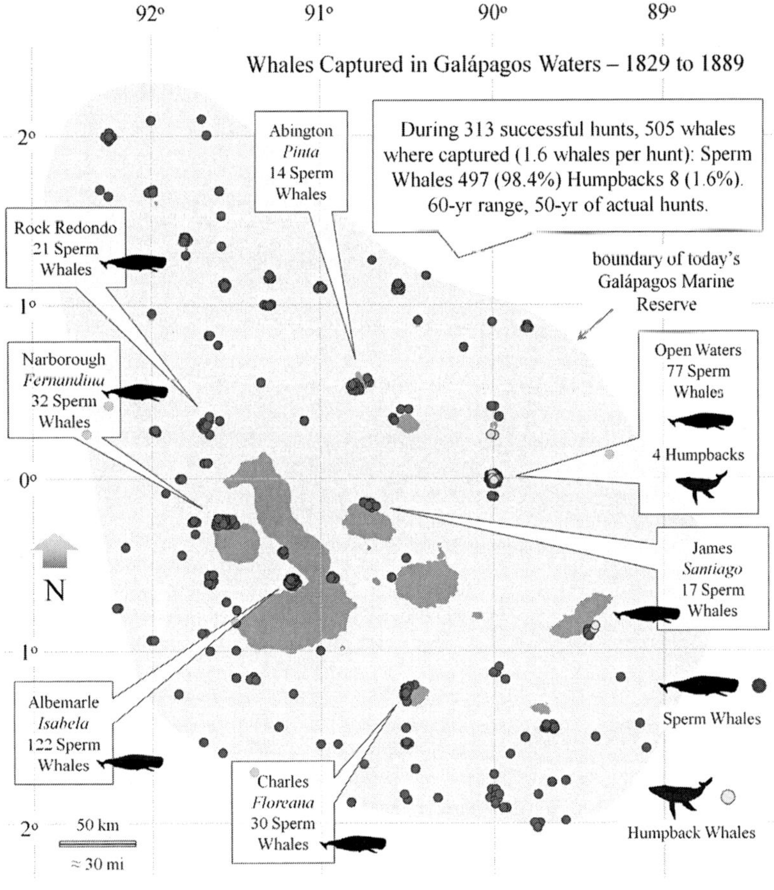

Fig. 4-19 Whales captured in Galápagos waters (1829 to 1889). During 313 successful hunts, 505 whales were captured (1.6 whales per hunt): sperm whales 497 (98.4%); humpbacks 8 (1.6%). Layout: GPC, data extracted from Whaling History (2017).

To zoom out and quantify whales hunted in Galápagos waters (Ground 38) and its neighbor Panama Bay (Ground 35), we relied again on the Woolmer (2013) and Halpin et al. (2009) dataset "Historical Distribution of Whales Shown by Logbook Records 1785-1913" (links in references and Appendix A). An online world map (based on Townsend's 1935 A-D Charts) allows users to mark/highlight areas on the map—with the computer cursor—for data access. We mark-highlighted the square shown in Fig. 4.20 and extracted the information. It corresponded to 1,433 whale hunts: 1,137 sperms (79.3%), and 296 humpbacks (20.7%).

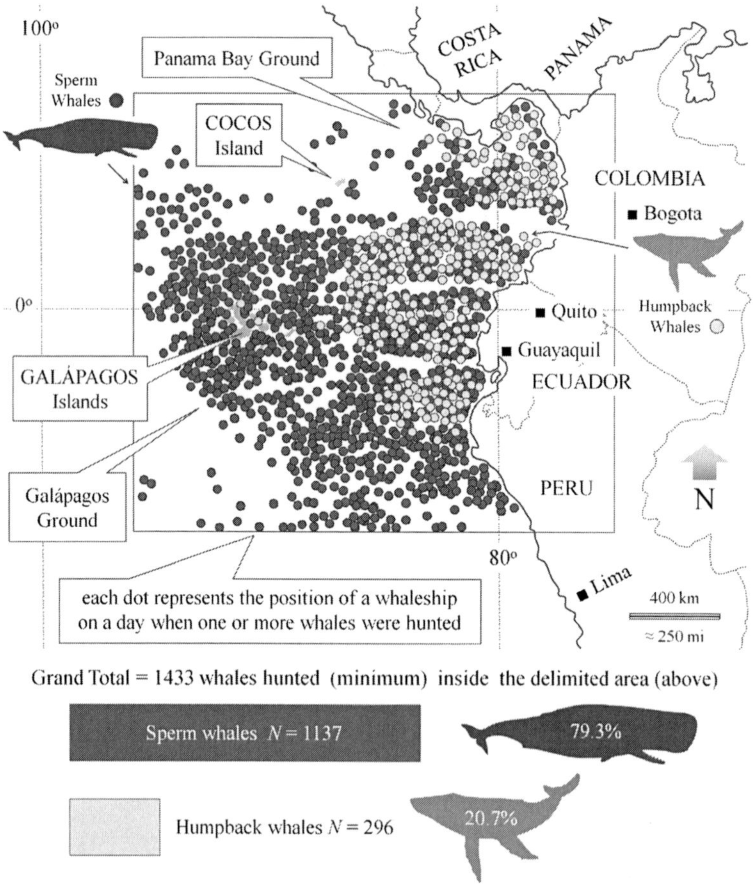

Fig. 4-20 Sperm and humpback whales hunted in Galápagos and Panama Bay (Grounds 38, 35), 1785-1913. Layout: GPC, data from Woolmer (2013), Halpin et al. (2009).

Note in Fig. 4.20 how sperm whales were hunted almost everywhere around the Galápagos and off the coasts of today's Costa Rica, Panama, Colombia, Ecuador, and Peru (keep in mind that back in 1785-1913 the borders between countries were different; Panama was still part of Colombia). By contrast, humpbacks were hunted close to the continental shores and rarely within Galápagos waters.

Did whaling in the Galápagos and surrounding waters have any impact on whale abundance? Yes. Researchers have examined whale-hunting practices in the Galápagos (and other grounds) from three informative perspectives: whale-sighting rate (number of whales seen as function of "whaling days" per whaleship); whale-catching rate (number of whales hunted as function of whaling days); and whale oil yields (number of whale oil barrels produced as function of whaling days) over time (Whitehead and Hope 1991). In Fig. 4.21, we introduce such approach. The sketch depicts sightings and/or catches (dots or diamonds on a map) of sperm whales in the Galápagos from 1830 to 1850. The data comes from 68 exemplar logbooks of whaling voyages (examined by Whitehead and Hope 1991). Note the "whole area", which covers 360,000 sq n mi (square nautical miles), ranging from 85°W to 95°W longitude, and 5°N to 5°S latitude. Within it, locate the "inner area" (islands inside), which covers 37,800 sq n mi (10.5% of the whole area). And notice the "outer area", which covers 322,200 sq n mi (89.5% of the whole area). As indicated in the dialog boxes in Fig. 4.21, the sightings were either "plural" (dark dots = more than one whale was observed) or "single" (white diamonds = single whales were observed).

In the "whole area", whaleships completed 7,017 whaling days, 657 sightings, and 548 catches. In the "inner area", whaleships completed 3,089 whaling days (44.0% of total), 284 sightings (43.2% of total), and 216 catches (39.4% of total). And in the "outer area", whaleships completed 3,928 whaling days (56.0% of total), 373 sightings (56.8% of total), and 332 catches (60.6% of total).

Despite the inner area covering only 10.5% of the whole-area surface, whaleships completed, in the former, 42.2% of the combined whaling days, sightings, and catches. By contrast, whaleships completed 57.8% of the combined whaling days, sightings, and catches in the outer area, which covered 89.5% of the whole-area surface (but see below).

The whale-sighting rate in the whole area (sightings 657/whaling days 7,017) was 0.093. In the inner area (284/3,089), the rate was 0.092; in the outer area (373/3,928), 0.095. The catch rate in the whole area (catches 548/whaling days 7,017) was 0.078. In the inner area (216/3,089), it was 0.069; in the outer area (332/3,928), 0.084. The outer area was higher!

It appears that whalers invested much effort to secure 39.4% of the total catches (216) in 10.5% of the whale-harvesting area (catch rate 0.069). In the outer area, however, they secured 60.6% of the total catches (332) in 89.5% of the whale-harvesting area (catch rate 0.084).

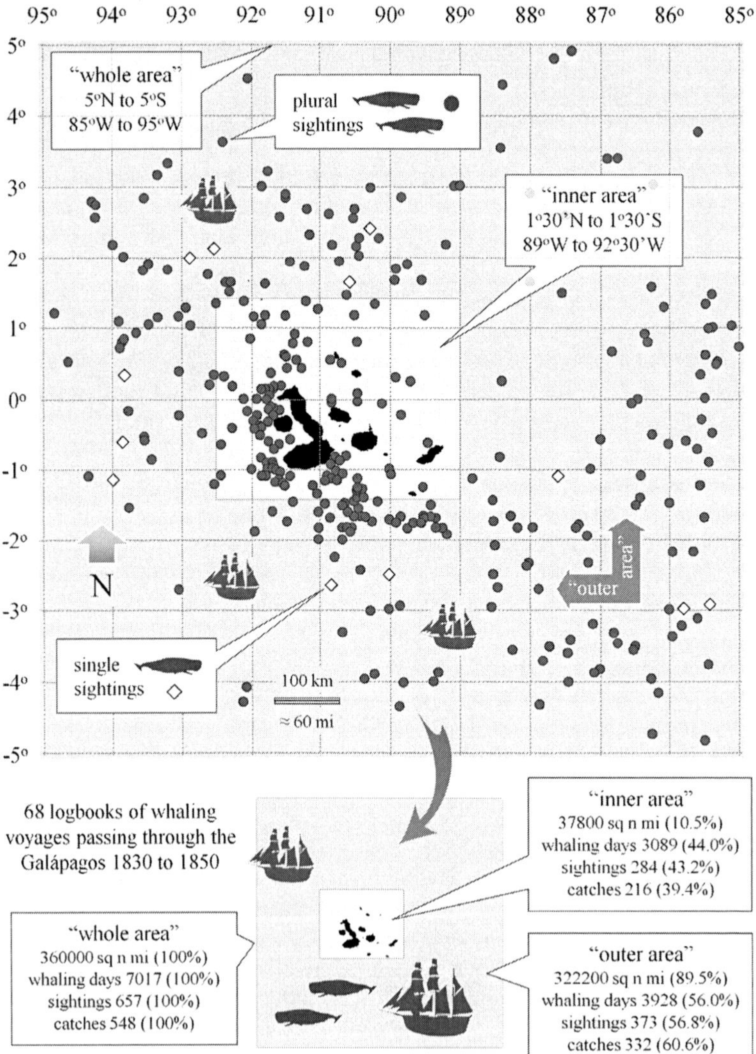

Fig. 4-21 Sightings and catches of sperm whales in the Galápagos, 1830 to 1850. Layout: GPC, redrawn and modified after Whitehead and Hope (1991).

Over time (1830 to 1850), sperm whale sighting rates, catching rates, and oil-yield rates declined in the Galápagos and surrounding waters (i.e. whole area), an indication of resource depletion (Fig. 4.22; data extracted from Whitehead and Hope 1991). Thus, at the peak of The Golden Age of Whaling (1840s-1850s), Galápagos "leviathans" were already declining!

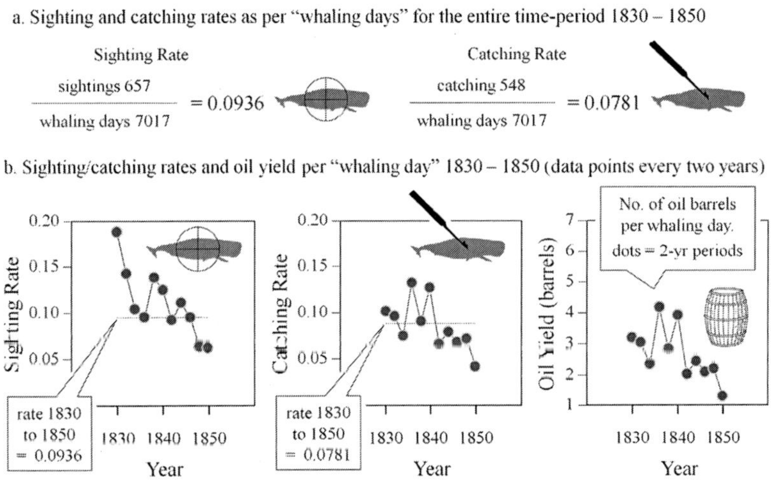

Fig. 4-22 Rates of sighting and catching sperm whales in the Galápagos, as well as oil yield from 1830 to 1850 (i.e. "whole area" in Fig. 4.26). Layout: GPC, redrawn and modified after Whitehead and Hope (1991).

In Fig. 4.23, we summarize the oil yields from sperm whales hunted in the Galápagos (1830 to 1850) as per whole, inner, or outer areas in Fig. 4.21 (after Whitehead and Hope 1991). On average, sperm whales hunted in the "whole area" (data available for *N*=205 individuals) yielded 31.22 oil barrels per whale. Those corresponding to the "inner area" (*N*=69) yielded 32.0 oil barrels per whale. And those caught in the "outer area" (*N*=136) yielded 30.82 oil barrels per whale. About half of the sperm whales hunted in the Galápagos produced ≈30 barrels of oil (range 0-9 to >100 barrels per individual), as shown in Fig. 4.23, bottom left. This indicates that they were, on average, smaller whales (likely females) than those caught, for example, in the Western North Pacific (same time period), which yielded ≈50 barrels per whale (*N*=540; many likely males). It is conceivable that the Galápagos ground was intrinsically less profitable for industrial whaling (due to ecological reasons, whale demographics or global distributions, female/male sperm-whale differential dispersal, or habitat preferences) than other grounds in the Pacific. But it is clear that sperm whales declined in the Galápagos from 1830 to 1850 (Fig. 4.22).

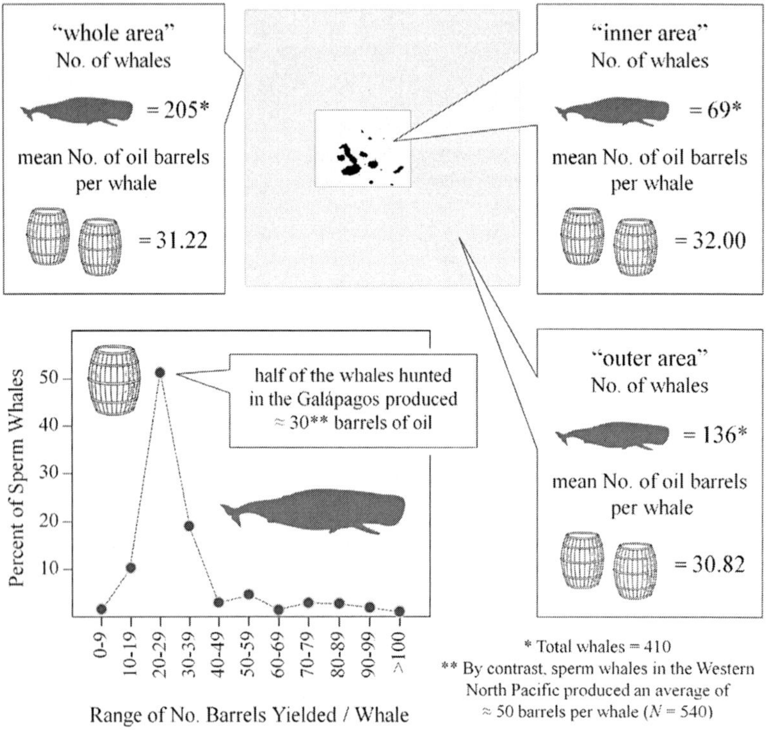

Fig. 4-23 Oil yields (mean values) from sperm whales hunted in the Galápagos, 1830 to 1850. Layout: GPC, redrawn and modified after Whitehead and Hope (1991).

What is the status of today's sperm whale population in Galápagos? Modern demographic trajectory models suggest that before intense industrial whaling took off (1700s), the world's sperm whale population was 1.11 million. By 1880, it had been reduced to 71% of its initial size. And by 1999, it was 32%. The current global estimate (2000s) is 360,000 individuals, one fifth of previously optimistic numbers (Whitehead 2002)!

Whaling appears to have completely ceased in the Galápagos even before the 1950s (the Galápagos National Park was established in 1959). It was considerably reduced by 1860 when global catches crashed (below; Smith et al. 2008). Some authors suggest that, prior to whaling ending in the waters close to the archipelago (late 1800s to early 1900s), up to 5,000 sperm whales were taken (Hope and Whitehead 1991). But sperm-whaling continued and later intensified (as a high-tech industry) in neighboring coastal Peru between 1957 and 1981 (Whitehead et al. 1997). Why is this relevant to today's Galápagos sperm whale population?

From a modern database of 3,737 high-quality photos of sperm whales found in the Galápagos between 1985 and 1995, researchers have estimated the number of whales present in the archipelago during that time (Fig. 4.24; Whitehead et al. 1997). [The data base contained 1,540 distinctive individuals as per tail-fluke features]. Note in Fig. 4.24b that by 1990 (the midpoint in the distribution), the estimated population had declined beyond 50% (from the initial ≈4,000 to ≈1,310 sperm whales), and it continued to decline to ≈429 individuals by 1995. The instantaneous population decline per year was 20% (e.g. if the population in 1985 was ≈4,000 individuals, by 1986 it had reduced to ≈3,200; by 1990 to ≈1,310; and by 1995 to ≈429; realize that these are our didactic estimates based on a 20% decline per year at an initial population of ≈4,000 individuals; in Whitehead et al.'s 1997 simulation, the numbers by 1990 were as low as 1,245 or even 1,035 whales; "by 1995 there were just a few hundred"!).

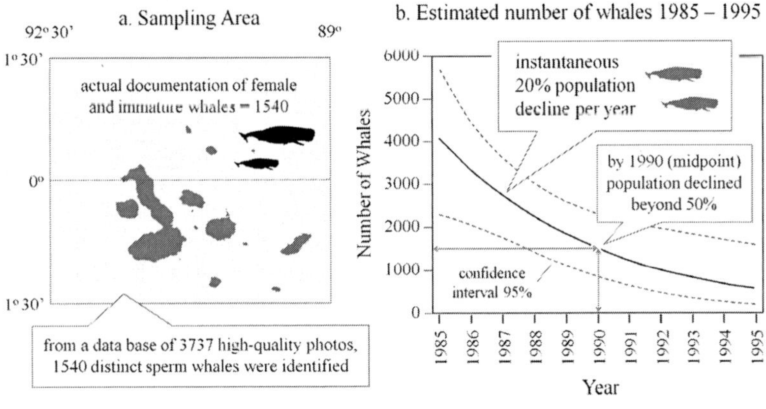

Fig. 4-24 Documented (a) and estimated (b) number of female and immature sperm whales present in the Galápagos, 1985-1995. About 80% of the sperm whale population was female and subadult. Layout: GPC, redrawn and modified after Whitehead et al. (1997).

If industrial whaling was no longer practiced in the Galápagos since the early 1900s (above), why did sperm whales continue to decline by the 1990s? It has been hypothesized that individuals from the sperm whale Galápagos population are dispersing away, filling in, and occupying available habitats elsewhere, particularly off the Peruvian shores, where hunting until 1981 depleted local populations (Whitehead et al. 1997). Thus, in a density-dependent habitat-selection scenario (rather than natural or accelerated mortality rate, unknown diseases, ocean pollution, sequels of damaging-to-the-whales industrial-fishing practices, or collisions with ships), sperm whales might be actively redistributing themselves into nearby available habitats (Whitehead et al. 1997). Although it is true that the rate of reproduction of Galápagos sperm whales is low (0.037 calves/individual/year, as supposed to 0.055 calves/individual/year to generate population growth), this might be associated with the scarcity of mature males rather than with females not reproducing or dying younger (>12-m/40-ft-long males are only 4% of the Galápagos population, significantly below the 15% reported in the 1830s-1850s whaling records). It appears that the sequels of intense whaling on past sperm whale populations (e.g. Peruvian sperm whale hunts up to 1981) can still be detected decades later in what one day was known as "Ground 38" (Whitehead et al. 1997).

The Whaling Paradox (1800s versus 1900s) and How Whales Learned to Outsmart Whalers

As indicated above, global whaling declined by the late 1800s. However, it recovered by the 1950s and reached historic catches—at rates never documented before—by the 1960s. Around 1830-1850, sperm whale abundance at two exemplar locations, Japan (Grounds 29, 30, 31; Fig. 4.13) and Galápagos (Ground 38; Fig. 4.13), plunged, as inferred from logbooks of American whaleships *encountering* cachalots (Fig. 4.25). If sperm whale populations had declined at previously productive Pacific grounds (e.g. Japan, Galápagos, and neighboring sites), how was it possible for whalers of the mid-1900s to encounter and catch whales at unprecedented rates in just recently depleted waters? [Keep in mind that whaling in the Galápagos ceased for good prior to the 1950s, above]. This decline in *encounter* rates during the 1800s versus the historic high *catches* of the 1900s was paradoxical (Smith et al. 2008).

In Fig. 4.25, we show the estimated removals (left axis) of sperm whales from 1750 to 2000 (i.e. Global, North, and South Pacific), as well as encounter rates per day in the mid-1800s (right axis). The dots (dark gray for Japan; light gray for Galápagos) correspond to encounter rates per day (which declined over time). The line corresponds to annual removals or actual catches of sperm whales (which increased threefold by the 1960s with respect to the 1800s). The "whaling paradox" 1800s (encounter rates) versus the 1900s (catching rates) emerges from such inconsistency! As shown in Fig. 4.25 (top), global catches of sperm whales in the 1800s peaked at <10,000/year; by 1960s, they reached >30,000 annually. In the North Pacific alone (Fig. 4.25, middle), catches of sperm whales in the 1800s were <2,000/year; by 1960s, they reached >15,000 annually. And in the South Pacific alone (Fig. 4.25, bottom), catches of sperm whales in the 1800s were <5,000/year; by 1960s, they reached >5,000 annually.

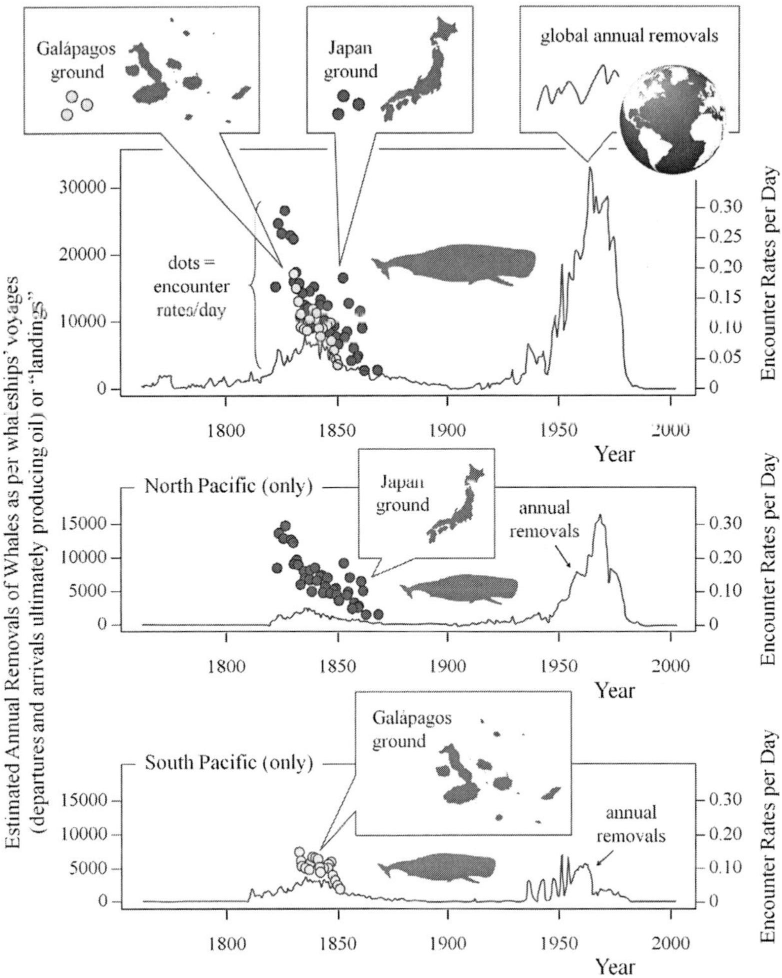

Fig. 4-25 Estimated removals of sperm whales (1750 to 2000): global, North Pacific, or South Pacific (left axis), and encounter rates per day in the mid-1800s (right axis). Layout: GPC, redrawn and modified after Smith et al. (2008).

Various scenarios have been proposed to explain the whaling paradox (Smith et al. 2008), including: (1) *Removal estimates*: The 1800s sperm-whale removals were substantially higher than estimated, or the 1900s removals were substantially lower than estimated. The reverse is conceivable. The point of this scenario is that the historical whaling record could be inaccurate, and the paradox apparent. (2) *Population structure*: Sperm whales belonged to distinct, isolated populations that were hunted and extracted from discrete grounds (easier to count and keep records), or they belonged to a single global population with high individual or group mobility and mixing (more difficult to count with accuracy). The latter suggests a potential "pooling fallacy" (consider repeated counts of the same whales as if they were independent from one another; Martin and Bateson 1993). Regardless of whales having clumped or widely spread spatial distributions, seasonal migrations made them travel long distances (and blend with other migrants). Whalers could have encountered them away from the usual grounds. (3) *Encounter rates proportional to whale abundance*: Whale-encounter rates derived from logbook data were reliable indicators (proxy) of whale abundance, or whale-encounter rates were unreliable to infer whale population status. Here, biology becomes relevant again, particularly prey behavior. Despite good whale-encounter data, whales could have learned—over time—to avoid whalers (predators) and become increasingly elusive. Plus, highly elusive groups could have retreated to remote locations. In other words, highly intelligent animals—all whales and particularly sperms—outsmarted human predators (more on this below)! (4) *Equilibrium population dynamics*: Environmental changes during centuries of whale exploitation (1700s to 1800s) could have altered the habitats in which whales occurred (i.e. the relative equilibrium prior to whaling). For example, differential hunting of whale species—by whalers—led to global disequilibrium in whale abundance, their natural foods or competitive interactions, and their subsequent fragmented and dissimilar recovery. Or climate-change impacts overlapped with the sequels of whaling (analysis of scenarios in Smith et al. 2008). Thus, these scenarios could, hypothetically, help explain the discrepancy between the decline in whale encounters during the 1800s and the historic high catches of the 1900s.

Did whales learn to outsmart whalers? Yes. Relying on a dataset of 78,000 whaleships' voyage days in the Northern Pacific (1841 to 1849), in which sperm whale sightings occurred on 2,400 (3.1%) of those days (data from Maury 1851; Smith et al. 2012), researchers have explored additional questions relevant to the drop in whale encounters during the 1800s: (1) Were earlier whalers more competent in finding and catching whales than later whalers? (2) Were vulnerable whales encountered and taken first (i.e.

the young, old, sick, foolhardy, belligerent, mothers protecting calves)? (3) Did whales learn to avoid whalers? In Fig. 4.26, we share the outcomes of such investigations (after Whitehead et al. 2021).

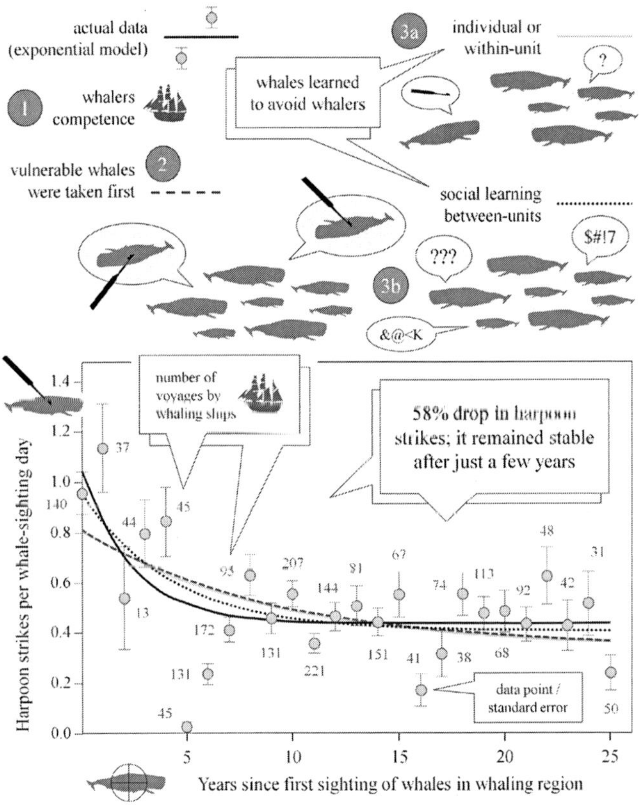

Fig. 4-26 Mean harpoon strikes per sperm whale sighting day as function of years since the whales were sighted for the first time in a region (Northern Pacific; 10°-50°N; 1841 to 1849). Three scenarios were evaluated: (1) whalers' differential competence over time (no evidence, not shown); (2) vulnerable whales were taken first; and (3) whales learned to avoid whalers. Black solid line=exponential model corresponding to the actual data points (gray circles and their standard errors). Dashed gray line=model under the assumption that whalers first took vulnerable whales. Solid gray line=model under the assumption that an individual whale within a group learned to avoid whalers. Dotted black line=model under the assumption that naïve social groups learned from experienced social groups to avoid whalers. The original dataset used to run the statistical models included 78,000 whaleships' voyage days; sperm whale sightings occurred on 2,400 (3.1%) voyage days (after Maury 1851 and Smith et al. 2012). Layout: GPC, redrawn and modified after Whitehead et al. (2021).

Note in Fig. 4.26 that mean values of harpoon strikes per sperm whale sighting day are plotted as function of years since the whales were sighted for the first time. The gray circles correspond to the actual data points (next to them is the number of voyages by whaling ships). The black solid line corresponds to the exponential model that fits these points. Just a few years after the first sighting of the whales took place in the whaling region (i.e. Northern Pacific), the harpoon strikes dropped by 58% and remained low for the next 20 years. Researchers found no evidence that earlier whalers were more competent in finding and catching whales than later whalers (Whitehead et al. 2021).

The dashed gray line in Fig. 4.26 corresponds to the trend (i.e. the mathematical model) under the assumption that whalers first took vulnerable whales. The trend is weak with respect to the actual data.

The solid gray line in Fig. 4.26 corresponds to the trend under the assumption that an individual whale—or within a group—learned to avoid whalers. The trend is also weak and overlaps with that of the "vulnerable whales" above.

Finally, the dotted black line in Fig. 4.26 corresponds to the trend under the assumption that naïve social groups learned from experienced social groups to avoid whalers. The trend fits the actual data more robustly than the other "causal models" (i.e. either "vulnerable whales" or "learning at the individual or within-group level"). Thus, social learning could have produced a noticeable decline in harpoon strike rate during the 1800s as whales learned from one another and in a social context to avoid whalers. We can speculate that highly intelligent, social animals like sperm whales communicated danger among groups, cooperated in, and coordinated fleeing behaviors (upwind), deep diving, and other predator avoidance tactics to escape human predators (Whitehead et al. 2021). Yes, the data suggests that sperm whales did learn to outsmart 1800s whalers!

Spatial and Seasonal Distribution of Whalers and Whales

During the late 1800s and early 1900s, thousands of whales continued to be hunted. This was the case despite the animals' behavioral ability to elude human predators. More recent assessments focused on whaleships' logbook data, 1780s-1920s, confirm this assertion (Smith et al. 2012). Whaling history specialists have combined two significant datasets: Townsend's 1930s and Maury's 1840s. We introduced Charles H. Townsend's work earlier in this chapter (Townsend 1935) and cited Maury's datasets numerous times (Maury 1851). Townsend compiled the locations of American whaling ships on the days when the whales were actually caught, i.e., only "positive" data. By contrast, Matthew Fontaine Maury—Officer and oceanographer of the United States Navy—extracted both "positive" data (as per actual sights and/or catches of whales) and "negative" data (as per "searches alone" of whales, but neither sights nor catches) from logbook records (review in Smith et al. 2012).

In Fig. 4.27, we show the daily locations of American whaling ships, sightings, and hunts of whales based on the combined work by Townsend (1930s) and Maury (1840s). This outstanding, open-access data collection comes from Smith et al. (2012) and represents about 10% of the American whaling voyages from 1780 to 1920. Note that Fig. 4.27a corresponds to data about five whale species: sperms (*Physeter macrocephalus*); rights (*Eubalaena* spp.); bowheads (*Balaena mysticetus*); humpbacks (*Megaptera novaeangliae*); and grays (*Eschrichtius robustus*). Fig. 4.27b corresponds only to sperms and grays.

In Fig. 4.28, we offer a close-up view of the Eastern Pacific, with the Galápagos (Ground 38), Panama Bay (Ground 35), Offshore (Ground 39), and Callao (Ground 52) whaling grounds in sight (i.e. a zoom-in view of one of Smith et al.'s maps). The 40x40-degree area (longitude 80°W to 120°W and latitude 20°N to 20°S) covers ≈18-million km^2 (≈7-million sq mi). In it, the daily locations (from 1780 to 1920) of American whaling ships, sightings, and hunts of sperm, humpback, and right whales are depicted. Spot the dialog box "*locations where whales were searched for but not sighted or caught*" (Fig. 4.28, top). The information on the map confirms that American whaleships looked for whales almost everywhere in the Galápagos, Panama Bay, Offshore, and Callao waters. In the bottom of Fig. 4.28, note three dialog boxes bringing the reader's attention to sites where sperm, humpback, or right whales were sighted (dots) or caught (+ symbols). Sperms were sighted or caught around the entire Galápagos, off the Northwestern coast of South America, and along the equator toward the West of the Galápagos Islands. Humpbacks were sighted or caught almost entirely along the Northwestern South American coast. And rights were rarely sighted or caught on these grounds.

What about the seasonal distribution of whaling? American vessels searched for whales in the Tropical Eastern Pacific [and in all oceans] year-round (Fig. 4.29). The searches increased from December to March (light gray dots in maps in Fig. 4.29). This coincided with El Niño Southern Oscillation (ENSO) phenomenon, which brought—still does—annual changes in surface sea-water temperature to the region, rainfall, and/or nutrient availability (details in Paz-y-Miño-C and Espinosa 2024b). Sights and catches of sperms, humpbacks, and rights (in that order) increased from December to March (Fig. 4.29). From May to September, searches,

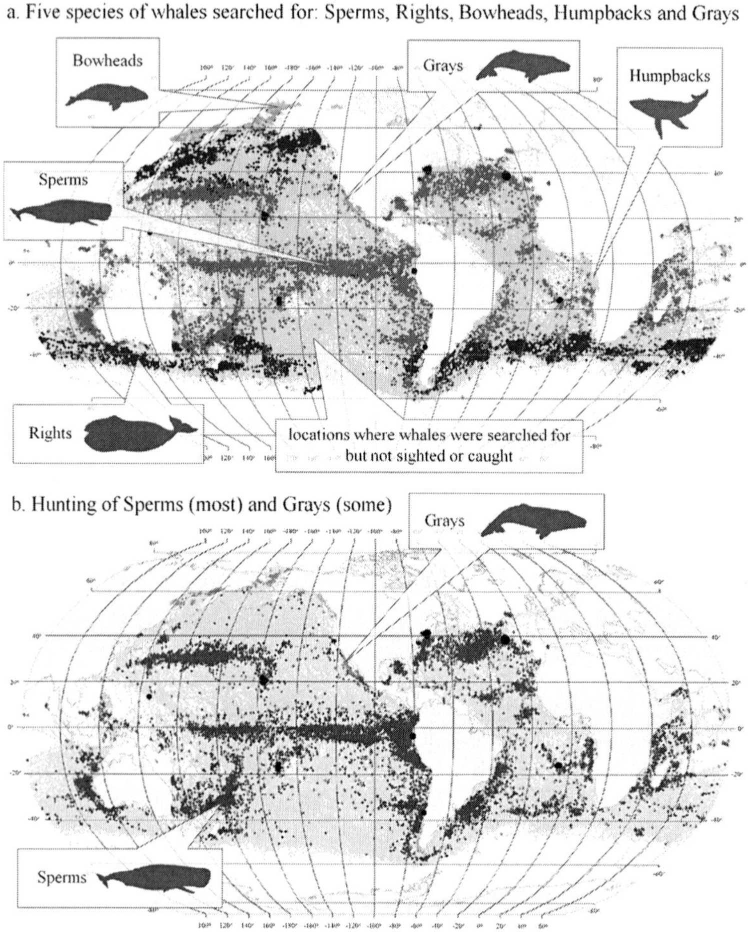

Fig. 4-27 Daily locations of American whaling ships, sightings, and hunts of whales as per logbook records from 1780 to 1920. Maps combine "positive" data (=actual sights and/or catches of whales; after Townsend 1930s) and "negative" data (=searches of whales but not sights or catches; after Maury 1840s, although Maury's compilations of whaleships' logbooks included both "positive" and "negative" data). American whalers preferentially hunted five species of whales. For details, see high-definition color maps in Smith et al. (2012); links in Appendix A. a. For didactic reasons, dialog boxes point at specific locations where whales of each of the five species were sighted and/or caught (these species had global distributions, although they varied in abundance and habitat preferences). b. Data restricted to sperms and grays to facilitate observation of the most commonly pursued and hunted species (i.e. sperms) from the 1780s to the 1920s. Maps extracted from open-access Smith et al. (2012).

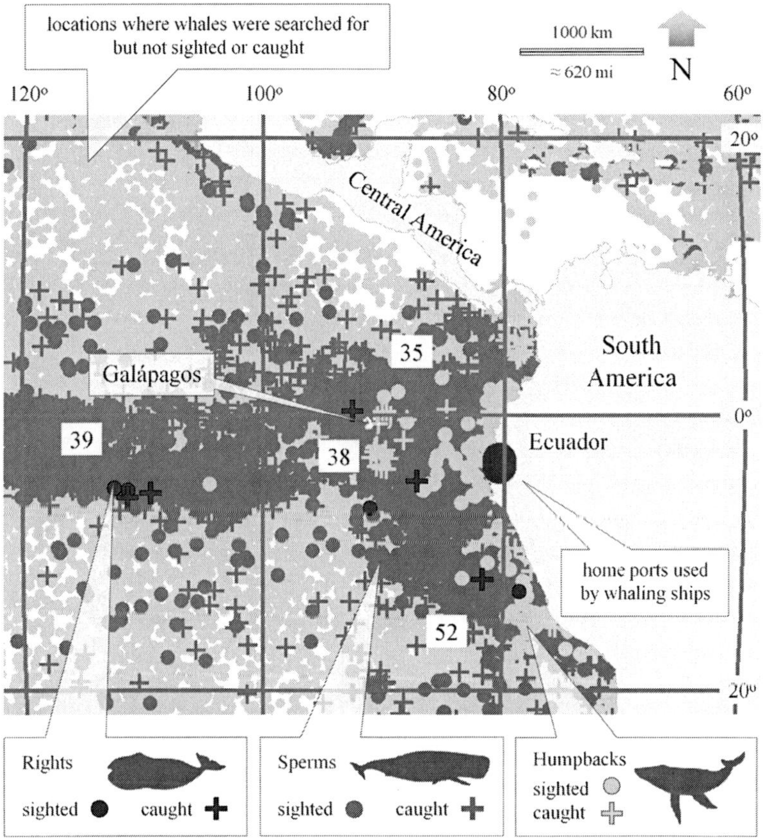

Fig. 4-28 Zoom in on the daily locations of American whaling ships, sightings, and hunts of whales (sperms, humpbacks, and rights) in the Eastern Pacific from 1780 to 1920 (Galápagos whaling Ground 38, Panama Bay Ground 35, Offshore Ground 39, and Callao Ground 52). The ocean area (40x40 degrees) between longitude 80°W to 120°W and latitude 20°N to 20°S corresponds to ≈18 -million km^2 (≈7-million sq mi). Layout: GPC. Background image extracted from open-access Smith et al. (2012); link in Appendix A.

sights, and catches of whales appear to have been concentrated along (not limited to) the equator. In fact, Ground 39, Offshore, remained productive for sperm-whale hunting the entire year (1780 to 1920). Sights and catches of humpbacks became more abundant from June to August (Fig. 4.29), but right whales were infrequently sighted or caught in the Tropical Eastern Pacific (Smith et al. 2012).

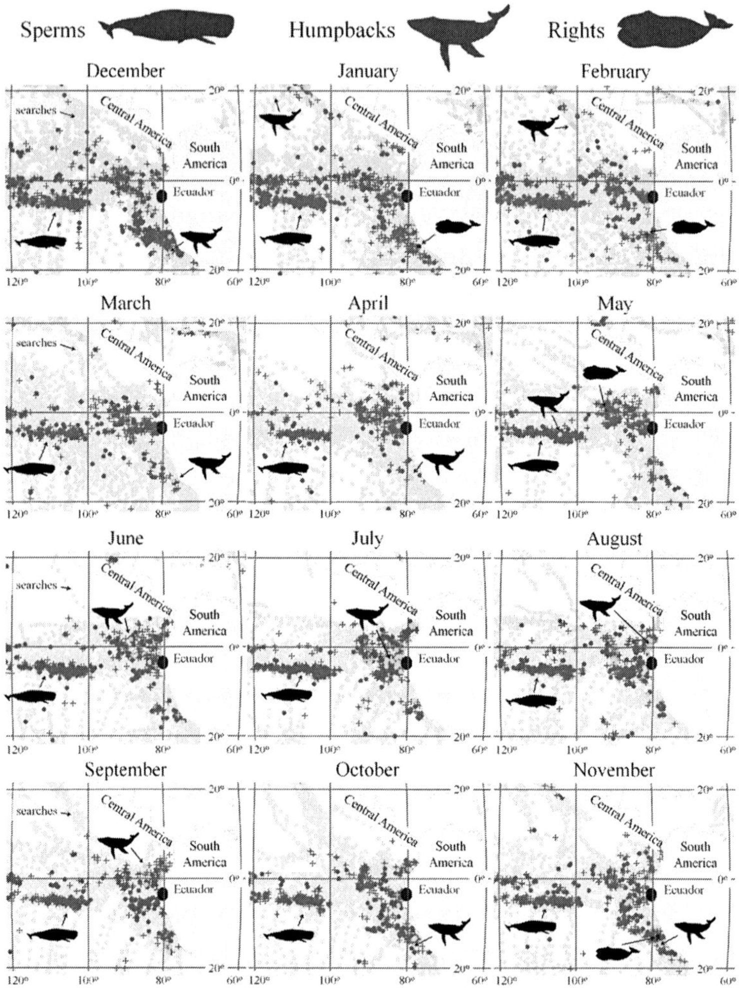

Fig. 4-29 Monthly (seasonal) locations of American whaling ships, sightings, and hunts of whales (sperms, humpbacks, and rights) in the Eastern Pacific from 1780 to 1920. The mapped area (60x40 degrees) corresponds to longitude 60°W to 120°W and latitude 20°N to 20°S. Various whaling grounds were located within this area: in the center Galápagos (Ground 38); to its North, Panama Bay (Ground 35); to the Southeast, Callao (Ground 52); to the South, Archer (Ground 51); to the West, Offshore (Ground 39); and next to it, farther West, "On the Line" (Ground 40); and distant to the Northwest of the Galápagos, "Southwest of Clipperton Island" (Ground 36; see details in Fig. 4.18). Layout: GPC. Background image extracted from open-access Smith et al. (2012); links to maps in Appendix A.

Realize that although American whalers dominated whaling at the peak of The Golden Age of Whaling (1840s-1850s), other open-boat whalers were active globally and in the Tropical Eastern Pacific. They too searched for whales in all oceans, chased them according to the whales' seasonal migrations, killed them, and consumed or commercialized their products. Technological advances in the 20th century (explosive, grenade-tipped harpoons, factory ships, whale-processing vessels, aircraft) made whaling even more efficient. The "slaughtering of the leviathans" was real.

Whale Falls, Whale Pumps, and Whale Absence

Prior to industrial whaling (1700s), the global whale populations totaled 4.05 million individuals; by the late 20th century, whale numbers dropped to 1.48 million (Smith et al. 2019). From 1900 to 1999, modern whalers caught 2,894,094 whales (9 species), of which 874,068 (30.2%) were fin whales (*Balaenoptera physalus*) and 761,523 (26.3%) were sperms (*Physeter macrocephalus*). The rest were blues (*B. musculus*; 379,185 or 13.1%), Seis (*B. borealis*; 291,540 or 10.1%), Minkes (*B. acutorostrata*; 283,905 or 9.8%), humpbacks (*Megaptera novaeangliae*; 249,433 or 8.6%), and three other species (Rocha et al. 2015).

Readers will be shocked to learn that from 1900 to 1962 (62 years), and then from 1962 to 1972 (10 years), the number of sperm whales taken from the world's oceans was ≈300,000 during each time-range (total ≈600,000). Compare these figures with the 360,000 sperm whales taken by open-boat whalers from 1712 to 1899 (177 years). Hunting-rate efficiency improved from 2,033 sperm whales/year (177-year period) to 4,838/year (62-year period) and to 30,000/year (10-year period; Rocha et al. 2015)!

The consumption of whale oil (as fuel or lubricant), spermaceti wax (for candles), bones, and baleen (for artifacts) declined gradually. They were replaced by petroleum oil derivatives (e.g. fuels or plastics). However, the "in-the-millions" hunts of whales (1900s) were, ironically, fueled by petroleum. Petroleum did not save whales (York 2017)!

Not only majestic animals of the sea were lost, but their ecological roles diminished (Roman et al. 2014). Whales and other megafauna in aquatic or terrestrial environments are central to transporting nutrients among habitats. The cycling of carbon, nitrogen, and phosphorus from the ocean depths to continental habitats depends on an "interlinked system" involving marine mammals, seabirds, fish (up-river migrants to spawn), and terrestrial fauna (Doughty et al. 2016). In Fig. 4.30, we sketch the role of whales in that system, with emphasis on the carbon cycle.

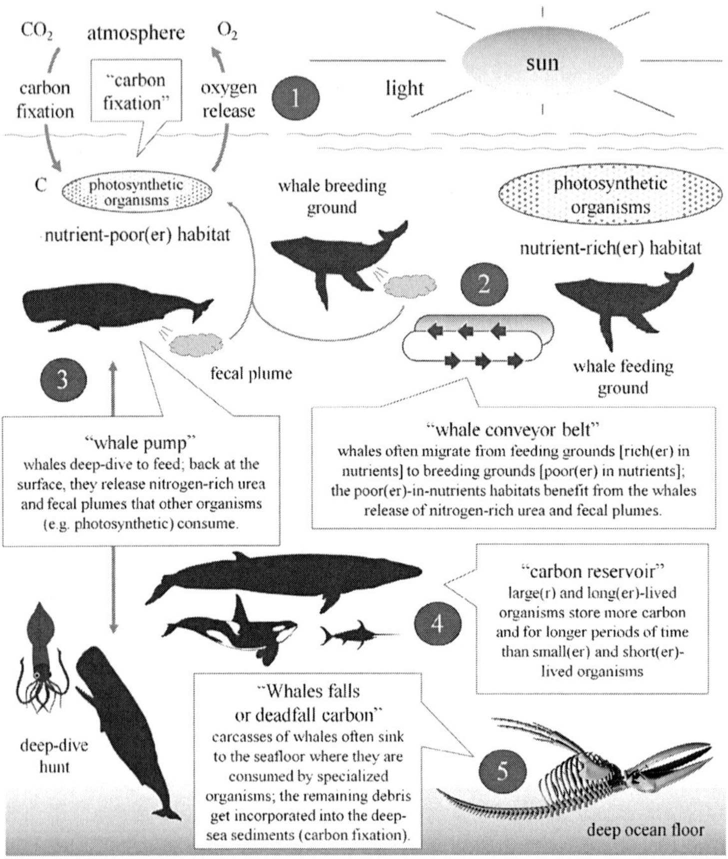

Fig. 4-30 Whales' role in the carbon cycle. Layout: GPC, redrawn and modified after Chami et al. (2019). For impact of whaling on the ocean C-cycle, see Pershing et al. (2010).

Fig. 4.30 summarizes components of the ocean carbon cycle in which whales are involved (Chami et al. 2019): (1) *Carbon fixation*: microscopic photosynthetic organisms fix carbon. These organisms are consumed by invertebrates; the latter are consumed by vertebrates (e.g. fish). Large concentrations of these life forms (by numbers and biomass) attract large consumers like baleen whales (e.g. see silhouette of a humpback whale at "feeding ground" in Fig. 4.30, top right). (2) *Whale conveyor belt*: whales, like the humpback silhouette on the sketch, often migrate from feeding grounds (rich in nutrients) to breeding grounds (poor in nutrients). The poor(er)-in-nutrients habitats benefit from the whales' release of nitrogen-rich urea (metabolite resulting from digesting proteins) and "fecal plumes" (see in Fig. 4.31 the release of a fecal plume by a blue whale). (3) *Whale pump*: some large-size predatory whales (i.e. sperm-whale silhouette in Fig. 4.30, top left) deep-dive to feed on large organisms (e.g. giant squids); back at the surface, these whales also release nitrogen-rich urea and fecal plumes that other life forms consume (Roman and McCarthy 2010). (4) *Carbon reservoir*: large(r)- and long(er)-lived organisms, such as whales, store more carbon for longer periods of time than small(er)- and short(er)-lived organisms. Readers should realize by now that industrial whaling did not only reduce global whale populations, but the burning of whale oil and spermaceti wax did accelerate the release of carbon, as CO_2, into the atmosphere (carbon that was previously fixed and contained in the soma of whales)! (5) *Whale falls or deadfall carbon*: carcasses of whales that naturally die often sink to the seafloor, where they are consumed by specialized organisms. The remaining debris gets incorporated into the deep-sea sediments (more carbon fixation). The sinking of whale carcasses, or "whale falls", secures nutrients for >100 species (whale-fall specialists) that reside in deep-sea habitats. Absence of whales could lead to the extinction of whale-fall specialists (Smith et al. 2019).

Fig. 4-31 Blue whale (*Balaenoptera musculus*) releasing a "fecal plume". The image, taken from a drone at about 50-m (≈160-ft) high, shows a blue whale releasing post-digestion matter. Courtesy of Ian Wiese (with permission), Geographe Bay, Western Australia.

The global decline of whale populations and marine mammals during the past few centuries has reduced the usual capacity of animals to transport *nutrients* (from richer to poorer habitats, above) to 5% its historic values. Phosphorus (proxy for other elements like nitrogen or iron) *upward* movement in the ocean is now down to 23% (in the past was 340-million kg of phosphorus, P, per year). The *lateral* diffusion of nutrients in the world's seas (by 13 species of whales) has been reduced to 10% in the North Pacific, 14% in the North Atlantic, and 2% in the Southern Ocean. Prior to industrial whaling, 11 marine mammals (6 whales) used to transport *vertically* (from deep to surface waters) 260-430 million kg of P per year; this has dropped to 54-110 million kg of P per year. Regionally, the vertical transport of P has plummeted to 34% its previous levels in the North Pacific, 28% in the North Atlantic, and 16% in the Southern Ocean (Doughty et al. 2016). Thus, the slaughtering of whales has deprived the biosphere of essential ecological functions, now compromised even further by the threats of climate change, which we discuss in Chapter Ten, *How Will Climate Change Affect Galápagos?*

Conclusions

Much happened in Galápagos before and after Charles Darwin's—and Captain Robert FitzRoy's—historical exploration of the archipelago in 1835. Three hundred years earlier (1535), the *yªs de galapagos* were accidentally discovered. Ecuador claimed jurisdiction over the *Archipiélago del Ecuador* in 1832. The United States established a military base on Baltra [Seymour] Island during World War II (1942-1946). Ecuador created the Galápagos National Park in 1959. UNESCO declared the islands a World Heritage Site in 1978. Soon after and coinciding with the tourism boom (1980s-2000s), the local population increased significantly from 5,545 in the 1980s to 28,583 by 2024. And tourism grew from 42,000 visitors in 1989 to 276,000 in 2018.

One distinctive human activity took place globally and affected the Galápagos' natural history prior to, during, and after Darwin's visit: whaling! In this chapter, we chose to discuss industrial whaling as it covered the world's oceans ("Global Whaling: 1760s to 1920s") and impacted the Galápagos Islands—also known as "Whaling Ground 38".

[We minimize citations below to abridge the text (sources for abridgment: this chapter and Paz-y-Miño-C and Espinosa 2024a,b). Each conclusion point summarizes topics substantiated by citations in this chapter. For comprehensive history of the Galápagos, see Luna-Tobar (1997); Idrovo (2005); and Woram (2005)].

(1) Industrial whaling became a profitable activity from the 1760s to the 1920s. The prosperity of the American colonies and the Industrial Revolution of the Young Republic (United States, 1820-1860) were linked to the slaughtering of whales. By the mid-1800s, whaling was the fifth-largest industry in America. New England was at the peak of The Golden Age of Whaling (sperm-whale oil trade), with ≈500 of the 730 American whaling ships—900 worldwide.

(2) Whales were hunted for their oil, spermaceti, meat, bones, baleen, teeth, and ambergris. Sperm-whale oil (waxy and thick) burned brightly and was ideal for lanterns; it was also used as a lubricant. Before petroleum oil products became popular (late 1800s onwards), almost all industrial machines or home instruments were lubricated with sperm whale oil. Spermaceti wax was shaped into candles. Whale bones and baleen were crafted into artifacts and utensils. In one of the most profitable years in whaling history, 1853, the sales of whale oil, spermaceti, and baleen in the United States totaled $11 million US (today's money, $413 US million).

(3) In Charles H. Townsend's 1935 work *The Distribution of Certain Whales as Shown by Logbook Records of American Whaleships*, he documented that, for 159 years (1761-1920), 744 New England and Yankee whaleships completed 1,665 voyages. They captured 53,877 whales, including: 36,908 sperm whales (*Physeter macrocephalus*); 5,114 bowheads (*Balaena mysticetus*); 2,118 Northern Rights (*Eubalaena glacialis*) in the Pacific Ocean; 1,685 Southern Rights (*E. australis*) in the Pacific Ocean; 35 Northern Rights in the Atlantic Ocean; 2,981 Southern Rights in the Atlantic Ocean; 1,596 Southern Rights in the Indian Ocean; 2,883 Humpbacks (*Megaptera novaeangliae*); and 557 Californian Grays (*Eschrichtius robustus*).

Townsend compiled detailed records from whaling grounds in the North Atlantic, South Atlantic, Pacific, and Indian oceans. In the Pacific, he noted the significance of various locations: Coast of Japan, Bonin Islands, Hawaiian Islands, Lower [Baja] California, off the shores of Central Mexico, as well as the grounds known as Panama, Galápagos, Offshore, On the Line, and others across the equator.

(4) In the archives of American whaling, Galápagos was cataloged as Ground 38. To its North was Panama Bay (Ground 35); to the Southeast, Callao (Ground 52); to the South, Archer (Ground 51); and to the West, Offshore (Ground 39). More distant and located to the Northwest of the archipelago was Ground 36, or "Southwest of Clipperton Island".

(5) Based on historic whaleship logbook compilations from 1785 to 1913, American vessels operating in the Galápagos and Panama Bay caught 1,433 whales, including 1,137 sperms, and 296 humpbacks. Modern

estimates, however, suggest that from the late 1800s to the early 1900s, up to 5,000 sperm whales alone were taken from the Galápagos, Panama Bay, and proximate waters. Note that sperm-whaling resumed and intensified (as a high-tech industry) in coastal Peru between 1957 and 1981.

(6) Prior to open-boat industrial whaling (1700s), the global whale populations totaled 4.05 million individuals; by the late 20th century, whale numbers dropped to 1.48 million. From 1900 to 1999, modern whalers caught 2,894,094 whales (9 species), of which 874,068 (30.2%) were fin whales (*Balaenoptera physalus*) and 761,523 (26.3%) were sperms (*Physeter macrocephalus*). The rest were blues (*B. musculus*; 379,185 or 13.1%), Seis (*B. borealis*; 291,540 or 10.1%), Minkes (*B. acutorostrata*; 283,905 or 9.8%), humpbacks (*Megaptera novaeangliae*; 249,433 or 8.6%), and three other species.

Shockingly, from 1900 to 1962 (62 years), and then from 1962 to 1972 (10 years), the number of sperm whales (i.e. the "whale of whales") taken from the world's oceans was ≈300,000 during each time-range (total ≈600,000). By contrast, from 1712 to 1899 (177 years), open-boat whalers caught 360,000 sperms. Thus, hunting rate efficiency improved from 2,033 sperm whales/year (177-year period) to 4,838/year (62-year period) and to 30,000/year (10-year period)!

The consumption of whale oil, spermaceti wax, bones, and baleen declined gradually. Whale products were replaced by petroleum oil derivatives (e.g. fuels or plastics). Ironically, petroleum fueled the "in-the-millions" hunts of whales in the 1900s. Petroleum did not save whales.

(7) If by the late 1800s sperm-whale populations (and other species) had declined at previously productive whaling grounds (e.g. Japan, Galápagos, and neighboring sites), how was it possible for whalers of the mid-1900s to encounter and catch whales at unprecedented rates in just-recently-depleted waters? This decline in encounter rates during the 1800s versus the historic high catches of the 1900s was called "the whaling paradox". A "social learning" hypothesis has been proposed to explain it. Scientists think that social learning apparently produced the noticeable decline in harpoon strike rate during the 1800s as whales learned, from one another and in a social context, to avoid whalers. It seems possible that highly intelligent social animals like sperm whales (and other species) communicated danger among groups and cooperated in and coordinated fleeing behaviors, deep diving, and other predator-avoidance tactics to escape human predators. If it is the case that whales outsmarted open-boat whalers of the 1800s, they succumbed to high-tech whalers of the 20th century (equipped with explosive or grenade-tipped harpoons, factory ships, whale-processing vessels, and aircraft).

(8) From a database of 3,737 high-quality photos of sperm whales found in the Galápagos from 1985 to 1995 (in it, 1,540 distinctive individuals), researchers determined that, by 1990, the local population had declined beyond 50% (from ≈4,000 to ≈1,310 individuals), and that it continued to decline to ≈429 individuals by 1995.

Twenty-seven species of Cetaceans (whales and dolphins) have been confirmed to occur in today's Galápagos (one third of the world's 87 species). The majority (21 species) are toothed-Cetaceans or Odontoceti (dolphins, orcas, sperm whales); the rest (6 species) are baleen-whales or Mysticeti (humpbacks, rights, blues). The most commonly observed Cetaceans in the Galápagos Marine Reserve nowadays are Bottle-nosed Dolphin (*Tursiops truncatus*), Bryde's Whale (*Balaenoptera edeni*), Common Dolphin (*Delphinus delphis*), Killer Whale or Orca (*Orcinus orca*), Humpback Whale (*Megaptera novaeangliae*), unidentified baleen species (*Balaenoptera* sp.), Short-finned Pilot Whale and/or unidentified pilot whale (*Globicephala macrorhynchus*/sp.), and Sperm Whale or Cachalot (*Physeter macrocephalus*).

(9) With whaling, not only majestic animals of the sea were lost, but their ecological roles diminished. Whales and other megafauna in aquatic or terrestrial environments were—still are—central to transporting nutrients among habitats. The cycling of organic carbon, nitrogen, and phosphorus from the ocean depths to the continental interiors depends on an "interlinked system" involving marine mammals, seabirds, fish (upriver migrants to spawn), and terrestrial fauna.

(10) The global decline of whale populations and marine mammals reduced the usual capacity of animals to transport nutrients (from richer to poorer habitats) to 5% its historic estimates. Phosphorus upward movement in the ocean is now down to 23%. The lateral diffusion of nutrients in the world's seas (by 13 species of whales) has been cut to 10% in the North Pacific, 14% in the North Atlantic, and 2% in the Southern Ocean. Some whales (sperms) deep-dive to feed on large organisms (giant squids); back at the surface, the whales release nitrogen-rich urea and fecal plumes that other life forms consume (whales act as "pumps", transporting nutrients from the bottom to the sea surface). "Whale falls" (i.e. sinking of whale carcasses after natural death) secure nutrients for >100 species (whale-fall specialists) that reside in deep-sea habitats. Thus, centuries of slaughtering whales have deprived the biosphere of ecological functions, now compromised even further by the threats of climate change.

Questions

The following questions (Q1–4) merit generous written answers. You might need to do additional reading or library work to enrich your answers (we suggest exploring Paz-y-Miño-C and Espinosa 2024a,b).

Question 1: (a) Why did global industrial whaling emerge? (b) Did "Whaling Ground 38" (i.e. the Galápagos) offer anything unique to the global industrial whaling? (c) Why did global industrial whaling come to an end? (d) Where is industrial whaling still being practiced? (e) What are the ecological and environmental impacts of today's industrial whaling?

Question 2: (a) What will be the impacts of climate change on the global populations of whales [and dolphins]? (b) Will climate change impact particular species of whales [and dolphins] more than others? (c) What will be the impacts of climate change on the populations of whales [and dolphins] that transit by or reside proximate to the Galápagos Marine Reserve (i.e. the Tropical Eastern Pacific)?

Question 3: In this chapter, we stated that "…among whalers, the 'whale of whales' was the Sperm Whale (*P. macrocephalus*), the most profitable catch for its quality of oil, spermaceti, and the infrequently found ambergris (after Dolin 2007)…" (a) What is ambergris, and how is it formed? (b) Why was ambergris highly valued during the Golden Age of Whaling? (c) What scientific evidence supports the notion that ambergris has unique chemical properties of lucrative value to any industry? (d) If ambergris chemical applications or benefits are real, can natural ambergris [which is infrequently found] be replaced by synthetic compounds?

Question 4: Examine Fig. 4.30 *Whales' role in the carbon cycle*. How would a sudden reduction in whales abundance, or their complete disappearance, affect the Galápagos and its Marine Reserve?

How to Answer the Questions Above

The questions above have various possible answers (A1–4). Here we suggest perspectives from which to address the queries.

Answer 1: (a) Industrial whaling overlapped with the industrial revolution. You need to examine why "burning whales" [their oil] became crucial for fast technological and economic development and, therefore, a profitable enterprise. Since "…New England and Yankee whalers dominated the chase, hunt, and processing of whales and their products…" (above; after Tower 1907; Moment 1957; Davis et al. 1997), you can also extrapolate the emergence of industrial whaling to the differential expansion of specific central economies in the world. (b) Think about the geographic position of "Whaling Ground 38" (i.e. the Galápagos) with respect to both the Tropical Eastern Pacific and the Pacific Ocean itself with its whaling grounds. Examine Figs. 4.10, 4.13, 4.14, and 4.20; they should give you hints about what "Whaling Ground 38" might have offered that was unique to the whalers' goals. Think also about Discussion 1 (below) and the provisioning role of the Galápagos Islands (via their human settlements) to whaling vessels during their long journeys—months and even years—before returning home; Galápagos became a must-stop! (c) Global industrial whaling ceased gradually, perhaps fast from a historical perspective. You need to explore what replaced industrial whaling (i.e. whales and their products) and how that replacement led to the emergence of new sources of energy with their own more advanced technologies. (d–e) You need to identify the countries that still sanction industrialized whaling and their justification for it (i.e. historical, cultural, and even scientific-research reasons, as some countries claim). International organizations (e.g. the International Whaling Commission) point out that both commercial and even subsistence whaling [by Indigenous peoples] need constant evaluation and extraction quotas to prevent unwanted ecological or environmental impacts. Your answer should be comprehensive.

Answer 2: (a–b) Climate change impacts [short-, medium-, and long-term] have been extensively discussed in the scientific literature (e.g. changes in sea water temperature, regional and global chemical cycles, and availability of nutrients for all sorts of organisms, including whales and dolphins; migration and reproduction patterns of ocean vertebrates; differential increase of environmental pollutants in oceanic habitats; and multiple other impacts). You need to identify how such changes in the ocean's environments will specifically affect cetacean populations. (c) In the case of Galápagos, keep in mind that seasonality drives nutrients' cycles and their availability in the islands and surrounding ocean. Particularly severe El Niño (hot/warm/rainy) or La Niña (cool/dry) years affect the entire Tropical Eastern Pacific. Now, as you learned from Figs. 4.11 and 4.12, Galápagos is rich in cetacean species (N=27). The Marine Reserve's whales and dolphins vary in body size, habitat preference, feeding habits, social behavior, migration patterns, and reproduction strategies. Thus, you should expect differential impact of climate change on Galápagos cetaceans. Consider that some species might be transient while others are philopatric [remain near particular habitats].

Answer 3: (a–d) Key to addressing Question 3 is to focus exclusively on scientific literature. Reject the misinformation about ambergris from the internet. Google Scholar is a good place to start. You will find plenty of scientific articles about "ambergris" here: *https://scholar.google.com/*.

Answer 4: Think of the whale's role in the carbon cycle (as per Fig. 4.30) and direct your answer exclusively to the Galápagos and its Marine Reserve (i.e. the local environments, habitats, and iconic organisms). Remember the concepts carbon fixation, whale conveyor belt, whale pump, carbon reservoir, and whale falls or deadfall carbon (Fig. 4.30).

Discussion Topics

The discussion topics below (D1–D2) should be addressed with your peers. We encourage college instructors to facilitate the dialogue.

Discussion 1: Whaling in the Galápagos was intricately linked to the extraction of the islands' giant tortoises. The reptiles survived for months on board whaling vessels; the seafarers later consumed their fresh meat. In the book *Arrivals of Life to the Galápagos* (Paz-y-Miño-C and Espinosa 2024a), we examine both the "Slaughtering of the Leviathans" (i.e. the whales during the whaling years, abridged in this volume) and the "Slaughtering of the '*galápagos*'" (the giant tortoises). Here is your DISCUSSION assignment: Identify the associations between whaling [as extractivism of natural resources] and the extraction of giant tortoises from the Galápagos. Realize that, up to this day, the depletion of giant tortoises' populations in the archipelago—and in some islands their extinctions—can be traced back to the historical unsustainable removal [extractivism] of the tortoises by seafarers (whalers, buccaneers, or privateers) and, to a lesser extent, consumption by the local people. [Note that since the creation of the Galápagos National Park (1959), giant tortoises have been protected by law. There is no noticeable extraction of tortoises for human consumption. However, the illegal trade of baby tortoises for the international markets is a current serious concern (we analyze it in Paz-y-Miño-C and Espinosa 2024a,b)].

Discussion 2: The "Whaling Paradox (1800s versus 1900s)" is, of course, paradoxical. As stated in this chapter, "…global whaling declined by the late 1800s. However, it recovered by the 1950s and reached historic catches—at rates never documented before—by the 1960s…" (Fig. 4.25). You know that several hypotheses have been proposed to explain the whaling paradox (i.e. H_1 removal estimates; H_2 population structure; H_3 encounter rates proportional to whale abundance; and H_4 equilibrium population dynamics). Another hypothesis examined in this chapter, which we can label H_5, is the "social learning proposal", in which the whaling paradox is explained by the learning ability of whales [intelligent social animals] to "outsmart whalers" (Fig. 4.26; Smith et al. 2012). H_5 is intriguing and, therefore, attractive to some scientists, predominantly cetacean behaviorists. Here is your DISCUSSION assignment: Compare the strengths and weaknesses of H_5, the "social learning proposal", to the other hypotheses (H_1 to H_4). Your goal should be to master the arguments favoring hypotheses H_1–H_4 as you evaluate them with respect to H_5. In the end, your discussion team should be able to rank all hypotheses (H_1–H_5) based on their rigor. [For this activity, your team might need a classroom with a writing board]. To close the discussion, ask your peers if the whaling paradox has already been explained by modern data or if more observations [which ones?] need to emerge to put the paradox to rest.

Suggested Observation Projects

The suggested observation projects below (P1–2) take into consideration that you will travel to Ecuador and the Galápagos legally; that you will visit the islands as a tourist [holding a Transit Control Card, a mandatory document issued by Ecuador and specifically by the Governing Council of Galápagos]; and that you will respect the regulations that protect the Galápagos National Park and the Marine Reserve (see complete list of rules in Appendix B). Only accredited scientists, academics, and/or naturalists who have obtained official permits can do research in Galápagos. You can take pictures, but flash photography is not permitted when taking photos of wildlife. Professional photography and videos recorded for commercial purposes must be authorized by the Galápagos National Park Directorate (GNPD). The ideas below are for enthusiastic college explorers interested in learning about the archipelago.

Project 1: When you visit the Galápagos and navigate around the islands, your tourist provider will likely take you to iconic locations within the Marine Reserve, the "must-see sites" or "must-stops". The probability of encountering cetaceans (whales and dolphins) will be high, and you will watch a few species (as per Figs. 4.11–4.12). —Save a copy of Fig. 4.19 [Whales captured in Galápagos waters (1829 to 1889)] on your cellphone, tablet, or laptop (internet access will be limited while at sea). During your navigation from one island to another, keep track of the specific locations where whales were hunted during the whaling years 1829–1889 (as per Fig. 4.19). Ask your naturalist guide to show you where Elizabeth Bay is located (Albemarle/Isabela Island as in Fig. 4.19). You will possibly spot whales and dolphins in this area, sometimes by the hundreds. Now, imagine whaling vessels of the 1800s capturing 122 sperm whales just there (Elizabeth Bay)! Seventy-seven sperm whales were caught in "open waters" of the archipelago (as also shown in Fig. 4.19); 32 in Northern Narborough/Fernandina Island; 30 near Charles/Floreana Island; 21 Rock

Redondo; 17 James/Santiago Island; and 14 Abington/Pinta Island. As we document in Fig. 4.19, at least 505 whales (497 sperms and 8 humpbacks) were taken and processed during a 60-year range [within what today is the Galápagos Marine Reserve]. Thus, your observation project is about realizing how the "slaughtering of the leviathans" materialized during the 1800s. As we stated earlier in this chapter, "…The [global] ecocide of the "leviathans" was on, but it was only acknowledged in the 20th century!"

Project 2: Prior to visiting the Galápagos, put together your own fully illustrated "Guide to the Whales and Dolphins of the Galápagos". Relying on Figs. 4.11 (Diversity of whales and dolphins, Cetaceans, in the Galápagos) and 4.12 (Most commonly observed whales and dolphins in the Galápagos), plus your own research, you can find images [best quality photos] of the 27 species that occur in the Marine Reserve. Build up a folding pocket guide [brochure] on which the images of the cetaceans, their common and scientific names, estimated body-size, and other useful natural history notes are summarized (keep it simple, good looking, and didactic). Laminate your pocket guide and bring it to the islands; enjoy it there. On the last page of the brochure, acknowledge the sources from which you obtained the information (keep in mind that the material found online is usually copyrighted and protected by law). Do you want to be bold and test your passion for marine life? Put together a "Guide to the Marine Mammals of the Galápagos".

References

Bannister, John L., Elizabeth A. Josephson, Randall R. Reeves, and Tim D. Smith. 2008. "There She Blew! Yankee Sperm Whaling Grounds 1760-1920". In *Oceans Past: Management Insights from the History of Marine Animal Populations*, edited by David J Srarkey, Poul Holm, and Michaela Barnard, 109-132. Sterling: Earthscan Research Editions.

Brown-Goode, George. 1887. *The Fisheries and Fishery Industries of the United States "Prepared through the Co-operation of the Commissioner of Fisheries and the Superintendent of the Tenth Census by George Brown Goode, Assistant Secretary of the Smithsonian Institution and a Staff of Associates"*. Washington: Government Printing Office.

Chami, Ralph, Thomas Cosimano, Connel Fullenkamp, and Sena Oztosun. 2019. "Nature's Solution to Climate Change". *Finance & Development*. Accessed June 13, 2022.
https://www.imf.org/external/pubs/ft/fandd/2019/12/natures-solution-to-climate-change-chami.htm.

Davis, Lance E., Robert E. Gallman, and Karin Gleiter. 1997. "Data Sets and Sources". In *In Pursuit of Leviathan: Technology, Institutions, Productivity, and Profits in American Whaling, 1816-1906*, edited by Lance E. Davis, Robert E. Gallman, and Karin Gleiter, 57-130. Chicago: The University of Chicago Press.

Denkinger, Judith, Javier Oña, Daniela Alarcón, Godfrey Merlen, Sandy Salazar, and Daniel M. Palacios. 2013. "From Whaling to Whale Watching: Cetacean Presence and Species Diversity in the Galapagos Marine Reserve". In *Science and Conservation in the Galapagos Islands: Frameworks & Perspectives*, edited by Stephen J. Walsh and Carlos F. Mena, 217-235. New York: Springer.

Dolin, Eric J. 2007. *Leviathan: The History of Whaling in America*. New York: W. W. Norton and Company Inc.

Doughty, Christopher E., Joe Roman, Søren Faurby, Adam Wolf, Alifa Haque, Elisabeth S. Bakker, Yadvinder Malhi, John B. Dunning, Jens-Christian Svenning. 2016. "Global Nutrient Transport in a World of Giants". *Proceedings of the National Academy of Sciences of the United States*, No. 113: 868-873.

Eguiguren, Ana, Enrico Pirotta, Maurício Cantor, Luke Rendell, and Hal Whitehead. 2019. "Habitat Use of Culturally Distinct Galápagos Sperm Whale *Physeter macrocephalus* Clans". *Marine Ecology Progress Series*, No. 609: 257-270.

Halpin, Patrick N., Andrew J. Read, Ei Fujioka, Ben D. Best, Ben Donnelly, Lucie J. Hazen, Connie Kot, Kim Urian, Erin LaBrecque, Andrew Dimatteo, Jesse Cleary, Caroline Good, Larry B. Crowder, and K. David Hyrenbach. 2009. "OBIS-SEAMAP: The World Data Center for Marine Mammal, Sea Bird, and Sea Turtle Distributions". *Oceanography*, No. 22:104-115.

Hohman, Elmo P. 1926. "Wages, Risk, and Profits in the Whaling Industry". *The Quarterly Journal of Economics*, No. 40: 644-671.

Hope, Patricia L., and Hal Whitehead. 1991. "Sperm Whales off the Galapagos Islands from 1830-50 and Comparisons with Modern Studies". *Reports of the International Whaling Commission*, No. 41: 273-286.

Idrovo, Hugo. 2005. *Galapagos: Footsteps in Paradise*. Quito: Libri Mundi.

Luna-Tobar, Alfredo. 1997. *Historia Política Internacional de las Islas Galápagos*. Quito: Ediciones Abya Yala.

Macleod, Ruairidh, Mikkel-Holger S. Sinding, Morten Tange Olsen, Matthew J. Collins, and Steven J. Rowland. 2020. "DNA Preserved in Jetsam Whale Ambergris". *Biology Letters*, No. 16: 20190819.

Martin, Paul, and Patrick Bateson. 1993. *Measuring Behavior: An Introduction Guide*. Cambridge: Cambridge University Press.

Maury, Mathew L. 1851. *Whale Chart*. Washington, DC: United States Bureau of Ordnance & Hydrography.

McGowen, Michael R. Michelle Spaulding, and John Gatesy. 2009. "Divergence Date Estimation and a Comprehensive Molecular Tree of Extant Cetaceans". *Molecular Phylogenetics and Evolution*, No. 53: 891-906.

Moment, David. 1957. "The Business of Whaling in America in the 1850's". *The Business History Review*, No. 31: 261-291.

Paz-y-Miño-C, Guillermo, and Avelina Espinosa. 2024a. *Arrivals of Life to the Galápagos: From Berlanga to Darwin, Melville, and World War II*. Newcastle upon Tyne: Cambridge Scholars Publishing.

—. 2024b. *A Synthesis of the Galápagos: Natural History, Human Conflict, and the Destiny of an Archipelago*. Newcastle upon Tyne: Cambridge Scholars Publishing.

Pershing, Andrew J., Line B. Christensen, Nicholas R. Record, Graham D. Sherwood, and Peter B. Stetson. 2010. "The Impact of Whaling on the Ocean Carbon Cycle: Why Bigger Was Better". *PLoS ONE*, No. 5: e12444.

Rocha, Robert C., Phillip J. Clapham, and Yulia V. Ivashchenko. 2015. "Emptying the Oceans: A Summary of Industrial Whaling Catches in the 20th Century". *Marine Fisheries Review*, No. 76: 37-48.

Roman, Joe, and James J. McCarthy. 2010. "The Whale Pump: Marine Mammals Enhance Primary Productivity in a Coastal Basin". *PLoS ONE*, No. 5: e13255.

—. James A. Estes, Lyne Morissette, Craig Smith, Daniel Costa, James McCarthy, J. B. Nation, Stephen Nicol, Andrew Pershing, and Victor Smetacek. 2014. "Whales as Marine Ecosystem Engineers". *Frontiers in Ecology and the Environment*, No. 12: 377-385

Smith, Tim D., Randall R. Reeves, Elizabeth A. Josephson, Judith N. Lund, and Hal Whitehead. 2008. "Sperm Whale Catches and Encounter Rates During the 19th and 20th Centuries: An Apparent Paradox". In *Oceans Past: Management Insights from the History of Marine Animal Populations*, edited by David J Srarkey, Poul Holm, and Michaela Barnard, 149-173. Sterling: Earthscan Research Editions.

—. Randall R. Reeves, Elizabeth A. Josephson, and Judith N. Lund. 2012. "Spatial and Seasonal Distribution of American Whaling and Whales in the Age of Sail". *PLoS ONE*, No. 7: e34905.

Smith, Craig R., Joe Roman, and J. B. Nation. 2019. "A Metapopulation Model for Whale-Fall Specialists: The Largest Whales Are Essential to Prevent Species Extinctions". *Journal of Marine Research*, No. 77: 283-302.

Stackpole, Edouard A. 1953. *The Sea-Hunters: The Great Age of Whaling*. New York: J. B. Lippincott Company.

Starbuck, Alexander. 1878. *History of the American Whale Fishery from Its Earliest Inception to the Year 1876*. Washington: Government Printing Office.

Tower, Walter S. 1907. *A History of the American Whale Fishery*. Philadelphia: The John C. Winston Company.

Townsend, Charles H. 1935. "The Distribution of Certain Whales as Shown by Logbook Records of American Whaleships". *Zoologica*, No. 19:1-50.

Whaling History. 2017. "Townsend's Whaling Charts". Accessed June 4, 2022. *https://whalinghistory.org/townsends-whaling-charts/*.

Whitehead, Hall. 2002. "Estimates of the Current Global Population Size and Historical Trajectory for Sperm Whales". *Marine Ecology Progress Series*, No. 242: 295-304.

—. and Patricia L. Hope. 1991. "Sperm Whalers off the Galápagos Islands and in the Western North Pacific, 1830 to 1850: Ideal Free Whalers?" *Ethology and Sociobiology*, No. 12: 147-161.

—. Jenny Christal, and Susan Default. 1997. "Past and Distant Whaling and the Rapid Decline of Sperm Whales off the Galápagos Islands". *Conservation Biology*, No. 11: 1387-1396.

—. Tim D. Smith, and Luke Rendell. 2021 "Adaptation of Sperm Whales to Open-Boat Whalers: Rapid Social Learning on a Large Scale?" *Biology Letters*, No. 17: 20210030.

Woolmer, Gillian. 2013. "Historical Distribution of Whales Shown by Logbook Records 1785-1913". Data from OBIS-SEAMAP. Accessed June 7, 2022. *https://www.gbif.org/dataset/01d9d254-b18c-458e-b6e1-fc569edb7f09*.

Woram, John. 2005. *Charles Darwin Slept Here*. Rockville Centre: Rockville Press Inc.

York, Richard. 2017. "Why Petroleum Did Not Save the Whales". *SOCIUS: Sociological Research for a Dynamic World*, No. 3: 1-13.

CHAPTER FIVE

WHY DID THE UNITED STATES ESTABLISH A MILITARY BASE IN GALÁPAGOS DURING WORLD WAR II?

> "When I look back on my visit to Galápagos, I know why every man there calls it 'The Rock.' To a geologist, I'm sure it would furnish several years of absorbing work, but to men establishing gun positions and defenses, building airfields and trying to find level space for a recreation field, it must be one of the most discouraging spots in the world. It is as though the earth had spewed forth rocks of every size and shape… [A]s one man said: 'You remove one rock, only to find two more...'"

These observations belong to Eleanor Roosevelt, the spouse of 32nd United States President Franklin D. Roosevelt (FDR). She visited Seymour South Island (Baltra or "The Rock") in March 1944, where the US had established, in 1942, a military base with the purpose of guarding the East Pacific during World War II (WWII) and, particularly, the nearby Panama Canal (located 1600 km/1000 mi Northeast). The US has had geopolitical control over the Canal since 1904 (right after Panama seceded from Colombia). Eleanor wrote her recollections about the Galápagos on April 4, 1944, once back in Washington, DC (she elaborated from her diary notes). In Box 5.1, we offer her complete assessment of "The Rock".

Eleanor's accounts about the Galápagos' desolate landscape and the boredom among the troops contrasted with Americans' perceptions of WWII, i.e., the daily routine at military bases overseas. In fact, two years before she arrived in Baltra, *Life* magazine had alarmed its readers when publishing, on March 2, 1942, the article "*How the US Must Fight for Its Life*". In it, the periodical speculated about "*Six Ways to Invade US. Axis Powers* [Japan and Nazi Germany] *Can Try It If They Combine Fleets to Win Sea Superiority*". The American psyche was sensitive to conspiracies about bringing the war to its soil; the attack on Pearl Harbor (Hawai'i) had just happened three months earlier, on December 7, 1941. An even bolder strike by Japan or Germany seemed conceivable in the minds of many (as per Paz-y-Miño-C and Espinosa 2024a).

Box 5-1 Eleanor Roosevelt's remarks on her visit to "The Rock" (Seymour Island, or Seymour South, South Seymour, Baltra), Galápagos, March 1944.

Washington, April 4, 1944 — *"When I look back on my visit to Galápagos, I know why every man there calls it 'The Rock'. To a geologist, I'm sure it would furnish several years of absorbing work, but to men establishing gun positions and defenses, building airfields and trying to find level space for a recreation field, it must be one of the most discouraging spots in the world. It is as though the earth had spewed forth rocks of every size and shape and, as one man said: 'You remove one rock, only to find two more underneath'.*

In between the rocks there is deep red dust, which permeates everything. A few cactus plants grow and also a few trees, which are easily blown over because they have no earth to root in.

All the water is either distilled from salt water or is brought in on a tank ship from another island. On the whole island, there are just two places with running water. Otherwise, tin basins or helmets are used for washing, and the regular set-up that men have at the front for showers is the order of the day. No luxurious living there! One boy in the dispensary, Corporal Edward Shwing, had spent much time putting up very good battle maps, covering every front, with colored pins showing the various activities. He told me that many boys came in to watch the pins change!

Galápagos is one of those places where 'going native' would be very easy. For that reason, during the day, men at work may wear as few clothes as they choose, but for evening inspection, every man must be in uniform. The Navy bars whites because they cannot be kept clean.

One of the most attractive places on the island is the Bluejackets Club, which the men have created themselves. Every bit of furniture and every decoration is their own handiwork. On the door of the club hangs a sign which reads: 'Bluejackets Club—women invited'. The joke is that there are no women there! American women in Ecuador have sent a few things like curtains and Ecuadorian straw mats to various service clubs on the island, but every day room and post exchange represents much work by the men themselves.

Commander Huffman has some pets—two goats, Blackie and Ruth, each with a painted green and red horn. And, in a little enclosure outside, are two prehistoric-looking iguanas. He was disappointed because I found these native pets interesting but not attractive!

The American man's sense of humor was evident everywhere. They held competition at the Navy base for the naming and the general appearance of the various quarters. The doctors' quarters were labeled 'Rock's Docs'. That won the prize.

You have to be deeply convinced that your job is an essential one in order to keep your balance and cheerfulness in these surroundings. I think perhaps it takes more fortitude and character to stand the loneliness and hardship of this kind of service without much excitement than the more active kind of service, though many women at home are probably happier with the knowledge that their men are not being sniped at by the enemy" (source 'Day by Day' archives from the Franklin D. Roosevelt Presidential Library; links in Appendix A). Note that Eleanor's husband, US President Franklin D. Roosevelt, had already visited Galápagos in 1938 (this Chapter).

By the 1950s, Eleanor Roosevelt hosted her long-time popular radio show "*Mrs. Eleanor Roosevelt's Own Program*". Above, she is interviewing Ecuador's President Galo Plaza Lasso, who was in Washington, DC, for an official visit to the United States, June 20-29, 1951. The event and image document a cordial interaction between the two. Image source: Harry S. Truman Library and Museum (public domain; see Appendix A for links).

The 13-page report in *Life* included maps and detailed plans (from 1 to 6) imagining the Japanese-German invasion. "*Plan Three*" referred to Galápagos and Ecuador (Fig. 5.1) as it called for a "*southern Pacific crossing by Japan*" reinforced by the Germans. *Life*'s conjecture presumed Japan's naval advantage over the US Fleet and assumed a "*surprise bombing on the Panama Canal, instantly followed by landings in Ecuador*". By 1942, the construction of two US Army-Navy bases was already planned or underway in Ecuador. One in Salinas, on the continental shores of the country, and the other in the Galápagos Islands (i.e. Seymour Island, or Seymour South, South Seymour, Baltra, or, ultimately, "The Rock"; Figs. 5.2 to 5.4).

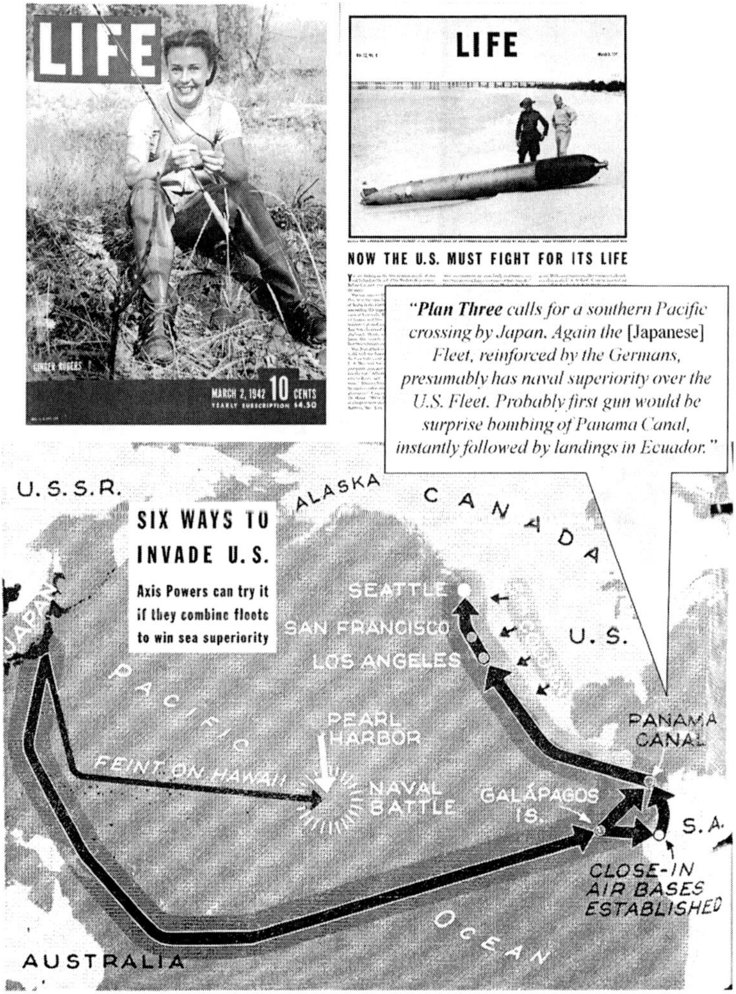

Fig. 5-1 *Life Magazine* of March 2, 1942: alarming article "*How the US Must Fight for Its Life: Six Ways to Invade US*" mentions Galápagos and Ecuador in "*Plan Three*" of a possible invasion (never materialized) by Japan and/or Nazi Germany. Note that in the dialog box, we have inserted the word [Japanese]. The actual caption by *Life* included an ethnic slur instead: "*Plan Three calls for a southern Pacific crossing by Japan. Again the* [Slur] *Fleet, reinforced by the Germans, presumably has naval superiority over the U.S. Fleet. Probably first gun would be surprise bombing of Panama Canal, instantly followed by landings in Ecuador*" (realize that by 1942, construction of two US Army-Navy bases was already conceived or underway in Ecuador: Salinas and Galápagos). Images links in Appendix A.

Seymour Island, Baltra or "The Rock"

In the past—and like other Galápagos Islands—Baltra had various names: Seymour Island, Seymour South, South Seymour, and, during WWII, "The Rock", as referred to by the Americans deployed to "Base Beta" (its code-name; jointly with "Alpha" in Nicaragua and "Gamma" in Salinas), run by the US Military from 1942 to 1946 (Idrovo 2005).

In Fig. 5.2, we show the "*Special Map, Galapagos Islands, Seymour Island*" put together by the Corps of Engineers, US Army, Panama Canal Department, in 1943. In Fig. 5.3, we share an aerial photo of Seymour taken by the US Air Force in 1959. And in Fig. 5.4, we outline a reconstruction of the road network on Seymour Island as built in the 1940s by the US Military. [Note that newer roads are also included]. For the sketch in Fig. 5.4, we relied on two sources: the 1943 US Army Map (as in Fig. 5.3) and modern Google-Earth images (aerial/satellite/road map). Readers should note in Fig. 5.4 the complexity of the 1940s road network. Nowadays, most of the old roads are not in use; only fragments remain. However, several of the contemporary roads follow the paths of the 1940s infrastructure. In Chapter Six, *Who Lives in Today's Galápagos Islands?*, we discuss the 2000s road system on Baltra Island (e.g. Fig. 6.6).

Baltra is a small island, a geologic "fragmented fragment" of Santa Cruz, located North with respect to the latter (Figs. 5.2 and 5.3, top right). Baltra's fragments are: North Seymour or Seymour North (or even "Little Seymour"); Seal Island or Mosquera Island (a tiny islet or what is left of the past continuity among land fragments); and South Seymour or Baltra, "The Rock" itself. Locate in Figs. 5.2 and 5.3 (bottom) the South Channel or Canal de Itabaca, which separates Baltra from Santa Cruz. Now, spot the two runways in Baltra: the main one to the Southwest of the island and the second to the East. The 1940s barracks and the complex network of roads were proximate to the main runway (still visible in Fig. 5.3, a photo from 1959, as well as in modern aerial/satellite images at Google-Earth; links in Appendix A). Interestingly, today's Seymour Galápagos Ecological Airport, or GPS, was built on top of the 1940s second runway (Fig. 5.4; see also Fig. 6.6 in Chapter Six)! The rest of the 1940s infrastructure, including the initial main runway, barracks, and small roads, is almost completely gone and of archeological value.

By 1943, there were 400 US officers, 4,500 military men, and 1,000 civilians living and/or working in Baltra. A year later, the US Department of State reduced the military presence to 1,900 US Army personnel and 450 US Navy marines (for ample historical accounts about Galápagos during WWII, see Luna-Tobar 1997; Idrovo 2005; Woram 2005).

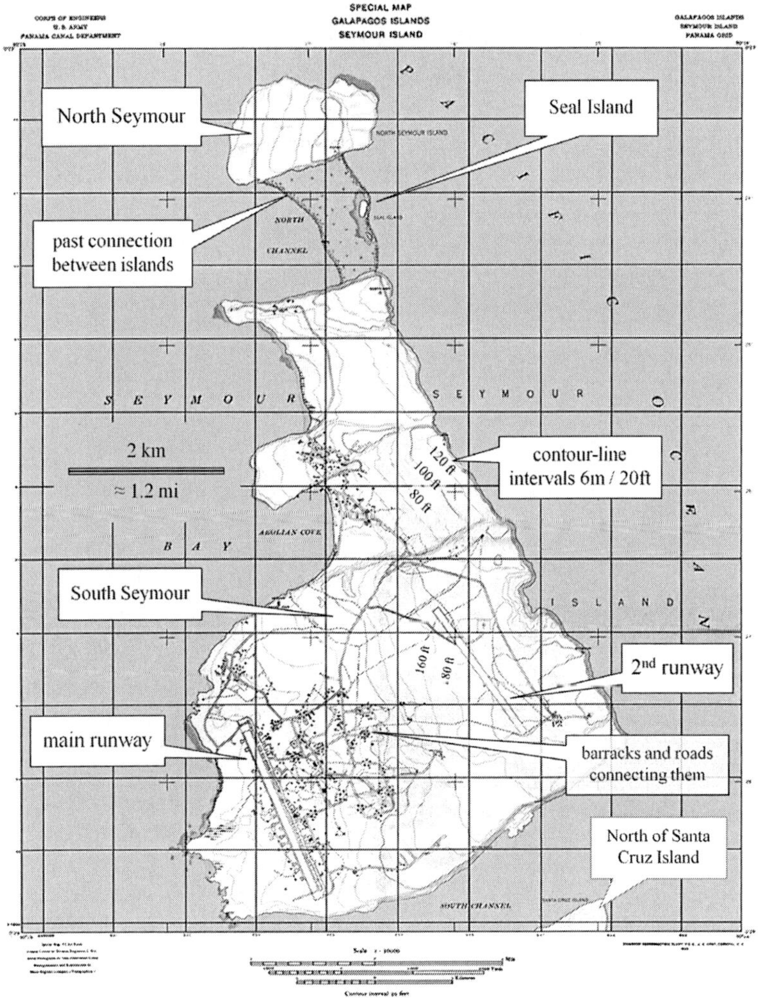

Fig. 5-2 "*Special Map, Galapagos Islands, Seymour Island*" by the Corps of Engineers, United States Army, Panama Canal Department, 1943 (courtesy historian John Woram).

In Fig. 5.5, we share images documenting Mrs. Eleanor Roosevelt's visit to "Base Beta" on Seymour Island in March 1944. She arrived from Panama via Salinas, Ecuador, together with Lieutenant General George Brett (Caribbean Defense Command), Admiral Harold Train (Panama Sea Frontier), and her secretary Malvina Thompson (Idrovo 2005).

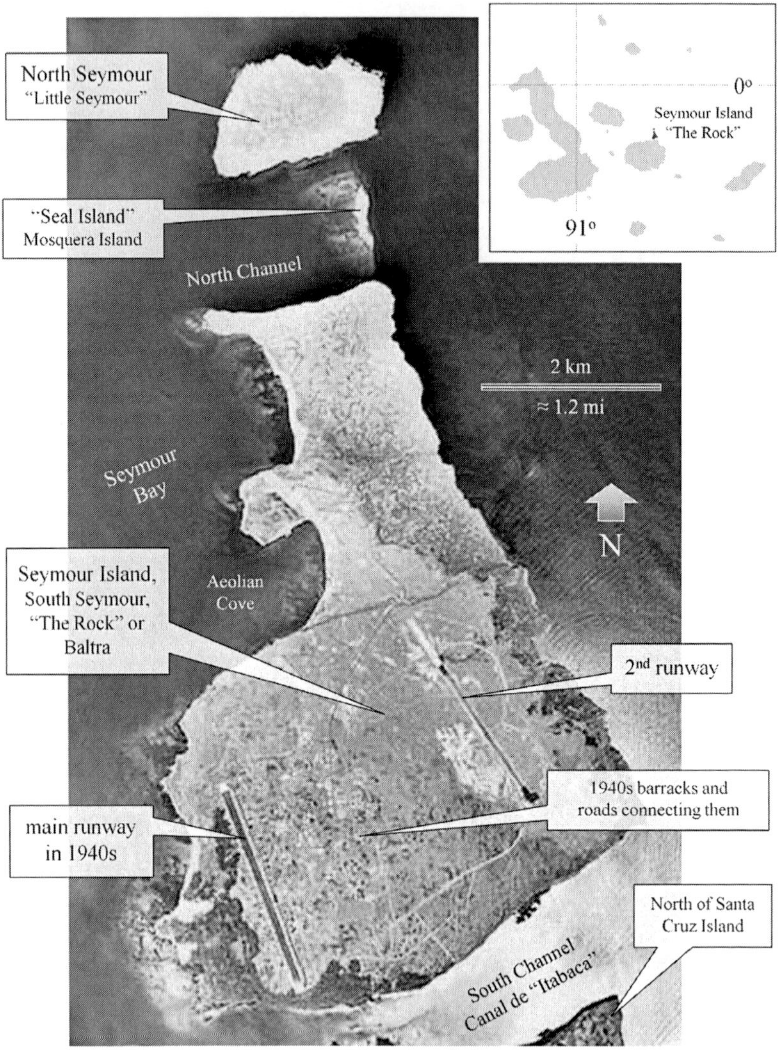

Fig. 5-3 Seymour Island, "The Rock" (or Seymour South, South Seymour, Baltra). Aerial photo by the United States Air Force, 1959 (www link in Appendix A).

There are hundreds, perhaps thousands, of photographs documenting the presence of the US military in Baltra during WWII. In Figs. 5.6 to 5.10, we show a sample of them. The images come from multiple sources, as detailed below. For www links see Appendix A.

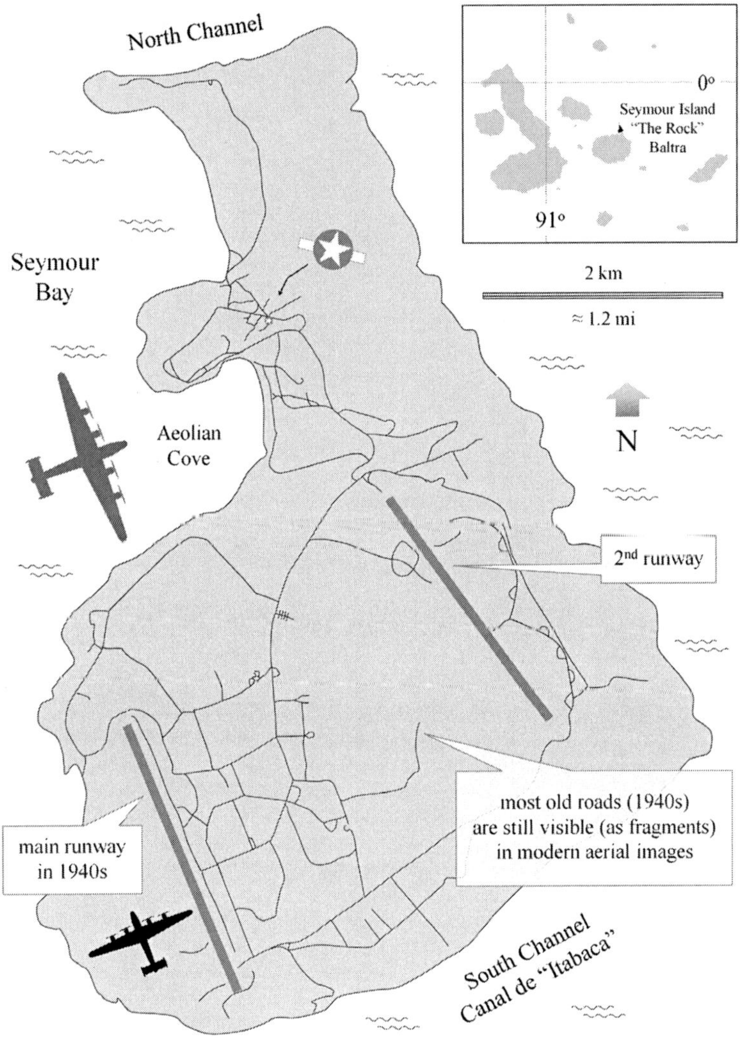

Fig. 5-4 Road network in Seymour Island (Baltra) built by the US Army in the 1940s plus newer roads. Reconstruction based on the 1943 US Army Map in Fig. 5.3 and Google-Earth images (aerial/satellite/road map). Note the complexity of the 1940s road network on the island. Most of the old roads are not in use nowadays; only fragments remain. Several of the contemporary roads follow the paths of the 1940s roads (for that, contrast this image with Fig. 6.6 in Chapter Six, which shows the 2000s roads). Silhouettes of planes are WWII B-24 Liberator bombers (not to scale). Design/sketch: G. Paz-y-Miño-C.

Fig. 5-5 Eleanor Roosevelt visiting the United States troops at "Base Beta" in Seymour Island or "The Rock" (i.e. Seymour South, South Seymour, Baltra), Galápagos (March 1944). Top: source "Day by Day" archives from the Franklin D. Roosevelt Presidential Library. Bottom: source National Archives and Records Administration via Wikimedia Commons. Both images ultimately come from the Corps of Engineers, US Army, Panama Canal Department. See Appendix A for links. In Box 5.1, we share Mrs. Roosevelt's summary-assessment of her visit to "The Rock", which she wrote back in Washington, DC, on April 4, 1944.

Fig. 5-6 United States Military at Baltra Island, Galápagos, during World War II. By 1943, there were 400 US officers, 4,500 military men, and 1,000 civilians living and/or working in Baltra. A year later, the US Department of State reduced the military presence to 1,900 US Army personnel and 450 US Navy marines (for ample historical accounts about Galápagos during WWII, see Luna-Tobar 1997; Idrovo 2005; Woram 2005). Top left: B-24 bomber flying over South Seymour, Baltra or "The Rock"; in the background, Daphne Major Island and Santiago (James) Island. Top right: "Knight Mare" B-24 bomber and crew. Middle right: the "Knight Mare" flying over the Northwestern side of North Seymour ("Little Seymour") and South Seymour (Baltra, "The Rock") islands; in the background, part of Santa Cruz (Indefatigable) Island can be spotted. Bottom right: aerial view of Northern Baltra where numerous buildings at "Base Beta" (the code name given to the Baltra military base) can be seen; the location of the "Five-Point Star" (roundabout) is pointed at and magnified in Fig. 5.4. Images come from multiple public-domain depositories, including Revisionismo Histórico del Ecuador (2022) and still shots from documentary films (links in Appendix A).

Fig. 5-7 Martin PBM Mariner flying boat, B-24 Liberator bombers flying over South Seymour/Baltra, and squadron on B-24, Baltra, Galápagos (1940s). Top: the Martin PBM Mariner (a patrol bomber flying boat) is flying South, on the Eastern side of Baltra Island; Daphne Major Island is visible in the background (left). Middle: B-24 bombers flying over Baltra (barracks and road network in the background); the main runway (see location with respect to the island in Figs. 5.2 to 5.4) appears on the bottom of the image. Bottom: +300 men in squadron with B-24 bomber. Images come from multiple public-domain depositories, including Revisionismo Histórico del Ecuador (2022) and documentary films (Appendix A).

Fig. 5-8 Daily life for US troops in South Seymour/Baltra, Galápagos (1940s). After visiting "Base Beta" in 1944, Mrs. Eleanor Roosevelt wrote (see full text in Box 5.1) "*You have to be deeply convinced that your job is an essential one in order to keep your balance and cheerfulness in these surroundings. I think perhaps it takes more fortitude and character to stand the loneliness and hardship of this kind of service without much excitement than the more active kind of service...*". Images come from multiple public-domain depositories, including Revisionismo Histórico del Ecuador (2022) and documentary films (Appendix A).

Fig. 5-9 US troops posing with wild and introduced animals, Galápagos (1940s). After visiting "Base Beta" in 1944, Mrs. Eleanor Roosevelt wrote (see full text in Box 5.1) "*Commander Huffman has some pets—two goats, Blackie and Ruth, each with a painted green and red horn. And, in a little enclosure outside, are two prehistoric-looking iguanas. He was disappointed because I found these native pets interesting but not attractive!*" Author John Woram (2005) has noted that "*this home* [Baltra] *so very far away from home* [was] *just 'The Rock', a term of endearment formerly reserved for Alcatraz, another prison watched over by gun-toting guards. But on this little rock in the Pacific, the guards* [were] *also the prisoners*". Images come from multiple public-domain depositories, including Revisionismo Histórico del Ecuador (2022) and still shots from documentary films (links in Appendix A).

Top: US officers and Ecuadorian Defense Minister meeting in Baltra, Dec. 1943

Lucius Drafts, George Brett and Alberto Romero (Ecuador's Defense Minister)

George Brett (second), Alberto Romero (third) and Lucius Drafts (fifth)

Bottom: Ecuadorian President José María Velasco Ibarra (fifth from the left) visiting Baltra (Sep. 1944) to assess the US military presence in the Galápagos

Fig. 5-10 Meeting at Baltra between United States officers and Ecuadorian Defense Minister Alberto Romero (December 1943) and arrival in Baltra of Ecuadorian President Velasco Ibarra (September 1944). In 1944, the US prepared a draft of a "*bilateral agreement for the* [extended] *leasing of* [military] *bases* [Galápagos and Salinas] *on Ecuadorian territory over... 99 years*". Ecuador did not accept the idea. In 1946, President Velasco Ibarra instructed the Ecuadorian Ambassador in Washington, Galo Plaza Lasso, to formally request the US "*withdrawal from Galápagos*" (history/context in Department of State 1967). The US officially returned Baltra to Ecuador on July 1, 1946. The last American personnel left Galápagos in 1948. Images come from public-domain depositories (links in Appendix A).

FDR in Galápagos: 1938

Prior to WWII or the events summarized in the images above (Figs. 5.6 to 5.10), or Eleanor Roosevelt's visit to "The Rock" in 1944, her husband, the 32nd US President Franklin D. Roosevelt (FDR), arrived in Galápagos—on an official *Inspection Cruise and Fishing Expedition on Board USS Houston*—in July 1938. The USS Houston departed from San Diego, California, on July 16, 1938 (Fig. 5.11). It navigated along the Baja California Peninsula (Mexico) and reached Socorro Island on July 20; then, Clipperton Island on July 21, and continued to Tower (Genovesa) Island in the "*Galápagos Group*", where both the USS Houston and its escort vessel, the USS McDougal, anchored off Darwin Bay on July 24 (Fig. 5.11). There are logbook records of FDR's activities in Galápagos (and the entire inspection cruise); the accounts, however, are centered on the fishing-expedition part of the journey, not geopolitics. Nowadays—and in retrospect—the latter is obvious. Here are the log excerpts:

> Sunday, July 24: "Poor boating conditions with a southeasterly wind of force four blowing, and the uninviting appearance of the terrain [Tower] led to... forego fishing in this area... At 1114 [11:14 am], after church services, the Presidential Detachment got underway... to [Sullivan] Bay, on the east coast of San Salvador (James) [Santiago] Island where the Detachment anchoring the President and his Party embarked in the fishing boats and for some three hours enjoyed excellent sport..."
>
> Monday, July 25: "In order to permit early arrival at Tagus Cove and to 'cross the line officially' [the equator] between 0830 and 0900 [8:30 and 9:00 am], the Houston and McDougal got underway at 0658 [6:58 am] and proceeded around the northern end of Albemarle [Isabela] Island... After anchoring at Tagus Cove..., at 1430 [2:30 pm], the President and all members of his party took to the boats and fished the smooth water of Tagus Cove and nearby strait. The results were excellent in a fishing sense with a large catch of Grouper, Mackerel and Tuna..."
>
> Tuesday, July 26: "...at 1016 [10:16 am], both vessels [Houston and McDougal] were underway and proceeding toward Elizabeth Bay on Albemarle [Isabela] Island. Narborough [Fernandina] Island was rounded during this passage affording a complete view of that bleak, volcanic bit of land... [A]fter anchoring in Elizabeth Bay at 1430 [2:30 pm]... all hands eagerly took to the boats for three hours of fishing... [T]he President caught a number of small fish of rare species as a result of his bottom fishing... Immediately after hoisting her boats (1823) [6:23 pm] the Houston got underway for Post Office Bay, on Charles [Floreana] Island, an easy overnight run."

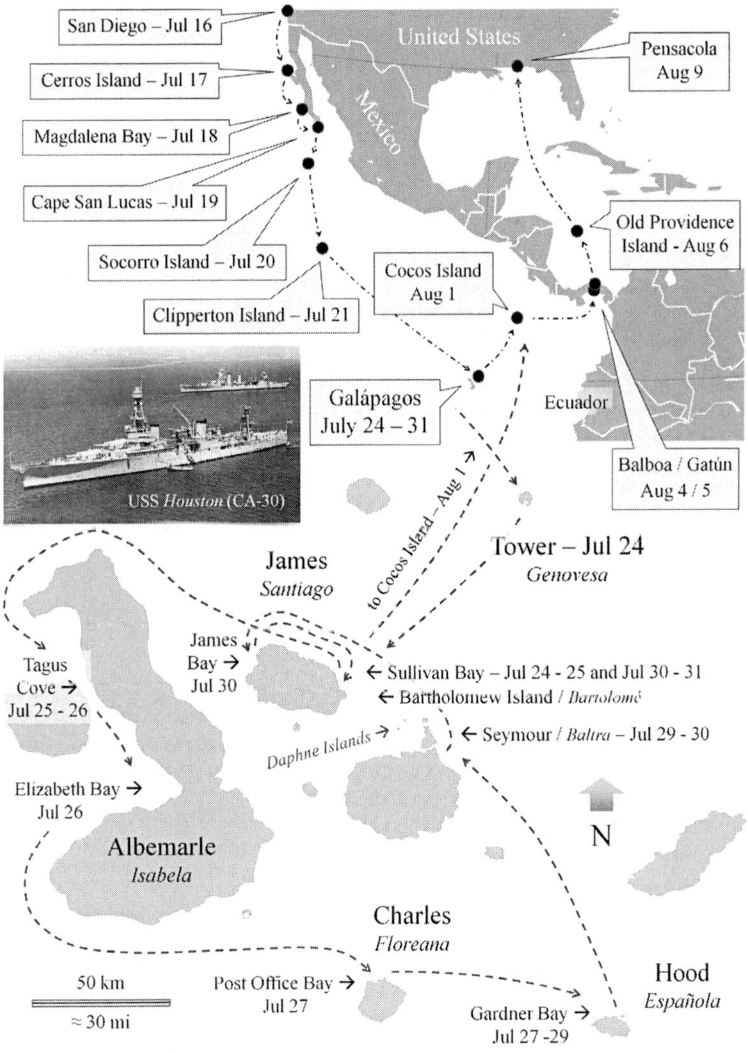

Fig. 5-11 President Franklin D. Roosevelt *Inspection Cruise and Fishing Expedition* on Board USS Houston (July 16 to August 9, 1938) included navigation to and in the Galápagos Islands, July 24-31, 1938 (see images in Fig. 5.12). Logbook descriptions of FDR's activities while in Galápagos waters (and the entire journey) exist. In this chapter, we summarize some of them (excerpts provided). Layout G. Paz-y-Miño-C after *Logs of the President's Trips: Inspection Cruise and Fishing Expedition on Board USS Houston, July 16 - August 9, 1938* (Grace Tully Archive of the FDR Presidential Library and Museum 2022; links Appendix A).

Wednesday, July 27: "...the Presidential Detachment anchored in Post Office Bay, Charles [Floreana] Island at 0725 [7:25 am]... [A] volunteer party of some fifty officers... put off in a motor launch and landed on a sandy beach. Supplies for Mr. Wittmer and family [Heinz, Margaret and son; German settlers who arrived in early 1930s], who reside on the Island, were taken ashore by this party... But on failing to contact him after an hour's hike magazines and newspapers were left on the trail... together with a letter stating the location of the supplies. Thereupon, the party returned to the beach and reembarked, arriving at the ship about 1100 [11:00 am]... [A]nother expedition of 'succor' [was] dispatched in a motor whaleboat to a point along the coast-line some four miles to the southward, where ship's aviators had noted two men and two women in the vicinity of miserable huts. It developed that these huts belonged to an Ecuadorian Indian and his companions. They were found to be sufficiently supplied with food, but additional delicacies were given them as well as to two Americans, a Mr. and Mrs. Conway, who later appeared on the scene with many letters for transmission to the United States... [T]he Houston at 1430 [2:30 pm] departed for Gardner Bay on [Española] (Hood) Island, anchoring there with the McDougal about 1800 [6:00 pm]. The Houston and McDougal remained at anchor overnight at Garner Bay."

Thursday, July 28: "The Presidential Detachment remained at anchor in Garden Bay throughout the day. Forenoon fishing of some two and a half hours was enjoyed by the President and members of his party. All hands attempted to catch large fish and avoid shoal Grouper waters... Several excellent specimens of the comparatively rare Golden Grouper were caught... At 1530 [3:30 pm], ...the fishermen were at the sport again. During the afternoon period the President's boat followed the northwestern coastline of the Island from Gardner Bay in the hopes of coming upon a school of large fish... The President's jaunt was unsuccessful with respect to large fish..."

Friday, July 29 [arguably, the most significant day]: "In order to reach Seymour [Baltra] Island off the North Coast of Indefatigable [Santa Cruz] Island at a seasonable hour the Houston and McDougal got underway at 0550 [5:50 am]. The two vessels anchored at 1055 [10:55 am]... just to the south'ard of the passage between Seymour and Indefatigable Islands [i.e. today's Itabaca Channel]. The President... embarked about 1130 [11:30 am] in a motor whaleboat for a fishing expedition around Daphne [Major and Minor] Islands some 4½ miles to westward of the Houston's anchorage. Indifferent fishing was found there... On the President's suggestion this fishing party proceeded thence to the windward (eastward) side of Seymour Island where, despite the rough seas, an hour and a half was spent in trying to land Tuna... The Presidential Detachment remained at anchor overnight off Seymour Island."

a. On board of the USS Houston, FDR poses with shark; Sullivan Bay, James Island / Santiago

b. The unmistakable Pinnacle Rock in Bartholomew Island / Bartolomé

c. View of Bartholomew / Bartolomé Island, Pinnacle Rock on the right

d. USS Houston's sailors with albatross at Gardner Bay, Hood Island / Española

Fig. 5-12 Snapshots of US President Franklin D. Roosevelt (FDR) inspection cruise and fishing expedition while in the Galápagos Islands, July 24-31, 1938. a-d. Images come from multiple public-domain depositories (links in Appendix A).

Saturday, July 30: "The President decided that a profitable forenoon might be spent on San Salvador (James) [Santiago] Island in search of the grave of Lieutenant [John S.] Cowan, USN[avy], attached to the [USS] Essex (Commodore David Porter, USN[avy]) and who, in 1813, as a result of a duel [with shipmate Lt. John M. Gamble], was killed and buried in the vicinity of James Bay. Accordingly, at 0530 [5:30 am]... the Houston and McDougal got underway from Seymour Island and Proceeded to an anchorage in James Bay which they reached about 0905 [9:05 am]. A landing party of some fifty officers and men disembarked on the beach about 0905 [9:05 am], and forming a skirmish line normal to the beach, swept the interior area from north to south for a distance of approximately one and a half

> miles and three hundred yards in depth in search of Lieutenant Cowan's grave. No conclusive evidence of its location was found... While the searching operations were in progress, the President... engaged in a bit of fishing which yielded a small number of Tuna, some Groupers and sundry smaller fishes... At 1250 [12:50 hrs], all partiers having returned to the ship and boats having been hoisted, the President Detachment departed for [Sullivan] Bay on the east coast of James [Santiago] Island, where the vessels anchored at 1500 [3:00 pm] for their second visit in a week to this area..."
>
> Sunday, July 31: "Arriving off the entrance to Darwin Bay, Tower [Genovesa] Island, at 0850 [8:50 am], it was decided that wind and sea conditions were such as to preclude the lowering of boats with any degree of comfort. So at 0910 [9:10 am], the Presidential Detachment proceeded for Cocos Island at 16 knots..."

Thus, on July 31, 1938, the USS Houston and escort vessel USS McDougal left the Galápagos for good. They arrived on Cocos Island (Costa Rica) on August 1. The logbook for that day read: *"What a welcome sight to eyes hardened by the almost universal desolation of the Galapagos was the island named 'Cocos' as the Presidential Detachment anchored... this morning in Chatham Bay"*. FDR's Inspection Cruise and Fishing Expedition continued. The ships crossed the Panama Canal (August 4-5) and reached Pensacola, Florida, on August 9 (Fig. 5.11).

Galápagos: Part of a Network of Air Military Bases

As indicated earlier, in 1942, the United States built "Base Beta" in Seymour (Baltra) Island, which the US Military run until 1946. By 1943, Galápagos became part of a large-scale plan—by the US—to establish a network of overseas air military bases as "*strategic frontier*" to safeguard its homeland (Pettyjohn 2012). In November 1943, the US Joint Chiefs of Staff prepared the document "*United States Military Requirements for Air Bases, Facilities, and Operating Rights in Foreign Territories*", which was approved by President Roosevelt, together with a map identifying +60 sites where access was necessary to construct the facilities (Fig. 5.13). The Galápagos Islands and Guayaquil [i.e. Salinas] in Ecuador were among such sites (Leffler 1984; Pettyjohn 2012). Washington, however, did not secure agreements with all potential host nations to build the military camps and faced local resistance or funding limitations (Pettyjohn 2012). But the bases in Galápagos and Salinas were—already—built.

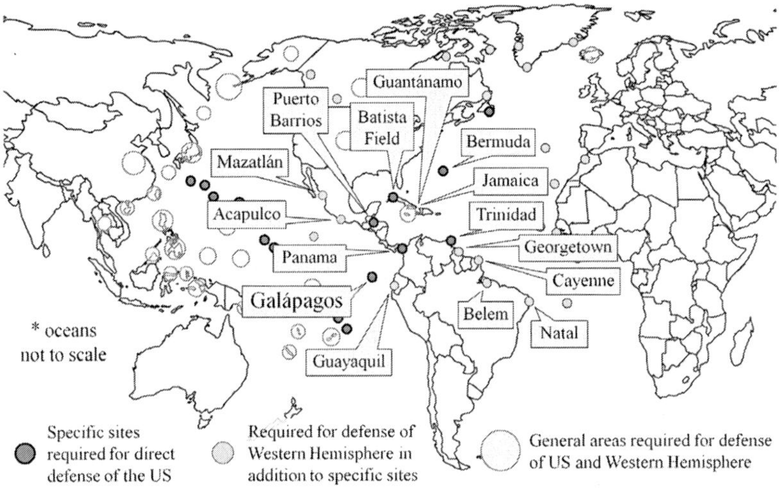

Fig. 5-13 Post World-War-Two Military Air Base Requirements [Western Hemisphere] by the United States, approved November 23, 1943. The dialog boxes highlight "*specific sites required for direct defense of the United States*" (10 sites, dark dots) or "*required for defense of the Western Hemisphere*" (5 sites, light dots), the latter in addition to specific sites. Clockwise (for practical reasons and not priority order): Galápagos (Ecuador); Panama; Acapulco and Mazatlán (Mexico); Puerto Barrios (Guatemala); Batista Field (nowadays San Antonio de los Baños, Cuba); Guantánamo (Cuba) and Jamaica both as a single "general area" required for defense of the US and Western Hemisphere (large circle); Bermuda; Trinidad; Georgetown (former British Guiana, nowadays Guyana); Cayenne (French Guiana); Belem and Natal (Brazil); and Guayaquil [Salinas] (Ecuador). Additional specific and/or required sites were depicted in 1943 maps as "*Western Hemisphere*" locations (here shown as dots, but without dialog boxes); they were in Canada and Greenland. The original 1943 map included air base requirements for Europe, Africa, and Asia-Pacific (most shown here). Oceans are not scaled. Figure layout: GPC, redrawn and modified after map "*Military Air Base Requirements-Period II*" (Peace Enforcement by Major Powers); enclosure "B" to Joint Chiefs of Staff memorandum to the President, November 15, 1943; subject "Post War Air Bases"–approved by the President [of the United States, Franklin D. Roosevelt], November 23, 1943. Source: Stacie L. Pettyjohn. 2012. U.S. Global Defense Posture, 1783-2011, Rand Project Air Force, RAND Corporation (links in Appendix A).

Ecuador's Position

Ecuador took legal possession of the Galápagos Islands as *Archipiélago del Ecuador* (Archipelago of Ecuador) on February 12, 1832. President Juan José Flores commissioned a civil and military expedition for this purpose. Thus, the prior condition of *res nullius*, in Latin, land "belonging to no one" (or "nobody's thing"), ended. At first, the islands became known as *Las Florianas*, and Charles Island itself, where the possession act materialized, *Floriana* (designations honoring President Flores). However, only "*Floreana*"—Charles Island—later remained (Luna-Tobar 1997; Idrovo 2005). But history is intricate, more so in geopolitics.

New-Orleans-born, in 1789, when Louisiana was a colony of Spain (before the Louisiana Purchase by the US of 1803), General-to-be José de Villamil arrived in Guayaquil (coastal Ecuador) in the early 1800s. By the 1830s, he was prosperous and well connected with local authorities. [Note that Ecuador had just left Greater Colombia in 1830, of which it had been a member since 1822, after becoming independent from Spain]. Villamil persuaded Ecuador's President Flores to claim possession of the Galápagos Islands. For that, Villamil would contribute with the creation of the *Galapagos Settlement Company* to secure Ecuadorian presence in the islands and work the land (agriculture and livestock). In October 1832, Villamil became Governor of the Galápagos Archipelago (Fig. 5.14). By 1835, Ecuador's Constitution recognized the *Archipiélago de Galápagos* as part of its territory (Luna-Tobar 1997; Idrovo 2005).

In 1840, Spain acknowledged—in an official document—Ecuador as a "*free nation, sovereign and independent*" and explicitly referred to the Galápagos Archipelago as part of Ecuador, as well as its provinces (i.e. the 1835 Constitution; Luna-Tobar 1997). Yes, Ecuador owns Galápagos!

Fig. 5-14 Tribute to General José de Villamil in Puerto Villamil, Isabela Island. In 1832, Villamil became the first Governor of the Galápagos. The *Archipiélago de Galápagos* was recognized as part of Ecuador in its Constitution of 1835. Photo: G. Paz-y-Miño-C.

Fifty-one years after Ecuador took possession of the Galápagos, the United States Senate, in 1883, declared the islands *res nullius*, "belonging to no one". Why? Two immediate reasons: geopolitics and the eventually-to-be-constructed Panama Canal (a plan under consideration). US Secretary of State James G. Blaine, during the administration of 21st US President Chester Allan Arthur, had previously commissioned, in 1881, George Earl Church, a Lieutenant Colonel with background in engineering and geography, plus extensive fieldwork in Mexico, Central America, and South America, to travel to Ecuador and gather information about the status of the Galápagos Islands. Church wrote in a report (mid-1881): "*In my opinion, Ecuador's occupation of the Galapagos does not exist to a sufficient degree to give them rights over other nations*". Church justified his assessment on the observation that José de Villamil's efforts to secure Ecuadorian presence in the islands [in other words, as per the *Galapagos Settlement Company* mentioned earlier] had failed and that after Villamil's passing (1866), the islands became practically abandoned by Ecuador (Luna-Tobar 1997; Idrovo 2005). Thus, Washington got from Church the report it was looking for. Two years later, the US Senate stated, "*there are serious doubts about Ecuadorian sovereignty in the islands*", and the archipelago was declared *res nullius*. In principle, it could be occupied or claimed by any nation. The construction of the Panama Canal was already envisioned and Galápagos—due to its proximity to Central America—was of major strategic and military defense value.

Via the Ministry of Foreign Affairs and its Secretary Antonio Flores, Ecuador expressed concern about the US position regarding Galápagos. *The New York Herald* published in May 1883 a letter by Mr. Flores, together with the newspaper's own analysis about potential consequences for the US if it disregarded Ecuador's sovereignty over the archipelago (Idrovo 2005). Ecuador acted fast and secured key international support.

Washington 1942: Arroyo del Río Meets FDR

Fifty-nine years after the US Senate had declared, in 1883, the Galápagos Islands *res nullius*, Ecuador's President Carlos Arroyo del Río was in Washington, DC, on an official visit to meet President Franklin D. Roosevelt, FDR (Fig. 5.15). Arroyo del Río arrived at Bolling Field on November 23, 1942. He was "*received with a 21-gun salute, playing of* [Ecuador's and the US'] *national anthems, and a motorcade down flag-decorated Constitution Avenue*" (as reported in the news). A series of adverse events preceded the encounter between the Ecuadorian President and FDR: In July 1941, half of Ecuador's territory (the Amazon) was invaded by its Southern neighbor, Peru. On January 29, 1942, Ecuador was compelled to sign a settlement accepting the loss of such territory (278,000 km^2/107,336 sq mi). The US, Argentina, Chile, and Brazil sanctioned the signing—in Brazil—of the *Protocol of Rio de Janeiro*, which granted Peru jurisdiction over the occupied land. [Up to 1998, Ecuador repudiated the Protocol for having been imposed under military occupation]. In addition, on December 7, 1941, Pearl Harbor, Hawai'i, was bombed by Japan (above). By December 15, 1941, the US Caribbean Defense Command was in urgent negotiations with Ecuador's Military to secure the construction of an air base in Salinas, "Base Gamma" (coastal Ecuador). On February 1942, the US Military began construction of "Base Beta" in Seymour Island (Baltra), Galápagos (Figs. 5.2 to 5.10), jointly with "Base Gamma". The rest is history (this chapter; Idrovo 2005).

Fig. 5-15 Ecuador's President Carlos Arroyo del Río with US President Franklin D. Roosevelt (FDR) after Arroyo del Río arrived at Bolling Field in Washington, DC, on November 23, 1942 (details of the trip in IBMC 1943). Arroyo del Río was "*received with a 21-gun salute, playing of* [Ecuador's and US'] *national anthems, and a motorcade down flag-decorated Constitution Avenue*" (photo credit Bettmann Collection, via Getty Images, with permission).

Arroyo del Río's November-1942 visit to Washington, DC, aimed at reassuring the alliance and cooperation between Ecuador and the United States. WWII was at its peak. His itinerary included stops in various Latin

American countries. On December 8, 1942, reporters in Havana, Cuba, asked him if Ecuador had requested compensation for granting rights to the US Military to operate air bases in his country. He told *El Mundo* that "*No deal has been made. It is not a matter of a loan, sale, lease or mortgage. The Galápagos Islands are and will always be Ecuadorian territory, subject to the laws of Ecuador and under its jurisdiction. The bases are temporary concession, valid while they are needed for the defense of the American continent and the Panama Canal, that Democracy is currently waging against totalitarian aggression*" (Idrovo 2005).

1944: United States Aims at Signing with Ecuador 99-year Lease for Military Bases in Galápagos and Salinas

By 1944, the United States had prepared a draft of a "bilateral agreement for the leasing of [military] bases on Ecuadorian territory over a period of 99 years. The principal points [were]: (1) the leasing of three islands (Baltra, North Seymour and Santa Cruz) in the Galápagos Archipelago and of a small area (Salinas) on the mainland; (2) the joint surveillance for security purposes of that part of the Archipelago not leased; (3) the payment of an annual rental to be decided upon in the course of negotiations…" [source Memorandum by the Acting Secretary of State to President Roosevelt, Washington, November 2, 1944; signed Edward R. Stettinius, Jr.]. Ecuador did not accept the idea, and two years later, in 1946, President Velasco Ibarra (Fig. 5.10 bottom) instructed the Ecuadorian Ambassador in Washington, Galo Plaza Lasso (photo in Box 5.1), to formally request the US "withdrawal from Galápagos at earliest possible moment". The US officially returned Baltra to Ecuador on July 1, 1946. All American personnel had left Galápagos by 1948.

Parallel to the 1944 discussions in Washington about a potential bilateral agreement for the leasing of military bases in Ecuador, a memorandum from President Roosevelt emerged on April 1 (Box 5.2). In it, FDR addressed the "*preservation of the fauna and flora of the Galapagos Islands for international park purposes*". He remarked, "*It is a new idea for which... there is no precedent... The whole conception is based on a different approach. It is a hemisphere matter... If I were handling this... I would interest some of the scientific societies, museums, etc. or other American Republics and have the initiative come from them rather than from us...*". FDR's initiative was adopted (see below) with the possible involvement of the Smithsonian Institution. Back in 1938, when FDR visited Galápagos (Figs. 5.11 and 5.12), he invited Waldo L. Schmitt, a scientist affiliated with the National Museum, as a guest in the *Inspection Cruise and Fishing*

Box 5-2 Memorandum from United States President Franklin D. Roosevelt (FDR) for the Secretary of State, Cordell Hull: Preservation of the Galápagos Islands for International Park Purposes, April 1, 1944.

The White House, Washington, April 1, 1944. Memorandum [from US President Franklin D. Roosevelt] for the [US] Secretary of State [Cordell Hull]: *I have your memorandum of March 30th in regard to the Galapagos Islands. I do not think that some of the people down the line have been realistic about the preservation of the fauna and flora of the Galapagos Islands for international park purposes. It is a new idea for which, of course, there is no precedent. I Think it has been viewed too much as a bilateral matter between the United States and Ecuador. The whole conception is based on a different approach. It is a hemisphere matter in which Brazil, Central America, Canada and the Straits of Magellan should be as much interested as Ecuador and the United States. If I were handling this matter, I would interest some of the scientific societies, museums, etc. or other American Republics and have the initiative come from them rather than from us. I have been at this for six or seven years and I would die happy if the State Department could accomplish some.......... on it! F.D.R.*

Expedition on Board USS Houston 16 July 1938 - 9 August 1938 (Fig. 5.11). From 1933 to 1935, Schmitt explored the Galápagos on three occasions. In 1941 and 1942, in collaboration with the US Navy, he assessed the possibility of establishing a research station in the Galápagos Islands (Smithsonian Institution Archives 2022).

In a memorandum of November 14, 1944, the US Acting Secretary of State, Edward R. Stettinius, Jr., wrote to the US Ambassador in Ecuador, Robert M. Scotten, that "*The* [US] *Department* [of State] *has discussed... the* [potential] *agreement on 'Scientific Study, Research and Wild Life Preservation in the Galápagos Archipelago' with a responsible official of the Smithsonian Institution and with the Joint Post War Committee of the Joint Chiefs of Staff. The former, in view of the plans which it had developed for a research station in the Archipelago prior to the present world conflict, would doubtless be the 'agency' appointed by this Government... President Roosevelt has expressed a deep interest in a wild life agreement, and it is hoped that the Ecuadorian Government will approve the consideration of this subject along with the matter of the* [military] *bases* [in Galápagos and Salinas]. *It is suggested that in the course of discussion you stress the international prestige which Ecuador would gain by having a station in the Galápagos such as that contemplated by the Smithsonian Institution*".

1959: Ecuador Creates the Galápagos National Park

The 1944 conversations between the United States and Ecuador about the potential establishment of a research station in the Galápagos (in partnership with the Smithsonian Institution), together with the 99-year leasing of territory for military bases in Galápagos and Salinas, did not prosper. After additional years of give and take between the two countries, entertaining multiple scenarios to prolong US presence in the archipelago, Baltra Island was ultimately returned to Ecuador in 1946. By 1948, all American personnel had left Galápagos.

In hindsight, Ecuador's lengthy process to colonize the Galápagos since it claimed sovereignty over the islands in 1832, plus the logistic and financial challenges to materialize their prosperity, led to the by-default preservation of the archipelago's unique environments and wildlife. [Nature does well when undisturbed]! A faster pace of development, with substantial transformation of the landscape (i.e. sizable airports, major roads, deep-keel ships in shallow reefs, large-scale housing, ubiquitous pollution), would have caused irreparable ecological damage (something we discuss in Paz-y-Miño-C and Espinosa 2024a,b).

Ecuador established the Galápagos National Park (GNP) in 1959. It covered 97% of the terrestrial ecosystems. The first offices of the Charles Darwin Research Station (CRDS) on Santa Cruz Island opened in 1964 (Fig. 5.16). Ecuador exerts sole sovereignty over the Galápagos (Fig. 5.17). The CDRS is the operative branch of the Charles Darwin Foundation (CDF), an international non-profit organization that sponsors science and conservation work on the islands. The mission of the CDF and its Research Station is to "*provide knowledge and assistance through scientific research and complementary actions to ensure the conservation of the environment and biodiversity in the Galápagos*". Their vision is to "*ensure a sustainable Galápagos by providing breakthrough research which informs conservation actions and inspires humanity to preserve this extraordinary archipelago and our planet as a whole*" (CDF 2020). Both CDF and CDRS work in partnership with the Galápagos National Park Directorate, GNPD (i.e. the governmental agency that manages the GNP). In 1978, UNESCO recognized Galápagos as a World Heritage Site.

The Galápagos Province (one of 24 provinces or administrative divisions in Ecuador) is administered by the "Special Government Council for Galápagos Affairs" (or *Consejo de Gobierno del Régimen Especial de Galápagos*). Its head leader (a president) ranks at the level of Secretary (or *Ministro*) and is appointed by Ecuador's President.

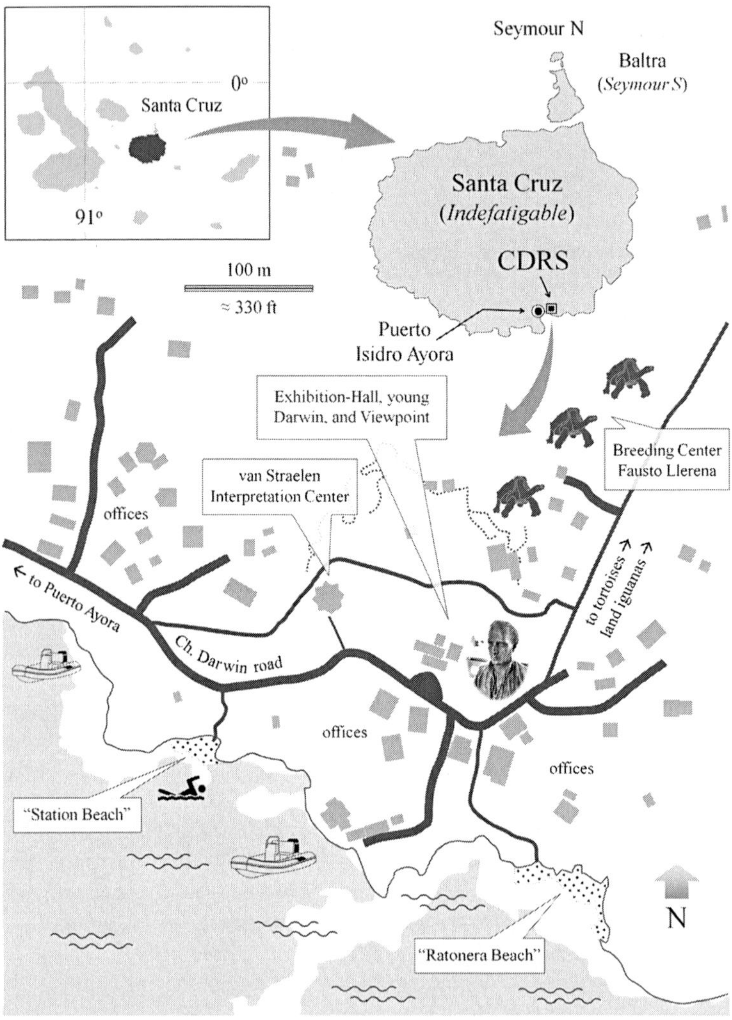

Fig. 5-16 Location of the Charles Darwin Research Station (CRDS) in Santa Cruz Island and summarized map of its facilities. The Galápagos National Park (GNP) was created in 1959. CDRS was established in 1964; CDRS is the operative branch of the Charles Darwin Foundation (CDF), an international non-profit organization that sponsors science and conservation work in the islands. CDF and CDRS work in partnership with the Galápagos National Park Directorate, GNPD (the governmental agency that manages GNP). One of the first and most successful conservation initiatives, the Giant Tortoises Captive Breeding and Reintroduction Program, was established at CDRS in 1965. Design/photo: G. Paz-y-Miño-C.

Fig. 5-17 Ecuador exerts sole sovereignty over the Galápagos Islands. The Galápagos Province (one of 24 provinces or administrative divisions in Ecuador) is administered by the Special Government Council for Galápagos Affairs. Ecuador's President appoints its head leader directly. Top left: Ecuador's flag in Puerto Villamil, Isabela Island (Albemarle) together with Navy flags. Top right: Galápagos Government Council's Building (*Consejo de Gobierno del Régimen Especial de Galápagos*) in Puerto Ayora, Santa Cruz Island (Indefatigable). Center left: main entrance to the Galápagos National Park in Santa Cruz. Center right: Puerto Ayora's harbor in Santa Cruz. Bottom left: Puerto Villamil on Isabela Island (Albemarle). Bottom right: the old-time emblem of the Charles Darwin Research Station (CDRS). Photos: G. Paz-y-Miño-C.

Besides the Galápagos National Park (6,937 km²/2,678 sq mi) and its Marine Reserve (141,100 km²/54,479 sq mi), Ecuador has numerous nature and wildlife reserves in its National System of Protected Areas (Fig. 5.18). As a "megadiversity" country, Ecuador ranks within the top 10 taxa-rich territories in the world (Mittermeier et al. 2005).

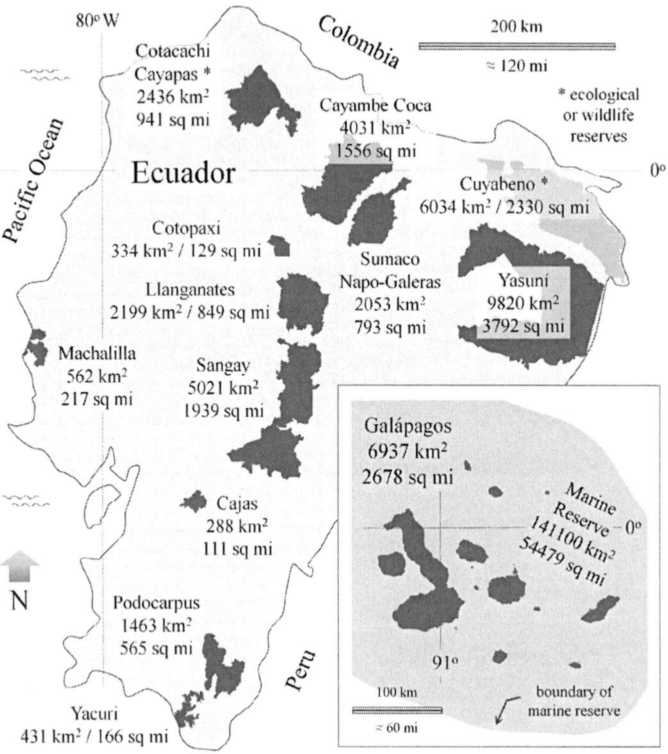

Fig. 5-18 National parks and exemplar protected areas in continental and insular Ecuador. Layout: G. Paz-y-Miño-C, redrawn/modified after *Mapa de Áreas Protegidas del Ecuador* (Sistema Nacional de Áreas Protegidas del Ecuador 2018; links in Appendix A).

The Galápagos Marine Reserve (GMR; Reck 2014), which was established in 1986, is proximate to various currently protected areas—by countries in the region—in the Tropical Eastern Pacific (Fig. 5.19), which include: Cocos Island National Park (Costa Rica), Coiba Island National Park (Panama), Malpelo Island Sanctuary (Colombia), and Ecuador's continental shores. In Paz-y-Miño-C and Espinosa (2024b), we discuss the

2021 "*Proposal for the Comprehensive Management of a Galápagos' Exclusive Economic Zone*" in the context of "*The Expansion of GMR*".

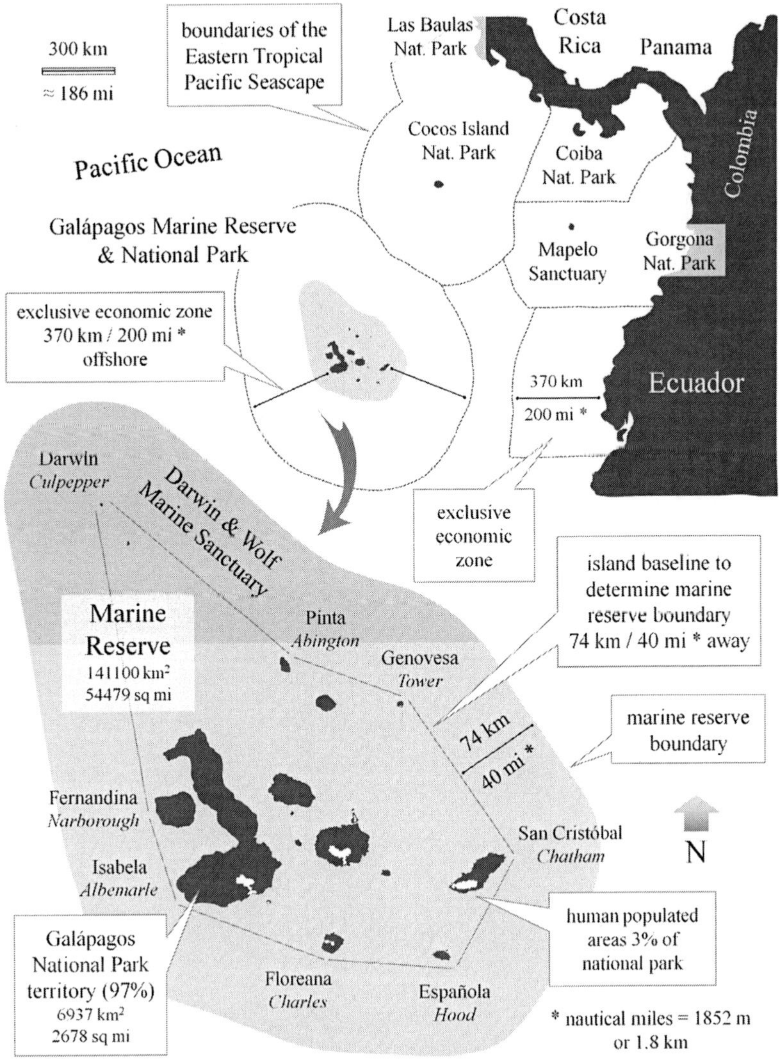

Fig. 5-19 Galápagos Marine Reserve (GMR) in the context of the Tropical Eastern Pacific. In 2021, Ecuador announced at the United Nations Climate Change Conference its commitment to protect 60,000 km^2 (23,166 sq mi) adjacent to the current GMR. Design: G. Paz-y-Miño-C.

2019: US Military Planes Back in Galápagos

In 2019, Ecuador's Ministry of Defense confirmed it had reached an agreement with the United States Military authorizing the landing of surveillance planes in Galápagos. It was part of a mutual effort to prevent drug cartels from using Ecuador's shores and maritime territory to smuggle banned substances into Central America, Mexico, the Caribbean, and ultimately the United States (Fig. 5.20). Ecuador approved routine arrivals of the Lockheed P-3 Orion and/or AWAC (i.e. AEW&C or Airborne Early

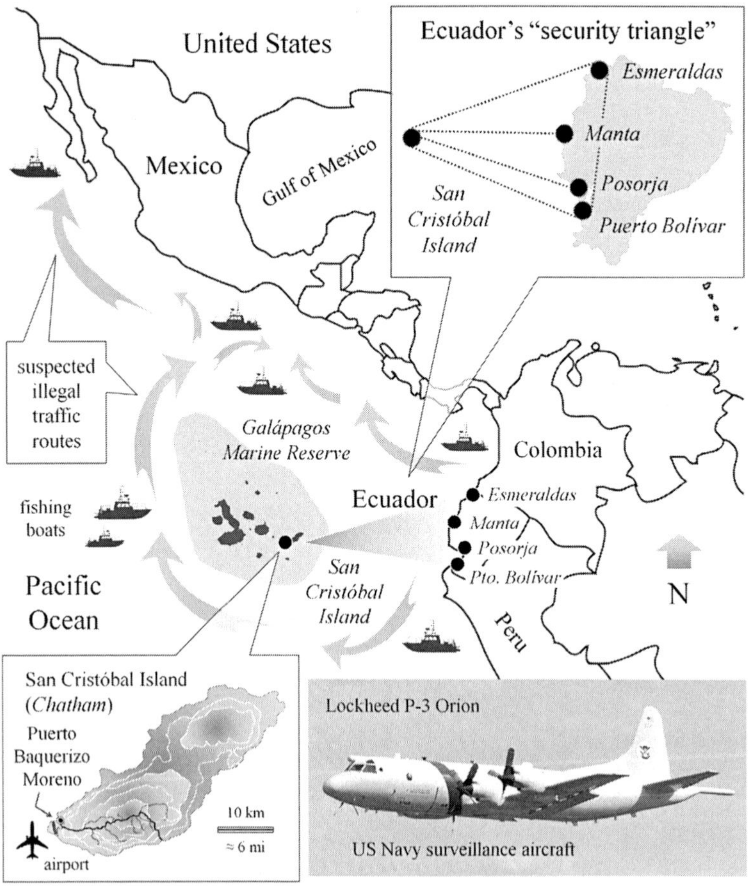

Fig. 5-20 Jurisdiction of the 2019 agreement Ecuador and the United States to Patrol the East Pacific: Galápagos and offshore Ecuador. Design: G. Paz-y-Miño-C. Photo-link Appendix A.

Warning and Control) planes at the San Cristóbal Island Airport, or SCY. In coordination with other strategic military posts on the continent, including Esmeraldas (Northern coast), Manta (Central coast), and Posorja and Puerto Bolívar (Southern coastal locations), Ecuador and the US would create a "Security Triangle"—from Galápagos to the coast of Ecuador—to patrol the area (Fig. 5.20, top right).

Ecuador's Minister of Defense, Oswaldo Jarrín, and US Admiral of the Southern Command, Craig S. Faller, announced the agreement during a press conference in Quito on April 25, 2019 (Fig. 5.21). Minister Jarrín stated that the international accord "*takes place under the doctrine of cooperative security among countries that face common threats: drug-trafficking, transnational organized crime, and illegal fishing*". [Note that large-scale international fishing in the Tropical Eastern Pacific is an ecological problem, plus illegal fishing within the Galápagos Marine Reserve has been documented in the recent past, together with the illegal trade of Galápagos wildlife; both seem associated with other unlawful activities in the region, as we discuss in Paz-y-Miño-C and Espinosa 2024a,b]. Admiral Faller added that "*these complex world challenges are better faced in co-operation*". The first Lockheed P-3 Orion plane—within the scope of

Fig. 5-21 Ecuador's [former] Minister of Defense, Oswaldo Jarrín, and [former] US Admiral of the Southern Command, Craig S. Faller, during a press conference in Quito on April 25, 2019, to announce the military cooperation Ecuador-United States to patrol the Galápagos and offshore Ecuador (photo: Rodrigo Buendía/AFP, via Getty Images, with permission).

the agreement—arrived in San Cristóbal on May 5, 2021 (previous arrivals occurred in an experimental phase). The US Ambassador to Ecuador, Michael J. Fitzpatrick, who flew in the Orion, was received by Minister Jarrín together with Ecuador's Commander of the Navy, Rafael Poveda, and local Galápagos authorities (El Comercio 2019, 2021; El Universo 2019; Paz-y-Miño-C and Espinosa 2024a).

The majority of the Ecuadorian people, however, did not adhere to the 2019 decision made by then President Lenín Moreno (in office 2017-2021) allowing US-Military planes in Galápagos, together with expanding the San Cristóbal runway and other infrastructure modifications to assist the aircraft. Due to the history of events at Baltra during WWII (this chapter) and Ecuador's recent experience of having hosted—from 1999 to 2009—a US military base in Manta, with apparently the same goal of deterring drug-trafficking, the public was skeptical about additional military agreements that might compromise the country's sovereignty or affect the environment in World Heritage Galápagos (as reported by BBC News Latin America 2019 and The Guardian 2019). [By 2025 when this book went to press, Ecuador and the United States had new officials as Minister of Defense or Commander of the US Southern Command, respectively].

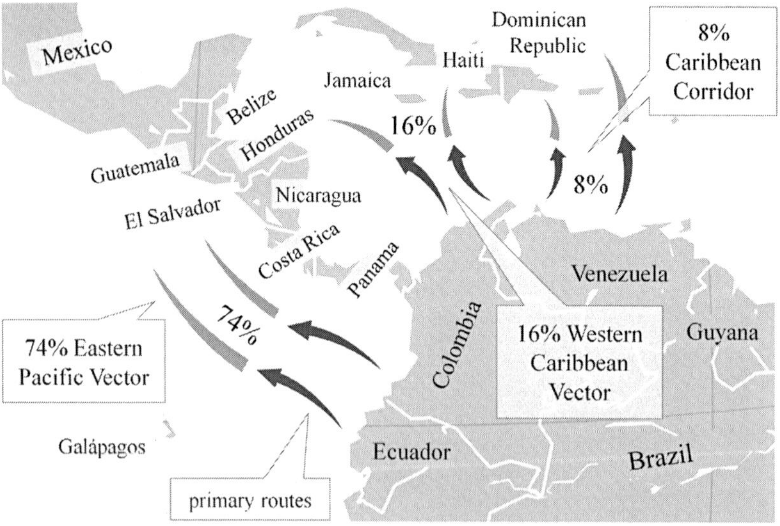

Fig. 5-22 The 2020 National Drug Threat Assessment released by the United States Department of Justice—Drug Enforcement Administration indicates that 74% of trafficking in cocaine from the Andean region (South America) to North America is currently shipped via Colombia (mainly) and Ecuador (partially). Layout: GPC, redrawn and modified after United Nations Office on Drugs and Crime (UNODC 2021).

However, drug smuggling via the waters of the Tropical Eastern Pacific is a problem. The *2020 National Drug Threat Assessment* released by the United States Department of Justice and Drug Enforcement Administration indicates that 74% of trafficking in cocaine from the Andean region (South America) to North America is shipped via Colombia (mainly) and Ecuador (partially). Both fronts are known as the "Pacific Vector" (Fig. 5.22). As shown in Fig. 5.22, other routes include the "Western Caribbean Vector", which facilitates the smuggling of 16% of the cocaine, and the "Caribbean Corridor", with 8% of shipments (United Nations Office on Drugs and Crime; UNODC 2021).

The future of the Security-Triangle collaboration between Ecuador- and US-Military (Fig. 5.21) seems uncertain; drug smuggling has affected the Tropical Eastern Pacific for decades. [Arguably, it started in the 1970s]. Ecuador's [former] Minister of Defense, Oswaldo Jarrín, pointed out, back in 2019, when announcing the collaboration, that "*the* [Lockheed P-3 Orion and AWAC] *aircraft are not here in Ecuador permanently; once they fulfil their mission, they will leave*" (El Universo 2019).

Conclusions

The Galápagos Archipelago belongs to Ecuador. The Republic of Ecuador claimed possession and jurisdiction over the *Archipiélago del Ecuador* (Archipelago of Ecuador) on February 12, 1832. Prior to that, the islands were "*res nullius*", they belonged to no one.

[Realize that, as we discuss in Chapters Two (*How Were the yas de galapagos Discovered?*) and Six (*Who Lives in Today's Galápagos Islands?*), Tomás de Berlanga and crew were the first humans to accidentally arrive (on March 10, 1535) in the later called "*yas de galapagos*" or "*isles of galapagos*". For 300 years, until Ecuador claimed the Galápagos Islands as its own (1832), no other republic or monarchy took possession of the islands].

By the end of the 1800s, the Galápagos Archipelago had gained geopolitical value due to its strategic location in the Tropical Eastern Pacific and with respect to the soon-to-be-constructed Panama Canal.

Ecuador was approached by England (early 1900s), France (1898), and the United States (1899) to negotiate the islands as either payment for Ecuador's debt with England (after England financed Ecuador's war of independence with Spain), or purchase an island for France, or lease one island for the United States. In 1911, the US proposed another lease—for 99 years, which was not accepted by Ecuador.

After the Japanese attack on Pearl Harbor in 1941, Ecuador signed an agreement with the United States for the establishment, in 1942, of US Military bases in Seymour (Baltra) Island (Galápagos) and Salinas (mainland Ecuador) to protect the Panama Canal and its surroundings.

By 1944, the United States had prepared a draft of a "*bilateral agreement for the leasing of* [military] *bases on Ecuadorian territory over a period of 99 years*". Ecuador did not accept the idea, and two years later, in 1946, Ecuador requested the US "*withdrawal from Galápagos*". The US officially returned Baltra to Ecuador on July 1, 1946. By 1948, all American personnel had left Galápagos.

Ecuador established the Galápagos National Park (GNP) in 1959. The first offices of the Charles Darwin Research Station (CRDS) on Santa Cruz Island opened in 1964. In 1978, UNESCO recognized Galápagos as a World Heritage Site.

The Galápagos Marine Reserve (GMR) was established in 1986. GMR is proximate to various currently protected areas in the Tropical Eastern Pacific, including Cocos Island National Park (Costa Rica), Coiba Island National Park (Panama), Malpelo Island Sanctuary (Colombia), and Ecuador's continental shores.

In 2019, Ecuador authorized the landing of US-Military surveillance planes in Galápagos as part of a mutual effort to prevent drug cartels from using Ecuador's shores and maritime territory to smuggle banned substances into Central America, Mexico, the Caribbean, and ultimately the United States. Ecuador's Military authorities indicated that the aircraft will not be in the country permanently and that once they fulfill their mission, they will leave. The Ecuadorian people expressed concern about present-day military agreements that might compromise the country's sovereignty or affect the environment in World Heritage Galápagos.

"The Rock" of the 1940s is nowadays a memory of WWII as well as a symbol of sovereignty—for Ecuador—with a complex past. Geopolitically, however, the destiny of the Tropical Eastern Pacific—and its multiple strategic locations, including the Galápagos Archipelago—will remain uncertain. Regional or global interests in this geographic space could manifest again in response to circumstantial change.

Questions

The following questions (Q1–3) merit generous written answers. You might need to do additional reading or library work to enrich your answers (we suggest exploring Paz-y-Miño-C and Espinosa 2024a,b).

Question 1: You have learned from this chapter that the Republic of Ecuador claimed possession and jurisdiction over the "*y^as de galapagos*" or "*isles of galapagos*" on February 12, 1832. Prior to that, the archipelago was "*res nullius*" ("belonging to no one"). In fact, for 300 years, from the time the islands were accidentally discovered in 1535—by Tomás de Berlanga and his crew—until Ecuador claimed possession and jurisdiction over them in 1832, no other republic or monarchy claimed the islands as their territory. (a) Why was—still is—the legal concept of "*res nullius*" central to Ecuador's ownership of the Galápagos Archipelago? (b) What is the difference between the concepts "*res nullius*" and "*terra nullius*"? (c) Does the concept "*terra nullius*" apply to the Galápagos? Why?

Question 2: The title of this chapter is "Why did the United States establish a military base in Galápagos during World War II?" (a) In hindsight, can you rank the geopolitical outcomes—for the United States—of establishing a military base in Galápagos during World War II? (b) How did Ecuador and the Galápagos benefit from the United States presence in Ecuador's territory during World War II? (c) From Ecuador's perspective, what were the adverse outcomes of the United States presence in Ecuador's territory during World War II?

Question 3: The Tropical Eastern Pacific (TEP)—where the Galápagos Archipelago and its Marine Reserve are located—is of great economic value, conservation/management priority, and geopolitical importance to multiple nations in Central, South, and North America. Explain why, from the economic standpoint, conservation/management needs, and geopolitical interests, the entire TEP [not only Galápagos and its Marine Reserve] requires a multi-nation comprehensive planning for its long-term preservation (i.e. functional ecosystems and their evolutionary processes).

How to Answer the Questions Above

The questions above have various possible answers (A1–3). Here we suggest perspectives from which to address the queries.

Answer 1: (a) Here you need to do some library research and learn the origin and meaning of the concept "*res nullius*" and how it has been used during history and in international law. Keep in mind that the principle of "*res nullius*" has been challenged numerous times as nations have disputed their authority over geographic space worldwide. (b–c) Both concepts, "*res nullius*" and "*terra nullius*", are related; however, over time, they have been used in distinctive ways as per international law. They should not be confused with one another. You need to establish if both concepts apply to the Galápagos case and justify your judgement.

Answer 2: (a) Keep in mind what were the geopolitical interests of the United States [both regionally in the Americas and worldwide] prior to, during, and after World War II. This will allow you to infer what the United States accomplished—and what it did not—when establishing military bases in Ecuador (Galápagos and Salinas) and worldwide (see Fig. 5.13 and text about it). (b–c) Ecuador's perspective should be at the center of your answers, not the United States' perspective. Realize that b and c are different questions.

Answer 3: Answering this question requires a regional viewpoint. Of course, Galápagos and its Marine Reserve are in the Tropical Eastern Pacific, but the region is large, with multiple players in Central, South, and North America [as well as in the Caribbean], as shown in Figs. 5.19, 5.20, and 5.22. The local fisheries are active in the countries' maritime jurisdiction, while global fisheries (many from Asia) operate—in principle—in international waters. Thousands of vessels navigate the area and cross the Panama Canal. And, as you learned in this chapter, drug cartels use the entire TEP shores to smuggle banned substances into Central America, Mexico, the Caribbean, and the United States (Fig. 5.22). Question 3 is complex and demands your meticulous work.

Discussion Topic

The discussion topic below (D1) should be addressed with your peers. We encourage college instructors to facilitate the dialogue.

Discussion 1: By December 2024, Ecuador's President Daniel Noboa authorized the United States to reinstate formal military presence in Galápagos. A bilateral agreement aiming at minimizing "drug trafficking and illegal fishing" around the Galápagos Marine Reserve would allow US vessels (ships, aircraft, and drones) and military personnel to patrol the area. San Cristóbal Island and its runway would become the center of operations. Environmental organizations in Ecuador, together with political leaders and the public, expressed concern about the return of the American military—after 80 years since World War II—to today's World Heritage Galápagos. Ecuador's fragile sovereignty over its territory and the archipelago were central to media debates criticizing President Noboa's announcement. Analysts, however, pointed out the difficulty for the country to face alone the challenges of growing drug trafficking and illegal fishing in both Ecuador's continental coasts and around the Galápagos Islands (Radio Pichincha 2024). Pundits highlighted that the United States geopolitical interests in the region (i.e. the entire Tropical Eastern Pacific) go beyond drug trafficking, illegal fishing, or migrant smuggling into the

US. The rising influence of China and Russia in Central and Latin America and in the Caribbean [i.e. trade, economic investments, cyber facilities, or space technology] is distressing to Washington, as summarized in a recent report by the United States Southern Command (2024). Thus, the US seeks to reinforce surveillance in the Tropical Eastern Pacific under the umbrella "control of illegal activities in the region" while guarding its geopolitical interests (as per Radio Pichincha 2024; teleSUR 2024).—Here is the scope of your discussion: forecast the future [short- and long-term] of the Galápagos Archipelago (its environments, ecological and evolutionary processes, its human-population presence, and their attitudes toward conservation and management of a World Heritage site), its Marine Reserve, and the neighboring protected areas in the Tropical Eastern Pacific (as per Fig. 5.19) under the conceivable scenario of mixed-global geopolitical interests and conflicts in the region.

Suggested Observation Projects

The suggested observation projects below (P1–2) take into consideration that you will travel to Ecuador and the Galápagos legally; that you will visit the islands as a tourist [holding a Transit Control Card, a mandatory document issued by Ecuador and specifically by the Governing Council of Galápagos]; and that you will respect the regulations that protect the Galápagos National Park and the Marine Reserve (see complete list of rules in Appendix B). Only accredited scientists, academics, and/or naturalists who have obtained official permits can do research in Galápagos. You can take pictures, but flash photography is not permitted when taking photos of wildlife. Professional photography and videos recorded for commercial purposes must be authorized by the Galápagos National Park Directorate (GNPD). The ideas below are for enthusiastic college explorers interested in learning about the archipelago.

Project 1: Together with your peers, college instructor, and a licensed naturalist guide sponsored by an authorized eco-tourism provider [there are plenty in Santa Cruz Island], you can explore the possibility of visiting Baltra Island. The one-day excursion could include sightseeing the Seymour [Baltra] Galápagos Ecological Airport (GPS). Remember that its runway rests on top of the "second airstrip" of the World War II years (see Figs. 5.2–5.4). Your eco-tour provider and naturalist guide might take you around in a vehicle and point at what is left of the 1940s roads, barracks, and the building blocks of the military base once run by the United States. You will be surprised by how much remains as an "archeological site". Compare your observations to the maps and aerial photo provided in Figs.

5.2–5.4. Google Maps will also give you a satellite view of Baltra Island ("Seymour South", "South Seymour", or "The Rock"), North Seymour ("Seymour North" or "Little Seymour"), and Seal Island or Mosquera Island (the tiny islet or what is left of the past continuity among land fragments, as we discussed in this chapter). [Obtain the Google Maps images in advance and save them on your cellphone, tablet, or laptop; remember that you will not have access to the internet while hiking in the area]. Your eco-tour provider will probably include a walk in North Seymour and snorkeling [or recreational diving if you are certified] in proximate sites to Baltra. Take notes and make sketches in your field book and get plenty of cellphone photos documenting the excursion. All these observations can be summarized in a slideshow presentation (for a 20-minute talk) to be shared with a general audience. Avoid slides with "text only" [they are often terrible]. Keep it rigorous yet fun and didactic.

Project 2: If you speak or understand Spanish, the documentary "The Rock: Galápagos During World War II" (*The Rock: Galápagos En La II Guerra Mundial*) might be both interesting and educational to watch (*https://www.youtube.com/watch?v=nYPibVzbPwA*). Even if you are not fluent in Spanish, the images and presentation of content, which include various segments in English, will be easy to follow. It is a 50-minute compilation of film clips from the 1940s showing routine activities by the American military in Seymour Island, images of the army base, interviews with veterans once deployed to "The Rock" [they narrate experiences from their homes in the United States], and a get-together while returning to Galápagos decades later. Their dialogue while stepping again on what is left of the World War II facilities is emotive. The film also addresses the views and memories of the local people, the *galapagueños*, those who worked at the military base or had some connection to it. A must watch. Organize a time to watch the film together with your peers and college instructor; discuss its content and remember Discussion 1 (above).

References

BBC News Latin America. 2019. "Ecuador 'Allows US Military Planes to Use Galapagos Island Airfield'". Accessed July 10, 2022. *https://www.bbc.com/news/world-latin-america-48663283*.

CDF. Charles Darwin Foundation. 2020. "Mission and Vision". Accessed July 10, 2022. *https://www.darwinfoundation.org/en/about*.

Department of State (United States of America). 1967. *Foreign Relations of the United States Diplomatic Papers 1944. Volume VII. The American Republics*. Washington DC: United States Government Printing Office.

El Comercio. 2019. “Defensa Trabaja en un ‘Triángulo de Seguridad’ Contra las Mafias”. Accessed July 11, 2022. *https://www.elcomercio.com/actualidad/defensa-triangulo-seguridad-mafias-narcotrafico.html*.

El Comercio. 2021. “El Avión Estadounidense Orión P3 Llegó a Galápagos para Reforzar las Operaciones Contra el Narcotráfico y la Pesca Ilegal”. Accessed July 11, 2022. *https://www.elcomercio.com /actualidad/seguridad/avion-estadounidense-orion-galapagos-operacio nes.html*.

El Universo. 2019. “El Avión Orión P-3 de Estados Unidos Opera Desde Galápagos por Primera Vez para Detectar Actividades de Narcotráfico y Pesca Ilegal”. Accessed July 11, 2022. *https://www.eluniverso.com/noticias/ecuador/avion-orion-p3-estados-unidos-galapagos-vigila-operaciones-ilicitas-narcotrafico-pesca-ilegal-nota/*.

Grace Tully Archive of the Franklin D. Roosevelt Presidential Library and Museum. 2022. “Log The Cruise 1938, Logs of the President’s Trips: Inspection Cruise and Fishing Expedition on Board USS Houston, July 16 - August 9, 1938”. Accessed July 12, 2022. *http://www.fdrlibrary.marist.edu/_resources/images/tully/7_03.pdf*.

IBMC International Business Machines Corporation. 1943. *Carlos A. Arroyo del Río Apostol del Panamericanismo*. New York: IMBC.

Idrovo, Hugo. 2005. *Galapagos: Footsteps in Paradise*. Quito: Libri Mundi.

Leffler, Melvyn P. 1984. “The American Conception of National Security and the Beginnings of the Cold War, 1945-48”. *American Historical Review*, No. 89: 349-357.

Life. 1942. “How the US Must Fight for Its Life”. Accessed July 10, 2022. *https://oldlifemagazine.com/march-02-1942-life-magazine.html*.

Luna Tobar, Alfredo. 1997. *Historia Política Internacional de las Islas Galápagos*. Quito: Ediciones Abya Yala.

Mittermeier, Russell A., Cristina Goettsch Mittermeier, Patricio Robles Gil, and Edward O. Wilson. 2005. *Megadiversity: Earth’s Biologically Wealthiest Nations*. México: CEMEX.

Paz-y-Miño-C, Guillermo, and Avelina Espinosa. 2024a. *Arrivals of Life to the Galápagos: From Berlanga to Darwin, Melville, and World War II*. Newcastle upon Tyne: Cambridge Scholars Publishing.

—. 2024b. *A Synthesis of the Galápagos: Natural History, Human Conflict, and the Destiny of an Archipelago*. Newcastle upon Tyne: Cambridge Scholars Publishing.

Pettyjohn, Stacie L. 2012. *U.S. Global Defense Posture, 1783-2011*. Santa Monica: RAND Corporation.

Radio Pichincha. 2024. “Qué Busca EE.UU. Con la Base Militar en Galápagos?” Accessed February 14, 2025. *https://www.radiopichincha.com/ee-uu-nueva-base-militar-galapagos/*.

Reck, Günther. 2014. "Development of the Galápagos Marine Reserve". In *The Galapagos Marine Reserve: A Dynamic Social-Ecological System*, edited by Judith Denkinger and Luis Vinueza, 139-158. New York: Springer.

Revisionismo Histórico del Ecuador. 2022. "Base Beta: Galápagos". Accessed July 12, 2022. *https://guerrade1941.blogspot.com/p/base-beta-galapagos.html*.

Sistema Nacional de Áreas Protegidas del Ecuador. 2018. "Mapa de Áreas Protegidas del Ecuador". Accessed July 7, 2022. *Foros Ecuador. http://www.forosecuador.ec/forum/ecuador/educaci%C3%B3n-y-ciencia/164669-51-%C3%A1reas-protegidas-del-ecuador-mapa-lista-de-zonas-naturales-protegidas*.

Smithsonian Institution Archives. 2022. "Record Unit 7231 Schmitt, Waldo L. (Waldo Lasalle), 1887-1977". Accessed July 6, 2022. *https://siarchives.si.edu/collections/siris_arc_217388*.

teleSUR. 2024. "What's Behind the New US Military Base in Galapagos". Accessed February 14, 2025. *https://www.telesurenglish.net/whats-behind-the-new-u-s-military-base-in-galapagos/*.

The Guardian. 2019. "Galápagos Islands: Outcry After Ecuador Allows US Military to Use Airstrip". Accessed July 10, 2022. *https://www.theguardian.com/world/2019/jun/17/galapagos-islands-ecuador-us-military-airstrip*.

United States Southern Command. 2024. *Statement of General Laura J. Richardson, Commander, United States Southern Command, Before the 118th Congress House Armed Services Committee*. Washington: United States Southern Command.

UNODC (United Nations Office on Drugs and Crime). 2021. *World Drug Report—Drug Market Trends: Cocaine Amphetamine-Type Stimulants*. Vienna: United Nations.

Woram, John M. 2005. *Charles Darwin Slept Here*. Rockville Centre: Rockville Press Inc.

Welcome to Galápagos
Seymour Galápagos Ecological Airport GPS
Baltra (*South Seymour*) Island

A giant tortoise welcomes you when arriving in the Galápagos.

proximate to Witch Hill (*Cerro Brujo*)
San Cristóbal (*Chatham*) Island

© GPC

Kicker Rock (*León Dormido*)
proximate to San Cristóbal (*Chatham*) Island

© GPC

Volcanic oceanic islands are always changing; they emerge, develop, and subside.

Kicker Rock (*León Dormido*)
proximate to San Cristóbal (*Chatham*) Island

This feature of Kicker Rock is evidence of volcanic islands' constant change.

Flamingo *Phoenicopterus*
Isabela (*Albemarle*) Island

Yellow warbler *Dendroica*
San Cristóbal (*Chatham*) Island

Yes, flamingos and yellow warblers often meet on the Galápagos' wetlands.

Red-billed Tropicbird *Phaethon*
Genovesa (*Tower*) Island

Nazca bobbies *Sula*
Genovesa (*Tower*) Island

Iconic Galápagos birds. Tropicbirds and boobies are typical shorebirds.

This majestic sea lion is resting upside-down. She is posing for the camera.

Giant tortoises *Chelonoidis*
San Cristóbal (*Chatham*) Island

© GPC

Giant tortoise *Chelonoidis*
Isabela (*Albemarle*) Island

© GPC

How many giant Galápagos tortoises can you count? Some are hiding.

Marine iguana *Amblyrhynchus*
Isabela (*Albemarle*) Island

Lava lizard *Microlophus*
Floreana (*Charles*) Island

No other place on Earth has marine iguanas and lava lizards living side by side.

Left or right leg? Anterior or posterior? How many scales? Look at those claws!

Medium ground finch (female) *Geospiza*
Santa Cruz (*Indefatigable*) Island

Medium ground finch (male) *Geospiza*
Santa Cruz (*Indefatigable*) Island

The world-famous Galápagos, or Darwin, finches. In truth, they are tanagers!

Looking at you. Galápagos pelicans are friendly; they often seem inquisitive.

Colors everywhere. Umbrellas in Santa Cruz and “the lobster” in San Cristóbal.

Primicias Cave (lava-tube cave or pyroduct)
Santa Cruz (*Indefatigable*) Island

© GPC

Remember to explore the lava tubes. Almost all islands have them.

Baby Sea Lion *Zalophus*
Rábida (*Jervis*) Island

Baby sea lion waiting for mother to arrive. She is probably hunting nearby.

Sally Lightfoot crab *Grapsus*
Santa Cruz (*Indefatigable*) Island
© GPC

Pyramid sea star *Pharia*
Rábida (*Jervis*) Island
© GPC

Zayapas on the shores, sea stars on the sea bottom. Galápagos is rich in marine life.

The splendor of a female cactus finch foraging on a prickly pear tree.

Picturesque Floreana and Isabela Islands. Their towns are always fun to sightsee.

Giant Prickly Pear *Opuntia*
Santa Cruz (*Indefatigable*) Island

Lava cactus *Brachicereus*
Genovesa (*Tower*) Island

Opuntia and *Brachicereus* cacti are iconic in the Galápagos. Look for them!

Puerto Villamil
Isabela (*Albemarle*) Island

Puerto Baquerizo Moreno
San Cristobal (*Chatham*) Island

Puerto Ayora
Santa Cruz (*Indefatigable*) Island

The beauty and charm of the Galápagos seaside towns.

A warm Galápagos sunset will likely be your forever memory of the islands.

CHAPTER SIX

WHO LIVES IN TODAY'S GALÁPAGOS ISLANDS?

> "…I let my thoughts stray back. We had left Amsterdam two months ago in July 1932. In Guayaquil, the chief port of Ecuador, we saw the mainland for the last time. Then we were a week on the seas to reach Chatham [San Cristóbal], the main island of the Galapagos group, where the Ecuadorian governor lived. We had paid to go as far as Floreana, but the captain of our primitive little ship declared he could go no further. I didn't know why… In this part of the world lots of things were different from the way they were back in Europe, and that was something you had to get used to. Right from the start of our new life I resolved to respect the customs and attitudes of these people, and on the whole I have managed it fairly well… We were stuck in [San Cristóbal] for three weeks, till the governor put at our disposal a fishing boat that was to take us to Floreana. 'Tomorrow we'll be there', the skipper promised proudly… His craft, twenty-five feet long and six feet wide, was so tightly crammed with her three-man crew, us three, our two dogs, and all the luggage, that it was almost impossible to move… We comforted ourselves with the thought that it would all be over in twelve hours. Water came splashing into the boat, and in a few minutes we were all drenched to the skin… We were miserably cold… Our teeth chattered in chorus. The night was pitch-black with no moon or stars. It was depressing and frightening to have only the endless expanses of the sea around me. [My husband] Heinz tried to cheer me up… 'it's really not so bad, it seems worse than it is because of the darkness. But that'll soon be over, and by dawn we'll be in sight of a journey's end, our island'…"—Margret Wittmer, 1959, Postlagernd Floreana [Floreana Adventure, 1961].

The Wittmers—Margret, her husband Heinz, son Harry, and their two dogs—arrived in Floreana Island, Galápagos, in August 1932, after leaving Germany for good. They were preceded by another German couple, Dore Strauch and Friedrich Ritter, who had abandoned their spouses, fled Berlin, and settled on the island a few years earlier. By 1930, the Galápagos population reached 571 inhabitants: San Cristóbal (*Chatham*) 389; Isabela (*Albemarle*) 155; Santiago (*James*) 17 salt mine workers, temporary residents; Santa Cruz (*Indefatigable*) 8; and Floreana (*Charles*) 2, Dore and Friedrich. When the Wittmers disembarked, pregnant Margret and her family added humanity to desolate Floreana. The second child, Rolf, was born Floreanan and Galapagean months later (for history and historicity of Galápagos, see Idrovo 2005).

Nowadays, it is hard to imagine that 100 years before the Wittmers immigrated to Galápagos, Floreana was the most populated island, with 80 houses and 120 people dedicated to agriculture, livestock, and provisioning the whaling vessels with water, giant tortoises for food, and firewood [we discussed whaling in Galápagos in Chapter Four], as described by José de Villamil, who in the early 1830s persuaded Ecuador's President, Juan José Flores, to claim possession of the Galápagos Islands. For that, Villamil would contribute with the creation of the *Galapagos Settlement Company* to secure Ecuadorian presence in the islands and work the land [we narrated these events in Chapter Five]. In 1832, President Flores instructed an official commission to travel to the islands and claim them as *Archipelago of Ecuador* (February 12, 1832). By October 1832, Villamil had become Governor of the Galápagos Archipelago (Luna-Tobar 1997).

The colonization of the islands began—arguably—right after they were discovered by Tomás de Berlanga in 1535 (Chapter Two). Thus, the Wittmers' terrifying navigation to Floreana, as narrated by Margret in her 1959 book *Postlagernd Floreana* [English edition *Floreana Adventure*; Wittmer (1961)], took place 100 years after Ecuador had claimed possession of the archipelago, and 400 years since the archipelago was first sighted. Much happened in between.

Margret's family settled in the highlands of Floreana and became legendary in the history of Galápagos' colonization during the 20th century. Her book alluded to the colorful life of another settler, Eloise Wagner de Bousquet—a self-proclaimed Austrian Baroness—and her three alleged servile lovers. A tale of intrigue, death, and obscurity around those who died. Journalists of the 1930s called it *The Galápagos Affair* and referred to Eloise as the *Empress of Galápagos*. In 2013, the film *The Galapagos Affair: Satan Came to Eden* was released. Yes, Galápagos was—still is—rich in accounts of immigrants from the Americas or Europe, memories of the "first settlers" as generations aged and vanished, stories of ambush privateers-buccaneers, adventures of whalers, diaries of explorers, and fables of pirates. For comprehensive compilations, see Luna-Tobar (1997), Idrovo (2005), and Woram (2005).

Below we summarize basic numbers about the Galápagos' population (demographics). It will help you understand the archipelago's history.

People Living in Galápagos: Census Facts

We advise readers to rely on official statistics for characterizations of Galápagos' demographics. In this section, we follow INEC's 2015 census (INEC=Ecuador's National Institute of Statistics and Census). The latest polling took place in 2022, but by the time we completed this chapter (February 2025), only preliminary results had been released (below).

A total of 25,244 people were living in Galápagos by 2015 (Fig. 6.1). The majority (62.2%) on Santa Cruz Island (*N*=15,701), followed by San Cristóbal (28.5%; *N*=7,199), and Isabela (9.3%; *N*=2,344). Floreana's residents (*N*≈100; 0.4%; perhaps 150 people) were included in San Cristóbal's total because Floreana's Parish (Parroquia Santa María) belongs to San Cristóbal's administrative jurisdiction (canton/county). Notice that INEC had previously projected Galápagos' population to reach 29,453 by 2015 and 33,042 by 2020 (more projections below).

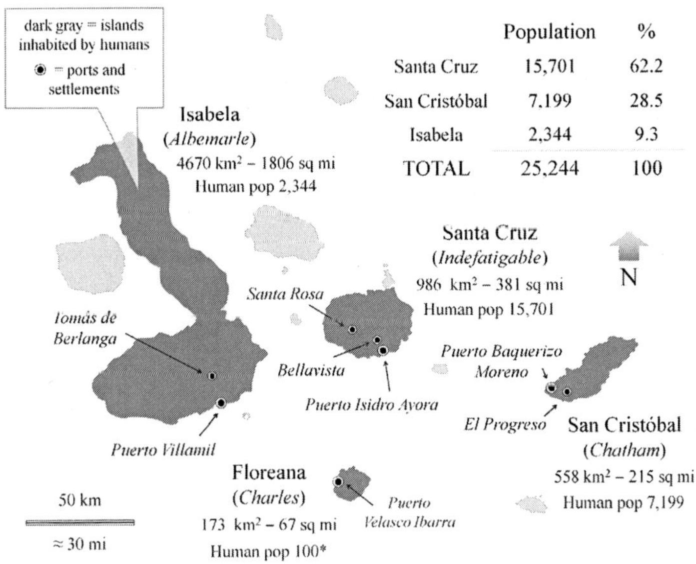

Fig. 6-1 Human population in the Galápagos Islands (as per 2015 census). 81.4% of the total population resides in urban/sub-urban areas (black circles). Census totals from INEC (Ecuador's National Institute of Statistics and Census 2015). INEC had previously projected Galápagos' population to reach 29,453 by 2015 and 33,042 by 2020. About 7,000 "Galápagos residents" (as per ID status) live elsewhere. *Floreana's residents (≈100; 0.4%; perhaps up to 150) are included in San Cristóbal's total because Floreana's Parish (Parroquia Santa María) belongs to San Cristóbal's administrative jurisdiction. Design: G. Paz-y-Miño-C.

The vast majority (81.4%) of the population resides in urban or suburban areas (black circles in Fig. 6.1). About 7,000 "Galápagos residents", as per ID status, live elsewhere (GAD-SC 2012; INEC 2020).

The inhabited islands differ substantially not only in population size but also in infrastructure. In Fig. 6.2, we sketch the main road networks, runways, and heliports in these islands, as well as the sites of iconic buildings like the Charles Darwin Research Station (CDRS) with its headquarters in Santa Cruz, or the Galápagos National Park with offices in the four islands. Note in Fig. 6.2 the location of the world famous giant tortoises breeding centers: Fausto Llerena (Santa Cruz), Galapaguera de Cerro Colorado or David Rodríguez/former Jacinto Gordillo (San Cristóbal), Arnaldo Tupiza (Isabela), and Asilo de la Paz, part of Floreana's Restoration Project (Floreana); the latter aims at recovering the ecological function of the tortoises—as top herbivores—since the local endemic species (*Chelonoidis elephantopus*) went extinct in mid-1800s (but see Miller et al. 2017). Tortoises from other islands were translocated to Floreana and are currently kept in fenced areas at Asilo de la Paz. [We discussed giant tortoises and the *Slaughtering of the "galápagos"* in *Arrivals of Life to the Galápagos*; Paz-y-Miño-C and Espinosa 2024a].

There are two sizeable airports in the archipelago: Seymour (Baltra) Galápagos Ecological Airport (GPS; see Fig. 6.5 for zoom in) and San Cristóbal Island Airport (SCY), both suitable for medium-size jet landings/take-offs (e.g. Airbus A31X/32X, capacity 140-170 passengers; Boeing 737-500 or 737-400, capacity 129-145), which arrive daily from continental Ecuador (via Quito-Guayaquil or Manta; the latter discontinued). A small runway exists in Isabela for inter-island flights, and only a heliport in Floreana. There is no direct international air travel; all commercial flights (or cargo ships) must arrive from continental Ecuador prior to biosafety inspection and fumigation to prevent introductions of invasive species (Toral-Granda et al. 2017). In 2002, ≈750 commercial flights landed in Galápagos (tourism); they increased to ≈2,200 in 2006, and by 2018 a total of 3,096 tourism-related flights arrived, as follows: 2,264 (73%) to Seymour/Baltra GPS and 832 (27%) to San Cristóbal SCY (as per Ecuador's Directorate General of Civil Aviation; Cruz and Causton 2008; MINTUR 2018). The main airlines operating in the archipelago [updated to 2023] included Avianca (former AeroGal), Latam (former LAN), Tame (ceased in 2020), Aeroregional, and Equair. The COVID-19 pandemic disrupted air travel and tourism in Galápagos from 2020 to 2022, but both recovered by mid-2023 (see Chapter Seven). The take-home message here is that infrastructure development (roads, runways, or heliports) correlates with population growth/presence.

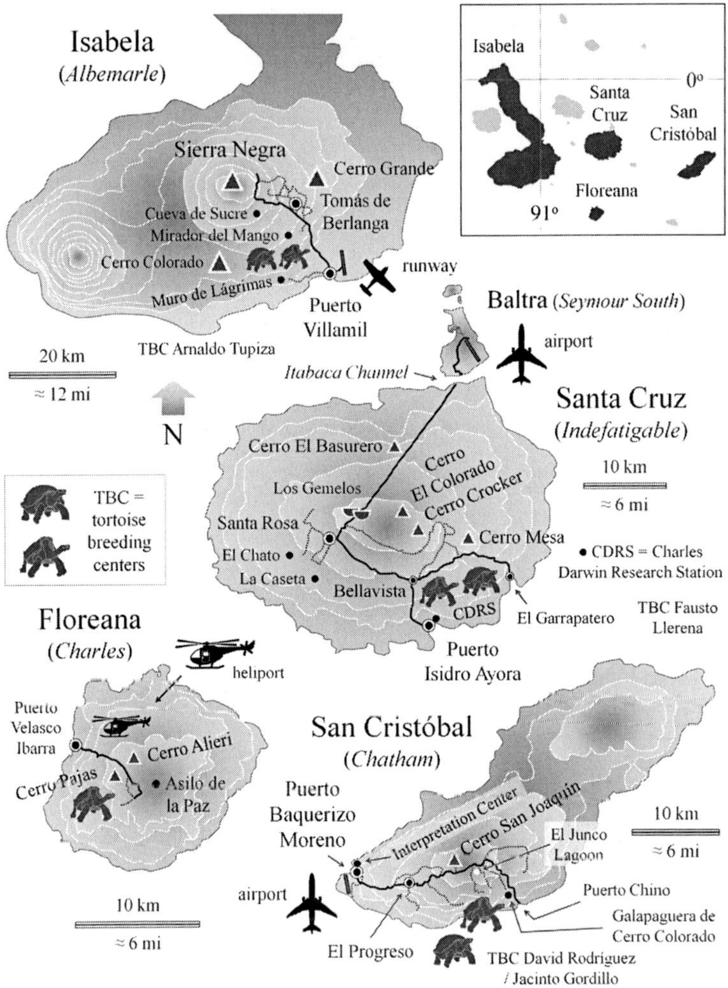

Fig. 6-2 Main road networks, runways, and heliports in the human-populated islands of the Galápagos. There are two sizeable airports: Seymour (Baltra) Galápagos Ecological Airport (GPS) and San Cristóbal Island Airport (SCY); a small runway in Isabela; and only a heliport in Floreana. There is no direct international air travel; all commercial flights (or cargo ships) must arrive from continental Ecuador prior biosafety inspection/fumigation to prevent introductions of invasive species (Toral-Granda et al. 2017). The Charles Darwin Research Station (CDRS) is in Santa Cruz (neighboring Puerto Isidro Ayora). The Galápagos National Park has offices on the four human-populated islands. Iconic giant tortoises' breeding centers exist on these islands (popular visitors' attraction). Design: G. Paz-y-Miño-C.

Puerto Isidro Ayora (population ≈12,000), Santa Cruz Island, is the largest settlement in Galápagos (Fig. 6.3); its urban infrastructure is the most developed (offices of the Governing Council of the Special Regime of Galápagos CGREG, NGOs, schools, hospitals, parks, hotels, and restaurants). The headquarters of Galápagos National Park, as well as the grounds of the Charles Darwin Research Station and the Fausto Llerena Giant Tortoises Breeding Center, are located to the East of the city.

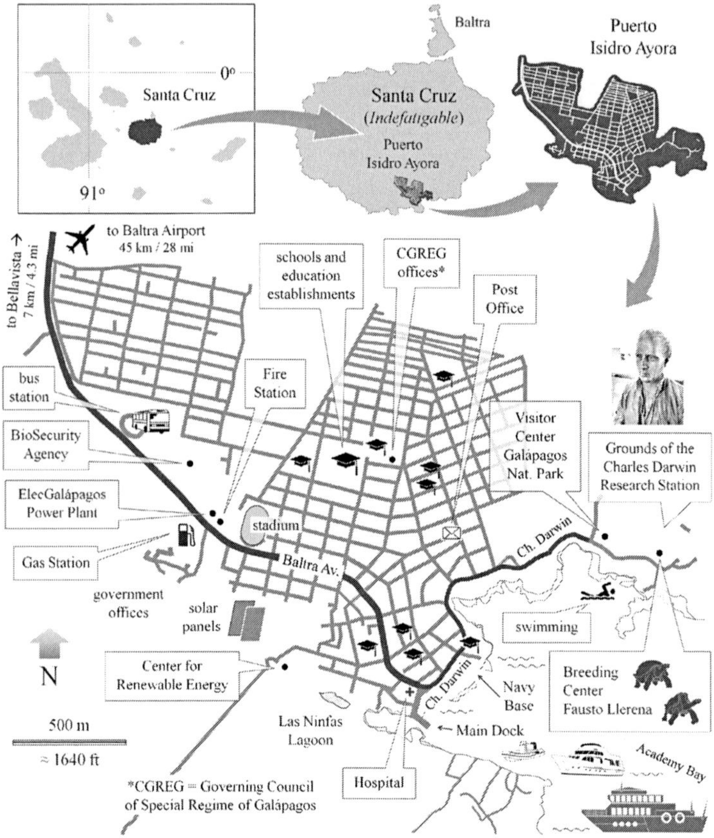

Fig. 6-3 Puerto Isidro Ayora (population ≈12,000), Santa Cruz Island (*Indefatigable*); the largest settlement in Galápagos (large town or small city). The headquarters of Galápagos National Park and its Visitors' Center, as well as the grounds of the Charles Darwin Research Station (CDRS, the operative branch of the Charles Darwin Foundation), and the Fausto Llerena Giant Tortoises Breeding Center are located to the East of the city. Design and photo of a young Charles Darwin statue at CDRS by G. Paz-y-Miño-C.

The rural settlements Santa Rosa (population ≈1,000) and Bellavista (population ≈2,400), both parishes under the administrative jurisdiction of Santa Cruz canton/county, are located in the highlands (Fig. 6.4): Santa Rosa elevation 350-m/1,150-ft; Bellavista elevation 245-m/800-ft. The popular Rancho Primicias and "El Chato", where giant tortoises can be observed roaming the farmlands, are located West of Santa Rosa.

North of Santa Cruz Island is Baltra (South Seymour), where the GPS airport is located (Fig. 6.5). [We discussed the significance of this island in Chapter Five, this volume; see also detailed analysis in *Arrivals of Life to the Galápagos*; Paz-y-Miño-C and Espinosa (2024a)].

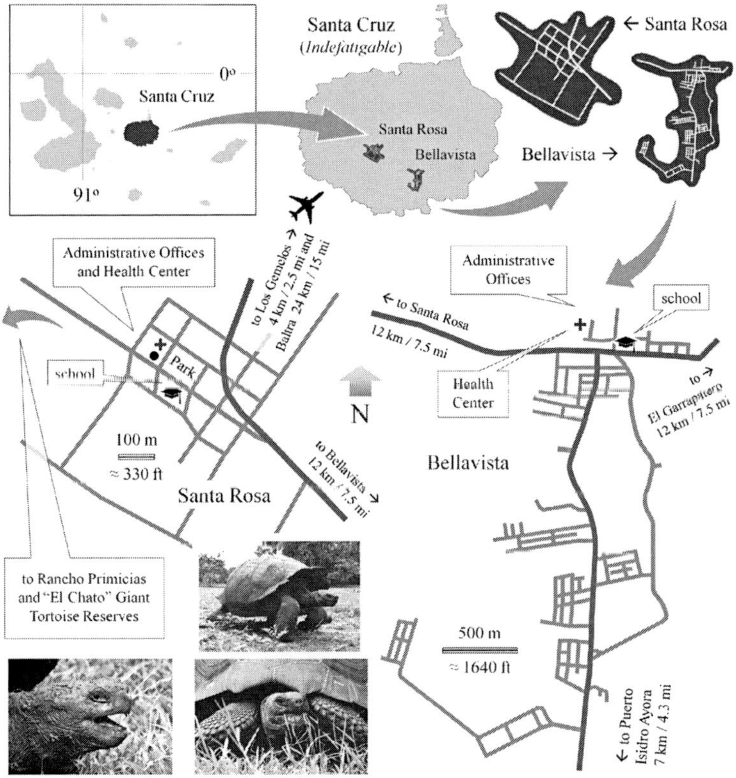

Fig. 6-4 Santa Rosa (population ≈1,000) and Bellavista (population ≈2,400), Santa Cruz Island (*Indefatigable*). Both rural settlements (and parishes under the administrative jurisdiction of Santa Cruz canton/county) located in the highlands: Santa Rosa elevation 350-m/1,150-ft; Bellavista elevation 245-m/800-ft. Rancho Primicias and "El Chato" Giant Tortoises Reserves located to the West of Santa Rosa. Design/photos: G. Paz-y-Miño-C.

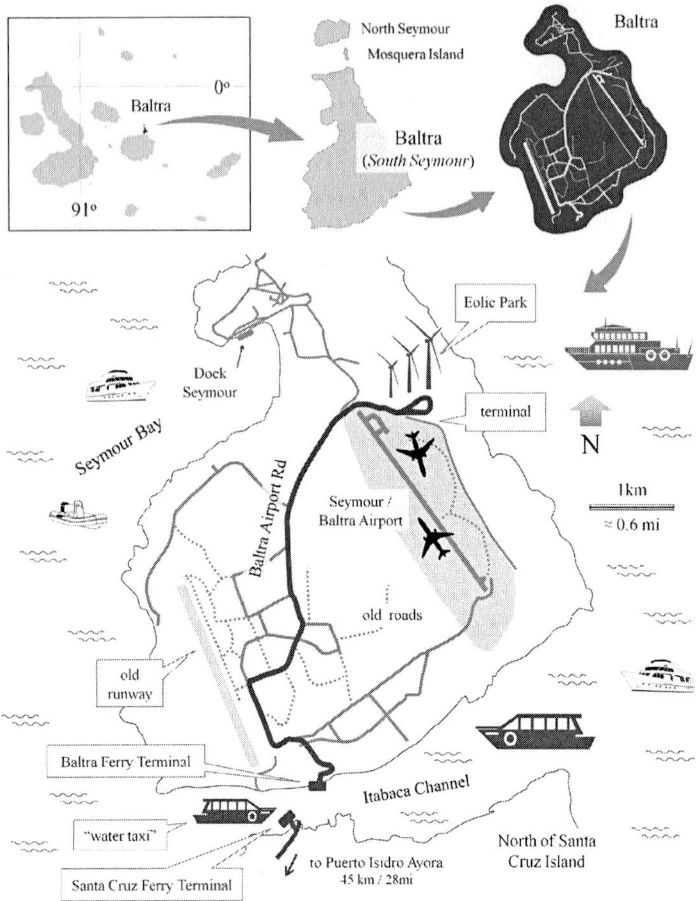

Fig. 6-5 Baltra Island (*South Seymour*). Spot two runways: a modern one in use and located to the East of the island; it corresponds to Seymour (Baltra) Galápagos Ecological Airport (GPS), which was built and re-built on top of what was left of a World War II military base airport (as discussed in Chapter Five, this volume, and also in *Arrivals of Life to the Galápagos*; Paz-y-Miño-C and Espinosa 2024a). An old, not-in-use runway is located to the Southwest; it was the main runway in the 1940s. By 1943, there were 400 United States officers, 4,500 military men, and 1,000 civilians living and/or working in Baltra. A year later, the US Department of State reduced the military presence to 1,900 US Army personnel and 450 US Navy marines. Nowadays, the Baltra population is restricted to staff working for the GPS airport, Ecuador's Army, and administrative offices. For historical accounts about Galápagos during WWII, see Luna-Tobar (1997), Idrovo (2005), and Woram (2005). Design: G. Paz-y-Miño-C.

The small town of Puerto Villamil (population ≈2,100) is the largest on Isabela Island (Fig. 6.6). The settlement has grown fast during the past few decades (government offices, schools, hospitals, parks, hotels, and restaurants). Puerto Villamil's beaches are among the most pristine in the human-populated islands of Galápagos, yet accessible to the public. The adjacent wetlands ("humedales") to the settlement are impressive, rich in mangroves and aquatic birds (flamingos, stilts, and gallinules).

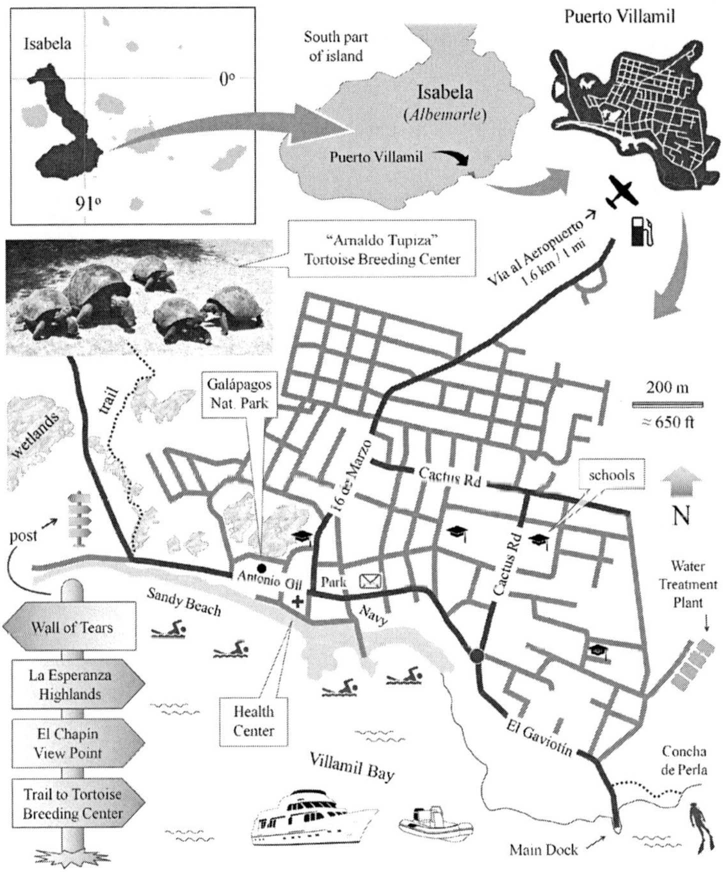

Fig. 6-6 Puerto Villamil (population ≈2,100), Isabela Island (*Albemarle*). Small town (and parish under the administrative jurisdiction of Isabela canton/county); its beaches are among the most pristine in the human-populated islands of the archipelago, yet accessible to the public. The adjacent wetlands (humedales) to the town are impressive, rich in mangroves and aquatic birds (flamingos, stilts, and gallinules). Design: G. Paz-y-Miño-C.

Tomás de Berlanga (population ≈200) is a rural settlement in the highlands of Isabela Island, elevation 160 m/530 ft (Fig. 6.7). It is proximate to the imposing caldera of Sierra Negra Volcano (elevation 1,124 m/3,688 ft). [We discussed Isabela's volcanoes in Chapter One; see Figs. 1.9, 1.10, 1.29]. Several picturesque neighborhoods (barrios) are typical of this area: La Esperanza (see close-up in Fig. 6.7), San Antonio de los Tintos, and Las Merceditas. The "Main Road" connects these sites with both Puerto Villamil in the lowlands and the Eastern edge of the Sierra Negra Volcano's caldera (Shelter Hike Volcano).

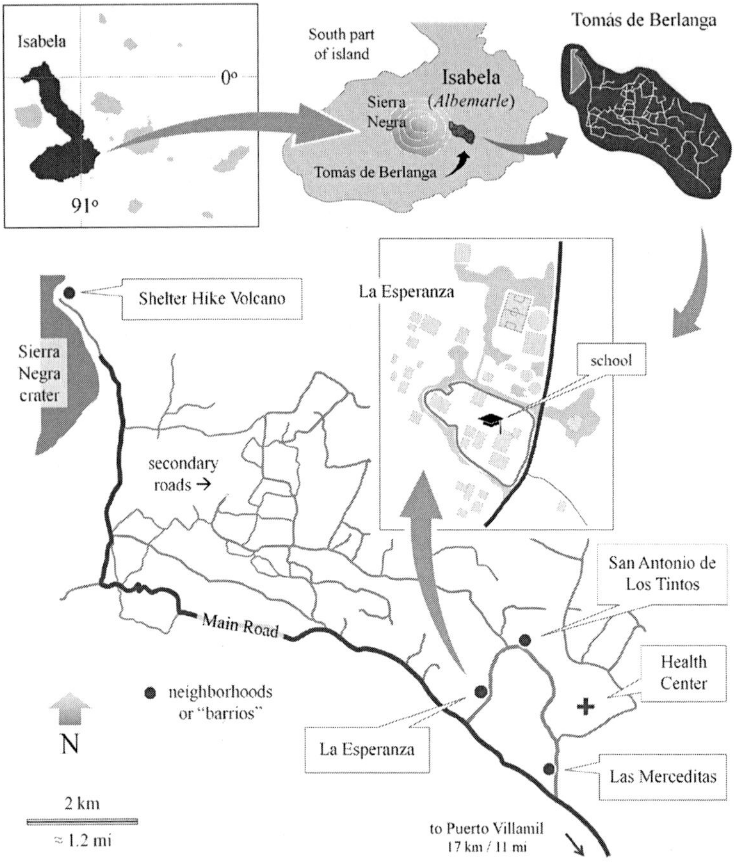

Fig. 6-7 Tomás de Berlanga (population ≈200), Isabela Island (*Albemarle*). Rural settlement (parish under the jurisdiction of Isabela canton/county) located in the highlands: elevation 160 m/530 ft (and up to Eastern edge of Sierra Negra caldera). Design: G. Paz-y-Miño-C.

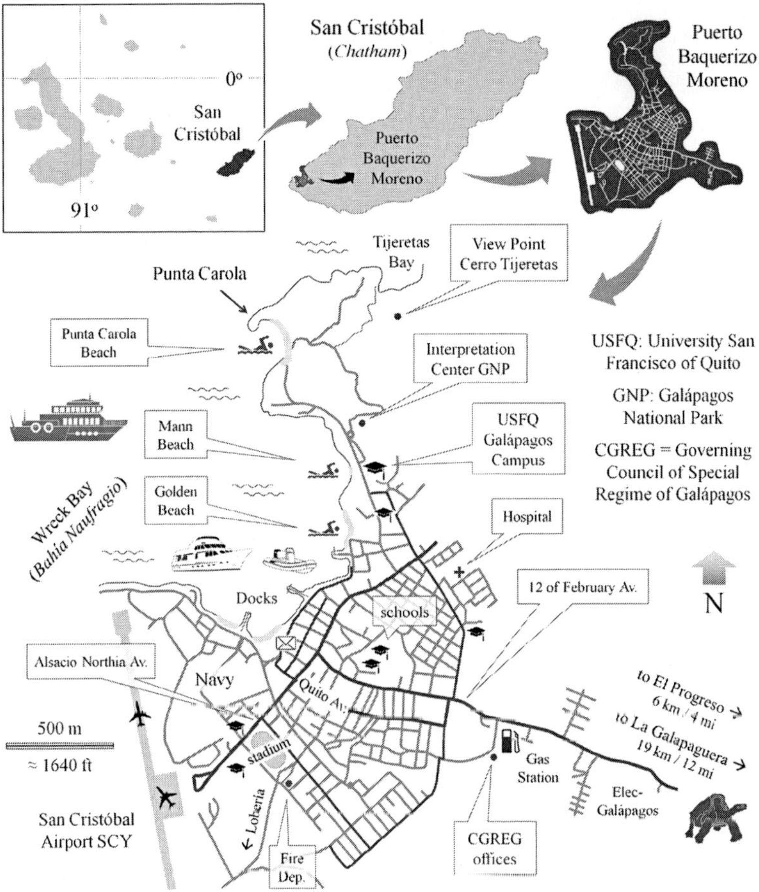

Fig. 6-8 Puerto Baquerizo Moreno (population ≈6,700), San Cristóbal Island (*Chatham*), the second largest human settlement in Galápagos (medium-size town or fast-developing small city). It is the capital of the Galápagos Province (Ecuador's administrative division in charge of the Galápagos Islands). CGREG= Governing Council. The San Cristóbal Island Airport (SCY) is located to the West. The Galápagos Campus of the University San Francisco of Quito (USFQ) and its Galápagos Science Center are in the Northern part of the town. The Interpretation Center GNP corresponds to facilities run by the Galápagos National Park (Centro de Interpretación Ambiental Gianni Arismendy). Design G. Paz-y-Miño-C.

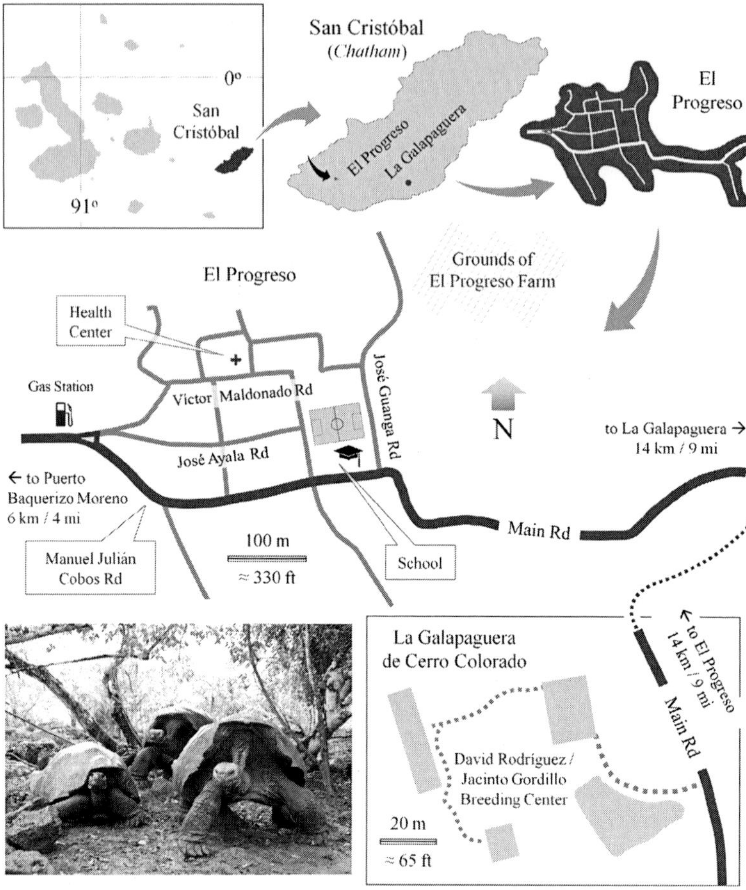

Fig. 6-9 El Progreso (population ≈700) and La Galapaguera, San Cristóbal Island (*Chatham*). Rural parish El Progreso is under the administrative jurisdiction of San Cristóbal canton/county; elevation 300 m/980 ft. Design/photo: G. Paz-y-Miño-C.

Puerto Baquerizo Moreno (population ≈6,700), San Cristóbal Island, is the second largest settlement in Galápagos (Fig. 6.8). As capital of the Galápagos Province (administrative division in charge of the Galápagos Islands), it is a fast-developing small city (CGREG's offices, NGOs', SCY airport, universities, schools, hospitals, parks, hotels, and restaurants). The elegant Galápagos campus of University San Francisco of Quito and its Galápagos Science Center are in the Northern part of the town.

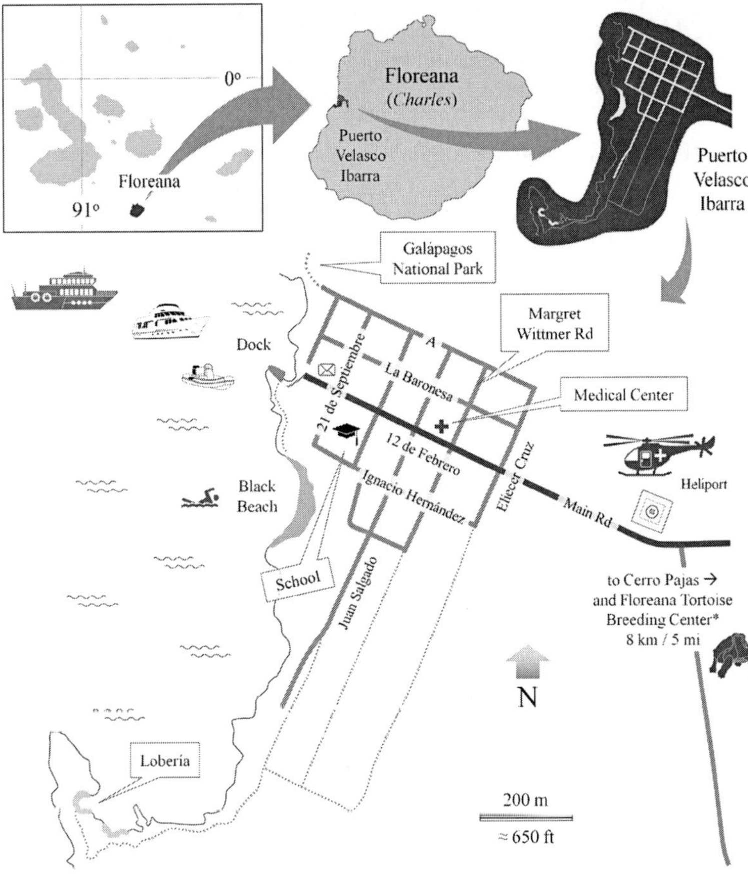

Fig. 6.10 Puerto Velasco Ibarra (population ≈100), Floreana Island (*Charles*). Floreana's Parish (Parroquia Santa María) belongs to San Cristóbal's (canton/county) administrative jurisdiction. Floreana's population is declining. Design: G. Paz-y-Miño-C.

El Progreso (population ≈700) and La Galapaguera are two well-visited sites in San Cristóbal Island (Fig. 6.9). Parish El Progreso (elevation 300-m/980-ft) is under the administrative jurisdiction of San Cristóbal canton/county. Like most settlements in the highlands of various islands, El Progreso consists of adjacent farms with centuries of human history (Luna-Tobar 1997; Idrovo 2005; Woram 2005). "La Galapaguera" refers to the giant tortoises' breeding center (galápago=tortoise).

Puerto Velasco Ibarra (population ≈100) is the main settlement in Floreana (Fig. 6.10). The population on the island is declining: 41 households (hogares) in 2017; by 2021, pollsters identified 37 (2.4 people per household; Vásconez et al. 2021). In 2021, Ecuador declared Floreana a Magic Town (Pueblo Mágico), acknowledging the island's past, from its discovery in 1535 (Tomás de Berlanga likely arrived to Floreana; Chapter Two) to whaling times (Chapter Four), to Ecuador's 1832 claim of the islands as Archipelago of Ecuador (Chapter Five), to Captain Robert FitzRoy and Charles Darwin visit in 1835 (Chapter Three), and to the settlers of the 1900s!

Colonization Led to Transformation of the Land

Where humans go, agriculture follows; 3.2% of the Galápagos islands is agricultural land, about 25,261 hectares/62,421 acres (Fig. 6.11). [As discussed in Chapter One, half of the archipelago (53%) is covered by native ecosystems; 43% corresponds to recent or old/er lava flows; 3% is agricultural land; and 1% is urban settings (Figs. 1.45-1.48)]. Santa Cruz Island has 11,439 ha/28,266 ac dedicated to agriculture; followed by San Cristóbal 8,341 ha/20,611 ac; Isabela 5,210 ha/12,874 ac; and Floreana 271 ha/669 ac (Fig. 6.11). Agricultural areas are surrounded by buffer zones (=transitions from farms to wilderness; 35,295 ha/87,215 ac), which are protected by law. Thus, agroecosystems (agricultural lands plus buffer zones) add up to 60,556 ha/149,637 ac (Fig. 6.11; Laso et al. 2020).

The most abundant and invasive taxa (as per vegetation cover) introduced by settlers to agricultural lands and—by default—to the islands include: *Psidium guajava* (common guava shrub/tree), *Pennisetum purpureum* (Napier/elephant grass), *Cedrella odorata* (Spanish/Cuban cedar tree), *Rubus niveus* (blackberry shrub), and *Cinchona pubescens* (red cinchona, quinine, or quina tree; Rivas-Torres et al. 2018; Laso et al. 2020). In *Arrivals of Life to the Galápagos* (Paz-y-Miño-C and Espinosa 2024a), we discuss how since the islands were discovered in 1535, about 1,579 alien species have been introduced, of which 1,476 are now established, with 868 of the latter (59%) naturalized (species that now spread and reproduce without active human involvement). The majority of introduced taxa are terrestrial plants (821 spp. introduced; 270 spp. naturalized), followed by terrestrial insects (545 spp. introduced; 467 spp. naturalized) and other groups: terrestrial invertebrates (68 spp. naturalized); pathogens (38 spp.); vertebrates (18 spp.); marine invertebrates (5 spp.); and marine plants (2 spp.). On average, 76% of all alien species have been introduced post-1970s, coinciding with the increase in human population growth, air and maritime travel practices, and tourism flow (factors associated with human connectivity and mobility in-and-out the archipelago; see below and Chapter Seven; Toral-Granda et al. 2017).

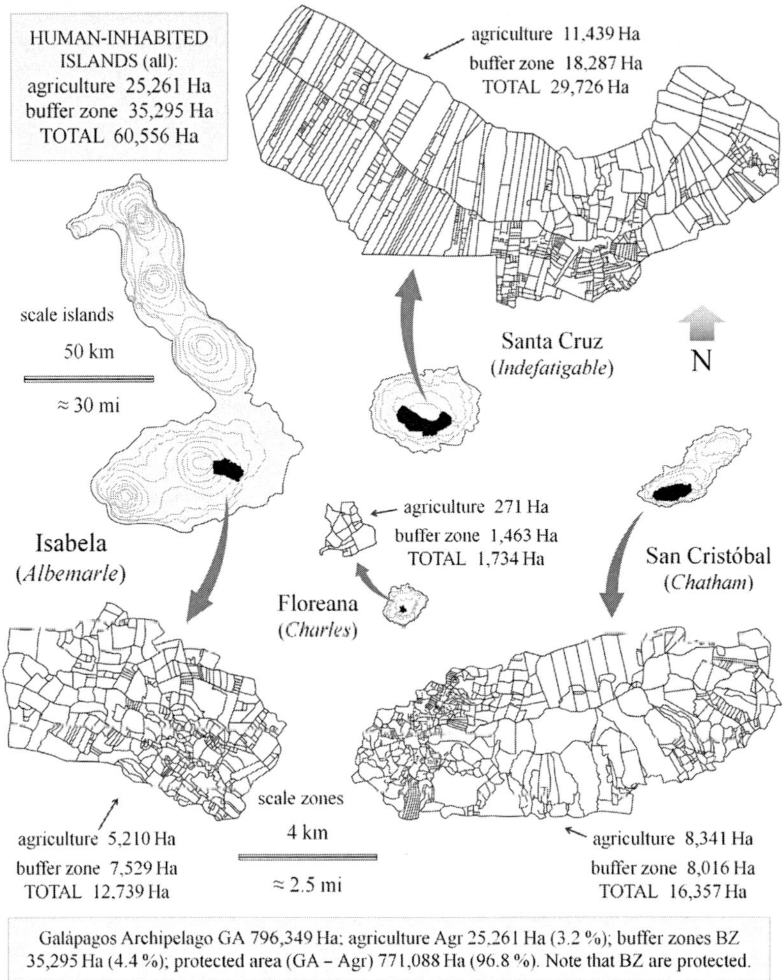

Fig. 6-11 Agroecosystems in the highlands of the human-inhabited Galápagos islands. Layout: GPC; data and agriculture/buffer-zone sketches after Laso et al. (2020).

Population Trends and Demographics

In Fig. 6.12a, we summarize outcomes of three censuses: 2001, 2010, and 2015. Note how although the number of inhabitants in the Galápagos has increased over the years, from 17,451 in 2001 to 23,046 in 2010, and to 25,244 in 2015, the percentage of people residing in each island (with respect to the grand total) has remained comparable over time; mean values: Santa Cruz 60.7%, San Cristóbal 30% [remember that Floreana's data is included in San Cristóbal's], and Isabela 9.3% (INEC 2015).

Now, examine Fig. 6.12b. From 1950 to 2015 (65 years), the Galápagos population increased 19-fold; the fastest rate of increase occurred between 1982 (4.0%) and 2001 (6.4%). Since 2001, the speed of population growth declined to 3.1% in 2010 and to 1.8% in 2015 (INEC 2015).

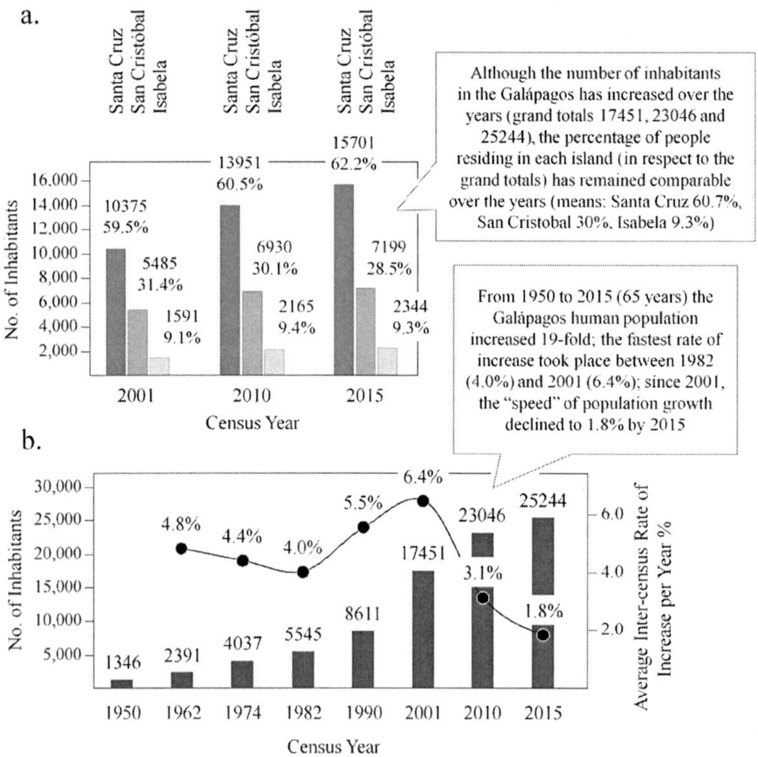

Fig. 6-12 a. Human population census 2001, 2010, 2015. b. Average inter-census rate of population increase per year 1950-2015 in the Galápagos Islands. Layout: GPC; data INEC (Ecuador's National Institute of Statistics and Census 2015).

The boost during the 1980s-1990s overlapped with the tourism boom (Chapter Seven). The decline by 2010/2015 corresponded to enhanced age-cohort-specific emigration (ages 15/17 to 24; below).

In Fig. 6.13a, we show an age-sex structure diagram that overlaps data from censuses 1990, 2001, 2010, and 2015. Note how during the four census events, the most prominent age-range was 15 to 64 years (67.2% of the population), followed by age 0 to 14 (29.0%), and age >65 (3.8%). The population decline within age-range 15/17 to 24 was evident across censuses (see light-gray dialog box in Fig. 6.13a) and associated with emigration of young people to pursue higher education or jobs elsewhere (mainland Ecuador). This is corroborated by Fig. 6.13b, which summarizes the age-sex structure of the population that migrated out of Galápagos after 2010 (a census year) and that by the time of the 2015 census had not yet returned (INEC 2015).

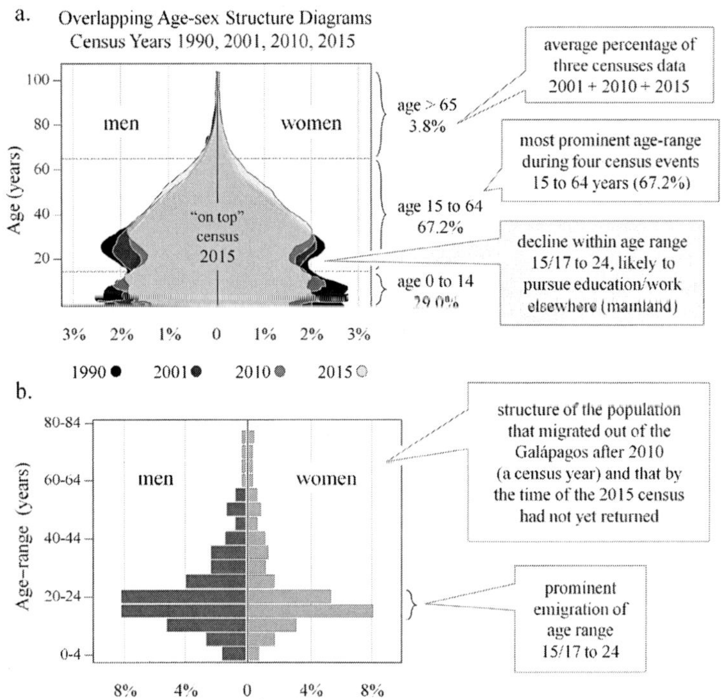

Fig. 6-13 Age-sex structure of the Galápagos Islands' human population. Note the decline within age range 15/17 to 24, likely to pursue education/work elsewhere. Layout: GPC; redrawn/modified after INEC (Ecuador's National Institute of Statistics and Census 2015).

In Fig. 6.14, we plot the last three censuses' data, i.e., age-sex structure 2001, 2010, and 2015, per human-populated island. As in the case of the entire Galápagos population (Fig. 6.13a), the age-range 15 to 64 years was the most prominent—regardless of census year—on Santa Cruz, San Cristóbal, or Isabela, followed by age 0 to 14, and age >65.

Where did the current Galápagos population come from? 62.5% immigrated from mainland Ecuador; 36.1% were born in Galápagos (natives of Galápagos Province); and 1.3% were international immigrants (Fig. 6.15a). If born in mainland Ecuador, 19.0% were from Guayas province; 12.1% Tungurahua; 6.1% Manabí; 5.8% Pichincha; 4.5% Loja; 2.2% Chimborazo; and 12.9% from other provinces (Fig. 6.15b). In other words,

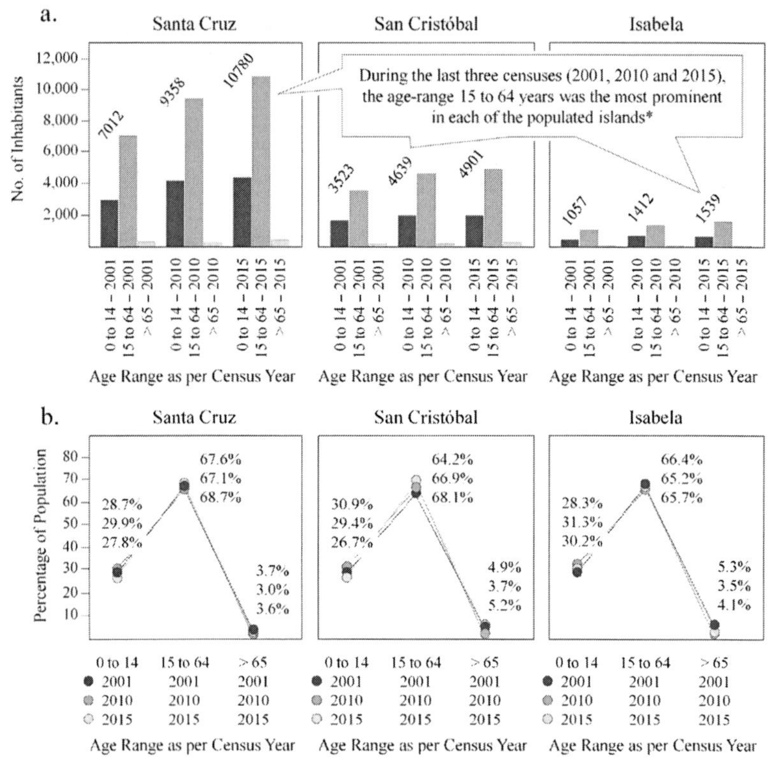

Fig. 6-14 Age-range distribution of the human population in the Galápagos Islands. Layout: GPC; redrawn and modified after INEC (Ecuador's National Institute of Statistics and Census 2015). *Floreana's data (≈100 people) included in San Cristobal's.

if visiting Galápagos and encountering locals at random while walking in the urban/sub-urban areas, chances are that 2 in every 3 people (particularly older adults) will be immigrants born in continental Ecuador (perhaps Guayas or Tungurahua), and 1 in every 3 people will be Galapagean (particularly younger adults and, of course, children).

Despite the decline in population growth rate, as per 2010 and 2015 censuses (Fig. 6.12b), the Galápagos population experienced a net gain of 2,576 individuals during that time. It was driven by higher immigration (3,368 people) than emigration (792 people; Fig. 6.16a). The majority of immigrants came from Guayas province (1,078 individuals; 32.0%), followed by "other provinces of mainland Ecuador" (*N*=592; 17.6%), Tungurahua (*N*=530; 15.7%), and Pichincha (*N*=476; 14.1%). The majority of emigrants were from Guayas (*N*=319; 40.3%), followed by Pichincha (*N*=167; 21.1%), and Tungurahua (*N*=98; 12.4%). Between 2010 and 2015, there was high mobility (in-and-out of the islands, although more "in" than "out") of immigrants/emigrants from Guayas, Tungurahua, and Pichincha. Note in Fig. 6.16a the particularly high immigration of people from other provinces (*N*=592; 17.6%) and other countries (*N*=222; 6.6%), totaling 814 (24.2% of all arrivals), which contrasted with 99 departures (12.5% of all emigrations); the net gain was 370 people.

When looking at emigration alone (*N*=792 individuals; Fig. 6.16b), it did increase from 4.8% in 2010 to 31.1% in 2015. If excluding mainland Ecuador's numbers and examining only the international emigrants (*N*=52), 23.1% were from the United States and a similar amount from "other countries" (Fig. 6.16c; INEC 2015).

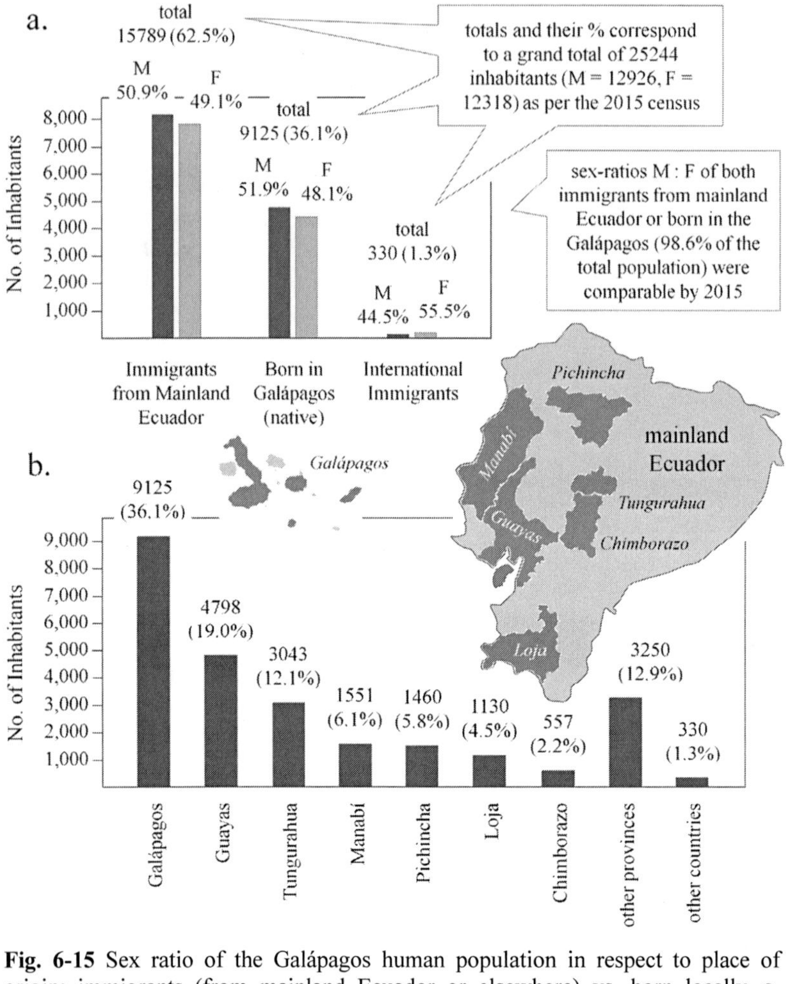

Fig. 6-15 Sex ratio of the Galápagos human population in respect to place of origin: immigrants (from mainland Ecuador or elsewhere) vs. born locally. a. Immigrants from mainland Ecuador, those born in the Galápagos, and immigrants from overseas; note that 2/3 people (62.5%) are immigrants born in continental Ecuador. b. Immigrants as per province of origin in mainland Ecuador, as well as from other countries. Layout: GPC; redrawn and modified after INEC (Ecuador's National Institute of Statistics and Census 2015).

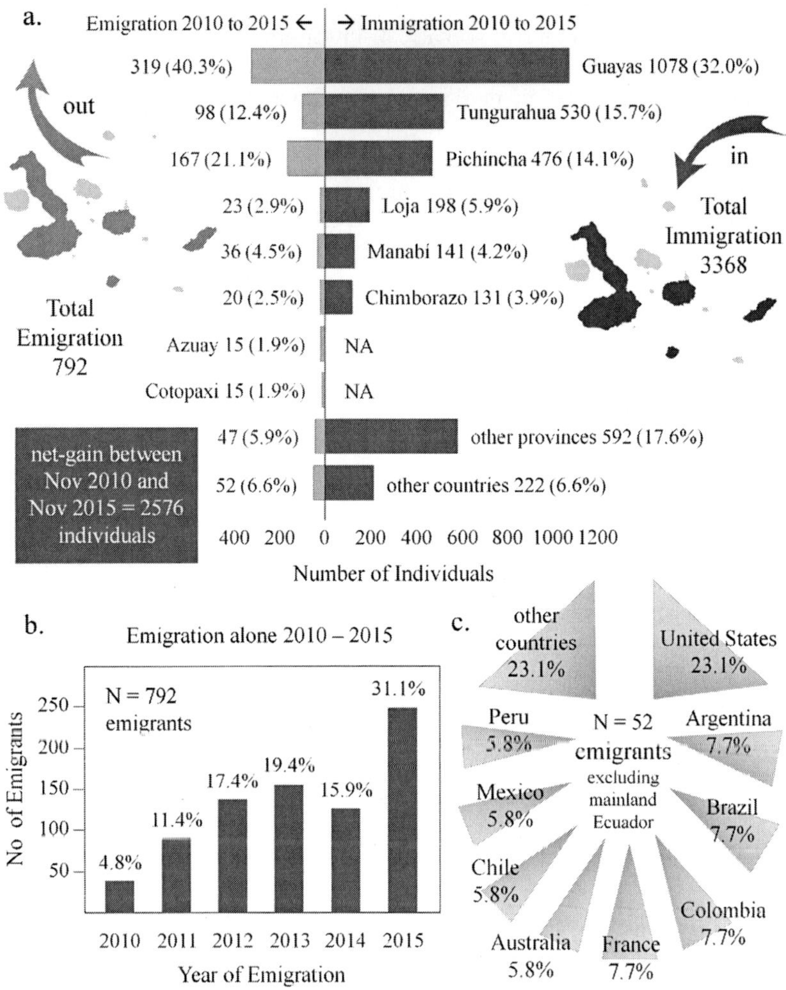

Fig. 6-16 Net gain in human population in the Galápagos Islands between November 2010 and November 2015 (census years). Layout: GPC; redrawn and modified after INEC (Ecuador's National Institute of Statistics and Census 2015).

Population Projections: from 2000 to 2049

In its *Plan Galápagos 2015-2020*, the Governing Council of the Special Regime of Galápagos (CGREG or *Consejo de Gobierno del Régimen Especial de Galápagos*) made projections about the future population size in the islands based on two scenarios: slower growth rate of 2.0% or faster growth rate of 3.75% per year (Fig. 6.17). The model relied on known data from 2000 to 2015 (some numbers shown above) and conceivable growth-rate circumstances. According to CGREG's estimates, by 2020, the population was expected to reach either 38,366 (slower growth, sg) or 42,614 (faster growth, fg). By 2033, the projection was to reach 49,759 (sg) or 69,386 (fg) people. And by 2049, the islands were expected to home 68,524 (sg) or 126,429 (fg) inhabitants!

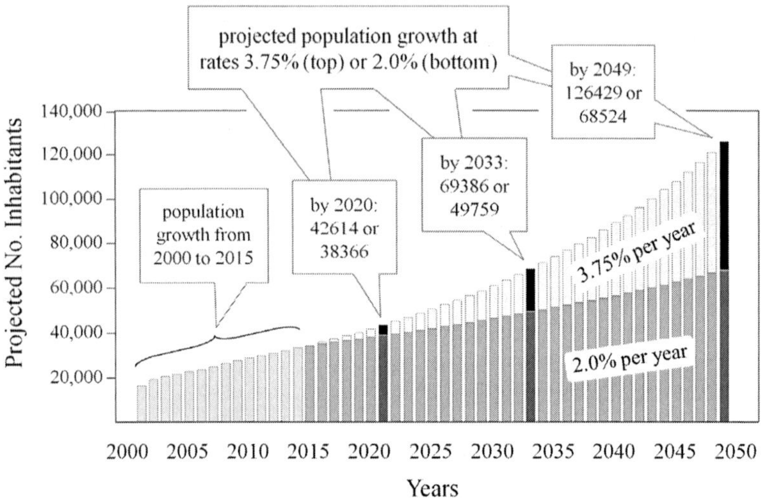

Fig. 6-17 Projected human population growth at different rates (2000-2050), Galápagos Islands. Layout: GPC; redrawn and modified after *Plan Galápagos 2015-2020* by Governing Council of Special Regime of Galápagos (CGREG 2016).

What is the carrying capacity of the Galápagos Islands in terms of human population? Short answer: we do not know. Long answer: the topic is complicated. In *Arrivals of Life to the Galápagos* (Chapter Eight, *Galápagos versus Other Archipelagos*; Paz-y-Miño-C and Espinosa 2024a), we discussed how smaller island groups [i.e. smaller than Galápagos' surface area 7,900 km^2/3,050 sq mi] hosted larger human populations: Mascarenes 2.19 million people (4,528 km^2/1,748 sq mi), Canaries 2.15 million (7,545 km^2/2,913 sq mi), Cape Verde 568 thousand (4,033 km^2/1,557 sq mi), Madeira & Selvagens 251 thousand (794 km^2/306 sq mi), Azores 243 thousand (2,332 km^2/900 sq mi), and Seychelles 99 thousand (235 km^2/90 sq mi). These archipelagos surpassed even Hawai'i's [the largest as per surface area 16,636 km^2/6,423 sq mi] human population of 1.47 million residents. The big difference between Galápagos and the other island groups was the amount of land (wilderness) protected by law: Galápagos 96%, Madeira & Selvagens 67%, Canaries 40%, Mascarenes 25%, Azores 20%, Seychelles 16%, Hawai'i 5%, and Cape Verde 0.2%. Another difference was tourism; while Galápagos received 271 thousand visitors in 2019, the Canaries got 13.1 million, Hawai'i 10.4 million, Mascarenes 1.38 million, Madeira & Selvagens 1.15 million, Azores 972 thousand, Cape Verde 758 thousand, and Seychelles 428 thousand (Paz-y-Miño-C and Espinosa 2024b). Thus, highly populated archipelagos, from smaller to larger than Galápagos in surface area (i.e. from Seychelles to Hawai'i), with hundreds of thousands to millions of inhabitants (i.e. from Seychelles to Mascarenes), and with hundreds of thousands to millions of visitors per year (i.e. from Seychelles to Canaries or Hawai'i), have lost much of their wilderness. But not Galápagos; at least not yet! [However, see discussions in both *Arrivals of Life to the Galápagos* (Paz-y-Miño-C and Espinosa 2024a) and *A Synthesis of the Galápagos* (Paz-y-Miño-C and Espinosa 2024b)].

As for the question above, what is the human population carrying capacity of the Galápagos Islands—or of any volcanic, distant archipelago? The query can be puzzling since the examples of highly populated islands mentioned earlier [with millions of inhabitants and/or visitors per year] correspond to destinations with large cities, considerable urban and sub-urban infrastructure, highways and road networks, vast resorts, hotels, golf courts, and year-round crowded beaches with water sports or recreation (e.g. windsurfing, kiteboarding, jet-skiing, parasailing, wakeboarding, waterskiing, bumper-tube riding, fly-boarding with hydro-flight aqua-jet propulsion, or jet-pack flying). Plus, luxurious shopping malls, assortment of businesses, restaurants, and day-and-night entertainment. These destinations, however, are supplied with almost all needs from the mainland (goods and services; imports are high and costly). None are self-sufficient. Neither is Galápagos, much less developed (Paz-y-Miño-C and Espinosa 2024a,b).

In this chapter, as well as in *Arrivals of Life to the Galápagos* (Paz-y-Miño-C and Espinosa 2024a), we have stated that "when humans arrive, transformations of the environment follow". Hawai'i (2.1 times larger than Galápagos and the largest island group among those discussed in Paz-y-Miño-C and Espinosa 2024a) was an iconic example of what could happen to any remote, volcanic, oceanic archipelago: habitat alteration, fragmentation, species extinctions, and entire ecosystem losses. "Lessons to never forget", we alerted. But also, opportunities to gain experience and—when possible—prevent comparable outcomes elsewhere. Just like Galápagos, the archipelagos mentioned above (Seychelles, Madeira & Selvagens, Azores, Cape Verde, Mascarenes, Canaries, or Hawai'i), once imposing and with matchless wilderness, started their journeys into the Anthropocene with very few human settlers. They were, too, discovered as uninhabited islands, although at different times during history, and gradually became colonized. But Galápagos can still be different long-term due to its unique cognitive awareness of 21st-century conservation challenges: it knows its past and current struggles, its significance to the world, and understands scientifically the better path to take to secure its forever persistence (as per Paz-y-Miño-C and Espinosa 2024a,b).

Preliminary Data of the 2022 Census

By the time we completed this chapter (mid 2025), comprehensive data of the 2022 population census had not been released by Ecuador's National Institute of Statistics and Census (INEC). In 2024, INEC distributed a flyer with summarized numbers (*Resultados Principales Galápagos*; INEC 2024). The overall 2022 statistics corroborated the trends we have discussed in this chapter, as follows: The total population in the archipelago reached 28,583 inhabitants. Thus, it did increase by 3,339 individuals since 2015 (for comparison, refer to Fig. 6.1). The majority (60.3%) resided on Santa Cruz Island (N=17,233), followed by San Cristóbal (29.0%; N=8,300) and Isabela (10.7%; N=3,050). [Remember that Floreana's population (≈100) is counted as part of San Cristóbal's]. All other statistics summarized in the INEC's 2024 flyer were in accordance with the trends reported and discussed in this chapter. For example, age and sex structure diagrams [men 50.3%, women 49.7%, were consistent with Fig. 6.13], or the proportion of the population residing in urban [79.6%] or suburban [20.4%] areas [were consistent with data for previous census provided above]. The INEC 2024 flyer also summarizes data on housing, access to energy, drinking water, and waste disposal. We did not discuss these topics in this chapter, but we do examine them, as per census (2001, 2010, and 2015), in *A Synthesis of the Galápagos* (Paz-y-Miño-C and Espinosa 2024b).

As pointed out earlier, INEC had projected the Galápagos' population to reach 29,453 by 2015 and 33,042 by 2020. The projections did not materialize. Although the number of residents in the islands has grown since 2015 (by 3,339 individuals), the rate ["speed"] of population increase went down to 1.6% in 2022. Remember that it was 6.4% in 2001, 3.1% in 2010, and 1.8% in 2015 (Fig. 6.12). [Note that we did not have access to complete statistics (e.g. fertility or mortality rates; differential trends in immigration versus emigration) by the time of completing this chapter]. Now, readers should keep in mind that the Galápagos population is growing! In seven years (from 2015 to 2022), 3,339 individuals were added to the count. Human population growth is always a concern at any World Heritage Site, including Galápagos.

Conclusions

We minimize citations in this section to abridge the text. Each conclusion point summarizes topics substantiated by citations in this chapter.

(1) The Galápagos human population increased from 17,451 inhabitants in 2001 to 23,046 in 2010 and to 25,244 in 2015. [By 2022, it reached 28,583]. Ecuador's National Institute of Statistics and Census (INEC) had projected the population to reach 29,453 by 2015 and 33,042 by 2020. The projections did not materialize.

(2) Results of the 2022 census were partially released by the time we wrote this chapter. Here we report [in brackets] the summarized data provided by INEC (2024). INEC's 2022 numbers were in accordance with the statistical trends reported and discussed in this chapter (i.e. 2001, 2010, and 2015).

(3) The majority (81.4%) of the population resided in urban or suburban areas. About 7,000 Galápagos residents, as per ID status, lived elsewhere.

(4) From 1950 to 2015, the population increased 19-fold. The fastest rate occurred between 1982 (4.0%) and 2001 (6.4%). By 2010, the speed of population growth declined to 3.1% and to 1.8% in 2015. [It was 1.6% in 2022]. The boost during the 1980s-1990s overlapped with the tourism boom. The decline by 2010/2015 corresponded to enhanced age-cohort-specific emigration (ages 15/17-24), presumably to pursue higher education or jobs elsewhere.

(5) Despite the decline in growth rate from 2010 to 2015, the population experienced a net gain of 2,576 individuals during that time. It was driven by higher immigration (3,368 people) than emigration (792 people). [Between 2015 and 2022, the population increased by 3,339 individuals].

(6) According to growth-projection models (at rates of 2.0% or 3.75% per year), by 2020, the population was expected to reach either 38,366 (slower growth, sg) or 42,614 (faster growth, fg). By 2033, the projection was to reach 49,759 (sg) or 69,386 (fg) people. And by 2049, the islands were expected to home 68,524 (sg) or 126,429 (fg) inhabitants!

(7) Where humans go, agriculture follows, and alien species are introduced. Indeed, 3.2% of the Galápagos Islands is agricultural land, about 25,261 hectares/62,421 acres. 76% of its alien species (N=1,579 total; terrestrial plants 821 taxa) were introduced post-1970s, coinciding with the increase in human population, air and maritime travel practices, and tour-

ism flow. The most abundant and invasive taxa introduced by settlers to agricultural lands and—by default—to the islands included *Psidium guajava*, *Pennisetum purpureum*, *Cedrella odorata*, *Rubus niveus*, and *Cinchona pubescens*.

(8) When humans arrive, transformations of the environment follow. In contrast to other island groups discussed in this chapter, Galápagos remains well preserved: half of the archipelago (53%) is covered by native ecosystems; 43% corresponds to recent or old/er lava flows; 3% is agricultural land; and 1% is urban settings. But just like Galápagos, the archipelagos mentioned above (Seychelles, Madeira & Selvagens, Azores, Cape Verde, Mascarenes, Canaries, or Hawai'i), once imposing and with matchless wilderness, started their journeys into the Anthropocene with very few human settlers. They were, too, discovered as uninhabited islands, although at different times during history, and gradually became colonized. [Now, as alerted by scholars, "a central focus on population is neither necessary nor sufficient to ensure biodiversity conservation across diverse ecological contexts" (Hughes et al. 2023)]. But Galápagos can still be different long-term due to its unique cognitive awareness of 21st-century conservation challenges: it knows its past and current struggles, its significance to the world, and understands scientifically the better path to take to secure its forever persistence.

Questions

The following questions (Q1–3) merit generous written answers. You might need to do additional reading or library work to enrich your answers (we suggest exploring Paz-y-Miño-C and Espinosa 2024a,b).

Question 1: In the section "Colonization Led to Transformation of the Land", we stated, "Where humans go, agriculture follows". In Fig. 6.11 [Agroecosystems in the highlands of the human-inhabited Galápagos islands], you saw that Santa Cruz, San Cristóbal, Isabela, and Floreana had "agroecosystems" of variable size (farms + buffer zones). As noted in Fig. 6.11, only 3.2% of the Galápagos terrestrial habitats corresponded to agricultural land, while 96.8% of the remaining archipelago was "national park" (protected by law). [Note that buffer zones (4.4%) are part of the protected land (as per Laso et al. 2020)]. (a) Why do scientists, managers, and decision-makers [e.g. officials overseeing national parks and protected areas] continue to warn us that introduced invasive species associated with the occurrence and/or expansion of agriculture are a "threat to global biodiversity", particularly to archipelagos like Galápagos? (b) How can the introductions of invasive species to—arguably—only 3.2% of the Galápagos terrestrial habitats be of significant impact to the entire archipelago? (c) Do introductions of invasive species affect different islands differently? If so, how? Why? (d) If the local human population continues to increase [which implies more traveling and transport in and out of the islands, more supply shipments from continental Ecuador, and more tourism], what predictions can you make about imminent new introductions of invasive species to the Galápagos environments? (e) What can be done long-term to mitigate the impacts of introduced invasive taxa to a World Heritage Site like Galápagos?

Question 2: Assume that the human population in the Galápagos Islands continues to grow at a slow rate, as reported for the last two censuses (2015 and 2022). [Remember that it was 6.4% in 2001, 3.1% in 2010, 1.8% in 2015, and 1.6 in 2022. From 2010 to 2015, the population increased by 2,576 individuals; and from 2015 to 2022, the net gain was 3,339 individuals]. (a) What will be the needs for the growing population in terms of food supply [a significant share comes from continental Ecuador], education, housing, transport (vehicles), access to energy, drinking water, or waste-disposal? (b) What will be the environmental impacts resulting from additional food-imports from the continent, new education facilities, houses and buildings, vehicles (from bicycles to scooters, motorcycles, trucks, buses, and boats), higher energy demands, increasing drinking-water needs, or generation of more waste? (c) Will such environmental impacts be restricted to the 3.2% of the terrestrial habitats which are not part of the Galápagos National Park (above)? Why? (d) Examine Fig. 6.13 [Age-sex structure of the Galápagos Islands' human population]. Remember that the decline within age range 15/17 to 24, which was observed during four censuses (1990, 2001, 2010, and 2015), was likely associated with the pursue of education and work outside Galápagos (emigration). If new education institutions (technical colleges or universities) are created in Galápagos [something already happening to meet the demand for local jobs], one can assume that more young people (ages 15/17 to 24) will stay in the islands. Thus, fresh higher-education opportunities and increasing local employment offer will likely reduce emigration. What will be the *demographic* consequences of that?

Question 3: When briefly comparing Galápagos to other archipelagos (i.e. Seychelles, Madeira & Selvagens, Azores, Cape Verde, Mascarenes, Canaries, or Hawai'i), we stated, "…highly populated archipelagos, from smaller to larger than Galápagos in surface area (i.e. from Seychelles to Hawai'i), with hundreds of thousands to millions of inhabitants (i.e. from Seychelles to Mascarenes), and with hundreds of thousands to millions of visitors per year (i.e. from Seychelles to Canaries or Hawai'i), have lost much of their wilderness…" And then, we emphasized, "But not Galápagos; at least not yet!" (a–b) What leads you to believe, or not, that Galápagos will be the exception and remain—for decades or centuries—as an exemplar World Heritage Site?

How to Answer the Questions Above

The questions above have various possible answers (A1–3). Here we suggest perspectives from which to address the queries.

Answer 1: (a–e) Think of the associations between human population presence/growth [demographics] and the arrival and proliferation of alien invasive species. You might choose to limit your answers to such a viewpoint. But keep in mind that experts agree that the main sources of alien species introductions to islands are multifactorial (i.e. trade and transport, stowaway, recreation and tourism, demography and migration, escape from captivity, and contaminants like microbes or insects; Lenzner et al. 2020). Likewise, factors facilitating the *establishment* and the *spread* of alien species on islands are numerous, including land use/cover change, overall biodiversity loss and degradation, climate change, stowaway, escape from captivity, and corridors (e.g. roads, trails; as per Lenzner et al. 2020; Paz-y-Miño-C and Espinosa 2024b). These hints might help.

Answer 2: Answering Q2 (a–d) requires combining your knowledge about Galápagos demographics and the environmental implications of having a gradually growing human population in the islands. The rate of increase [the "speed"] of such growth can be higher [6.4% in 2001] or lower [1.6% in 2022] and have differential impacts [as per time scale] on the local environments. You should be able to envision the needs of a growing human population and the consequences of such growth for both the islands' residents themselves and the World Heritage Site they ought to protect (as per UNESCO's expectations of a WHS).

Answer 3: You should answer Q3 from opposing perspectives. (a) Yes, Galápagos will be the exception and remain—for decades or centuries—as an exemplar World Heritage Site (explain why and document rationale). (b) Galápagos will not be the exception and remain—for decades or centuries—as an exemplar World Heritage Site (explain why and document rationale). Now, archipelagos share similarities in origin (e.g. oceanic or continental), history of colonization, environmental resilience, culture and traditions of local populations, levels of socio-economic development, and conservation commitment. Thus, you could make inferences and predictions based on similarities among archipelagos as well as Galápagos' uniqueness. Some library work will be needed. Our Galápagos books (Paz-y-Miño-C and Espinosa 2024a,b) will certainly help; however, you will need to go beyond them.

Discussion Topics

The discussion topics below (D1–D2) should be addressed with your peers. We encourage college instructors to facilitate the dialogue.

Discussion 1: Conservation scientists often remind us that the most effective preservation and management of protected areas is accomplished when the local population [i.e. those living nearby or within the very ecosystems that nations and international organizations aim to safeguard] participates in all aspects of the planning and execution of conservation programs. Investigate and discuss why, despite the observable conservation accomplishments in the Galápagos (1960s onwards), it continues to be a challenge to bring the local residents together and fully commit—*in praxis*—to the *forever* preservation of the Galápagos National Park and the Galápagos Marine Reserve. Keep in mind that the local population, which continues to grow [this chapter], has its own subsistence needs from employment to housing, purchasing food, health care, education, recreation, transport, access to energy, drinking water, or waste-disposal. Although tourism is central to the Galápagos economy (Chapter Seven), so are subsistence fishing and agriculture to the local population.

Discussion 2: Climate change is expected to have multiple impacts on the fragile ecosystems of islands (both terrestrial and aquatic). Discuss how climate change will influence the *demographics* of the Galápagos population. Note that extreme weather (severe hot-rainy and cool-dry seasons) and sea-level rise will bring stress to human settlements (i.e. cities, towns, rural villages, and farmers' houses and storage facilities). Frequent rainstorms will affect buildings, harbors, homes, and infrastructure connectivity (runways, roads, and bridges); sewage pipes will break and lead to contamination of the drinking water. Food supply to the islands, either coming from the continent (e.g. cargo ships) or being distributed within the archipelago (i.e. small boats), will be affected by extreme weather (e.g. more turbulent seas will lead to longer journeys, extra use of fuel, and higher market prices for goods and supplies). Remember that your discussion should be about the impacts of climate change on Galápagos' *demographics* (i.e. human population as per Figs. 6.12–6.14, 6.16–6.17). Here are some readings that will help you enrich your knowledge and discussion: Paz-y-Miño-C and Espinosa (2024b), Nguyen et al. (2023), Hauer et al. (2020), and Zickgraf (2020).

Suggested Observation Project

The suggested observation project below (P1) takes into consideration that you will travel to Ecuador and the Galápagos legally; that you will visit the islands as a tourist [holding a Transit Control Card, a mandatory document issued by Ecuador and specifically by the Governing Council of Galápagos]; and that you will respect the regulations that protect the Galápagos National Park and the Marine Reserve (see complete list of rules in Appendix B). Only accredited scientists, academics, and/or naturalists who have obtained official permits can do research in Galápagos. You can take pictures, but flash photography is not permitted when taking photos of wildlife. Professional photography and videos recorded for commercial purposes must be authorized by the Galápagos National Park Directorate (GNPD). The ideas below are for enthusiastic college explorers interested in learning about the archipelago.

Project 1: Reconnaissance of the urban and suburban areas in the human-populated islands of the Galápagos—call it *A Town Tour* or *A Series of Town Tours* or even *A One-Town-A-Day Experience*. This activity should be done in a group of 3–4 people. Never do it alone! For it, you can adopt two approaches: bikes and/or a taxi ride. Bikes can be rented for a few dollars a day, together with protective gear, at the local towns, including Puerto Ayora (Santa Cruz Island), Puerto Baquerizo Moreno (San Cristóbal), Puerto Villamil (Isabela), or Puerto Velasco Ibarra (Floreana). Taxis can be contacted daily and prearranged for your journey. Some drivers might look puzzled about what your group is requesting. They may ask, "Do you want a town tour"? Explain that your team is interested in learning about the picturesque Galápagos towns. Your hotel can also assist and get in touch with a driver. If your team feels more comfortable hiring a licensed naturalist guide for this purpose, it will add confidence to your journey. Be smart and discuss exactly what your group wants and agree on fees. Here is the activity: Bike around and/or ask your taxi driver to show you as much as possible of the local towns. If biking, you will be in control of the ride, time allocation to each stop, or route change when necessary. If you hire a taxi, there will be less flexibility, although your team will move faster from one point to another. Thus, we recommend combining bike rides with taxi rides. One whole-day excursion per town will probably be enough. To plan the scouting, refer to the maps in Figs. 6.3–6.4, 6.6–6.10. In addition, get hard copies of satellite images—from Google Maps—of the human settlements on the islands; do it in advance (access to the internet will be limited while outdoors). Bring with you a notebook and a cell phone camera. Your team's goals should be: (i) *In situ* reconnaissance of the Galápagos towns (Puerto Ayora, Puerto Baquerizo Moreno, Puerto Villamil, and Puerto Velasco Ibarra) and their proximate suburban areas (as in Figs. 6.3–6.4, 6.6–6.10). (ii) Become knowledgeable about the Galápagos urban and suburban settlements. (iii) Integrate what you have learned from this chapter about the Galápagos' demographics with the "on the ground reality" of the human-inhabited islands. And (iv) enjoy the picturesque character of the Galápagos towns. Ask the locals for feedback and directions, e.g., where are the schools, hospitals, fire stations, solar-panel fields, gas stations, government offices, runways, outdoor markets, heliports, power plants, stores, loading docks, stadiums, recycling facilities, coffee shops, waste disposal fields, parks, churches, restaurants, neighborhoods, monuments, recreation sites, fish markets, gyms, pharmacies, and all that comes to mind? You will be overwhelmed by the amount of information you can gather. Plus, it will be an excellent way to really get to know the towns. Master their settings by intelligently exploring them. Take notes and photos, document the location of all identified sites on your maps, and organize the information thoroughly. Upload your observations to an Excel file and turn the data into descriptive statistics [i.e. totals of this or that and the percentage representation of your items; be creative]. Put together a slideshow presentation (for a 50-minute talk) to be shared with a general audience. Avoid slides with "text only" [they are often terrible]. Keep it rigorous yet fun and didactic. You can also upload amateur videos to the web and share them with the world. [But remember, professional photography and videos recorded for commercial purposes must be authorized by the GNPD]. Remember to observe the law, be respectful, and be safe!

References

CGREG (Consejo de Gobierno del Régimen Especial de Galápagos). 2016. Plan Galápagos 2015-2020: Plan de Desarrollo Sustentable y Ordenamiento Territorial del Régimen Especial de Galápagos. Puerto Baquerizo Moreno: CGREG.

Cruz, David, and Charlotte Causton. 2008. "Air Traffic to Galapagos is Increasing". In Galapagos Report 2006-2007, edited by Charles Darwin Foundation (CDF), Galápagos National Park (GNP), and Galápagos National Institute (INGALA), 48-54. Puerto Ayora: CDF, GNP and INGALA.

GAD-SC (Gobierno Autónomo Decentralizado de Santa Cruz). 2012. "Sistema de Acentamientos Humanos". In Plan de Desarrollo y Ordenamiento Territorial del Cantón Santa Cruz, edited by GAD-SC, 215-341. Puerto Ayora: GAD-SC.

Hauer, Mathew E., Elizabeth Fussell, Valerie Mueller, Maxine Burkett, Maia Call, Kali Abel, Robert McLeman, and David Wrathall. 2020. "Sea-Level Rise and Human Migration". Nature Reviews Earth & Environment, No. 1: 28-39.

Hughes, Alice C., Kévin Tougeron, Dominic A. Martin, Filippo Menga, Bruno H.P. Rosado, Sebastian Villasante, Shweta Madgulkar, Fernando Gonçalves, Davide Geneletti, Luisa Maria Diele-Viegas, Sebastian Berger, Sheila R. Colla, Vitor de Andrade Kamimura, Holly Caggiano, Felipe Melo, Marcelo Guilherme de Oliveira Dias, Elke Kellner, and Edivando Vitor do Couto. 2023. "Smaller Human Populations Are Neither a Necessary Nor Sufficient Condition for Biodiversity Conservation". Biological Conservation, No. 277: 109841.

Idrovo, Hugo. 2005. Galapagos: Footsteps in Paradise. Quito: Libri Mundi.

INEC (Instituto Nacional de Estadísticas y Censos). 2015. Análisis de Resultados Definitivos Censo de Población y Vivienda Galápagos 2015. Quito: INEC.

—. 2020. "¿Cómo Crecerá la Población en Ecuador?" Accessed March 30, 2023.
https://www.ecuadorencifras.gob.ec/documentos/web-inec/Poblacion_y_Demografia/Proyecciones_Poblacionales/presentacion.pdf.

—. 2024. Resultados Principales Galápagos. Quito: INEC.

Laso, Francisco J., Fátima L. Benítez, Gonzalo Rivas-Torres, Carolina Sampedro, and Javier Arce-Nazario. 2020. "Land Cover Classification of Complex Agroecosystems in the Non-Protected Highlands of the Galapagos Islands". Remote Sensing, No. 12: 65.

Lenzner, Bernd, Guillaume Latombe, César Capinha, Céline Bellard, Franck Courchamp, Christophe Diagne, Stefan Dullinger, Marina Golivets, Severin D. H. Irl, Ingolf Kühn, Brian Leung, Chunlong Liu, Dietmar Moser, Núria Roura-Pascual, Hanno Seebens, Anna Turbelin, Patrick Weigelt, and Franz Essl. 2020. "What Will the Future Bring for Biological Invasions on Islands? An Expert-Based Assessment". Frontiers in Ecology and Evolution, No. 8: 280.

Luna-Tobar, Alfredo. 1997. Historia Política Internacional de las Islas Galápagos. Quito: Ediciones Abya Yala.

Miller, Joshua M., Maud C. Quinzin, Nikos Poulakakis, James P. Gibbs, Luciano B. Beheregaray, Ryan C. Garrick, Michael A. Russello, Claudio Ciofi, Danielle L. Edwards, Elizabeth A. Hunter, Washington Tapia, Danny Rueda, Jorge Carrión, Andrés A. Valdivieso, and Adalgisa Caccone. 2017. "Identification of Genetically Important Individuals of the Rediscovered Floreana Galápagos Giant Tortoise (Chelonoidis elephantopus) Provides Founders for Species Restoration Program". Scientific Reports, No. 7: 11471.

MINTUR. 2018. Observatorio de Turismo de Galápagos, Estadísticas del Turismo en Galápagos. Puerto Ayora: Ministerio de Turismo del Ecuador.

Nguyen, Trung Thanh, Ulrike Grote, Frank Neubacher, Dil B. Rahut, Manh Hung Do, and Gokul P. Paudel. 2023. "Security Risks from Climate Change and Environmental Degradation: Implications for Sustainable Land Use Transformation in the Global South". Current Opinion in Environmental Sustainability, No. 63: 101322.

Paz-y-Miño-C, Guillermo, and Avelina Espinosa. 2024a. Arrivals of Life to the Galápagos: From Berlanga to Darwin, Melville, and World War II. Newcastle upon Tyne: Cambridge Scholars Publishing.

—. 2024b. A Synthesis of the Galápagos: Natural History, Human Conflict, and the Destiny of an Archipelago. Newcastle upon Tyne: Cambridge Scholars Publishing.

Rivas-Torres, Gonzalo, Fátima L. Benítez, Danny Rueda, Christian R. Sevilla, and Carlos F. Mena. 2018. "A Methodology for Mapping Native and Invasive Vegetation Coverage in Archipelagos: An Example from the Galápagos Islands". Progress in Physical Geography, No. 42: 83-111.

Toral-Granda, M. Verónica, Charlotte E. Causton, Heinke Jäger, Mandy Trueman, Juan Carlos Izurieta, Eddy Araujo, Marilyn Cruz, Kerstin K. Zander, Arturo Izurieta, and Stephen T. Garnett. 2017. "Alien Species Pathways to the Galapagos Islands, Ecuador". PLoS ONE, No. 12: e0184379.

Vásconez, Sigrid, Edwin Naula, and Juan Carlos Izurieta. 2021. Análisis Socioeconómico del Proyecto para la Restauración Ecológica de la Isla Floreana. Puerto Ayora: Dirección del Parque Nacional Galápagos.

Wittmer, Margret. 1961. Floreana Adventure. New York: E. P. Dutton & Co., Inc.

Woram, John M. 2005. Charles Darwin Slept Here. Rockville Centre: Rockville Press Inc.

Zickgraf, Caroline. 2020. "Climate Change, Slow Onset Events and Human Mobility: Reviewing the Evidence". Current Opinion in Environmental Sustainability, No. 50: 21-30.

CHAPTER SEVEN

WHY IS NATURE TOURISM ESSENTIAL TO GALÁPAGOS?

> Our current generations could be the last to explore and sightsee what is left of pristine nature; its fragmented habitats; its surviving ecosystems. Soil you can still cultivate; air you can still breathe; water you can still drink. Its native peoples and their cultures; its biodiversity; its calm weather and pleasant seasons. The imminent Anthropocene, however, will bring a contrasting experience to our descendants. Climate change and its sequels will be their foremost challenge. In hindsight, and as both biologists and global travelers, we have had the privilege to explore and sightsee what was left to us as "pristine nature". —GPC and AE.

Forty years ago, as high school graduates, we visited Galápagos for the first time. We were among ≈20,000 yearly tourists at that time. In retrospect, exploring and sightseeing on the islands turned us into biologists. Few places on our planet can have such an impact. Later in our lives, we have tried to pass on our passion for Galápagos to students.

Galápagos' subsistence is powered by tourism. In 2018—a record-high year before the COVID-19 pandemic, which plummeted global travel and collapsed exploring or sightseeing the islands—275,817 visitors arrived in the archipelago. Each day, 756 new people walked the towns, hiked the volcanoes, or navigated the marine reserve; each week, 5,304 people; and each month, 22,985! As shown in Fig. 7.1, 76% of the travelers arrived via Seymour-Baltra-GPS and 24% via San Cristóbal-SCY airports. By 2019, total arrivals dropped to 271,238 visitors (negative 4,579 with respect to the previous year). By 2020, a COVID-19 pandemic year, arrivals decreased to 72,519 visitors (3.8 times below 2018). By 2021, arrivals had partially recovered to 136,336 visitors (49.4% with respect to 2018). And in 2022, they reached 267,688 visitors (97% with respect to 2018). An expected post-pandemic boost occurred in 2023, with an all-time high of 329,477 arrivals! During 2024, the numbers decreased to 279,277 visitors, which suggested tourism trends were normalizing after the COVID-19 pandemic (data Galápagos National Park Directorate; El Comercio 2025).

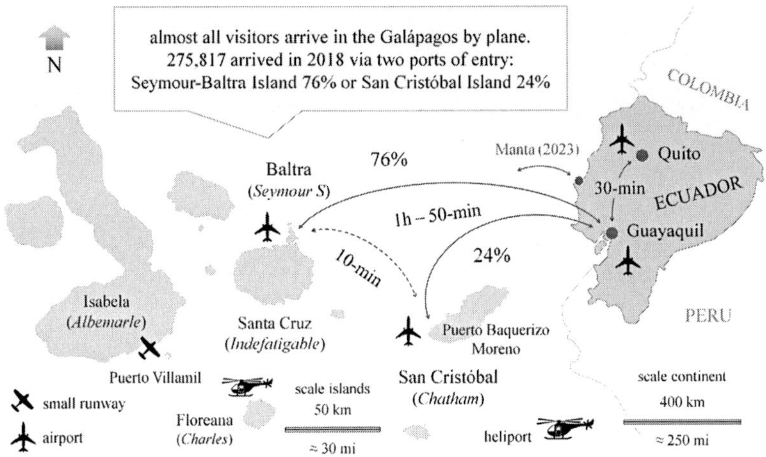

Fig. 7-1 Total number of visitors arriving in Galápagos in 2018: 275,817 (76% arrived via Seymour-Baltra-GPS and 24% via San Cristóbal-SCY airports). By 2019, total arrivals dropped to 271,238 visitors. By 2020, a COVID-19 pandemic year, arrivals decreased to 72,519 visitors. By 2021, arrivals recovered to 136,336 visitors. And in 2022, they reached 267,688 visitors. An expected post-pandemic boost occurred in 2023, with a record of 329,477 arrivals. In 2024, the numbers decreased to 279,277 visitors, suggesting tourism trends were normalizing after the COVID-19 pandemic (data Galápagos National Park Directorate; El Comercio 2025). Commercial flights usually arrive in the archipelago (GPS and/or SCY) via Quito UIO–Guayaquil GYE (30-min flight with stop in Guayaquil GYE) continuing to Galápagos GPS/SCY (1h-50-min) or direct flights Guayaquil GYE-Galápagos GPS/SCY (1h-50-min). In 2023, commercial flights Manta MEC-Galápagos GPS/SCY (1h-25-min) were approved to land 2-times/week [but later discontinued]. Layout: GPC, after Dirección del Parque Nacional Galápagos and Observatorio de Turismo de Galápagos (2018), Dirección del Parque Nacional Galápagos (2019, 2020, 2021, 2022), and CGREG (2023a). For location of runways in the human-populated islands of Galápagos, see Fig. 6.2 in Chapter Six.

In this chapter, we rely on data from 2018/2019 as the latest to discuss Galápagos' tourism (i.e. exemplar years in history under the regular and expected tourism growth rate of the recent past). COVID-19 years 2020 to 2022, and even 2023 (a record-high recovery year), were unusual since tourism collapsed during the pandemic. Note that 2024 was, apparently, a "return to normal year". Below, we do contrast 2018/2019 vs. the unusual years but focus our attention on 2018/2019 for extrapolations: from tourism trends to most visited sites per island (Figs. 7.3-7.9), to revenue generated by tourism, and to administrative expenses by CGREG, the government agency that manages the islands (Figs. 7.10-7.13).

Fig. 7-2 The "expedition experience" while visiting Galápagos (small groups aiming at low environmental impact). Photos: G. Paz-y-Miño-C.

The "expedition experience" while visiting Galápagos is a popular format to learn about the archipelago (Fig. 7.2). Groups of maximum 16 travelers accompanied by a naturalist guide sightsee the land or seascapes. Exploring various islands requires navigation in boats or larger vessels. Visitors hike along trails (lowlands or highlands of the islands), stop by interpretation centers (e.g. facilities around the Charles Darwin Research Station in Santa Cruz Island or the Gianni Arismendy Environmental Interpretation Center in San Cristóbal Island) or giant tortoises' captive-breeding centers in the human-populated islands [for locations of environmental interpretation and tortoises' captive-breeding centers, see Figs. 6.2-6.10 in Chapter Six], or snorkel and spend time at the beach. All group activities within Galápagos National Park are supervised by a licensed naturalist, although tourists can freely visit the towns and/or travel on their own to selected iconic sites (e.g. nature trails, some volcanoes' tops, lava caves, lagoons, recreational beaches, private farms, and coffee plantations; as per Paz-y-Miño-C and Espinosa 2024a,b).

Tourism's Trends

In Fig. 7.3, we summarize arrivals of visitors to Galápagos during the past three decades (1990s, 2000s, and 2010s). From 1989 to 2018, the number of visitors to Galápagos grew more than six times: 42,000 to 276,000 (rounded numbers; 6.71% compound annual growth rate for 30 years). Note the impact of the COVID-19 pandemic in 2020 (73,000 visitors), its halfway recovery by 2021 (136,000 visitors), and its 97% recovery by 2022 (268,000 visitors). [Growth rates not in figure: 1980s=9.16%; 1990s=4.46%; 2000s=9.59%; and 2010s=3.04%].

How did local tourism (i.e. from Ecuador) versus international tourism compare? In 2018, visitors from Ecuador and the United States comprised 62% of all arrivals to the islands (Fig. 7.4 top). By 2019, the share of both countries remained comparable (62%). During the COVID-19 pandemic's most severe years (2020, 2021), the share of both countries changed: 66% in 2020; 84% in 2021. And in 2022, it was 75%. In 2018, >50% of visitors from United States, United Kingdom, Canada, and Australia stayed on board vessels while exploring Galápagos. By contrast, >50% of visitors from Ecuador, Germany, Argentina, France, Netherlands, and Spain stayed at hotels or lodges (Fig. 7.4 bottom). A comparable trend of staying "on board of vessels" vs. "at hotels or lodges" remained for these countries up to 2022. From 1990s to 2000s to 2010s, the number of visitors staying at hotels or lodges increased significantly, as did the availability of hotels and lodges in the islands (DPNG and OTG 2018; DPNG 2019, 2020,

2021, 2022). In 2019, 72% of visitors stayed at hotels or lodges, while 28% stayed on board vessels (ESPOL 2021).

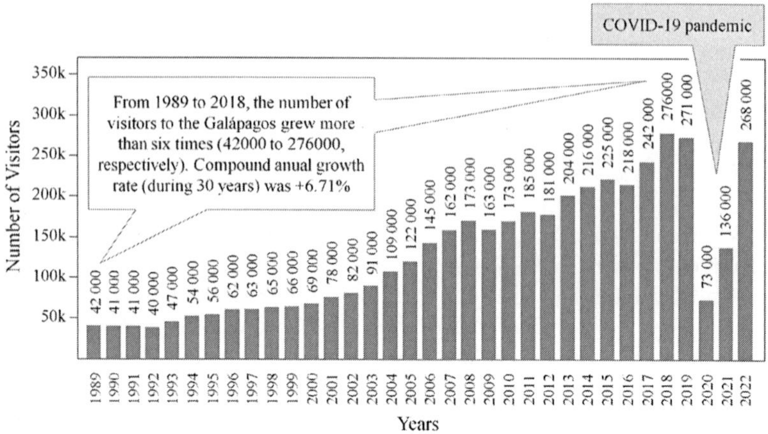

Fig. 7-3 Arrivals of visitors to Galápagos during the past three decades (rounded numbers). Note the impact of the COVID-19 pandemic. Layout: GPC; redrawn and modified after Dirección del Parque Nacional Galápagos and Observatorio de Turismo de Galápagos (2018) and Dirección del Parque Nacional Galápagos (DPNG 2019, 2020, 2021, 2022). As noted in this chapter, 329,477 visitors arrived in 2023 (a record high year) and 279,277 in 2024 (a "return to normal year"; data Galápagos National Park Directorate; El Comercio 2025).

How did Galápagos tourism differ with respect to other protected areas in Ecuador? Although by 2015 Galápagos received more visitors (>200,000) than other protected areas in Ecuador, its tourism compound annual growth rate was the lowest (3.72%) during a 9-year period (2007 to 2015; Fig. 7.5). In other words, tourism at protected areas in continental Ecuador grew at a faster pace than in Galápagos, as follows: Cotopaxi National Park 4.63% (2007-2015); Cotacachi-Cayapas Ecological Reserve 5.83%; Yasuní National Park 6.40%; Cuyabeno Wildlife Reserve 9.97%; Cajas National Park 13.77%; Cayambe-Coca Ecological Reserve 17.85%; and Machalilla National Park 20.24% (Izurieta 2017).

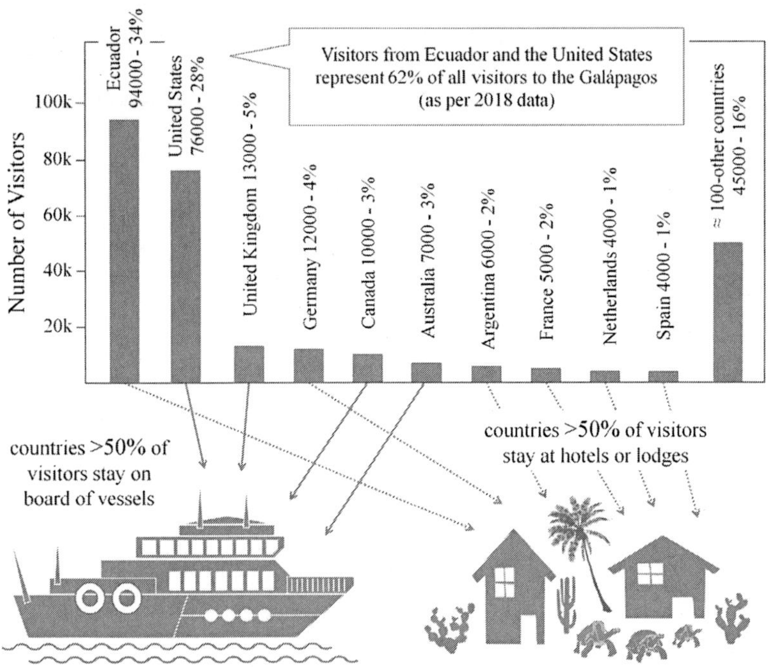

Fig. 7-4 Visitors to the Galápagos Islands: top 10 countries. In 2018, visitors from Ecuador and the United States represented 62% of all arrivals to the islands (N_T=275,817 total visitors; 34% Ecuador; 28% US). By 2019, the share of both countries (62%) remained comparable (N_T=271,238; 33% Ecuador; 29% US). During the COVID-19 pandemic's most severe years (2020, 2021), the share of both countries changed: in 2020 was 66% (N_T=72,519; 42% Ecuador; 24% US); in 2021 was 84% (N_T=136,336; 59% Ecuador; 25% US). In 2022, a recovery year, was 75% (N_T=267,688; 46% Ecuador; 29% US). Note how, in 2018, >50% of visitors from United States, United Kingdom, Canada, and Australia stayed on board vessels while exploring Galápagos. By contrast, >50% of visitors from Ecuador, Germany, Argentina, France, Netherlands, and Spain stayed at hotels or lodges. A comparable trend of staying "on board of vessels" vs. "at hotels or lodges" remained for these countries up to 2022. From 1990s to 2000s to 2010s, the number of visitors staying "at hotels or lodges" has increased significantly, as well as the availability of hotels/lodges. Layout: GPC; data from Dirección del Parque Nacional Galápagos and Observatorio de Turismo de Galápagos (2018), and Dirección del Parque Nacional Galápagos (2019, 2020, 2021, 2022).

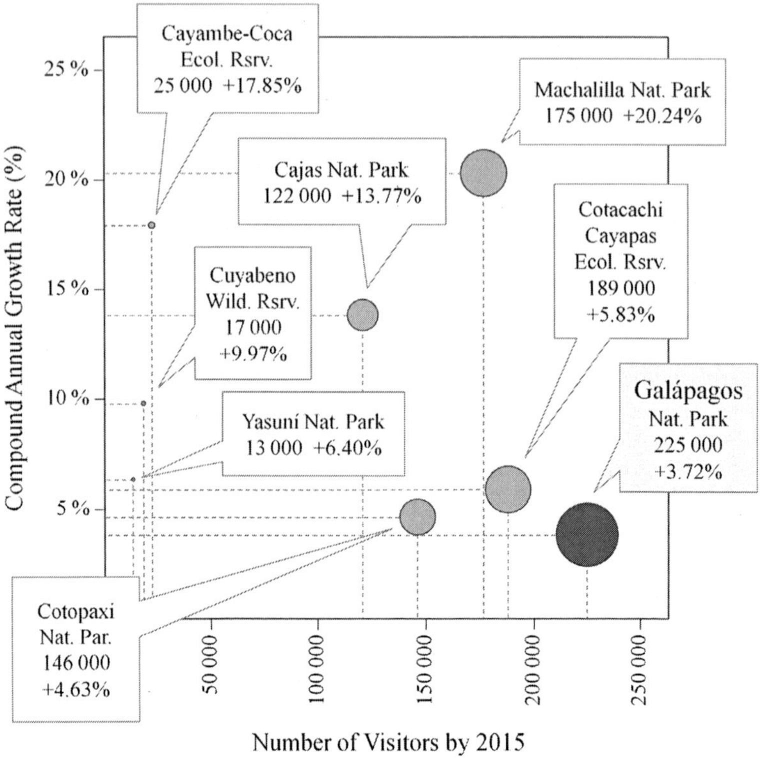

Fig. 7-5 Compound annual growth rate of visitors to the Galápagos National Park in comparison to other protected areas in Ecuador (2007-2015). Although by 2015 Galápagos received more visitors (>200,000) than other protected areas in Ecuador, its tourism compound annual growth rate was the lowest. [For map of national parks and protected areas in continental and insular Ecuador, see Fig. 5.18; for map of the Galápagos Marine Reserve in the context of the Tropical Eastern Pacific, see Fig. 5.19; both figures in Chapter Five]. Note in this figure that the value of +3.72% for Galápagos differs with respect to +6.71% growth rate in Fig. 7.3; this is because the former refers to a 9-year period (2007-2015), while the latter to a 30-year period (1989-2018). Layout: GPC; redrawn and modified after Izurieta (2017).

How old were the tourists visiting Galápagos? Women or men? In 2018, the average age of Ecuador's visitors to Galápagos was 34 years old (54% women; 46% men), while other countries was 45 y/o (55% W; 45% M). Note that 43% of the Ecuadorian tourists were 21-40 y/o, while the age distribution of other countries was bimodal (Fig. 7.6).

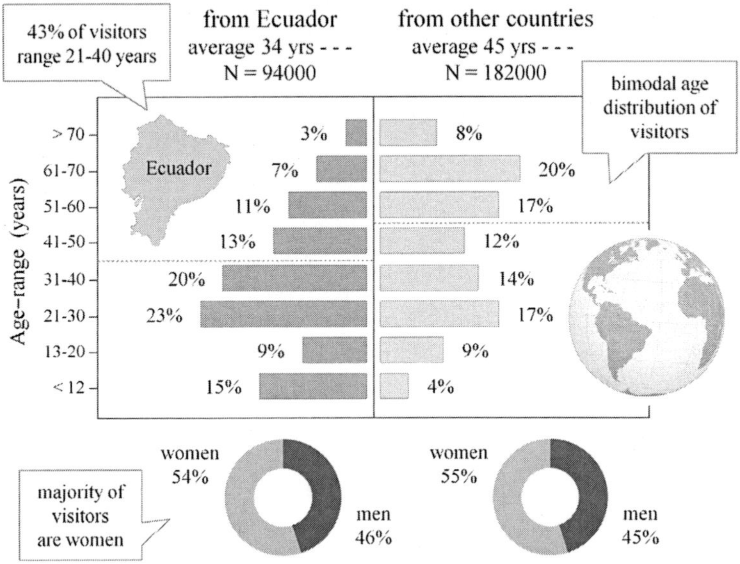

Fig. 7-6 Age and sex distribution of visitors to the Galápagos Islands: Ecuador vs. other countries of origin; data 2018. Note that in 2019, average age of Ecuador's visitors was 34 years old (55% women; 45% men), while other countries was 42 y/o (54% W; 46% M). In 2020, average age of Ecuador's visitors was 34 y/o (54% women; 46% men), while other countries was 48 y/o (53% W; 47% M). In 2021, average age of Ecuador's visitors was 35 y/o (55% women; 45% men), while top nine other countries was 41 y/o (48% W; 52% M). And in 2022, average age of Ecuador's visitors was 36 y/o, while other countries was 45 y/o (no data available on sex ratio). Layout: GPC; redrawn and modified after Dirección del Parque Nacional Galápagos and Observatorio de Turismo de Galápagos (2018) and Dirección del Parque Nacional Galápagos (2019, 2020, 2021, 2022).

Is Galápagos tourism affected by seasonality? As shown in Fig. 7.7a, visitors' monthly arrivals in Galápagos fluctuate during a given year, apparently coinciding with: (i) summer in the Northern Hemisphere (source of the majority of international visitors in July-August); (ii) Ecuadorian tourism to the islands, which tends to increase in July-August-September (vacation time in part of the country); and (iii) local seasonality (e.g. less arrivals in September during the islands' cool-dry season and more arrivals in February-March during the hot-rainy season). [We discussed Galápagos seasonality in Chapter One, Figs. 1.43-1.44]. Note in Fig. 7.7a that during three exemplar years (2016, 2017, 2018), there was a significant increase in visitors' monthly arrivals. Notably, during 2018, the

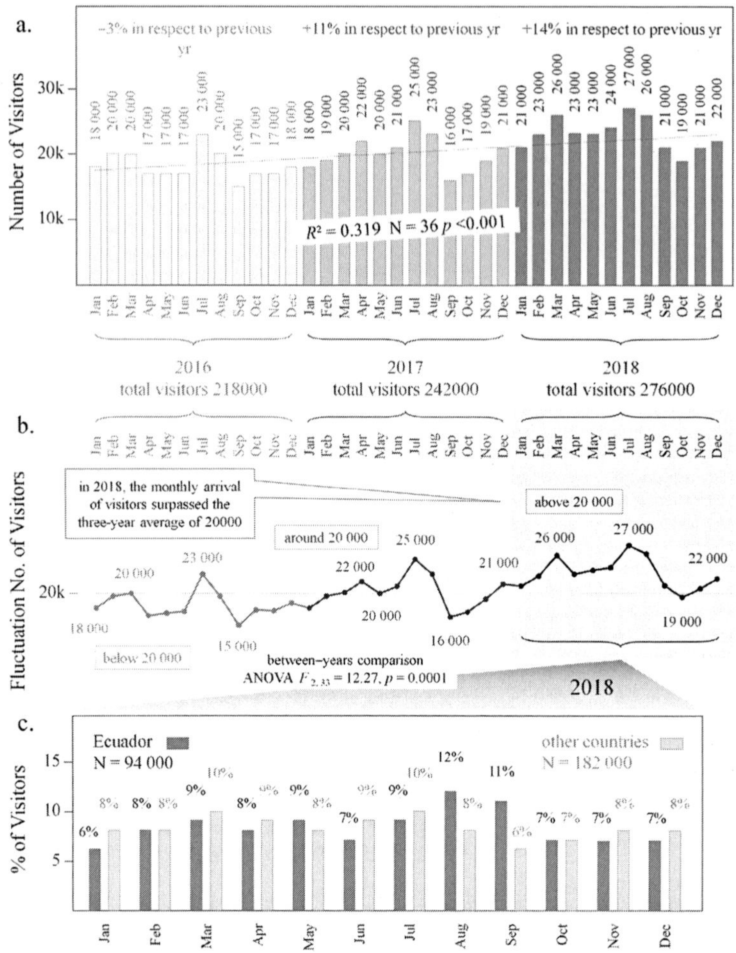

Fig. 7-7 Fluctuations of visitors' monthly arrivals in the Galápagos Islands from 2016 to 2018. a. For three years (2016, 2017, 2018), there was a significant increase (as per trend) in visitors' monthly arrivals to the archipelago (R^2=0.319, N=36, p<0.001). b. During 2018, the monthly arrival of visitors surpassed the three-year average of 20,000 (2016 was a below-average year; 2017 around average; and 2018 above average). Both 2016 and 2017 were statistically different from 2018, but not from one another; thus, 2018 contributed most of the difference (ANOVA $F_{2,33}$=12.27, p=0.0001; Holm-Sidak post-hoc test for pairwise year comparisons, p<0.05). c. Ecuador's vs. other countries' monthly visitors' fluctuation during 2018. Layout: GPC; redrawn and modified after Dirección del Parque Nacional Galápagos and Observatorio de Turismo de Galápagos (2018).

monthly arrival of travelers surpassed the three-year average of 20,000 (Fig. 7.7b); 2016 was a below-average year; 2017 was around average; and 2018 was above average. Both 2016 and 2017 were statistically different from 2018 but not from one another; thus, 2018 contributed most of the difference (Fig. 7.7b). Ecuador's vs. other countries' monthly visitors to Galápagos did fluctuate during 2018, but with minor differences between the two, except for August-September when Ecuadorian travelers surpassed other countries' (Fig. 7.7c).

While staying in Galápagos, where do visitors go? Almost everywhere (see Fig. 7.8; as per Paz-y-Miño-C and Espinosa 2024b).

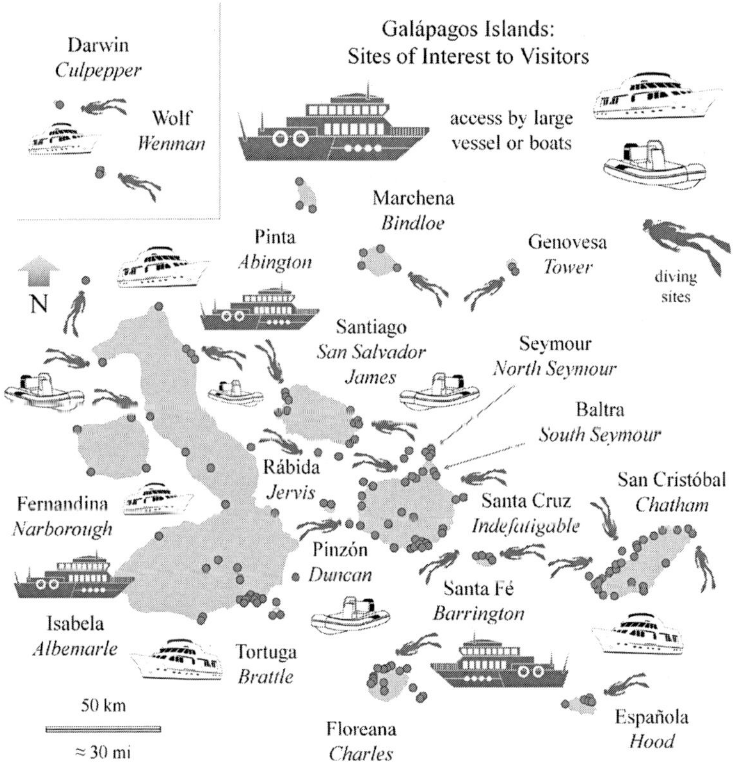

Fig. 7-8 Sites of interest to visitors to the Galápagos. The human-populated islands (Santa Cruz, San Cristóbal, Isabela, and Floreana) offer the most-visited sites (inland or coastal). Layout: GPC; redrawn/modified after Ministerio de Turismo del Ecuador et al. (2018) with data input from Charles Darwin Foundation and World Wildlife Fund Ecuador (CDF and WWF-Ecuador 2018); popular diving sites after Rosenberg and Sarbone (2004).

In Fig. 7.8, we depict some popular sites of interest to visitors while exploring/sightseeing the islands. The human-populated islands (Santa Cruz, San Cristóbal, Isabela, and Floreana) offer the most-visited sites (complete analysis in Paz-y-Miño-C and Espinosa 2024b).

Tourism Under Diverse Growth-Rate Scenarios

In Fig. 7.9, we sketch the projected number of visitors to the Galápagos Islands under diverse growth-rate scenarios (emphasis 2012 to 2033; after Walsh and Mena 2016 with input from Fig. 7.3 as well as reports from the Galápagos National Park Directorate). Examine Fig. 7.9 as follows: (1) Locate the actual tourism growth trend from 1979 to 2012. (2) Spot the diverse growth rate scenarios departing from 2012 (solid/dashed/dotted lines) and reaching 2033. (3) Realize that the gray dots correspond to actual

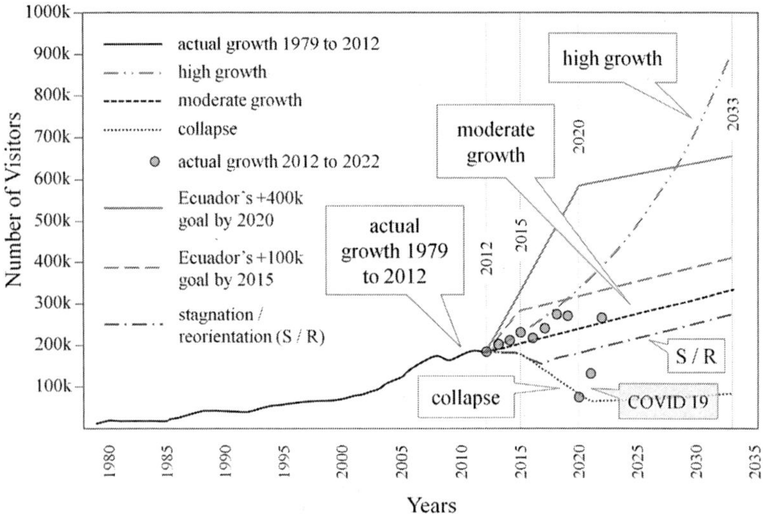

Fig. 7-9 Projected number of visitors to the Galápagos Islands under diverse growth-rate scenarios (emphasis 2012 to 2033). Gray dots correspond to actual growth during eleven consecutive years (2012-2022), including the most severe COVID-19-impacted years (2020 and 2021); the latter matched the growth-rate scenario under a possible "collapse" in tourism flow, particularly in 2021, but tourism recovered much faster than suggested by the model (dotted line). Layout GPC; redrawn and modified after Walsh and Mena (2016). Data on actual growth of tourism from 2012 to 2022 comes from Fig. 7.3 (after Dirección del Parque Nacional Galápagos and Observatorio de Turismo de Galápagos 2018; Dirección del Parque Nacional Galápagos 2019, 2020, 2021, 2022).

data on tourists' arrivals from 2012 to 2022 (as per Fig. 7.3). (4) Under high-growth scenario, tourists' arrivals to Galápagos could reach +300k [thousand] by 2020 and +850k by 2033. (5) Under moderate growth scenario, arrivals could reach ≈250k by 2020 and ≈350k by 2033. (6) Ecuador's goal to reach +400k visits by 2020 did not materialize (i.e. the actual numbers even before the COVID-19 pandemic did not match such goal). (7) Ecuador's goal of reaching, by 2015, +100k visits over the 181,000 registered in 2012 did not materialize either, although by 2018, the actual numbers reached 276,000 visitors [but for the next three years, the COVID-19 pandemic disrupted such a growing trend]. (8) A tourism collapse scenario (dotted line) did match with accuracy the sequels of the COVID-19 pandemic; by 2020, only 73,000 visitors arrived; but tourism recovered much faster than suggested by the model: 136,000 visitors arrived in 2021 and 268,000 in 2022 [as indicated above, 329,477 visitors arrived in 2023 (a record high year) and 279,277 in 2024 (a "return to normal year"; data Galápagos National Park Directorate; El Comercio 2025)]. (9) The stagnation/reorientation scenario (S/R) has not materialized thus far. (10) Despite the impact of the COVID-19 pandemic, the actual arrivals from 2012 to 2018, plus the fast recovery by 2022 and subsequent 2023/2024, suggest Galápagos' tourism is growing at a faster rate than a "moderate-growth scenario" and just below the rate of a "high-growth scenario"; the latter needs confirmation during the coming years (Paz-y-Miño-C and Espinosa 2024b).

Tourism's Revenue: Galápagos in the Context of Ecuador

How profitable is Galápagos' tourism? How does it contrast with respect to nature-oriented tourism in other protected areas of Ecuador? By 2019 [before the COVID-19 pandemic and a comparable year to 2018 as per tourism-related growth trends], Ecuador's gross domestic product (GDP) was \$108.1 billion USD. The country ranked 8th in Latin America and 69th worldwide (Fig. 7.10 top). Tourism was Ecuador's 4th source of income, with \$2.28 billion USD, after crude oil (\$7.78 billion USD); shrimp, fish, and fish products (\$4.19 billion USD); and bananas and plantains (\$3.29 billion USD). Tourism's revenue with respect to the previous year grew from \$0.49 billion USD (0.8%) in 2006 to \$2.27 billion USD (12.9%) in 2018, and to \$2.28 billion USD (0.4%) in 2019. It collapsed to \$0.70 billion USD (-69.2%) in 2020 (Fig. 7.10 bottom; data from MINTUR 2017; ASOBANCA 2019; IMF 2020; ESPOL 2021; CEIC 2023; OEC 2023; World's Top Exports-Ecuador 2023).

In 2019, the Galápagos National Park alone generated $356 million USD, ≈60% of the profit produced by all protected areas in Ecuador (Fig. 7.10, bottom right). [For map of national parks and exemplar protected areas in continental and insular Ecuador, see Fig. 5.18 in Chapter Five]. Examine Fig. 7.11; it summarizes the number of businesses, their offer, and jobs directly created by tourism in Galápagos (up to 2019): 63% of tourism-related businesses (*N*=874) were hotels, lodges, and travel agencies; the rest were restaurants and bars (15%), small boats for transportation (12%), boats and ships with on-board-staying while exploring/sightseeing the archipelago (9%), and transportation on land (1%; MINTUR and OTG 2019, 2020, 2021). In terms of employment, 3,955 jobs were directly created by Galápagos' tourism, 60% of them at hotels, lodges, and with on-board staying boats and ships. [We discuss cruises below]. An important component of the tourism workforce was the naturalists; about 817 were licensed by the Galápagos National Park by 2019; however, not all were active or employed concurrently. [The boats and ships with on-board staying (cruises) employed 394 naturalists (48% of the total licensed ones) during 2019; ESPOL (2021)].

Note that a vital source of income for the Galápagos National Park (including the offices of the Galápagos National Park Directorate, GNPD/DPNG), the Governing Council of Special Regime of Galápagos (CGREG), the local municipalities (Santa Cruz, San Cristóbal/Floreana, and Isabela Islands), and parishes (administrative sub-governing sections of each municipality) comes from an "entrance fee" applied to all *visitors* to the archipelago. The fee varies as per nationality, age, residency status, or transient-visitor status (10 categories exist; DPNG 2025a). Foreign adult tourists pay $200 USD (minors, $100 USD). Foreign adult tourists from the Andean Community of Nations or MERCOSUR pay $100 USD (minors, $50 USD). Foreign students enrolled in Ecuadorian education institutions pay $50 USD. Ecuadorian adult tourists or foreign residents in Ecuador pay $30 USD (minors, elderly, or disabled, $15 USD). "Transient visitors", considered as such by the Governing Council of Special Regime of Galápagos, pay $15 USD (DPNG 2025a).

In 2024, the "entrance fee" generated $21.1 million USD. About 50% of the revenue regularly goes to the Galápagos National Park itself and the Galápagos National Park Directorate (i.e. conservation programs, infrastructure, education and outreach activities, and salaries = jobs); 25% goes to the local municipalities; 20% to the Governing Council of Special Regime of Galápagos; and 5% to the parishes (DPNG 2025b). Thus, there is a direct benefit connection between the "entrance fee" to the National Park and *in situ* conservation and administration work.

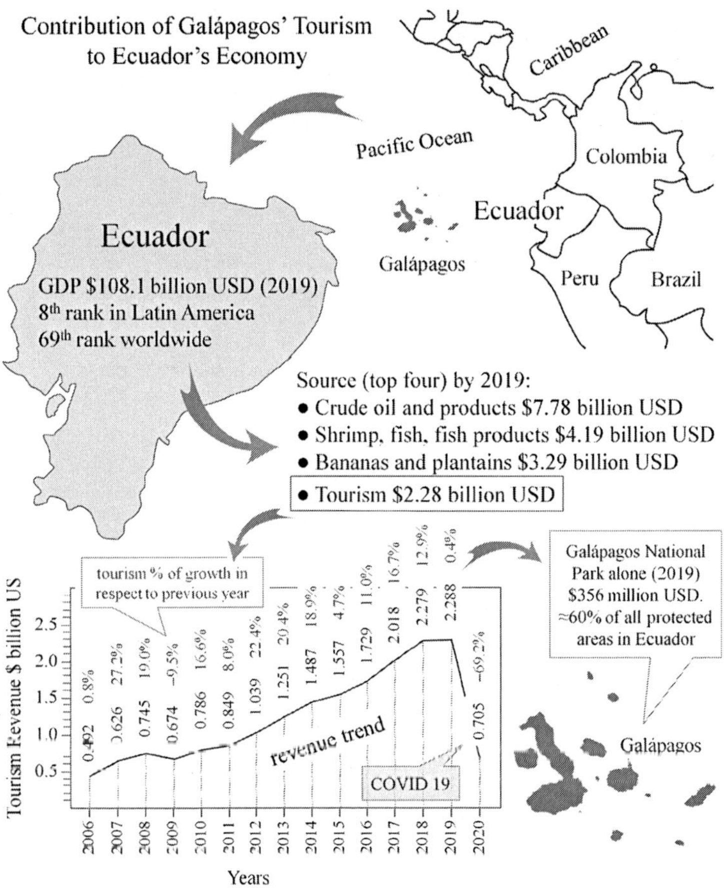

Fig. 7-10 Contribution of Galápagos' tourism to Ecuador's economy (2019 as last exemplar year before the COVID-19 pandemic; 2019 was a +0.4% tourism growth year with respect to 2018 and, thus, comparable to the latter in other tourism growth trends shown in this figure). Tourism's revenue ($2.28 billion USD) was Ecuador's fourth source of income in 2019, after crude oil ($7.78 billion USD), shrimp, fish, and fish products ($4.19 billion USD), or bananas and plantains ($3.29 billion USD). The Galápagos National Park alone generated $356 million USD (≈60% of the profit produced by all protected areas in Ecuador). [As shown in Figs. 7.11-7.12, the cruise companies (i.e. the "stay on board of vessels" modality of exploring and sightseeing the islands) captured $266 million USD (75%) of the total sales; these companies (*N*=76) comprised only 9% of all Galápagos' tourism businesses (N_{total}=874); they also created 45% of the jobs directly associated with tourism]. Layout: GPC; data from MINTUR (2017), ASOBANCA (2019), IMF (2020), ESPOL (2021), CEIC (2023), OEC (2023), and World's Top Exports-Ecuador (2023).

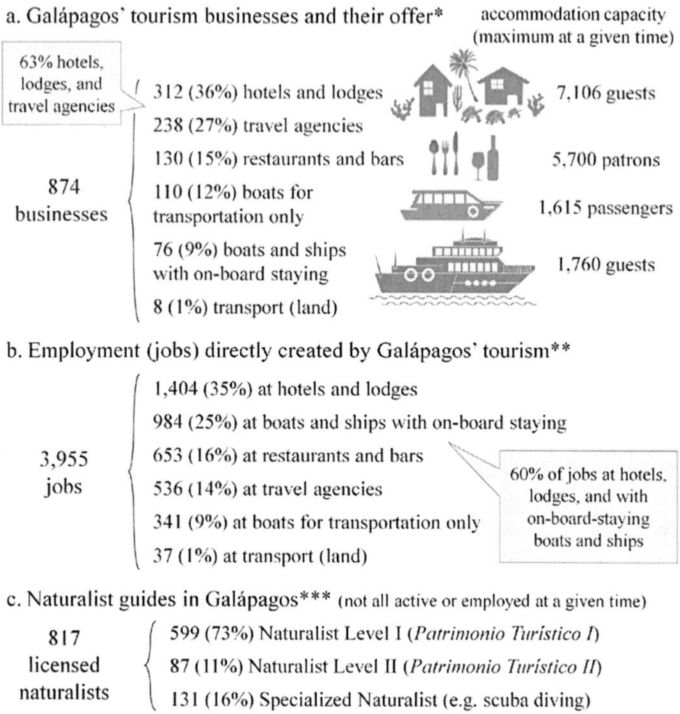

Fig. 7-11 Tourism businesses, their offer, and jobs directly created by tourism in Galápagos 2019-2020 (data MINTUR and OTG). *63% of tourism-related businesses were hotels, lodges, and travel agencies. **60% of jobs were offered by hotels, lodges, or with-on-board-staying boats and ships. ***Licenses to naturalists were granted by the Galápagos National Park (GNP); applicants to the GNP accreditation attended a training course and passed a basic test (professional development workshops were—still are—offered periodically to licensed naturalists); not all licensed naturalists were/are active/employed at a given time. In principle, Naturalist Level I required basic language skills, i.e., English proficiency level B1 (according to the Common European Framework of Reference for Languages); Naturalist Level II required English proficiency level B2 and an additional language at level B1; and Specialized Naturalist required English proficiency level B2 plus "adventure specialization", e.g., scuba diving. [During 2018, one-third of the naturalists (29%) had 1-5 years of experience in the job (mean=14 years)]. By 2019-2020, 73% of the naturalists were men (mean 47 years old) and 27% were women (mean 43 years old); 57% of the naturalists had completed high school, 32% had college or professional degrees, 5% had a technical degree, 5% had a graduate degree or equivalent, and 1% had completed only elementary school; among those with degrees, 51% had certification in tourism, biology/environment, and/or business. Layout: GPC after MINTUR, OTG (2019, 2020, 2021).

The cruise companies were central to Galápagos' tourism. In Fig. 7.12, we summarize the contribution of stay-on-board-of-vessels tourism to Ecuador's and Galápagos' economies: (1) As we learned earlier (Fig. 7.10 bottom right), Galápagos tourism generated \$356 million USD as direct revenue in 2019; (2) of such profits, \$108 million USD (30%) were spent in Galápagos, while \$248 million USD (70%) in continental Ecuador; (3) 75% of the total revenue, i.e., \$266 million USD in sales, originated in "cruises" or the "stay-on-board-of-vessels" modality while exploring and sightseeing the islands; (4) before/after visiting Galápagos, cruises' passengers spent \$133 million USD in continental Ecuador (hotels, tours, and air-tickets); in addition, they paid \$1.5 million USD in airport fees (Galápagos Transit Control Card); (5) cruises' passengers also spent \$4.1 million USD in purchases at stores while in Galápagos (adding to the expenses made *in situ*); (6) cruises companies paid \$24 million USD in taxes to the State of Ecuador (\$20 million USD) or the Galápagos Province (\$2 million USD) [in Fig. 7.13 we summarize the allocation of tax-generated funds by Galápagos' tourism]; and (7) the stay-on-board-of-vessels tourism created 2,646 jobs (45% of all jobs directly created by Galápagos' tourism), more than half of them (N=1,407, 53%) employed permanent residents in the islands, i.e., ≈12% of Galápagos' economically active population (N≈12,000; Fig. 7.12; after ESPOL 2021; with input from MINTUR and OTG 2019, 2020, 2021).

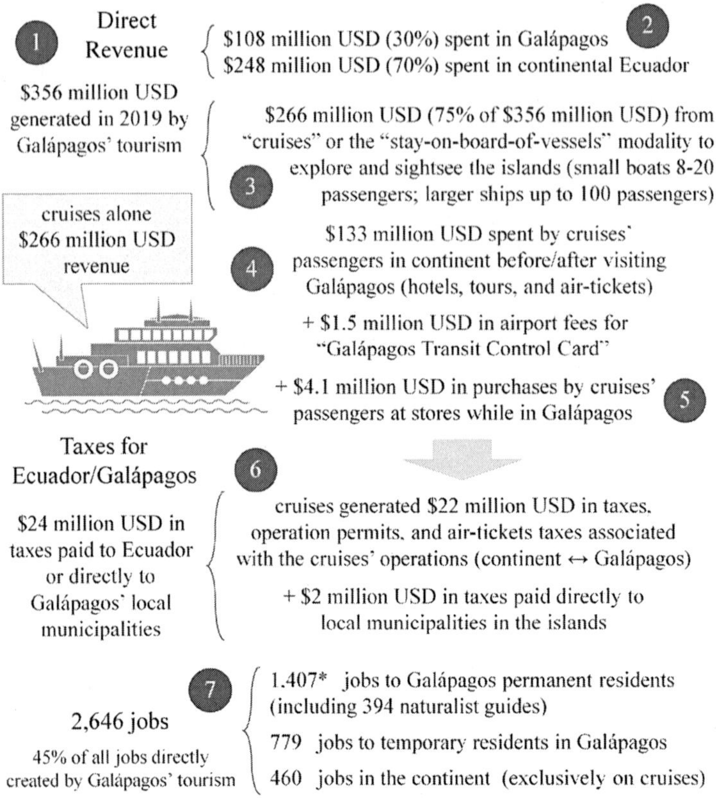

Fig. 7-12 Contribution of the stay-on-board-of-vessels tourism (cruises) to Ecuador's and Galápagos' economy. *=12% of the economically active population (*N*=12,000). Layout GPC after ESPOL (2021) with input from MINTUR and OTG (2019, 2020, 2021).

Where did taxes go? As indicated in Fig. 7.13, during fiscal year 2018, Ecuador's government allocated nationally $810 million USD to the governing offices of 24 provinces (mean $34 million USD; range $11 to $150 million USD). The Galápagos Province—via its Governing Council of Special Regime of Galápagos (CGREG)—received/spent $20.6 million USD (2.5%). In 2019, CGREG's operative/spent budget was $28.1 million USD; in 2020, it reached $24.0 million USD; in 2021, was $21.9 million USD; and in 2022, was $25.3 million USD (data from Ecuador's Ministry of Economy and Finance as well as CGREG; Ministerio de Economía y Finanzas del Ecuador 2018; CGREG 2019, 2020a, 2021, 2022, 2023b).

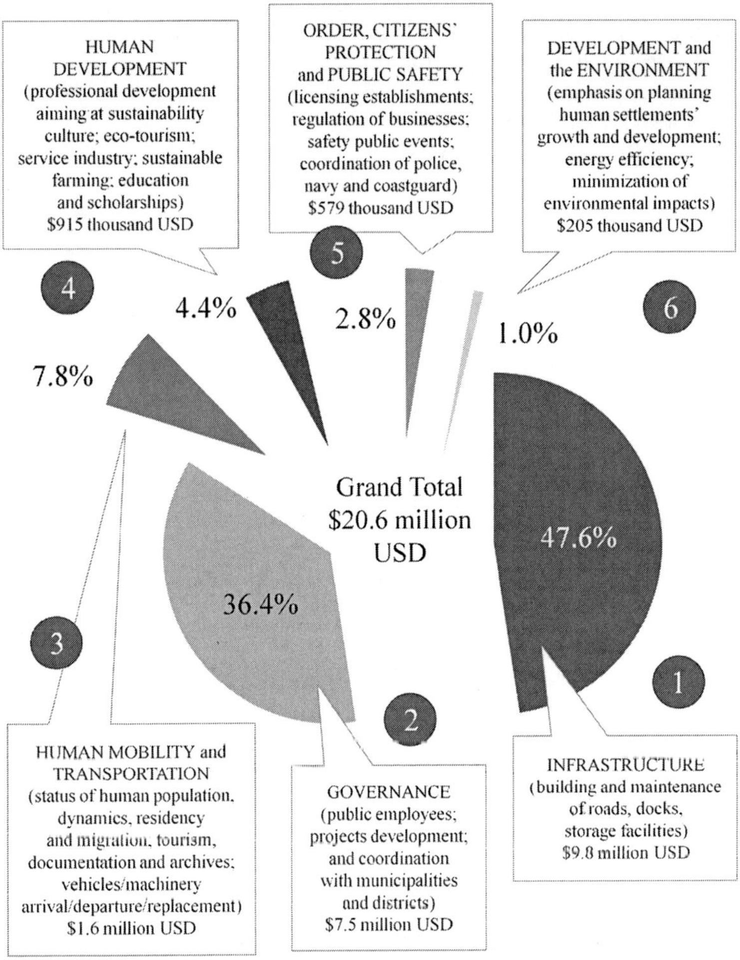

Fig. 7-13 Operative budget available to the Governing Council of Special Regime of Galápagos (CGREG) in 2018. During fiscal year 2018, Ecuador's government allocated $810 million USD to the governing offices of 24 provinces (mean $34 million USD; range $11 to $150 million USD). Galápagos received/spent $20.6 million USD (2.5%). In 2019, CGREG's operative/spent budget was $28.1 million USD; in 2020, it reached $24.0 million USD; in 2021, was $21.9 million USD; and in 2022, was $25.3 million USD. [In 2019, Galápagos' per capita GDP was $7,592 USD (population as per 2015 census=25.2k), while for the entire country of Ecuador was $6,318 USD (population 17.3 million; CGREG 2020b)]. Layout: GPC; data from Ecuador's Ministry of Economy and Finance (Ministerio de Economía y Finanzas del Ecuador 2018) and CGREG (2019; 2020a; 2021; 2022; 2023b).

Follow this sequence in Fig. 7.13: (1) about half of CGREG's 2018 operative/spent budget ($9.8 million USD, 47%) went to infrastructure (building and maintenance of roads, docks, and storage facilities); (2) $7.5 million USD (36.4%) went to governance (public employees; projects' development; and coordination with local municipalities and districts in the human populated islands: Santa Cruz, San Cristóbal, Isabela, Floreana); (3) $1.6 million USD (7.8%) was spent in human mobility and transportation (status of human population, dynamics, residency and migration; tourism, documentation and archives; and vehicles/machinery arrival/departure/replacement); (4) $915 thousand USD (4.4%) were spent in human development (professional development aiming at sustainability culture; eco-tourism; service industry; sustainable farming; and education and scholarships); (5) $579 thousand USD (2.8%) went to law and order, citizens' protection, and public safety (licensing establishments; regulation of businesses; safety public events; and coordination of police, navy and coastguard); and (6) $205 thousand USD (1%) went to development and the environment (emphasis on planning human settlements' growth and development; energy efficiency; and minimization of environmental impacts; Ministerio de Economía y Finanzas del Ecuador 2018; CGREG 2019, 2020a, 2021, 2022, 2023b).

Conclusions

We omit citations in this section to abridge the text. Each conclusion point summarizes topics substantiated by citations in this chapter.

(1) Galápagos' subsistence is powered by tourism.

(2) In 2018—a record-high year before the COVID-19 pandemic—275,817 visitors arrived in the archipelago. By 2019, total arrivals dropped to 271,238 visitors. By 2020, arrivals decreased to 72,519 visitors (3.8 times below 2018). By 2021, arrivals had partially recovered to 136,336 visitors (49.4% with respect to 2018). And by 2022, they reached 267,688 visitors (97% with respect to 2018). An expected post-pandemic boost occurred in 2023, with an all-time record high of 329,477 arrivals. During 2024, the numbers decreased to 279,277 visitors, suggesting tourism trends were normalizing after COVID-19.

(3) The "expedition experience" while visiting Galápagos is a popular format to learn about the archipelago. Groups of maximum 16 travelers accompanied by a naturalist guide sightsee the land or seascapes. Visitors hike along trails, stop by interpretation centers or giant tortoises' captive-breeding centers, or snorkel and spend time at the beach. Exploring various islands requires navigation in boats or larger vessels (cruises).

(4) All group activities within Galápagos National Park are supervised by a licensed naturalist, although tourists can freely visit the towns and/or travel on their own to selected sites (nature trails, volcanoes' tops, lava caves, lagoons, recreational beaches, private farms, coffee plantations).

(5) From 1989 to 2018, the number of visitors to Galápagos grew more than six times: 42,000 to 276,000 (rounded numbers; 6.71% compound annual growth rate for 30 years).

(6) In 2018, visitors from Ecuador and the United States comprised 62% of all arrivals to the islands.

(7) In 2018, >50% of visitors from United States, United Kingdom, Canada, and Australia stayed on-board-of-vessels while exploring Galápagos. By contrast, >50% of visitors from Ecuador, Germany, Argentina, France, Netherlands, and Spain stayed at hotels or lodges.

(8) By 2019, 72% of all visitors stayed at hotels or lodges, while 28% stayed on board of vessels.

(9) In 2018, the average age of Ecuador's visitors to Galápagos was 34 years old (54% women; 46% men), while other countries was 45 y/o (55% W; 45% M). 43% of the Ecuadorian tourists were 21-40 y/o, while the age distribution of other countries was bimodal (peak at 21-30 y/o with 17% of travelers and peak at 61-70 y/o with 20% of travelers).

(10) Scientists have projected the number of visitors to the Galápagos under diverse growth-rate scenarios. Under a moderate-growth scenario, visitors' arrivals could reach ≈250,000 by 2020 or ≈350,000 by 2033. Under a high-growth scenario, arrivals could reach +300,000 by 2020 or +850,000 by 2033.

(11) By 2019, tourism was Ecuador's 4th source of income, with $2.28 billion USD, after crude oil ($7.78 billion USD); shrimp, fish, and fish products ($4.19 billion USD); and bananas and plantains ($3.29 billion USD). Tourism's revenue with respect to the previous year grew from $0.49 billion USD (0.8%) in 2006 to $2.27 billion USD (12.9%) in 2018, and to $2.28 billion USD (0.4%) in 2019. It collapsed to $0.70 billion USD (-69.2%) in 2020 due to the COVID-19 pandemic.

(12) In 2019, the Galápagos National Park alone generated $356 million USD, ≈60% of the profit produced by all protected areas in Ecuador.

(13) There is a direct benefit connection between the "entrance fee" [for tourists] to the Galápagos National Park and *in situ* conservation and administration work. In 2024, the fee generated $21.1 million USD; half of it went to the National Park itself (conservation programs, infrastructure, education and outreach activities, and salaries = jobs).

(14) In 2018-2019, 63% of Galápagos' tourism-related businesses (*N*=874) were hotels, lodges, and travel agencies; the rest were restaurants

and bars (15%), small boats for transportation (12%), boats and ships with on-board stays while exploring/sightseeing the archipelago (9%), and transportation on land (1%).

(15) An estimated total of 3,955 jobs were directly created by Galápagos' tourism in 2018-2019, 60% of them at hotels, lodges, and on-board boats and ships. An important component of the tourism workforce was the naturalist guides; about 817 were licensed by the Galápagos National Park by 2019; however, not all were active or employed concurrently. Note that the boats and ships with on-board stays (cruises) employed 394 naturalists (48% of the total licensed ones) during 2019.

(16) Cruise companies were central to Galápagos' tourism. By 2019, \$266 million USD in sales (75% of the total \$356 million USD revenue mentioned above) originated in "cruises" or the "stay-on-board-of-vessels" modality while exploring and sightseeing the islands. Before/after visiting Galápagos, cruise passengers spent \$133 million USD in continental Ecuador (hotels, tours, and air tickets); in addition, they paid \$1.5 million USD in airport fees and also spent \$4.1 million USD in purchases at stores while in Galápagos (*in-situ* expenses).

(17) Cruise companies paid \$24 million USD in taxes to the State of Ecuador (\$20 million USD) or the Galápagos Province (\$2 million USD). They created 2,646 jobs (45% of all jobs directly created by Galápagos' tourism); more than half of them (N=1,407, 53%) employed permanent residents in the islands, i.e., ≈12% of Galápagos' economically active population (N≈12,000).

(18) Where did taxes go? During fiscal year 2018, the Galápagos Province—via its Governing Council of Special Regime of Galápagos (CGREG)—received/spent \$20.6 million USD (2.5%). In 2019, CGREG's operative/spent budget was \$28.1 million USD; in 2020, it reached \$24.0 million USD; in 2021, was \$21.9 million USD; and in 2022, was \$25.3 million USD.

Questions

The following questions (Q1–2) merit generous written answers. You might need to do additional reading or library work to enrich your answers (we suggest exploring Paz-y-Miño-C and Espinosa 2024a,b).

Question 1: To better understand the environmental benefits or impacts of Galápagos' tourism and place it in a global context, you need to compare it with tourism models at other archipelagos, particularly volcanic ones; for example, Hawai'i, Cape Verde, Canaries, Madeira & Selvagens, Azores, Marquesas, Juan Fernández, Seychelles, or Mascarenes (Paz-y-

Miño-C and Espinosa 2024a). (a) How does Galápagos tourism differ with respect to other archipelagos? (b) How do human population size [local inhabitants], amount of land (wilderness) protected by law, or number of visitors arriving yearly correlate with the quality of tourism offered and/or the local environmental impacts [from low/manageable impact to high/unmanageable impact]? (c) How does differential socio-economic development among archipelagos influence the quality of tourism offered, from eco-friendly to eco-unfriendly? (d) Based on your comparisons among archipelagos' tourism practices, can you make projections about how Galápagos tourism will look like in the future? (e) Is there anything that Galápagos must do to secure eco-friendly tourism in the future [keep in mind that Galápagos is a World Heritage Site, as per UNESCO 1978; see UNESCO (2025), World Heritage Convention, Galápagos Islands]?

Question 2: Ecotourism was conceptualized in the 1980s as "an alternative to other forms of tourism" [e.g. outdoor recreation], to "ensure a positive feedback loop between tourism and conservation", and to "benefit wildlife and biodiversity, create incentives to protect landscapes, and support local communities" (Stronza et al. 2019). Researchers have alerted, however, that unrestricted [not carefully regulated] tourism could lead to environmental harm and socio-economic struggle (i.e. higher per-capita impact on fragile habitats and an increased cost of living for the local population; Burbano et al. 2022) or that tourism can become eco-trapped by unattainable island sustainability, i.e., the seeming harmony between economic prosperity and conservation by advertising conventional tourism or outdoor recreation as "ecotourism" (Grydehøj and Kelman 2017; Kelman 2021; Stronza et al. 2019). (a) Is Galápagos' nature tourism an exemplar of ecotourism? Why? (b) Can Galápagos attain long-term "island sustainability"? How? (c) What can Galápagos do to avoid becoming "eco-trapped" by unattainable island sustainability?

How to Answer the Questions Above

The questions above have various possible answers (A1–2). Here we suggest perspectives from which to address the queries.

Answer 1: (a–e) Key to answering this question is to do library work and make *reasonable* comparisons—in the context of *tourism*—between Galápagos and other archipelagos. Keep in mind that Galápagos is a UNESCO World Heritage Site (WHS), and other archipelagos might not be. [World Heritage Sites must meet requirements to remain as WHS, e.g., long-term protections and management plans. WHS can lose their designation if they fail to meet requirements]. At the same time, realize that seemingly extreme comparisons between highly populated, wealthy, and socio-economically developed archipelagos versus Galápagos are valid and belong in your answer. Such comparisons can help you envision how Galápagos could change over time. Thus, be judicious when addressing what Galápagos must do to secure eco-friendly tourism in the future.

Answer 2: (a–c) Three concepts should be clear to you prior to answering Q2: ecotourism, island sustainability, and the eco-trap. Each concept, however, has arguable definitions. Academics and conservationists have debated for decades about the rigor, applicability, or actual value of these concepts. Yet, the scientific literature and nature-conservation documents continue to rely on these concepts to advance tourism management policies and actions. When answering Q2, your goals should be to first characterize the type(s) of tourism offered in Galápagos. Second, address how the archipelago can attain long-term "sustainability" [keeping in mind that tourism shall continue to be essential to the local economy]. And third, determine if Galápagos is destined, or not, to become "eco-trapped". If it is, then you should outline what needs to be done to reverse such a course. [Remember to explore Grydehøj and Kelman (2017), Stronza et al. (2019), Kelman (2021), Burbano et al. (2022), and Paz-y-Miño-C and Espinosa (2024a,b)].

Discussion Topic

The discussion topic below (D1) should be addressed with your peers. We encourage college instructors to facilitate the dialogue.

Discussion 1: Here you need to learn a few more facts about Galápagos conservation in order to discuss them—in the context of tourism—with your peers and instructor. Examine Fig. 7.14. It summarizes the estimated annual number of visitors to "public-use areas" (PUA) within 13 "high-ecological-value areas" (HEVA 1-13) in the Galápagos terrestrial and aquatic ecosystems (after Escobar-Camacho et al. 2021). HEVA cover 23% of the archipelago (14,715 km^2/5,682 sq mi; terrestrial ecosystems 20%; marine ecosystems 3%). [Examples of HEVA include high-primary-productivity or nutrient-rich environments; high-biodiversity sites; sheltering, foraging, or breeding grounds for endemic or transient taxa; among other ecosystem services]. Researchers have identified four "drivers of change" (i.e. natural- or human-induced stressors) affecting the high-ecological-value areas. *Unsustainable tourism* and *local population growth* ranked second among the four main drivers of change! [Remember that we discussed human population growth in Chapter Six]. Climate change ranked first; overfishing and illegal undeclared and unregulated fishing ranked third; and inva-

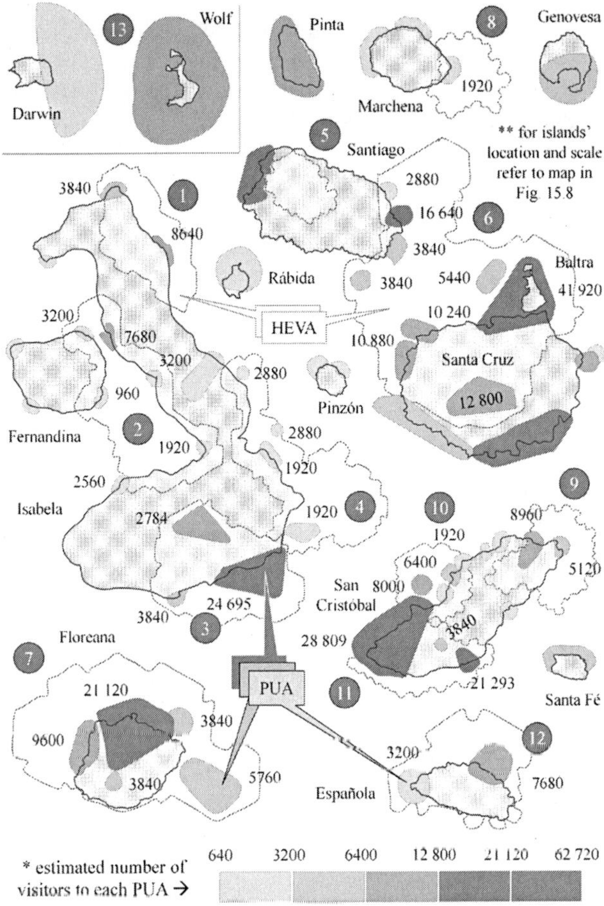

Fig. 7-14 Estimated annual number of visitors to public use areas (PUA) within 13 high-ecological value areas (HEVA 1-13) in the Galápagos terrestrial and aquatic ecosystems. *Gray-scaled color codes correspond to number of visitors (range) per PUA. Numbers in the "hundreds" or "thousands" next to each PUA correspond to annual visitors as per 2018. **For islands' actual location and scale, refer to map in Fig. 7.8. HEVA: 1. Corridor Wolf Volcano, Punta Albemarle, and Cape Marshall; 2. Bolivar Channel and Elizabeth Bay South; 3. Corridor Sierra Negra Volcano-Isabela South; 4. Corridor Cartago Bay-San Luis seabed; 5. Santiago Island highlands; 6. Conservation area Santiago-Santa Cruz; 7. Floreana and islets; 8. Marchena coral remnants; 9. Corridor la Galapaguera-Punta Pitt; 10. León Dormido (Kicker Rock); 11. Corridor El Junco and Southern seabeds; 12. Española and Gardner Islands; 13. Darwin and Wolf Islands. Layout: GPC; redrawn and modified after Escobar-Camacho et al. (2021); for expanded analysis, see Paz-y-Miño-C and Espinosa (2024b).

sive species ranked fourth. [We discuss climate, overfishing, and invasive species in Chapters Eight, Nine, and Ten of this volume; see also comprehensive analysis in Paz-y-Miño-C and Espinosa 2024b]. At variable temporal or spatial scales, the drivers of change [factors] can interact with one another and magnify their impact on HEVA (Escobar-Camacho et al. 2021). Note in Fig. 7.14 bottom that the grayscale color-codes correspond to number of visitors (range) per public-use areas (PUA). Numbers in the "hundreds" or "thousands" next to each PUA indicate annual visitors during 2018 [an exemplar year for examining tourism for reasons explained earlier]. HEVA include: (1) Corridor Wolf Volcano, Punta Albemarle, and Cape Marshall; (2) Bolivar Channel and Elizabeth Bay South; (3) Corridor Sierra Negra Volcano-Isabela South; (4) Corridor Cartago Bay-San Luis seabed; (5) Santiago Island highlands; (6) Conservation area Santiago-Santa Cruz; (7) Floreana and islets; (8) Marchena coral remnants; (9) Corridor la Galapaguera-Punta Pitt; (10) León Dormido (Kicker Rock); (11) Corridor El Junco and Southern seabeds; (12) Española and Gardner Islands; and (13) Darwin and Wolf Islands (full descriptions in Escobar-Camacho et al. 2021). Thus, public-use areas (PUA) overlap with high-ecological-value areas (HEVA). Visits to PUA range from hundreds to thousands of people each year (640 to 62,720 tourists, as in Fig. 7.14)! Here is your DISCUSSION: Since *tourism* and *local population growth* have been identified as drivers of change [stressors] on HEVA, *how exactly* would *tourism*, combined with *local population growth*, threaten the fragile Galápagos ecosystems? How could such impacts [i.e. if tourism and local population growth become unsustainable] be enhanced by the interaction with other environmental stressors: climate change, overfishing, and introduced invasive species? What can be done to minimize the impacts of *tourism* on the Galápagos Archipelago and its Marine Reserve (from short-term to long-term)? Your discussion should be informed rather than opinion based. We suggest exploring Escobar-Camacho et al. (2021) and Paz-y-Miño-C and Espinosa (2024b). Additional library work will be needed.

Suggested Observation Projects

The suggested observation projects below (P1–P2) take into consideration that you will travel to Ecuador and the Galápagos legally; that you will visit the islands as a tourist [holding a Transit Control Card, a mandatory document issued by Ecuador and specifically by the Governing Council of Galápagos]; and that you will respect the regulations that protect the Galápagos National Park and the Marine Reserve (see complete list of rules in Appendix B). Only accredited scientists, academics, and/or naturalists who have obtained official permits can do research in Galápagos. You can take pictures, but flash photography is not permitted when taking photos of wildlife. Professional photography and videos recorded for commercial purposes must be authorized by the Galápagos National Park Directorate (GNPD). The ideas below are for enthusiastic college explorers interested in learning about the archipelago.

Project 1: When you visit Galápagos, keep track of all the sites you explore and plot them on a hard or electronic copy of the maps on Fig. 7.14. This will summarize your exploration of the islands' ecosystems [terrestrial and aquatic] as per "public-use areas" (PUA in Fig. 7.14) within the 13 "high-ecological-value areas" (HEVA in Fig. 7.14). Put together a slideshow presentation (for a 20-minute talk) about "PUA and HEVA in Galápagos" to be shared with a general audience. Connect your story of exploring the archipelago's ecosystems with the "public-use areas" and in the context of the "high-ecological-value areas". Examine the potential impact that your visit had, together with that of others [i.e. in the hundreds or thousands of tourists every year], upon HEVA. Keep in mind that, as we stated above in reference to Fig. 7.14, "…Researchers have identified four 'drivers of change' (i.e. natural- or human-induced stressors) affecting the high-ecological-value areas. *Unsustainable tourism* and *local population growth* ranked second among [them]…". Discuss this point in your presentation. Avoid slides with "text only" [they are often terrible]. Keep it rigorous yet fun and didactic. And remember to consult the literature (Escobar-Camacho et al. 2021; Paz-y-Miño-C and Espinosa 2024b).

Project 2: The International Ecotourism Society defines ecotourism as "responsible travel to natural areas that conserves the environment, sustains the well-being of the local people, and involves interpretation and education" (TIES 2015). TIES lists eight principles of ecotourism: (i) Minimize physical, social, behavioral, and psychological impacts. (ii) Build environmental and cultural awareness and respect. (iii) Provide positive experiences for both visitors and hosts. (iv) Provide direct financial benefits for conservation. (v) Generate financial benefits for both local people and private industry. (vi) Deliver memorable interpretative experiences to visitors that help raise sensitivity to host countries' political, environmental, and social climates. (vii) Design, construct, and operate low-impact facilities. (viii) Recognize the rights and spiritual beliefs of the Indigenous People in your community and work in partnership with them to create empowerment. Here is your OBSERVATION PROJECT: First, and prior to field-data collection, build an Excel file (or any spreadsheet) in which you list [in the first column] TIES' principles of ecotourism.

Second, create row cells on your spreadsheet to be filled with the names of each tourism activity in which you participate while visiting the archipelago. Third, go down the list of principles of ecotourism and record on your spreadsheet the activities that meet the principles [when "non applicable", write N/A]. Now you are ready to assess and quantify (i.e. descriptive statistics from an ecotourism perspective) your exploration of the Galápagos terrestrial and marine ecosystems. Put together a slideshow presentation (for a 20-minute talk) about "Assessment of the Ecotourism Offer in Galápagos" to be shared with a general audience. Avoid slides with "text only" [they are often terrible]. Keep it rigorous yet fun and didactic.

References

ASOBANCA. 2019. *Boletín Macroeconómico: Enero 2019*. Quito: Asociación de Bancos del Ecuador.

Burbano, Diana V., Juan Carlos Valdivieso, Juan Carlos Izurieta, Thomas C. Meredith, and Diego Quiroga Ferri. 2022. "'Rethink and Reset' Tourism in the Galapagos Islands: Stakeholders' Views on the Sustainability of Tourism Development". *Annals of Tourism Research Empirical Insights*, No. 3: 100057.

CDF (Charles Darwin Foundation) and WWF-Ecuador (World Wildlife Fund Ecuador). 2018. *Atlas de Galápagos, Ecuador: Especies Nativas e Invasoras*. Quito: CDF and WWF-Ecuador.

CEIC (Global Economic Data). 2023. "Ecuador Tourism Revenue Growth 2009-2020". Accessed May 4, 2023. *https://www.ceicdata.com/en/indicator/ecuador/tourism-revenue-growth*.

CGREG (Consejo de Gobierno del Régimen Especial de Galápagos). 2019. *Rendición de Cuentas 2018: Informe Preliminar*. Puerto Ayora: CGREG.

—. 2020a. *Rendición de Cuentas Ene-Dic 2019*. Puerto Ayora: CGREG.

—. 2020b. *Información Base para el Diálogo: Información para la Elaboración Participativa del Plan de Reactivación de Galápagos y Plan Galápagos 2030*. Puerto Ayora: CGREG.

—. 2021. *Rendición de Cuentas 2020*. Puerto Ayora: CGREG.

—. 2022. *Rendición de Cuentas 2021*. Puerto Ayora: CGREG.

—. 2023a. Pleno del Consejo de Gobierno de Galápagos Habilita al Aeropuerto de Manta para Cubrir la Ruta Hacia estas Islas. Accessed May 4, 2023.

https://www.gobiernogalapagos.gob.ec/pleno-del-consejo-de-gobierno-de-galapagos-habilita-al-aeropuerto-de-manta-para-cubrir-la-ruta-hacia-estas-islas/.

—. 2023b. *Rendición de Cuentas 2022*. Puerto Ayora: CGREG.

DPNG, and OTG (Dirección del Parque Nacional Galápagos, and Observatorio de Turismo de Galápagos). 2018. *Informe Anual de Visitantes a las Áreas Protegidas de Galápagos del Año 2018*. Puerto Ayora: Dirección del Parque Nacional Galápagos and Observatorio de Turismo de Galápagos.

DPNG (Dirección del Parque Nacional Galápagos). 2019. *Informe Anual de Visitantes a las Áreas Protegidas de Galápagos del Año 2019*. Puerto Ayora: Dirección del Parque Nacional Galápagos.

—. 2020. *Informe Anual de Visitantes a las Áreas Protegidas de Galápagos del Año 2020*. Puerto Ayora: Dirección del Parque Nacional Galápagos.

—. 2021. *Informe Anual de Visitantes a las Áreas Protegidas de Galápagos del Año 2021*. Puerto Ayora: Dirección del Parque Nacional Galápagos.

—. 2022. *Informe Anual de Ingreso de Visitantes a las Áreas Protegidas de Galápagos del Año 2022*. Puerto Ayora: Dirección del Parque Nacional Galápagos.

—. 2025a. "Entry Fee for Conservation of the Protected Areas of Galapagos". Accessed March 20, 2025. *https://galapagos.gob.ec/2025/files/tasa_esp_ing.pdf.*

—. 2025b. "Tributo de Ingreso al Parque Nacional Galápagos". Accessed March 20, 2025. *https://galapagos.gob.ec/tributo-de-ingreso/.*

El Comercio. 2025. "Estas Son las Cifras del Turismo en Galápagos en 2024 ¿Bajó o Subió el Número de Visitantes?" Accessed March 19, 2025. *https://www.elcomercio.com/actualidad/ecuador/cifras-turismo-galapagos-2024-numero-visitantes.html.*

Escobar-Camacho, Daniel, Paulina Rosero, Mauricio Castrejón, Carlos F. Mena, and Francisco Cuesta. 2021. "Oceanic Islands and Climate: Using a Multi-criteria Model of Drivers of Change to Select Key Conservation Areas in Galapagos". *Regional Environmental Change*, No. 21: 47.

ESPOL (Escuela Superior Politécnica del Litoral). 2021. *Contribución Económica del Turismo de Crucero Navegable en el Archipiélago de Galápagos y el País*. Guayaquil: ESPOL Centro de Investigaciones Económicas.

Grydehøj, Adam, and Kelman, Ilan. 2017. "The Eco-Island Trap: Climate Change Mitigation and Conspicuous Sustainability". *Area*, No. 49: 106-113.

IMF (International Monetary Fund). 2020. *Ecuador IMF Country Report No. 20/325: First Review Under the Extended Arrangement Under the Extended Fund Facility and Request for Modification of Quantitative Performance Criteria*. Washington DC: IMF Publication Services.

Izurieta, Juan Carlos. 2017. "Behavior and Trends in Tourism in Galapagos between 2007 and 2015". In *Galapagos Report 2015-2016*, edited by Galápagos National Park Directorate (GNPD), Governing Council of Special Regime of Galápagos (GCREG), Charles Darwin Foundation (CDF), and Galápagos Conservancy (GC), 83-89. Puerto Ayora: GNPD, GCREG, CDF, and GC.

Kelman, Ilan. 2021. "Critiques of Island Sustainability in Tourism". *Tourism Geographies*, No. 23: 397-414.

Ministerio de Economía y Finanzas del Ecuador. 2018. "Acuerdo Ministerial 0115 del 27 de Septiembre, 2018". Accessed May 5, 2023. *https://www.finanzas.gob.ec/wp-content/uploads/downloads/2018/10/Acuerdo-GAD.pdf*.

Ministerio de Turismo del Ecuador, Ministerio de Defensa del Ecuador, Comando Conjunto de las Fuerzas Armadas del Ecuador, Ejército Ecuatoriano, Instituto Geográfico Militar del Ecuador, Ministerio del Ambiente del Ecuador, Parque Nacional Galápagos, and Fundación Charles Darwin. 2018. *Galápagos: Mapa Ecoturístico*. Quito: MTE, MDE, CCFAE, EE, IGME, MAE, PNG, and FCD.

MINTUR (Ministerio de Turismo del Ecuador). 2017. *Boletín de Estadísticas Turísticas 2012-2016*. Quito: Ministerio de Turismo del Ecuador.

MINTUR, and OTG (Ministerio de Turismo del Ecuador and Observatorio de Turismo de Galápagos). 2019. *Estadísticas del Turismo en Galápagos en 2018*. Galápagos: MINTUR and OTG.

—. 2020. *Informe de Capacidad Instalada Galápagos Año 2019*. Galápagos: MINTUR and OTG.

—. 2021. *Informe de Capacidad Instalada Galápagos Año 2020*. Galápagos: MINTUR and OTG.

OEC (The Observatory of Economic Complexity). 2023. "Ecuador". Accessed May 3, 2023. *https://oec.world/en/profile/country/ecu*.

Paz-y-Miño-C, Guillermo, and Avelina Espinosa. 2024a. *Arrivals of Life to the Galápagos: From Berlanga to Darwin, Melville, and World War II*. Newcastle upon Tyne: Cambridge Scholars Publishing.

—. 2024b. *A Synthesis of the Galápagos: Natural History, Human Conflict, and the Destiny of an Archipelago*. Newcastle upon Tyne: Cambridge Scholars Publishing.

Rosenberg, Steve, and Ellen I. Sarbone. 2004. *Galapagos Islands: The Diving Guide*. Dunedin: Cruising Guide Publications, Inc.

Stronza, Amanda L., Carter A. Hunt, and Lee A. Fitzgerald. 2019. "Ecotourism for Conservation?" *Annual Review of Environment and Resources*, No. 44: 229-253.

TIES (The International Ecotourism Society). 2015. "What Is Ecotourism?" Accessed March 24, 2025. *https://ecotourism.org/what-is-ecotourism/.*

UNESCO. 2025. "World Heritage Convention, Galápagos Islands". Accessed March 21, 2025. *https://whc.unesco.org/en/list/001/.*

Walsh, Stephen J., and Carlos F. Mena. 2016. "Interactions of Social, Terrestrial, and Marine Sub-systems in the Galapagos Islands, Ecuador". *Proceedings of the National Academy of Sciences of the United States*, No. 113: 14536-14543.

World's Top Exports-Ecuador. 2023. "Ecuador's Top 10 Exports". Accessed May 3, 2023. *https://www.worldstopexports.com/ecuadors-top-10-exports/.*

CHAPTER EIGHT

WHAT IS THE CONSERVATION VALUE OF THE GALÁPAGOS?

In 1978, the United Nations Educational, Scientific and Cultural Organization (UNESCO) designated Galápagos a World Heritage Site. UNESCO describes Galápagos' "Outstanding Universal Value" as follows: "…This archipelago and its immense marine reserve is known as the unique 'living museum and showcase of evolution'. Its geographical location at the confluence of… ocean currents makes it one of the richest marine ecosystems in the world. Ongoing seismic and volcanic activity reflects the processes that formed the islands. These processes, together with the… isolation of the islands, led to the development of [matchless] plant and animal life—such as marine iguanas, flightless cormorants, giant tortoises, huge cacti, endemic trees and the many different subspecies of mockingbirds and finches…" (UNESCO WHC 2025a).

Before we elaborate on the concept of "conservation value", readers might ask, what is a World Heritage Site? "…World Heritage is the designation for places on Earth that are of outstanding universal value to humanity and, as such, have been inscribed on the World Heritage List to be protected for future generations to appreciate and enjoy…" Galápagos is one of 1,223 World Heritage Sites [952 cultural, 231 natural, and 40 mixed properties; list update October 2024; UNESCO WHC 2025b]. The World Heritage Convention (WHC, an international treaty) highlights four criteria (VII, VIII, IX, and X) that warrant Galápagos' designation as a World Heritage Site. Below, we summarize UNESCO WHC's rationale.

According to Criterion VII of the WHC, "…The [Galápagos] Marine Reserve is an underwater wildlife spectacle with abundant life ranging from corals to sharks to penguins to marine mammals. No other site in the world can offer the experience of diving with such a diversity of marine life forms that are so [friendly to] human beings that they accompany divers. The diversity of underwater geomorphological forms is an added value to the site, producing a unique display, which cannot be found anywhere else in the world…" (UNESCO WHC 2025a).

Criterion VIII states, "…The archipelago's geology begins at the sea floor and emerges above sea level where biological processes continue. Three major tectonic plates—Nazca, Cocos, and Pacific—meet at the [bottom] of the ocean, which is of significant geological interest. In comparison with most oceanic archipelagos, the [Galápagos] are very young, with the largest and youngest islands, Isabela and Fernandina, with less than one million years of existence, and the oldest islands, Española and San Cristóbal, somewhere between three to five million years. The site demonstrates the evolution of the younger volcanic areas in the [West] and the older islands in the [East]. Ongoing geological and geomorphological processes, including recent volcanic eruptions, small seismic movements, and erosion, provide key insights to the puzzle of the origin of the [Galápagos] Islands. Almost no other site in the world offers protection of such a complete continuum of geological and geomorphological features…" (UNESCO WHC 2025a).

Criterion IX states, "…The origin of the flora and fauna of the [Galápagos] has been of great interest to people ever since the publication of the *Voyage of the Beagle* by Charles Darwin in 1839. The islands constitute an almost unique example of how ecological, evolutionary, and biogeographic processes influence the flora and fauna on both specific islands [and] the entire archipelago. Darwin's finches, mockingbirds, land snails, giant tortoises, and a number of plant and insect groups represent some of the best examples of adaptive radiation, which still continues today. Likewise, the Marine Reserve, situated at the confluence of… major eastern Pacific currents and influenced by climatic phenomena such as El Niño, has had major evolutionary consequences and provides important clues about species evolution under changing conditions. The direct dependence on the sea for much of the islands' wildlife (e.g. seabirds, marine iguanas, sea lions) is abundantly evident and provides an inseparable link between the terrestrial and marine worlds…" (UNESCO WHC 2025a).

And Criterion X states, "…The islands have relatively high species diversity for such [a] young oceanic [archipelago] and contain emblematic taxa [like] giant tortoises and land iguanas, the most northerly species of penguin in the world, flightless cormorants, [and] the historically important Darwin's finches and [Galápagos] mockingbirds. Endemic flora [like] the giant daisy trees *Scalesia*... and many other genera have also radiated on the islands… [The] native flora [comprises] about 500 vascular plant species, of which about 180 are endemic. Examples of endemic and threatened [lineages] include 12 native terrestrial mammal species (11 endemic, with 10 threatened or extinct) and 36 reptile species [all endemic and most considered threatened/extinct], including the only marine iguana in the world. Likewise, the marine fauna has an unusually high level of diversity and endemism, with 2,909 marine species identified with 18.2% endemism. [High-profile] marine species include sharks, whale sharks, rays, and cetaceans. The interactions between the marine and terrestrial biotas (e.g. sea lions, marine and terrestrial iguanas, and seabirds) are also exceptional. Recent exploration of [deep-sea] communities continues to produce new additions to science…" (UNESCO WHC 2025a).

In summary, Criterion VII is about the spectacular underwater wildlife and seascape features of the Galápagos. Criterion VIII highlights its plate tectonics, volcanism, and ongoing geomorphological processes. Criterion IX focuses on the ecological, evolutionary, and biogeographic [processes and] characteristics of terrestrial and marine environments. And Criterion X underlines the local biodiversity and endemism and the conservation status of the flora and fauna [more recent data in Tables 8.1-8.7, below]. Thus, according to the UNESCO World Heritage Convention, these are the "outstanding universal values" of the Galápagos, which should be "protected for future generations to appreciate and enjoy".

Today, managers of the Galápagos National Park and its Marine Reserve, together with the Galápagos people, the country of Ecuador, and the world—since Galápagos, as a World Heritage Site, belongs to all of us—are responsible for protecting the archipelago's "outstanding universal values". Now, we ought to ask, what are we really safeguarding? In our view, three areas require particular examination: (1) biodiversity, (2) endemism, and (3) ecosystem functions. We will discuss them next.

1 Biodiversity: Exemplar Terrestrial and Marine Taxa

In Chapter One, we briefly addressed the question, where did the Galápagos' organisms [biodiversity] come from? We discussed three working hypotheses explaining the origins of the archipelago's biota [Fig. 1.35, Chapter One]. *Hi* overwater dispersal from mainland—in which organisms flew, swam, or were carried to the islands by ocean currents [Fig. 1.36, Chapter One]; *Hii* stepping-stone dispersal from mainland (a variant of *Hi*)—in which organisms dispersed via the now-submerged paleo islands, or seamounts, and made it to the current archipelago after "stepping" from older to younger seamounts [when they were above-sea-level volcanoes]; and *Hiii* metapopulation vicariance—in which groups of organisms originated *in situ* during the breakup of widespread ancestral populations (in the deep geological past; reviews in Grehan 2001; Heads and Grehan 2021; Paz-y-Miño-C and Espinosa 2024a). We advise readers to reexamine the section "Origins of the Galápagos Biota" in Chapter One. Below we will refer to it numerous times.

The vicariance model (*Hiii*) brings perspective to our understanding of the origins of the Galápagos biodiversity. Because the current islands are recent (youngest Fernandina = 0.032-0.06 Ma; oldest San Cristóbal = 2.4-4 Ma and Española = 3-3.5 Ma), biogeographic relations tend to be dated according to the age of the present archipelago. Such approach disregards that today's islands were preceded by earlier ones (now seamounts) and that ancestors of some contemporary taxa likely "hovered around" and diversified *in situ* as new islands continued to emerge. Thus, although many taxa—perhaps the majority—arrived via dispersal from proximate or distant mainland, not all did (Heads and Grehan 2021).

In Fig. 8.1, we illustrate some of the taxa discussed below, and in Table 1.3 [Chapter One], we list examples of species colonizations (terrestrial and marine) of the archipelago, including birds, reptiles, mammals, arthropods, and plants. Scientists rely on genetic markers (e.g. mtDNA mitochondrial DNA; rRNA ribosomal RNA; nDNA nuclear DNA; rDNA ribosomal DNA; cpDNA chloroplast DNA) to reconstruct phylogenies, infer relationships among members of diverse taxa and their closest relatives on the mainland (e.g. Central/South/North America, the Caribbean or elsewhere), and deduce the number of successful colonizations of the islands. Although the majority of taxa seem to have colonized the archipelago only once (and dispersed among islands afterwards), others did it two, three, or multiple times. That is the case of lava lizards or carpenter bees (2 colonizations), leaf-toed geckos (3 colonizations), and common Guava or humans (multiple colonizations; Table 1.3, Chapter One). Note in Table 1.3 [Chapter One] that many of such colonizations occurred within the time frame of existence of the current islands (up to 4 Ma). But other taxa seem to be older than the present archipelago: mockingbirds (1.6-5.5 Ma), land and marine iguanas (5.85-11.06 Ma), lava lizards (1.69-4.54 Ma), leaf-toed geckos (3.03-13.8 Ma), land snakes (4.4-6.9 Ma), weevils (8.6-11.5 Ma), and sand spiders (15 Ma; Table 1.3, Chapter One). Thus, these "older-than-the-islands taxa" must have arrived at paleo islands (now seamounts) and continued to disperse and colonize newer islands in a "stepping-stone" fashion (*Hi* and *Hii*, above), or, alternatively, their ancient lineages "hovered around" and diversified *in situ* (*Hiii*) as tectonics modified and shaped the landscape above the archipelago's hotspot. Consequently, the Galápagos biogeographic affinities result from complex interactions among multiple variables: geological/tectonic past; proximate or distant relations with the mainland; ocean currents (seasonality); patterns of organisms' colonizations of the islands; and *in situ* differentiation. Below, we provide examples.

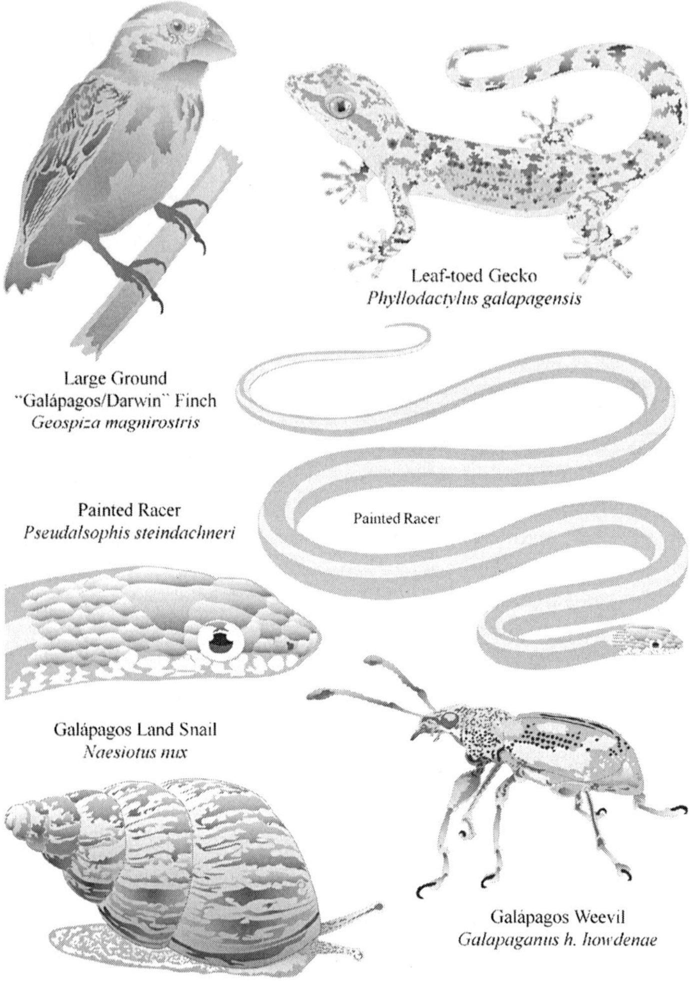

Fig. 8-1 Exemplar taxa discussed in this chapter (additional examples are listed in Table 1.3 and/or illustrated in Fig. 1.36, Chapter One). Three main working hypotheses (among others) explain the origins of the Galápagos biota: *Hi* overwater dispersal from mainland; *Hii* stepping-stone dispersal from mainland (a variant of *Hi*); and *Hiii* metapopulation vicariance. Sketches: G. Paz-y-Miño-C (computer-generated based on multiple www sources; see Appendix A).

Galápagos' Finches or Darwin's Finches

In Chapter Three, we indicated that recent taxonomic/biogeographic analyses place the Galápagos/Darwin's finches within the tanagers family Thraupidae, subfamily Coerebinae—not finches at all—and with biogeographic links to the Caribbean or—less likely—South America (as per Sato et al. 2001; Funk and Burns 2018; see Figs. 3.16-3.19, Chapter Three). This is consistent with pattern No. 7 of biogeographic affinities in Fig. 1.37, Chapter One (i.e. South America and Caribbean; association with tectonic structures: Andean uplift, Romeral fault system, and Southern Caribbean transform margin; as per Grehan 2001; Heads and Grehan 2021; Paz-y-Miño-C and Espinosa 2024a). [For essential terminology used in this chapter, see Box 8.1].

In Fig. 8.2, we illustrate the biogeographic regions in Central/South America and the Caribbean (*N* = 8: Central America; Pacific Coast of South America; Andes; Northern SA; Eastern SA; Caribbean Islands; Galápagos; Cocos Island) to which the Galápagos'/Darwin's finches' geographic affinities (i.e. origin) can be attributed (after Funk and Burns 2018). Keep in mind that, based on genetic data, 15 species of "finches" are recognized (14 in Galápagos and 1 in Cocos Island; Burns et al. 2014; Lamichhaney et al. 2015). And in Fig. 8.3, we summarize the outcomes of a biogeographic analysis called Dispersal Extinction Cladogenesis (DEC+J+X model; Funk and Burns 2018), which aims at reconstructing the evolution of a taxon geographic range (i.e. the finches) that is consistent with the topology of a robust phylogenetic tree. For that, the biogeographic model DEC+J+X relied on a previously generated phylogeny of the tanagers (Thraupidae) based on six genetic markers (mtDNA gene regions cyt *b* and ND2; and nuclear loci RAG1, ACO1-I9, MB-I2, and FGB-I5; details in Burns et al. 2014). The DEC+J+X model has various assumptions (consistent with *Hi* and *Hii*, above), including: (1) each diversification (speciation) event in a lineage is accompanied by a change in the geographic range distribution of the lineage; (2) when jump-dispersal events occur (J in model), the newly formed lineage disperses into an area outside the distribution of its ancestor; and (3) the likelihood of a dispersal event between two regions co-depends on the distance that separates them (X in model; higher likelihood to proximate regions). In other words, a quantitative, analytical approach to "telling a biogeography story" that makes sense when interpreting the diversification pattern (topology) of a phylogeny! DEC+J+X suggests (with 99.5% confidence) that the Galápagos/Darwin finches have biogeographic affinity with the Caribbean tanagers (Fig. 8.3). Start by spotting the Galápagos/Darwin finches clade in Fig. 8.3 (arrow pointing at shaded rectangle; black circles = 15 spe-

Box 8.1 Essential terminology for this chapter.

Bayesian and Maximum Likelihood Methods. Statistical tools used by researchers to assess the *probability of correctness* of the phylogenetic trees they generate [via computer algorithms] and the *likelihood of their results* based on the available data. Phylogenetic trees are explanatory hypotheses. Researchers use statistics to evaluate the strength of the data supporting their hypotheses.

Dispersal Extinction Cladogenesis. Model used to infer a taxon's or lineage's historical geographic range. Biogeographers rely on DEC to hypothesize the origin and adaptive radiation of taxa/lineages. DEC methods use phylogenies.

Ecosystem Functions. *Functional benefits* that ecosystems provide to their organisms (micro/macro life) and the planet, including chemical or biological cycles. Humans benefit from ecosystems (clean air, fishing, or nature tourism).

Emblematic Taxa. = *flag species* or *ambassador species* are central to conservation campaigns. *In situ* protection of ET leads to preservation of their habitats. Giant tortoises, marine iguanas, or Galápagos'/Darwin's finches are charismatic and emblematic; they bring attention to Galápagos protection.

Endemism. In biology, endemism refers to *uniqueness in occurrence*, e.g., taxon or entire lineages with restricted or localized geographic distribution. Endemism and rareness (=infrequent occurrence) of taxa are typical of oceanic islands.

Factorial Correspondence Analysis. Statistical method used to *uncover* [*factors associated with*] *phylogenetic relationships* among taxa/lineages as per multiple traits (molecular, anatomical). Researchers *combine FCA with other methods* to explore evolutionary, biogeographic or ecological relations among taxa/lineages.

Founder-effect Differentiation. It occurs when a small group of individuals establishes [founder effect] a *new population*, which, over time, can *evolve* and *differentiate independently*. Pink iguanas in Galápagos are an example of FED.

Genetic Distance. It often refers to the *amount of genetic dissimilarity* (GDis) between or among taxa or populations within taxa. E.g., in this chapter, we refer to higher dissimilarity [=genetic distance] between or among species or their populations as a function of geographic separation [or geographic distance].

Genetic Diversity. The *amount of genetic variation* within or across populations. In this chapter, we refer to GDiv as a function of population size, island size, genetic drift, or gene flow. [The terms heterozygosity H_e (=allele variability) or allelic richness *AR* (=number of alleles) are also used in the context of GDiv].

Haplotype-network Analyses. Graphical representations of the relationships among genetic markers [haplotypes]. Scientists often use HNA to assess genetic variation as a function of geographic distributions of populations or species.

Heterozygosity. A *metric of genetic diversity*. H provides information about a population's genetic structure and hints about its evolutionary past. H refers to the *number* and *frequency* of alleles in a population. H values range from 0.0 (no heterozygosity) to 1.0 (high number of alleles among those being investigated). H is the proportion of heterozygotes with respect to the sampled genotypes.

Hybridization Speciation. The *emergence of a new lineage* after the *cross-reproduction between* or among members of *distinctive species*. New species can emerge from such events if the hybrid descendants are fertile and reproduce.

Jump-dispersal Events. JDE often involve newly formed lineages [themselves resulting from hybridization] that *disperse* into *areas outside* the *geographic range* of the *parental species*. Thus, JDE can lead to genetic isolation and the emergence of new species. E.g., JDE likely occurred in Galápagos' finches.

Maximum Clade Credibility Trees. Researchers rely on computer programs to generate phylogenies, which often result in many tree topologies (=hypotheses). Scientists then use *serial statistics* to identify *probabilistically* even more *robust trees*. After *cumulative sets of analyses*, they can ultimately *generate MCCT*.

Maximum Likelihood vs. Maximum Parsimony. *Statistical methods* to *optimize phylogenetic reconstructions*. ML helps identify the *most probable* phylogenetic *tree*. MP *minimizes* the *number of steps* needed to match the data to the tree.

Morphospecies. Organisms assigned to the *category of species* based entirely on their *morphological traits* (often due to lack of robust genetic data).

Phylogenetic Chronogram. A tree-like diagram [a variant of a phylogenetic tree] in which the *length of* its *branches* corresponds to estimated *time* (i.e. thousands or millions of years before present). The "variable time" is inferred from estimates of mutation rate across taxa (e.g. mtDNA or nDNA sequences).

Polyploidy. P refers to organisms carrying *multiple copies* of their *chromosomes*. P is common in plants but not exclusive to them. Galápagos *Opuntia* prickly-pear trees are polyploid. P can offer evolutionary information about lineages.

Reticulated Evolution. RE refers to *processes* and *patterns*. The patterns result from events during the evolutionary past (recent or long ago). E.g., the species complex of the Galápagos'/Darwin's finches or the *Opuntia* prickly pears include instances of hybridization across ancestors (or polyploidy in the case of *Opuntia*). When the origin of present traits (genetic or morphological) is assessed, their placement on a phylogeny can reveal a *network* of past events (e.g. multiple hybridizations or polyploidy) leading to today's patterns. [Some patterns might be apparent; researchers, however, look for shared ancestry].

Fig. 8-2 Biogeographic regions used to assess the possible origin of the Galápagos/Darwin finches. Nowadays, 15 species of "finches" are recognized (14 in Galápagos and 1 in Cocos Island). Layout: GPC, redrawn and modified after Funk and Burns (2018).

cies in the genera *Platyspiza*, *Geospiza*, *Camarhynchus*, *Pinaroloxias*, and *Certhidea*). Then, locate the Cocos-Island finch (*Pinaroloxias inornata*), which likely originated in a jump-dispersal event from the Galápagos to Cocos Island (dark gray circle). Under this perspective, it is plausible that the entire Galápagos/Darwin finches clade "originated" from a jump-dispersal event from the Caribbean to the Galápagos (Funk and Burns 2018). This inference is based on the current geographic distributions of the taxa outside the Galápagos/Darwin finches clade, all in the "Caribbean Islands" (gray circles): *Tiaris olivaceus*, *Loxigilla portoricensis*, *L. violacea*, *Melopyrrha nigra*, *Euneornis campestris*, *Loxipasser anoxanthus*, *T. canorus*, *Melanospiza richardsoni*, *T. bicolor*, *L. barbadensis*, and *L. noctis*

(Fig. 8.3). Now, locate in the phylogeny the immediate sister taxa to the Galápagos/Darwin finches clade, i.e., *T. fuliginosus* and *T. obscurus*. Their current distributions in the Andes (for both species) as well as Northern and Eastern South America (for *T. fuliginosus*) indicate that another jump-dispersal event probably occurred, which led to the recolonization of South America (Andes, Northern/Eastern SA) by the ancestor of *T. fuliginosus* and *T. obscurus*, while the ancestor of the Galápagos/Darwin finches remained in the archipelago (Fig. 8.3; Funk and Burns 2018).

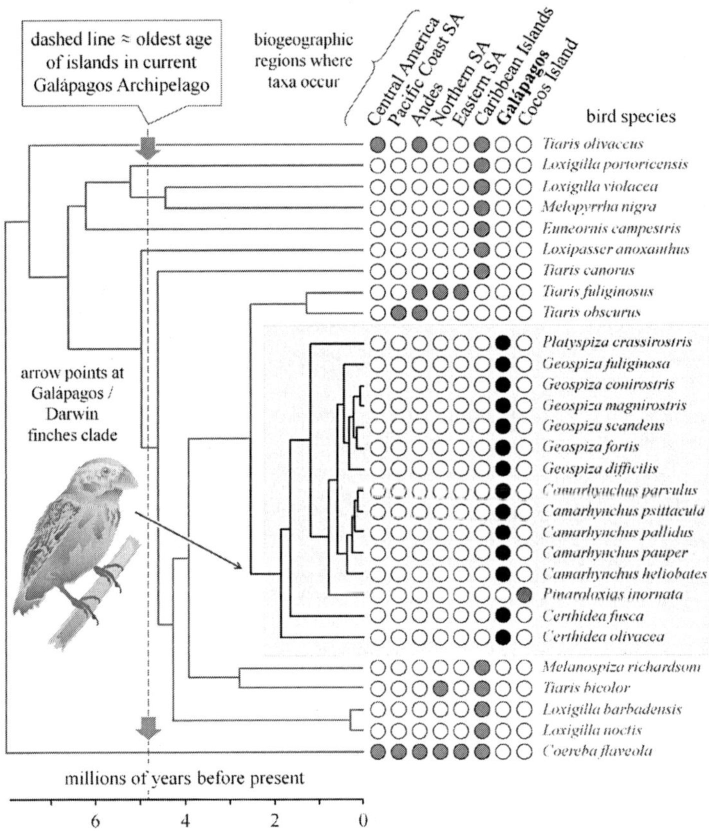

Fig. 8-3 Suspected biogeographic affinities of the Galápagos'/Darwin's finches. A Dispersal Extinction Cladogenesis analysis (DEC+J+X model applied to eight biogeographic regions) suggests the "finches" have affinity with the Caribbean tanagers (finches' clade originated ≈2 Ma). Layout: GPC, redrawn and modified after Funk and Burns (2018).

Galápagos Penguin

As the only penguin to reach the equator, the Galápagos penguin (*Spheniscus mendiculus*) is one of 19 living species distributed almost entirely in the Southern Hemisphere. Breeding colonies of Galápagos penguins occur on the shores of Fernandina, Isabela, Santiago, and Floreana, some in Seymour, Baltra and Santa Cruz (Fig. 8.4a). Its closest phylogenetic relatives are the Humboldt (*S. humboldti*), Magellanic (*S. magellanicus*), and African/Cape or South African (*S. demersus*) penguins (Fig. 8.4b; after Gavryushkina et al. 2017). Ancestors of Galápagos penguins seem to have reached the archipelago recently, 1.5 Ma. Biogeographically, *S. mendiculus* has affinity with coastal South America (West and South), Southern Africa, and New Zealand (Gavryushkina et al. 2017; Ramos et al. 2018; Pelegrín and Acosta-Hospitaleche 2022). This is in partial agreement with pattern No. 9 of biogeographic affinities in Fig. 1.37, Chapter One (i.e. Ecuador-Peru-Chile; association with tectonic structures: East Pacific Rise; igneous plateaus; Grehan 2001; Heads and Grehan 2021; Paz-y-Miño-C and Espinosa 2024a).

Giant Tortoises

Sixteen species of Galápagos giant tortoises (12 living, 3 extinct, 1 undescribed extinct) are/were distributed in 11 islands of the archipelago (Fig. 8.5a). In Fig. 8.5b, we sketch a simplified phylogenetic reconstruction of the Galápagos giant tortoises and their closest living relatives in Western South America: *C. chilensis*, *C. carbonaria*, and *C. denticulata*. Scientists are confident that ancestors of the Galápagos clade were linked to South America. This is in agreement with pattern No. 10 of biogeographic affinities in Fig. 1.37, Chapter One (i.e. Chile and Argentina; association with tectonic structure: East Pacific Rise; igneous plateaus; Grehan 2001; Heads and Grehan 2021).

In Fig. 8.6, we show the estimated times of taxa divergence in the Galápagos giant tortoises based on a phylogenetic reconstruction (mtDNA sequences of living and museum specimens; after Poulakakis et al. 2012 and various other studies listed in Fig. 8.6). Ancestors of the *Chelonoidis* Galápagos clade were likely present in the archipelago as far back as 3.2 Ma (Fig. 8.7); taxa divergence within the islands began 1.74 Ma (Fig. 8.6). Three clades within the Galápagos *Chelonoidis* are recognizable: (1) Eastern/Northern (divergence time 0.82 Ma); (2) Western/Northern (0.40 Ma); and (3) Central/South/Western (1.26 Ma; Fig. 8.6).

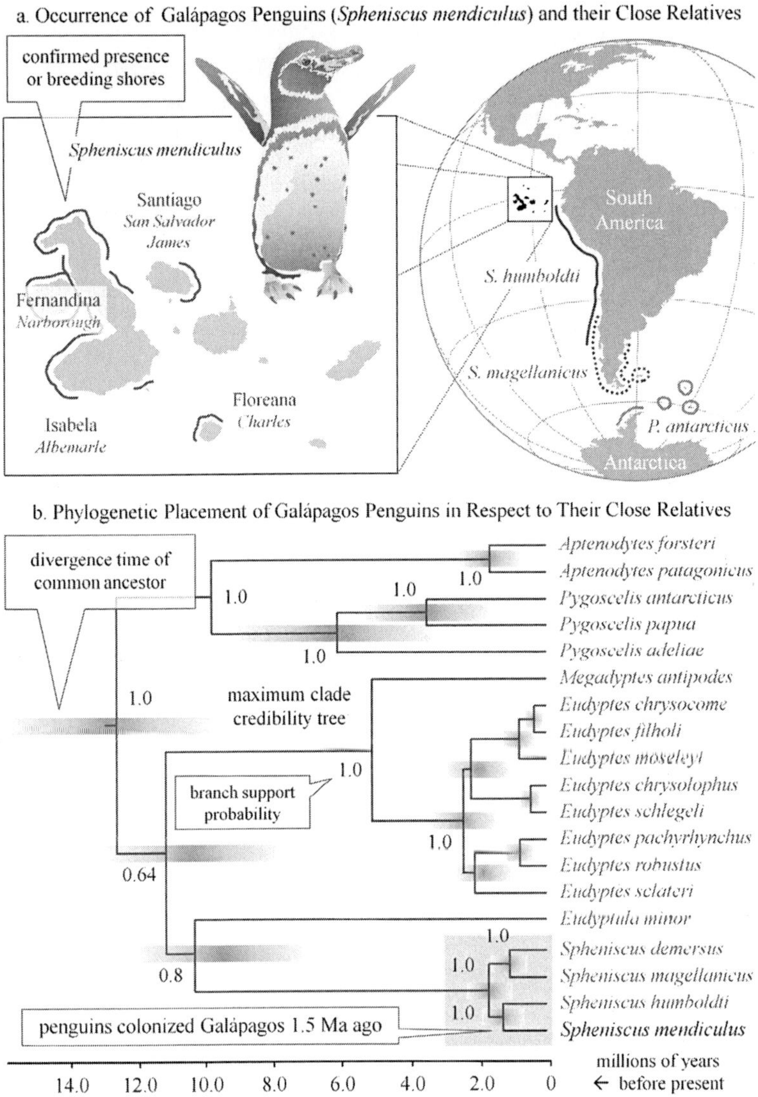

Fig. 8-4 Distribution (a) and phylogenetic relations (b) of the Galápagos penguin (*Spheniscus mendiculus*). Layout: GPC. a. Modified after Ramos et al. (2018), Fundación Charles Darwin and WWF-Ecuador (2018). b. Maximum clade credibility tree (based on nDNA RAG-1 gene; mtDNA 12S, 16S, CO1, cyt *b* genes); redrawn after Gavryushkina et al. (2017). See all-inclusive phylogeny (+fossils) in Pelegrín and Acosta-Hospitaleche (2022).

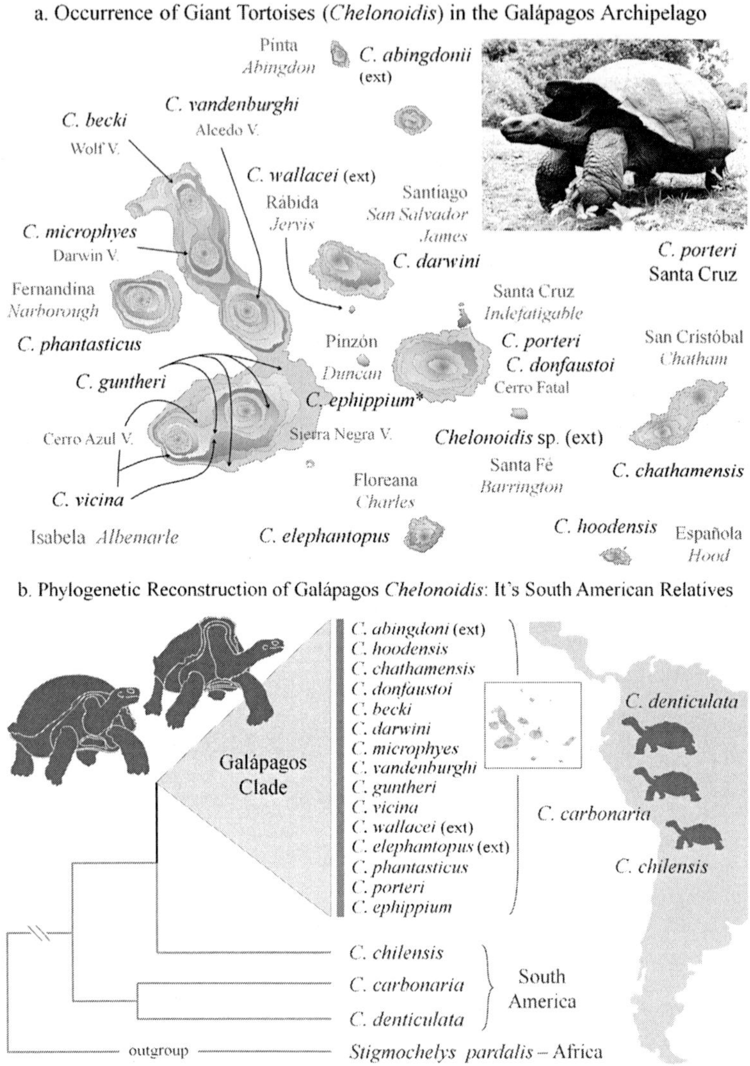

Fig. 8-5 Distribution (a) and phylogenetic relations (b) of the Galápagos giant tortoises (*Chelonoidis* spp.). The close phylogenetic relation to *C. chilensis*, *C. carbonaria* and *C. denticulata* suggests Galápagos tortoises share biogeographic affinity with South America. Layout: GPC. a. Data from Poulakakis et al. (2008, 2012, 2015); Gaughran et al. (2017); Miller et al. (2017, 2018). b. 50% majority rule consensus tree (16S rRNA, cyt *b* mtDNA) for Maximum Parsimony (Santa Fé excluded); modified after Caccone et al. (1999).

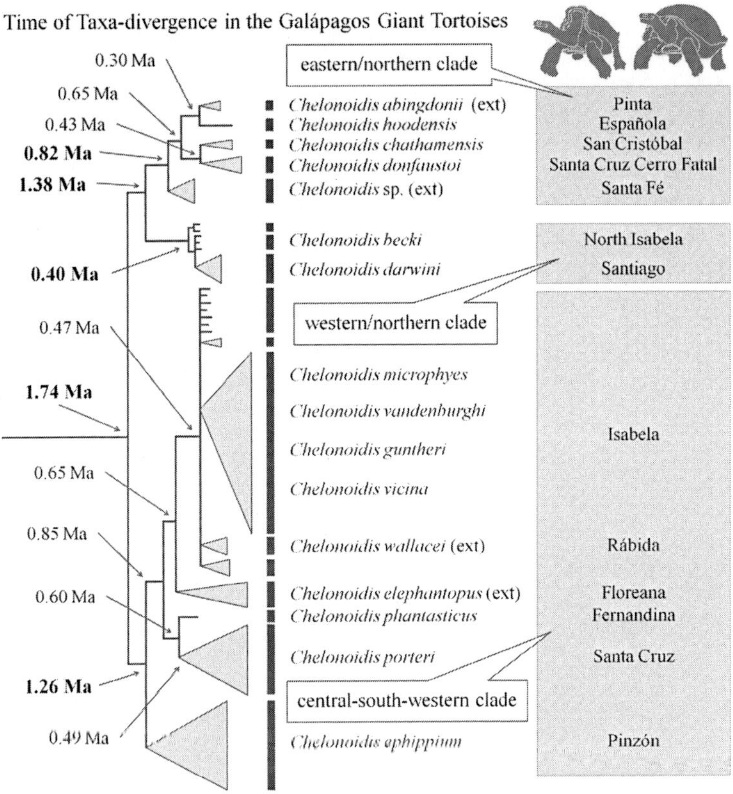

Fig. 8-6 Time of taxa divergence in the Galápagos giant tortoises. Bayesian phylogenetic reconstruction based on mtDNA sequences (control region CR and cyt *b*) of living and museum specimens. Ancestors of the *Chelonoidis* Galápagos clade were likely present in the archipelago 3.2 Ma (Fig. 8.7); taxa divergence within *Chelonoidis* spp. began 1.74 Ma. Layout: GPC, after Poulakakis et al. (2008, 2012, 2015); Gaughran et al. (2017); Miller et al. (2017, 2018); ext=extinct on island. Top right=silhouettes of morphotypes saddleback- and dome-shaped shell/carapace. Some scientists contend that the Galápagos tortoises belong to separate species, or subspecies, or to the genera *Geochelone* or *Testudo* (summary of dispute in Poulakakis et al. 2015).

Both the Eastern/Northern and Western/Northern clades shared a common ancestor 1.38 Ma. The Eastern/Northern clade is/was distributed in Pinta (*C. abingdonii* ext), Española (*C. hoodensis*), San Cristóbal (*C. chathamensis*), Santa Cruz (*C. donfaustoi*), and Santa Fé (*Chelonoidis* sp. undescribed and ext) islands (Fig. 8.6 top; see map in Fig. 8.5a).

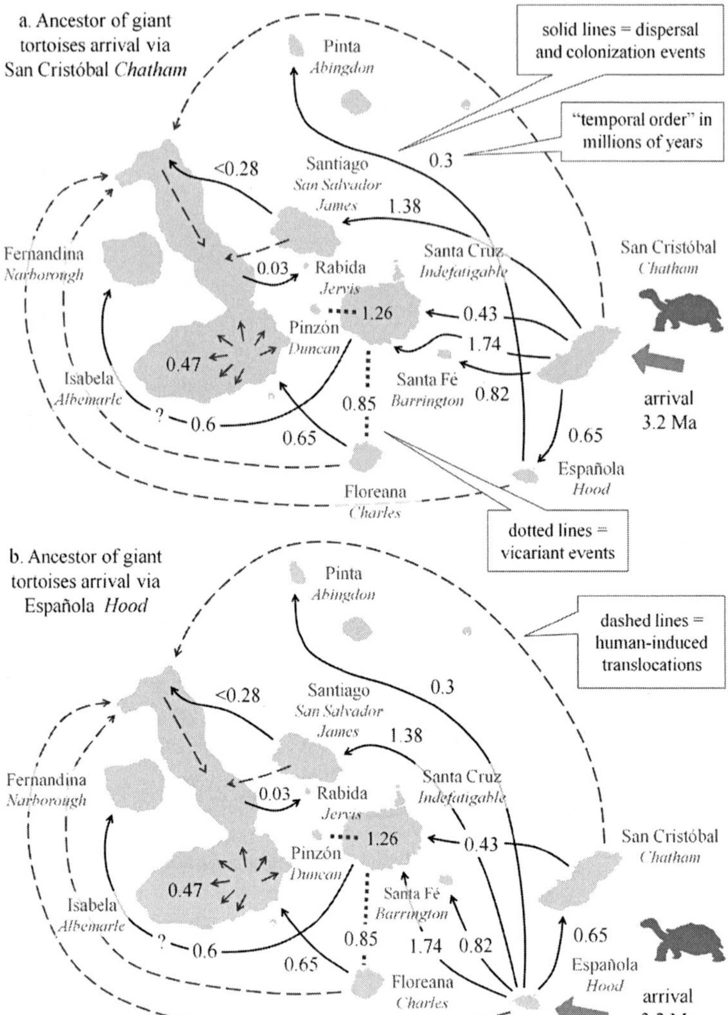

Fig. 8-7 Scenarios of possible arrivals to the Galápagos Archipelago by ancestors of the giant tortoises. Time of arrival set to 3.2 Ma, via San Cristóbal (a) or Española (b) islands (oldest islands in the archipelago). Arrow lines = hypothetical dispersal and/or colonization paths with possible times of occurrence (Ma). Dotted lines = vicariance events involving Santa Cruz, Floreana, or Rábida islands. Dashed lines = suspected translocations of tortoises among islands by humans during the past few hundred years (1535 onwards; see Chapter Two). Layout: GPC, redrawn and modified after Poulakakis et al. (2012).

The Western/Northern clade is distributed in North Isabela (Wolf Volcano, *C. becki*) and Santiago (*C. darwini*) islands (Fig. 8.6, center right; map in Fig. 8.5a). The Central/South/Western clade is quite intricate and distributed on various islands. Isabela alone has four species from this group (divergence during last 0.47 Ma): Darwin Volcano (*C. microphyes*); Alcedo Volcano (*C. vandenburghi*); slopes of Sierra Negra and Cerro Azul volcanoes (*C. guntheri*); and slopes of Cerro Azul and Sierra Negra volcanoes (*C. vicina*; Fig. 8.6 center; map in Fig. 8.5a). Rábida (0.47 Ma; *C. wallacei* ext); Floreana (0.65 Ma; *C. elephantopus* ext); Fernandina (0.60 Ma; *C. phantasticus*); Santa Cruz (0.49 Ma; *C. porteri*); and Pinzón (1.26 Ma; *C. ephippium*; Fig. 8.6 bottom; map in Fig. 8.5a).

Realize that identification of new lineages of Galápagos giant tortoises continues. Researchers have inferred their existence from genetic samples of living and museum specimens (e.g. Santa Cruz, Chiari et al. 2009 and Poulakakis et al. 2015; Floreana, Miller et al. 2017; San Cristóbal, Jensen et al. 2022; Fernandina, Tapia 2022).

In Fig. 8.7, we offer two hypothetical scenarios of "arrivals" to the Galápagos Archipelago by ancestors of the giant tortoises. The time of arrival and subsequent dispersal—via San Cristóbal (a) or Española (b) islands—is set to 3.2 Ma, which matches the age of the oldest islands (as indicated earlier, divergence of the *Chelonoidis* Galápagos lineage began 1.74 Ma; Fig. 8.6). Both scenarios are complex and include: diverse inter-island dispersal and colonization paths; differential timing of their occurrence; noticeable vicariance events involving Santa Cruz, Floreana, or Rábida (among other islands); and suspected translocations of tortoises—by humans—between islands since the archipelago was discovered in 1535 (see Chapter Two). Keep in mind that two central factors characterized the patterns of arrival, colonization, and dispersal of not only the giant tortoises but many of the taxa discussed in this chapter: ocean currents (i.e. Fig. 1.32) and tectonics and volcanism (Chapter One). Thus, imagine the relevance of the constantly changing archipelago configuration during the past 3 Ma, as depicted in Fig. 8.8, for the arrival, colonization, and dispersal of ancestors and descendants of the tortoises.

Land and Marine Iguanas

Land iguanas (*Conolophus* spp.) occur in seven of the Galápagos Islands (Fernandina, Isabela, Santiago, Baltra, Seymour, Santa Cruz, and Santa Fé). Marine iguanas (*Amblyrhynchus* sspp.) are present in the entire archipelago. Both taxa have close relatives in Central America (Fig. 8.9).

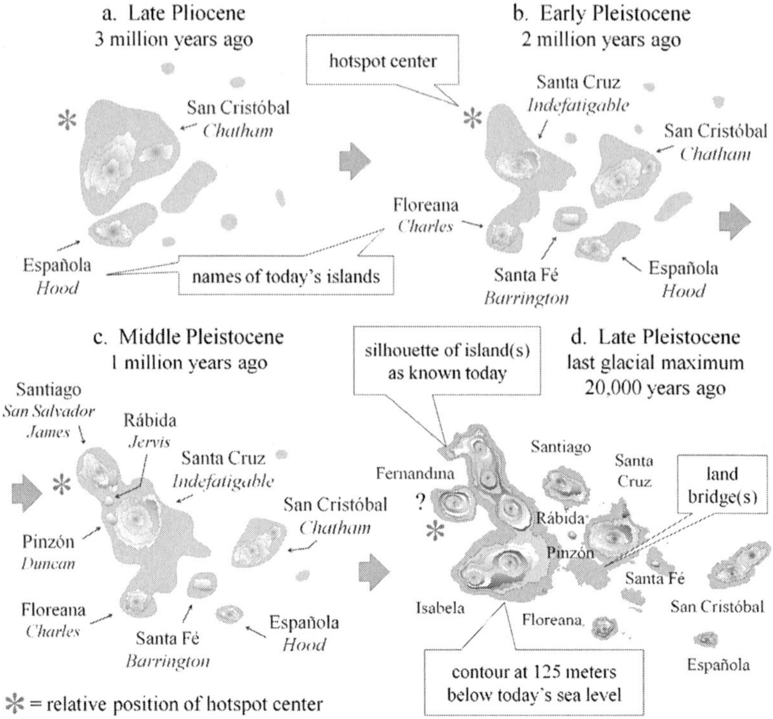

Fig. 8-8 Island-configuration model of the Galápagos during the Pliocene to Pleistocene (the idea is to bring context to the colonization, dispersal, and distribution of the Galápagos giant tortoises in the current archipelago). Layout: GPC. a-d. Redrawn and modified after Poulakakis et al. (2012). d. Redrawn and modified after Geist et al. (2014). See also Chapter One: Figs. 1.8, 1.11, and 1.12.

The phylogenetic split between Galápagos iguanas and their sister taxon *Ctenosaura* (Central America) apparently took place 8.26 Ma (Fig. 9.8b). Divergence between *Conolophus* and *Amblyrhynchus* dates back to 4.52 Ma ("older than the islands", above). And noticeable divergence within marine iguanas occurred in the last 0.23 Ma (Fig. 9.8b; MacLeod et al. 2015). This is consistent with pattern No. 1 of biogeographic affinities in Fig. 1.37, Chapter One (i.e. Central and West Pacific; structures: East Pacific Rise; Western Pacific plate margin; igneous plateau translation; Grehan 2001; Heads and Grehan 2021).

Take a closer look at the distribution of the three species of Galápagos land iguanas in Fig. 8.10. The most common, *Conolophus subcristatus*, occurs in Fernandina, Isabela, Santiago, Baltra, Seymour, and Santa Cruz.

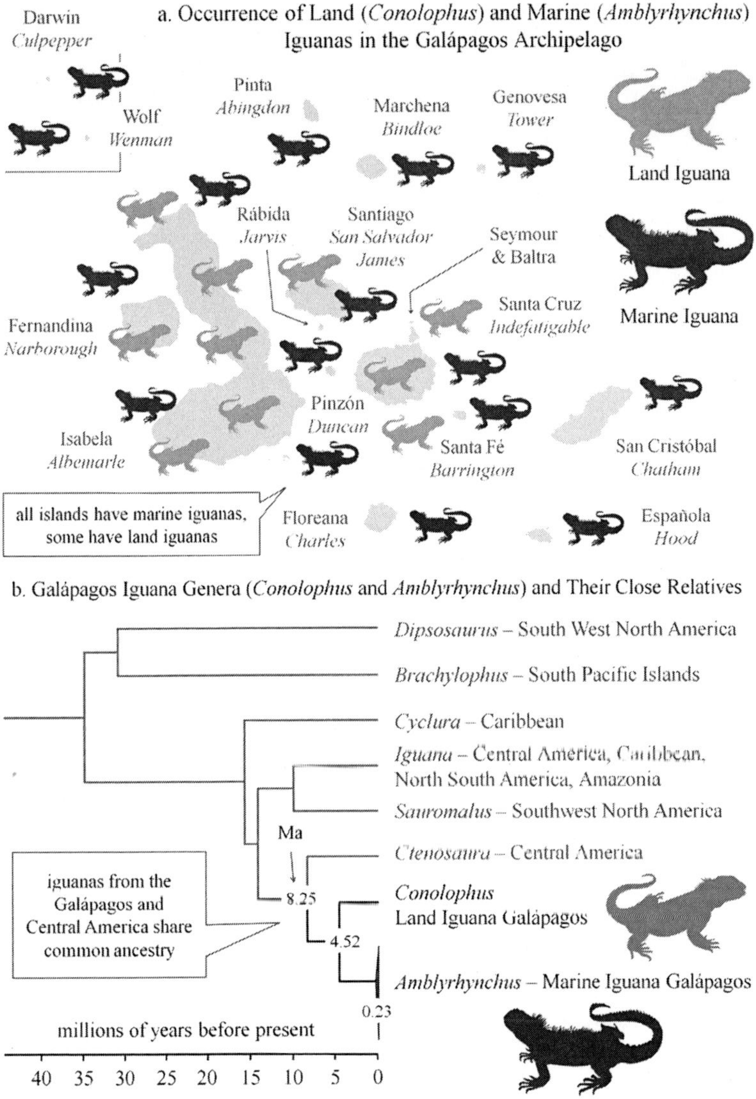

Fig. 8-9 Distribution (a) and phylogenetic relations (b) of the Galápagos land (*Conolophus*) and marine (*Amblyrhynchus*) iguanas. Layout: GPC. a. Species occurrence follows Miralles et al. (2017) and Arteaga et al. (2019). b. Phylogenetic reconstruction of genera relations (Bayesian Inference) based on nDNA genes (concatenated alignment of RAG1, BDNF, R35, NKTR sequences). Redrawn and modified after MacLeod et al. (2015).

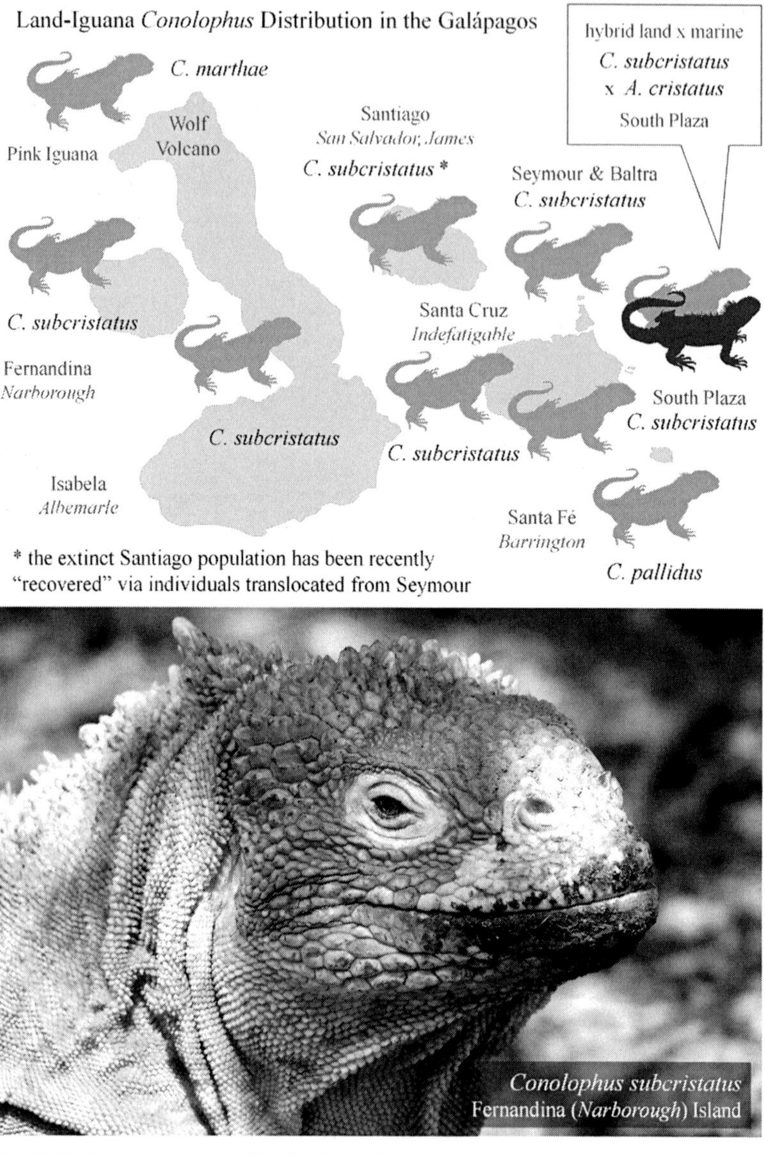

Fig. 8-10 Occurrence and distribution of land iguanas (*Conolophus* spp.) in the Galápagos Archipelago. Layout/photo: GPC; data comes from Arteaga et al. (2019).

The Pink Iguana (*C. marthae*) is restricted to the slopes of Wolf Volcano in Isabela; and the Santa Fé Iguana (*C. pallidus*) is unique to the island. The common ancestor of Pink iguanas and the other two species of land iguanas (*C. subcristatus* and *C. pallidus*) dates back to 1.52 Ma. And the common ancestor of *C. subcristatus* and *C. pallidus* dates back to 0.29 Ma (MacLeod et al. 2015). Note that the extinct land iguana of Santiago (*C. subcristatus* ext) has been "recovered" by translocations from Seymour Island (*C. subcristatus*; Williams 2019). Hybridization between land and marine iguanas (*Conolophus subcristatus* x *Amblyrhynchus cristatus*) does occur, as observed in South Plaza Island (dialog box in Fig. 8.10 top right; Rassmann et al. 2009).

In Fig. 8.11, we show the distribution of 11 subspecies of marine iguanas in the archipelago (after Miralles et al. 2017). And in Fig. 8.12a, we summarize the timeline of lineage diversification of all Galápagos iguanas (after MacLeod et al. 2015). Examine closely the phylogenetic reconstruction of the 11 marine-iguana lineages in Fig. 8.12b (analysis based on nuclear DNA markers; Miralles et al. 2017). Spot the monophyletic groups with their probability support values (%). They correspond to genetically distinctive clusters from: Española (100% *A. c. venustissimus*); Santa Fé (83% *A. c. trillmichi*); Pinta (100% *A. c. sielmanni*); Marchena (100% *A. c. hayampi*); Genovesa (100% *A. c. nanus*); San Cristóbal North (100% *A. c. godzilla*); San Cristóbal South (100% *A. c. mertensi*); Wolf (100% *A. c. jeffreysi*); Isabela and Fernandina (each 100% *A. c. cristatus*).

Note that the Floreana marine iguanas cluster together with Española (as *A. c. venustissimus*) and both branch off as sister taxa of the Santa Fé iguanas (*A. c. trillmichi*). Note also that the Pinta, Marchena, and Genovesa clusters belong together to a well-supported (100%) monophyletic group. Although the populations from Isabela and Fernandina (*A. c. cristatus*) are only moderately supported (75%; Fig. 8.12b). Interestingly, marine iguanas from Santiago (*A. c. wikelskii*), Santa Cruz (*A. c. hassi*), and even Floreana (above) do not form independent groups; in fact, Santiago's and Santa Cruz's clusters branch off paraphyletically. Something comparable occurs with Rock Redonda and Darwin Islands specimens (*A. c. jeffreysi*; Fig. 8.12b; Miralles et al. 2017).

The San Cristóbal North (*A. c. godzilla*) and South (*A. c. mertensi*) clusters are noteworthy (both occur in the same island). They are distinctive individually (each 100% branch support probability) and robustly linked as sister taxa (100%). The Northern lineage is called the Punta Pitt population; the Southern lineage, La Lobería (map in Fig. 1.27, Chapter One). A subpopulation of Española migrants/hybrids (*A. c. venustissimus*) exists on the Eastern shores of San Cristóbal (see * in Fig. 8.11; MacLeod et al. 2015; Miralles et al. 2017).

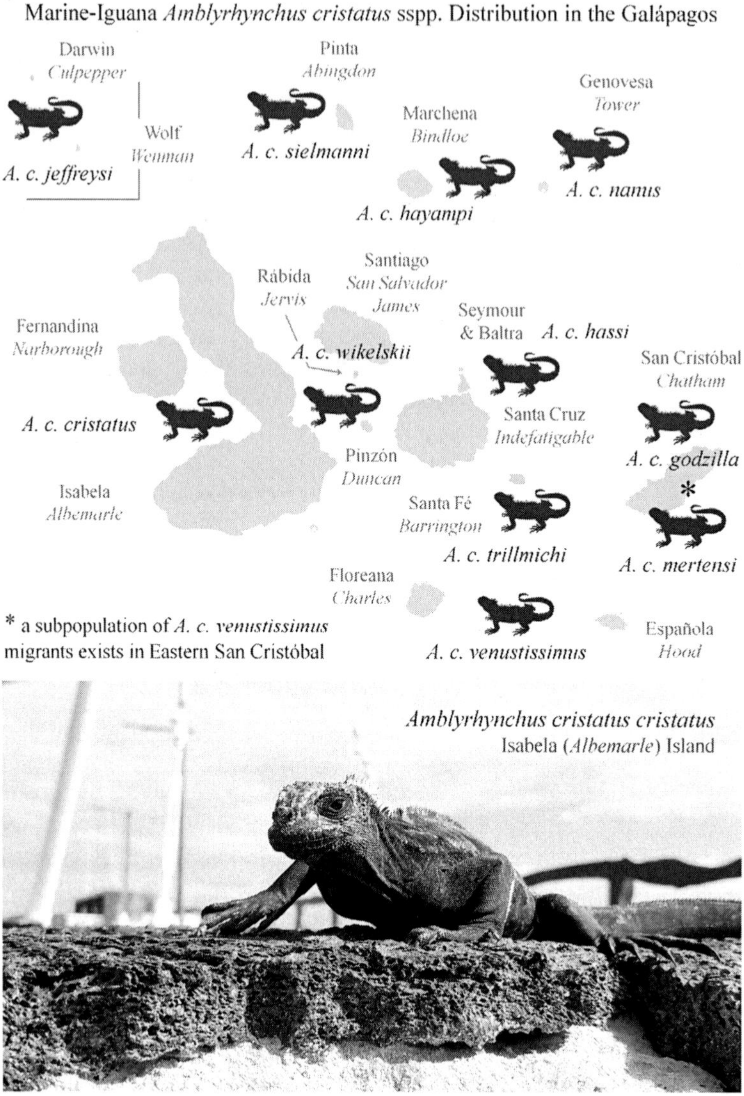

Fig. 8-11 Occurrence and distribution of marine iguanas (*Amblyrhynchus cristatus* sspp.) in the Galápagos Archipelago. There are 11 subspecies currently recognized as per nDNA markers (phylogeny in Fig. 8.12b). San Cristóbal has two distinct lineages (*A. c. godzilla* and *A. c. mertensi*). *A subpopulation of Española migrants/hybrids (*A. c. venustissimus*) exists in Eastern San Cristóbal. Layout and photo: GPC, after Miralles et al. (2017).

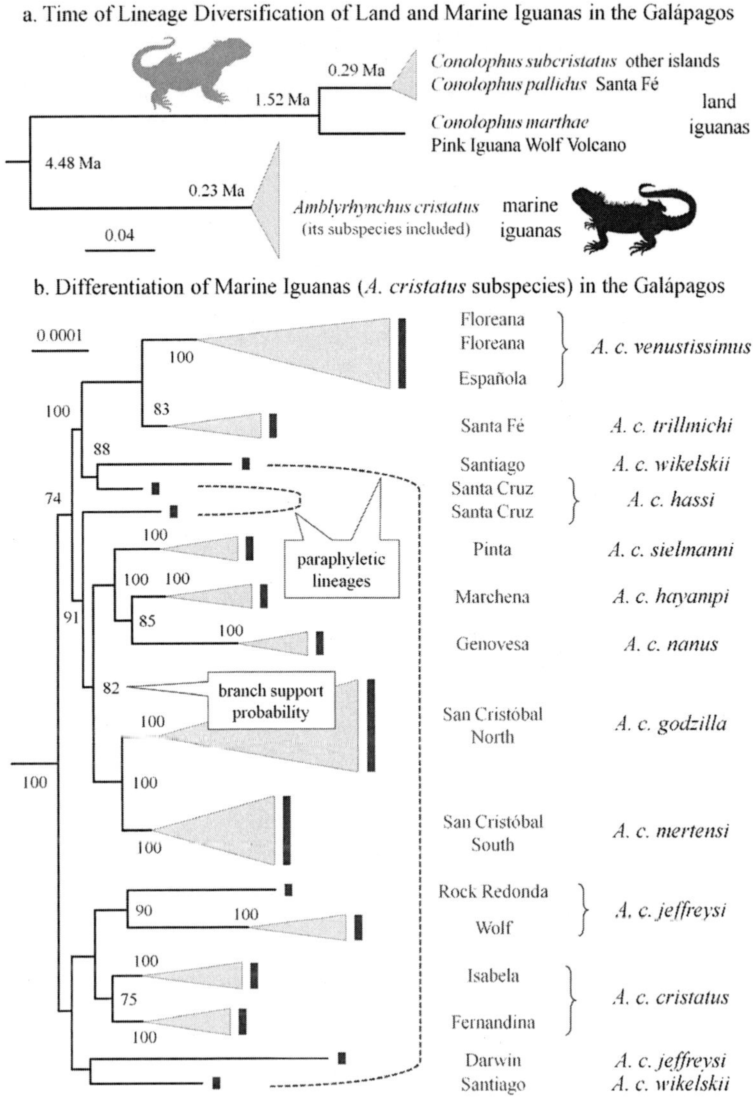

Fig. 8-12 Timeline of lineage diversification between land and marine iguanas (a) as well as genetic differentiation patterns within marine iguanas (b) in the Galápagos Archipelago. Layout: GPC. a. Maximum parsimony credibility tree based on mtDNA; simplified after MacLeod et al. (2015). b. Maximum likelihood phylogeny based on restriction site-associated nDNA sequencing RADseq; simplified after Miralles et al. (2017).

Lava Lizards

There are 9 species of lava lizards (*Microlophus*) in the Galápagos (Figs. 8.13-8.14). They are distributed in 14 islands (and surrounding islets): Fernandina and Isabela (*M. albemarlensis*); Pinta (*M. pacificus*); Marchena (*M. habeli*); Santiago and Rábida (*M. jacobi*); Santa Cruz, Baltra, and Seymour (*M. indefatigabilis*); Pinzón (*M. duncanensis*); San Cristóbal (*M. bivittatus*); Floreana (*M. grayii*); Española (*M. delanonis*); and Santa Fé (*M. barringtonensis*; Benavides et al. 2009; Arteaga et al. 2019). Their closest living relatives reside in Northwestern South America (Ecuador and Peru), but the story is intricate (Fig. 8.15).

Take a close look at the phylogenetic chronogram in Fig. 8.15. Two independent radiation events, "Western" or "Eastern" clades, seem to have given origin to the current species of lizards in the Galápagos. Both events are traceable to a common ancestor on the continent (3.69-4.54 Ma; center-left dialog box in Fig. 8.15). The Western radiation clade apparently spanned 1.69-3.69 Ma and gave origin to 7 of the 9 species in the islands (center-top dialog box in Fig. 8.15). The Eastern radiation clade dates back to 2.09-2.8 Ma and gave origin to 2 lineages: *M. bivittatus* in San Cristóbal and *M. habeli* in Marchena (center-bottom dialog box in Fig. 8.15; Kizirian et al. 2004; Benavides et al. 2009). The phylogenetic relations among species of *Microlophus* in the Galápagos and their sister taxa in Northwestern South America (Fig. 8.15 bottom right) are in agreement with pattern No. 9 of biogeographic affinities in Fig. 1.37, Chapter One (i.e. Ecuador-Peru-Chile; association with tectonic structures: East Pacific Rise; igneous plateaus; Grehan 2001; Heads and Grehan 2021).

In Fig. 8.16, we summarize three "phases" that explain the process of colonization and radiation of lava lizards to/within the archipelago (after Benavides et al. 2009). First Phase or "Eastern" and "Western" colonization (>2.2 Ma; Fig. 8.16a): South American ancestors of lava lizards arrived—separately and independently—in what today are the oldest Galápagos islands, i.e., Española and San Cristóbal. The Española population gave origin to the "Western" colonization and radiation (as in Fig. 8.16b), while the San Cristóbal population gave origin to the "Eastern" colonization and radiation, which eventually reached Marchena island (Fig. 8.16c).

Second Phase or "Western" colonization and radiation (1.0-2.2 Ma; Fig. 8.16b): lava lizards from Española colonized Santa Cruz, then Pinzón (closer to Santa Cruz than to other larger islands), and Floreana. Santa Cruz and Floreana Islands were colonized almost simultaneously.

Third Phase or "Eastern" and "Western" colonization and radiation—continued (<1.0 Ma; Fig. 8.16c): as younger islands continued to emerge, lizards from San Cristóbal arrived in Marchena (part of the "Eastern" colonization), while Santa Cruz's individuals dispersed to Santa Fé;

Fig. 8-13 Male and female lava lizards (*Microlophus albemarlensis*) from Isabela Island. Wetlands Trail (Sendero de Humedales), South Isabela. Photos: G. Paz-y-Miño-C.

Pinzon's lizards reached Santiago. Note that Isabela was likely colonized from Floreana; later, Fernandina and Pinta received immigrants from Isabela (part of the "Western" colonization).

Fig. 8-14 Distribution of Lava-Lizard species (*Microlophus*) in the Galápagos Archipelago. Layout/photos: GPC, redrawn and modified after Benavides et al. (2009). *Some authors consider Santa Fé lizards a separate species (*M. barringtonensis*; Arteaga et al. 2019).

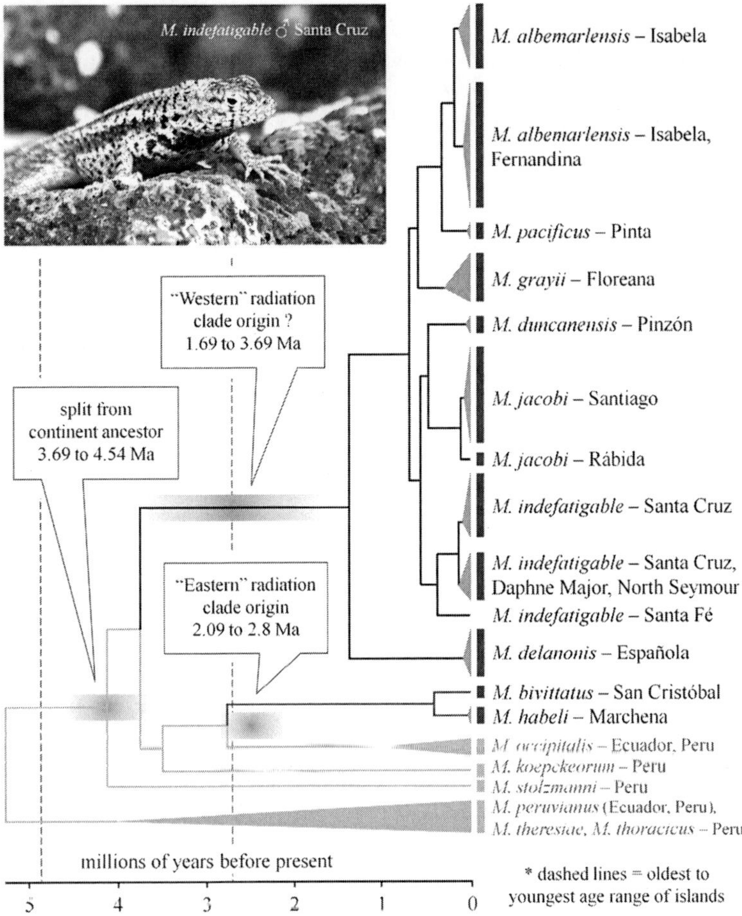

Fig. 8-15 Phylogenetic chronogram of *Microlophus* species (Lava Lizards) with respect to their radiation and occurrence in the Galápagos Archipelago. Lineage relatedness based on mtDNA and nDNA sequences (Bayesian and Maximum Likelihood methods). Layout/photo: GPC, redrawn and modified after Benavides et al. (2009).

The key points to keep in mind are that the current Galápagos Islands were colonized twice by ancestors of lava lizards (Figs. 8.15-8.16). Both the "Western-" and "Eastern-radiation" Galápagos clades share a common ancestor (3.69-4.54 Ma) with Peruvian *M. stolzmanni*. The "Western" clade (7 species), in turn, shares a common ancestor (≈1.69-3.69 Ma) with Galápagos' (2 species: San Cristóbal's and Marchena's, part of the "Eastern" clade) and the South American species *M. occipitalis* (Ecuador,

Peru), *M. koepckeorum*, and *M. stolzmanni* (both Peru). And the "Eastern" clade, which resulted from a second colonization and radiation (2.08-2.09 Ma), is sister taxa to *M. occipitalis* (Ecuador, Peru; Fig. 8.15).

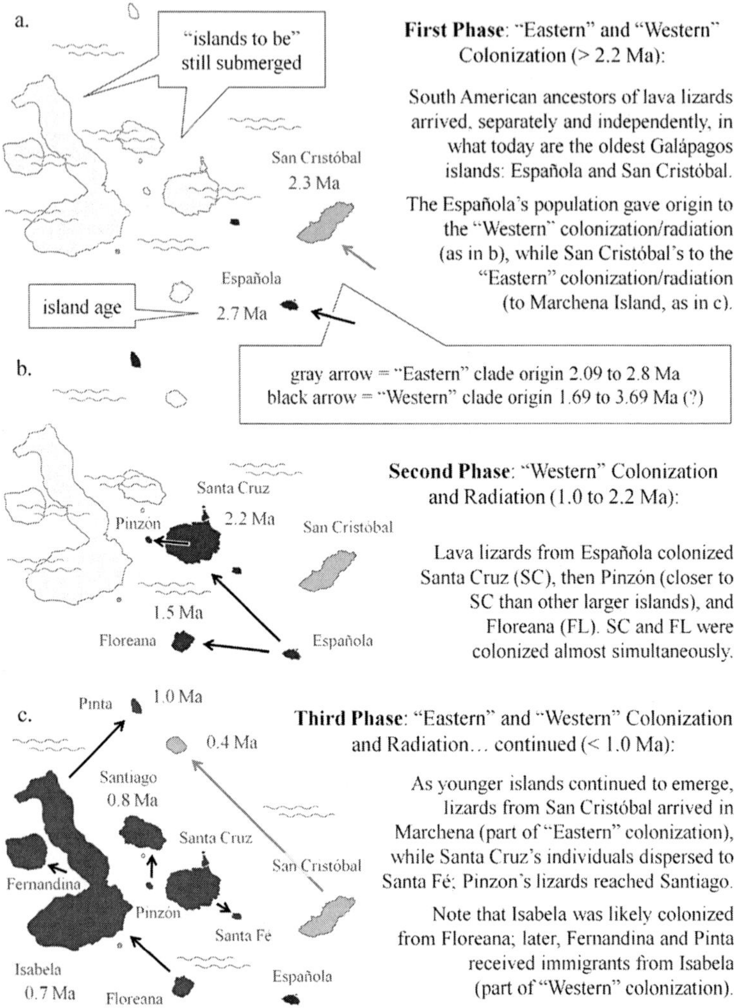

Fig. 8-16 "Eastern" and "Western" colonization/radiation of Lava Lizards (*Microlophus*) to/within the Galápagos Archipelago (for didactic purpose a-c = First to Third "phases"). Layout: GPC, redrawn and modified after Benavides et al. (2009). Note that >2.2 Ma Western/Central islands were not fully formed (see Fig. 8.8 and also Chapter One).

Leaf-toed Geckos

Ten lineages of leaf-toed geckos (*Phyllodactylus*) exist in the Galápagos (Fig. 8.17). They are distributed in 11 islands (and surrounding islets): Wolf (*P. gilberti*); Isabela (*P.* species 1); Santiago (*P.* species 2); Santa Cruz, Baltra, and Seymour (*P. galapagensis*); Pinzón (*P. duncanensis*); Santa Fé (*P. barringtonensis*); San Cristóbal (*P. darwini* and *P. leei*); Floreana (*P. baurii*); and Española (*P. gorii*; Torres-Carvajal et al. 2014, 2016). In addition, there are four introduced species into human-inhabited areas (*Hemidactylus frenatus*, *Lepidodactylus lugubris*, *Phyllodactylus reissii*, and *Gonatodes caudiscutatus*; Arteaga et al. 2019), but we will not discuss them here.

The phylogenetic analysis summarized in Fig. 8.17b suggests three events of colonization by *Phyllodactylus'* ancestors arriving in Galápagos, all with sister taxa in Northwestern South America (Ecuador, Peru). The oldest event took place 5.49 Ma and involved the subsequent diversification into 8 Galápagos taxa (consistent with the "older-than-the-islands" scenario, above; see gray background rectangle in Fig. 8.17, bottom right). A more recent colonization of San Cristóbal Island took place 3.03 Ma (*P. darwini* with sister taxa *P. leoni* from the Andes and South Ecuador). The most recent colonization (Wolf Island) occurred 0.69 Ma (*P. gilberti* with sister taxa *P. reisii* from the Pacific Coast of Northern South America; dialog boxes in Fig. 8.17b; Torres-Carvajal et al. 2016). Three independent, asynchronous colonizations indeed!

The phylogenetic relations among species of *Phyllodactylus* in the Galápagos and their sister taxa in Northwestern South America (Fig. 8.17b) are in agreement with pattern No. 5 of biogeographic affinities in Fig. 1.37, Chapter One (i.e. Western Americas and West Indies or "Caribbean", but disjunction in Central America; association with structures: Rifting of North and South America; formation of the Caribbean Plateau at the Galápagos hotspot; Grehan 2001; Heads and Grehan 2021).

Land Snakes or "Racer Snakes"

Nine species of land snakes or "racer snakes" (*Pseudalsophis*) exist in Galápagos (Fig. 8.18). They are distributed in 11 islands (and neighboring islets): Española (*P. hoodensis*); Santa Cruz (*P. steindachneri*); Santiago (*P. hephaestus*); Pinzón (*P. slevini*); Isabela and Tortuga (*P. darwini*); San Cristóbal (*P. biserialis*); Isabela, Fernandina and Tortuga (*P. occidentalis*); Seymour and Santa Fé (*P. dorsalis*); and Santiago and Rábida (*P. thomasi*). Their closest phylogenetic relative, *P. elegans*, is distributed in Northwestern South America (Zaher et al. 2018; Arteaga et al. 2019).

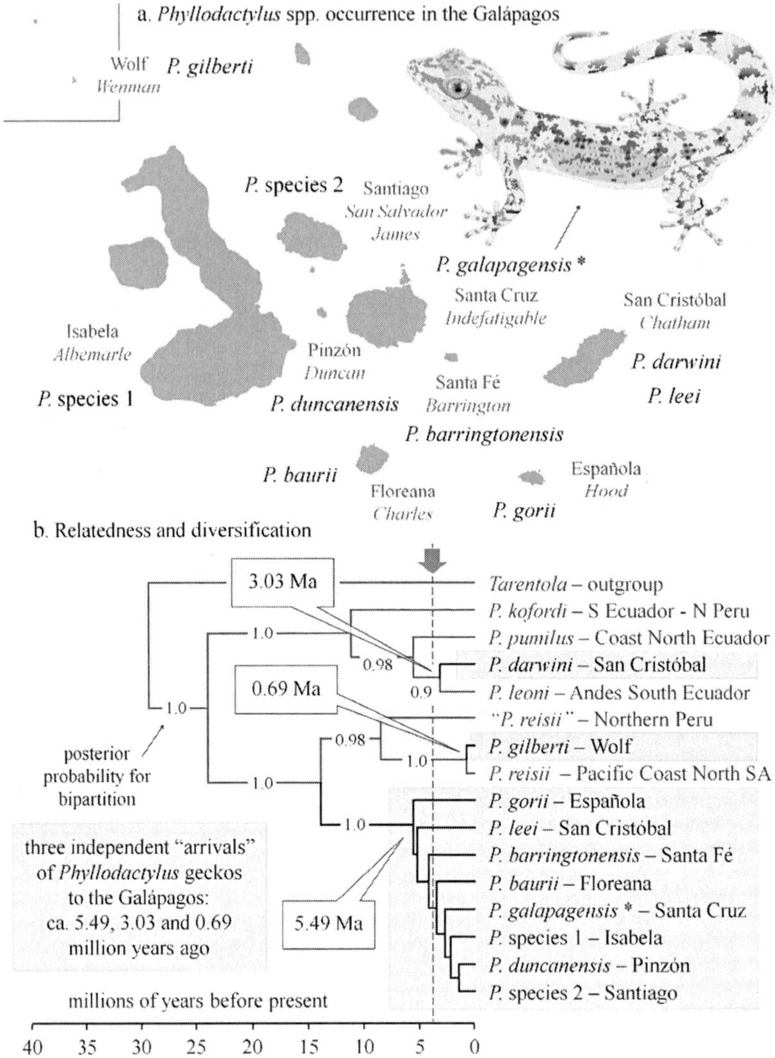

Fig. 8-17 Distribution (a), relatedness and diversification (b) of Galápagos' Leaf-toed Geckos (*Phyllodactylus*). Maximum Clade Credibility Tree (based on nDNA and mtDNA sequences). Layout: GPC, redrawn after Torres-Carvajal et al. (2014, 2016). *Digital sketch of *P. galapagensis* (exemplar *galapagensis* clade) after Arteaga et al. (2019).

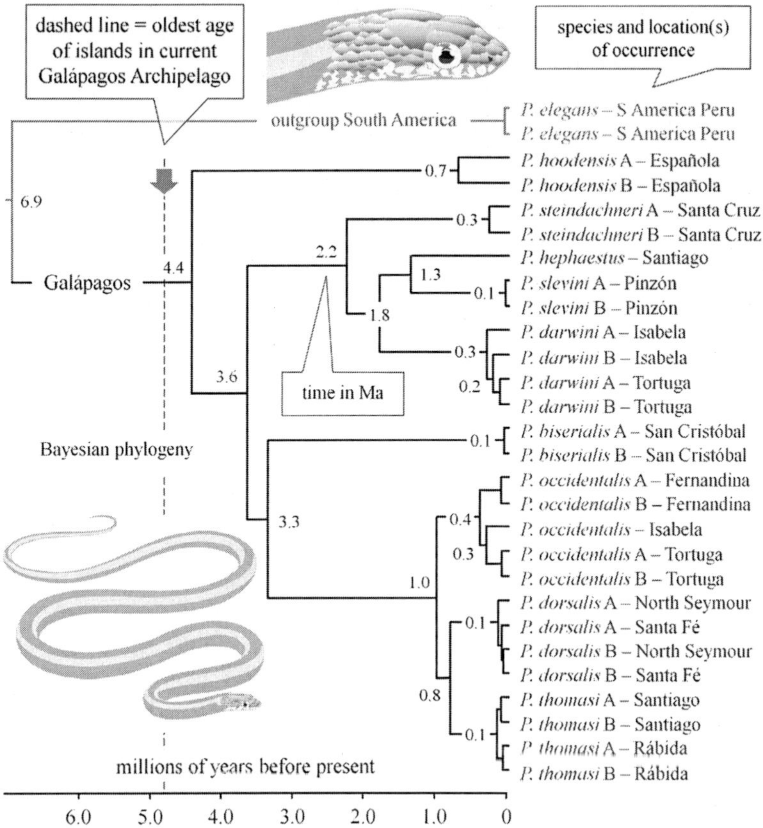

Fig. 8-18 Land-snakes (*Pseudalsophis*) species radiation in the Galápagos Archipelago. Phylogenetic reconstruction of *Pseudalophis* (calibrated Bayesian tree) based on mtDNA and nDNA sequences corresponding to two individual samples (A and B) per island, as well as the outgroup (*P. elegans*). Layout: GPC; phylogeny redrawn and modified after Zaher et al. (2018). Digital sketches of *P. steindachneri* after Arteaga et al. (2019).

Common ancestry between the Galápagos *Pseudalsophis* spp. and South American relatives dates back to 6.9 Ma (Fig. 8.18). The Galápagos lineages seem to have originated 4.4 Ma with significant diversification in the last 1-2 Ma (Fig. 8.18; Zaher et al. 2018). It is conceivable that ancestors of the Galápagos racer snakes inhabited paleo-islands, which are now submerged seamounts (above). This is in agreement with pattern No. 9 of biogeographic affinities in Fig. 1.37, Chapter One (i.e. Ecuador-Peru-

Chile; association with tectonic structures: East Pacific Rise; igneous plateaus; Grehan 2001; Heads and Grehan 2021).

Sea Lions and Fur Seals

Galápagos sea lions (*Zalophus wollebaeki*) and fur seals (*Arctocephalus galapagoensis*) belong to separate lineages of carnivore-marine mammals, the subfamilies cf. Otariinae and cf. Arctocephalinae, respectively (ancestor dates back to 8 Ma). Sea lions breeding colonies are distributed in the entire archipelago (Figs. 8.19-8.20a), while fur seals are noticeably located toward the West (Fig. 8.20b; for population genetics assessment, see Chaves et al. 2022). It appears that a "Western [sub]population" of sea lions, which is located around Fernandina Island and its proximate waters to Northern Isabela, might be a separate lineage or a "subspecies or species in the making" (see dialog box in Fig. 8.20a top left; Wolf et al. 2008).

The Galápagos sea lions' close phylogenetic relatives (as cf. Otariinae) are distributed in the North Pacific; its sister taxa is the California sea lion (*Z. californianus*; Fig. 8.21 top). The common ancestor of *Z. wollebaeki* and *Z. californianus* dates back to at least 2.3 Ma (Wolf et al. 2007); but the colonization and dispersal within the Galápagos islands seem recent, <1.0 Ma (dialog box Fig. 8.21 top). Both Galápagos and California sea lions are sister taxa to the already extinct (1970s) Japanese sea lion (*Z. japonicus*; Wolf et al. 2007). Such phylogenetic relations suggest a Northern arrival and later colonization of sea lions to/within the Galápagos (after Churchill et al. 2014; Berta et al. 2018). This is consistent with pattern No. 1 of biogeographic affinities in Fig. 1.37, Chapter One (i.e. Central and West Pacific; structures: East Pacific Rise; Western Pacific plate margin; igneous plateau translation; Grehan 2001; Heads and Grehan 2021).

By contrast, the Galápagos fur seals' closest phylogenetic relative (within the cf. Arctocephalinae) is the South American fur seal (*A. australis*). Note in Fig. 8.21 (bottom) that both *A. galapagoensis* and *A. australis* are sister taxa to the New Zealand fur seal (*A. forsteri*). These three lineages diverged <1.0 Ma (Fig. 8.21, bottom right). The distribution of other members of cf. Arctocephalinae includes: South Africa, Australia, and New Zealand (*A. pusillus*); Southern Ocean (*A. tropicalis*, *A. gazella*); South America (*A. philippii*); and only one of them in the North Pacific (*A. townsendi*; Fig. 8.21 bottom). Thus, a Southern arrival and later colonization of fur seals to/within the Galápagos is suspected (after Churchill et al. 2014; Berta et al. 2018). This is in agreement with pattern No. 10 of biogeographic affinities in Fig. 1.37, Chapter One (i.e. Chile and Argentina;

association with tectonic structure: East Pacific Rise; igneous plateaus; Grehan 2001; Heads and Grehan 2021).

As readers might notice, while Otariinae roam in the Northern oceans, the Arctocephalinae roam in the South. Both distinctive distributions.

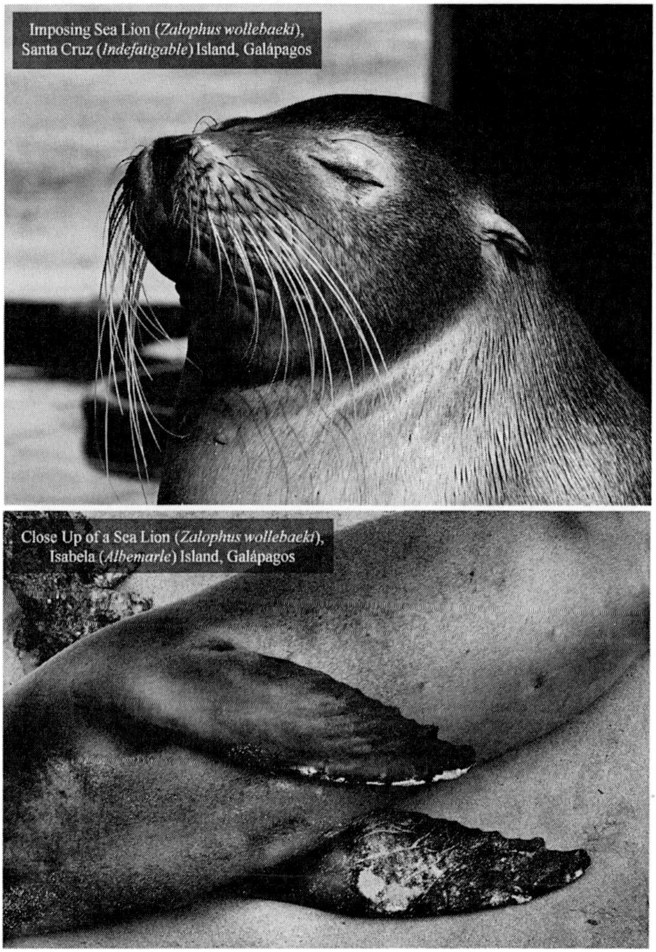

Fig. 8-19 Imposing sea lion (*Zalophus wollebaeki*) from Santa Cruz Island (top) and detail of a juvenile individual from Isabela Island (bottom). Galápagos sea lions' close phylogenetic relatives (within the subfamily Otariinae) are distributed in the North Pacific. Its sister taxa is the California sea lion (*Z. californianus*). The common ancestor of both Galápagos and California sea lions dates back to <1.0 Ma (Fig. 8.21). Photos: G. Paz-y-Miño-C.

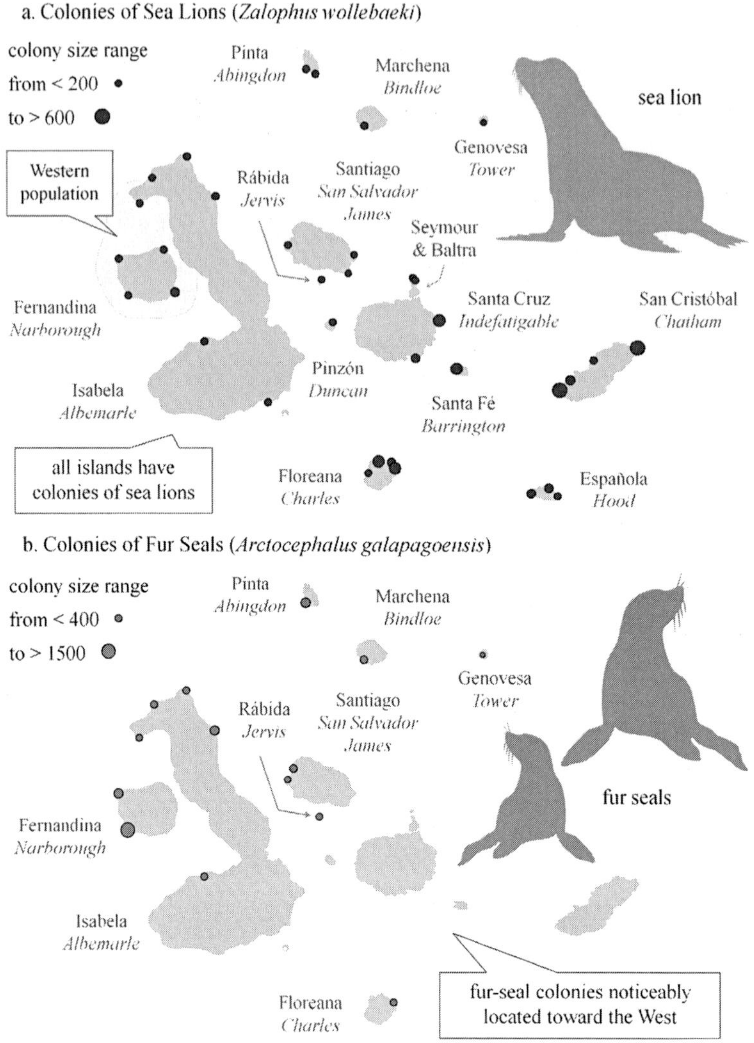

Fig. 8-20 Breeding colonies of sea lions (*Zalophus wollebaeki*) and fur seals (*Arctocephalus galapagoensis*) in the Galápagos Archipelago. Sea lions' colonies are distributed in the entire archipelago (a), while fur seals are noticeably located toward the West (b), although colonies in the Northern islands (Pinta and Marchena) are sizable. The "Western population" of sea lions (dialog box top left) might be a separate lineage or a "subspecies or species in the making" (Wolf et al. 2008). Layout/sketches GPC; data from Fundación Charles Darwin (FCD) and World Wildlife Fund (WWF) Ecuador (2018); and Chaves et al. (2022).

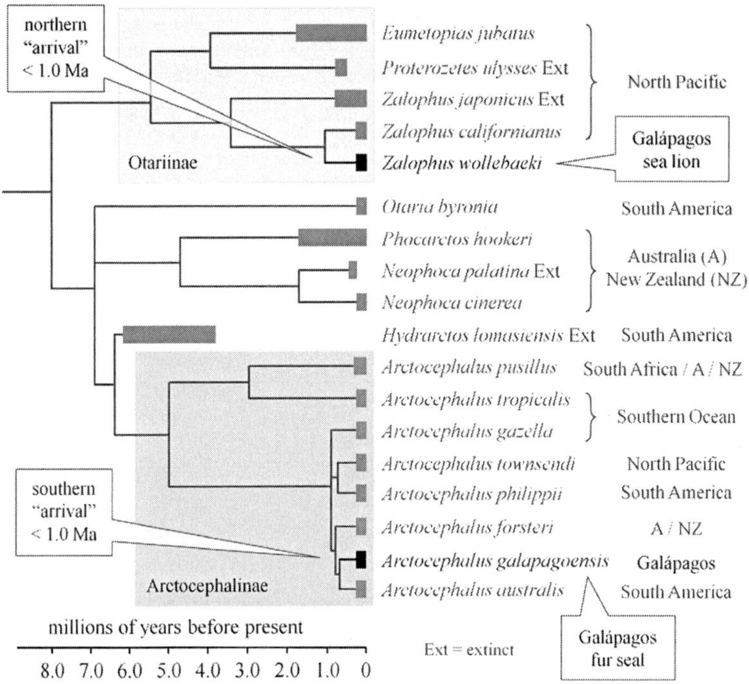

Fig. 8-21 Phylogenetic placement of Galápagos sea lions (*Zalophus wollebaeki*) and fur seals (*Arctocephalus galapagoensis*) with respect to their relatives and biogeographic distributions. Background tree based on combined evidence (morphology, mtDNA and nDNA sequences). Layout: GPC, redrawn/modified after Churchill et al. (2014), Berta et al. (2018).

"Flightless" Weevils

At least 10 species of "flightless" weevils (*Galapaganus*) occur in the Galápagos (Fig. 8.22). They are distributed in all main islands (and surrounding islets): *G. darwini* (Darwin and Wolf); *G. conwayensis* (Pinta, Isabela, and Santa Cruz); *G. crockeri* (Marchena and Genovesa); *G. blairi* (Santiago); *G. williamsi* (Isabela); *G. galapagoensis* (San Cristóbal); *G. collaris* (San Cristóbal and Floreana); *G. vandykei* (San Cristóbal, Floreana and Española); *G. caroli* (Floreana); and the introduced lineage from continental Ecuador, *G. h. howdenae* (Santa Cruz). The common ancestor of *Galapaganus* and the mainland taxa (e.g. *G. h. howdenae*) dates back to 8.6-11.5 Ma; this links the archipelago's species to Northwestern South America (Fig. 8.23; Sequeira et al. 2008).

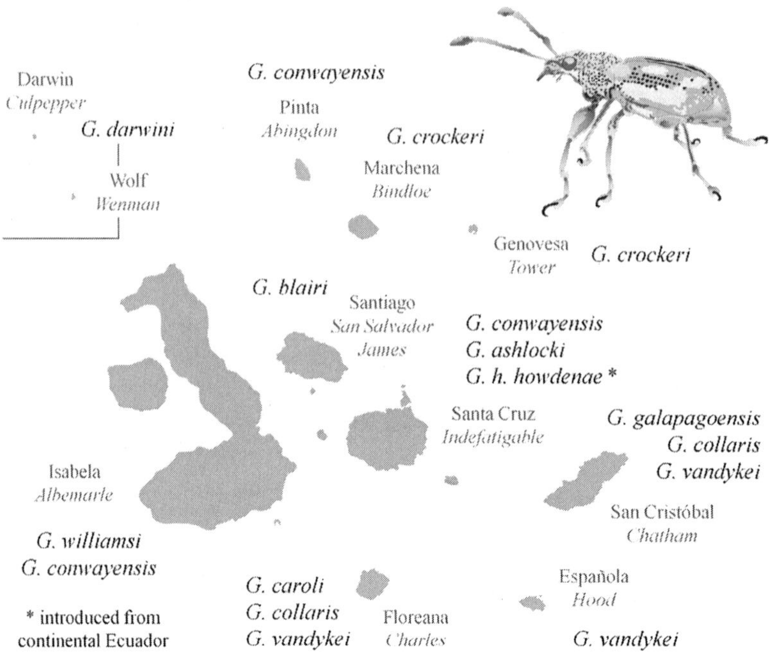

Fig. 8-22 Occurrence of "flightless" weevils (*Galapaganus* spp.) in the Galápagos. Layout/digital sketch: GPC; redrawn and modified after Sequeira et al. (2008).

Note in Fig. 8.23 that the Galápagos clade alone dates back to >8 Ma ("older than the islands"; above and Sequeira et al. 2000), and that both the current pattern of island distribution (Fig. 8.22) plus phylogenetic branching and sister-taxa connections to Northwestern South America (Fig. 8.23) suggest a complex sequence of inter-island colonization and *in situ* differentiation of *Galapaganus*. This is consistent with the metapopulation vicariance model (*Hiii* above), in which groups of organisms originated *in situ* during the breakup of widespread ancestral populations (in the deep geological past), and particularly with pattern No. 9 of biogeographic affinities in Fig. 1.37, Chapter One (i.e. Ecuador-Peru-Chile; association with tectonic structures: East Pacific Rise; igneous plateaus; Grehan 2001; Heads and Grehan 2021).

Galapaganus flightless weevils are likely exemplars of biogeographic processes and patterns common to other groups of terrestrial invertebrates in the Galápagos (i.e. ancient lineages). Of the 2,000 species of insects and 350 species of land arthropods known to occur in the archipelago, 486 are beetles. Flightlessness is frequent on islands (Peck 2006).

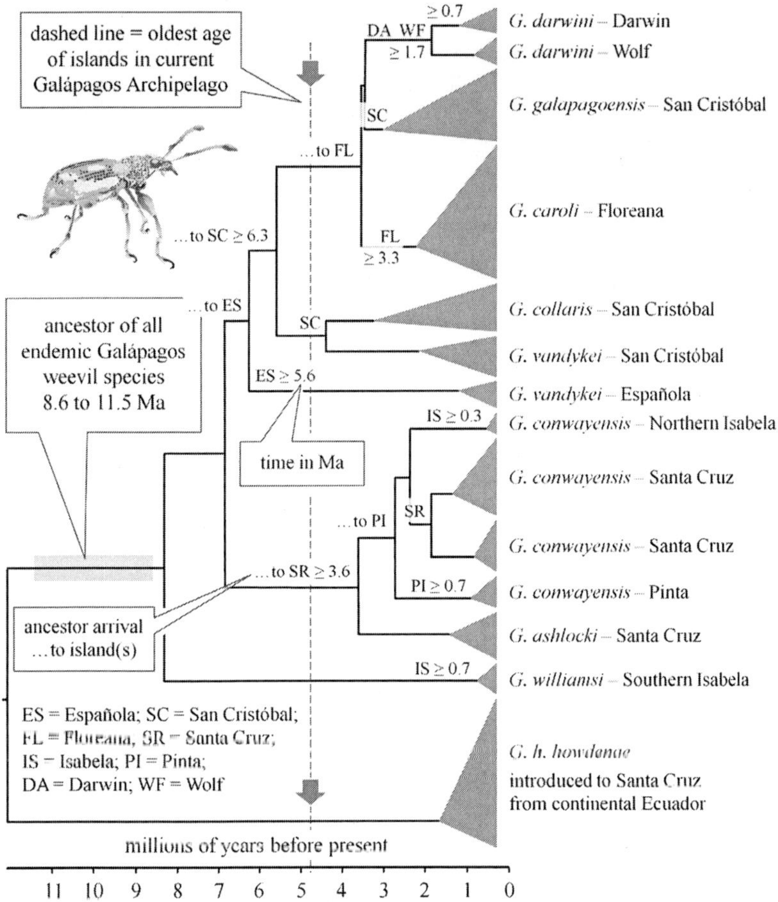

Fig. 8-23 Colonization and diversification time-range of the Galápagos' "flightless" weevil (*Galapaganus* spp.) as per mtDNA phylogeny and archipelago's age. Layout/digital sketch: GPC; simplified after Sequeira et al. (2008).

Carpenter Bees

A single species of carpenter bee (*Xylocopa darwini*) is distributed on nine islands in the Galápagos (Fig. 8.24a). Mitochondrial DNA (mtDNA) haplotype-network analyses suggest that two independent colonizations (<1.5 Ma) by mainland ancestors of *X. darwini* gave origin to today's populations in the archipelago (Fig. 8.24b; Vargas et al. 2015).

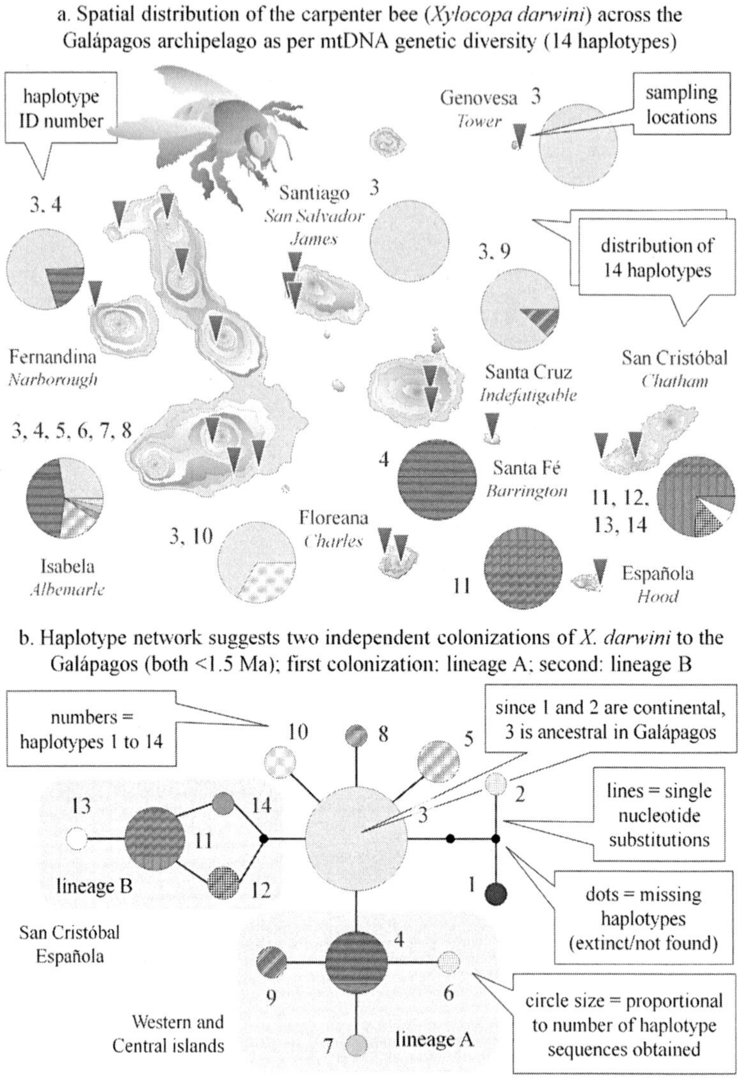

Fig. 8-24 Spatial distribution of carpenter bees (*Xylocopa darwini*) in the Galápagos (a), and mtDNA haplotype network of various populations used as proxy to infer patterns of colonization/dispersal within the archipelago (b). Haplotypes 1 and 2 come from continental species (Ecuador and Argentina) far-distantly related to *X. darwini*. Layout/digital sketch: GPC; redrawn and modified after Vargas et al. (2015).

The haplotype network shown in Fig. 8.24b is quite revealing. Realize that 14 mtDNA haplotypes (numbers 1 to 14) have been used by scientists to infer patterns of colonization-dispersal (Vargas et al. 2015). Haplotypes 1 and 2 belong to continental species (*Neoxylocopa* sp. from coastal Ecuador and *X. ordinaria* from Argentina) far-distantly related to *X. darwini*. Since haplotype 3 is linked to them and to all others in the network, 3 is considered "ancestral" in the Galápagos (Fig. 8.24b). In fact, haplotype 3 is present on six of the nine islands (Isabela, Fernandina, Santiago, Santa Cruz, Floreana, and Genovesa).

Note in Fig. 8.24b that lineage A (first colonization event) shares haplotypes 4, 6, 7, and 9, which are distributed in Isabela, Fernandina, Santa Cruz, and Santa Fé. Lineage B (second colonization) shares haplotypes 11-14, which are restricted to San Cristóbal and Española. Keep in mind that inter-island connectivity during the last glacial maximum (Fig. 8.8d) likely influenced the colonization and dispersal of carpenter bees and other "recent arrivals" to the Galápagos.

Sand Spiders

The sand spider *Sicarius utriformis* is widely distributed in the Galápagos. There are 21 species in the genus *Sicarius* in South and Central America, with a common ancestor dating back to >50 Ma. In fact, the South and Central American species share deep ancestry (>80 Ma) with the South African genus *Hexophthalma* (Fig. 8.25a; Magalhaes et al. 2021).

Morphological and molecular data (mtDNA, nDNA, rRNA) suggest the Galápagos species is sister taxa to the Peruvian *S. peruensis* (with a common ancestor 13-15 Ma, mean 8.8 Ma), which indicates biogeographic affinity with Western South America (Fig. 8.25a). The close phylogenetic relation between Galápagos and Peruvian sand spiders has robust statistical support (72-97% confidence; see thick arrow in Fig. 8.25a).

In Fig. 8.25b, we sketch three biogeographic scenarios for the "arrival"—or colonization—of sand spiders to/within the Galápagos. Scenario 1: arrival/colonization to/within a recently originated archipelago (3-5 Ma). Scenarios 2 and 3: arrival/colonization to/within paleo-islands currently submerged (15 Ma). Scenario 1 is consistent with *Hi* overwater dispersal from mainland (above). Scenarios 2 and 3 are consistent with *Hii* stepping-stone dispersal from mainland (above). Realize that metapopulation vicariance (*Hiii* above), in which sand spiders originated *in situ* during the breakup of widespread ancestral populations (deep ancestry), is a possibility (as per Grehan 2001; Heads and Grehan 2021).

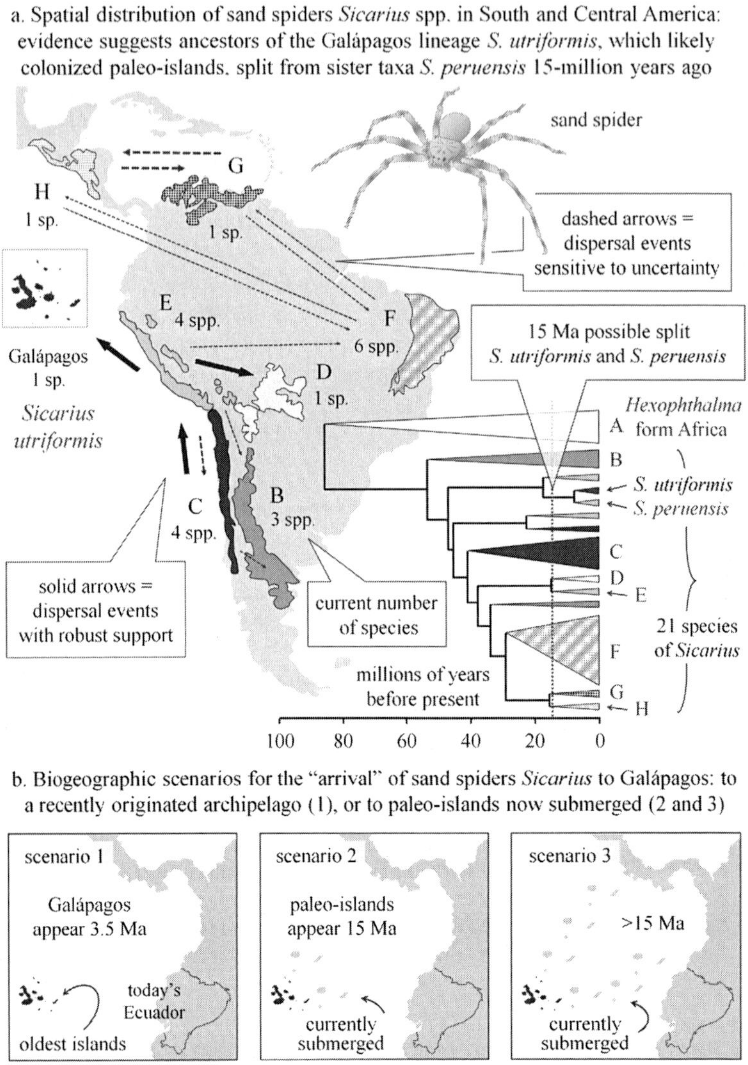

Fig. 8-25 Biogeographic affinities of Galápagos sand spiders (*Sicarius utriformis*) inferred from the distribution of other *Sicarius* species in South/Central America and their phylogenetic relations (a), as well as scenarios for the colonization/dispersal of sand spiders in the archipelago (b). Maximum clade credibility (MCC) tree based on morphology and molecular data (mtDNA, nDNA, rRNA sequences; Magalhaes et al. 2019). Layout/digital sketch: GPC; a-b redrawn and modified after Magalhaes et al. (2021).

Land Snails

At least 50 species of *Naesiotus* land snails exist in the Galápagos (Fig. 8.26). They are broadly distributed in the archipelago (Fig. 8.27a; Phillips et al. 2020). Although their biogeographic relations remain uncertain, *Naesiotus* spp. are likely connected to South America (Parent 2008).

Look at the phylogenetic reconstruction of the Galápagos *Naesiotus* lineages in Fig. 8.26. Within-island differentiation is typical and has probably resulted from habitat heterogeneity (Parent 2012). The common ancestor of all Galápagos taxa dates back to >3.0 Ma. Out of the 51 species accounted for in the phylogeny, half occur in San Cristóbal (17), Floreana (6), and Española (2), the oldest islands. They form a monophyletic group with 2-3 Ma ancestry (Fig. 8.26, top right). Species from all other islands (Santiago, Santa Cruz, Isabela, Fernandina, Pinta, and Rábida) cluster in a monophyletic group of their own, with 2.5-3 Ma ancestry (Fig. 8.26 bottom). Note that each caldera volcano in Isabela or Fernandina Islands has 1-2 species (see close-up view of phylogeny in Fig. 8.26, top left); this diversification is recent and likely took place within <1.0 Ma (Phillips et al. 2020).

In Fig. 8.27a, we sketch a suspected colonization sequence of ancestors of *Naesiotus* to the Galápagos, with inter-island colonization events indicated by arrows (after Parent and Crespi 2006; Phillips et al. 2020). Merge the information provided in Fig. 8.27ab. Realize that the colonization of the current archipelago is traceable—first—to the oldest islands of San Cristóbal (2.4-4.0 Ma), Española (3.0-3.5 Ma), and Floreana (1.5-2.3 Ma). Back-and-forth snail exchange between San Cristóbal and Española, or San Cristóbal and Floreana, likely occurred (two-way arrows in Fig. 8.27a). Inter-island colonization also took place between Santa Cruz (1.1-2.3 Ma) and Santiago (0.8-1.4 Ma). Snails in Rábida and Pinta (0.7-0.8 Ma) probably came from Santiago, while Isabela's (0.5-0.8 Ma) seem to have originated in both Santiago and Santa Cruz (but keep in mind that Isabela's volcanoes gave origin to their own and/or neighbor-volcano lineages). The most recent snail inter-island colonization and/or exchange apparently involved Isabela and Fernandina (0.03-0.06 Ma).

When plotting *Naesiotus* clade age as function of island or volcano-in-island age, as in Fig. 8.27c, the statistical correlation between variables is sustained (R^2=0.65; p≤0.01). Thus, the notion that taxa colonization and subsequent within-island diversification take place in archipelagos, from older to younger islands (or individual volcanoes since they function as islands in the landscape), seems to apply to *Naesiotus* in the Galápagos (i.e. "progression rule"; Shaw and Gillespie 2016; Phillips et al. 2020).

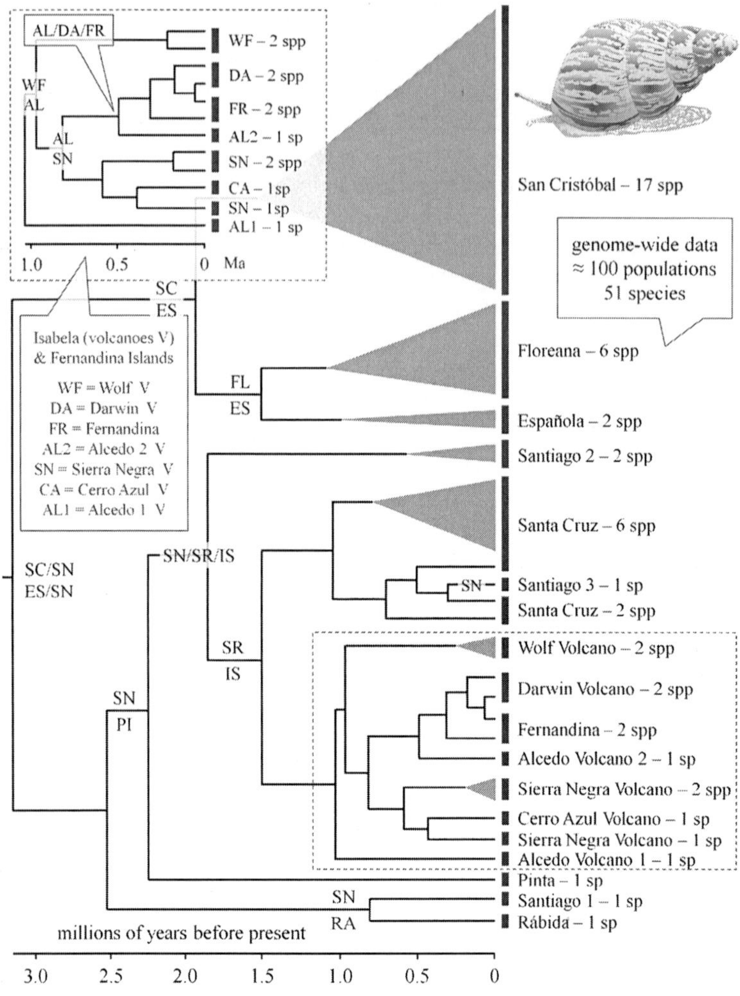

Fig. 8-26 Phylogenetic reconstruction of Galápagos land-snail lineages (*Naesiotus*) as per nDNA and time of taxa divergence in the archipelago (Bayesian inference methods). The common ancestor of all the Galápagos taxa dates back to >3.0 Ma. Out of the 51 species accounted for in the phylogeny, half occur in San Cristóbal (*N*=17), Floreana (*N*=6), and Española (*N*=2), the oldest islands. Note that each caldera volcano in Isabela or Fernandina has 1-2 species (close-up view of phylogeny top left); this diversification is recent and took place within <1.0 Ma. Within-island differentiation is typical and has probably resulted from habitat heterogeneity (i.e. *in situ* diversification per island/caldera-volcano). Layout/digital sketch: GPC, redrawn and modified after Phillips et al. (2020).

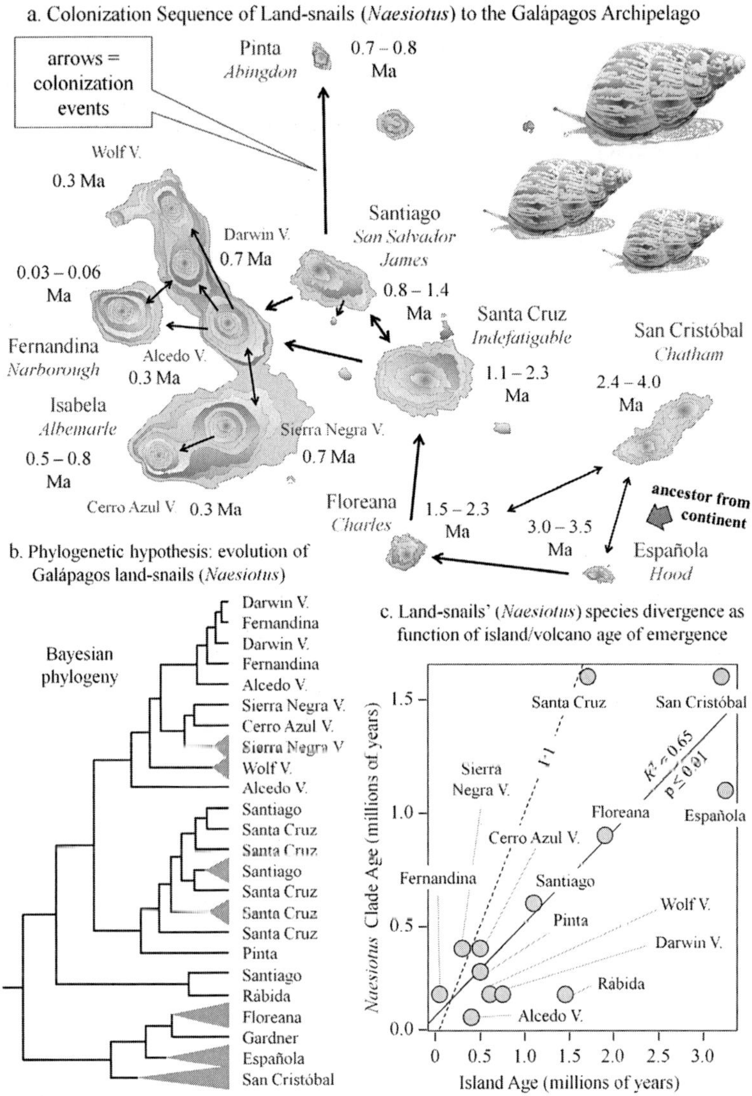

Fig. 8-27 Suspected colonization sequence of ancestors of *Naesiotus* to the Galápagos (a) inferred from phylogeny (b) and island age (c). Layout/digital sketch: GPC, redrawn and modified after: a. Parent and Crespi (2006), Phillips et al. (2020); min-max emergence ages of islands: Geist et al. (2014); Isabela volcanoes' ages: Naumann et al. (2002), Parent and Crespi (2006). b-c. Phylogeny and plot: Phillips et al. (2020).

Snapdragon Shrub

Three subspecies of snapdragon (*Galvezia leucantha* sspp.) exist in the Galápagos; they are distributed on four islands: *G. l. leucantha* (Isabela and Fernandina), *G. l. porphyrantha* (Santiago), and *G. l. pubescens* (Rábida; Fig. 8.28a; after Guzmán et al. 2017). Their closest phylogenetic relatives (*G. fruticosa* and *G. elisensii*) occur in South America (Northwestern Peru). The estimated divergence of the Galápagos *Galvezia* lineage dates back to 0.09-0.66 Ma (see dialog box in Fig. 8.28b).

The phylogenetic relations among the subspecies of *Galvezia leucantha* in the Galápagos and their sister taxa in Northwestern South America (Northwestern Peru and more distant relatives in Ecuador, e.g. *G. grandiflora*; see Fig. 8.28b bottom) are in agreement with pattern No. 9 of biogeographic affinities in Fig. 1.37, Chapter One (i.e. Ecuador-Peru-Chile; association with tectonic structures: East Pacific Rise; igneous plateaus; Grehan 2001; Heads and Grehan 2021).

Interestingly, the snapdragon shrubs of the Galápagos are pollinated by insects, particularly the carpenter bee, *Xylocopa darwini* (above). By contrast, the mainland relatives are pollinated by birds, particularly hummingbirds. A shift from bird-mediated to insect-mediated pollination has apparently taken place once *Galvezia leucantha*'s ancestors diverged in the archipelago (Guzmán et al. 2017).

Prickly-Pear Trees

Opuntia prickly-pear trees are the most conspicuous life form of the Galápagos flora, particularly in lowland arid habitats (e.g. deciduous forests and shrublands). They are distributed throughout the archipelago (Figs. 8.29-8.30). Traditionally, 6 morphospecies have been recognized, including 11 subspecies (varieties) within them (Fig. 8.29a; after Wiggins and Porter 1971 with input from Helsen et al. 2009): (1) *Opuntia galapageia* (Santiago, Rábida, Pinzón, Pinta, and Eastern Isabela); its subspecies include *O. g. galapageia*, *O. g. macrocarpa*, and *O. g. profusa*. (2) *Opuntia echios* (Santa Cruz, Baltra, Seymour, and Southern Isabela); its subspecies include *O. e. echios*, *O. e. barringtonensis*, *O. e. gigantea*, *O. e. intermis*, and *O. e. zacana*. (3) *Opuntia megasperma* (San Cristóbal, Española, and Floreana); its subspecies include *O. m. megasperma*, *O. m. mesophytica*, and *O. m. orientalis*. (4) *Opuntia insularis* (Fernandina, Northern and Central Isabela). (5) *Opuntia saxicola* (Southern Isabela). And (6) *Opuntia helleri* (Marchena, Genovesa, Darwin, and Wolf).

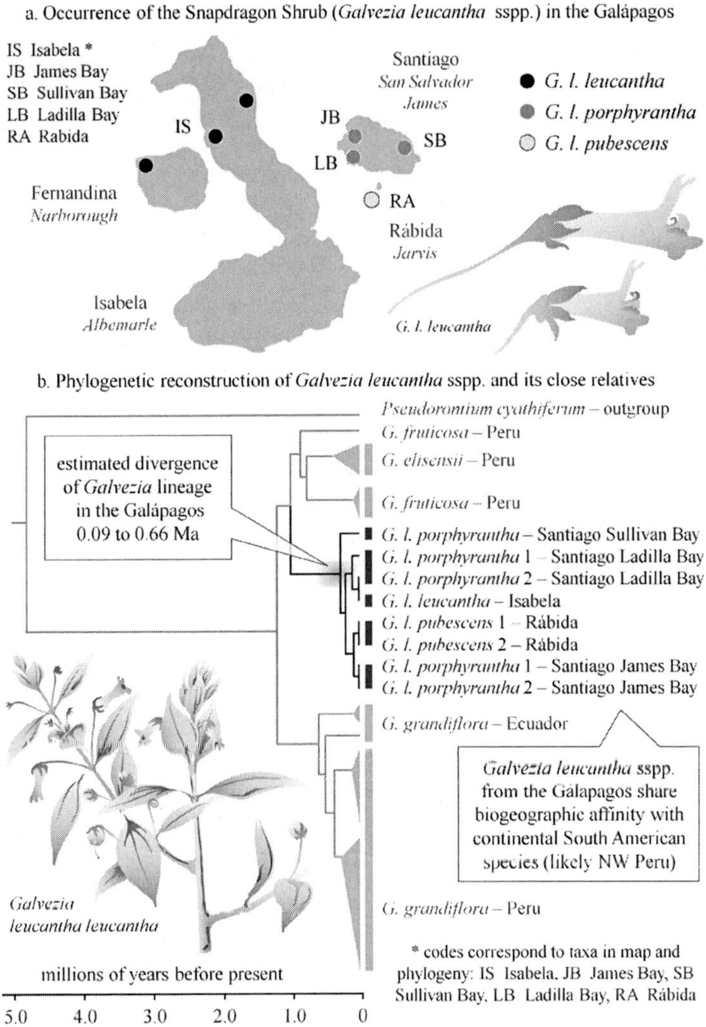

Fig. 8-28 Occurrence of the Snapdragon shrub (*Galvezia leucantha* sspp.) in the Galápagos (a); and phylogeny showing its close and distant relatives (b). There are three subspecies of snapdragon in the islands: *G. l. leucantha* (Isabela and Fernandina), *G. l. porphyrantha* (Santiago), and *G. l. pubescens* (Rábida). The Galápagos lineages share biogeographic affinity with continental South America (NW Peru). Maximum clade credibility (MCC) tree based on nDNA and cpDNA sequences (Bayesian inference). Layout: GPC; digital sketches after Wiggins and Porter (1971); a-b redrawn and modified after Guzmán et al. (2017).

Phylogenetic and biogeographic analyses suggest ancestors of *Opuntia* originated in Southern or Southwestern South America during the late Miocene (10-5 Ma; Arakaki et al. 2011). The early lineages expanded to the Central Andean Valleys and Western North America. Interestingly, the North American populations diverged further (last 5 Ma) and dispersed back South, toward what today is Central America and the Caribbean, and eventually "returned" to South America (Majure et al. 2012). The North American radiation was significant. It often gave origin to taxa via the merging of ancestor lineages (reticulated evolution) and polyploidy. Because the closest relatives of the Galápagos' *O. galapageia*, *O. echios*, and *O. megasperma* (Fig. 8.29b) are currently distributed in South America (South-Southeast; South; South-Southwest; West; Northwest), as noted in the phylogeny in Fig. 8.29b, it is suspected that their ancestors originated in Western South America and likely colonized the archipelago during just 1 Ma (Helsen et al. 2009; Majure et al. 2012). This is consistent with pattern No. 7 of biogeographic affinities in Fig. 1.37, Chapter One (i.e. South America and Caribbean; association with tectonic structures: Andean uplift, Romeral fault system, and Southern Caribbean transform margin; as per Grehan 2001; Heads and Grehan 2021).

Note that it has been a challenge to differentiate genetically among the *Opuntia* lineages (as sketched in the abridged phylogeny in Fig. 8.29b; for complete analysis, see Majure et al. 2012). In fact, as shown in the phylogenetic reconstruction in Fig. 8.31, low genetic variability among populations is ubiquitous (as per 240 individual samples collected at 15 locations in diverse islands; Helsen et al. 2009). This contrasts with the classically rich morphospecies attributions (as per Wiggins and Porter 1971; see dialog box in Fig. 8.31, top right). For example, there is little genetic differentiation between *O. echios* and *O. galapageia* (arguably two distinguishable morphospecies), although both branch as monophyletic groups A and B (B within A) in Fig. 8.31. The same occurs with *O. megasperma*, *O. galapageia*, and *O. insularis*, which seem genetically proximate to one another regardless of their morphological differences (Fig. 8.31, bottom right). Thus, no clear correlation appears to exist between morphospecies distinctiveness and genetic differentiation. Yet, inter-island gene flow and long-distance dispersal are both limited (Helsen et al. 2009). It may be the case that the genetic markers used for the phylogenetic reconstruction (i.e. cpDNA and nrDNA sequences) lack robustness to produce clear taxonomic groups. But it seems more likely that the *Opuntia* lineages in question (not only in the Galápagos but elsewhere) have indeed low genetic divergence and high anatomical variation (Helsen et al. 2009; Majure et al. 2012).

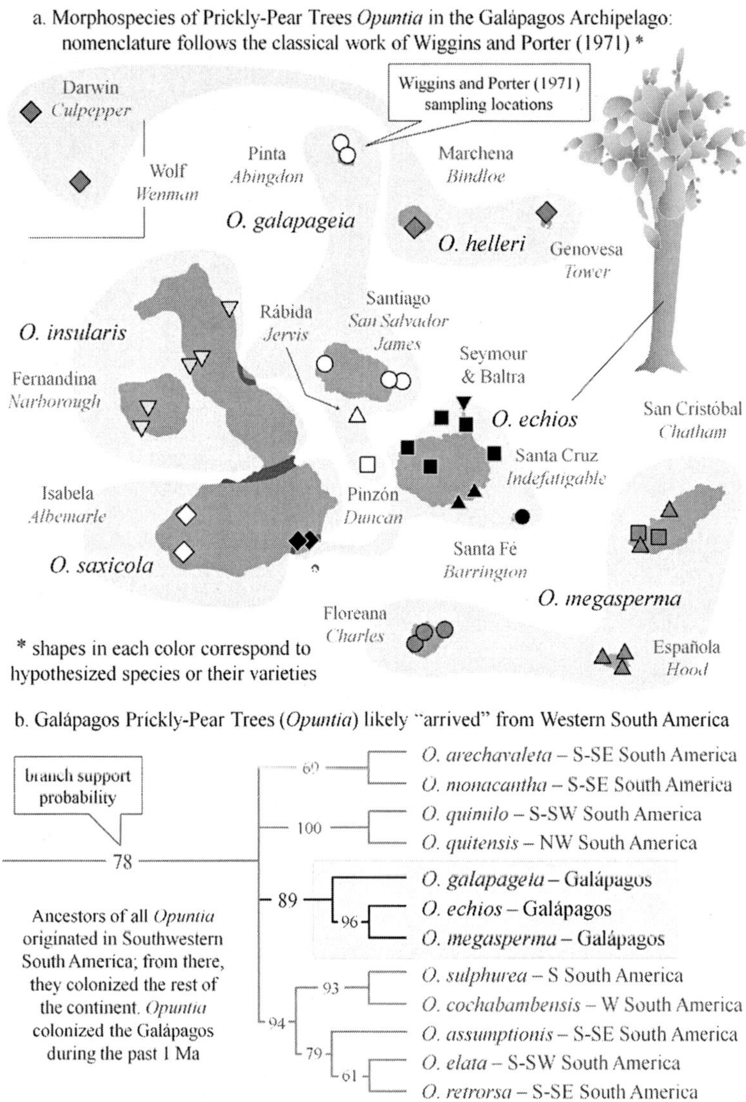

Fig. 8-29 Occurrence of six recognized morphospecies of Galápagos prickly pears (*Opuntia*) in the archipelago (a), and phylogenetic relation of the Galápagos lineages with other *Opuntia* species from South America (b). Layout: GPC. a. Modified after Wiggins and Porter (1971) with input from Helsen et al. (2009). b. Randomized Axelerated Maximum Likelihood tree (nDNA, cpDNA) extracted/zoomed-in from Majure et al. (2012).

Fig. 8-30 Majestic specimens of Galápagos prickly-pear trees (*Opuntia*). Top: Isabela Island (trail Muro de Lágrimas). Bottom: Santa Cruz Island (at Charles Darwin Research Station). *Opuntia* cacti are the most conspicuous life form of the Galápagos flora, particularly in lowland arid habitats (e.g. deciduous forests and shrublands, see "Vegetation Cover" section in Chapter One, Figs. 1.45-1.47). Six morphospecies of Galápagos *Opuntia* have been recognized, including 11 (or 14) subspecies. Photos: G. Paz-y-Miño-C.

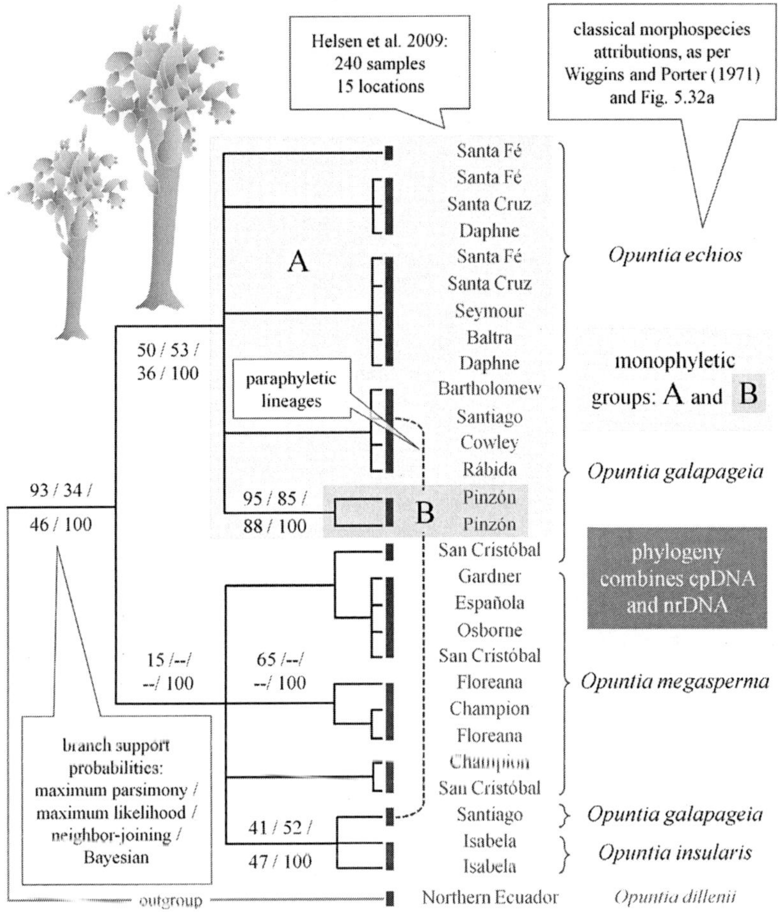

Fig. 8-31 Phylogenetic reconstruction of prickly-pear (*Opuntia*) specimens as per 15 locations in the Galápagos where biological samples ($N = 240$) were collected. Note how the low genetic variability among samples (assessed by cpDNA and nrDNA sequences; left) contrasts with the classically rich morphospecies attributions by Wiggins and Porter (1971; right). Layout: GPC, redrawn and modified after Helsen et al. (2009).

Scalesia: Darwin's Giant Daisies

Fifteen species of *Scalesia* are distributed on 11 islands in the Galápagos (Fig. 8.32). The highest diversity (6 spp.) occurs in Santa Cruz (*S. affinis*, *S. aspera*, *S. crockeri*, *S. helleri*, *S. pedunculata*, *S. retroflexa*); followed by

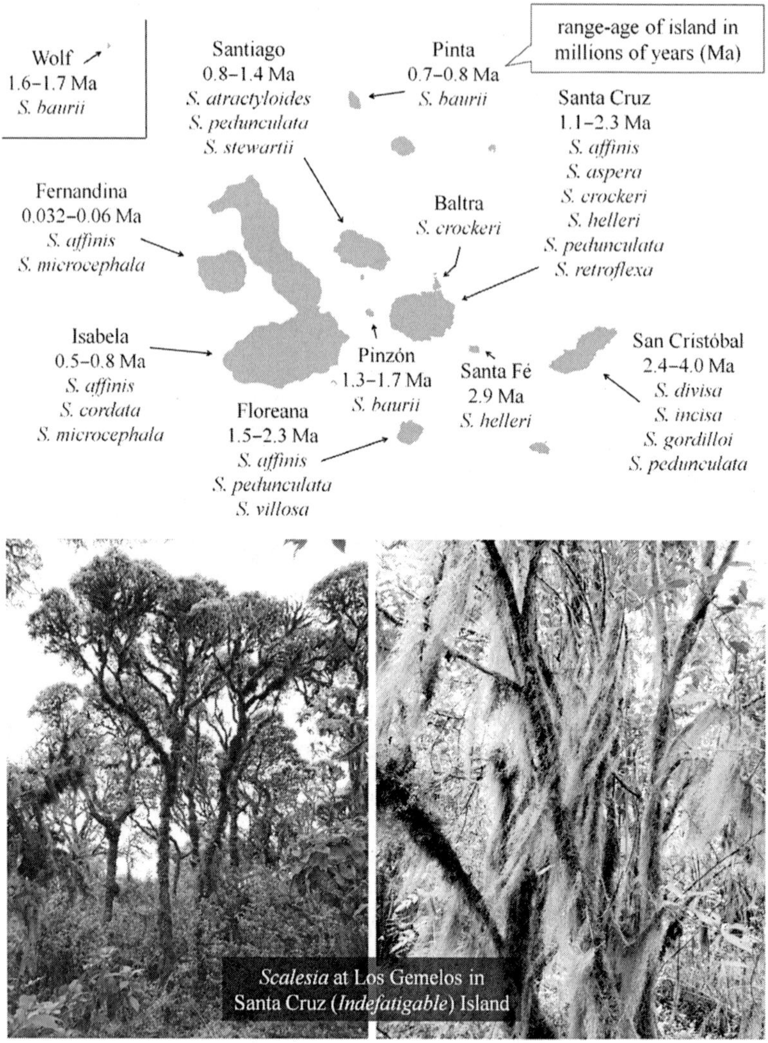

Fig. 8-32 Diversity of "Darwin's Giant Daisies" (*Scalesia*) in the Galápagos islands. Layout and photos: G. Paz-y-Miño-C; redrawn and modified after Fernández-Mazuecos et al. (2020).

San Cristóbal (4 spp.: S. divisa, S. incisa, S. gordilloi, S. pedunculata); Santiago (3 spp.: S. atractyloides, S. pedunculata, S. stewartii); Floreana (3 spp.: S. affinis, S. pedunculata, S. villosa); Isabela (3 spp.: S. affinis, S.

cordata, S. microcephala); Fernandina (2 spp.: S. affinis, S. microcephala); Pinta (S. baurii); Baltra (S. crockeri); Pinzón (S. baurii); Santa Fé (S. helleri); and Wolf (S. baurii). There is no statistical association between island age and Scalesia species richness in the archipelago ("progression rule", above; Fernández-Mazuecos et al. 2020).

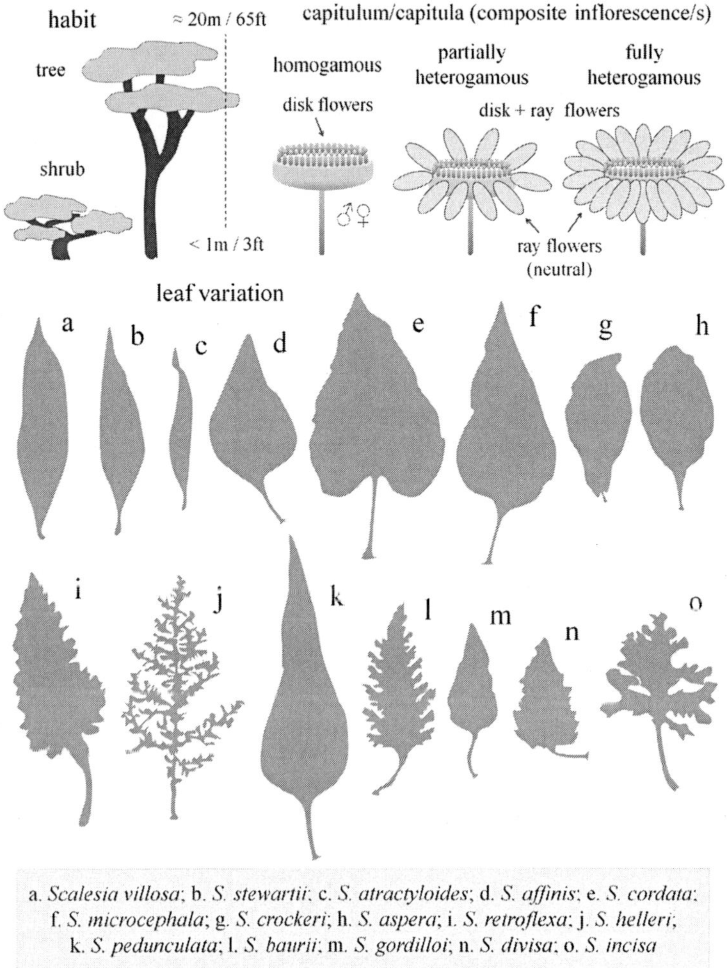

Fig. 8-33 Morphological variation (habit, capitula and leaves) among the species of "Darwin's Giant Daisies" (*Scalesia*) in the Galápagos. Layout: GPC, redrawn and modified after Fernández-Mazuecos et al. (2020) with input from Wiggins and Porter (1971).

Scalesia lineages exhibit considerable anatomical variation (Fig. 8.33 top), as per habit (tree or shrub), capitula or composite inflorescences (homogamous or partially-or-fully heterogamous), and leaves (Fig. 8.33). Note in Fig. 8.33 that the homogamous inflorescence has male and female disk flowers, while the heterogamous capitulum has a combination of disk and ray flowers. Ray flowers, a few or numerous per capitulum, are neutral or infertile in *Scalesia* (Fig. 8.33; after Fernández-Mazuecos et al. 2020 with input from Wiggins and Porter 1971).

The *Scalesia* genus is sister taxa to the Western South American genera *Syncretocarpus*, *Heiseria*, and *Pappobolus*, distributed in Ecuador and Peru; their last common ancestor dates back to ≈3 Ma (Fig. 8.34). These four taxa have distant phylogenetic relatives in Mesoamerica (today's West-Southern Mexico and Northern Central America), e.g., *Aldama*, from which they split about 3-4 Ma (Fig. 8.34 bottom). This is consistent with pattern No. 4 of biogeographic affinities in Fig. 1.37, Chapter One (i.e. Southwestern Mexico; association with tectonic structure: Guerrero terrane; as per Grehan 2001; Heads and Grehan 2021).

Scalesia radiation within the Galápagos took place during ≈1 Ma (Fig. 8.34 bottom). Although widely distributed across habitats, these trees or shrubs are particularly abundant in the highlands (180-550-m/600-1800-ft elevation; Wiggins and Porter 1971; McMullen 1999). Examine the phylogenetic reconstruction of the 15 *Scalesia* species in Fig. 8.35. To the right of the phylogeny, we have sketched the species' habit (tree or shrub), capitula (homogamous, partially heterogamous, or fully heterogamous), and leaf variation (after Fernández-Mazuecos et al. 2020). Note that the most recently diverged lineages (*S. pedunculata*, *S. gordilloi*, *S. divisa*, and *S. incisa*) occur on one of the oldest islands, San Cristóbal (2.4-4.0 Ma; Fig. 8.35 bottom). Interestingly, the youngest islands of Fernandina (0.032-0.06 Ma) and Isabela (0.5-0.8 Ma) possess comparably diverged (as per pattern of differentiation over time) *Scalesia* lineages: *S. affinis*, *S. cordata*, and *S. microcephala* (Fig. 8.35 center). By contrast, the middle-age island of Santiago (0.8-1.4 Ma) has two of the older lineages (*S. stewartii* and *S. atractyloides*; Fig. 8.35 top). Thus, there is no association between island age and lineage-diversification pattern. This adds to the lack of correspondence between island age and *Scalesia* species richness, as indicated above ("progression rule"). If anything, within-island lineage differentiation (i.e. speciation) is typical of the Galápagos *Scalesia* rather than between-island differentiation (Fernández-Mazuecos et al. 2020).

When looking at the anatomical variation among the 15 species as function of their phylogenetic divergence (Fig. 8.35), it becomes evident that convergence in habit, capitula, or leaf morphology is common (i.e.

species with similar phenotypes occur on different islands). Selection has likely encouraged the emergence of comparable anatomy under redundant habitat and ecological challenges (Fernández-Mazuecos et al. 2020).

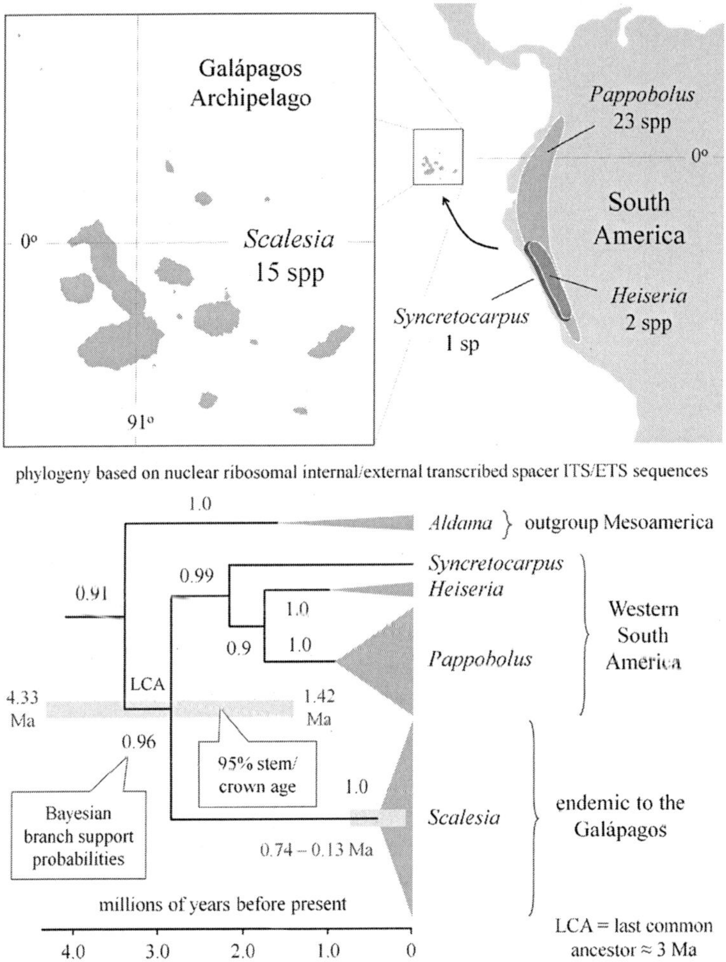

Fig. 8-34 Phylogenetic relations and likely origin of "Darwin's Giant Daisies" (*Scalesia*) in the Galápagos. The 15 species in the archipelago have close relatives in Western South America (LCA ≈3 Ma). *Scalesia* radiation in Galápagos 1 Ma. Time-calibrated phylogenetic relations based on nrDNA sequences (Bayesian posterior probabilities methods). Layout: GPC, redrawn and modified after Fernández-Mazuecos et al. (2020).

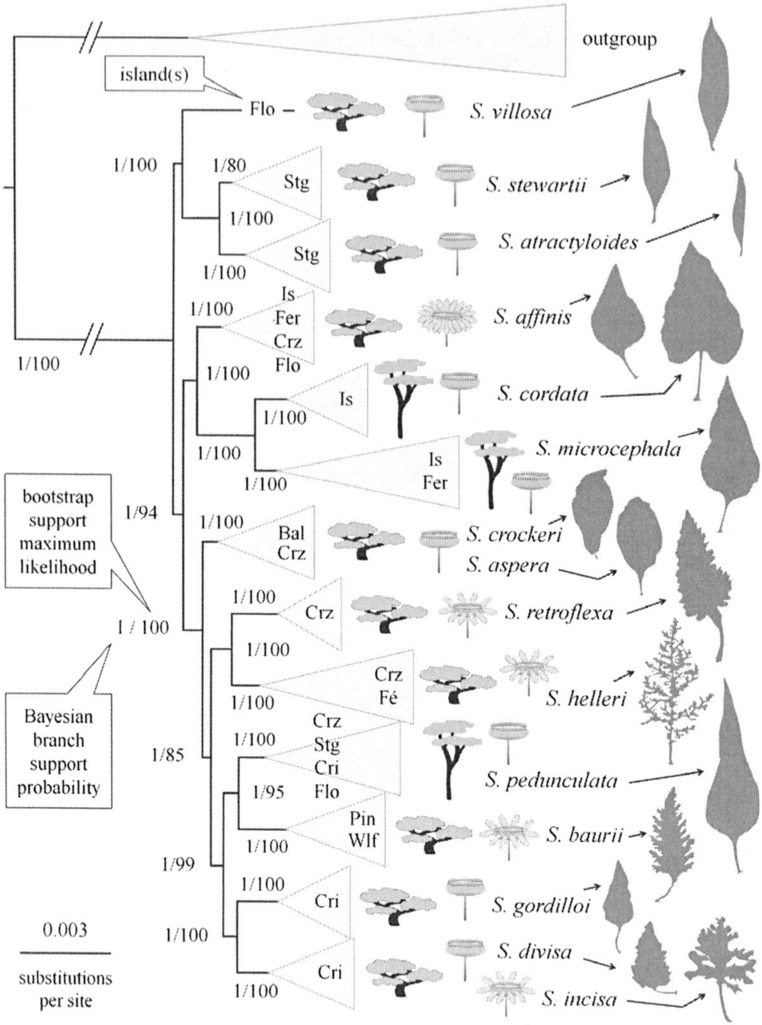

Fig. 8-35 Phylogenetic reconstruction of 15 species of "Darwin's Giant Daisies" (*Scalesia*) distributed in diverse islands. Flo = Floreana; Stg = Santiago; Is = Isabela; Fer = Fernandina; Crz = Santa Cruz; Bal = Baltra; Fé = Santa Fé; Cri = San Cristóbal; Pin = Pinta; Wlf = Wolf. Majority-rule consensus tree (Bayesian analysis based on concatenated DNA sequences of 9,363 loci). Note that convergence in habit, capitula, or leaf morphology is common (i.e. species with similar phenotypes occur on different islands). Layout: GPC, redrawn and modified after Fernández-Mazuecos et al. (2020).

2 Endemism: A Glimpse of Galápagos' Endemics

Going back to the questions, what is the conservation value of the Galápagos? And what are we really protecting? Here we discuss endemics! As exemplar volcanic archipelago, Galápagos is rich in endemic taxa. In Table 8.1, we list species numbers and proportions of endemics for some terrestrial and marine organisms (compilation of 20+ studies), including plants, mammals, birds, reptiles, insects, spiders, land snails, fishes, mollusks, corals (counting gorgonians and sea anemones hydroids), echinoderms (sea stars, sea cucumbers, sea urchins), bryozoans (moss animals), and polychaete (bristle worms). Endemics range from 8.3% in corals, gorgonians and sea anemones hydroids (169 species; Chiriboga et al. 2011) to 75% in reptiles (56 species; Jiménez-Uzcátegui et al. 2013) and to 100% in *Naesiotus* land snails (45 species; Phillips et al. 2020). More than half (53.9%) of the beetles are endemic (493 species; Peck 2006), as are 51.6% of the spiders (159 species; Buchholz et al. 2020). Half (50.0%) of the breeding and migrant birds are endemic (88 species; Wiedenfeld 2006). One-third (32.1%) of the flies (305 species of diptera), 35.5% of all plants (642 species), and 45.7% of the angiosperms in the lowlands (313 species) are endemic (Wiggins and Porter 1971; Sinclair 2011; Vargas et al. 2014). And close to one-third (28.1%) of all mammals (57 terrestrial and marine species) are unique to Galápagos (Jiménez-Uzcátegui and Snell 2014).

In the sections below, we discuss case studies on: (i) endemism and seed-dispersal in Galápagos' angiosperms; (ii) endemism in Galápagos' birds; (iii) endemism in iconic Galápagos' vertebrates; (iv) endemism, landscape, and adaptive radiation in Galápagos'/Darwin's finches; (v) endemism and hybridization in Galápagos mockingbirds; (vi) endemic island rodents; (vii) endemism in Galápagos' beetles; (viii) endemism in Galápagos' spiders; and (ix) endemism in Galápagos' marine organisms. Readers should realize that many more examples of natural-history studies involving Galápagos endemic taxa (i.e. the essence of the archipelago) are discussed in Paz-y-Miño-C and Espinosa (2024a,b).

Endemism and Seed-Dispersal in Galápagos' Angiosperms

Botanists have assessed endemism of flowering plants (Angiosperms) in the lowlands (below 300-400 m/1000-1300 ft elevation) of 12 Galápagos islands with surface area >10 km^2/4 mi^2 (Fig. 8.36 top). The researchers relied on a database of 403 species of native angiosperms available from the Charles Darwin Foundation (details in Vargas et al. 2014). Of this total

Table 8.1 Species numbers and proportion of endemics for exemplar taxa (terrestrial and marine) in the Galápagos Archipelago

environments and taxa	total spp	endemics	%	source
terrestrial environments				
plants	642	228	35.5	A
angiosperms in lowlands	313	143	45.7	B
hornworts and liverworts	150	15	10.0	C
Pertusaria lichens	24	7	29.2	D
mammals (includes cetaceans)	57	16	28.1	E
birds (including vagrants)	178	45	25.3	F
birds (breeding and migrants only)	88	44	50.0	G
reptiles	56	42	75.0	H
coleoptera (insects)	493	266	53.9	I
diptera (insects)	305	98	32.1	J
lice (insects)	96	17	17.7	K
ants (insects)	28	4	14.3	L
odonata (insects)	10	1	10.0	M
spiders	159	82	51.6	N
Naesiotus land snails	45	45	100.0	O
marine environments				
fishes	444	41	9.2	P
marine mollusks in shallow waters	640	115	18.0	Q
corals, gorgonians, sea anemones hydroids	169	14	8.3	R
echinoderms (stars, cucumbers, urchins)	198	17	8.6	S
bryozoans (moss animals)	184	34	18.5	T
polychaete (bristle worms)	192	50	26.0	U

endemics are quite common in the Galápagos; range from 8.3% to >75% of the species in the listed taxa

A = Wiggins and Porter (1971); B = Vargas et al. (2014); C = Gradstein and Ziemmeck (2011); D = Bungartz et al. (2015); E = Jiménez-Uzcátegui and Snell (2014); F = Jiménez-Uzcátegui et al. (2012); G = Wiedenfeld (2006); H = Jiménez-Uzcátegui et al. (2013); I = Peck (2006); J = Sinclair (2011); K = Palma and Peck (2013); L = Wauters et al. (2016); M = Lorenzo-Carballa et al. (2021); N = Buchholz et al. (2020); O = Phillips et al. (2020); P = Grove and Lavenberg (1997); Q = Kay (1991); R = Chiriboga et al. (2011); S = Maluf (1991); T = Banta (1991); U = Blake (1991)

313 species (78%) were distributed in the lowlands. The botanists focused their attention on this subset of taxa (N=313) for meaningful comparisons among islands (i.e. only lowland habitats were examined since islands varied widely in altitude). One of the goals of the study was to contrast seed-dispersal strategies among endemic versus non-endemic angio-

sperms. As summarized in Fig. 8.36 bottom, 161 species (76 endemics, 85 non-endemics) belonged to a variety of seed-dispersal syndromes: dispersal by wind (anemochory), dispersal inside animals (endozoochory), dispersal attached to animals (epizoochory), or drifters on seawater (thalassochory). By contrast, 152 species (67 endemics, 85 non-endemics) relied on other means of dispersal not directly associated with inter-island colonization (i.e. medium-distance dispersal or MDD); the investigators' main interest was MDD (Vargas et al. 2014).

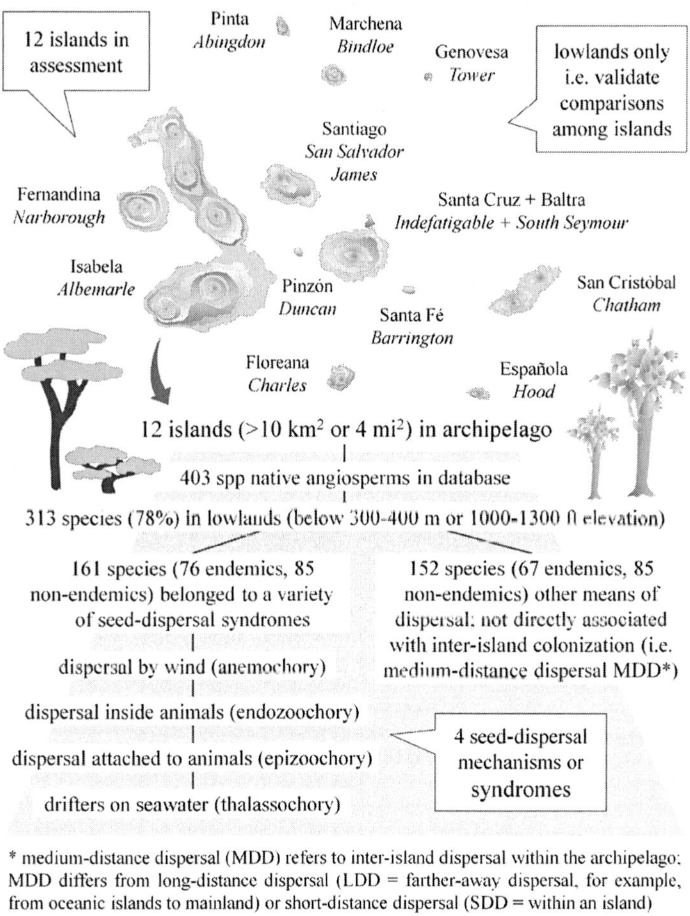

Fig. 8-36 Native endemic and non-endemic angiosperms in the Galápagos Archipelago: emphasis on seed-dispersal mechanisms in the lowlands. Layout: GPC; redrawn and modified after Vargas et al. (2014).

In Fig. 8.37a, we plot the occurrence of endemic and non-endemic angiosperms (all natives) in the lowlands of the 12 sampled islands (numbers 1 to 12). Note how the cumulative number of endemics found per island significantly decreased as more islands were added to the count (left histogram plot; regression line r_p= –0.755; p=0.004), while there was no statistically significant trend for the cumulative number of non-endemics (right; regression line r_p= –0.193; p=0.547; Vargas et al. 2014). Realize that such trends are often observed in nature when cumulatively accounting for taxa (endemics or non-endemics) because a greater number of taxa tend to be distributed widely and in various locations (e.g. islands in an archipelago), while a small number of taxa tend to be confined to just a few locations (e.g. a few islands). The latter pattern was particularly accentuated among endemics, perhaps due to their restricted occurrence only in Galápagos, which was not the case among non-endemics (broadly distributed elsewhere). By looking at the data summarized in Fig. 8.37b, it becomes evident that endemics relied on wind (anemochory; 26 species, 18.2%) or inside-animal-gut transport (endozoochory; 29 species, 20.3%) strategies for inter-island seed dispersal or medium-distance dispersal (MDD), while non-endemics relied on drifting on seawater (thalassochory; 47 species, 27.6%; Vargas et al. 2014) [these mechanisms were likely used differentially to reach the archipelago in the past when first colonizing it; Paz-y-Miño-C and Espinosa (2024b)].

The botanists concluded that Angiosperms with specialized strategies for inter-island dispersal had successfully colonized more islands within the archipelago than species lacking such specializations (Vargas et al. 2014). This can be inferred from the data summarized in Fig. 8.37b.

a. Occurrence of endemic and non-endemic angiosperms (all natives) in the lowlands (below 300-400 m or 1000-1300 ft elevation) of the Galápagos Archipelago (12 islands, each >10 km² or 4 mi²)

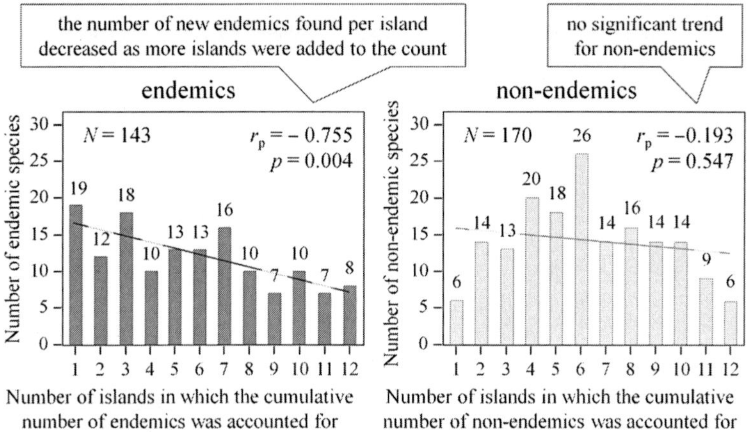

b. Seed-dispersal strategies between endemic vs. non-endemic angiosperms in the lowlands (below 300-400 m or 1000-1300 ft elevation) of the Galápagos Archipelago (12 islands, each >10 km² or 4 mi²) – contingency analyses (*G* test) of rounded % values

	endemics	%	non-endemics	%	*G* test	*p*
total number species	143	100	170	100		
with dispersal specialization	76	53.1	85	50.0	$G_1 = 0.567$	0.452
without dispersal specialization	67	46.9	85	50.0		
* anemochory	26	**18.2**	4	2.4	$G_1 = 57.24$	**<0.001**
* endozoochory	29	**20.3**	22	12.9	$G_1 = 4.39$	**0.036**
* epizoochory	19	13.3	24	14.1	$G_1 = 0.063$	0.802
* thalassochory	13	9.1	47	**27.6**	$G_1 = 20.09$	**<0.001**

endemics relied on wind (anemochory) or inside-animal-gut transport (endozoochory) strategies for inter-island seed dispersal vs. non-endemics relied on drifting on seawater (thalassochory) as strategy for seed dispersal

* note that 11 endemics and 12 non-endemics had more than one dispersal specialization, which was accounted for in the subtotals of anemo-, endozoo-, epizoo- or thalassochory

Fig. 8-37 Occurrence of endemic and non-endemic angiosperms (a) and their seed-dispersal strategies in the lowlands of the Galápagos (b). Note in b how endemics relied on wind (anemochory 18.2%) or inside-animal-gut transport (endozoochory 20.3%) strategies for inter-island seed dispersal, while non-endemics relied on drifting on seawater (thalassochory 27.6%). Layout: GPC; redrawn and modified after Vargas et al. (2014).

Endemism in Galápagos' Birds

The grand total of bird species in Galápagos is 163 (Fig. 8.38 top). This includes vagrants (57 species) and taxa of suspected presence (18 species). The number of resident bird species, migrants, or introduced is 88; half of them (50.0%) are endemic species (N=29) or subspecies (N=15); 12 species come yearly to breed in the archipelago; 27 species are migrants from

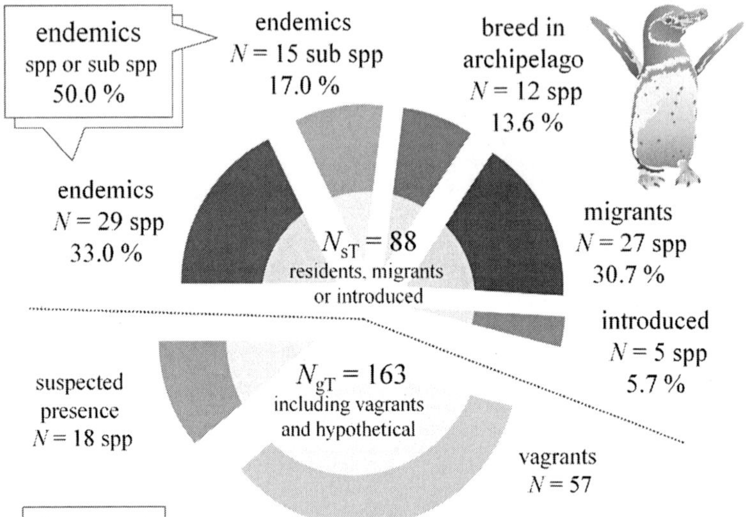

frequented habitat	terrestrial	marine	coasts and lagoons		total
endemic species	22	5	2		29
endemic subspecies	4	6	5		15
breed in archipelago	2	5	5		12
migrants	4	3	20		27
introduced	5	0	0		5
				sub total = (N_{sT} residents, migrants or introduced)	88
vagrants	14	15	28		57
				sub total =	145
suspected presence	2	11	5		18
probable grand total N_{gT} in the archipelago				grand total =	163

Fig. 8-38 Birds in the Galápagos Archipelago: species, endemics, and other counts. Layout/digital sketch: G. Paz-y-Miño-C; data from Wiedenfeld (2006).

both hemispheres that come to Galápagos during the winter season in their breeding grounds; and 5 species are introduced (Fig. 8.38 top). Of the endemic species (*N*=29), 22 live in terrestrial, 5 in marine, and 2 in coastal/lagoon habitats, while of the endemic subspecies (*N*=15), 4 live in terrestrial, 6 in marine, and 5 in coastal/lagoon habitats (Fig. 8.38 bottom; Wiedenfeld 2006).

Endemism in Iconic Galápagos' Vertebrates

The stories of six exemplar Galápagos species can help us illustrate how selection (*ns*), gene flow (*gf*), genetic drift (*gd*), mutation (*mu*), and inbreeding (*in*) have contributed to shaping the evolution of vertebrate taxa in the archipelago. They are all endemic: flightless cormorants (*Phalacrocorax harrisi*), pink land iguanas (*Conolophus marthae*), marine iguanas (*Amblyrhynchus cristatus* sspp.), Galápagos hawks (*Buteo galapagoensis*), giant tortoises (*Chelonoidis* spp.), and Galápagos'/ Darwin's finches (4-5 genera). In Fig. 8.39a,b,c—8.39d,e,f (continued), we summarize how such factors (*ns*, *gf*, *gd*, *mu*, *in*) have interacted and were likely and characteristically associated with these species' endemism (after Hedrick 2019). [We alert readers that Hedrick's (2019) work was not focused on endemism but rather on the overall evolution of these vertebrates in Galápagos and their population genetics; we have chosen to use Hedrick's perspective to link it to endemism in these species].

In Fig. 8.39a-f, we summarize, in text, the distribution of the six endemic vertebrates in the archipelago and provide statements on how natural selection, gene flow, genetic drift, mutation, or inbreeding were possibly associated with their evolution and endemism. In Table 8.2, we share data on these vertebrates' species richness, estimated time of colonization ("arrival to" or "adaptive radiation" in Galápagos), number of islands/islets of occurrence, and probable population size. And in Table 8.3, we rank (as very high, high, low, very low, etc.) the relative importance of *ns*, *gf*, *gd*, *mu*, and *in* in the evolution of these vertebrates (information extracted from Hedrick 2019).

Flightless cormorants (*Phalacrocorax harrisi*) are currently distributed in Fernandina and Isabela. Their ancestors colonized Galápagos 2.0-2.4 Ma. Their estimated population size is 2,200 individuals (Fig. 8.39a and Table 8.2). Natural selection likely drove flightlessness in nutrient-rich environments (fueled by upwelling to the West of Fernandina and Isabela); positive selection fixed gene variants (e.g. *Cux1*) associated with flightlessness; and selection also favored facultative sequential polyandry in which females mate serially with various males, while males become primary providers to the sequential offspring (males 40% larger than females; males also choose mates; Fig. 8.39a; Table 8.3; Hedrick 2019).

a. Flightless cormorants (*Phalacrocorax harrisi*): distributed in Fernandina and Isabela

selection (diet): drove flightlessness in nutrient-rich environments (upwelling in western islands); the only flightless cormorant among 40 world species

mutation: positive selection fixed gene variants (e.g. *Cux1*) associated with flightlessness

selection (behavior): favored facultative sequential polyandry; females mate serially with various males; males become primary providers to the sequential offspring; males are 40% larger than females; males too choose mates

bottlenecks (genetic drift): relatively small population sizes in either Fernandina or Isabela and low genetic diversity suggest bottleneck events occurred in the recent past; currently, limited gene flow assumed to take place between islands

b. Pink land iguanas (*Conolophus marthae*): restricted to Wolf Volcano, North Isabela

bottlenecks (genetic drift): the current population is very small (about 192 individuals); the existence of a single type of mtDNA in the population indicates both low genetic variation and past bottlenecks; very low reproductive success hints to possible inbreeding depression

founder-effect: founder-effect differentiation and subsequent speciation are possible

c. Marine iguanas (*Amblyrhynchus cristatus* sspp.): distributed in the entire archipelago

selection (habitat, diet): drove adaptations to marine habits and algae-feeding in shores; marine iguanas evolved from terrestrial ancestors and are the sole marine species in the world

mutation: positive selection possibly fixed genes (unknown) associated with deep diving and diet; excess salt accumulated in body through diet is expelled out via salt-glands in nostrils ("salt sneezing")

gene flow: inter-island movement and hybridization among populations can occur

Fig. 8.39 a,b,c A close look at exemplar endemic vertebrates in the Galápagos; evolutionary factors likely and characteristically associated with their endemism: flightless cormorants; pink land iguanas; and marine iguanas (examples continue on Fig. 8.39d,e,f). Layout: GPC; adapted from Hedrick (2019).

Pink land iguanas (*Conolophus marthae*) are restricted to Wolf Volcano in Northern Isabela (see map of island in Fig. 1.10, Chapter One). Pink iguanas diverged from land iguanas 1.5 Ma (ancestors of all iguanas colonized Galápagos 4.5 Ma). [See occurrence and distribution of land iguanas in Fig. 8.10]. The estimated population of pink iguanas is 192 individuals (Fig. 8.39b and Table 8.2). The existence of a single type of mtDNA in the population indicates both low genetic variation and past bottlenecks; very low reproductive success suggests possible inbreeding depression. Pink iguanas likely originated via founder-effect differentiation and subsequent speciation (Table 8.3; Hedrick 2019). They are unique among the already-unique Galápagos iguanas.

Marine iguanas (*Amblyrhynchus cristatus* sspp.) are distributed throughout the entire Galápagos archipelago. [We discussed them in the context of colonization of the islands in Figs. 8.9, 8.11-8.12]. Marine and land iguanas are sister taxa; their ancestors colonized Galápagos 4.5 Ma. *Amblyrhynchus cristatus* occurs in 16 islands with an estimated population of 37,000 to 280,000 individuals (Fig. 8.39c and Table 8.2). Natural selection likely drove adaptations to marine habits and algae-feeding on shores. Positive selection possibly fixed genes (unknown) associated with deep diving and diet; excess salt accumulated in body through diet is expelled out via salt-glands in the nostrils ("salt sneezing"). Inter-island movement and hybridization among subspecies populations can occur (Table 8.3; Hedrick 2019).

No long ago, Galápagos hawks (*Buteo galapagoensis*) were distributed on most islands; nowadays, they are absent from Santa Cruz, San Cristóbal, and Floreana (human-populated islands; Fig. 8.39d). They are the most recently arrived vertebrate in the archipelago (0.125 Ma; Table 8.2). Relatively small populations of *B. galapagoensis* exist (400-500 adults, 300-400 juveniles; total ≈1,000) in the hawk-inhabited islands (Fernandina, Isabela, Santiago, Pinzón, Santa Fé, Española, plus islets; Table 8.2). Major-histocompatibility-complex (MHC) and mtDNA studies suggest overall low genetic variability (likely due to founder effect and genetic drift). There is little migration (gene flow) between island populations. Selection has favored cooperative polyandry in some islands (i.e. one female with several unrelated males) or strict monogamy (e.g. Española; Fig. 8.39d; Table 8.3; Hedrick 2019).

Giant tortoises (*Chelonoidis* spp.) are distributed across the larger islands of the archipelago and their volcanoes, or some of the smaller islands; a few species became recently extinct (Hedrick 2019). [We discussed giant tortoises in diverse contexts of their colonization of the archipelago; Figs. 8.5-8.7]. Ancestors of these tortoises colonized Galápagos 3.0-4.0

d. Galápagos hawks (*Buteo galapagoensis*): in the past, distributed in most islands; nowadays absent from Santa Cruz, San Cristóbal and Floreana

founder effect (genetic drift): most recently arrived vertebrate in the archipelago (125k yr ago); relatively small populations in the hawk-inhabited islands (400-500 adults, 300-400 juveniles); MHC and mtDNA studies suggest overall low genetic variability

gene flow: apparently little migration between island populations

selection (behavior): favored cooperative polyandry in some islands (one female with several unrelated males; all provide to the nest), or strict monogamy (i.e. in Española)

e. Giant tortoises (*Chelonoidis* spp.): species distributed across larger islands and their volcanoes, or some of the smaller islands; a few species recently extinct

selection (habitat, diet): drove adaptations to rocky volcanic-island environments; seasonal altitudinal migration of large individuals (seeking food-vegetation in highlands) during dry months in the lowlands (small individuals stay in the lowlands)

selection (volcanoes): in Isabela island, each volcano homes a distinctive species, suggesting habitat isolation (in highlands) drives population differentiation and, over time, leads to further isolation in respect to neighboring populations; large population size in various volcanoes (healthy) suggests little inbreeding

dome and saddleback morphotypes: trait not fully understood*; (dome = ancestral; saddleback = more recent, might have evolved multiple times); dome-shape tortoises feed well on easy-to-reach vegetation (highlands); long-necked saddlebacks feed well on taller vegetation (lowlands); genetic basis suspected for trait

f. Finches (4-5 genera): 14-17 species distributed in all islands

selection (habitat, diet): drove quick adaptive radiation of ancestors into diverse habitats and feeding habits, from lowlands to highlands, and from seeds to invertebrates; seasonality strongly associated with finch survival; beak morphology and function varies across species (genes known, e.g. *HMGA2*) and correlates with variation in food abundance

gene flow: frequent among some species and in specific habitats

hybridization speciation: documented in Daphne Major island

Fig. 8.39 d,e,f (continued from Fig. 8.39a,b,c) A close look at exemplar endemic vertebrates in the Galápagos; evolutionary factors likely and characteristically associated with their endemism: Galápagos hawks; giant tortoises; and Galápagos'/Darwin's finches. *For discussion about dome and saddleback morphotypes in giant tortoises, see Paz-y-Miño-C and Espinosa (2024a) and references therein (i.e. Hunter et al. 2013, 2020; Chiari et al. 2017; Chiari 2021). Layout: GPC; adapted from Hedrick (2019).

Table 8.2 Species richness, time of colonization, number of islands/islets of occurrence, and population size of exemplar vertebrate endemics (birds or reptiles) in the Galápagos Archipelago (modified after Hedrick 2019)

	No. species	colonization (millions yrs)	No. islands and/or islets	population size * (individuals)
flightless cormorants	1	2.0 - 2.4	2	2200
pink iguanas	1	4.5 **	1	192
marine iguanas	1	4.5	16	37 000 - 280 000
Galápagos hawks	1	0.125	8	1000
giant tortoises	11	3.0 - 4.0	6	26 700 - 35 400
finches	14 or 17 ***	2.3 - 2.6	14 islands 18 islets	1 000 000 - 2 400 000

* all population sizes are estimates based on diverse census methods; ** pink iguanas diverged from land iguanas 1.5 million years ago (4.5 million years refers to the arrival of the ancestors of all land iguanas); *** 14 species of finches are often recognized in the Galápagos, 3 other species (total 17) include taxa acknowledged by some researchers, plus the Cocos finch (*Pinaroloxias inornata*), from the Cocos Island (750 km/460 mi to the Northeast)

Table 8.3 Relative importance* of population genetics factors in the evolution of exemplar vertebrate endemics in the Galápagos Archipelago (after Hedrick 2019)

	selection	gene flow	genetic drift	mutation	inbreeding and/or inbreeding depression
flightless cormorants	very high	low	high	high	high
pink iguanas	unknown	none	high	unknown	very high
marine iguanas	very high	some	some	possible	unknown
Galápagos hawks	unknown	low	high	low	high
giant tortoises	high	very low	low	possible	unknown
finches	very high	high	low	high	high (some populations)

* relative importance of factors was inferred from natural history, phylogenetics, population genetics and demographics (when available), but not actually tested as one factor vs. another

Ma; the current populations range from 26,700 to 35,400 individuals (Table 8.2). Natural selection likely favored adaptations to rocky volcanic-island environments; seasonal altitudinal migration of large individuals (seeking food-vegetation in highlands) during dry months in the lowlands (note that small individuals stay in the lowlands; for details on migration patterns in *Chelonoidis*, see Bastille-Rousseau et al. 2016, 2019). For example, in Isabela, each volcano homes a distinctive species, suggesting that habitat isolation (in highlands) drives population differentiation and, over time, leads to further isolation with respect to neighboring populations; large population size in various volcanoes (healthy status) suggests little inbreeding (Fig. 8.39e; Table 8.3; Hedrick 2019).

The iconic Galapagos'/Darwin's finches (4-5 genera; 14-17 species) are distributed on all islands (Fig. 8.40). Their ancestors colonized the archipelago 2.3-2.6 Ma. Their all-inclusive populations reach between 1,000,000 and 2,400,000 individuals (Fig. 8.39f; Table 8.2; Hedrick 2019). [We already discussed "finches" in Chapter Three, Figs. 3.16-3.19; and in this chapter, their origins and distribution, Figs. 8.2-8.3]. Natural selection likely favored quick adaptive radiation of their ancestors into diverse habitats and feeding habits, from lowlands to highlands and from seeds to invertebrates. Seasonality is strongly associated with finch survival. Beak morphology and function vary across species (genes known, e.g. *HMGA2*) and correlate with variation in food abundance (Fig. 8.39f). Gene flow occurs frequently among some species and in specific habitats, and hybridization speciation has been documented on Daphne Major Island (Fig. 8.39f; Table 8.3; Hedrick 2019).

Endemism, Landscape, and Adaptive Radiation in Galápagos'/Darwin's Finches

Let us look at the Galápagos/Darwin finches' endemism in more detail. What happens to populations of endemic taxa—like the finches—that go through adaptive radiation and generate new lineages that further evolve in a scattered archipelago landscape? This question has been addressed by contrasting populations of cactus finches (*Geospiza scandens* and *G. conirostris*), sharp-beaked ground finches (*Geospiza difficilis*), and warbler finches (*Certhidea olivacea* and *C. fusca*; Petren et al. 2005). As discussed in previous chapters, there are phylogenetic reconstructions of the finches based on various methodologies, e.g., autosomal genomes (Chapter Three, Fig. 3.17), biogeographic analysis and mtDNA+nDNA (this chapter, Fig. 8.3), or whole genomes (Lamichhaney et al. 2018; for review, see Paz-y-Miño-C and Espinosa 2024a,b). The finches' distribution in the archipelago is well known (Fig. 8.40), as is their natural history, from ground to tree finches and to the unique woodpecker finch (*Cactospiza pallida*), which uses and modifies tools (Tebbich et al. 2012)!

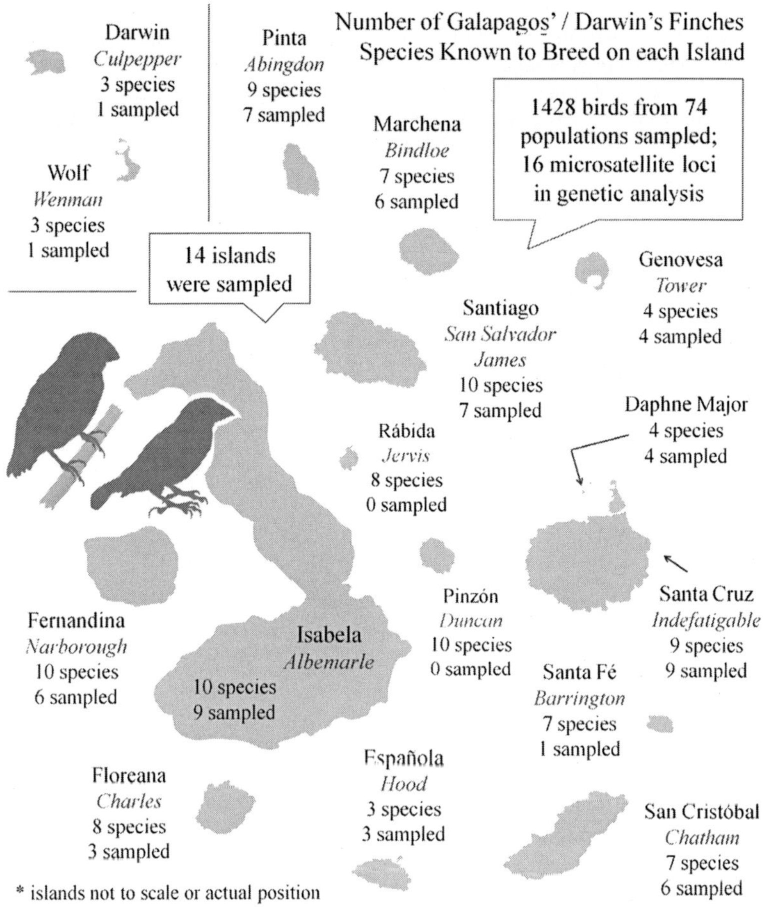

Fig. 8-40 Number of Galapagos'/Darwin's finches species known to breed on each island. Across 14 islands, 1,428 birds from 74 populations were sampled for a genetic assessment of island populations (e.g. heterozygosity as function of island area; inter-populations' genetic distances as function of geographic distances; as well as beak morphology and genetic variation among populations: cactus finches vs. warbler finches; or sharp-beaked ground finches vs. warbler finches); 16 microsatellite loci were used in the genetic analysis. Layout: GPC; redrawn and modified after Petren et al. (2005).

Table 8.4 Species of Galápagos' / Darwin's finches which genetic diversity has been assessed (16 microsatellite loci) in various populations and islands, with the purpose of establishing the interplay among selection, gene flow, genetic drift or hybridization during adaptive radiation (modified after Petren et al. 2005)

genus / species	Dar	Wol	Pin	Mar	Gen	Fer	Isb	Snt	Dph	Scrz	Sfe	Stbl	Flo	Esp	total
Geospiza															
G. fuliginosa			X				X	X	X	X		X	X	X	8
G. fortis			X	X			X	X	X	X		X			7
G. magnirostris				X	X		X	X	X	X					6
G. scandens *			X	X				X	X	X		X			6
G. conirostris *					X									X	2
G. difficilis **	X	X	X		X	X						X			6
Camarhynchus															
C. parvulus						X	X			X			X		4
C. psittacula			X	X		X	X			X					5
C. pauper													X		1
Cactospiza															
C. pallida						X	X	X		X		X			5
C. heliobates							X								1
Platyspiza crassirostris			X	X		X	X	X		X					6
Certhidea (2 spp) ***			X	X	X	X	X	X		X	X	X		X	10
spp sampled in island	1	1	7	6	4	6	9	7	4	9	1	6	3	3	
number of spp in island	3	3	9	7	4	10	10	10	4	9	7	7	8	3	

* *G. scandens* and *G. conirostris* are cactus finches; ** *G. difficilis* is the sharp-beaked ground finch; *** *Certhidea* includes two species (2 spp) of warbler finches *C. olivacea* and *C. fusca*; the three groups (i.e. cactus finches, sharp-beaked ground finch, and warbler finches) are the most genetically divergent of the Galápagos'/Darwin's finches, they belong to morphologically and ecologically similar allopatric populations; abbreviations of islands: Dar = Darwin; Wol = Wolf; Pin = Pinta; Mar = Marchena; Gen = Genovesa; Fer = Fernandina; Isb = Isabela; Snt = Santiago; Dph = Daphne Major; Scrz = Santa Cruz; Sfe = Santa Fé; Stbl = San Cristóbal; Flo = Floreana; Esp = Española; total = number islands sampled; spp = species

In Fig. 8.40, we summarize the number of Galápagos'/Darwin's finches species known to breed on each island. Researchers sampled for genetic assessment 1,428 birds from 74 populations across 14 islands. Their purpose was to characterize populations' heterozygosity as function of island area; inter-populations' genetic distances as function of geographic distances; as well as beak morphology and genetic variation across populations. In addition, the biologists wanted to establish the interplay among selection, gene flow, genetic drift, or hybridization during adaptive radiation. Sixteen microsatellite loci were used in that analysis (Table 8.4; Petren et al. 2005). In Fig. 8.41, we explain the methodological approach, and in Figs. 8.42-8.43, we summarize the results.

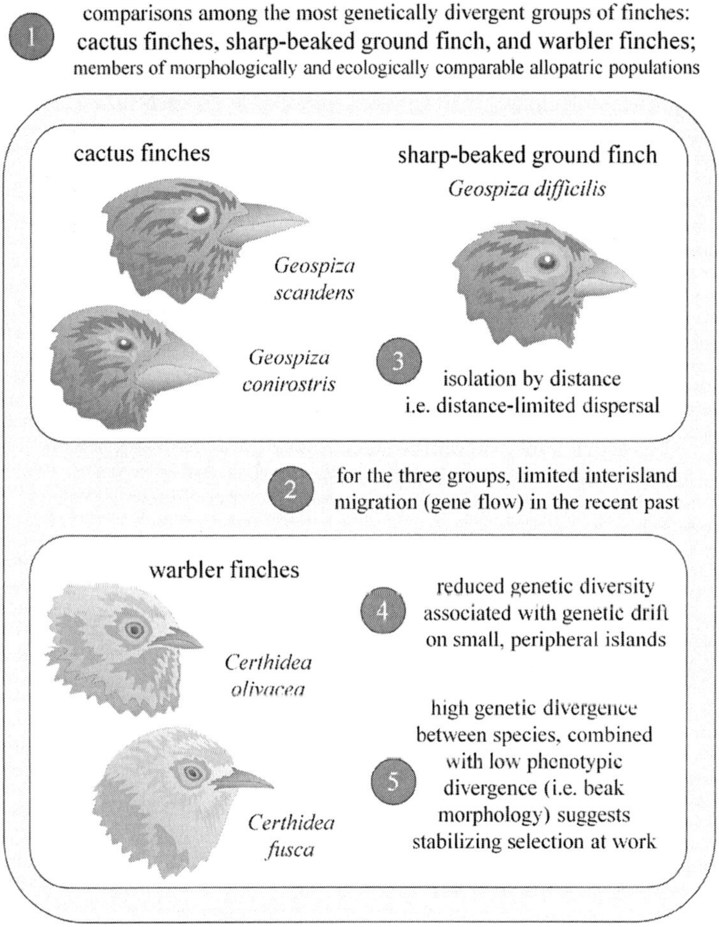

Fig. 8-41 Analysis of genetic divergence (16 microsatellite loci) of 74 populations of Galápagos'/Darwin's finches (*N*=1,428 birds) in 14 islands suggests differential effect of geographic features of the archipelago's landscape on the genetic structure of the finches' populations. Layout: GPC; after Petren et al. (2005).

Follow the numeric sequence in Fig. 8.41: (1) The researchers compared among the most genetically divergent groups of finches: cactus finches, sharp-beaked ground finches, and warbler finches, all members of morphologically and ecologically comparable allopatric populations. (2) For the three groups, researchers knew there was limited interisland migration (gene flow) in the recent past. (3) Isolation by distance (i.e. distance-limited dispersal) was typical of the populations of cactus finches and sharp-beaked ground finches. (4) There was reduced genetic diversity associated with genetic drift on small, peripheral islands, particularly for warbler finches. And (5) there was high genetic divergence between species (warbler finches) combined with low phenotypic divergence (beak morphology), suggesting stabilizing selection at work (Petren et al. 2005).

Bring your attention to Fig. 8.42, in which we show the association between genetic diversity and island size (Fig. 8.42a) as well as genetic distance and geographic distance (Fig. 8.42b) in populations of cactus finches (*Geospiza scandens*, *G. conirostris*), sharp-beaked ground finches (*G. difficilis*) or warbler finches (*Certhidea olivacea*, *C. fusca*). Note on the left plot that genetic diversity across populations (assessed as per heterozygosity H_O) was lower in smaller islands than in larger islands; the trend was statistically significant only in warbler finches *C. olivacea* and *C. fusca* (together $p \leq 0.01$); when *C. fusca* was analyzed alone, $p \leq 0.001$. Thus, in *C. olivacea* and *C. fusca*, reduced genetic diversity was likely driven by genetic drift acting on small, peripheral islands (in agreement with point 4 in Fig. 8.41). Now, as shown on the right plot, genetic distance across populations was greater as geographic distance increased; the trend was statistically significant in *G. difficilis* ($p \leq 0.05$) and *G. scandens* (only when analyzed alone, $p \leq 0.01$); when *C. fusca* was analyzed alone, $p=0.07$. Thus, in *G. difficilis* and *G. scandens*, reduced gene flow due to isolation-by-distance (rather than reduced population size) drove the association between genetic distance and geographic separation (in agreement with point 3 in Fig. 8.41; Petren et al. 2005).

Let us examine Fig. 8.43, in which beak morphology variation (Fig. 8.43a: beak depth x beak length) and genetic variation (Fig. 8.43b: pairwise microsatellite distance) among populations of cactus finches (*Geospiza scandens*, *G. conirostris*), sharp-beaked ground finches (*G. difficilis*) or warbler finches (*C. olivacea*, *C. fusca*) are plotted. Note on the left plot how cactus finches had larger beaks (as per depth and length) than the sharp-beaked ground finches (intermediate size) and warbler finches (smaller size); the beak morphology of *C. olivacea* and *C. fusca* (depth x length) was similar, not identical. Now, as shown on the right plot, allopatric populations of *C. olivacea* and *C. fusca* were distinctive as per genetic distances; notice in Fig. 8.43a, however, little beak morphology differentiation. Thus, high genetic divergence combined with low phenotypic divergence suggested stabilizing selection at work (in agreement with point 5 in Fig. 8.41; Petren et al. 2005).

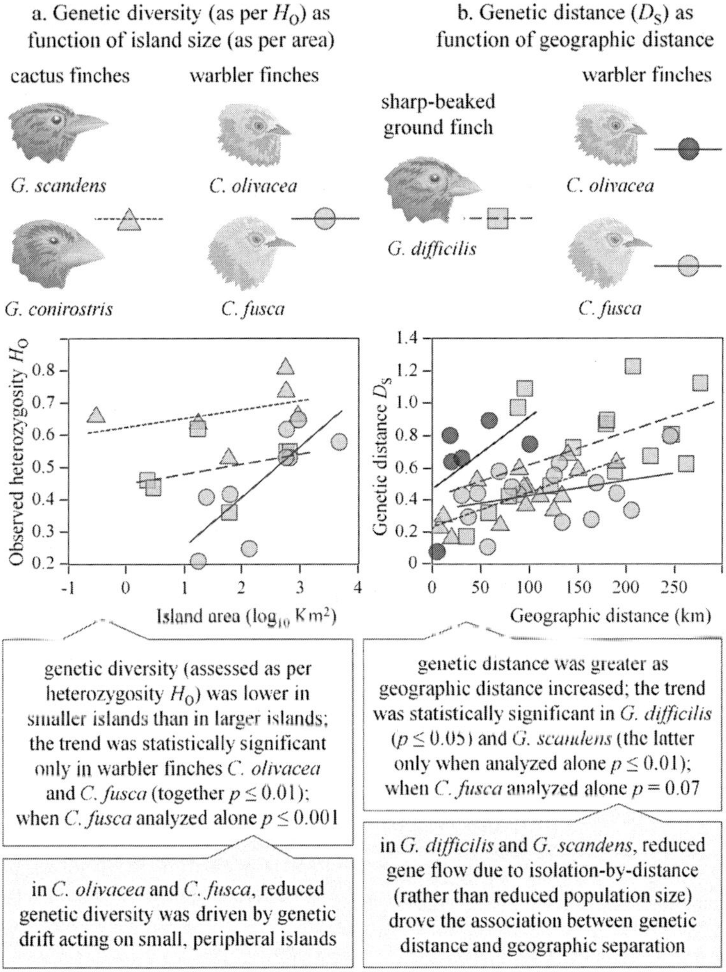

Fig. 8-42 Impact of island size/area (left) or separation by distance (right) on the genetic diversity or genetic distance in populations of cactus finches (*Geospiza scandens*, *G. conirostris*), sharp-beaked ground finches (*G. difficilis*) or warbler finches (*Certhidea olivacea*, *C. fusca*). Layout: GPC; redrawn and modified after Petren et al. (2005).

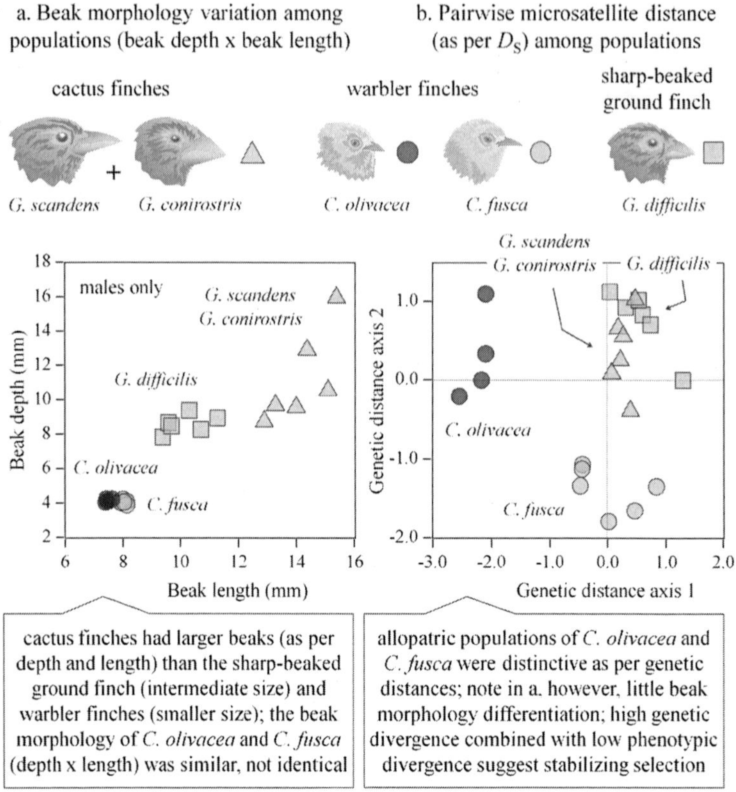

Fig. 8-43 Beak morphology (a) and genetic (b) variation among populations of cactus finches (*Geospiza scandens*, *G. conirostris*), sharp-beaked ground finches (*G. difficilis*) or warbler finches (*Certhidea olivacea*, *C. fusca*). Layout: GPC; redrawn and modified after Petren et al. (2005).

In synthesis, genetic diversity among exemplar populations of Galápagos'/Darwin's finches was lower in smaller islands than in larger islands; such trend was statistically significant in warbler finches (*C. olivacea* and *C. fusca*) and was likely driven by genetic drift (Fig. 8.42a; number 4 in Fig. 8.41). Genetic distance among finch populations was greater as geographic distance increased; in the case of *G. difficilis* (sharp-beaked ground finch) and *G. scandens* (cactus finch), reduced gene flow due to isolation-by-distance (not small population size) drove the association between genetic distance and geographic separation (Fig. 8.42b; number 3 in Fig. 8.41). Cactus finches (*G. scandens*, *G. conirostris*) had larger beaks than sharp-beaked ground finches (*G. difficilis*; intermediate size) and warbler finches (smaller size; Fig. 8.43a). The beak morphology of *C. olivacea* and *C. fusca* (warbler finches) was statistically similar, not identical (Fig. 8.43a); the high genetic divergence between these two species, combined with low phenotypic divergence, suggested that stabilizing selection drove the evolution of their phenotypes (Fig. 8.43a-b; number 5 in Fig. 8.41; Petren et al. 2005).

Endemism and Hybridization in Galápagos Mockingbirds

Archipelagos are ideal to study genetic diversity as function of small population size, genetic drift, or gene flow (i.e. variables associated with endemism); the endemic Galápagos mockingbirds are a great system to test such associations in a historical context. Ancestors of cucubes colonized the archipelago 1.6-5.5-Ma (Table 1.3, Chapter One). There are currently four species (but see below): *Mimus parvulus*, the most common and distributed in the Central, Western, and Northern Islands; *Mimus melanotis* from San Cristóbal; *Mimus trifasciatus* from Floreana and its satellite islands Gardner and Champion; and *M. macdonaldi* from Española and its satellite island Gardner (Fig. 8.44).

Researchers conducted a spatio-temporal population genetic analysis to explore the role of genetic drift or gene flow in mockingbirds' evolution. They relied on contemporary blood samples (years 2003 to 2008) of 543 birds from 19 populations on 14 islands and historic museum tissue samples (1899, and from the 1905-1906 Expedition to Galápagos by the California Academy of Sciences) of 349 specimens from 13 islands. A total of 16 microsatellite loci were used in the analysis, which was restricted to 12 islands from which both contemporary and historic samples were available (Fig. 8.44; Hoeck et al. 2010).

In Fig. 8.45, genetic diversity is plotted as function of island size in contemporary and historic populations of Galápagos mockingbirds; measures of genetic diversity were expected heterozygosity H_e (left axis) and allelic richness *AR* (right axis). Note how genetic diversity (either H_e or *AR*) increased significantly with island size (area). No statistical difference was found for contemporary vs. historic samples. In addition, genetic diversity (H_e or *AR*) decreased significantly with island age (trend not shown); and allelic richness (*AR*) alone decreased significantly with island isolation (trend not shown; Fig. 8.45; Hoeck et al. 2010). In other words, mockingbird populations on larger islands were more genetically diverse than those on smaller islands. And separation-by-distance (geographic isolation) led to less genetic diversity in populations (as assessed by allelic richness *AR*).

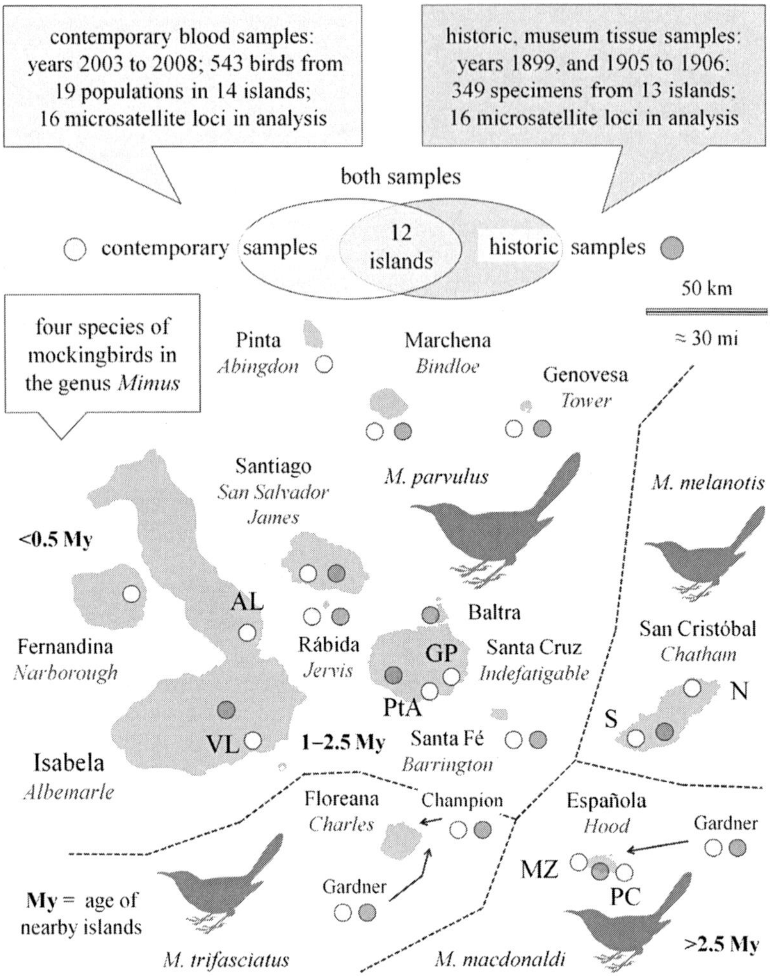

Fig. 8-44 Spatio-temporal population genetic analysis of Galápagos mockingbirds (*Mimus* spp.): contemporary (blood) and historic (tissue) samples were used in microsatellite analyses to test the role of genetic drift or gene flow in mockingbirds' evolution. Relevant sampling sites included: Isabela AL=Alcedo and VL=Puerto Villamil; Santa Cruz PtA=Puerto Ayora and GP=Galapaguera; San Cristóbal N=North and S=South; Española PC=Punta Cevallos and MZ=Suárez. Historic tissue samples came from museum collections of the California Academy of Sciences (a few samples from 1899; most from CAS Expedition to Galápagos, 1905-1906). Layout: GPC; redrawn and modified after Hoeck et al. (2010).

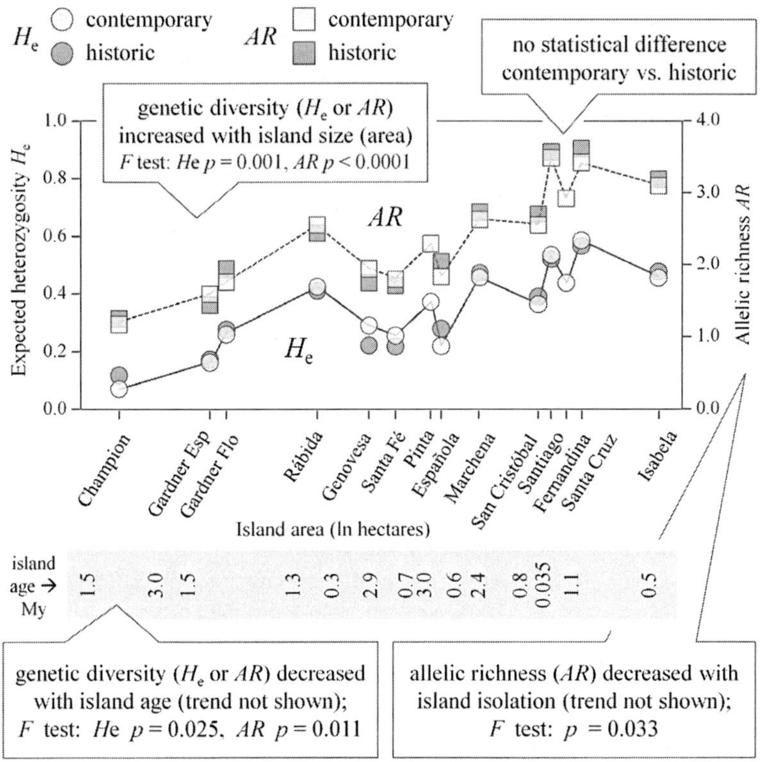

Fig. 8-45 Genetic diversity as function of island size in contemporary and historic populations of Galápagos mockingbirds (*Mimus* spp.): measures of genetic diversity were expected heterozygosity H_e and allelic richness AR (information on the effects of island age and geographic isolation is also provided). Layout: GPC; redrawn and modified after Hoeck et al. (2010).

When comparing change in gene frequencies (assessed as per F_{ST} a statistical estimate of population differentiation over time) among the 12 populations for which both contemporary and historic samples were available, researchers found that mockingbirds in larger islands showed less genetic differentiation—over time—than those in smaller islands (Fig. 8.46a). This indicated that genetic drift was—as expected—stronger in smaller than larger islands/populations during the past 100 years (Hoeck et al. 2010). This was corroborated by the observation that smaller islands had smaller estimated effective population size (N_e) than larger islands (Champion had the lowest N_e=43 breeding individuals, while Isabela the highest N_e=1,591 breeding individuals; Fig. 8.46b; Hoeck et al. 2010).

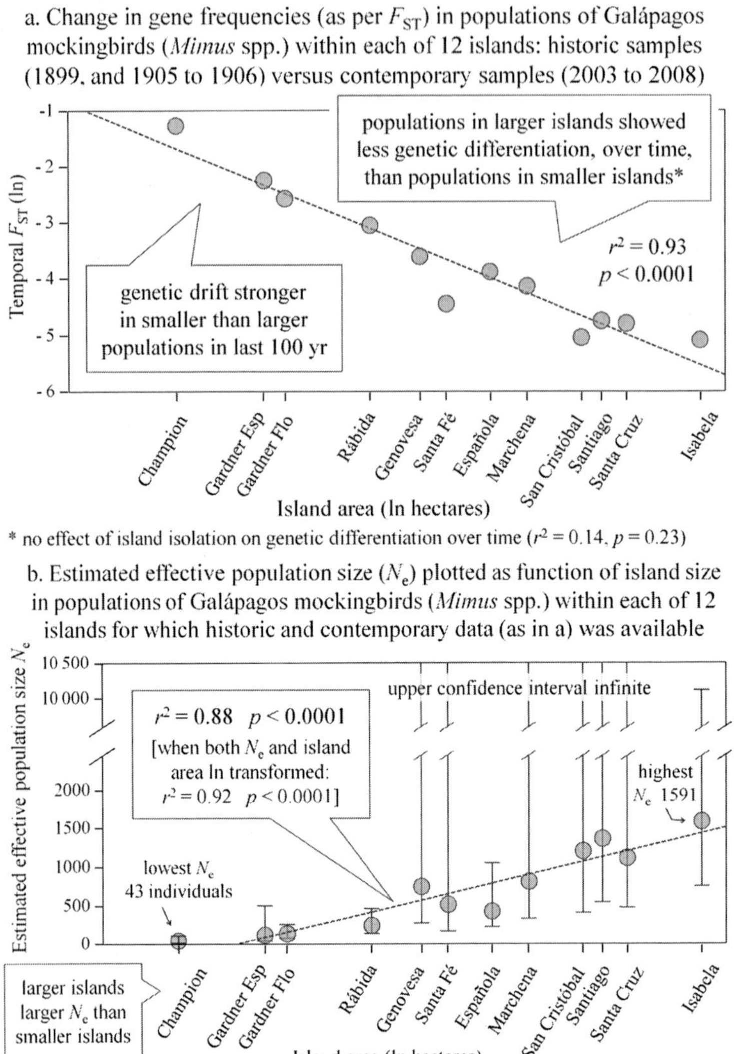

Fig. 8-46 a. Change in gene frequencies (as per fixation index F_{ST} a statistical estimate of population differentiation, polymorphisms) in populations of Galápagos mockingbirds (*Mimus* spp.) within each of 12 islands: historic versus contemporary samples. b. Estimated effective population size (N_e) plotted as function of island size in populations of Galápagos mockingbirds within each of 12 islands for which historic and contemporary data were available. Layout: GPC; redrawn and modified after Hoeck et al. (2010).

Researchers concluded that genetic drift had modulated genetic variance in Galápagos mockingbird populations (within/between) and species! Genetic diversity did not change significantly during the past 100 years, at least at the archipelago scale, but genetic drift did cause loss of genetic variability in small populations. Galápagos mockingbird populations were (are) evolving in partial isolation (Hoeck et al. 2010).

How do endemic taxa generate new lineages (e.g. subspecies, species) in an archipelago like Galápagos? The same team of investigators and collaborators has suggested one possible scenario: introgressive hybridization (Nietlisbach et al. 2013). Their observations come from the Genovesa Island cucubes (*Mimus parvulus bauri*; Fig. 8.47). [Realize that hybridization between/among lineages of various Galápagos birds, including finches, can be intricate; for discussion, see Paz-y-Miño-C and Espinosa 2024b]. Let us follow the story in Fig. 8.47: (1) Genovesa mockingbirds *M. p. bauri* are genetically quite similar to other *M. parvulus* from the Central, Western, and Northern islands of the archipelago (analysis based on 26 autosomal microsatellites; for species' distribution map, refer to Fig. 8.44). (2) Genovesa mockingbirds, however, are of mixed genetic ancestry; they share mitochondrial haplotypes with *M. melanotis* (San Cristóbal) and *M. macdonaldi* (Española), which indicates that the three taxa had a common female ancestor in the past; in addition, some autosomal loci (*FIB7*, *TGF*) of the San Cristóbal and Española populations are detectable in the Genovesa subspecies. (3) Natural history analyses suggest that at least one female from San Cristóbal—the closer island—dispersed to Genovesa 124,000 to 6,000 years ago; the immigrant female(s) introduced her (their) mtDNA into the Genovesa population, as well as distinctive nuclear loci (*FIB7*, *TGF*). (4) Over time, the three taxa differentiated morphologically while living on separate, distant islands. (5) Scientists hypothesized that the Genovesa subspecies was the product of introgressive hybridization (IH) among non-sister taxa. [IH involves the transfer of genetic material from one species to another and the fixation of some transferred haplotypes in the descendants' (hybrid) genome]. This helps explain why various phylogenetic reconstructions based on mtDNA loci have placed Genovesa mockingbirds proximate to the San Cristóbal and Española populations (e.g. Fig. 8.48). Thus, Genovesa mockingbirds might be in the process of becoming a separate species (Fig. 8.47; after Nietlisbach et al. 2013). As mentioned above, Galápagos mockingbird populations are evolving in relative isolation. Genetic drift, small population size, separation by distance (geographic isolation), and limited gene flow are shaping their genetic structure (Hoeck et al. 2010).

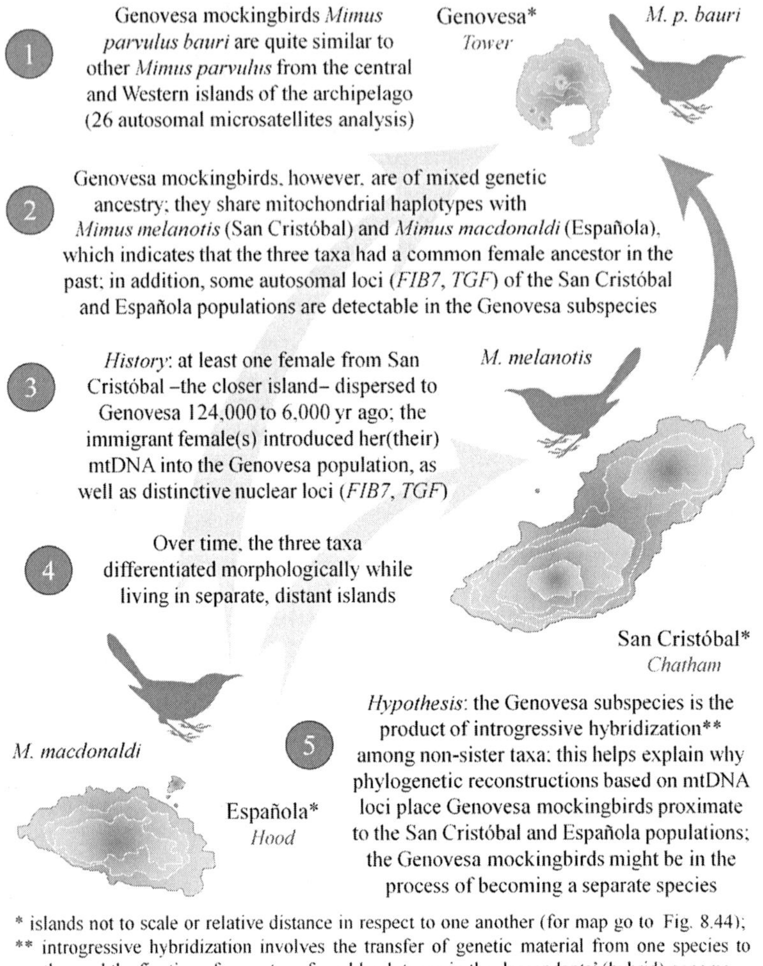

Fig. 8-47 The hybrid ancestry of the Genovesa mockingbirds (*Mimus parvulus bauri*) illustrates the intricacy of the genetic makeup of some Galápagos endemics. Layout and sketch: G. Paz-y-Miño-C; rationale after Nietlisbach et al. (2013).

Examine Fig. 8.48, which shows a phylogenetic reconstruction of Galápagos mockingbirds (*Mimus* spp.) based on the mitochondrial gene *ND2* (NADH dehydrogenase subunit 2). Note how the Genovesa mockingbirds cluster with San Cristóbal and Española species (top right dialog box; after Nietlisbach et al. 2013). The topology of this reconstruction, however, differs

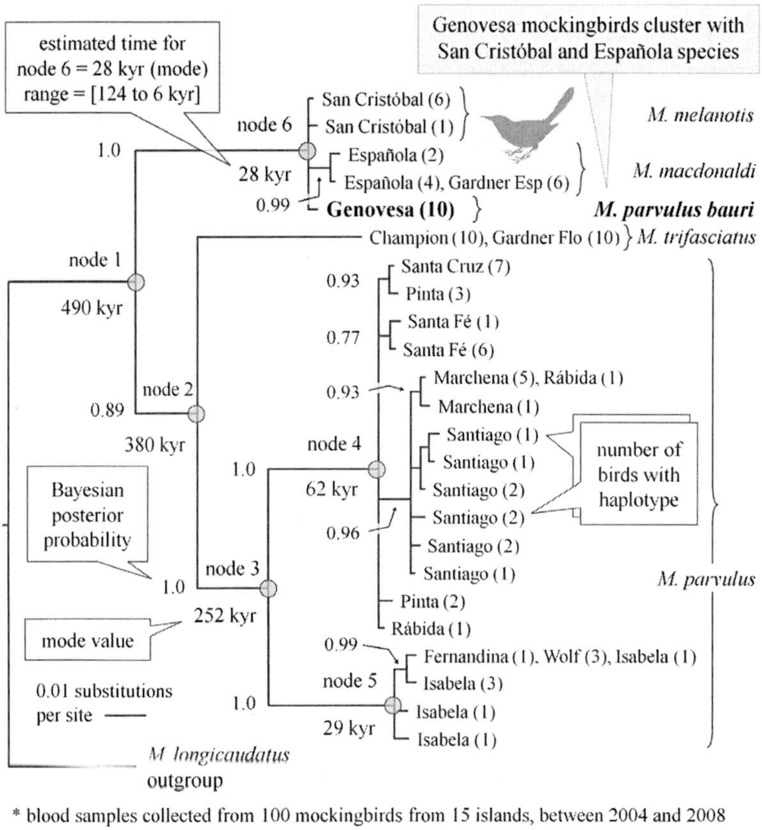

Fig. 8-48 Phylogenetic reconstruction of Galápagos mockingbirds (*Mimus* spp.) based on mitochondrial gene *ND2* (NADH dehydrogenase subunit 2) sequences*. Layout: GPC; redrawn and modified after Nietlisbach et al. (2013).

from that shown in Fig. 8.49, which is based on 16 autosomal microsatellites (contemporary and historic samples; Hoeck et al. 2010). On the latter phylogeny, Genovesa mockingbirds cluster within populations of *M. parvulus*!

A closer genetic analysis of Genovesa mockingbirds, based on autosomal nuclear DNA genes (*FIB7* and *TGF*), revealed that *M. p. bauri* (Genovesa, largely a *M. parvulus* genetic background) had nuclear features common to San Cristóbal (*M. melanotis*) and Española (*M. macdonaldi*) populations (Fig. 8.50a; Nietlisbach et al. 2013). Note in the median-joining network plots of Fig. 8.50a that for *FIB7* (beta-fibrinogen gene), the Genovesa haplo-

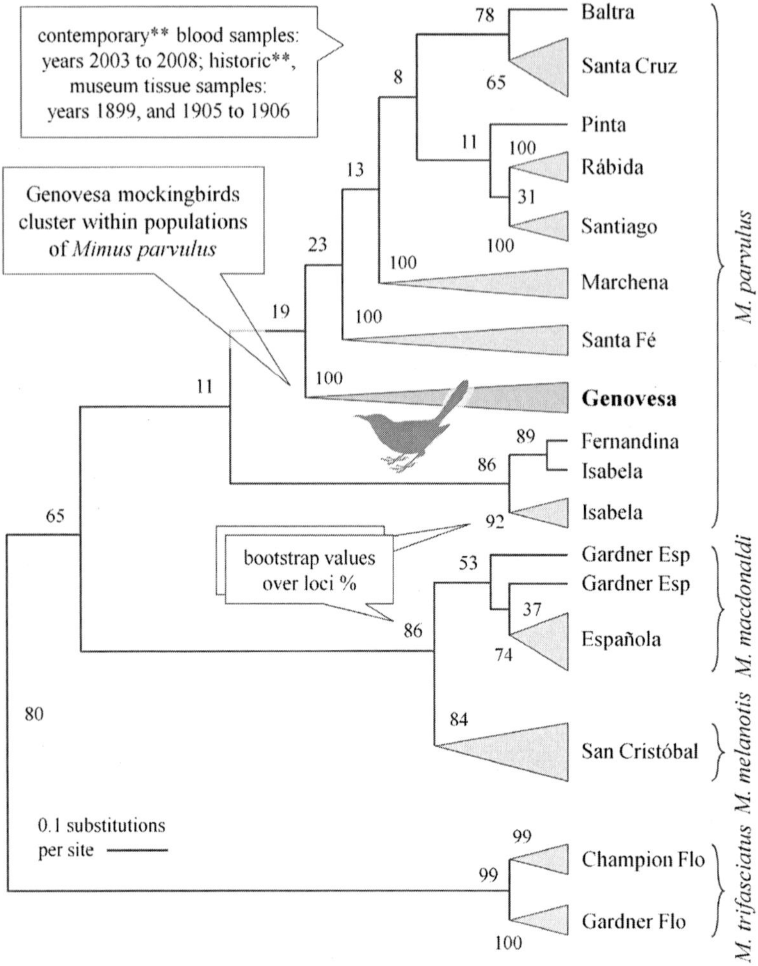

* UPGMA tree (unweighted pair group method with arithmetic mean) based on Nei's Ds (standard genetic distances); 1000 bootstrap resamplings among loci; ** contemporary samples: 543 birds, 19 populations, 14 islands; historic samples: 349 specimens, 13 islands

Fig. 8.49 Phylogenetic reconstruction (unrooted UPGMA* tree) of Galápagos mockingbirds (*Mimus* spp.) based on 16 autosomal microsatellites (contemporary and historic samples). Layout: GPC; redrawn and modified after Hoeck et al. (2010).

type is shared with *M. parvulus* and *M. macdonaldi* (thick black arrows on the right-side haplotype pies). Likewise, for *TGF* (transforming growth factor β2), the Genovesa mocking bird haplotype is shared with *M. parvulus*,

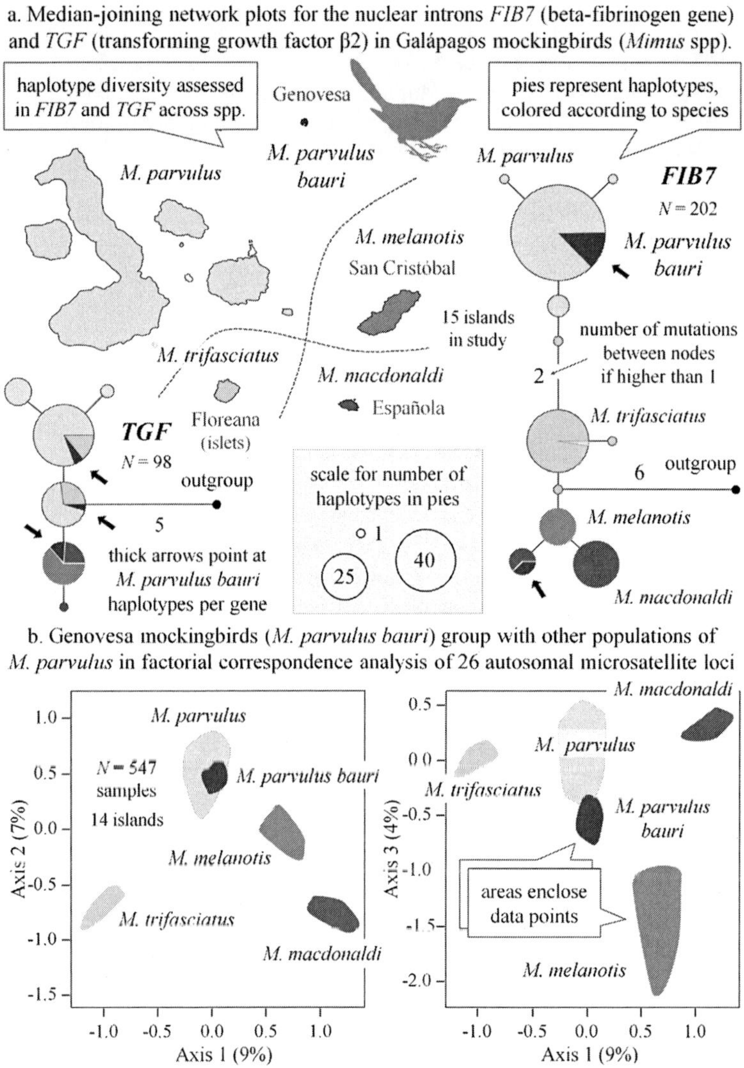

Fig. 8-50 a. Median-joining network plots for the nuclear introns *FIB7* (beta-fibrinogen gene) and *TGF* (transforming growth factor β2) in Galápagos mockingbirds (*Mimus* spp). b. Genovesa mockingbirds (*M. parvulus bauri*) group with other populations of *M. parvulus* in factorial correspondence analysis of 26 autosomal microsatellite loci. Genovesa mockingbirds are likely the result of introgressive hybridization among non-sister species. Layout: GPC; redrawn and modified after Nietlisbach et al. (2013).

M. trifasciatus, and *M. macdonaldi* (thick black arrows on the left-side haplotype pies of Fig. 8.50a). In addition, as shown in the factorial correspondence analysis of 26 autosomal microsatellite loci depicted in Fig. 8.50b, the Genovesa mockingbirds' group with other populations of *M. parvulus*. Thus, researchers think that Genovesa mockingbirds are the result of introgressive hybridization among non-sister species (i.e. mix ancestry) and that they might be undergoing speciation considering that their morphology (as per wing, beak, and tarsus measurements) has also become characteristically distinctive with respect to other *M. parvulus* populations in Galápagos (Nietlisbach et al. 2013).

Endemic Island Rodents

Introduced rodents often infest islands. Ship rats, Norway rats, and house mice are a global conservation problem and a threat to archipelagos' species (local plants and animals), including their own endemic rodents. Of the 2,217 species of rodents documented worldwide, 388 species (17.5%) are restricted to islands (Fig. 8.51). Of the 10 families occurring in islands, Muridae (258 species; 66.5% of total) and Sciuridae (54 species; 13.9% of total) are dominant (Fig. 8.51a). And of the 127 island-rodent genera, 83 (65.4%) are endemic and comprise 197 endemic species (50.8%; Fig. 8.51a; Amori et al. 2008).

Examine Fig. 8.51b, in which we show rodent-genera endemism index (*EI* = proportion of endemic species belonging to endemic genera) plotted as function of oceanic or continental islands' area (after Amori et al. 2008). The association between *EI* and island area is weak, but island isolation (oceanic vs. continental) is a stronger predictor of *EI*. Rate of endemism is higher in oceanic islands (dark-gray circles; mean *EI*=70%; range 7–100%) than in continental islands (light-gray circles; mean *EI* =22%; range 10–65%). Galápagos' rodents' *EI*=67%.

Except for the introduced invasive species (*Rattus rattus*, *R. norvegicus*, and *Mus musculus*), all other rodent species in Galápagos are endemic. [Note that the genus *Nesoryzomys* in the Cricetidae family is endemic]. Here are the taxa: *Nesoryzomys swarthi* (Santiago Rice Rat); *N. fernandinae* (small Fernandina Rice Rat or Fernandina Galápagos Mouse); *N. narboroughii* (large Fernandina Rice Rat or Heller's Galápagos Mouse); and *Oryzomys bauri* or "*Aegialomys galapagoensis*" (Galapagos' *Aegialomys*; Santa Fé Rice Rat or Galápagos Rice Rat; Swash and Still 2005; Harper and Carrión 2011; Tirira et al. 2020). In addition, the Galápagos

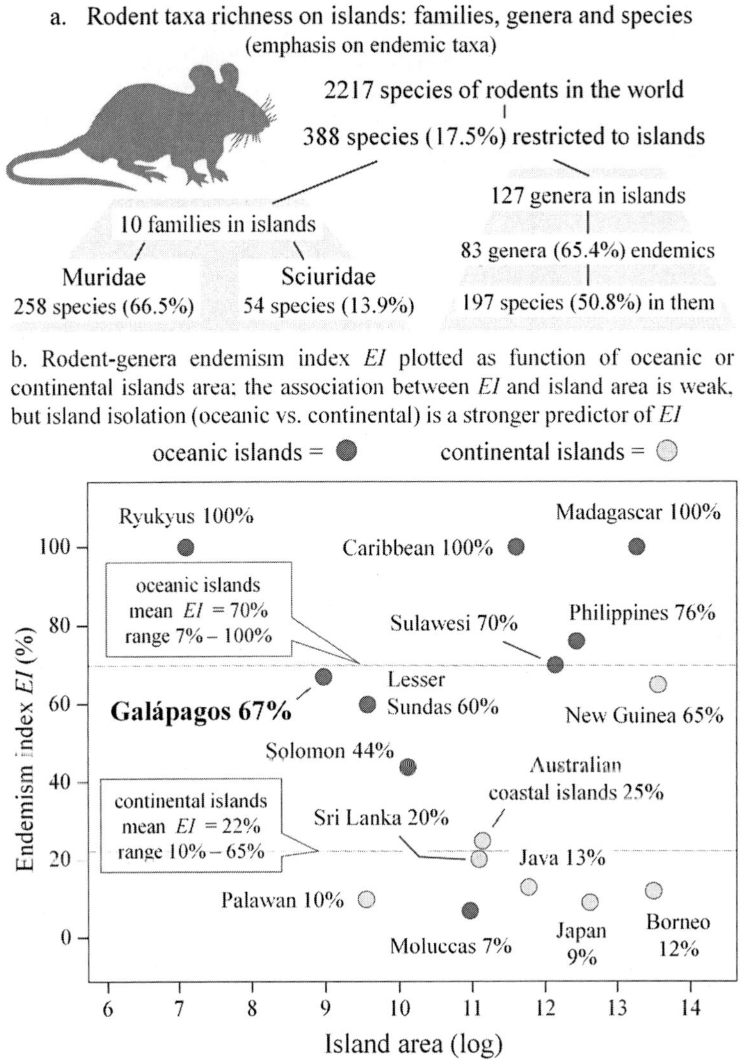

Fig. 8-51 a. Rodent taxa richness on islands: families, genera, and species (emphasis on endemic taxa). b. Rodent-genera endemism index (*EI*) plotted as function of oceanic or continental island area; the association between *EI* and island area is weak, but island isolation (oceanic vs. continental) is a stronger predictor of *EI*. In Galápagos (*EI*=67%), 4 extant and 3 extinct rodent taxa have been reported (see text). Layout: GPC; redrawn and modified after Amori et al. (2008) with additional data from Tirira et al. (2020).

Giant Rat (*Megaoryzomys curioi*) became extinct during the late Pleistocene to early Holocene (see fossils' presence in Fig. 1.20, Chapter One); *Nesoryzomys darwini* (Darwin's *Nesoryzomys* or Darwin's Galápagos Mouse) did it during the past 150 years; and *N. indefessus* (Santa Cruz Galápagos Mouse) vanished perhaps more recently (Amori et al. 2008; Tirira et al. 2020).

Endemism in Galápagos' Beetles

In Fig. 8.52, we summarize beetle species richness and endemics in the Galápagos Archipelago (17 exemplar islands). And in Table 8.5, we provide the number of genera and species per island (26 total islands), with emphasis on natives and single-island endemics (after Peck 2006). Of the 493 species documented to occur in Galápagos, 266 (54%) are endemic. [Note that 104 species (39%) are single-island endemics not shared with other islands, while 162 (61%) species are "Galápagos endemics" found on more than one island]. Larger and/or human-populated islands have more species, the latter likely because they have been more exhaustively surveyed, as follows: Santa Cruz 258 species, 27 endemics; Isabela 205 species, 24 endemics; San Cristóbal 153 species, 17 endemics; Santiago 148 species, 14 endemics; and Floreana 141 species, 5 endemics. All other islands reported in Fig. 8.52 or Table 8.5 have ≤87 species (range: Pinta 87 species to Beagle 1 species) and few endemics (range: Española 4 species to various islands 0 endemics; Peck 2006).

Seven genera are endemic to Galápagos: *Stomion* (family Tenebrionidae, darkling beetles) with 11 species; *Neoryctes* (Scarabaeidae, scarabs or scarab beetles) 4 species; *Mystroceridius* (Carabidae, ground beetles) 2 species; *Galapagodacnum* (Hydrophilidae, water scavenger beetles) 1 species; *Pinostygus* (Staphylinidae, rove beetles) 1 species; *Docema* (Chrysomelidae, leaf beetles) 1 species; and *Galapagonotus* (Curculionidae, snout beetles or true weevils) 1 species (Peck 2006).

The number of native genera per island is notable and suggests moderately rich phylogenetic beetle diversity in both single islands and the entire archipelago (i.e. considering the isolation of Galápagos with respect to the mainland): Santa Cruz 186 native genera; Isabela 158; San Cristóbal 123; Santiago 119; and Floreana 114. All other islands reported in Table 8.5 have ≤76 native genera (range: Pinta 76 genera to Beagle 1 genus; data extracted from Peck 2006).

Of the 167 world beetle families, 56 (34%) occur in Galápagos. Staphylinidae (rove beetles) with 52 genera and 78 species; Carabidae (ground beetles) 15 genera and 44 species; Tenebrionidae (darkling beetles) 16 ge-

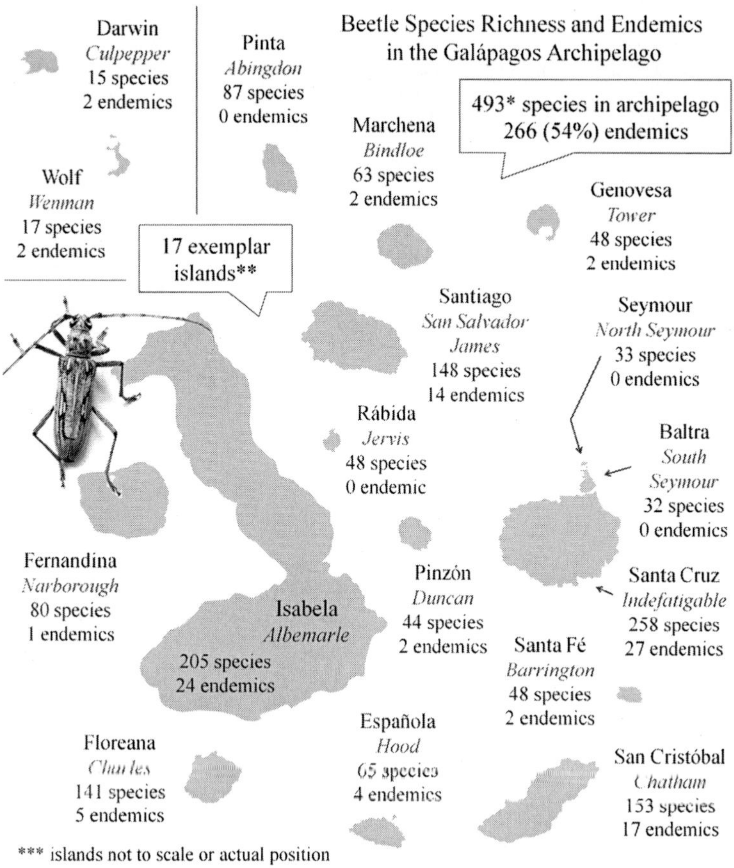

* Peck (2006) reported 493 species of beetles in the Galápagos, including 266 endemics (53.9%); in this Figure we provide total numbers of species per island (17 exemplars), which can be shared among various islands (thus, the summation surpasses 493); however, the numbers of endemics are single-island endemics (not shared with other islands), in this case 104, as accounted for in the Figure (note that 162 other endemics are found in more than one island, data not shown); ** in his assessment, Peck (2006) covers 26 islands (see Table 8.5 for complete list) from which we have selected the 17 exemplars; *** map not to scale or actual position of islands in the archipelago (for scaled map go to Fig. 8.44)

Fig. 8-52 Beetle species richness and endemics in the Galápagos Archipelago. Of the 493 species documented to occur in Galápagos, 266 (54%) are endemic. Of the 266 endemics, 104 (39%) are single-island endemics (not shared with other islands), while 162 (61%) are "'Galápagos endemics" found on more than one island. Figure layout and photo of longhorn-beetle from Isabela Island (*Eburia lanigera*, endemic to Galápagos and with wide distribution across islands): G. Paz-y-Miño-C; data extracted from Peck (2006).

Table 8.5 Beetle diversity in the Galápagos: number of genera* and species* per island (emphasis on natives and single-island endemics)

island	native genera	native species	endemic* species	%
Santa Cruz**	186	258	27	10.5
Isabela**	158	205	24	11.7
San Cristóbal**	123	153	17	11.1
Santiago	119	148	14	9.5
Floreana**	114	141	5	3.5
Pinta	76	87	0	0.0
Fernandina	72	80	1	1.3
Española	55	65	4	6.2
Marchena	56	63	2	3.2
Rábida	46	48	0	0.0
Genovesa	44	48	2	4.2
Santa Fé	40	48	2	4.2
Pinzón	42	44	2	4.5
Seymour North	31	33	0	0.0
Baltra	28	32	0	0.0
Plaza Sur	17	17	0	0.0
Bartolomé	16	17	0	0.0
Wolf	17	17	2	11.8
Champion	16	16	0	0.0
Darwin	15	15	2	13.3
Caamaño	9	10	0	0.0
Eden	8	8	0	0.0
Gardner Flo	7	7	0	0.0
Daphne Major	3	4	0	0.0
Tortuga	3	3	0	0.0
Beagle	1	1	0	0.0

* Peck (2006) reported 493 species of beetles in the Galápagos, including 266 endemics (53.9%); note that the numbers of native genera or species can be shared among islands, however, the numbers of endemics are single-island endemics (not shared with other islands), in this case 104, as accounted for on the Table (note that 162 other endemics are found in more than one island, data not shown); ** gray area includes the top five islands, as per beetle species richness, with >100 native species; four of these islands (Santa Cruz, Isabela, San Cristóbal, Floreana) are human populated, i.e. better known beetle diversity than other islands

nera and 57 species; and Curculionidae (snout beetles or true weevils) 38 genera and 70 species. All other families (*N*=52) are represented by ≤13 genera (range: Cerambycidae or longhorn beetles 13 genera to various families 1 genus) or ≤14 species (range: Cerambycidae 14 species to various families 1 species). [Note that 110 species in 26 families have been introduced in the archipelago; data from Peck 2006)].

Endemism in Galápagos' Spiders

Of the 159 spider species found in Galápagos, 82 (52%) are endemic (Fig. 8.53). In Table 8.6, we summarize data on the top 10 families and their species richness in the archipelago, and in Table 8.7, the numbers of families and species per island, with emphasis on endemics, natives, and introduced taxa (after Buchholz et al. 2020).

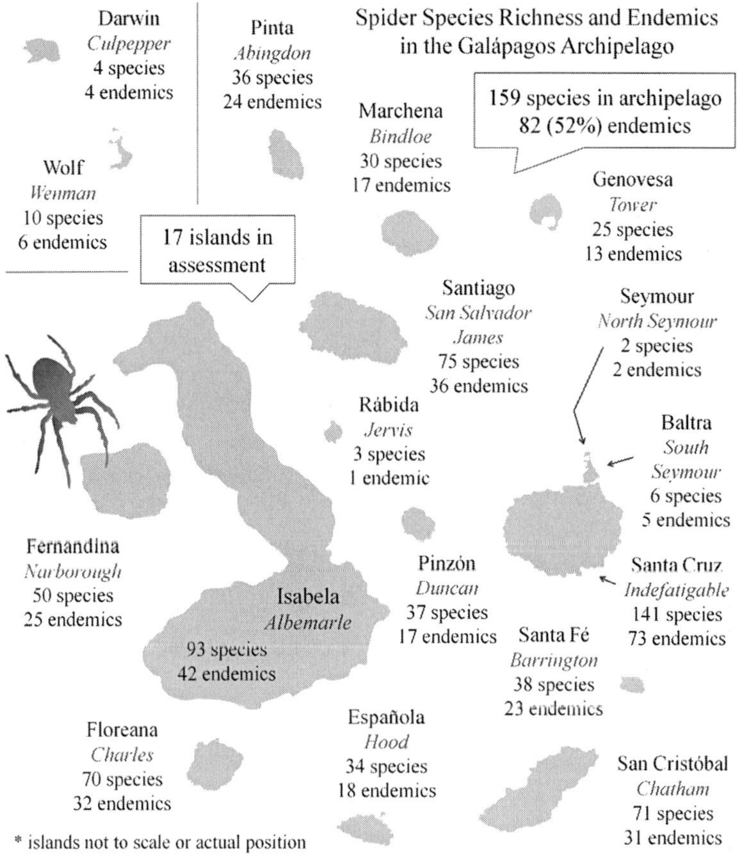

Fig. 8-53 Spider species richness and endemics in the Galápagos Archipelago. Layout: G. Paz-y-Miño-C; data extracted from Buchholz et al. (2020).

Table 8.6 Spiders in the Galápagos: top 10 families and species richness (emphasis on endemics, natives and introduced species in the islands)*

families	species	endemic spp	native spp	introduced spp
Theridiidae	24	4	16	4
Salticidae	15	9	3	3
Araneidae	14	6	8	0
Oonopidae	14	5	9	0
Pholcidae	14	12	0	2
Linyphiidae	11	5	6	0
Gnaphosidae	9	7	2	0
Lycosidae	7	7	0	0
Tetragnathidae	5	1	4	0
Zoridae	4	4	0	0
sub total top 10 families	117 (74%)	60 (38%)	48 (30%)	9 (6%)
24 other families	42 (26%)	22 (14%)	13 (8%)	7 (4%)
Total 34 families	159 (100%)	82 (52%)	61 (38%)	16 (10%)

* Data extracted from Buchholz et al. (2020)

Similar to beetles (above), spiders' diversity is higher on larger and/or human-populated islands, which are more exhaustively surveyed by arachnologists (Fig. 8.53). [Larger islands also vary in altitudinal habitat diversity]. The top ten dominant spider families in Galápagos account for 117 (74%) of the species (N_T=159), 60 of the endemic species (N_T=82), 48 of the native species (N_T=61), and 9 of the introduced species (N_T=16; Table 8.6). The richest spider family is Theridiidae (tangle-web spiders), with 24 species, 4 endemics; followed by Salticidae (jumping spiders), 15 species, 9 endemics; Araneidae (orb-weaver spiders), 14 species, 6 endemics; Oonopidae (goblin spiders), 14 species, 5 endemics; and Pholcidae (cellar spiders or daddy-long-legs spiders), 14 species, 12 endemics. All other spider families have ≤11 species (Table 8.6; data from Buchholz et al. 2020).

Santa Cruz Island has the highest so-far-documented spider diversity, with 141 species, 73 endemics; seconded by Isabela, 93 species, 42 endemics; Santiago, 75 species, 36 endemics; San Cristóbal, 71 species, 31 endemics; Floreana, 70 species, 32 endemics; and Fernandina, 50 species, 25 endemics. All other islands have ≤38 species (Table 8.7).

Table 8.7 Spider diversity in the Galápagos: numbers of families* and species* per island (emphasis on endemics, natives and introduced)**

island	families	species	endemics	%	natives	%	introduced	%
Santa Cruz	32	141	73	52	52	37	16	11
Isabela	29	93	42	45	37	40	14	15
Santiago	27	75	36	48	31	41	8	11
San Cristóbal	24	71	31	44	28	39	12	17
Floreana	23	70	32	46	23	33	15	21
Fernandina	20	50	25	50	22	44	3	6
Santa Fé	23	38	23	61	11	29	4	11
Pinta	18	36	24	67	9	25	3	8
Española	16	34	18	53	12	35	4	12
Pinzón	18	34	17	50	13	38	4	12
Marchena	16	30	17	57	11	37	2	7
Genovesa	12	25	13	52	7	28	5	20
Wolf	9	10	6	60	4	40	0	-
Baltra	4	6	5	83	0	-	1	17
Darwin	3	4	4	100	0	-	0	-
Rábida	3	3	1	33	2	67	0	-
North Seymour	2	2	2	100	0	-	0	-

Data extracted from Buchholz et al. (2020). *As shown in Table 8.6, a total of 159 species [in 34 families] of spiders have been found in the Galápagos Archipelago (endemics = 82, natives = 61, introduced = 16). **Note in this table (gray area) that for islands with 30 or more species (from Santa Cruz to Marchena in the list) about half of the species are endemics (mean = 52%, range 44% to 67%); about one in every three species (mean 36%) are natives (range 25% to 44%); introduced species are more common in the human populated islands (Santa Cruz = 16 species, Isabela = 14 species, San Cristóbal = 12 species, and Floreana = 15 species). Santiago Island was human-populated in the past.

Spiders are efficient at long-distance dispersal (e.g. via ballooning or kiting, typical of spiderlings, which release silk threads and become lifted by winds; spiders can travel thousands of kilometers/miles). About one-quarter (24%) of the Galápagos' spider species are cosmopolitan (global distribution), while 26% are confined to the Americas. Still, the majority of Galápagos spiders are distributed on only 1-2 islands; very few species are found on >10 islands; and no species has been documented to occur on all islands (Buchholz et al. 2020). [Keep in mind that taxa surveying remains limited to the most accessible and/or human-populated islands; Isabela, the largest island, likely remains unexplored].

Endemism in Galápagos' Marine Organisms

In Fig. 8.54, we give readers an idea about endemism in some groups of Galápagos' marine organisms (data extracted from Hickman 2009, with input from Bustamante et al. 2002). Of the 780 species of mollusks (snails, clams, squids, and octopuses) estimated to occur in the Galápagos Marine Reserve, 141 (18.1%) are endemic. They are followed by fishes (bony and cartilaginous), 447 species, 51 (11.4%) endemics; soft bottom meiofauna (small multicellular organisms living proximate to or on/in the seafloor), 390 species (endemics unidentified); algae, 333 species, 130 (39.0%) endemics; echinoderms (sea stars, brittle stars, sea urchins, sea cucumbers), 200 species, 34 (17.0%) endemics; polychaetes (marine segmented worms, bristle worms), 192 species, 50 (26.0%) endemics; bryozoans (sea moss animals), 184 species, 34 (18.5%) endemics; brachyurans (true crabs), 120 species, 23 (19.2%) endemics; hydroids (jellyfishes), 96 species, 14 (14.6%) endemics; caridea (true shrimp) and stenopods (shrimp-like crustaceans), 65 species, 10 (15.4%) endemics; amphipods (crustaceans), 50 species, 19 (38.0%) endemics; opisthobranchs (sea slugs or sea hares), 49 species, 18 (36.7%) endemics; ahermatypic corals (non-matrix building corals), 43 species, 17 (39.5%) endemics; marine mammals (whales, dolphins, porpoises, sea lions, fur seals), 24 species, 2 (8.3%) endemics; marine birds (shore/coastal birds), 19 species, 5 (26.3%) endemics; and barnacles (cirripede crustaceans), 18 species, 4 (22.2%) endemics.

Of the grand total 3,089 species accounted for in Fig. 8.54, about one in every five (569 species, 18.3%) are Galápagos endemics. Keep in mind that earlier in this chapter, we discussed exemplar endemic marine taxa in the archipelago. For example, the biogeographic affinities of Galápagos penguins (*Spheniscus mendiculus*; Fig. 8.4), marine iguanas (*Amblyrhynchus cristatus*; Figs. 8.9-8.11), sea lions (*Zalophus wollebaeki*; Figs. 8.19-8.21), and fur seals (*Arctocephalus galapagoensis*; Figs. 8.20-8.21). In Chapter Eight of our book *Arrivals of Life to the Galápagos* (Paz-y-Miño-C and Espinosa 2024a), we examine [although indirectly focused on endemism or endemics] fish assemblages in tropical islands' reefs, with emphasis on the Tropical Eastern Pacific (TEP) and Galápagos; deep-ocean fishes and invertebrates in TEP; and coral species richness/dissimilarity across Pacific reefs. In Chapter Nine of the same volume, we evaluate the extinction risk of endemic taxa, including the endangered Galápagos Flamingo (*Phoenicopterus ruber glyphorhynchus*), vulnerable Galápagos Cormorant (*Phalacrocorax harrisi*), endangered Galápagos Penguin, endangered Blue-footed Booby (*Sula nebouxii excisa*),

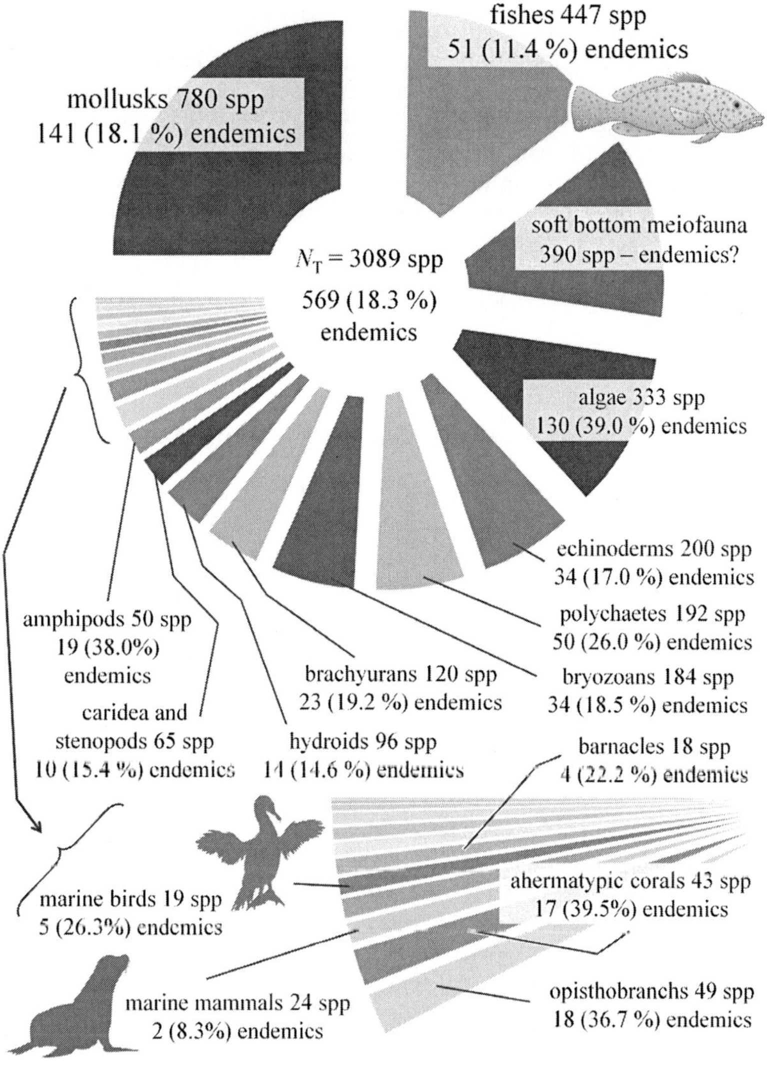

Fig. 8-54 Marine organisms in the Galápagos (exemplars): species' and endemics' counts. Layout and sketches GPC; data extracted from Hickman (2009) with input from Bustamante et al. (2002).

vulnerable Lava Gull (*Leucophaeus [Larus] fuliginosus*), near threatened Magnificent Frigatebird (*Fregata magnificens magnificens*), vulnerable American Oystercatcher (*Haematopus palliatus galapagensis*), and the near threatened Lava or Galápagos Heron (*Butorides sundevalli*). In our book *A Synthesis of the Galápagos* (Paz-y-Miño-C and Espinosa 2024b), we discuss various topics associated with the archipelago's endemic taxa, including the population status (rookeries) of Galápagos sea lions and their pups; the ability of marine iguanas to body-shrink and survive severe El Niño events; a cormorant that became flightless; oceanic bioregions and the cycles of life; the islands of Darwin and Wolf as source of coral larvae in TEP; interactions between coral reefs and sharks; and thirty-five species of sharks and counting. [More about some of these themes below]. We encourage readers to study these topics—directly or indirectly related to endemism in Galápagos' marine organisms—and integrate the information therein with this chapter (e.g. Fig. 8.54)!

3 Ecosystem Functions: A Glimpse of Functional Benefits

In Box 8.1, we defined ecosystem functions [as part of the conservation value of the Galápagos] in the context of the *functional benefits* that ecosystems provide to their organisms (from microbes to macro life) and the planet, including chemical or biological cycles. We indicated that humans benefit from ecosystems by having access to clean air, fishing, or participating in nature-oriented tourism. Our definition was, of course, abridged. Ecosystem functions and benefits can include anything and everything ecosystems do! From primary production (via photosynthesis or chemosynthesis) to nutrient cycling and energy flow (consumption of nutrients by all trophic levels) to ecological relationships (predation, competition, mutualism, and other across-organisms interactions such as pollination, ectoparasite removal, or hosting microbiomes) to large-scale natural transformations of the environment by organisms' behavior (from microbes to invertebrates, large vertebrates, or huge-tree forests, and from the ocean floor to the highest mountains) and to the cognitive interactions among organisms (e.g. highly social species) that result in substantial habitat modifications. Thus, ecosystem functions and benefits can emerge from a wide range of processes in the ecosystem. Now, from a historically simplistic viewpoint, "ecosystem benefits" have been attributed to the for-profit extraction of goods/services from the environment, e.g., natural resources (timber, petroleum, wildlife, or groundwater), pristine areas for urban development (rich-soil valleys or seashores for mega resorts), or massive outdoor recreation (conventional tourism). Although these activities rely on ecosystem outputs (i.e. what nature provides), they disregard the long-term environmental impacts of merely profiting from nature. [For a review on the concept of "ecosystem services", not to be confused with "ecosystem functions" (our emphasis here), see Balvanera et al. (2017). The distinction is intricate since both concepts overlap. *Ecosystem services* refer to direct or indirect *benefits* that *humans receive from the environment*: provisioning (resources extracted from nature), regulating (gains resulting from legislating the use of the ecosystem), supporting (natural processes that sustain the ecosystem and keep it running), and cultural (traditions, sense of identity, knowledge inspired in nature and used in environmental interpretation or ecotourism, as well as the aesthetic value of the ecosystem; Balvanera et al. (2017). Readers should keep in mind that the terms *ecosystem functions* and *ecosystem services* have distinctive meanings. In this chapter, we focus on ecosystem functions and benefits in the Galápagos].

Below, we discuss five examples of ecosystem functions and benefits in the Galápagos, which help us address two key questions in the chapter: what is the conservation value of the archipelago and its marine reserve, and what are we really safeguarding? We examine altitudinal migration of giant tortoises [remember that these reptiles are emblematic species, above; migration behavior is essential to their survival and reproduction]. We describe how tiger sharks come to hunt in the Galápagos Marine Reserve while their favorite prey, the green sea turtles, come to nest [i.e. predator-prey interactions]. We highlight the importance of the islands of Wolf and Darwin as "sources of coral larvae" for the Galápagos marine ecosystems and the entire Tropical Eastern Pacific [i.e. biological processes of local and regional relevance, likely global]. We address the interactions between coral reefs and sharks [i.e. reefs as exemplar ecosystems and providers of habitat, nutrients, and shelter for sharks; the latter as exemplar mesopredators and top/apex predators]. And we close by underlining the Galápagos' impressive shark diversity, with its "thirty-five species of sharks and counting" [i.e. sharks as both essential to Galápagos' marine-species assemblages and attractive predators to explorers of nature (eco-diving and ecotourism) and conservationists].

Altitudinal Migration of Giant Tortoises

Yes, in Galápagos, giant tortoises migrate! Not all tortoises do it, nor do they migrate on all islands. For decades, farmers in the agricultural lands, park rangers familiar with the habits of local fauna, and some naturalists knew that giant tortoises traveled to the highlands during the cool season (when rains were scarce in the lowlands). Scientific evidence of such migrations has been recently produced, but the story is intricate. [Content below abridged from Paz-y-Miño-C and Espinosa (2024b)].

Researchers tracked the movements of 50 GPS-tagged individual tortoises (i.e. equipped with Global Positioning System devices) during 2009-2012 at contrasting sites as per island, vegetation cover, and altitude (Fig. 8.55; after Bastille-Rousseau et al. 2016): (1) Santa Cruz (altitude 864 m/2,835 ft): 13 individuals (7 females, 6 males) of *Chelonoidis donfaustoi* in the Eastern slopes of the island (Cerro Fatal; between 200-400-m/656-1,312-ft elevation); and 15 individuals (8 females, 7 males) of *C. porteri*

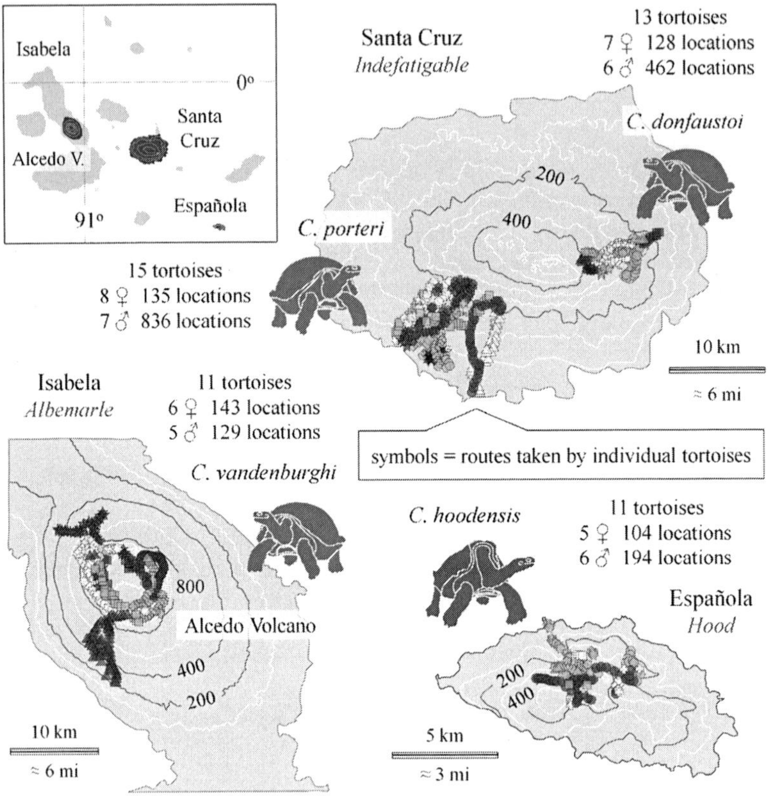

Fig. 8.55 Movement patterns of Galápagos tortoises (*Chelonoidis*) in Santa Cruz, Isabela (Alcedo Volcano), and Española Islands. Fifty GPS-tagged individuals were tracked during 2009-2012. Note that the Santa Cruz and Isabela species corresponded to "dome" morphotype individuals, while the Española species was "saddleback". For distribution of all giant tortoise species across the archipelago, phylogenetic relations, time of divergence among taxa, and scenarios of possible arrivals to the archipelago by ancestors of giant tortoises, see Figs. 8.5-8.7. Layout: GPC; redrawn and modified after Bastille-Rousseau et al. (2016).

in the Southwestern part of the island (La Reserva; between 0-200-m/0-656-ft elevation). (2) Alcedo Volcano in Isabela (altitude 1,130 m/3,710 ft): 11 individuals (6 females, 5 males) of *C. vandenburghi* around the caldera and Western slopes of the volcano (between 200-800-m/656-2,624-ft elevation). And (3) Española (altitude 206 m/676 ft): 11 individuals (5 females, 6 males) of *C. hoodensis* in the Northern-central part of the island (particularly between 200-400-m/656-1,312-ft elevation). The symbols in Fig. 8.55 correspond to routes taken by individual tortoises during the tracking period. The number of GPS tracking locations per species is given as, e.g., Santa Cruz, *C. donfaustoi*, 13 tortoises, 7 ♀ 128 locations, and 6 ♂ 462 locations (top right in Fig. 8.55).

By looking at Fig. 8.55, readers can infer that giant tortoises of the four species (at all tracking sites) did move and travel across the landscape; they followed redundant paths (i.e. natural trails likely built by themselves while moving about). Scientists examined if such movements correlated with various variables: species (*C. donfaustoi*, *C. porteri*, *C. vandenburghi*, *C. hoodensis*), tortoises' size (body mass inferred from curved carapace length), sex (female, male), tracking site (as per island), probability of movement (likelihood to initiate locomotion after remaining inactive at a location), directional persistence (overall movement direction as per GPS-tag position), hourly movement (distance), daily movement (distance), monthly movement (distance), and time of the year movement (to infer seasonality in movement patterns; Bastille-Rousseau et al. 2016).

Chelonoidis vandenburghi (Alcedo Volcano, Isabela) showed noticeably higher values across the studied variables (above) during January or July. This was consistent with seasonal movement patterns in this species (i.e. migration). The other species showed differential but weaker statistical consistency across variables to infer clear seasonal movement patterns (data not shown here). However, by simply examining the routes taken by the tortoises in Santa Cruz (much less in the smaller, lower-elevation Española), it was obvious that they did travel across altitudinal gradients during the study period (Fig. 8.55). Note that larger, heavier tortoises were more active and "motivated" (i.e. bouts of intense locomotion) to move and cover long distances, although smaller individuals could reach comparable distances during longer periods of time if motivated to move (data in Bastille-Rousseau et al. 2016). With this knowledge in hand, researchers suspected that body mass had something to do with the tortoises' altitudinal and seasonal movements. Since vegetation biomass varied seasonally, particularly in the lowlands of the larger islands (Isabela, Santa Cruz) with respect to the highlands (i.e. during the cool season, the lowlands become dry while rainfall remains steady in the highlands; see Figs. 1.43-1.44 in

Chapter One), scientists hypothesized that Galápagos giant tortoises migrated altitudinally and seasonally to compensate for nutrients' scarcity (deficit) in the lowlands during the "cool/dry" season (Yackulic et al. 2017). The biologists explained that larger tortoises were more sensitive than smaller ones to spatial and temporal variation in food abundance due to body size effects on foraging and metabolism. Thus, to achieve energetic balance, such sensitivity in larger individuals (i.e. an incentive to track resources) triggered migratory behavior: the forage hypothesis (Yackulic et al. 2017).

In Fig. 8.56, we synthesize the essence of the forage hypothesis. Note in Fig. 8.56a how during the "hot/rainy" season, both large(r) and small(er) tortoises roam in the lowlands of Santa Cruz, where food is plentiful. Keep in mind that the tortoises nest in the lowlands. During the "cool/dry" season, when food becomes scarce in the lowlands, the large(r) tortoises (>80-kg/>176-lbs in weight) migrate to the highlands (as indicated by the arrow), where they compensate for the nutrient deficit in the lowlands. Fig. 8.56b shows how an exemplar 240-kg/529-lbs male of *C. porteri* from Southwestern Santa Cruz migrated altitudinally from/back the highlands to the lowlands. This individual remained in the highlands (at +300-m/+980-ft altitude) during the peak of the "cool/dry" season, from August to October and up to January. It descended to the lowlands in just 9 days, where it remained during the peak of the "hot/rainy" season, from February to March, and up to July, when it returned to the highlands in just 16 days (Yackulic et al. 2017).

Look at Fig. 8.56c [a model] in which the response (i.e. sensitivity) of small(er) or large(r) tortoises to seasonal changes in vegetation forage quantity is plotted. The ratio of energy surplus/deficit (i.e. energy balance) for 20-kg/44-lbs or 200-kg/440-lbs tortoises is plotted as per Normalized Difference Vegetation Index (NDVI), i.e., a proxy for seasonal variation in vegetation forage quantity based on satellite imagery tracking actual changes in seasonal vegetation cover. See how the dotted-line trend for the small(er) tortoise is either zero or positive as NDVI values increase. For the large(r) tortoise, by contrast, the solid-line trend switches abruptly from negative to positive values as NDVI increases. Large(r) tortoises are simply more sensitive to seasonal changes in vegetation forage quantity than small(er) tortoises (Yackulic et al. 2017)!

Fig. 8.56d [a model] shows that small(er) tortoises can afford to stay in the lowlands during the entire "cool/dry" season (June-December) since the energy balance (surplus/deficit) practically overlaps between lowlands and highlands. Although there is evidently more vegetation forage quantity available for the tortoises in the lowlands than in the highlands during the

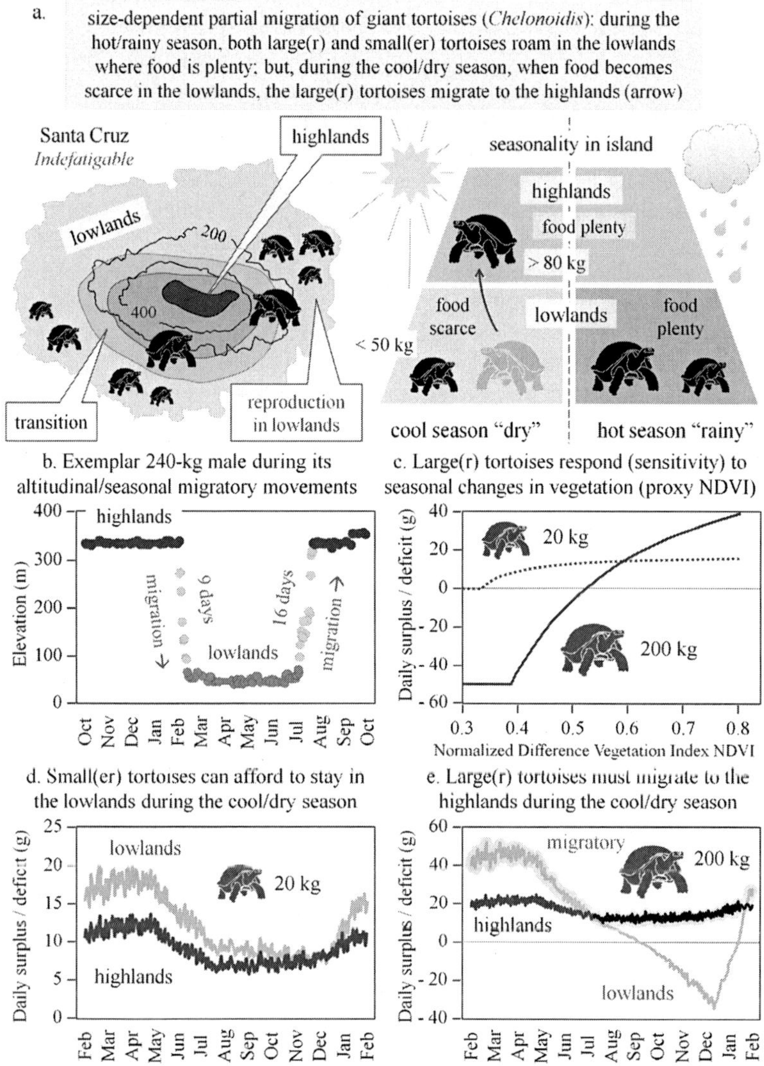

Fig. 8-56 Altitudinal migration of Galápagos tortoises on Santa Cruz Island. The forage hypothesis suggests that larger tortoises are more sensitive to spatio-temporal variation in food abundance due to body size effects on foraging/metabolism. To achieve energetic balance, larger individuals must track resources, which triggers migratory behavior. Layout: GPC; redrawn and modified after Yackulic et al. (2017).

"hot/rainy" season (January-May). By contrast, as shown in Fig. 8.56e [a model], large(r) tortoises must migrate to the highlands during the "cool/dry" season to keep up with their energy balance (surplus/deficit) demands! Note in Fig. 8.56e how the lowlands offer plenty of vegetation for the large(r) tortoises to forage during the "hot/rainy" season but not during the "cool/dry" season (Yackulic et al. 2017). [For hypotheses-based, detailed analyses of the factors involved in Galápagos tortoises' migrations, see Blake et al. 2013; Bastille-Rousseau et al. 2019]. Natural selection has likely favored geno-phenotypes with the ability to sense and respond to seasonal variation in resources' abundance, track the spatial distribution and availability of such resources, and engage in energy-efficient altitudinal movement patterns to maximize intake of resources. Ultimately, migrations seem to be driven by lowland vegetation dynamics.

Going back to the questions: what is the conservation value of the archipelago, and what are we really protecting? In the case of the giant tortoises, an emblematic species and top herbivore in the Galápagos, their *in-situ* preservation will only be possible if the islands' ecosystems remain functional. By safeguarding emblematic species and their habitats, we also perpetuate ecological and evolutionary processes.

Tiger Sharks Come to Hunt, Green Sea Turtles to Nest

Where prey go, predators follow. That is the case of green sea turtles (*Chelonia mydas*) and tiger sharks (*Galeocerdo cuvier*), both cosmopolitan species (Fig. 8.57a). On Galápagos' crowded shores [full of wildlife], tiger sharks come to hunt while sea turtles nest (although both species are present year-round). Between 2014 and 2015, biologists tagged 20 tiger sharks (13 females, 7 males) in the vicinity of two green-sea turtles' sandy-beach nesting sites: Isabela-South (Is) in Southern Isabela Island and Bachas-Salinas (BS) in Santa Cruz/Baltra Islands. A third study site, Cerro Ballena (CB), also in Southern Isabela but in a rocky area, was selected due to the offshore presence of juvenile tiger sharks (Fig. 8.57b). The researchers monitored the sharks' movement patterns (via satellite tracking of tags and baited underwater video) within the Galápagos Marine Reserve (GMR 133,000 km^2/51,000 sq mi) and the proximate ocean (coastal Central and South America; Fig. 8.57b; Acuña-Marrero et al. 2017). [Keep in mind that tiger sharks and green sea turtles are exemplars of the ecological relationships between organisms (in this case predator-prey interactions), including the ecological and evolutionary processes that sustain them, being protected by the Galápagos Archipelago's status as a National Park, Marine Reserve, and World Heritage Site (above)].

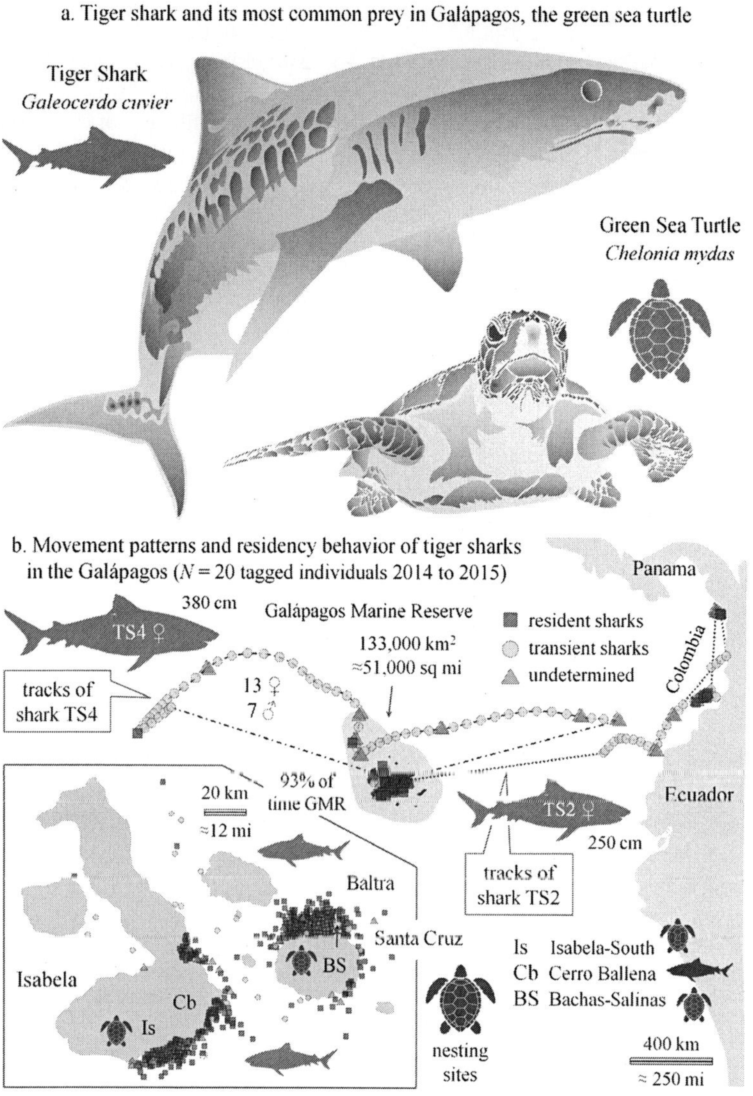

Fig. 8-57 a. Tiger shark (*Galeocerdo cuvier*) and its most common prey in the Galápagos Marine Reserve (GMR 133,000 km²/51,000 sq mi), the green sea turtle (*Chelonia mydas*). b. Movement patterns and residency behavior of tiger sharks in GMR ($N = 20$ tagged individuals; satellite tracking and underwater video 2014 to 2015). a-b. Layout/digital sketches: G. Paz-y-Miño-C. b. Redrawn and modified after Acuña-Marrero et al. (2017).

In Fig. 8.57b, we summarize the movement patterns and residency behavior of the tagged sharks during the 2-year tracking period (after Acuña-Marrero et al. 2017). Look how sharks spent 93% of their time within the Galápagos Marine Reserve, i.e., they behaved as "resident sharks" during 2014-2015 (dark-gray squares in Fig. 8.57b, bottom left). Two females (TS2 and TS4), however, behaved in a mixed manner. TS2 (250 cm/100 in long) remained as resident within GMR during April 2014. It later switched to "transient" (as represented by the light-gray circles in Fig. 8.57b) and swam toward coastal Ecuador (900 km/560 mi), where it continued offshore-North (additional 800 km/500 mi; coastal Colombia and proximate to Panama); the journey took about three months. TS2 was later detected again (by underwater receivers) within GMR, nearby her tagging location at Bachas-Salinas. Returning to Bachas-Salinas coincided with the onset of *C. mydas* nesting (Acuña-Marrero et al. 2017). For comparison, follow TS4's tracks on Fig. 8.57b, a much larger shark (380 cm/150 in long). TS4 remained within GMR from January to July 2014. Her satellite signal reappeared a couple of months later, about 1,000 km/620 mi West of her tagging location (Bachas-Salinas). TS4 swam East, passed by GMR, and approached coastal Ecuador (900 km/560 mi). Her signal was lost again (October 2014) until it reappeared in November 2014 (beginning of green sea turtle nesting season). TS4 returned to Bachas-Salinas (Acuña-Marrero et al. 2017). Both tiger sharks, TS2 and TS4, came back to hunt green sea turtles!

In Fig. 8.58, we sketch the collective presence of various tiger sharks (i.e. pooled tracking data from numerous individuals) at the three study sites Isabela-South (Is), Cerro Ballena (CB), and Bachas-Salinas (BS). Tracking information is organized according to Core Range (CR, where the collective presence of sharks was highest, usually close to the shoreline at each study site; these areas are highlighted dark-gray in Fig. 8.58); Activity Space (AS, the broad range of presence where sharks swam about within GMR; these areas are highlighted light-gray in Fig. 8.58); and intermediate space (between CR and AS; highlighted intermediate-gray-color in Fig. 8.58). Activity Space (AS) is all inclusive of CR and intermediate space (Acuña-Marrero et al. 2017). The sketches in Fig. 8.58 correspond to: (a) Large-size sharks >300 cm/>120 in total length; collective presence of 3 individuals; CR = 422 km^2/163 sq mi; AS = 3,324 km^2/1,283 sq mi. (b) Medium-size sharks 200-300 cm/80-120 in total length; collective presence of 10 individuals; CR = 406 km^2/157 sq mi; AS = 7,088 km^2/2,737 sq mi. (c) Small-size sharks <200 cm/<80 in total length; presence of 1 individual; CR = 179 km^2/69 sq mi; AS = 782 km^2/302 sq mi. (d) Collective presence of 11 sharks during nesting season of green sea turtles; CR = 356 km²/137 sq mi; AS = 6,500 km²/2,510 sq mi. (e) Collective presence of 6 sharks during non-nesting season of green sea turtles; CR = 1,046 km²/404 sq mi; AS = 3,827 km²/1,478 sq mi (complete data in Acuña-Marrero et al. 2017).

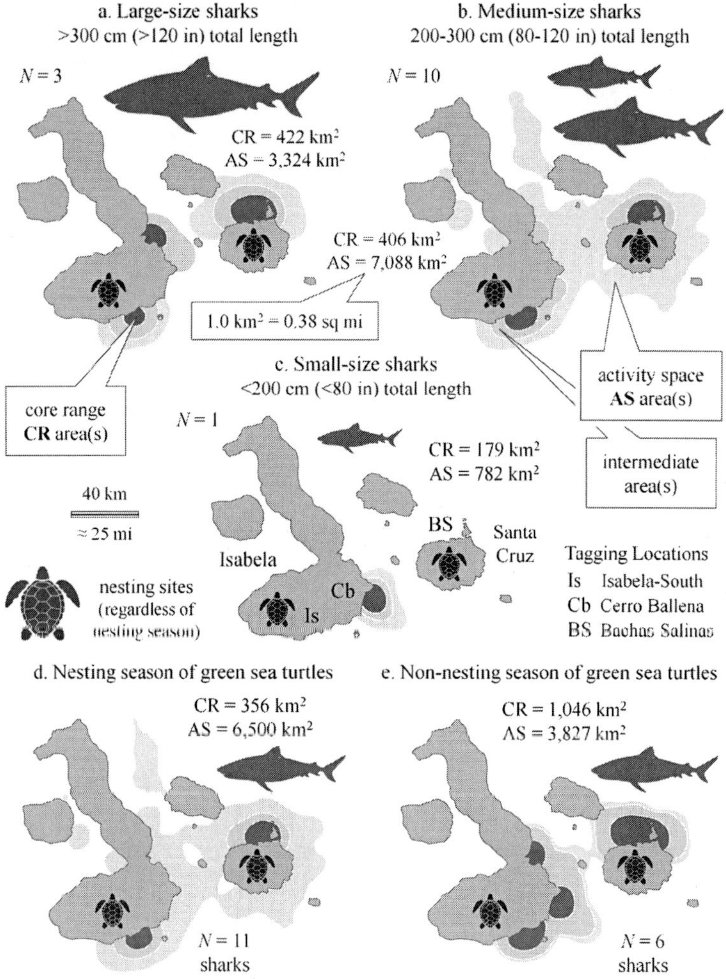

Fig. 8-58 Collective core range (CR) and activity space (AS) areas used by tiger sharks in Galápagos. Study sites included Isabela-South (Is), Cerro Ballena (CB), and Bachas-Salinas (BS). Shark size (a-c) and green-sea turtle (prey) nesting season (d) or non-nesting season (e). Layout: GPC; redrawn and modified after Acuña-Marrero et al. (2017).

Fig. 8.59 shows a zoom-in view of the 12-hour tracking time of tiger-shark positions at Isabela South (Is), Cerro Ballena (CB), and Bachas-Salinas (BS). The sketches correspond to the 3 large-, 10 medium- and 1 small-

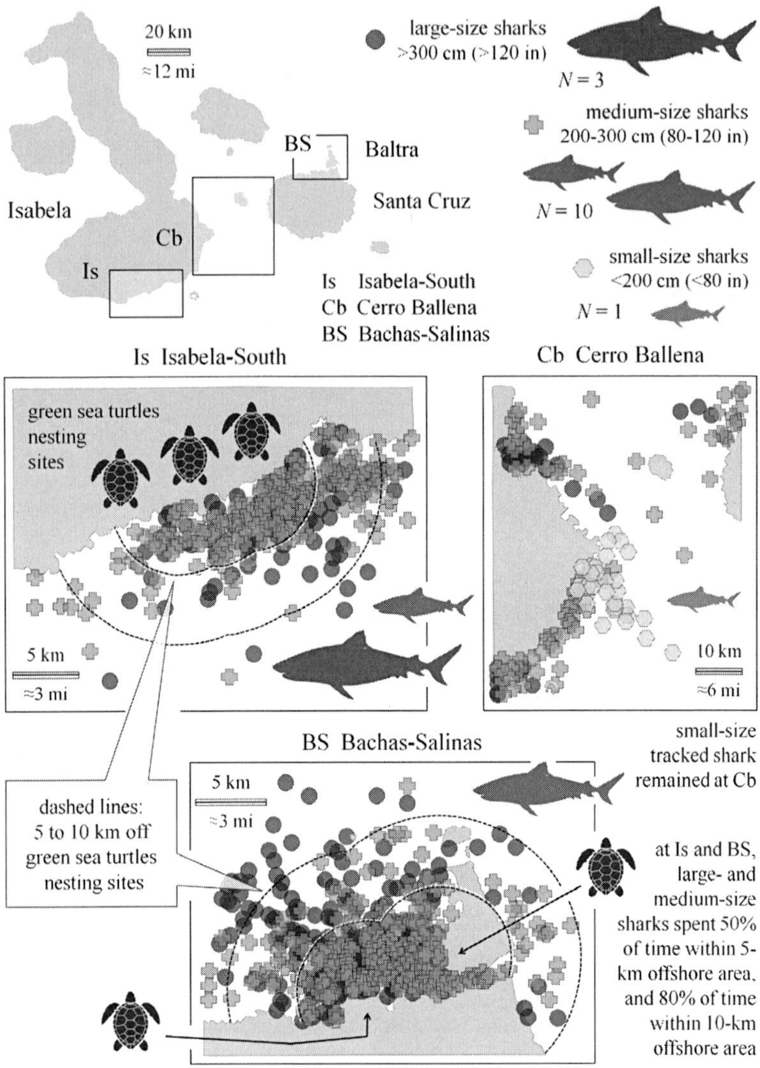

Fig. 8-59 Twelve hours of tracking tiger-shark positions at three offshore study sites in the Galápagos: Isabela South (Is), Cerro Ballena (CB), and Bachas-Salinas (BS). Layout GPC; redrawn and modified after Acuña-Marrero et al. (2017).

size shark discussed in Fig. 8.58 (as per Core Range CR, Activity Space AS, and intermediate space collective presence; Fig. 8.58). As highlighted in Fig. 8.59 top, tracking locations of large-size sharks during the 12-hour observation period are marked by dark-gray circles; medium-size sharks by gray-crosses; and the small-size shark by light-gray hexagons. Concentration of large- and medium-size sharks was particularly high proximate to the green-sea turtle nesting sites at Isabela South (center-left map in Fig. 8.59) and Bachas-Salinas (bottom map in Fig. 8.59). In both maps, spot the dashed lines marking 5 to 10 km (3.1 to 6.2 mi) offshore the green-sea turtle nesting sites. Large- and medium-size sharks spent 50% of their time within the 5-km/3.1-mi offshore area and 80% of time within the 10-km/6.2-mi offshore area. Note how the small-size shark was tracked exclusively close to Cerro Ballena (center-right map in Fig. 8.59). At this site, large- and medium-size sharks were also present, but at much lower density than at Is or BS. Based on this and previous observations of small tiger sharks nearby Cerro Ballena, scientists inferred that juveniles frequent this rocky area for daytime protection while foraging elsewhere by night (Acuña-Marrero et al. 2017).

In synthesis, the data presented in Figs. 8.57-8.59 suggests that, when resident, large- or medium-size tiger sharks remain within the Galápagos Marine Reserve (i.e. collectively, the tracked individuals spent 93% of their time within GMR; sharks were philopatric). When transient, exemplar sharks (females TS2 and TS4) traveled thousands of kilometers (miles) away from GMR yet returned seasonally to the very sites where they were tagged, i.e., nearby green sea turtles' nesting sites (e.g. Isabela-South, Bachas-Salinas). When juvenile, tiger sharks seem to find protection in the rocky shores during the day (e.g. Cerro Ballena), while foraging elsewhere by night. In adulthood, they will hunt green sea turtles!

Darwin and Wolf: Source of Coral Larvae

Scuba divers often characterize Galápagos' sea waters as murky. In our experience, visibility while diving or snorkeling is typically within 10-20 m/33-66 ft, although it can be as good as 30 m/100 ft. Turbulence and upwelling (seasonally enhanced by currents and winds) can set in motion sediments and microorganisms. Yes, Galápagos shores and the open sea are regularly crowded not only with sea lions, marine iguanas, green sea turtles, or tiger sharks, but also with minute life, e.g., coral larvae, fish fry, multiple invertebrates' larvae, algae, and microbes. [We discuss microbes of the Galápagos in our book *A Synthesis of the Galápagos* (Paz-y-Miño-C and Espinosa 2024b)].

The shallow depths of the archipelago are usually occupied by coral communities, structurally built by calcium-carbonate-fixing organisms, the "corals" (i.e. colonial invertebrates in the class Anthozoa). [In our book *Arrivals of Life to the Galápagos* (Paz-y-Miño-C and Espinosa 2024a), we introduced readers to coral reef communities and other invertebrates in the Galápagos and other comparable oceanic archipelagos]. In his 1842 book *The Structure and Distribution of Coral Reefs*, Charles Darwin hypothesized that "*...these wonderful formations* [the rim-shaped reefs in the periphery of some volcanic islands]... *have been raised from the summits of extinct volcanoes...*" under the premise that—as we know nowadays—volcanic islands are destined to sink (we discussed Galápagos' volcanic islands' ontogeny in Chapter One of this volume). Darwin referred specifically to the stunning atolls (ring-shaped islands with their shallow and surrounding circular reefs) in the Central-, Western-Pacific, and Indian Oceans, but his observations that islands' seaboards were the substrate on which coral colonies grew applied elsewhere.

Reefs are scattered across the Galápagos' shores. In Table 8.8, we list 22 species of corals identified at 13 island locations during an 18-year assessment (2000-2018; data from Riegl et al. 2019). The Northwestern islands of Darwin and Wolf have the highest coral species diversity (20 and 19 species, respectively, and statistically different from the other islands). All other islands have about half or less than half the number of taxa in Darwin and Wolf, as follows: Floreana 13; Santiago 10; Española 9; Bartolomé 8; Baltra and Santa Cruz 7; San Cristóbal 6; Marchena 4; Isabela East 4; Santa Fé 3; Isabela West 2; and Fernandina 1.

[Keep in mind that the coral reef communities of the islands of Darwin and Wolf, as well as corals in the entire Galápagos Archipelago, are exemplars of microbial life assemblages, primary production, nutrient cycling, and energy flow (involving multiple trophic levels, from microbes to invertebrates to algae and to large vertebrates like sea turtles, sharks, and whales), whose ecological and evolutionary processes are being protected by the Galápagos Archipelago's status as a National Park, Marine Reserve, and World Heritage Site (above). Coral reef communities are part of the broad conservation value of the Galápagos and its Marine Reserve (this chapter)].

Table 8.8 Coral Species Identified in Galápagos (2000 to 2018; after Riegl et al. 2019)

main reef builders belong to *Pocillopora*, *Pavona* and *Porites*	Darwin	Wolf	Marchena	Santiago	Bartolomé	Baltra/Sta Cruz	Santa Fé	San Cristóbal	Española	Floreana	Isabela East	Isabela West	Fernandina
Psammocora stellata	•	•	d	•	•	•	•	d	•	d	d		
Pocillopora verrucosa	•	•	•	•	•	•		•	•	•			
Pocillopora damicornis	•	•	•	•							d		
Pocillopora meandrina	•	•		•					•	•			
Pocillopora capitata	•	•											
Pocillopora eydouxi	•	•											
Pocillopora effusus	•	•			•					•			
Pocillopora inflata	•	•		•					•	•			
Pocillopora ligulata	•	•											
Pocillopora woodjonesi	•	•											
Pavona varians	•	•		•	•	•	•	•		•			
Pavona chiriquiensis	•	•		•	•	•	•	•	•	•			
Pavona clavus	•	d		d	d	d		d	•	d	•		
Pavona gigantea	•	•		•	•	•		•	•	•		•	d
Pavona maldivensis	•	•											
Gardineroseris planulata	•	•				nlr				•			
Leptoseris scabra	•	•											
Leptoseris sp. 1	•	•											
Cycloseris curvata	nlr								nlr	cvr			
Diaseris distorta										d			
Porites lobata	d	d	•	•	•	•		nlr	•	•	•	d	
Porites evermanni **	•								•				
TOTAL*** spp. / island	20	19	4	10	8	7	3	6	9	13	4	2	1

* d = numerically dominant species in island reef; nlr = no longer recorded in island reef; cvr = currently very rare in island reef; ** *Porites evermanni* and *P. lobata* are quite similar species and difficult to tell apart by morphology (i.e. approach used by Riegl et al. 2019), thus, it is likely that part of the *P. lobata* identifications correspond to *P. evermanni*; *** species richness (i.e. mean values: islands' sampled sites/surveyed years) was higher at Darwin and Wolf than at other islands (ANOVA $_{11,15}$ $F = 80.4$, $p < 0.001$); Darwin's species richness (alone) was higher than at other islands ($p < 0.001$) except Wolf

The main reef-builder genera listed in Table 8.8 (i.e. corals that contribute most to the accumulation of calcium carbonate via excretion) are *Pocillopora*, *Pavona*, and *Porites* (after Riegl et al. 2019). [Corals proliferate by "budding" (polyps that grow large and split) or translocation of coral fragments from one place to another by strong currents].

In Fig. 8.60a, we show coral reef species-composition similarity among the 13 sampled Galápagos islands. The map is based on the data plotted in Fig. 8.60b (we do this for didactic reasons), which depicts species presence or absence similarity across the 13 samples. Examine Fig. 8.60b. Spot the five groups of islands as per distances among sampling sites, species occurrence, and richness (Coordinates 1, 2, 3). Darwin and Wolf form a distinctive group (highest species richness; range 19-20 spp.). Floreana and Española belong to another group (second highest species richness; range 9-13 spp.). Santiago, Bartolomé, Baltra/Santa Cruz, San Cristóbal, and Santa Fé cluster together (range 3-10 spp.). Marchena and Isabela East also form a group (4 spp.), as well as Fernandina and Isabela West (range 1-2 spp.; Riegl et al. 2019). In other words, the coral-reef communities within the Galápagos Archipelago are idiosyncratic as per sampling site (the five groups of islands), species occurrence (where the taxa happen to be found), and species richness (diversity)!

What environmental or biological factors influence the Galápagos coral reef communities? Fig. 8.61 helps us address this question (data from Riegl et al. 2019). When both environmental and biological variables are combined in a single analysis, three main clusters of island reefs become identifiable: Northwestern islands (Darwin and Wolf); Western islands (Fernandina and Isabela West); and Central, South, and Southeastern islands, which all cluster together (Baltra/Santa Cruz, Isabela East, Española, Santiago/Bartolomé, San Cristóbal, Floreana, Santa Fé, and Marchena). [Realize that Marchena belongs to the Northern Galápagos Volcanic Province; see Chapter One, Figs. 1.5 to 1.7; this may also explain its peculiar position in Fig. 8.61]. These three island clusters appear as gray-shaded areas within Fig. 8.61, which we suggest examining as follows: (1) Locate the arrows (vectors) departing from the zero point (centroid) in Fig. 8.61; they correspond to abiotic factors influencing coral reef community structure and frame-building (the reef's structural development), i.e., pH, long-term minimum temperature, phosphate (PO_4), and nitrate (NO_3) levels (both essential nutrients). Arrows depict direction and strength (as per vector length) of each abiotic factor. For example, warmer sea-water temperature is characteristic of Darwin and Wolf, the most diverse coral-reefs; all other islands are exposed to cooler waters. By contrast, nitrate-(NO_3)-rich waters are typical of the Eastern side of the archipelago where coral-reef species composition is idiosyncratic to the is-

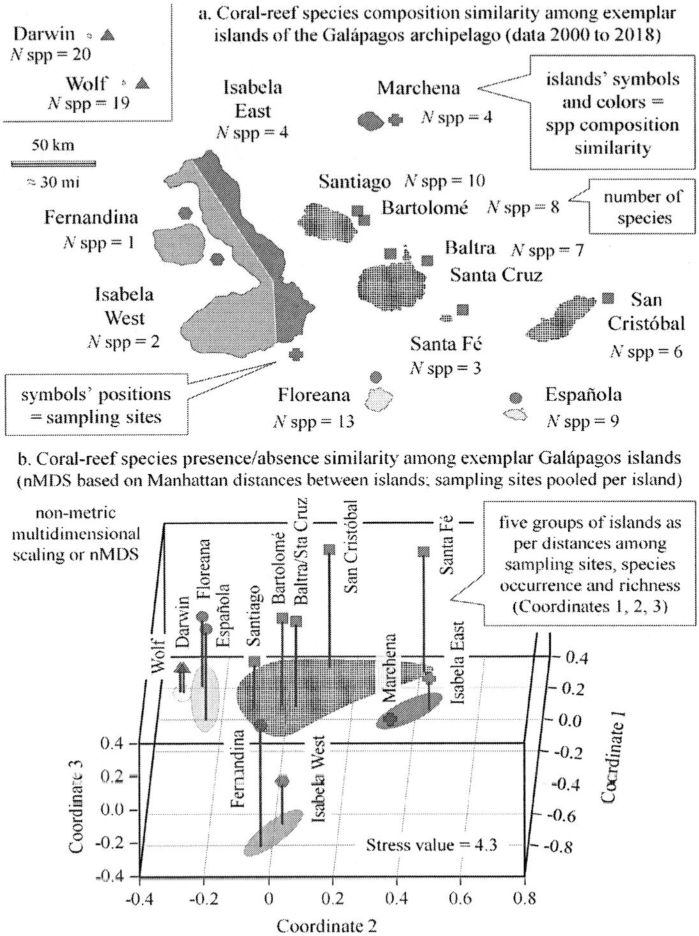

Fig. 8-60 Coral-reef species composition similarity (a) determined as per species presence or absence similarity (b) among 13 exemplar Galápagos islands. The top figure (map) is based on the data from the bottom figure (plot). Sampling sites correspond to Darwin: Wellington Reef, Anchorage Bay, Hidden Reef; Wolf: Shark Bay and The Anchorage; Marchena: Punta Espejo; Santiago: Sullivan Bay; Bartolomé: North-shore, about 1-km East of Pinnacle Rock; Baltra/Santa Cruz: Western entry to Itabaca Channel, Eastern part of Itabaca Channel; Santa Fé: Northeast Anchorage/Barrington Bay (inside Islote Black); San Cristóbal: Punta Pitt, Manzanillo; Española: Gardner Island, Osborn (Gayle) and Xarifa (Tortuga) Island, Gardner Bay; Floreana: Corona del Diablo, Tres Cuevitas; Isabela East: Concha de Perla lagoon, Caseta; Isabela West: Urbina Bay; Fernandina: Punta Espinosa. Layout: GPC; redrawn and modified after Riegl et al. (2019).

lands therein (Table 8.8). (2) The abiotic factors influence multiple coral reef characteristics, including the number of species per site, median colony size, and thickness of the reefs' structural framework. For example, *Pocillopora, Pavona* and *Porites* main builders of reefs) occupy different positions with respect to the centroid of the plot and the direction and reach of vectors. (3) Coral communities are also influenced by biotic factors, including recruitment frequency (how often corals produce larvae that add up to the population dispersal or recovery, locally or elsewhere; more on this point below) and skeletal density (how thick/thin corals deposit skeletal material; Riegl et al. 2019).

Why do coral-reef communities in Darwin and Wolf differ with respect to the rest of the archipelago (Table 8.8; Figs. 8.60 and 8.61)? Readers know that Galápagos' seasonal weather runs the cycles of life (Chapter One; Figs. 1.33-1.34 and 1.43-1.44). Extreme weather (El Niño or La Niña) affects coral communities. The Northwestern islands have higher coral taxa diversity, which covers more seafloor surface, has larger colonies, and recovers better and faster from extreme El Niño (i.e. abrupt changes in sea water temperature, and acidity levels, chemical imbalance) than the rest of the islands (Riegl et al. 2019). But most importantly, Darwin and Wolf have better coral-reef larval connectivity than the other islands, which brings us back to the "murky waters" mentioned earlier.

When reproducing sexually, corals spawn billions of larvae or gametes that are taken by ocean currents and dispersed either proximally or thousands of kilometers/miles away. In Fig. 8.62, we show two biophysical models depicting coral-reef larval dispersal connectivity around the Galápagos and in the context of the Tropical Eastern Pacific (TEP; Riegl et al. 2019). Note in Fig. 8.62a how coral larvae originated in Darwin and Wolf disperse widely to the Northwest, North and Northeast, and reach reefs in Cocos Island, Gulf of Chiriquí, and the islands of Coiba, Malpelo and Gorgona on the Western coasts of Central and South America. Thus, Darwin and Wolf act as sources of coral larvae for TEP (Fig. 8.62a). By contrast, coral larvae originated in the Western, Central, and Southeastern islands of the Galápagos mix primarily *in situ*, reach Darwin and Wolf, and can also disperse to the West, Northwest, Northeast, and even Southeast toward Isla de la Plata (coastal Ecuador).

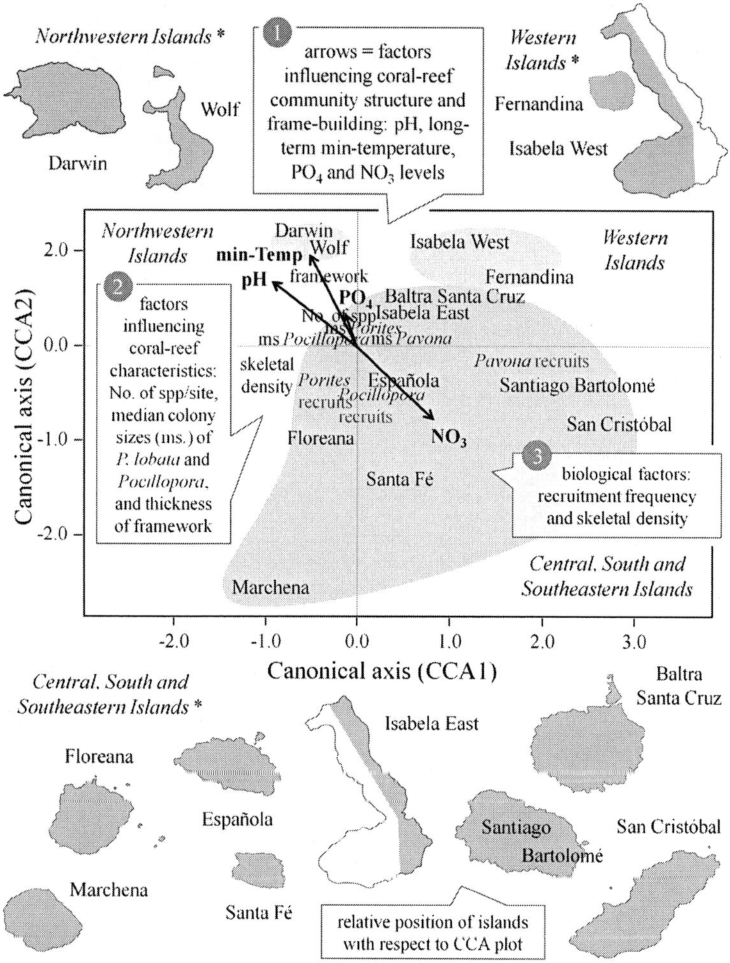

* three clusters of coral-reefs/islands can be characterized as function of environmental and biological factors; islands' sketches not to scale with respect to one another; note that Marchena is located rather North with respect to the central islands of the archipelago (see Chapter One)

Fig. 8-61 Environmental and biological factors influencing coral reef species richness and reef-building dynamics in the Galápagos Archipelago. Canonical correspondence analysis (CCA) triplot as per island-reef (sampling sites), biological variables (larvae recruitment, skeletal density), and environmental variables (pH, long-term minimum temperature, phosphate and nitrate levels) characterizing clusters. Three clusters of island-reefs are evident: (i) Northwestern; (ii) Western; and (iii) Central, South, and Southeastern, which all cluster together. Layout: GPC; redrawn and modified after Riegl et al. (2019).

a. Expected settlement density of coral-reef larvae released from Darwin and Wolf; note how larvae originated in Darwin and Wolf disperse widely to the Northwest, North and Northeast, and can reach reefs in Cocos Island, Gulf of Chiriquí, and the islands of Coiba, Malpelo and Gorgona in the Western coasts of Central and South America*

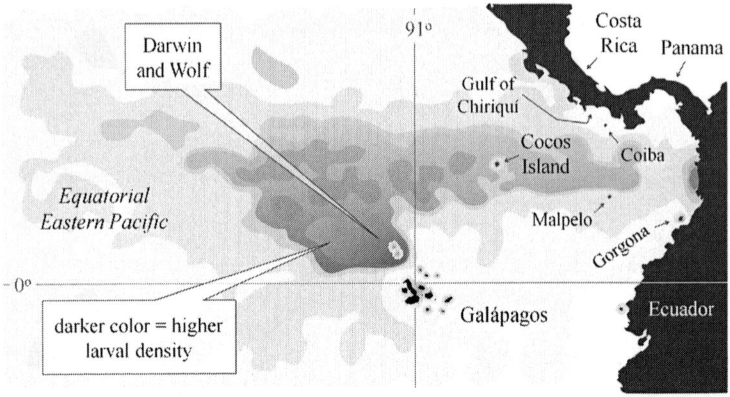

b. By contrast in respect to a (above), coral larvae originated in the Western, Central and Southeastern islands of the Galápagos mix *in situ*, reach Darwin and Wolf, disperse to the West, Northwest, Northeast, and even Southeast toward Isla de la Plata*

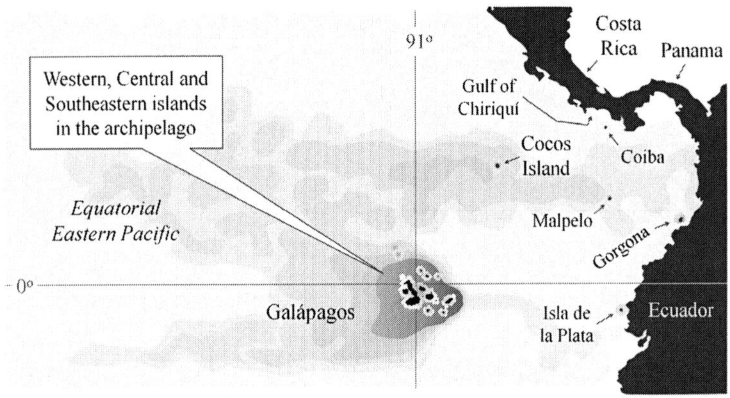

* together a and b help us understand how the dynamics of coral larval-dispersal connectivity contribute to the high abundance and coral species richness in Darwin and Wolf

Fig. 8-62 Biophysical model (simulation) depicting coral-reef larval dispersal connectivity around the Galápagos and in the context of the Tropical Eastern Pacific. Multiple variables contribute to Darwin's and Wolf's robust coral reef communities (as per taxa diversity, seafloor cover, colony size, or recovery from extreme El Niño events), from geographic location (far Northwest) to abiotic, environmental, and biotic factors. Broad coral larval connectivity (including input from other islands in the archipelago) benefits Darwin and Wolf. Layout: GPC; redrawn and modified after Riegl et al. (2019).

In consequence, multiple variables contribute to Darwin's and Wolf's robust coral communities (as per taxa diversity, seafloor cover, colony size, or recovery from extreme weather), from geographic location (far Northwest) to abiotic, environmental, and biotic factors. Broad coral larval connectivity (input from other islands) benefits Darwin and Wolf.

Interactions between Coral Reefs and Sharks

Coral-reef ecosystems, which include the coral-reef communities (the Anthozoan invertebrates that build the physical reef), are complex and among the most diverse environments on the planet. About 3,000 species (≈20% endemic) exist in the Galápagos Marine Reserve, including mollusks (snails, clams, squids, octopuses), fishes (bony, cartilaginous), echinoderms (sea stars, brittle stars, sea urchins, sea cucumbers), crustaceans (barnacles, shrimp, crabs, lobsters), polychaetes (marine segmented worms, bristle worms), and other invertebrates, vertebrates, and algae (Bustamante et al. 2000). Fishes (450 species) are the most conspicuous vertebrates in the Galápagos' coral reefs and—among them— sharks (cartilaginous fishes) arguably the most fascinating. [We discuss fish and invertebrate species assemblages in the Tropical Eastern Pacific in *Arrivals of Life to the Galápagos* (Paz-y-Miño-C and Espinosa 2024a), with emphasis on deep-sea waters and shallow reefs; for a comprehensive review of Galápagos fishes, see Grove and Lavenberg (1997)].

Fig. 8.63 summarizes ecological interactions between coral reefs and sharks (after Roff et al. 2016). Reefs provide habitat, nutrients, and shelter for sharks. Sharks act as transient apex predators or reef-associated mesopredators. (1) Apex predators like Scalloped Hammerheads (*Sphyrna lewini*) or Tiger Sharks (*Galeocerdo cuvier*) are usually large (adults), transient (regularly traveling from reefs to continental shores to pelagic areas), and have extensive home ranges (dispersal range >1,000 km/>620 mi). They can be omnivorous, hunt large prey, including other sharks, and have few wild predators. These sharks shape, directly or indirectly, the population structure and species composition of coral reef communities. Their "nutrient cycling" (nc in Fig. 8.63) or prey consumption-digestion-and-excretion can lead to nutrient flux (transport) across distant habitats (coastal areas to open sea, back and forth). As facultative scavengers, apex predators can consume carcasses of large fish, rays, or marine mammals and nutrient-cycle them within or between reefs. (2) Reef-associated mesopredators like Galápagos Sharks (*Carcharhinus galapagensis*) are often smaller, resident (dispersal range <50 km/<30 mi), and hunt smaller prey (although less omnivorous than apex predators). Depending on body-size range, lar-

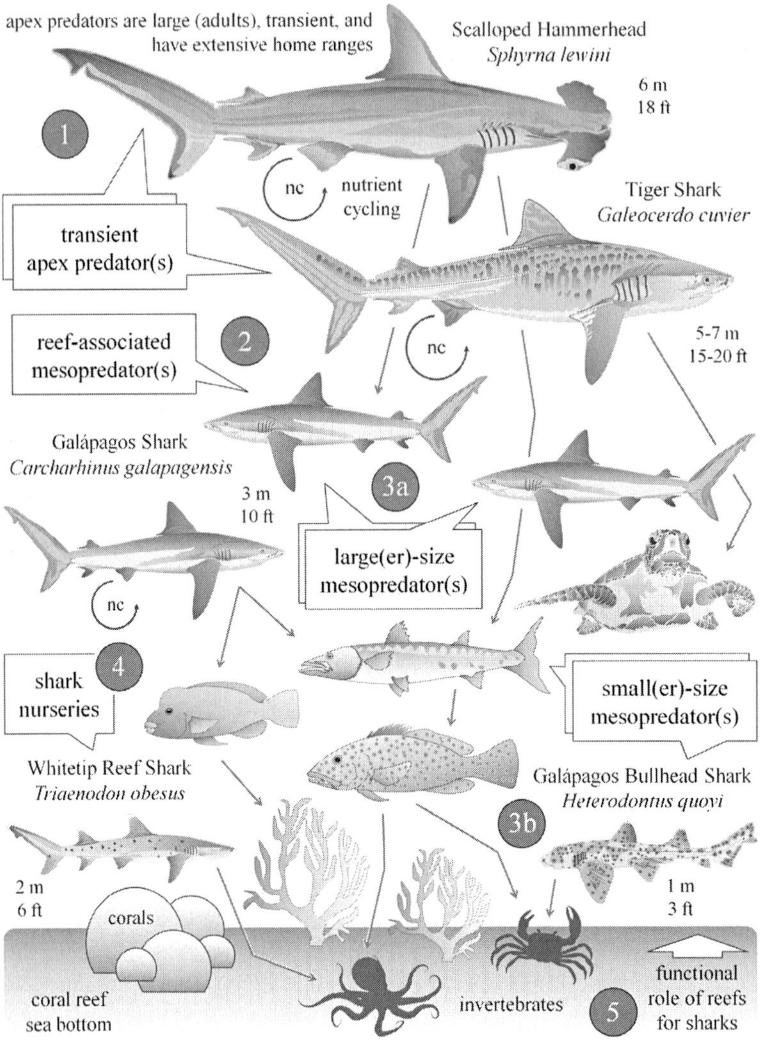

Fig. 8-63 Ecological interactions between coral reefs and sharks: reefs as habitats/nutrients/shelter providers; sharks as transient apex predators or reef-associated mesopredators. Layout: G. Paz-y-Miño-C; redrawn and modified after Roff et al. (2016). [For discussion on fish and invertebrate species assemblages in Galápagos and the Tropical Eastern Pacific, with emphasis on deep-sea waters and shallow reefs, see Paz-y-Miño-C and Espinosa (2024a). For a comprehensive review of Galápagos fishes, see Grove and Lavenberg (1997)].

ger-size mesopredators (3a), like the very Galápagos sharks (when larger), or smaller-size mesopredators (3b), like Galápagos Bullhead Sharks (*Heterodontus quoyi*), can occupy lower trophic levels and predate upon smaller fishes or invertebrates (crabs, octopuses). (4) Some species, like the Whitetip Reef Shark (*Triaenodon obesus*), rely on coral-reef habitats for reproduction and/or foraging for juveniles (shark nurseries). Thus, (5) the bottom-up functional role of reefs for sharks can be diverse, including not only feeding grounds or nurseries for juveniles but also places to rest or hide from predators (caves, tidal pools) or ecto-parasite removal by "cleaner fishes" (Roff et al. 2016).

The Galápagos Marine Reserve (GMR) is quite rich in sharks and rays (also cartilaginous fishes). The world's largest reef-fish biomass (17.5 tons per hectare) has been found around Darwin and Wolf, and a substantial portion of it (73%) corresponds to sharks (hammerheads *S. lewini* 48%; Galápagos *C. galapagensis* 20%; blacktips *C. limbatus* 5%; Salinas-de-León et al. 2016). Likewise, the world's largest population of Giant Oceanic Manta Rays (*Mobula birostris*) has been documented for coastal Ecuador (2,800 identified individuals; estimated abundance 22,300; Harty et al. 2022) located just 926 km/575 mi East from Galápagos.

What about regional species richness? The Galápagos (GAL) Marine Biogeographic Province—which engulfs GMR—has half the species of sharks (*N*=33) and rays (*N*=23) with respect to the entire Tropical Eastern Pacific (TEP) Marine Biogeographic Province (sharks=59; rays=53). [Note that TEP is vast and expands from coastal Mexico to Central America and Northern Peru]. And in Tropical America (from latitude 23° North to 23° South), only the Tropical Northwestern Atlantic (TNWA, the whole Caribbean) has more species of sharks (*N*=92) and rays (*N*=68) than TEP (Figs. 8.64 and 8.65a; Carrillo-Briceño et al. 2018)!

In Fig. 8.65b-d, we show topologies summarizing shark and ray species-similarity analyses among the six marine biogeographic provinces of Tropical America (after Carrillo-Briceño et al. 2018). Note how provinces in the Eastern Pacific (EP) and Western Atlantic (WA) cluster separately when both shark and ray species are analyzed together (8.65b); sharks alone (8.65c); or rays alone (8.65d). In other words, the cartilaginous-fish communities on the Eastern Pacific or Western Atlantic are distinctive. They likely became idiosyncratic after the formation of the Panama Isthmus in Central America (3.5 Ma) separated the ocean into today's Eastern Pacific and Western Atlantic (above; Carrillo-Briceño et al. 2018). Note as well in Fig. 8.65b-d how Galápagos (GAL) and the Tropical Eastern Pacific (TEP) share close species similarity for sharks and rays together

(8.65b); sharks alone (8.65c); or rays alone (8.65d), as should be expected from their biogeographic proximity.

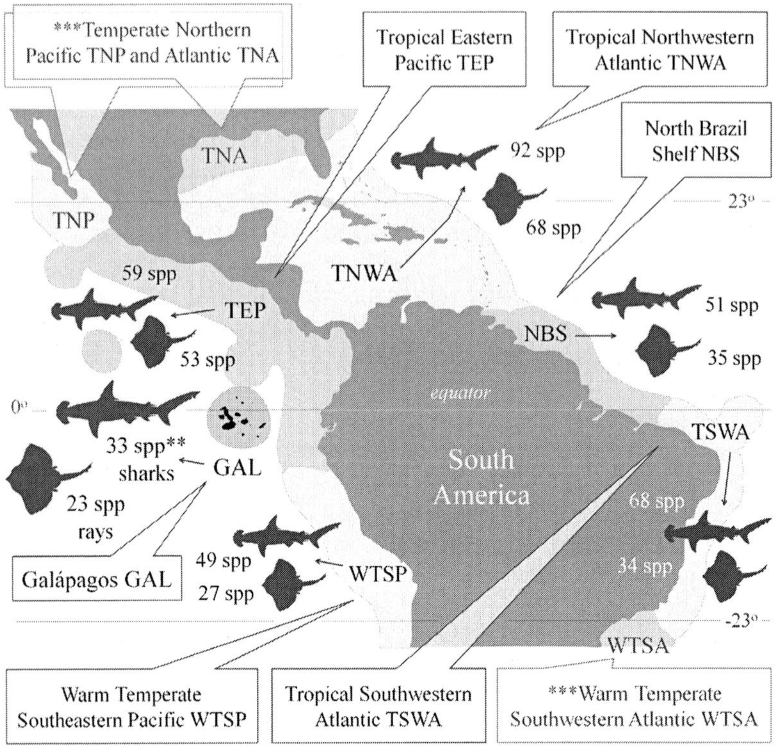

* Tropical America expands from latitude 23° North to 23° South; ** Carrillo-Briceño et al. (2018) report 33 species of sharks for the Galápagos (GAL) biogeographic province; however, in recent years, at least 35 species have been identified in the Galápagos Marine Reserve; *** the Tropical America biogeographic provinces –from the shark/ray species-richness perspective (Pacific: TEP, GAL, WTSP; Atlantic: TNWA, NBS, TSWA), overlap in continuity with the neighboring temperate provinces (TNP, TNA, WTSA; spp data not shown)

Fig. 8-64 Shark and ray species richness in the Galápagos (GAL) Marine Biogeographic Province with respect to other marine provinces in tropical America* (latitude 23° North to 23° South). Layout: GPC; redrawn and modified after Carrillo-Briceño et al. (2018).

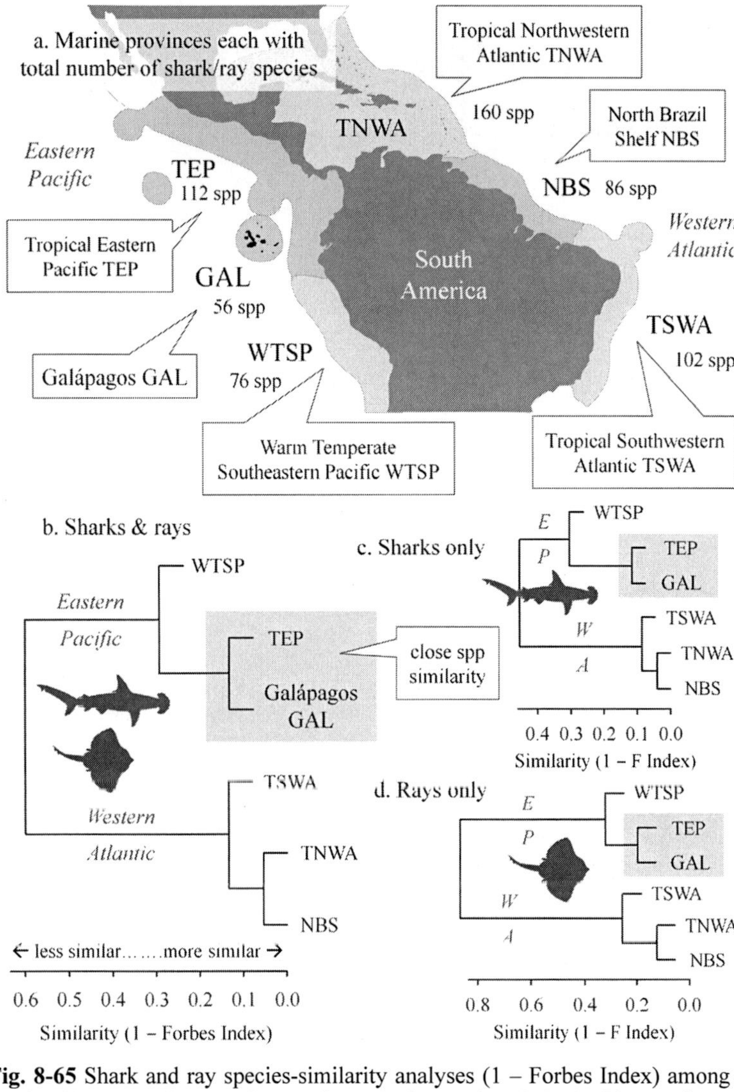

Fig. 8-65 Shark and ray species-similarity analyses (1 – Forbes Index) among six marine biogeographic provinces of Tropical America, latitude 23° North to 23° South (provinces as in a). Note how the Eastern Pacific (EP) and Western Atlantic (WA) provinces cluster separately when both shark and ray species are analyzed together (b); sharks alone (c); or rays alone (d). Note as well how Galápagos (GAL) and the Tropical Eastern Pacific (TEP) share close species similarity for sharks and rays together (b); sharks alone (c); or rays alone (d). Layout: GPC; redrawn/modified after Carrillo-Briceño et al. (2018).

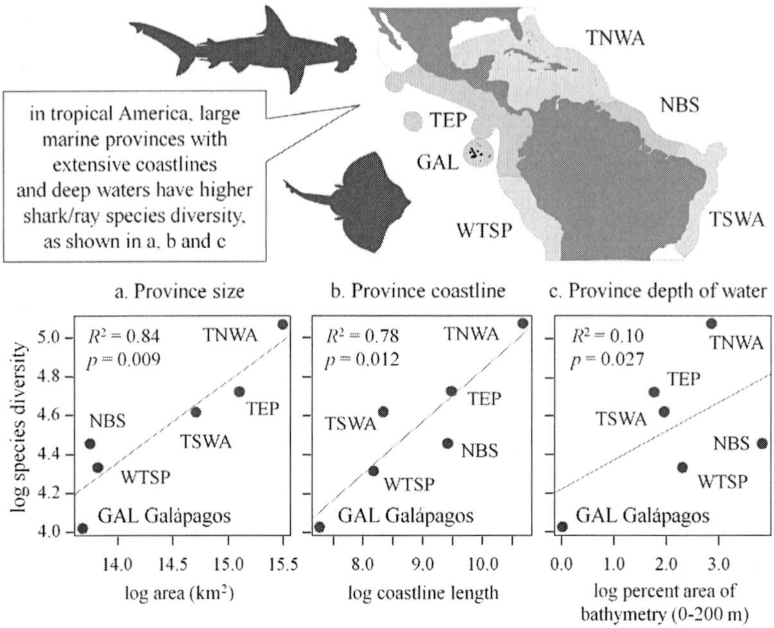

* Eastern Pacific provinces: TEP = Tropical Eastern Pacific; GAL = Galápagos; WTSP = Warm Tropical Southeastern Pacific; Western Atlantic provinces: TNWA = Tropical Northwestern Atlantic; NBS = North Brazilian Shelf; TSWA = Tropical Southwestern Atlantic

Fig. 8-66 Shark and ray species diversity in marine biogeographic provinces* of Tropical America is associated with the provinces' size (a), coastline length (b), or depth of water (c). Large provinces with extensive coastlines and deep waters have higher shark/ray species diversity. Layout: GPC; redrawn and modified after Carrillo-Briceño et al. (2018).

Why are the cartilaginous-fish communities in the marine biogeographic provinces of Tropical America distinctive as per species richness? Figs. 8.66 and 8.67 help us address this question. The plots in Fig. 8.66 suggest that large marine provinces (Fig. 8.66a) with extensive coastlines (Fig. 8.66b) and deep waters (Fig. 8.66c) have higher shark and ray species diversity. Compare/contrast Galápagos (GAL) with the other provinces in Fig. 8.66a-c. GAL is the least rich (sharks=33; rays=23; total=56), while Tropical Northwestern Atlantic (TNWA) is the richest (sharks=92; rays=68; total=160). The other provinces rank between GAL and TNWA (total species WTSP=76; NBS=86; TSWA=102; TEP=112; Fig. 8.66; after Carrillo-Briceño et al. 2018). Now, bring your attention to Fig. 8.67. In it, we summarize some physical factors (habitat and habit) associated with shark and ray species diversity in the marine biogeographic provinces of Tropical America: Eastern Pacific (with emphasis on Galápagos) versus Western Atlantic. Each province is unique and offers a characteristic share of habitat (benthic/demersal, benthopelagic, pelagic) and habit (coastal, oceanic, bathyal) types to sharks and rays (Carrillo-Briceño et al. 2018). For example, the Tropical Eastern Pacific (TEP) and Tropical Northwestern Atlantic (TNWA) are particularly rich in ray species that live or feed close to the sea bottom (benthic/demersal) near the coast (coastal/offshore; Fig. 8.67). Comparably, both TEP and TNWA offer plenty of habitats near the sea bottom and also in mid-waters close to the surface (benthopelagic), which are associated with their high shark species diversity (Fig. 8.67). Galápagos (GAL) has its own uniqueness of habitats and habits for sharks and rays to live and feed in. GAL is rich in benthopelagic and coastal areas where the majority of its shark and ray species can be found (Fig. 8.67a,c, center). Thus, in Tropical America, large marine biogeographic provinces with extensive coastlines and deep waters, rich in habitat and habit types offshore and toward the open sea, have higher shark and ray species diversity (comprehensive analysis in Carrillo-Briceño et al. 2018).

In synthesis, coral-reef ecosystems, which include coral-reef communities, are complex and among the most diverse environments on the planet. Fishes are the most conspicuous vertebrates in coral reefs and—among them—sharks arguably the most fascinating. Reefs provide habitat and feeding grounds for sharks, together with nurseries for juveniles, places to rest or hide from predators (caves, tidal pools), or removal of ecto-parasites by "cleaner fishes". Sharks act as transient apex predators or reef-associated mesopredators. The Galápagos Marine Reserve (GMR) is rich in sharks and rays (both cartilaginous fishes). The world's largest reef-fish biomass (17.5 tons per hectare) has been found around Darwin and Wolf; 73% of it corresponds to sharks. The world's largest population of Giant Oceanic Manta Rays (*Mobula birostris*) has been documented in coastal Ecuador, just 926 km/575 mi East of GMR. The Galápagos (GAL) Marine Biogeographic Province—which engulfs GMR—has half the species of sharks (N=33) and rays (N=23) with respect to the entire Tropical Eastern Pacific (TEP) Marine Biogeographic Province (sharks=59; rays=53). GAL and TEP share close species similarity for sharks and rays together, sharks alone, or rays alone. With respect to other marine biogeographic provinces in Tropical America, GAL also offers unique habitats and habits for sharks and rays to live and feed in. GAL is rich in benthopelagic and coastal areas where the majority of its shark and ray species can be found.

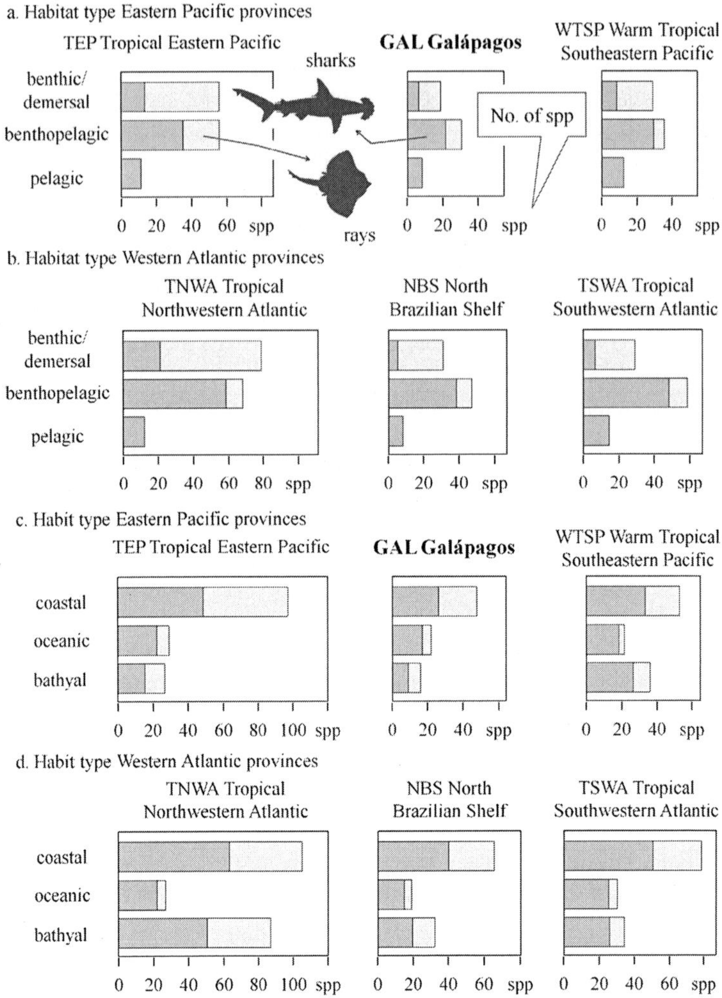

Fig. 8-67 Physical factors (habitat and habit) associated with shark and ray species diversity in marine biogeographic provinces of Tropical America: Eastern Pacific (emphasis Galápagos) vs. Western Atlantic. Habitat type: benthic/demersal (sea bottom); benthopelagic (near sea bottom and also mid-waters close to the surface); pelagic (open ocean). Habit type: coastal (offshore); oceanic (away from coastline); bathyal (between continental shelf and the deep ocean). Note how Galápagos (GAL) is rich in benthopelagic and coastal areas where the majority of its shark and ray species can be found. Layout: GPC; redrawn and modified after Carrillo-Briceño et al. (2018).

Thirty-Five Species of Sharks and Counting

As noted in Table 8.9, in 2020, two additional species of sharks (numbers 34 and 35) were added to the Galápagos' list of cartilaginous fishes (Buglass et al. 2020; Dando 2020). Requiem sharks (Carcharhinidae) are represented taxonomically by 11 species, followed by cat/ground sharks (Scyliorhinidae) with 4 species, hammerheads (Sphyrnidae) with 4 species, and eleven other families: Triakidae (3 spp.), Alopiidae (2 spp.), Lamnidae (2 spp.), Hexanchidae (2 spp.), Echinorhinidae (1 sp.), Heterodontidae (1 sp.), Centrophoridae (1 sp.), Rhincodontidae (1 sp.), Etmopteridae (1 sp.), Dalatiidae (1 sp.), and Odontaspididae (1 sp.).

Table 8.9 Species of Sharks in the Galápagos Marine Reserve (after Dando 2020)

	Species	Family		Species	Family
1	*Alopias pelagicus*	Alopiidae	19	*Rhincodon typus*	Rhincodontidae
2	*Alopias superciliosus*	Alopiidae	20	*Apristurus kampae*	Scyliorhinidae
3	*Carcharhinus albimarginatus*	Carcharhinidae	21	*Apristurus stenseni*	Scyliorhinidae
4	*Carcharhinus altimus*	Carcharhinidae	22	*Bythaelurus giddingsi*	Scyliorhinidae
5	*Carcharhinus falciformis*	Carcharhinidae	23	*Galeus sp.*	Scyliorhinidae
6	*Carcharhinus galapagensis*	Carcharhinidae	24	*Sphyrna lewini*	Sphyrnidae
7	*Carcharhinus limbatus*	Carcharhinidae	25	*Sphyrna zygaena*	Sphyrnidae
8	*Carcharhinus longimanus*	Carcharhinidae	26	*Sphyrna mokarran*	Sphyrnidae
9	*Carcharhinus plumbeus*	Carcharhinidae	27	*Sphyrna tiburo*	Sphyrnidae
10	*Galeocerdo cuvier*	Carcharhinidae	28	*Centroscyllium nigrum*	Etmopteridae
11	*Nasolamia velox*	Carcharhinidae	29	*Isistius brasiliensis*	Dalatiidae
12	*Prionace glauca*	Carcharhinidae	30	*Mustelus mento*	Triakidae
13	*Triaenodon obesus*	Carcharhinidae	31	*Triakis maculata*	Triakidae
14	*Echinorhinus cookei*	Echinorhinidae	32	*Mustelus albipinnis*	Triakidae
15	*Heterodontus quoyi*	Heterodontidae	33	*Odontaspis ferox*	Odontaspididae
16	*Centrophorus squamosus*	Centrophoridae	34*	*Notorynchus cepedianus*	Hexanchidae
17	*Carcharodon carcharias*	Lamnidae	35*	*Hexanchus griseus*	Hexanchidae
18	*Isurus oxyrinchus*	Lamnidae			

Alopiidae: thresher sharks; Carcharhinidae: requiem sharks; Echinorhinidae: prickly sharks; Heterodontidae: bullhead sharks; Centrophoridae: squaliform sharks; Lamnidae: mackerel sharks; Rhincodontidae: whale sharks; Scyliorhinidae: cat/ground sharks; Sphyrnidae: hammerhead sharks; Etmopteridae: lantern sharks; Dalatiidae: kitefin sharks; Triakidae: hound/ground sharks; Odontaspididae: sand tiger sharks; Hexanchidae: cow sharks. * Numbers 34 and 35 were added to the list in 2020 (Buglass et al. 2020).

Relying on baited remote underwater stereo-video systems (BRUVs), scientists have monitored taxa occurrence and estimated relative abundance of coastal sharks in the Galápagos. In Fig. 8.68, we summarize their findings, which correspond to 8 species with robust data (after Acuña-Marrero et al. 2018). Note the higher abundance of Scalloped Hammerheads (*Sphyrna lewini*) on the Northwestern islands of Darwin and Wolf (4 individuals per 90-min video recording) relative to the other islands. By contrast, the Central, East, and Southeastern islands are notable for their higher species richness (particularly around Santiago SA-North, SA-South; Santa Cruz SCrz-N-S-E-W; and Islets ISL). Overall, Fig. 8.68 suggests differential taxa occurrence and relative abundance of the 8 shark species across the video-recorded sampling locations (N=40-45) in the entire archipelago. This is corroborated by the data in Fig. 8.69, in which we show the number of shark sightings (No. individuals) and relative abundance (% of total) per family and species (N_{Total}=877 sharks) based on recordings at the BRUVs. Four species were the most commonly sighted: Galápagos Shark (*Carcharhinus galapagensis*) 334 sightings (38.0%); Scalloped Hammerhead (*Sphyrna lewini*) 207 sightings (23.6%); Blacktip Shark (*Carcharhinus limbatus*) 117 sightings (13.3%); and Whitetip Reef Shark (*Triaenodon obesus*) 72 sightings (8.2%; Fig. 8.69a; data from Acuña-Marrero et al. 2018). Now, look at Fig. 8.69b; it summarizes the events when diverse species were co-detected during the 629 BRUVs recording efforts and the level of similarity (%) among co-detection events as per sighting location and depth. Ten species of sharks are included in the non-metric multi-dimensional scaling MDS plot of Fig. 8.69b, each represented by a symbol matching the taxonomic families to which the sharks belong: circles for Carcharhinidae (6 spp.); squares for Sphyrnidae (1 sp.); crosses for Triakidae (2 spp.); and hexagons for Heterodontidae (1 sp.). Bring your attention to the large ellipse at the center of the plot. It includes four species: *Carcharhinus galapagensis* (semipelagic), *Carcharhinus limbatus* (semipelagic), *Sphyrna lewini* (semipelagic), and *Triaenodon obesus* (benthic). These species were co-detected during 20% of the 629 BRUVs recording efforts (=similarity in spatial occurrence); these taxa are also the most abundant in Galápagos (as shown in Fig. 8.69a). Co-occurrences of three species (*C. galapagensis*, *C. limbatus*, and *T. obesus*) were documented during 40% of the BRUVs. Benthic taxa like *Triakis maculata* and *Mustelus albipinnis* were co-detected during 20% of the BRUVs (right ellipse in Fig. 8.69b). Researchers interpret these co-occurrences as spatial inter-species associations [perhaps ecological, behavioral, or environmental in nature] in the surveyed coastal habitats (co-detections as per sampling location and depth in the water column: semipelagic/pelagic=offshore and open ocean; or benthic=close to the sea bottom; Acuña-Marrero et al. 2018). Some sharks were often detected solely with conspecifics: *Carcharhinus altimus* (semipelagic), *C. falciformis* (semipelagic), *Heterodontus quoyi* (benthic), and *Galeocerdo cuvier* (semipelagic; small ellipses in Fig. 8.69b).

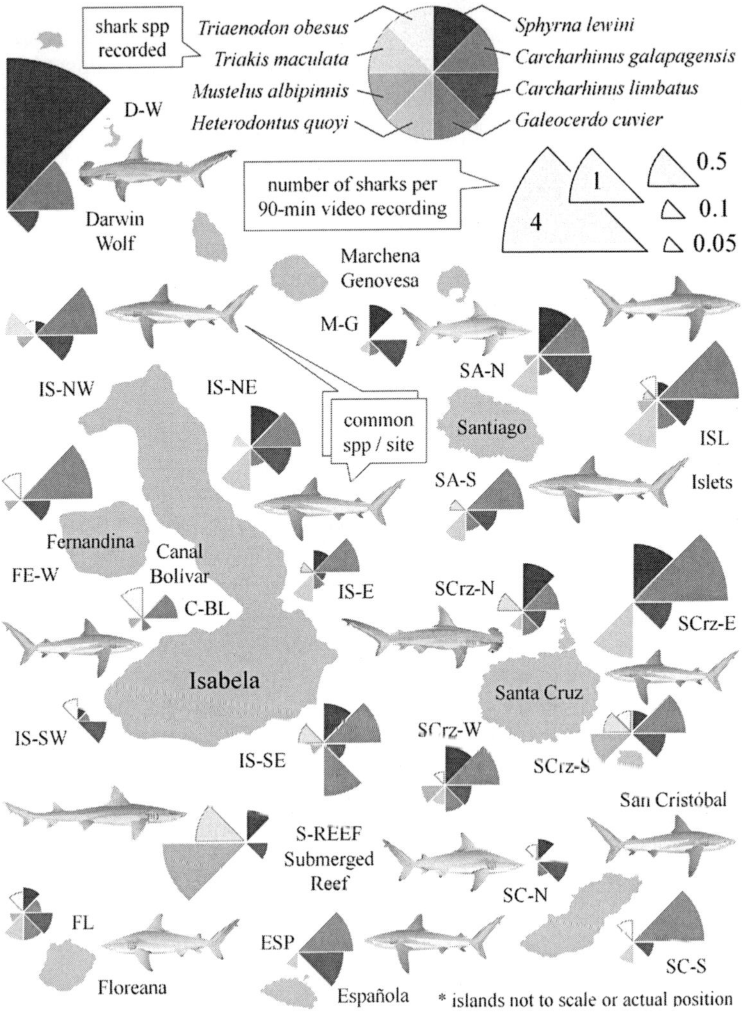

Fig. 8-68 Occurrence and abundance of coastal sharks in the Galápagos Archipelago: data based on baited remote underwater stereo-video systems (*N*=629 stereo BRUVs; *N*=40-45 locations). Each video-sampling site is unique as per shark species abundance or species composition. Note the higher abundance of Scalloped Hammerheads (*Sphyrna lewini*) on the Northwestern islands of Darwin and Wolf (4 individuals per 90-min video recording) relative to other islands. By contrast, the Central, East, and Southeastern islands have higher species richness (particularly around Santiago SA-North, SA-South; Santa Cruz SCrz-N-S-E-W; and Islets ISL). Layout: GPC; redrawn/modified after Acuña-Marrero et al. (2018).

To close this section, in Figs. 8.70 to 8.77, we offer a graphic summary of the 35 species of sharks so far documented in the Galápagos Marine Reserve. For each species, we include the conservation status category by 2025, according to the International Union for Conservation of Nature and Natural Resources IUCN (see asterix symbol * below). As noted earlier, two species (*Notorynchus cepedianus*, *Hexanchus griseus*) were added to the Galápagos list in 2020 (Table 8.9; Buglass et al. 2020). Our computer-generated sketches below were inspired by various sources, including Grove and Lavenberg (1997), Jiménez-Prado and Béarez (2004), Ebert (2016), Dando (2020), Cantú and Levi-Méndez (2022), and Shark Research Institute (2022). For a complete, public-friendly assessment of the Galápagos Sharks, we recommend *Sharks of the Galápagos Marine Reserve* (*Tiburones de la Reserva Marina de Galápagos*) by the Charles Darwin Foundation (CDF 2020). For natural-history and conservation-status assessments of the iconic Whale Sharks (*Rhincodon typus*) in GMR and the Tropical Eastern Pacific (i.e. population dynamics, satellite-monitored movement patterns, reproductive behavior and seasonality, microbiome sampling), we suggest Acuña-Marrero et al. (2018) and Green et al. (2020, 2021).

[*By 2025, the IUCN conservation status category of the 35 species of sharks found in the Galápagos Marine Reserve was: 5 species were considered to be critically endangered CR (1 of them with populations "largely depleted"), 8 endangered E (1 of them with populations "largely depleted"), 8 vulnerable V (1 of them with populations "moderately depleted"), 4 near threatened NT (1 of them with populations "largely depleted"), 7 least concern LC, 2 data deficient DD, and 1 n/a or non-applicable (the latter known at the genus sp. level). All this information is presented in Figs. 8.70 to 8.77. And yes, the conservation value of the Galápagos Marine Reserve can also be treasured for its "shark-diversity currency"; sharks as mesopredators and top/apex predators are essential to ecosystem functions (see Fig. 8.63)].

a. Shark Sightings in the Galápagos Archipelago: relative shark abundance (%) based on recordings at 629 baited remote underwater stereo video systems BRUVs (N = 877 sharks)

Family / habitat	Species	No. individuals	% of total
Carcharhinidae ●		**585**	**66.70**
semipelagic	*Carcharhinus galapagensis*	334	38.08
	Carcharhinus limbatus	117	13.34
	Carcharhinus altimus	6	0.68
	Carcharhinus falciformis	6	0.68
	Galeocerdo cuvier	42	4.79
benthic	*Triaenodon obesus*	72	8.21
	unidentified sp	8	0.91
Sphyrnidae ■		**207**	**23.60**
semipelagic	*Sphyrna lewini*	207	23.60
Triakidae ✚		**61**	**6.96**
benthic	*Mustelus albipinnis*	24	2.74
	Triakis maculata	36	4.10
	unidentified sp	1	0.11
Heterodontidae ⬢		**24**	**2.74**
benthic	*Heterodontus quoyi*	24	2.74

b. Shark-species Co-occurrence (co-detection of species during 629 BRUVs recording efforts) and Levels of Similarity among Co-occurrence Clusters, Galápagos Archipelago

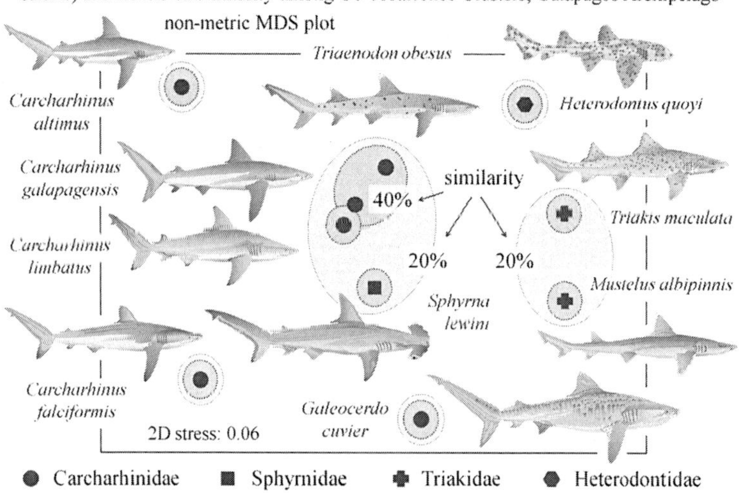

Fig. 8-69 a. Shark sightings in the Galápagos Archipelago (data extracted from Acuña-Marrero et al. 2018). b. Shark-species spatial co-occurrence in the Galápagos Marine Reserve (similarity of co-occurrence as per sampling location in Fig. 8.68 and position in the water column, i.e., pelagic or benthic). Researchers interpret these co-occurrences as spatial inter-species associations [perhaps ecological, behavioral, or environmental in nature]. Layout: GPC; redrawn and modified after Acuña-Marrero et al. (2018).

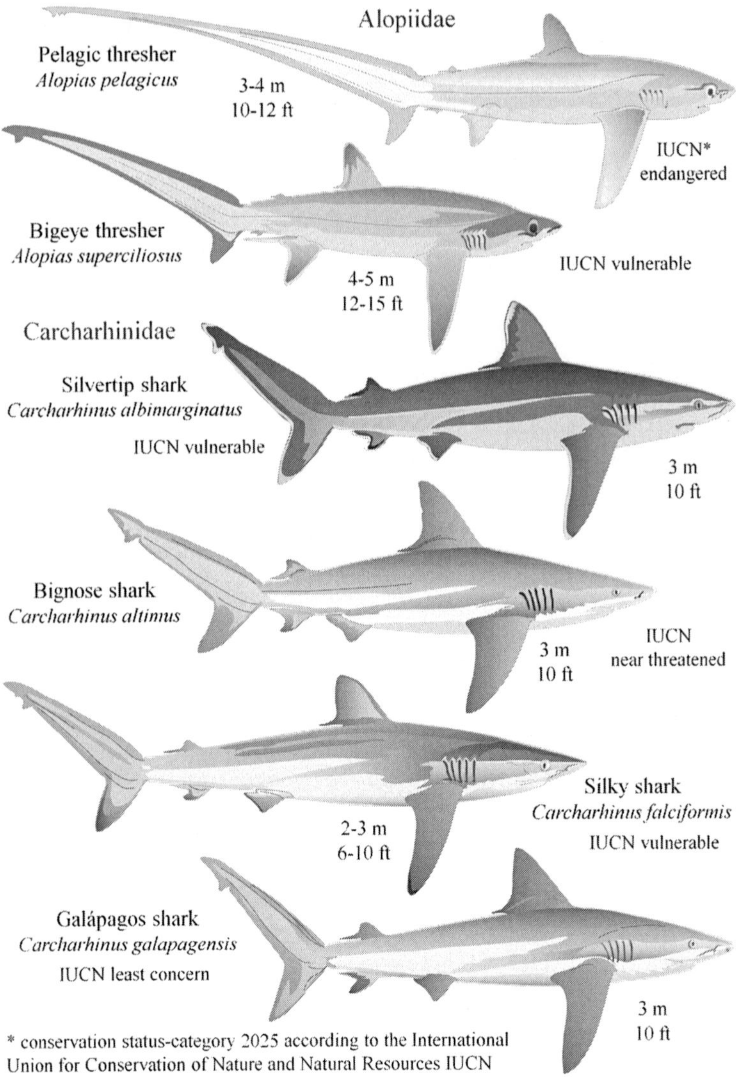

Fig. 8-70 Alopias pelagicus; Alopias superciliosus; Carcharhinus albimarginatus; Carcharhinus altimus; Carcharhinus falciformis; Carcharhinus galapagensis (all differential global distribution). Layout, computer-generated sketches: G. Paz-y-Miño-C; redrawn and modified after Grove and Lavenberg (1997); Jiménez-Prado and Béarez (2004); Ebert (2016); Dando (2020); Cantú and Levi-Méndez (2022); Shark Research Institute (2022).

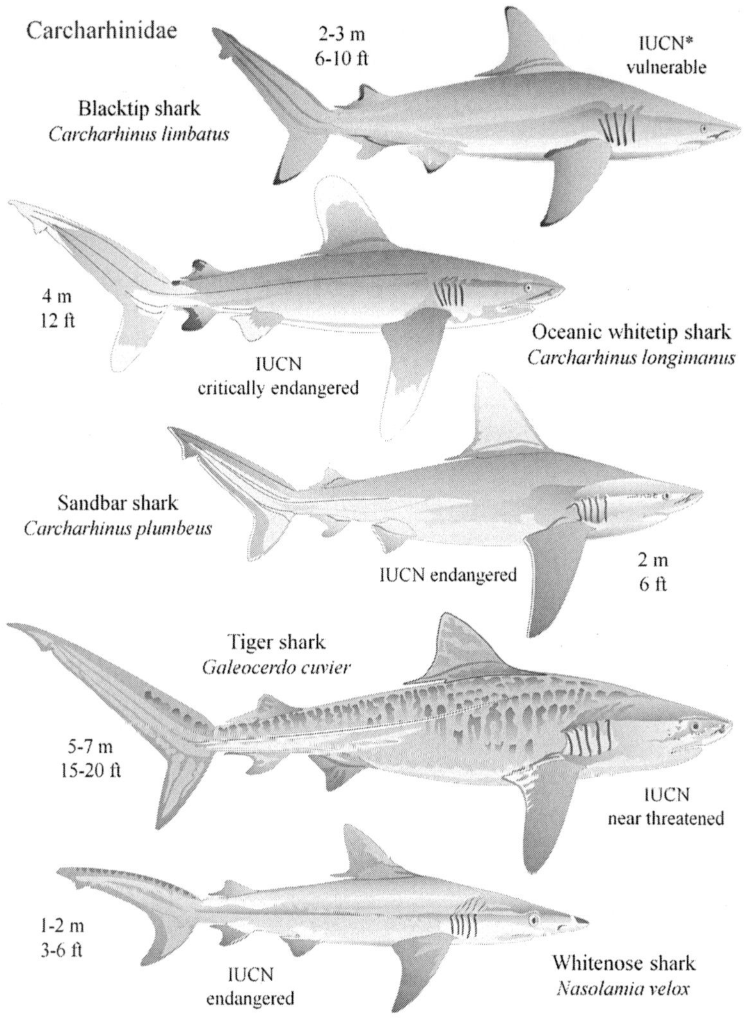

Fig. 8-71 *Carcharhinus limbatus*; *Carcharhinus longimanus*; *Carcharhinus plumbeus*; *Galeocerdo cuvier* (all differential global distribution); *Nasolamia velox* (Tropical Eastern Pacific). Layout, computer-generated sketches: G. Paz-y-Miño-C; redrawn and modified after Grove and Lavenberg (1997); Jiménez-Prado and Béarez (2004); Ebert (2016); Dando (2020); Cantú and Levi-Méndez (2022); Shark Research Institute (2022).

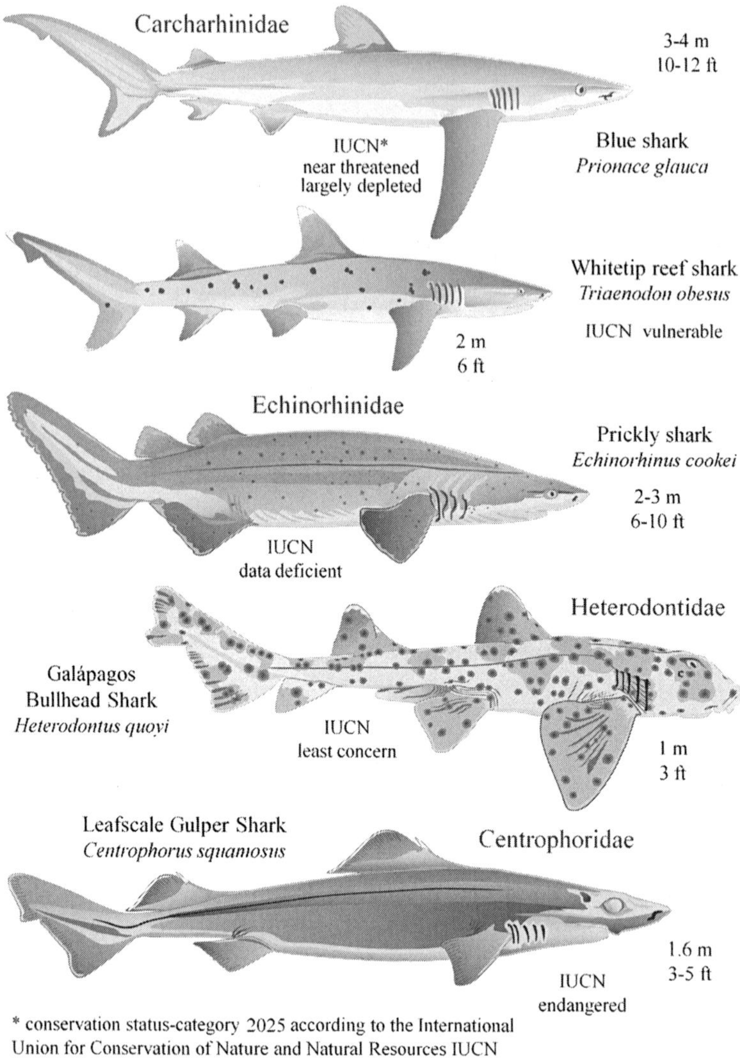

Fig. 8-72 *Prionace glauca* (global distribution); *Triaenodon obesus* (Pacific Ocean); *Echinorhinus cookei* (scattered distribution Pacific Ocean); *Heterodontus quoyi* (Tropical Eastern Pacific); *Centrophorus squamosus* (scattered global distribution). Layout, computer-generated sketches: G. Paz-y-Miño-C; redrawn and modified after Grove and Lavenberg (1997); Jiménez-Prado and Béarez (2004); Ebert (2016); Dando (2020); Cantú and Levi-Méndez (2022); Shark Research Institute (2022).

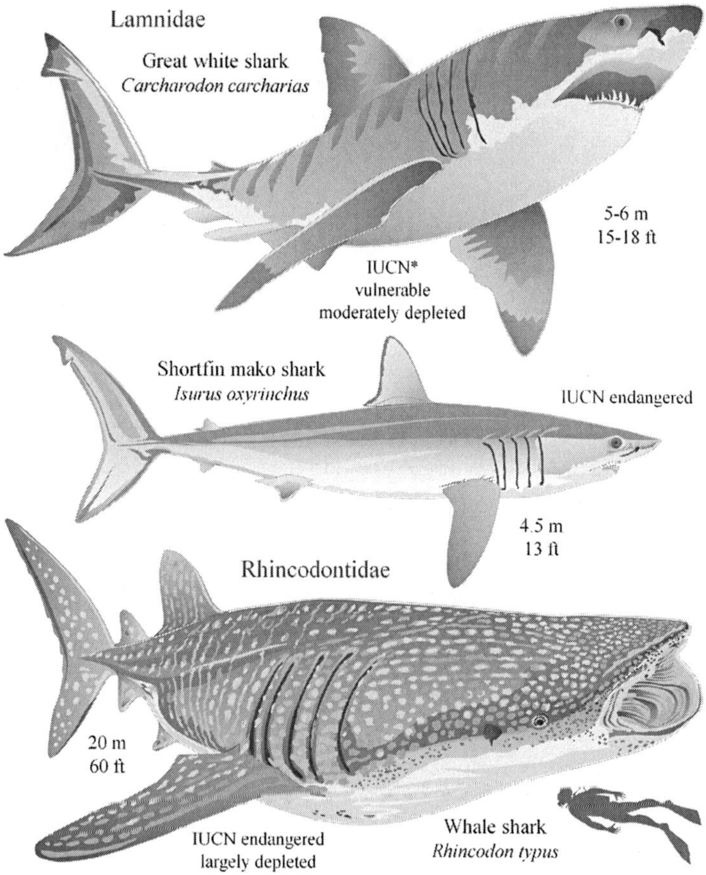

Fig. 8-73 *Carcharodon carcharias*; *Isurus oxyrinchus*; *Rhincodon typus* (all global distribution). The *Galápagos Whale Shark Project* (science, conservation, and education) monitors presence/abundance of whale sharks within the Galápagos Marine Reserve (GMR); maintains photo records of all possible whale sharks for individual identification (about 600 individual sharks are in database); tracks the movement patterns of satellite-tagged individuals within GMR and Tropical Eastern Pacific, including diving behavior; and obtains blood samples to assess health/reproductive condition of sharks or stress levels (Green et al. 2020, 2021). During two monitoring seasons (July-October in both 2011 and 2012), scientists tagged 39 female sharks around Darwin Island (most 8-12 m/26-40 ft long). Sharks tagged in July usually traveled West (the open sea); those tagged in September-October traveled East (toward coastal South America). Traveling range after being

tagged was 50-2,750 km/30-1,700 mi; median daily movement 40 km/25 mi. Five of the tagged sharks returned to Darwin Island more than a month after being tagged (Hearn et al. 2016). Highest abundance of whale sharks around Darwin Island (monitored 2007-2013) occurs during July-December (cool/dry season). Whale sharks have been detected frequently at 20-30-m/65-100-ft depth; water temperature 24-25 °C/75-77 °F. Researchers think that Darwin Island's surroundings (perhaps the entire GMR) are an important stopover for whale sharks during ocean travel (Acuña-Marrero et al. 2014). Layout, computer-generated sketches: G. Paz-y-Miño-C; redrawn and modified after Grove and Lavenberg (1997); Jiménez-Prado and Béarez (2004); Ebert (2016); Dando (2020); Cantú and Levi-Méndez (2022); Shark Research Institute (2022).

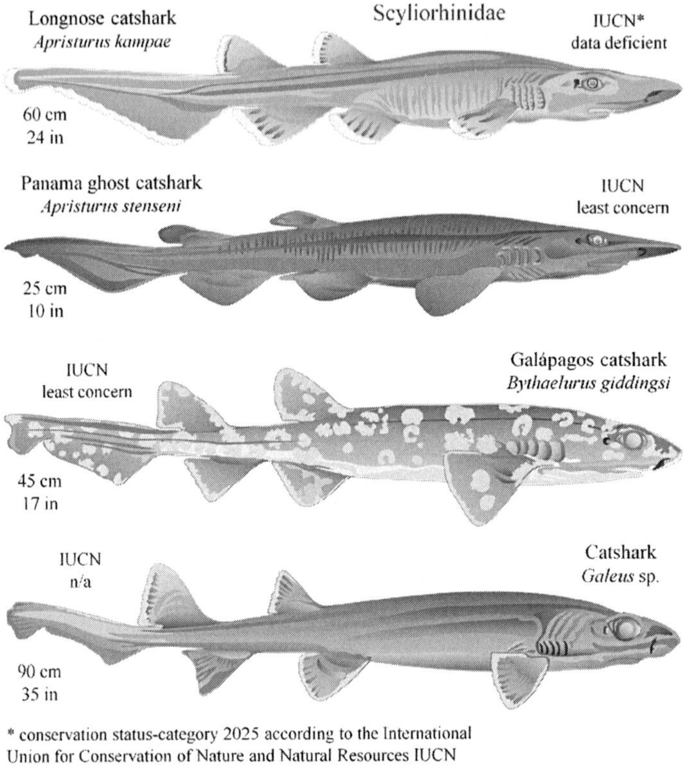

Fig. 8-74 *Apristurus kampae* (scattered distribution Eastern Pacific Ocean); *Apristurus stenseni* (Tropical Eastern Pacific); *Bythaelurus giddingsi* (Tropical Eastern Pacific); *Galeus* sp. (insufficient data). Layout, computer-generated sketches: G. Paz-y-Miño-C; redrawn and modified after Grove and Lavenberg (1997); Jiménez-Prado and Béarez (2004); Ebert (2016); Dando (2020); Cantú and Levi-Méndez (2022); Shark Research Institute (2022).

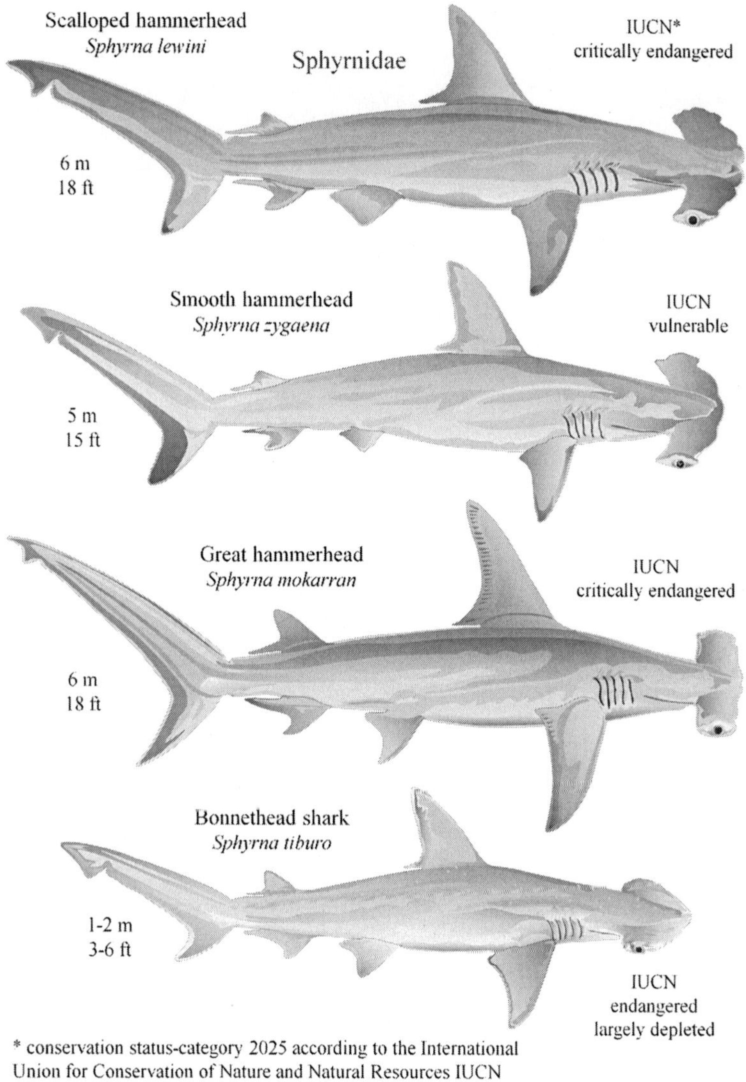

Fig. 8-75 *Sphyrna lewini*; *Sphyrna zygaena*; *Sphyrna mokarran* (all differential global distribution); *Sphyrna tiburo* (Tropical Eastern Pacific and Tropical Western Atlantic). Layout, computer-generated sketches: G. Paz-y-Miño-C; redrawn and modified after Grove and Lavenberg (1997); Jiménez-Prado and Béarez (2004); Ebert (2016); Dando (2020); Cantú and Levi-Méndez (2022); Shark Research Institute (2022).

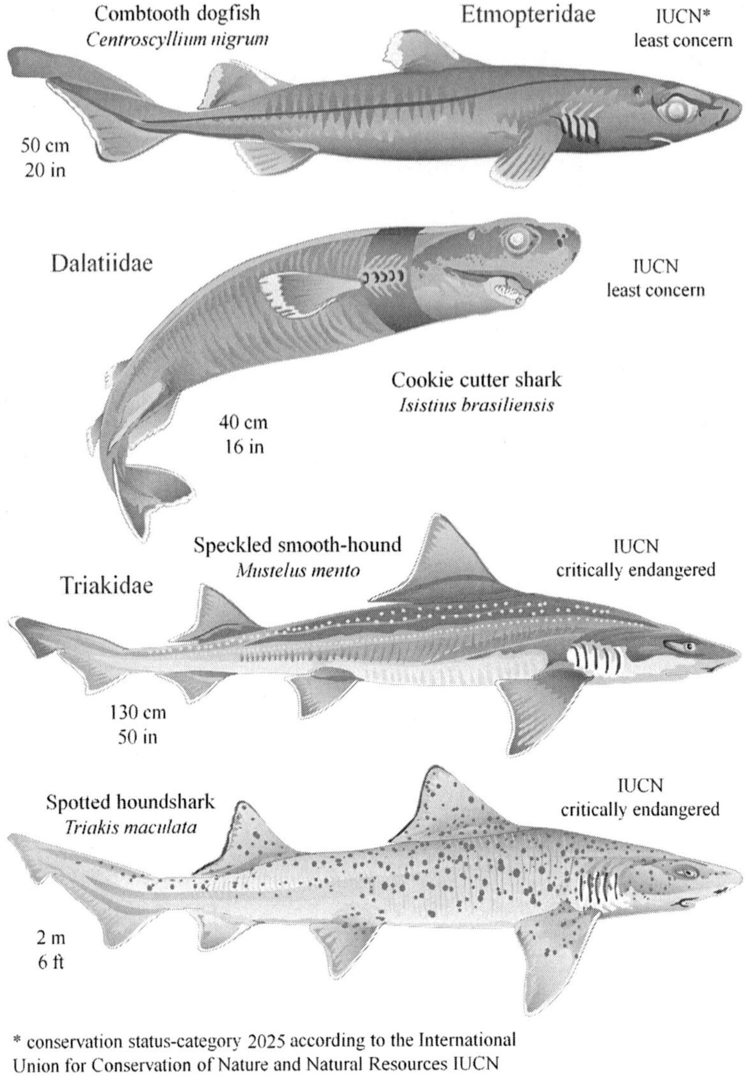

Fig. 8-76 *Centroscyllium nigrum* (Eastern Pacific); *Isistius brasiliensis* (scattered global distribution); *Mustelus mento* (Eastern Pacific); *Triakis maculata* (Tropical Eastern Pacific). Layout, computer-generated sketches: G. Paz-y-Miño-C; redrawn and modified after Grove and Lavenberg (1997); Jiménez-Prado and Béarez (2004); Ebert (2016); Dando (2020); Cantú and Levi-Méndez (2022); Shark Research Institute (2022).

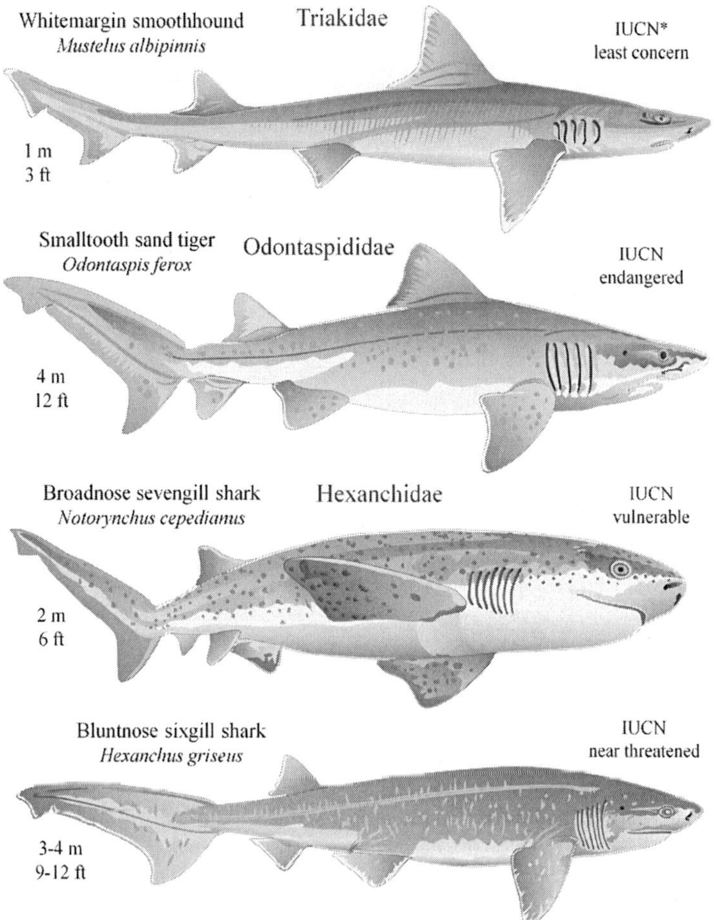

Fig. 8-77 *Mustelus albipinnis* (Tropical Eastern Pacific); *Odontaspis ferox* (scattered global distribution); *Notorynchus cepedianus* (scattered global distribution); *Hexanchus griseus* (scattered global distribution). Both *N. cepedianus* and *H. griseus* were video recorded for the first time in Galápagos in Sep-Oct 2019 (sandy/rocky substrates) via free-falling autonomous deep-ocean cameras (Buglass et al. 2020). *Notorynchus cepedianus* was spotted North of Floreana at 210-m/690-ft depth. *Hexanchus griseus* was seen North of Española at 420-m/1,380-ft depth. Layout, computer-generated sketches: G. Paz-y-Miño-C; redrawn and modified after Grove and Lavenberg (1997); Jiménez-Prado and Béarez (2004); Ebert (2016); Dando (2020); Cantú and Levi-Méndez (2022); Shark Research Institute (2022).

Conclusions

What is the conservation value of the Galápagos? By preserving the Galápagos Archipelago and its Marine Reserve, what are we really safeguarding?

(1) In 1978, the United Nations Educational, Scientific and Cultural Organization (UNESCO) designated Galápagos a World Heritage Site.

(2) "…World Heritage is the designation for places on Earth that are of outstanding universal value to humanity and, as such, have been inscribed on the World Heritage List to be protected for future generations to appreciate and enjoy…"

(3) UNESCO describes Galápagos' "Outstanding Universal Value" as follows: "…This archipelago and its immense marine reserve is known as the unique 'living museum and showcase of evolution'. Its geographical location at the confluence of… ocean currents makes it one of the richest marine ecosystems in the world. Ongoing seismic and volcanic activity reflects the processes that formed the islands. These processes, together with the… isolation of the islands, led to the development of [matchless] plant and animal life—such as marine iguanas, flightless cormorants, giant tortoises, huge cacti, endemic trees and the many different subspecies of mockingbirds and finches…"

(4) The World Heritage Convention (WHC, an international treaty) highlights four criteria (VII, VIII, IX, and X) that warrant Galápagos' designation as a World Heritage Site. Criterion VII is about the spectacular underwater wildlife and seascape features of the Galápagos. Criterion VIII highlights its plate tectonics, volcanism, and ongoing geomorphological processes. Criterion IX focuses on the ecological, evolutionary, and biogeographic [processes and] characteristics of terrestrial and marine environments. And Criterion X underlines the local biodiversity and endemism and the conservation status of the flora and fauna.

(5) When addressing the question, "What are we really safeguarding?", we indicated that, in our view [this chapter], three areas required particular examination: biodiversity, endemism, and ecosystem functions.

(6) In the section on *biodiversity*, we asked, where did the Galápagos' organisms [biodiversity] come from? We discussed three working hypotheses explaining the origins of the archipelago's biota. *Hi* overwater dispersal from mainland—in which organisms flew, swam, or were carried to the islands by ocean currents; *Hii* stepping-stone dispersal from mainland (a variant of *Hi*)—in which organisms dispersed via the now-submerged paleo islands, or seamounts, and made it to the current archipelago after "stepping" from older to younger seamounts [when they were above-sea-level volcanoes]; and *Hiii* metapopulation vicariance—in which groups of

organisms originated *in situ* during the breakup of widespread ancestral populations (in the deep geological past). We discussed 17 exemplar terrestrial and marine taxa, including Galápagos' Finches or Darwin's Finches, Galápagos Penguins, Giant Tortoises, Land and Marine Iguanas, Lava Lizards, Leaf-toed Geckos, Land Snakes or "Racer Snakes", Sea Lions and Fur Seals, "Flightless" Weevils, Carpenter Bees, Sand Spiders, Land Snails, Snapdragon Shrub, Prickly-Pear Trees, and *Scalesia* Darwin's Giant Daisies. These taxa represent both the unique Galápagos biodiversity, which we all aim at valuing and protecting, as well as the essence of the archipelago's natural history.

(7) In the section on *endemism*, we discussed case studies on: (i) endemism and seed-dispersal in Galápagos' angiosperms; (ii) endemism in Galápagos' birds; (iii) endemism in iconic Galápagos' vertebrates; (iv) endemism, landscape, and adaptive radiation in Galápagos'/Darwin's finches; (v) endemism and hybridization in Galápagos mockingbirds; (vi) endemic island rodents; (vii) endemism in Galápagos' beetles; (viii) endemism in Galápagos' spiders; and (ix) endemism in Galápagos' marine organisms. By valuing and protecting endemic taxa [i.e. species and lineages with restricted or localized geographic distribution], we are bound to preserve their habitats and the ecological and evolutionary processes that sustain endemism or give origin to it.

(8) In the section on *ecosystem functions*, we stated that ecosystem functions and benefits can include anything and everything ecosystems do! From primary production to nutrient cycling and energy flow to ecological relationships to large-scale natural transformations of the environment by organisms' behavior and to the cognitive interactions among organisms that result in substantial habitat modifications. Ecosystem functions and benefits can emerge from a wide range of processes in the ecosystem. In the context of ecosystem functions and benefits, we examined altitudinal migration of giant tortoises [migration behavior is essential to their survival and reproduction]. We described how tiger sharks come to hunt in the Galápagos Marine Reserve while their favorite prey, the green sea turtles, come to nest [i.e. predator-prey interactions]. We highlighted the importance of the islands of Wolf and Darwin as "sources of coral larvae" for the Galápagos marine ecosystems and the entire Tropical Eastern Pacific [i.e. biological processes of local and regional relevance, likely global]. We addressed the interactions between coral reefs and sharks [i.e. reefs as exemplar ecosystems and providers of habitat, nutrients, and shelter for sharks; the latter as exemplar mesopredators and top/apex predators]. And we closed by underlining the Galápagos' impressive shark diversity, with its "thirty-five species of sharks and counting" [i.e. sharks as both essential

to Galápagos' marine-species assemblages and attractive predators to explorers of nature (ecodiving and ecotourism) and conservationists]. By valuing and protecting ecosystem functions and benefits, we perpetuate—by default—the long-term survival of the Galápagos Archipelago and its Marine Reserve.

(9) The value or values of the Galápagos Archipelago and its Marine Reserve will only be forever preserved [as a World Heritage Site] under sound sustainable development practices leading to limiting human population expansion on the islands, sponsoring solely eco-friendly tourism, and securing revenue fairness for the local residents.

(10) Today, managers of the Galápagos National Park and its Marine Reserve, together with the Galápagos people, the country of Ecuador, and the world—since Galápagos, as a World Heritage Site, belongs to all of us—are responsible for protecting the archipelago's "outstanding universal value".

Questions

The following questions (Q1–2) merit generous written answers. You might need to do additional reading or library work to enrich your answers (we suggest exploring Paz-y-Miño-C and Espinosa 2024a,b).

Question 1: In the introduction to this chapter, we indicated that the UNESCO's World Heritage List included 1,223 World Heritage Sites, i.e., 952 cultural, 231 natural, and 40 mixed properties (the latter both cultural and natural sites). Galápagos is a "natural" site. (a) What attributes should Galápagos have to also qualify for the designation "cultural" World Heritage Site? (b) According to the World Heritage Convention (an international treaty), World Heritage Sites ought to be "protected for future generations to appreciate and enjoy". In simple words, be protected forever. What should Galápagos [its residents and local institutions], the country of Ecuador [its sovereign nation and the branches of government], and the world [all of us and the international organizations to which global citizens belong] do to ensure its forever preservation? (c) How feasible is the forever preservation of the Galápagos Islands and Marine Reserve?

Question 2: As explained in this chapter, endemism (=uniqueness in occurrence) and rareness (=infrequent occurrence) of taxa are typical of oceanic islands. Now, scientists refer to islands as either "Darwinian islands" (DIs, under the assumption that Charles Darwin envisioned islands, particularly volcanic ones, to form *de novo*) or "fragment islands" (FIs; perhaps because Alfred Russell Wallace, 1880, co-discoverer of evolution by means of natural selection, noticed that many islands formed as fragments of continents; Gillespie and Roderick 2002; Graham et al. 2017). (a) Examine Fig. 8.78. In it we explain the types of endemism that can emerge from DIs (neo-endemics) or FIs (paleo-endemics). (b) What is the conservation value of Darwinian islands versus fragment islands? (c) What difference does it make to preserve the biodiversity, endemism, and ecosystem functions of Darwinian islands versus fragment islands? Note that we are *not* asking you to choose or favor one type of islands versus the other; that would be silly. We are asking you to characterize the value

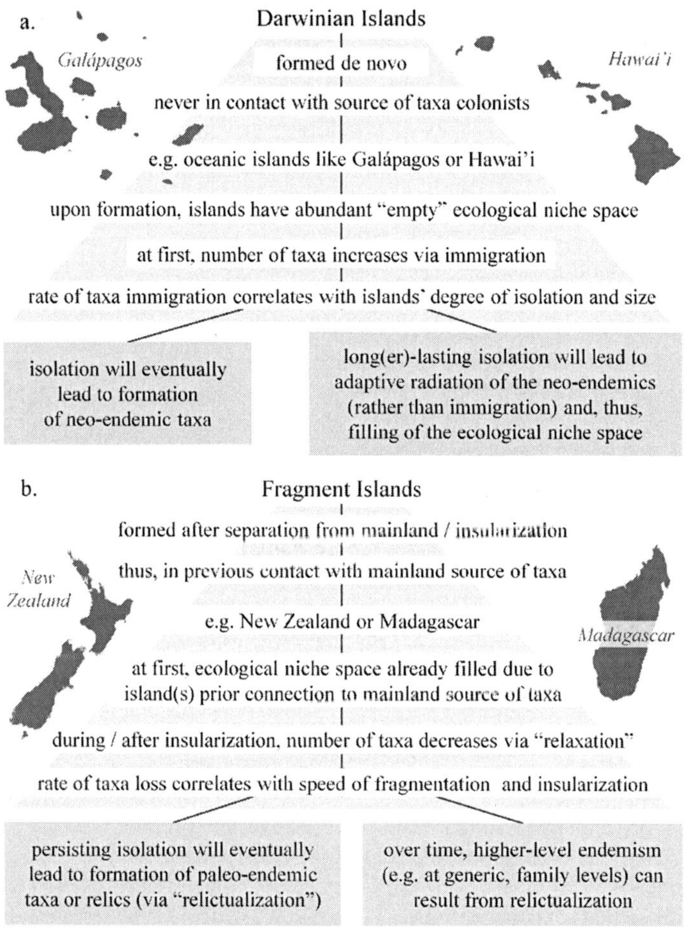

Fig. 8-78 Endemism on "Darwinian" (a) versus "fragment" (b) islands. Design: GPC; adapted from/modified after Gillespie and Roderick (2002); Graham et al. (2017). For discussion, see Paz-y-Miño-C and Espinosa (2024b).

of preserving Darwinian islands versus fragment islands in a comparative manner, as well as to characterize the distinctive outcomes of preserving—forever—the biodiversity, endemism, and ecosystem functions of Darwinian islands versus fragment islands!

How to Answer the Questions Above

The questions above have various possible answers (A1–2). Here we suggest perspectives from which to address the queries.

Answer 1: (a) You need to know the difference between "cultural" and "natural" World Heritage Sites and why some sites qualify for both designations (i.e. mix properties). (b–c) The key principle to answer b–c is "forever preservation" and the actions needed to ensure it by local, regional, and global participation of citizens and organizations. You might want to explore cases in which the status of World Heritage Site has ended (i.e. "delisted" sites). Such cases illustrate that much needs to be done to maintain good standing ("listed") as a World Heritage Site.

Answer 2: (a–c) Keep in mind that endemism is the confinement of taxa to a geographic location (e.g. islands). You need to explore the types of endemism that can emerge not only in Darwinian islands versus fragment islands, but endemism in general. Several concepts exist, e.g., phylogenetic endemism, neo-endemism, paleo-endemism, mix-endemism, or super-endemism (Veron et al. 2019; Paz-y-Miño-C and Espinosa 2024b). Understanding these concepts is key to answering Q2.

Discussion Topics

The discussion topics below (D1–D2) should be addressed with your peers. We encourage college instructors to facilitate the dialogue.

Discussion 1: You have learned in this chapter that emblematic taxa (=flag species or ambassador species) are central to conservation campaigns. You know that giant tortoises, marine iguanas, or Galápagos'/Darwin's finches are charismatic and emblematic; thus, they bring attention to Galápagos protection. You also understand that *in situ* conservation of emblematic taxa leads to preservation of their habitats. Your DISCUSSION should be based on the following questions: (a) What criticism do you have about the concept of emblematic taxa? (b) Would less charismatic taxa, yet highly endangered, become neglected if conservation priorities and efforts were focused mainly or exclusively on flag species? (c) Would the extinction of emblematic taxa affect conservation programs in the habitats where they used to exist, thus leading to abandoning all other species in those habitats? (d) Can you put together a list of charismatic Galápagos taxa based entirely on your preference? Should they all be considered emblematic? Were they all large animals? What about invertebrates? What about plants? Do they all require equal protection? (e) Now, what do you really think about the conservation rigor of the concept of emblematic species?

Discussion 2: The conservation value of the Galápagos is often discussed from the perspective of its terrestrial environments. Your DISCUSSION should be based on the following questions: (a) What is the conservation value of the Galápagos Marine Reserve? (b) How is the conservation value of the Marine Reserve linked to the conservation value of the terrestrial environments? (c) Why is the conservation value of the Galápagos Marine Reserve often discussed in a regional [Tropical Eastern Pacific] and global context [planet Earth]?

Suggested Observation Projects

The suggested observation projects below (P1–P2) take into consideration that you will travel to Ecuador and the Galápagos legally; that you will visit the islands as a tourist [holding a Transit Control Card, a mandatory document issued by Ecuador and specifically by the Governing Council of Galápagos]; and that you will respect the regulations that protect the Galápagos National Park and the Marine Reserve (see complete list of rules in Appendix B). Only accredited scientists, academics, and/or naturalists who have obtained official permits can do research in Galápagos. You can take pictures, but flash photography is not permitted when taking photos of wildlife. Professional photography and videos recorded for commercial purposes must be authorized by the Galápagos National Park Directorate (GNPD). The ideas below are for enthusiastic college explorers interested in learning about the archipelago.

Project 1: Prepare a short survey with questions to be asked to as many people as you run into while visiting the Galápagos (i.e. locals, other tourists, some naturalist guides, diverse people of all ages [adults], and anyone you think could answer your friendly questions). For example, you could ask, *Do you know that Galápagos is a World Heritage Site? Do you know what a World Heritage Site is? Do you know about other World Heritage Sites besides Galápagos?* You could ask locals, *What does it mean for you to live in a World Heritage Site like Galápagos? What do you personally do to ensure that Galápagos continues to be a World Heritage Site?* You can write your own questions, but keep in mind that surveying should be friendly, short, and for personal interest only. Organize your data in an Excel file and rely on descriptive statistics to summarize the outcomes. [As stated above, only accredited scientists, academics, and/or naturalists who have obtained official permits can do research in Galápagos]. Put together a slideshow presentation (for a 20-minute talk) about "Peoples' Level of Awareness that Galápagos is a World Heritage Site: A Friendly Survey" to be shared with a general audience. Avoid slides with "text only" [they are often terrible]. Keep it rigorous yet fun and didactic.

Project 2: The top 10 biggest natural World Heritage Sites include French Austral Lands and Seas (France), Phoenix Islands Protected Area (Kiribati), Papahānaumokuākea (United States), Great Barrier Reef (Australia), Galápagos (Ecuador), Kluane/Wrangell-St Elias/Glacier Bay/Tatshenshini-Alsek (United States and Canada), Lake Baikal (Russia), Aïr and Ténéré Natural Reserves (Niger), Tassili n'Ajjer (Algeria), and Central Amazon Conservation Complex (Brazil). Mark these 10 sites on a world map (electronic image file), label them, and provide their surface area. Build an Excel file in which you compare these sites as per biodiversity richness, endemism, and ecosystem functions being protected [do library work]. Make the Galápagos Archipelago and its Marine Reserve the center of your attention when comparing among sites. Highlight Galápagos' *uniqueness* with respect to the other natural World Heritage Sites at the same time that you honor the "outstanding universal value" of all sites. Put together a slideshow presentation (for a 20-minute talk) about "Galápagos: One of the Top 10 World Heritage Sites" to be shared with a general audience. Avoid slides with "text only" [they are often terrible]. Keep it rigorous yet fun and didactic.

References

Acuña-Marrero, David, Jesús Jiménez, Franz Smith, Paul F. Doherty, Alex Hearn, Jonathan R. Green, Jules Paredes-Jarrín, and Pelayo Salinas-de-León. 2014. "Whale Shark (*Rhincodon typus*) Seasonal Presence, Residence Time and Habitat Use at Darwin Island, Galapagos Marine Reserve". *PLoS ONE*, No. 9: e115946.

—. Adam N. H. Smith, Neil Hammerschlag, Alex Hearn, Marti J. Anderson, Hannah Calich, Matthew D. M. Pawley, Chris Fischer, and Pelayo Salinas-de-León. 2017. "Residency and Movement Patterns of an Apex Predatory Shark (*Galeocerdo cuvier*) at the Galapagos Marine Reserve". *PLoS ONE*, No. 12: e0183669.

—. Adam N. H. Smith, Pelayo Salinas-de-León, Euan S. Harvey, Matthew D. M. Pawley, and Marti J. Anderson. 2018. "Spatial Patterns of

Distribution and Relative Abundance of Coastal Shark Species in the Galapagos Marine Reserve". *Marine Ecology Progress Series*, No. 593: 73-95.

Amori, Giovanni, Spartaco Gippoliti, and Kristofer M. Helgen. 2008. "Diversity, Distribution, and Conservation of Endemic Island Rodents". *Quaternary International*, No. 182: 6-15.

Arakaki, Mónica, Pascal-Antoine Christin, Reto Nyffeler, Anita Lendel, Urs Eggli, R. Matthew Ogburn, Elizabeth Spriggs, Michael J. Moore, and Erika J. Edwards. 2011. "Contemporaneous and Recent Radiations of the World's Major Succulent Plant Lineages". *Proceedings of the National Academy of Sciences of the United States*, No. 108: 8379-8384.

Arteaga, Alejandro, Lucas Bustamante, Jose Vieira, Washington Tapia, and Juan M. Guayasamin. 2019. *Reptiles of the Galápagos*. Quito: Tropical Herping.

Balvanera, Patricia, Sandra Quijas, Daniel S. Karp, Neville Ash, Elena M. Bennett, Roel Boumans, Claire Brown, Kai M.A. Chan, Rebecca Chaplin-Kramer, Benjamin S. Halpern, Jordi Honey-Rosés, Choong-Ki Kim, Wolfgang Cramer, Maria José Martínez-Harms, Harold Mooney, Tuyeni Mwampamba, Jeanne Nel, Stephen Polasky, Belinda Reyers, Joe Roman, Woody Turner, Robert J. Scholes, Heather Tallis, Kirsten Thonicke, Ferdinando Villa, Matt Walpole, and Ariane Walz. (2017). "Ecosystem Services". In *The Geo Handbook On Biodiversity Observation Networks*, edited by Michele Walters and Robert J. Scholes, 39-78. Cham, Switzerland: Springer Nature.

Banta, William C. 1991. "The Bryozoa of the Galápagos". In *Galápagos Marine Invertebrates: Taxonomy, Biogeography, and Evolution in Darwin's Islands*, edited by Mathew J. James, 371-389. New York and London: Plenum Press.

Bastille-Rousseau, Guillaume, Charles B. Yackulic, Jacqueline L. Frair, Freddy Cabrera, and Stephen Blake. 2016. "Allometric and Temporal Scaling of Movement Characteristics in Galapagos Tortoises". *Journal of Animal Ecology*, No. 85: 1171.1181.

—. Charles B. Yackulic, James P. Gibbs, Jacqueline L. Frair, Freddy Cabrera, and Stephen Blake. 2019. "Migration Triggers in a Large Herbivore: Galápagos Giant Tortoises Navigating Resource Gradients on Volcanoes". *Ecology*, No. 100: e02658.

Benavides, Edgar, Rebecca Baum, Heidi M. Snell, Howard L. Snell, and Jack W. Sites. 2009. "Island Biogeography of Galápagos Lava Lizards (Tropiduridae: *Microlophus*): Species Diversity and Colonization of the Archipelago". *Evolution*, No. 63: 1606-1626.

Berta, Annalisa, Morgan Churchill, and Robert W. Boessenecker. 2018. "The Origin and Evolutionary Biology of Pinnipeds: Seals, Sea Lions, and Walruses". *Annual Review of Earth and Planetary Sciences*, No. 46: 203-228.

Blake, James A. 1991. "The Polychaete Fauna of the Galápagos Islands". In *Galápagos Marine Invertebrates: Taxonomy, Biogeography, and Evolution in Darwin's Islands*, edited by Mathew J. James, 75-96. New York and London: Plenum Press.

Blake, Stephen, Charles B. Yackulic, Fredy Cabrera, Washington Tapia, James P. Gibbs, Franz Kümmeth, and Martin Wikelski. 2013. "Vegetation Dynamics Drive Segregation by Body Size in Galapagos Tortoises Migrating Across Altitudinal Gradients". *Journal of Animal Ecology*, No. 82: 310-321.

Buchholz, Sascha, Léon Baert, Jacqueline Rodríguez, Charlotte E Causton, and Heinke Jäger. 2020. "Spiders in Galapagos–Diversity, Biogeography and Origin". *Biological Journal of the Linnean Society*, No. 130: 41-48.

Buglass, Salome, Shannon Nagy, David Ebert, Paulina Sepa, Alan Turchik, Katherine L. C. Bell, Fernando Rivera, and Jonatha Giddens. 2020. "First Records of the Seven-Gilled *Notorynchus cepedianus* and Six-Gilled *Hexanchus griseus* Sharks (Chondrichthyes: Hexanchiformes: Hexanchidae) Found in the Galápagos Marine Reserve". *Journal of Fish Biology*, No. 97: 926-929.

Bungartz, Frank, John A. Elix, Alba Yánez-Ayabaca, and Alan W. Archer. 2015. "Endemism in the Genus *Pertusaria* (Pertusariales, lichenized Ascomycota) from the Galapagos Islands". *Telopea*, No. 18: 325-369.

Burns, Kevin J., Allison J. Shultz, Pascal O. Title, Nicholas A. Mason, F. Keith Barker, John Klicka, Scott M. Lanyon, and Irby J. Lovette. 2014. "Phylogenetics and Diversification of Tanagers (Passeriformes: Thraupidae), the Largest Radiation of Neotropical Songbirds". *Molecular Phylogenetics and Evolution*, No. 75: 41-77.

Bustamante, Rodrigo, Ken J. Collins, and Robert Bendsted-Smith. 2000. "Biodiversity Conservation in the Galápagos Marine Reserve". *Bulletin de l'Institut Royal des Sciences Naturelles de Belgique*, No. 70: 31-38.

—. Gerald M. Wellington, George M. Branch, Graham J. Edgar, Priscilla Martinez, Fernando Rivera, F. Cliff Smith, and Jon Witman. 2002. "Outstanding Marine Features of Galápagos". In *A Biodiversity Vision for the Galapagos Islands*, edited by Robert Bensted-Smith, 60-71. Puerto Ayora: Charles Darwin Foundation and World Wildlife Fund.

Caccone, Adalgisa, James P. Gibbs, Valerio Ketmaier, Elizabeth Suatoni, and Jeffrey R. Powell. 1999. "Origin and Evolutionary Relationships of Giant Galápagos Tortoises". *Proceedings of the National Academy of Sciences of the United States*, No. 96: 13223-13228.

Cantú, Juan Carlos, and Raziel Levi-Méndez. 2022. "Sharks of Mexico and Central America–Pacific Ocean". Accessed November 24, 2022. *https://www.defenders.org/sites/default/files/publications/Sharks-of-Mexico-and-Central-America-Pacific-Ocean-in-English.pdf.*

Carrillo-Briceño, Jorge Domingo, Juan D. Carrillo, Orangel Antonio Aguilera, and Marcelo R. Sanchez-Villagra. 2018. "Shark and Ray Diversity in the Tropical America (Neotropics)–An Examination of Environmental and Historical Factors Affecting Diversity". *PeerJ*, No. 6: e5313.

CDF (Charles Darwin Foundation). 2020. [*Sharks of the Galápagos Marine Reserve*] *Tiburones de la Reserva Marina de Galápagos*. Puerto Ayora, Santa Cruz, Galápagos: Charles Darwin Foundation.

Chaves, Jaime A., Fernando Lopes, Daniela Martínez, Dario F. Cueva, Gabriela I. Gavilanes, Sandro L. Bonatto, Larissa Rosa de Oliveira, and Diego Páez-Rosas. 2022. "Population Genetics and Phylogeography of Galapagos Fur Seals". *Frontiers in Genetics*, No. 13: 725772.

Chiari, Ylenia, Chaz Hyseni, Tom H. Fritts, Scott Glaberman, Cruz Marquez, James P. Gibbs, Julien Claude, and Adalgisa Caccone. 2009. "Morphometrics Parallel Genetics in a Newly Discovered and Endangered Taxon of Galápagos Tortoise". *PLoS ONE*, No. 4: e6272.

—. Arie van der Meijden, Adalgisa Caccone, Julien Claude, and Benjamin Gilles. 2017. "Self-righting Potential and the Evolution of Shell Shape in Galápagos Tortoises". *Scientific Reports*, No. 7: 15828.

—. 2021. "Morphology". In *Galapagos Giant Tortoises*, edited by James P. Gibbs, Linda J. Cayot, and Washington Tapia, 139-156. London: Academic Press.

Chiriboga, Angel, Diego Ruiz, and Stuart Banks. 2011. "CDF Checklist of Galapagos Corals, Gorgonians, Sea Anemones Hydroids". In *Charles Darwin Foundation Galapagos Species Checklist*, edited by Frank Bungartz, Henri Herrera, Patricia Jaramillo, Nathalia Tirado, Gustavo Jímenez-Uzcategui, Diego Ruiz, Anne Guézou, and Frauke Ziemmeck, 1-21. Puerto Ayora: Charles Darwin Foundation.

Churchill, Morgan, Robert W. Boessenecker, and Mark T. Clementz. 2014. "Colonization of the Southern Hemisphere by Fur Seals and Sea Lions (Carnivora: Otariidae) Revealed by Combined Evidence Phylogenetic and Bayesian Biogeographical Analysis". *Zoological Journal of the Linnean Society*, No. 172: 200-225.

Dando, Marc. 2020. "Guía de Especies de Tiburones Registradas en Galápagos". In *Tiburones de la Reserva Marina de Galápagos*, edited by

Charles Darwin Foundation, 42-75. Puerto Ayora, Santa Cruz, Galápagos: Charles Darwin Foundation.

Ebert, David A. 2016. *Deep-Sea Cartilaginous fishes of the Southeastern Pacific Ocean*. Rome: Food and Agriculture Organization (FAO) of the United Nations.

Fernández-Mazuecos, Mario, Pablo Vargas, Ross A. McCauley, David Monjas, Ana Otero, Jaime A. Chaves, Juan Ernesto Guevara Andino, and Gonzalo Rivas-Torres. 2020. "The Radiation of Darwin's Giant Daisies in the Galápagos Islands". *Current Biology*, No. 30: 4989-4998.

Fundación Charles Darwin (FCD), and World Wildlife Fund (WWF) Ecuador. 2018. *Atlas de Galápagos, Ecuador: Especies Nativas e Invasoras*. Quito: FCD y WWF-Ecuador.

Funk, Erik R., and Kevin J. Burns. 2018. "Biogeographic Origins of Darwin's Finches (Thraupidae: Coerebinae)". *The Auk Ornitological Advances*, No. 135: 561-571.

Gaughran, Stephen J., Maud C. Quinzin, Joshua M. Miller, Ryan C. Garrick, Danielle L. Edwards, Michael A. Russello, Nikos Poulakakis, Claudio Ciofi, Luciano B. Beheregaray, and Adalgisa Caccone. 2017. "Theory, Practice, and Conservation in the Age of Genomics: The Galápagos Giant Tortoise as a Case Study". *Evolutionary Applications*, No. 11:1084-1093.

Gavryushkina, Alexandra, Tracy A. Heath, Daniel T. Ksepka, Tanja Stadler, David Welch, and Alexei J. Drummond. 2017. "Bayesian Total-Evidence Dating Reveals the Recent Crown Radiation of Penguins". *Systematic Biology*, No. 66: 57-73.

Geist, Dennis, Howard Snell, Heidi Snell, Charlotte Goddard, and Mark D. Kurz. 2014. "A Paleogeographic Model of the Galápagos Islands and Biogeographical and Evolutionary Implications". In *The Galapagos: A Natural Laboratory for the Earth Sciences*, edited by Karen S. Harpp, Eric Mittelstaedt, Noémi d'Ozouville, and David W. Graham, 145-166. Washington DC: American Geophysical Union and John Wiley & Sons Inc.

Gillespie, Rosemary G., and George K. Roderick. 2002. "Arthropods on Islands: Colonization, Speciation, and Conservation". *Annual Review of Entomology*, No. 47: 595-632.

Gradstein, S. Robert, and Fraunke Ziemmeck. 2011. "CDF Checklist of Galapagos Horn-Liverworts". In *Charles Darwin Foundation Galapagos Species Checklist*, edited by Frank Bungartz, Henri Herrera, Patricia Jaramillo, Nathalia Tirado, Gustavo Jímenez-Uzcategui, Diego Ruiz, Anne Guézou, and Frauke Ziemmeck, 1-16. Puerto Ayora: Charles Darwin Foundation.

Graham, Natalie R., Daniel S. Gruner, Jun Y. Lim, and Rosemary G. Gillespie. 2017. "Island Ecology and Evolution: Challenges in the Anthropocene". *Environmental Conservation*, No. 44: 323-335.

Green, Jonathan R., Sofía M. Green, and Alex Hearn. 2020. *Galápagos Whale Shark Project: Field Report 2020*. Puerto Ayora: Galápagos Whale Shark Project.

—. Sofía M. Green, Alex Hearn, and Jenny Waack. 2021. *Galápagos Whale Shark Project: Field Report 2021*. Puerto Ayora: Galápagos Whale Shark Project, Galápagos National Park Directorate, Georgia Aquarium.

Grehan, John. 2001. "Biogeography and Evolution of the Galapagos: Integration of the Biological and Geological Evidence". *Biological Journal of the Linnean Society*, No. 74: 267-287.

Grove, Jack S., and Robert J. Lavenberg. 1997. *The Fishes of the Galápagos Islands*. Stanford: Stanford University Press.

Guzmán, Beatriz, Ruben Heleno, Manuel Nogales, Walter Simbaña, Anna Traveset, and Pablo Vargas. 2017. "Evolutionary History of the Endangered Shrub Snapdragon (*Galvezia leucantha*) of the Galápagos Islands". *Diversity and Distributions*, No. 23: 247-260.

Harper, Grant A., and Victor Carrión. 2011. "Introduced Rodents in the Galápagos: Colonisation, Removal and the Future". In *Island Invasives: Eradication and Management*, edited by C. Richard Veitch, Mick N. Clout, and David R. Towns, 63-66. Gland, Switzerland: International Union for Conservation of Nature.

Harty, Kanina, Michel Guerrero, Anna M. Knochel, Guy M. W. Stevens, Andrea Marshall, Katherine Burgess, and Joshua D. Stewart. 2022. "Demographics and Dynamics of the World's Largest Known Population of Oceanic Manta Rays *Mobula birostris* in Coastal Ecuador". *Marine Ecology Progress Series*, No. 700: 145-159.

Heads, Michael, and John R. Grehan. 2021. "The Galápagos Islands: Biogeographic Patterns and Geology". *Biological Reviews*, No. 96: 1160-1185.

Hearn, Alex R., Jonathan Green, Marlon H. Román, David Acuña-Marrero, Eduardo Espinoza, and A. Peter Klimley. 2016. "Adult Female Whale Sharks Make Long-Distance Movements Past Darwin Island (Galapagos, Ecuador) in the Eastern Tropical Pacific". *Marine Biology*, No. 163: 214.

Hedrick, Phillip W. 2019. "Galapagos Islands Endemic Vertebrates: A Population Genetics Perspective". *Journal of Heredity*, No. 110: 137-157.

Helsen, Philippe, Robert A. Browne, David J. Anderson, Peter Verdyck, and Stefan van Dongen. 2009. "Galápagos' Opuntia (prickly pear)

Cacti: Extensive Morphological Diversity, Low Genetic Variability". *Biological Journal of the Linnean Society*, No. 96: 451-461.

Hickman, Cleveland P. 2009. "Evolutionary Responses of Marine Invertebrates to Insular Isolation in Galapagos". *Galapagos Research*, No. 66: 32-42.

Hoeck, Paquita E. A., Jennifer L. Bollmer, Patricia G. Parker, and Lukas F. Keller. 2010. "Differentiation with Drift: a Spatio-Temporal Genetic Analysis of Galápagos Mockingbird Populations (*Mimus* spp.)". *Philosophical Transactions of the Royal Society B*, No. 365: 1127-1138.

Hunter, Elizabeth A., James P. Gibbs, Linda J. Cayot, and Washington Tapia. 2013. "Equivalency of Galápagos Giant Tortoises Used as Ecological Replacement Species to Restore Ecosystem Functions". *Conservation Biology*, No. 27: 701-709.

—. James P. Gibbs, Linda J. Cayot, Washington Tapia, Maud C. Quinzin, Joshua M. Miller, Adalgisa Caccone, and Kevin T. Shoemaker. 2020. "Seeking Compromise Across Competing Goals in Conservation Translocations: The Case of the 'Extinct' Floreana Island Galapagos Giant Tortoise". *Journal of Applied Ecology*, No. 57: 136-148.

Jensen, Evelyn L., Maud C. Quinzin, Joshua M. Miller, Michael A. Russello, Ryan C. Garrick, Danielle L. Edwards, Scott Glaberman, Ylenia Chiari, Nikos Poulakakis, Washington Tapia, James P. Gibbs, and Adalgisa Caccone. 2022. "A New Lineage of Galapagos Giant Tortoises Identified from Museum Samples". *Heredity*, No. 128: 261-270.

Jiménez-Prado, Pedro, and Philippe Béarez. 2004. *Marine Fishes of Continental Ecuador*. Quito: SIMBIOE/NAZCA/IFEA.

Jiménez-Uzcátegui, Gustavo, David A. Wiedenfeld, F. Hernán Vargas, and Howard L. Snell. 2012. "CDF Checklist of Galapagos Birds". In *Charles Darwin Foundation Galapagos Species Checklist*, edited by Frank Bungartz, Henri Herrera, Patricia Jaramillo, Nathalia Tirado, Gustavo Jímenez-Uzcategui, Diego Ruiz, Anne Guézou, and Frauke Ziemmeck, 1-33. Puerto Ayora: Charles Darwin Foundation.

—. Cruz Márquez, and Howard L. Snell. 2013. "CDF Checklist of Galapagos Reptiles". In *Charles Darwin Foundation Galapagos Species Checklist*, edited by Frank Bungartz, Henri Herrera, Patricia Jaramillo, Nathalia Tirado, Gustavo Jímenez-Uzcategui, Diego Ruiz, Anne Guézou, and Frauke Ziemmeck, 1-39. Puerto Ayora: Charles Darwin Foundation.

—. and Howard L. Snell. 2014. "CDF Checklist of Galapagos Mammals". In *Charles Darwin Foundation Galapagos Species Checklist*, edited by Frank Bungartz, Henri Herrera, Patricia Jaramillo, Nathalia Tirado,

Gustavo Jímenez-Uzcategui, Diego Ruiz, Anne Guézou, and Frauke Ziemmeck, 1-18. Puerto Ayora: Charles Darwin Foundation.

Kay, E. Alison. 1991. "The Marine Mollusks of the Galápagos: Determinants of Insular Marine Faunas". In *Galápagos Marine Invertebrates: Taxonomy, Biogeography, and Evolution in Darwin's Islands*, edited by Mathew J. James, 235-252. New York and London: Plenum Press.

Kizirian, David, Adrienne Trager, Maureen A. Donnelly, and John W. Wright. 2004. "Evolution of Galapagos Island Lava Lizards (Iguania: Tropiduridae: *Microlophus*)". *Molecular Phylogenetics and Evolution*, No. 32: 761-769.

Lamichhaney, Sangeet, Jonas Berglund, Markus Sällman Almén, Khurram Maqbool, Manfred Grabherr, Alvaro Martinez-Barrio, Marta Promerová, Carl-Johan Rubin, Chao Wang, Neda Zamani, B. Rosemary Grant, Peter R. Grant, Matthew T. Webster, and Leif Andersson. 2015. "Evolution of Darwin's Finches and their Beaks Revealed by Genome Sequencing". *Nature*, No. 518: 371-375.

—. Fan Han, Matthew T. Webster, Leif Andersson, B. Rosemary Grant, and Peter R. Grant. 2018. "Rapid Hybrid Speciation in Darwin's Finches". *Science*, No. 359: 224-228.

Lorenzo-Carballa, María Olalla, Rosser W. Garrison, Andrea C. Encalada, and Adolfo Cordero-Rivera. 2021. "Darwin Returns to the Galapagos: Genetic and Morphological Analyses Confirm the Presence of *Tramea darwini* at the Archipelago (Odonata, Libellulidae)". *Insects*, No. 12: 21.

MacLeod, Amy, Ariel Rodríguez, Miguel Vences, Pablo Orozco-terWengel, Carolina García, Fritz Trillmich, Gabriele Gentile, Adalgisa Caccone, Galo Quezada, and Sebastian Steinfartz. 2015. "Hybridization Masks Speciation in the Evolutionary History of the Galapagos Marine Iguana". *Proceedings of the Royal Society B*, No. 282: 2015.0425.

Magalhaes, Ivan L. F., Danilo M. Neves, Fabrício R. Santos, Teofânia H. D. A. Vidigal, Antônio D. Brescovit, and Adalberto J. Santos. 2019. "Phylogeny of Neotropical *Sicarius* Sand Spiders Suggests Frequent Transitions From Deserts to Dry Forests Despite Antique, Broad-Scale Niche Conservatism". *Molecular Phylogenetics and Evolution*, No. 140: 106569.

—. Adalberto J. Santos, and Martín J. Ramírez. 2021. "Incorporating Topological and Age Uncertainty into Event-Based Biogeography of Sand Spiders Supports Paleo-Islands in Galapagos and Ancient Connections among Neotropical Dry Forests". *Diversity*, No. 13: 418.

Majure, Lucas C., Raul Puente, M. Patrick Griffith, Walter S. Judd, Pamela S. Soltis, and Douglas E. Soltis. 2012. "Phylogeny of *Opuntia* s.s.

(Cactaceae): Clade Delineation, Geographic Origins, and Reticulate Evolution". *American Journal of Botany*, No. 99: 847-864.

Maluf, Linda Yvonne. 1991. "Echinoderm Fauna of the Galápagos Islands". In *Galápagos Marine Invertebrates: Taxonomy, Biogeography, and Evolution in Darwin's Islands*, edited by Mathew J. James, 345-367. New York and London: Plenum Press.

McMullen, Conley K. 1999. *Flowering Plants of the Galápagos*. Ithaca: Cornell University Press.

Miller, Joshua M., Maud C. Quinzin, Nikos Poulakakis, James P. Gibbs, Luciano B. Beheregaray, Ryan C. Garrick, Michael A. Russello, Claudio Ciofi, Danielle L. Edwards, Elizabeth A. Hunter, Washington Tapia, Danny Rueda, Jorge Carrión, Andrés A. Valdivieso, and Adalgisa Caccone. 2017. "Identification of Genetically Important Individuals of the Rediscovered Floreana Galápagos Giant Tortoise (*Chelonoidis elephantopus*) Provides Founders for Species Restoration Program". *Scientific Reports*, No. 7: 11471.

—. Maud C. Quinzin, Danielle L. Edwards, Deren A. R. Eaton, Evelyn L. Jensen, Michael A. Russello, James P. Gibbs, Washington Tapia, Danny Rueda, and Adalgisa Caccone. 2018. "Genome-Wide Assessment of Diversity and Divergence Among Extant Galapagos Giant Tortoise Species". *Journal of Heredity*, No. 109: 611-619.

Miralles, Aurélien, Amy Macleod, Ariel Rodríguez, Alejandro Ibáñez, Gustavo Jiménez-Uzcategui, Galo Quezada, Miguel Vences, and Sebastian Steinfartz. 2017. "Shedding Light on the Imps of Darkness: An Integrative Taxonomic Revision of the Galápagos Marine Iguanas (genus *Amblyrhynchus*)". Zoological Journal of the Linnean Society, No. 118: 678-710.

Naumann, Terry, Dennis Geist, and Mark Kurz. 2002. "Petrology and Geochemistry of Volcán Cerro Azul: Petrologic Diversity among the Western Galápagos Volcanoes ". *Journal of Petrology*, No. 43: 859-883.

Nietlisbach, Pirmin, Peter Wandeler, Patricia G. Parker, Peter R. Grant, B. Rosemary Grant, Lukas F. Keller, and Paquita E. A. Hoeck. 2013. "Hybrid Ancestry of an Island Subspecies of Galápagos Mockingbird Explains Discordant Gene Trees". *Molecular Phylogenetics and Evolution*, No. 69: 581-592.

Palma, Ricardo L., and Stewart B. Peck. 2013. "An Annotated Checklist of Parasitic Lice (Insecta: Phthiraptera) from the Galápagos Islands". *Zootaxa*, No. 3627: 1-87.

Parent, Christine E. 2008. "Diversification on Islands: Bulimulid Land Snails of Galápagos". PhD diss., Simon Fraser University.

—. 2012. "Biogeographical and Ecological Determinants of Land Snail Diversification on Islands". *American Malacological Bulletin*, No. 30: 207-215.

—. and Bernard J. Crespi. 2006. "Sequential Colonization and Diversification of Galapagos Endemic Land Snail Genus *Bulimulus* (Gastropoda, Stylommatophora)". *Evolution*, No. 60: 2311-2328.

Paz-y-Miño-C, Guillermo, and Avelina Espinosa. 2024a. *Arrivals of Life to the Galápagos: From Berlanga to Darwin, Melville, and World War II*. Newcastle upon Tyne: Cambridge Scholars Publishing.

—. 2024b. *A Synthesis of the Galápagos: Natural History, Human Conflict, and the Destiny of an Archipelago*. Newcastle upon Tyne: Cambridge Scholars Publishing.

Peck, Stewart B. 2006. *The Beetles of the Galápagos Islands, Ecuador: Evolution, Ecology, and Diversity (Insecta: Coleoptera)*. Ottawa: NRC Research Press.

Pelegrín, Jonathan S., and Carolina Acosta-Hospitaleche. 2022. "Evolutionary and Biogeographical History of Penguins (Sphenisciformes): Review of the Dispersal Patterns and Adaptations in a Geologic and Paleoecological Context". *Diversity*, No. 14: 255.

Petren, Kenneth, Peter R. Grant, B. Rosemary Grant, and Lukas F. Keller. 2005. "Comparative Landscape Genetics and the Adaptive Radiation of Darwin's Finches: The Role of Peripheral Isolation". *Molecular Ecology*, No. 14: 2943-2957.

Phillips, John G., T. Mason Linscott, Andrew M. Rankin, Andrew C. Kraemer, Nathaniel F. Shoobs, and Christine E. Paren. 2020. "Archipelago-Wide Patterns of Colonization and Speciation Among an Endemic Radiation of Galápagos Land Snails". *Journal of Heredity*, No. 111: 92-102.

Poulakakis, Nikos, Scott Glaberman, Michael Russello, Luciano B. Beheregaray, Claudio Ciofi, Jeffrey R. Powell, and Adalgisa Caccone. 2008. "Historical DNA Analysis Reveals Living Descendants of an Extinct Species of Galápagos Tortoise". *Proceedings of the National Academy of Sciences of the United States*, No. 105: 15464-15469.

—. Michael Russello, Dennis Geist, and Adalgisa Caccone. 2012. "Unravelling the Peculiarities of Island Life: Vicariance, Dispersal and the Diversification of the Extinct and Extant Giant Galápagos Tortoises". *Molecular Ecology*, No. 21: 160-173.

—. Danielle L. Edwards, Ylenia Chiari, Ryan C. Garrick, Michael A. Russello, Edgar Benavides, Gregory J. Watkins-Colwell, Scott Glaberman, Washington Tapia, James P. Gibbs, Linda J. Cayot, and Adalgisa Caccone. 2015. "Description of a New Galapagos Giant Tortoise

Species (*Chelonoidis*; Testudines: Testudinidae) from Cerro Fatal on Santa Cruz Island". *PLoS ONE*, No. 10: e0138779.

Ramos, Barbara, Daniel González-Acuña, David E. Loyola, Warren E. Johnson, Patricia G. Parker, Melanie Massaro, Gisele P. M. Dantas, Marcelo D. Miranda, and Juliana A. Vianna. 2018. "Landscape Genomics: Natural Selection Drives the Evolution of Mitogenome in Penguins". *BMC Genomics*, No. 19: 53.

Rassmann, Kornelia, Fritz Trillmich, and Diethard Tautz. 2009. "Hybridization Between the Galápagos Land and Marine Iguana (*Conolophus subcristatus* and *Amblyrhynchus cristatus*) on Plaza Sur". *Journal of Zoology*, No. 242: 729-739.

Riegl, Bernhard, Matthew Johnston, Peter W. Glynn, Inti Keith, Fernando Rivera, Mariana Vera-Zambrano, Stuart Banks, Joshua Feingold, and Peter J. Glynn. 2019. "Some Environmental and Biological Determinants of Coral Richness, Resilience and Reef Building in Galápagos (Ecuador)". *Scientific Reports*, No. 9: 10322

Roff, George, Christopher Doropoulos, Alice Rogers, Yves-Marie Bozec, Nils C. Krueck, Eleanor Aurellado, Mark Priest, Chico Birrell, and Peter J. Mumby. 2016. "The Ecological Role of Sharks on Coral Reefs". *Trends in Ecology & Evolution*, No. 31: 395-407.

Salinas-de-León, Pelayo, David Acuña-Marrero, Etienne Rastoin, Alan M. Friedlander, Mary K. Donovan, and Enric Sala. 2016. "Largest Global Shark Biomass Found in the Northern Galápagos Islands of Darwin and Wolf". *PeerJ*, No. 4: e1911.

Sato, Akie, Herbert Tichy, Colm O'hUigin, Peter R. Grant, B. Rosemary Grant, and Jan Klein. 2001. "On the Origin of Darwin's Finches". *Molecular Biology and Evolution*, No. 18: 299-311.

Sequeira, Andrea S., Analía A. Lanteri, M. Amalia Scataglini, Viviana A. Confalonieri, and Brian D. Farrell. 2000. "Are Flightless *Galapaganus* Weevils Older than the Galápagos Islands They Inhabit?". *Heredity*, No. 85: 20-29.

—. Analía A. Lanteri, L. Roque Albelo, Sayantani Bhattacharyam, and Manisha Sijapati. 2008. "Colonization History, Ecological Shifts and Diversification in the Evolution of Endemic Galápagos Weevils". *Molecular Ecology*, No. 17: 1089-1107.

Shark Research Institute. 2022. "Learn About Some of the 400+ Species of Sharks". Accessed November 24, 2022. *https://www.sharks.org/species*.

Shaw, Kerry L., and Rosemary G. Gillespie. 2016. "Comparative Phylogeography of Oceanic Archipelagos: Hotspots for Inferences of Evolutionary Process". *Proceedings of the National Academy of Sciences of the United States*, No. 113: 7986-7993.

Sinclair, Bradley J. 2011. "CDF Checklist of Galapagos Flies". In *Charles Darwin Foundation Galapagos Species Checklist*, edited by Frank Bungartz, Henri Herrera, Patricia Jaramillo, Nathalia Tirado, Gustavo Jímenez-Uzcategui, Diego Ruiz, Anne Guézou, and Frauke Ziemmeck, 1-34. Puerto Ayora: Charles Darwin Foundation.

Swash, Andy, and Rob Still. 2005. *Birds, Mammals, and Reptiles of the Galápagos Islands*. New Heaven and London: Yale University Press.

Tapia, Washington. 2022. "Previewing: Expedition Fernandina". Galapagos Conservancy. Accessed May 7, 2022. *https://www.galapagos.org/blog/previewing-the-fernandina-expedition/*.

Tebbich, Sabine, Irmgard Teschke, Erica Cartmill, and Sophia Stankewitz. 2012. "Use of a Barbed Tool by an Adult and a Juvenile Woodpecker Finch (*Cactospiza pallida*)". *Behavioural Processes*, No. 89: 166-171.

Tirira, Diego G., Jorge Brito, Santiago F. Burneo, and Comisión de Diversidad de la Asociacioón Ecuatoriana de Mastozoología. 2020. *Mammals of Ecuador: Updated Check List*. Quito: Asociación Ecuatoriana de Mastozoología.

Torres-Carvajal, Omar, Charles W. Barnes, María J. Pozo-Andrade, Washington Tapia, and Gabriela Nicholls. 2014. "Older than the Islands: Origin and Diversification of Galápagos Leaf-toed Geckos (Phyllodactylidae: *Phyllodactylus*) by Multiple Colonizations. *Journal of Biogeography*, No. 41: 1883-1894.

—. Andrea Rodríguez-Guerra, and Jaime A. Chaves. 2016. "Present Diversity of Galápagos Leaf-toed Geckos (Phyllodactylidae: *Phyllodactylus*) Stems from Three Independent Colonization Events". *Molecular Phylogenetics and Evolution*, No. 103: 1-5.

UNESCO WHC (World Heritage Convention). 2025a. "Galápagos Islands". Accessed March 27, 2025. *https://whc.unesco.org/en/list/1/*.

—. 2025b. "World Heritage List". Accessed March 27, 2025. *https://whc.unesco.org/en/list*.

Vargas, Pablo, Manuel Nogales, Patricia Jaramillo, Jens M. Olesen, Anna Traveset, and Ruben Heleno. 2014. "Plant Colonization Across the Galápagos Islands: Success of the Sea Dispersal Syndrome". *Botanical Journal of the Linnean Society*, No. 174: 349-358.

—. Beatriz Rumeu, Ruben H. Heleno, Anna Traveset, and Manuel Nogales. 2015. "Historical Isolation of the Galápagos Carpenter Bee (*Xylocopa darwini*) Despite Strong Flight Capability and Ecological Amplitude". *PLoS ONE*, No. 10: e0120597.

Veron, Simon, Thomas Haevermans, Rafaël Govaerts, Maud Mouchet, and Roseli Pellens. 2019. "Distribution and Relative Age of Endemism Across Islands Worldwide". *Scientific Reports*, No. 9: 11693.

Wauters, Nina, Wouter Dekoninck, Frederik Hedrick, and Henri W. Herrera. 2016. "Habitat Association and Coexistence of Endemic and Introduced Ant Species in the Galápagos Islands". *Ecological Entomology*, No. 41: 40-50.

Wiedenfeld, David A. 2006. "Aves, The Galapagos Islands, Ecuador". *Check List*, No. 2: 1-27.

Wiggins, Ira L., and Duncan M. Porter. 1971. *Flora of the Galápagos Islands*. Stanford: Stanford University Press.

Williams, Ted. 2019. "Recovery: Victories in Galápagos National Park". *The Nature Conservancy: Cool Green Science*. Accessed May 7, 2022. *https://blog.nature.org/science/2019/06/25/recovery-victories-in-galapagos-national-park/*.

Wolf, Jochen B. W., Diethard Tautz, and Fritz Trillmich. 2007. "Galápagos and Californian Sea Lions Are Separate Species: Genetic Analysis of the Genus *Zalophus* and Its Implications for Conservation Management". *Frontiers in Zoology*, No. 4: 20.

—. Chris Harrod, Sylvia Brunner, Sandie Salazar, Fritz Trillmich, and Diethard Tautz. 2008. "Tracing Early Stages of Species Differentiation: Ecological, Morphological and Genetic Divergence of Galápagos Sea Lion Populations". *BMC Evolutionary Biology*, No. 8: 150.

Yackulic, Charles B., Stephen Blake, and Guillaume Bastille-Rousseau. 2017. "Benefits of the Destinations, Not Costs of the Journeys, Shape Partial Migration Patterns". *Journal of Animal Ecology*, No. 86: 972-982.

Zaher, Hussam, Mario H. Yánez-Muñoz, Miguel T. Rodrigues, Roberta Graboski, Fabio A. Machado, Marco Altamirano-Benavides, Sandro L. Bonatto, and Felipe G. Grazziotin. 2018. "Origin and Hidden Diversity within the Poorly Known Galápagos Snake Radiation (Serpentes: Dipsadidae)". *Systematics and Biodiversity*, No. 16: 614-642.

CHAPTER NINE

WHAT ARE THE MAIN CONSERVATION PROBLEMS IN GALÁPAGOS?

Galápagos has multiple conservation-related problems. As we discuss in our books, *Arrivals of Life to the Galápagos* and *A Synthesis of the Galápagos* (Paz-y-Miño-C and Espinosa 2024a,b), despite decades of legal protection efforts of terrestrial and marine environments (1960s onwards, after the creation of the Galápagos National Park and later the Galápagos Marine Reserve), restoration of historically altered habitats (e.g. successful eradication of large introduced herbivores in some islands, followed by reintroduction of native giant tortoises or land iguanas), and carefully regulated use of the land and local resources (e.g. agricultural areas, artisanal fishing, extraction of sea cucumbers and spiny lobsters), 21st-century Galápagos is destined to be affected by: (i) the impacts of climate change (=a human-induced phenomenon) on the local habitats, the people living on the islands, and their means of subsistence (agriculture, artisanal fishing, and tourism). [We examine this topic in Chapter Ten]. (ii) Large-scale predatory fishing in the vicinity of the Galápagos Marine Reserve (particularly by international fisheries); this chapter. (iii) Macro- and microplastics in the ocean, permeating all trophic levels (globally and in Galápagos); this chapter. (iv) Unmanageable local population growth in the islands of Santa Cruz, San Cristóbal, and Isabela. [We examine this topic in Chapter Six]. (v) Increasing eco unfriendly tourism. [We examine this topic in Chapter Seven; note that 4 and 5 correlate with introductions of invasive species, as discussed in this chapter]. (vi) Wildlife trade (e.g. baby giant tortoises, which are smuggled for the global market; this chapter). (vii) Illegal drug trafficking in the Tropical Eastern Pacific that has targeted the archipelago as a steppingstone on its way to international consumers (a serious, growing problem, as discussed in Chapter Five). And (viii) geopolitical instability in the region (part of it derived from drug trafficking; discussed in Chapter Five). These conservation-related problems pose unprecedented threats to the Galápagos, and all entail human conflict with nature (Paz-y-Miño-C and Espinosa 2024a,b)!

Below, we discuss case studies regarding iconic conservation problems in the Galápagos. Some are local problems, others are regional or global, and affect the archipelago directly or indirectly: (1) Introduced Species: 1535 Onwards; (2) Invasive Species at World Heritage Sites; (3) The Global Decline of Shark Populations; (4) The *Fu Yuan Yu Leng 999* and the Shark-Fin Market; (5) Tuna Fisheries Around the Galápagos Marine Reserve; (6) The Decline of Sea Lions and Fur Seals; (7) Wildlife Trade: Tortoises and Iguanas for Sale; (8) Wildlife Roadkills: *Homo* the Driver in Galápagos; (9) Oceans' Microplastics Reach Galápagos; (10) Microplastics in the Guts and Tissues of Marine Life: Coastal Ecuador and Galápagos; (11) Macroplastics Present in All Galápagos Islands; and (12) Plastics Ingested and Uptaken by All Trophic Levels.

1 Introduced Species: 1535 Onwards

Since the discovery of the Galápagos Archipelago in 1535 (Chapter Two), its connectivity with the mainland—facilitated by human presence, transformation of the land, trading, or tourism—has increased significantly (Fig. 9.1a). Such connectivity has caused or expedited the introduction, accidental or intentional, of "alien species" (Fig. 9.1b; after Toral-Granda et al. 2017). Invasive alien species are one of the most serious threats to biodiversity, particularly in archipelagos (Lenzner et al. 2020).

As depicted in Fig. 9.1a, the connectivity and mobility within/between the Galápagos Islands and continental Ecuador is obvious. The arrival of "alien species" has long been associated with human presence and trading, something summarized in Fig. 9.1b. Note that since the islands were discovered +490 years ago, about 1,579 alien species have been introduced, of which 1,476 are now established, with 868 of the latter (59%) naturalized (i.e. species that now spread and reproduce without active human involvement). The majority of introduced taxa are terrestrial plants (821 spp. introduced; 270 spp. naturalized), followed by terrestrial insects (545 spp. introduced; 467 spp. naturalized) and other groups: terrestrial invertebrates (68 spp. naturalized); pathogens (38 spp.); vertebrates (18 spp.); marine invertebrates (5 spp.); and marine plants (2 spp.). Note as well in Fig. 9.1b bottom-right that, on average, 76% of all alien species have been introduced post-1970s, coinciding with the gradual increase in human population growth, air and maritime travel practices (flights, boats, cargo ships), and tourism flow (i.e. factors associated with human connectivity and mobility in and out of the archipelago; Toral-Granda et al. 2017; Paz-y-Miño-C and Espinosa 2024b).

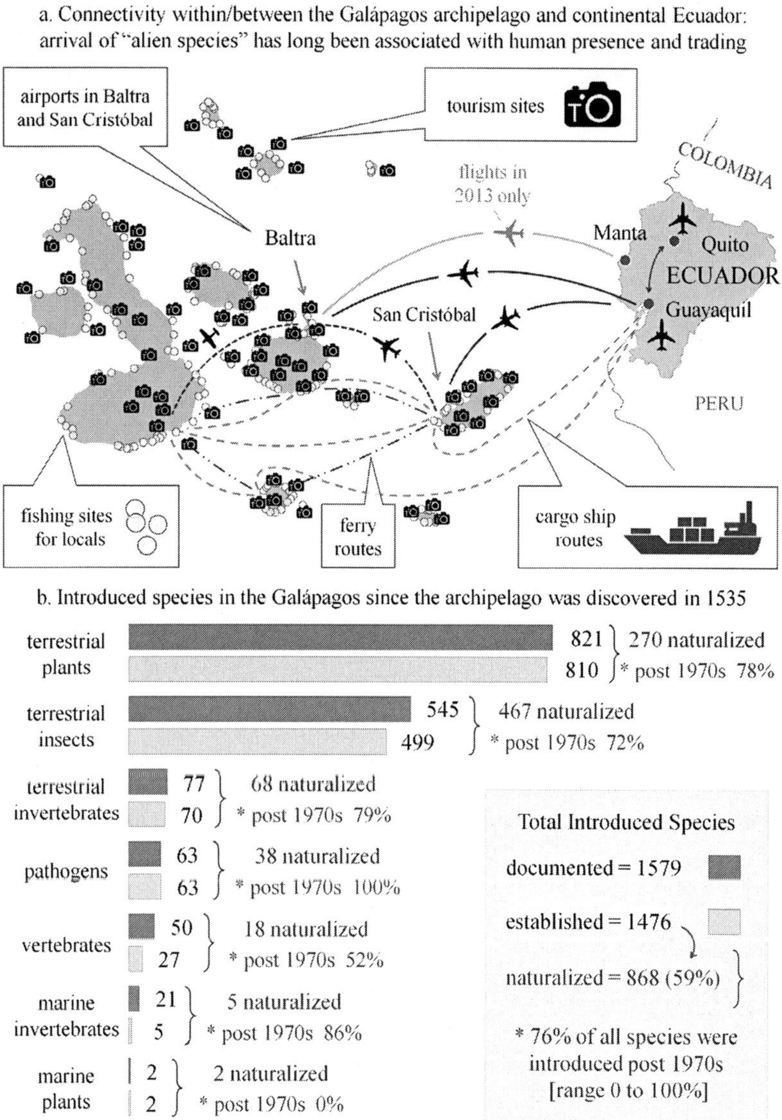

Fig. 9-1 Connectivity and mobility within/between the Galápagos Islands and continental Ecuador (a) and introduced species in the archipelago since 1535 (b). Layout: GPC. a. Redrawn and modified after Toral-Granda et al. (2017). b. Data extracted from Toral-Granda et al. (2017). For assessment of marine bio-invasions alone, see Carlton et al. (2019).

2 Invasive Species at World Heritage Sites

Environmental scientists have characterized the "threat level from invasive species" at 241 UNESCO's World Heritage Sites (WHS), including 7 exemplar case studies (Fig. 9.2 top): Galápagos Islands (Ecuador), Doñana National Park (Spain), Serengeti National Park (Tanzania), Vredefort Dome (South Africa), Aldabra Atoll (Seychelles), Keoladeo National Park (India), and Kakadu National Park (Australia; data extracted from Shackleton et al. 2020). Note in Fig. 9.2-top how the threat level from invasive species is coded as: very high (circles), high (squares), low or very low (crosses), and deficient data or not assessed (triangles). The exemplar WHS case studies are represented by stars. As stated in Fig. 9.2-top, about 300 species of invasive alien species are a threat to about 50 (21%) of the 241 WHS! The Galápagos Islands are threatened by 60 invasive species (second highest after Doñana's 75; Fig. 9.2, bottom), the threat level is "very high", and the management efforts to reduce the threat are considered "moderately effective (some concern)". [Realize that "very high" threat level refers to "*invasive alien species have or are likely to induce irreversible changes to community structure and ecosystem services with no likelihood of them returning to their original state resulting in irreversible damage to the values and integrity of the site even with effective management*", while "moderately effective (some concern)" management efforts to reduce the threat correspond to "*management is taking place but is not effectively reducing the threat and impact of IAS* (invasive alien species)—*this could affect the values and integrity of the site in the long term—adaptive management could potentially improve the situation. There is funding* (for management programs), *but it might not be adequate or sustained in the long run*" (Shackleton et al. 2020)]. In Table 9.1, we list the most threatening invasive alien species to the 7 exemplar WHS (after Shackleton et al. 2020). In the case of Galápagos there are 10 species: *Cedrela odorata* (tree), *Felis catus* (cat), *Rattus norvegicus* (brown rat), *Rattus rattus* (black rat), *Philornis downsi* (fly), *Polistes versicolor* (wasp), *Psidium guajava* (shrub/tree), *Solenopsis geminata* (ant), *Rubus niveus* (shrub), and *Wasmannia auropunctata* (ant). [Keep in mind that earlier in this chapter, we reported 1,579 introduced taxa in the Galápagos, as per another study (Toral-Granda et al. 2017)].

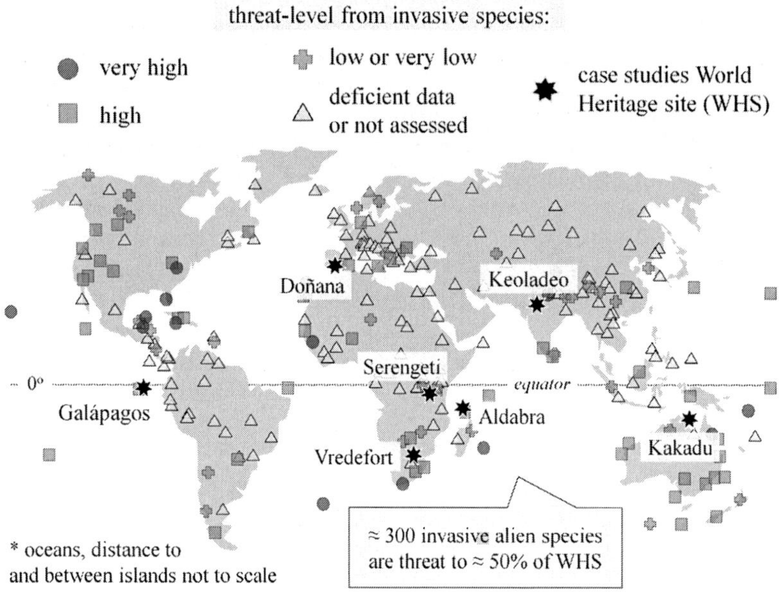

WHS	No. invasive spp	threat level	management to reduce threat
Galápagos	60**	very high	moderately effective (some concern)
Doñana	75	high	moderately effective (some concern)
Serengeti	23	high	uncertain long term (serious concern)
Vredefort	44	moderate	sustainable long term (effective)
Aldabra Atoll	5	high	sustainable long term (effective)
Keoladeo	14	high	uncertain long term (serious concern)
Kakadu	60	high	moderately effective (some concern)

** number corresponds to invasive alien species; note that in other sections of this chapter, we report 1579 total introduced taxa (terrestrial, marine) in the Galápagos (Toral-Granda et al. 2017)

Fig. 9-2 Threat level from invasive species at world heritage sites (*N*=241), including seven exemplar case studies: Galápagos Islands (Ecuador), Doñana National Park (Spain), Serengeti National Park (Tanzania), Vredefort Dome (South Africa), Aldabra Atoll (Seychelles), Keoladeo National Park (India), and Kakadu National Park (Australia). Layout: GPC; redrawn and modified after Shackleton et al. (2020).

The Galápagos Archipelago has been transformed since its discovery to the Anthropocene [=arguably, a geological time in which humans' actions have impacted, uniquely and distinctly, Earth's environments]. As summarized in Fig. 9.3a, iconic historical events include: (1) 1535: Tomás de Berlanga accidental arrival in what is today known as the Galápagos Archipelago, its pristine environments, with no prior human settlements. (2) 1835: Robert FitzRoy (Captain HMS Beagle) and Charles Darwin explored the Galápagos (globalism ongoing); local population about 1,000 settlers; Earth's atmospheric CO_2 concentration <300 ppm. And (3) 2000s: the Galápagos become world-connected; 25,000 inhabitants and about 250,000 tourists per year; Earth's atmospheric CO_2 concentration >400 ppm; extreme weather, i.e., El Niño and La Niña more frequent (after Salinas-de-León et al. 2020). [We discussed the discovery of the archipelago in Chapter Two, *How Were the yªs de galapagos Discovered?*; the Beagle's navigation around Galápagos, as well as FitzRoy's and Darwin's whereabouts, in Chapter Three, *Why Was Charles Darwin's Visit to Galápagos Significant?*; the colonization of the islands in Chapter Six, *Who Lives in Today's Galápagos Islands?*; and modern tourism in Chapter Seven, *Why Is Nature Tourism Essential to Galápagos?*].

Table 9.1 Most threatening invasive alien species to exemplar World Heritage sites (WHS) (assessment by Shackleton et al. 2020)

WHS	most threatening invasive alien species
Galápagos	*Cedrela odorata* (tree), *Felis catus* (cat), *Rattus norvegicus* (brown rat), *Rattus rattus* (black rat), *Philornis downsi* (fly), *Polistes versicolor* (wasp), *Psidium guajava* (shrub/tree), *Solenopsis geminata* (ant), *Rubus niveus* (shrub), *Wasmannia auropunctata* (ant); $N = 10$*
Doñana	*Azolla filiculoides, Eucalyptus camaldulensis, Carpobrotus edulis, Cyprinus carpio, Eriocheir sinensis, Linepithema humile, Micropterus salmoides, Procambarus clarkii*; $N = 8$
Serengeti	*Opuntia stricta, Lantana camara, Parthenium hysterophorus, Pistia stratiotes, Opuntia monacantha, Chromolaena odorata*; $N = 6$
Vredefort	*Arundo donax, Cestrum laevigatum, Cyprinus carpio, Eichhornia crassipes, Eucalyptus spp., Gleditsia triacanthos, Micropterus salmoides, Myriophyllum aquaticum, Opuntia spp., Tamarix ramosissima*; $N = 10$
Aldabra Atoll	*Rattus rattus, Felis catus, Casuarina equisetifolia, Icerya seychellarum, Stachytarpheta jamaicensis*; $N = 5$
Keoladeo	*Bos taurus, Clarisa gariepinus, Eichhornia crassipes, Lantana camara, Parapoynx diminutalis, Paspalum distichum, Prosopis juliflora*; $N = 7$
Kakadu	*Andropogon gayanus, Salvinia molesta, Mimosa pigra, Hymenachne amplexicaulis, Pennisetum polystachion, Themeda quadrivalvis, Jatropha gossypiifolia*; $N = 7$

* number corresponds to most threatening invasive alien species; note that in other sections of this chapter, we report 1579 total introduced taxa in the Galápagos (Toral-Granda et al. 2017)

Look at Fig. 9.3b. The plot shows both the cumulative number of residents and tourists in the Galápagos from 1830s to 2000s (after Schiller et al. 2015). Spot the steep increase in cumulative number of residents and tourists, i.e., where both trend lines overlap, from 1980s to 2000s. [See also Fig. 6.17 in Chapter Six and Fig. 7.3 in Chapter Seven]. Now, explore Fig. 9.3c, which highlights the steep increase of invasive alien species 1950s onwards; "invasives" augmented acutely during the 1980s to 2000s (coinciding with both accelerated human population growth and tourism boom). Finally, note in Fig. 9.3d the abrupt decline of landings of reef finfishes captured in Galápagos from 1950s to 1990s. [For the latter, see also Figs. 9.5-9.7 and 9.11-9.13, this chapter]. The overall message from Fig. 9.3 is clear-cut: since its discovery in 1535 to the 2000s, Galápagos has been exposed to incremental human presence (a sequel of invasive global human expansion), particularly from 1980s to 2000s. Invasive species have been introduced by humans—accidentally or deliberately—mainly from 1950s onwards. Landings of reef finfishes have abruptly declined, at least from 1950s to 1990s (but see "spillover effect" below, as in Figs. 9.11-9.13), suggesting overfishing has depleted finfishes. And, during the recent past (20th century), Earth's atmospheric CO_2 concentration has reached >400 ppm; extreme weather events (El Niño and La Niña) have become more frequent (Schiller et al. 2015; for comprehensive analysis of this topic, see Paz-y-Miño-C and Espinosa 2024b).

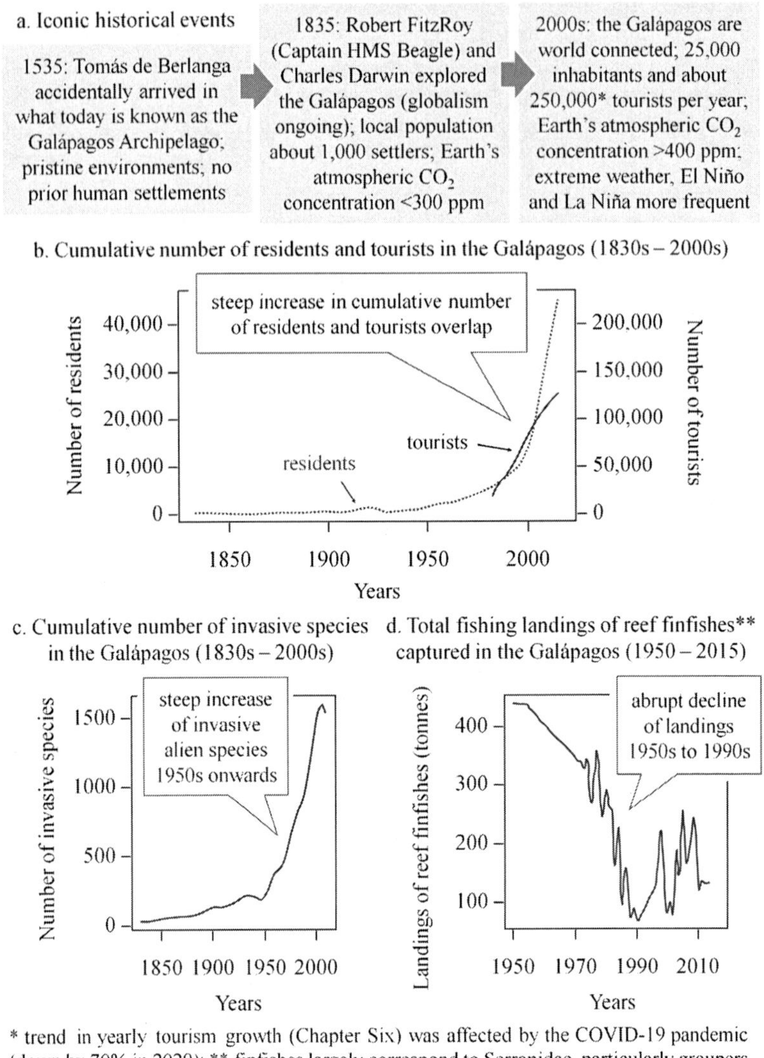

Fig. 9-3 Transformations in the Galápagos since its discovery to the Anthropocene [a geological time in which humans' actions have impacted, uniquely and distinctly, Earth's environments]. Layout: GPC; redrawn/modified after Salinas-de-León et al. (2020). **Data in b (1830s-2000s) and d (=landings of reef finfishes) with input from Schiller et al. (2015).

3 The Global Decline of Shark Populations

The waters around Darwin (*Culpepper*) and Wolf (*Wenman*), the Northwestern islands in the Galápagos Marine Reserve, have the largest shark biomass in the world. Scientists have sampled sharks' and other fishes' abundance at seven locations proximate to these islands (Fig. 9.4a; after Salinas-de-León et al. 2016). In Fig. 9.4b, we highlight the contrasting fish biomass between Darwin and Wolf (Galápagos) versus 9 comparable "no-take" (=no intense fishing) marine reserves: Cocos Island (off Costa Rica), Chagos Archipelago (Southern Indian Ocean), Kingman Reef (North Pacific), Cabo Pulmo National Park (Gulf of California), Farquhar Atoll (Indian Ocean, North of Madagascar), Palmyra Atoll (North Pacific), Northwestern Hawaiian Islands (Pacific Ocean, Hawai'i), Pitcairn Islands (South Pacific), and Salas & Gomez Islands (South Pacific, off Chile). Of the total fish biomass estimated around Darwin and Wolf (17.5 tons/hectare), 75% (13.1 t/ha) corresponded to apex predators: 48% scalloped hammerhead sharks (*Sphyrna lewini*), 20% Galápagos sharks (*Carcharhinus galapagensis*), 5% blacktip sharks (*Carcharhinus limbatus*), and 2% jacks and groupers. The remaining 25% (4.4 t/ha) of the total biomass were "other fishes". Note in Fig. 9.4b how the other exemplar marine reserves had less than half the fish biomass at Darwin and Wolf (Salinas-de-León et al. 2016). Why is this relevant? Because global shark populations are declining. Sharks are indicators of the health and wellbeing of reef communities. Thus, sharks and their habitats (e.g. Galápagos) are central to marine environments' conservation.

According to the Living Planet Index or LPI (=quantitative approach to inferring the global population status of a species based on regional, annual assessments of its abundance), the global presence of 18 shark and ray species has declined by 71% since 1970 (Fig. 9.5; Pacoureau et al. 2021). Twelve of such 18 monitored species occur in the Galápagos Marine Reserve. As shown in Fig. 9.5a, populations of both sharks and rays in the Pacific Ocean have declined from 1970 onwards. Their habitats have also declined in tropical and temperate regions (Fig. 9.5b). Sharks and rays of all sizes (particularly large and medium ones) have become less abundant (Fig. 9.5c). And sharks and rays of all generation cohorts have become less abundant (Fig. 9.5d; data from Pacoureau et al. 2021).

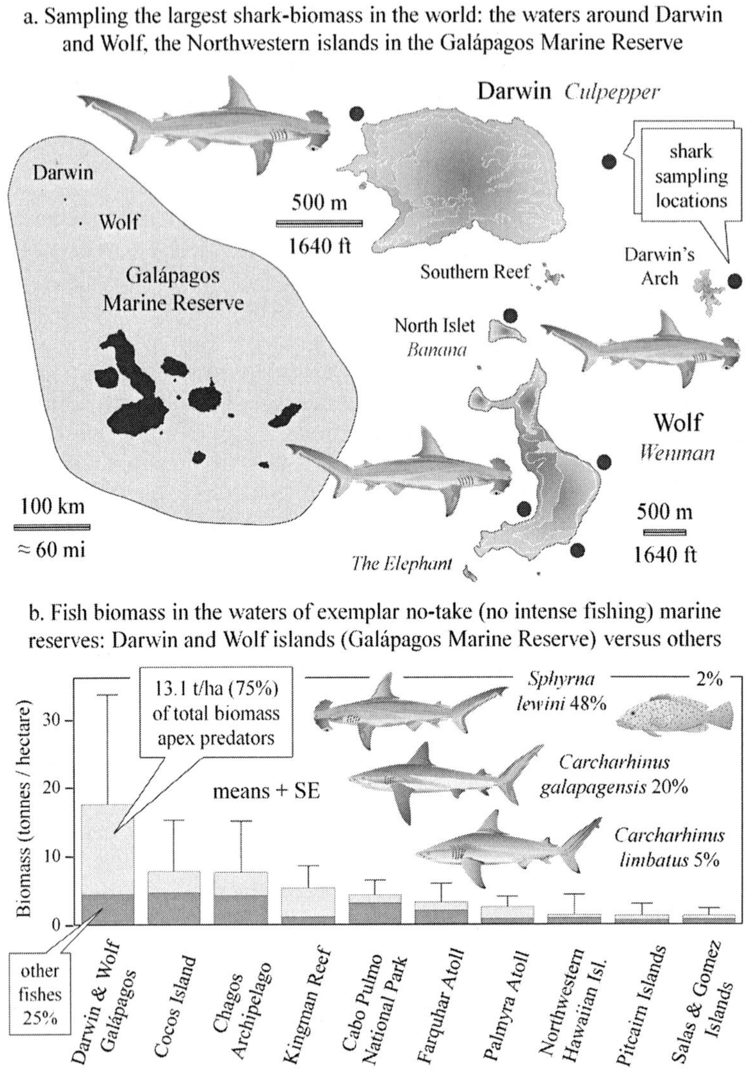

Fig. 9-4 a. Shark-biomass sampling locations around Darwin (*Culpepper*) and Wolf (*Wenman*) Islands, Galápagos (sampling relied on stereo-video surveying during 2013-2014). b. Fish biomass of apex predators and other fishes around Darwin and Wolf Islands in comparison to exemplar "no-intense-fishing" marine reserves. Of the total fish biomass estimated around Darwin and Wolf (17.5 tons/hectare), 75% (13.1 t/ha) corresponded to apex predators. Layout: GPC; redrawn and modified after Salinas-de-León et al. (2016).

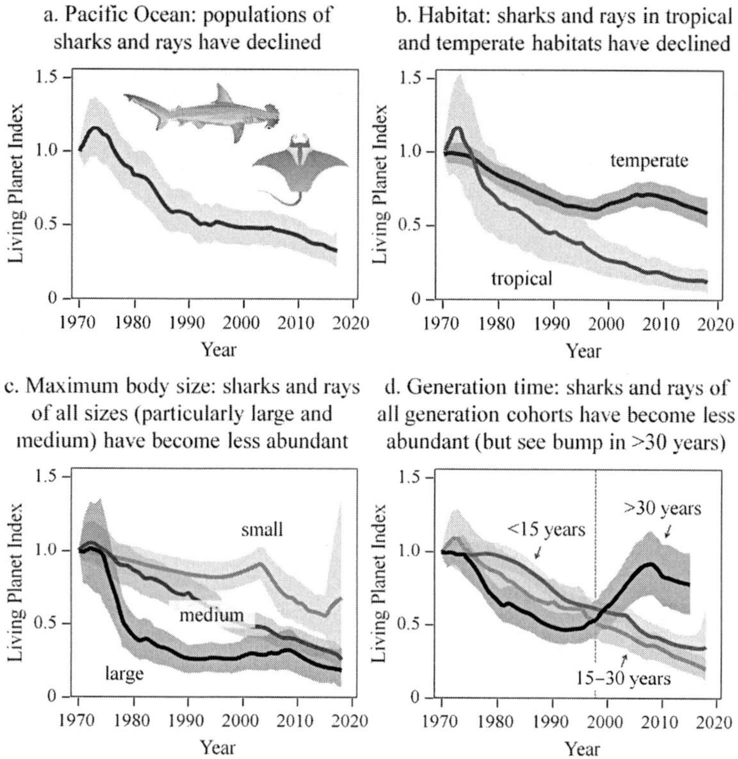

* Living Planet Index (LPI) is a quantitative approach to inferring the global population status of a species based on regional, annual assessments of its abundance (Pacoureau et al. 2021); 12 of the 18 species in the assessment occur in the Galápagos Marine Reserve

Fig. 9-5 Global changes in population abundance of 18 oceanic species of sharks and rays: variations as per Living Planet Index* (data range: 1970-onwards). Layout: GPC; redrawn and modified after Pacoureau et al. (2021).

In Fig. 9.6, we depict global fluctuations in 18 oceanic species of sharks and rays (1970-onwards). Note how almost all populations have declined globally (as per Living Planet Index, LPI, values <1.0). The 12 species occurring in the Galápagos are sketched to the left of the plots; they include 10 sharks and 2 rays: oceanic whitetip shark (*Carcharhinus longimanus*), silk shark (*Carcharhinus falciformis*), blue shark (*Prionace glauca*), great hammerhead (*Sphyrna mokarran*), scalloped hammerhead (*Sphyrna lewini*), smooth hammerhead (*Sphyrna zygaena*), bigeye thresher (*Alopias superciliosus*), pelagic thresher (*Alopias pelagicus*), great white shark (*Carcharodon carcharias*), shortfin mako (*Isurus oxyrinchus*), pelagic stingray (*Pteroplatytrygon violacea*), and giant manta (*Manta birostris*). Their most recent LPI values range from 0.14 (decline) in giant manta to 1.29 (increase) in smooth hammerhead (Fig. 9.6).

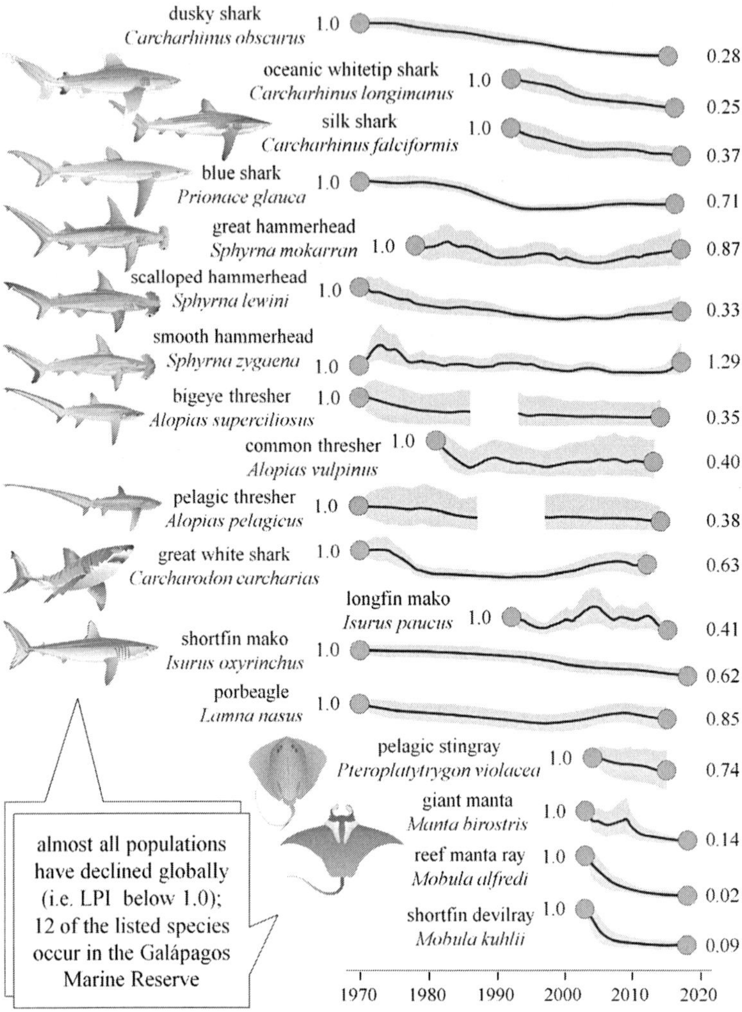

Fig. 9-6 Living-Planet-Index (LPI) fluctuations in 18 oceanic species of sharks and rays: global data (1970-onwards); species occurring in the Galápagos are sketched. Layout: GPC; redrawn and modified after Pacoureau et al. (2021). Note that we discussed Galápagos' sharks and rays in Chapter Eight, see Figs. 8.64-8.67; see also *Thirty-Five Species of Sharks and Counting*, Figs. 8.68-8.77 and Table 8.9.

What caused this global decline in sharks and rays? An 18-fold increase in relative fishing pressure, as underscored in Fig. 9.7 (after Pacoureau et al. 2021). Note in Fig. 9.7a how fishing with longlines and seine nets (which catch most of the oceanic sharks) has increased from 1970

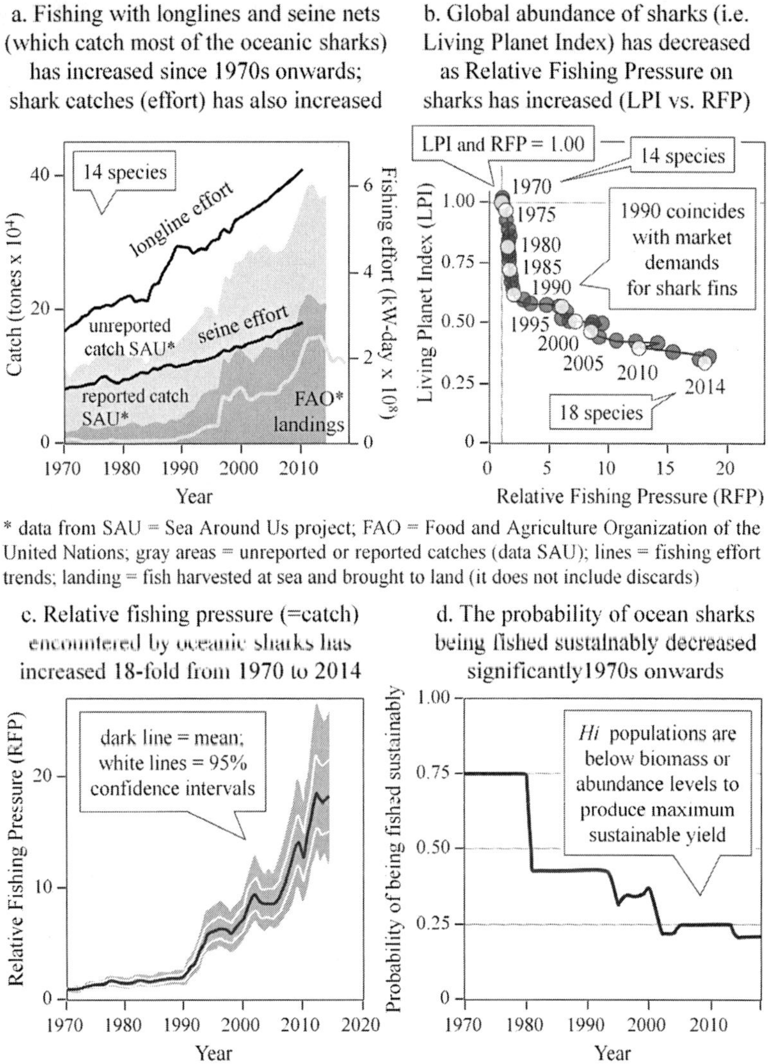

Fig. 9-7 Overfishing, the main threat to oceanic sharks: global assessment. Layout: GPC; redrawn and modified after Pacoureau et al. (2021).

onwards; shark catches (effort) have also increased. The global abundance of sharks (as per Living Planet Index) has decreased as Relative Fishing Pressure (RFP) on sharks has increased (LPI vs. RFP in Fig. 9.7b). As shown in Fig. 9.7b, the relationship between LPI and RFP was assumed to be (for methodological reasons) 1:1 in 1970 (LPI and RFP=1.00; 14 species monitored). By 1990, coinciding with the increased market demand for shark fins, LPI dropped to ≈0.60 and continued to decline to ≈0.30 by 2014 as RFP increased (18 species monitored). Likewise, the relative fishing pressure (RFP=catch) encountered by oceanic sharks increased 18-fold from 1970 to 2014 (Fig. 9.7c). And the probability of ocean sharks being fished sustainably decreased significantly from 1970 onwards, suggesting that current populations are below biomass or abundance levels to produce maximum sustainable yield (Fig. 9.7d; Pacoureau et al. 2021).

4 The *Fu Yuan Yu Leng 999* and the Shark-Fin Market

In August 2017, Ecuadorian authorities detained the Chinese Reefer *Fu Yuan Yu Leng 999*. The ship, a refrigerated cargo-transporter of perishable goods, was navigating without authorization in the Galápagos Marine Reserve, just 19 km (12 mi) off Punta Pitt, Northeastern San Cristóbal (*Chatham*) Island (for map of island, see Fig. 1.27 in Chapter One). The vessel contained 572 tons of fish and fish products, yet only +1 of its 6 freezers was full. In other words, the *Fu Yuan Yu Leng 999* had plenty of space for additional cargo. Both the fish catch and the ship were confiscated by authorities (Bonaccorso et al. 2021).

As detailed in Fig. 9.8, the reefer contained 7,639 sharks (juveniles and adults=7,207; unborn=432); 2,114 bony fishes; and 537 sacks with shark fins (cuts). From a sample of 929 juvenile/adult sharks (12% of the total cargo), researchers identified morphologically and genetically-barcoded 12 species (Bonaccorso et al. 2021): 27.7% silky sharks (*Carcharhinus falciformis*); 20.2% oceanic whitetip sharks (*Carcharhinus longimanus*); 13.1% pelagic threshers (*Alopias pelagicus*); 13.1% scalloped hammerheads (*Sphyrna lewini*); 11.7% blue sharks (*Prionace glauca*); 7.2% smooth hammerheads (*Sphyrna zygaena*); 5.8% bigeye threshers (*Alopias superciliosus*); and 1% other species, including shortfin mako sharks (*Isurus oxyrinchus*); grey reef sharks (*Carcharhinus amblyrhynchos*; rare in Galápagos); tiger sharks (*Galeocerdo cuvier*); a whale shark (*Rhincodon typus*); and a common smooth-hound (*Mustelus mustelus*; apparently, not reported as a Galápagos' species).

Species		No.**	%
silky shark *Carcharhinus falciformis*		257	27.7
oceanic whitetip shark *Carcharhinus longimanus*		188	20.2
pelagic thresher *Alopias pelagicus*		122	13.1
scalloped hammerhead *Sphyrna lewini*		122	13.1
blue shark *Prionace glauca*		109	11.7
smooth hammerhead *Sphyrna zygaena*		67	7.2
bigeye thresher *Alopias superciliosus*		54	5.8
shortfin mako shark *Isurus oxyrinchus*		4	0.4
grey reef shark *Carcharhinus amblyrhynchos*	rare in Galápagos	2	0.2
tiger shark *Galeocerdo cuvier*		2	0.2
whale shark *Rhincodon typus*		1	0.1
common smooth-hound *Mustelus mustelus*	not in Galápagos	1	0.1

* the *Fu Yuan Yu Leng 999* contained in storage 572 tons of frozen fish: 7639 sharks (juveniles/adults = 7207; unborn = 432), 2114 bony fishes, and 537 sacks with shark fins (cuts); ** from a sample of 929 juvenile/adult sharks (12% of the total cargo), researchers identified morphologically and genetically-barcoded 12 species (Bonaccorso et al. 2021)

Fig. 9-8 Shark species identified in the cargo of the Chinese reefer vessel *Fu Yuan Yu Leng 999** after being detained by Ecuadorian authorities in August 2017; the ship was navigating without authorization in the Galápagos Marine Reserve. Layout: GPC; data comes from Bonaccorso et al. (2021).

Navigation records indicated that, prior to being seized, the *Fu Yuan Yu Leng 999* was active in the West and Northwest of the Galápagos Marine Reserve (latitude 2-7°N; longitude 120-110°W). Later, the vessel entered GMR and was intercepted proximate to San Cristóbal Island (Bonaccorso et al. 2021). Authorities suspected the ship [not a fishing vessel] procured its cargo from other crafts in the surroundings of GMR [the species listed in Fig. 9.8 occurred in the region].

Researchers alerted that the *Fu Yuan Yu Leng 999* "exemplif[ied] how industrial fisheries [were] likely depleting shark populations in the Eastern Tropical Pacific Ocean" (Bonaccorso et al. 2021). In fact, sharks—and particularly their fins—were, still are, the target of fishing efforts (below).

In related research and to determine the potential source populations supplying the global trade of sharks (their fins), scientists obtained shark-fin samples from 75 randomly selected Hong Kong fish markets (2014-2015). The samples were genetically compared to an already-existing database [compiled by the scientists] of 9 probable source populations (shaded areas in Fig. 9.9): Western Pacific, Central Pacific, Eastern Pacific,

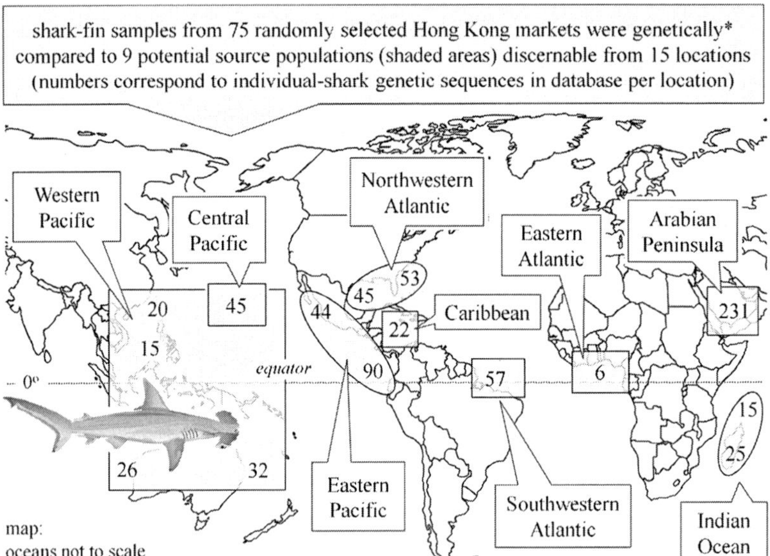

* genetic stock identification (GSI) was used to infer the probable origin of mix-population samples of scalloped-hammerhead-shark fins purchased at 75 Hong Kong fish markets (from 2014 to 2015); the technique relies on comparing/contrasting genetic markers (e.g. mitochondrial DNA sequences, control region mtCR) from individual samples of unknown origin to a known database of genetic sequences, which population affinity had previously been established; this figure shows such database plotted on a map

Fig. 9-9 Potential source populations supplying the global trade of the critically endangered scalloped hammerhead shark (*Sphyrna lewini*). Layout: GPC; redrawn and modified after Fields et al. (2020).

Northwestern Atlantic, Eastern Atlantic, Caribbean, Southwestern Atlantic, Arabian Peninsula, and Indian Ocean (Fields et al. 2020). Relying on genetic stock identification (GSI) techniques, biologists compared and contrasted genetic markers (mitochondrial DNA sequences) from the shark-fin individual samples of unknown origin (i.e. from the 75 Hong Kong markets) to the known database (i.e. the 9 probable shark-source populations depicted in Fig. 9.9). Their analysis focused on the cosmopolitan yet critically endangered scalloped hammerhead shark (*Sphyrna lewini*), the fourth most common species in the market samples (Fields et al. 2020). The results are summarized in Fig. 9.10. Note how 61.4% of the shark fins from the Hong Kong markets likely originated in the Eastern Pacific, where the Galápagos Marine Reserve is located! 15.5% of the fins were traceable to the Western Pacific; 8.7% to the Indian Ocean; 5.8% to the Eastern Atlantic; 5.0% to the Northwestern Atlantic; and 3.7% to the Caribbean.

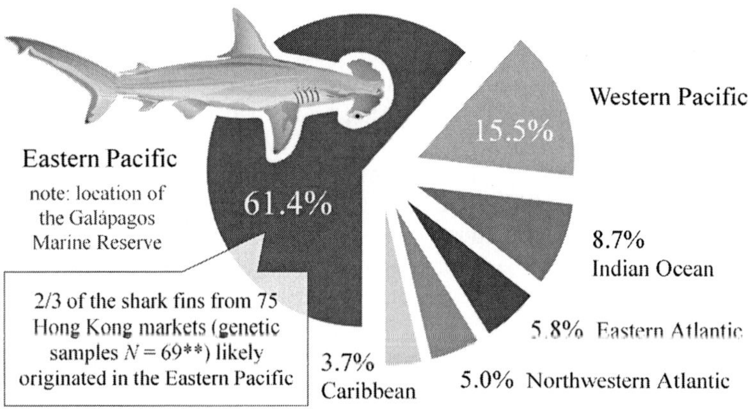

* the estimated contribution (%) of each Potential Source Population (PSP) supplying the global trade of scalloped-hammerhead-shark fins was generated using SPAM (Statistics Program for Analyzing Mixtures); thus, each value is an estimate (standard errors: Eastern Pacific SE = 7.08%; Western Pacific SE = 5.89%; Indian Ocean SE = 3.39%; Eastern Atlantic SE = 2.81%; Northwestern Atlantic SE = 2.87%; Caribbean SE = 2.53%); ** Fields et al. (2020) obtained 4800 shark-fin trimmings from 75 randomly selected Hong Kong markets (from 2014 to 2015) of which 196 (4.1%) were genetically confirmed to be *Sphyrna lewini* (i.e. scalloped hammerhead sharks); 69 of the trimmings had haplotypes identical to those in the authors' database and were inferred to belong to six PSPs (as shown in figure); *S. lewini* was the fourth most common species in the 4800 samples, after blue sharks (*Prionace glauca*), silky sharks (*Carcharhinus falciformis*), or members of the blacktip-shark species complex (*C. limbatus*, *C. amblyrhynchoides*, *C. leiodon*, *C. tilstoni*)

Fig. 9-10 Possible origin of scalloped-hammerhead-shark fins (*Sphyrna lewini*) purchased at 75 randomly selected Hong Kong fish-markets*. 2/3 of shark fins (61.4%) likely originated in the Tropical Eastern Pacific. Layout GPC; redrawn/modified after Fields et al. (2020).

Even if illegal fishing of sharks within GMR were rare or non-existent [the *Fu Yuan Yu Leng 999* case suggests otherwise], shark populations proximate to the Galápagos Archipelago are being hunted, as inferred from the shark-fin trade (Fig. 9.10). The habitat connectivity between GMR and the surrounding ocean is such that sharks spared within GMR can be captured just outside the reserve. Thus, the 2021 commitment for the protection of the Tropical Eastern Pacific by Ecuador, Costa Rica, Panama, and Colombia is essential (Paz-y-Miño-C and Espinosa 2024b).

5 Tuna Fisheries Around the Galápagos Marine Reserve

The fishing industry regularly monitors fisheries' activities worldwide, including around the Galápagos Marine Reserve (GMR is a non-commercial fishing zone) and within the Galápagos Exclusive Economic Zone (GEEZ). In Figs. 9.11-9.13, we summarize some relevant observations. Researchers gathered historical records on tuna catches, fishing effort, and catches per unit of fishing effort and plotted the data on a 1x1 degree latitude-longitude grid covering GMR and GEEZ. The grid expanded from 6ºN to 6ºS and from 85ºW to 97ºW (Fig. 9.11; Boerder et al. 2017). Examine in Fig. 9.11 the location of 49 fishing vessels within the grid—tracked from 2011 to 2015 (via automatic identification systems). Note their almost ubiquitous presence around GMR, their navigation within GEEZ (i.e. its West part), and activity outside GEEZ. Locate the fisheries' "hotspots" area (darker gray-color grid) adjacent to the West of GMR, where the vessels frequently aggregated during the 5-year monitoring period. These "hotspots" coincided with the locations of highest tuna catches, fishing effort, and catches per unit of fishing effort, as summarized in Fig. 9.12. [Readers should realize that these "hotspots" overlapped with the region of *seasonal upward surges of cold, rich-in-nutrients deep-sea waters* (to the West of the Galápagos Archipelago), i.e., *the 'upwelling' that fuels the biological cycles of microorganisms (primary production), which, in turn, sustain all other life forms in the islands* (Figs. 7.1-7.2, above). And, as previously discussed in this chapter, *ultimately, the upwelling nourishes sea and land organisms, from micro to macro life*. Thus, Fig. 9.11 suggests how tuna fisheries concentrate their catches, fishing effort, and catches per unit of fishing effort proximate to the nutrient-rich areas of the archipelago (=frequented by pelagic predators)].

Let us examine Fig. 9.12 in more detail. It depicts tuna fishing trends around GMR from 2011 to 2015 (spatial hotspot analysis). Notice how catches, fishing effort, and catches per unit of fishing effort were all concentrated to the West of GMR. Statistically, the hotspots received more fishing intensity than other locations (Boerder et al. 2017). The researchers identified both "hotspots" (with the most fishing intensity in respect to the mean value) and "coldspots" (with the least fishing intensity in respect to the mean value) of tuna fishing activity by the vessels. "Hot" to the West of GMR (=black- and dark-gray-color range in Fig. 9.12); "cold" to the Southeast (=light-gray-color range in Fig. 9.12).

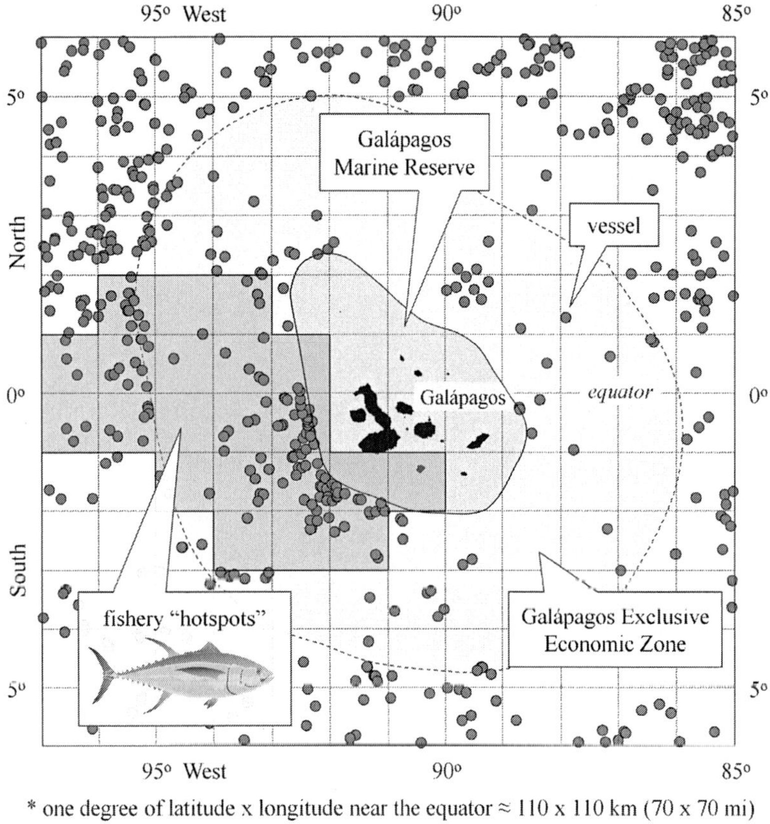

Fig. 9-11 Tuna fisheries around the Galápagos Marine Reserve (GMR=no fishing zone) and its exclusive economic zone: 49 vessels tracked from 2011 to 2015 (automatic identification systems). Layout: GPC; redrawn and modified after Boerder et al. (2017).

Bring your attention to Fig. 9.13. It shows fishing trends for tuna *within* the hotspots versus other areas around GMR, from 1990 to 2015 (data from Boerder et al. 2017). Upon the enforcement of GMR in 2002, the trend for tuna catches and fishing effort (hotspots versus other areas outside GMR) significantly and progressively diverged by 2015 (i.e. more catches and fishing effort at hotspots than at other areas). Notably, from 2002

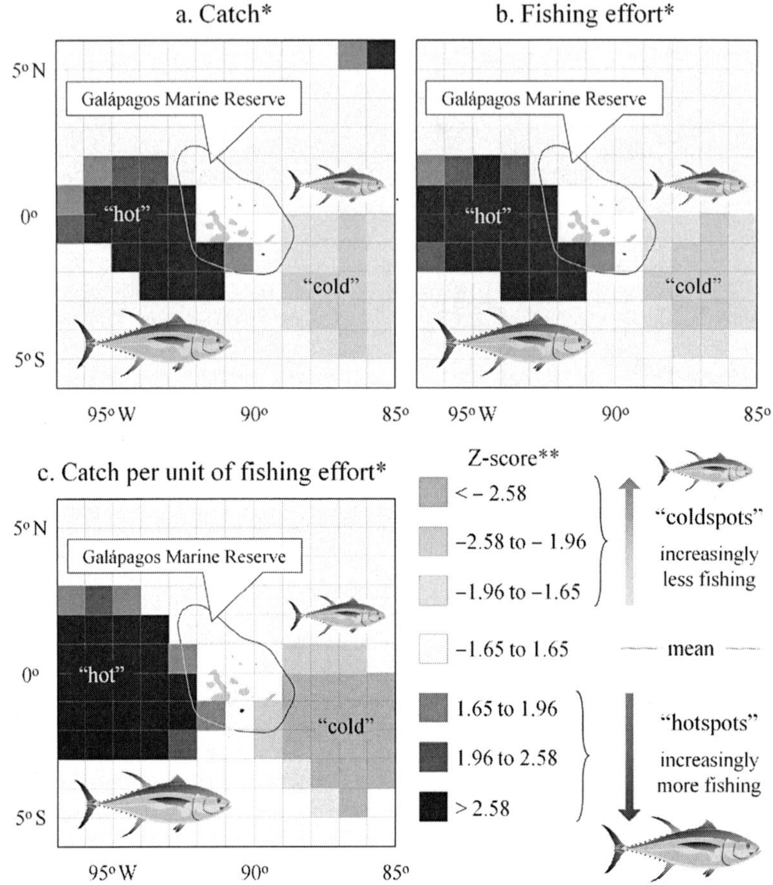

* catch, fishing effort, and catches per unit of fishing effort were all concentrated to the West ("hotspots") of the Galápagos Marine Reserve; the hotspots received more fishing intensity than other locations (statistically as per z-scores >1.65); ** z-scores are statistical indicators of how far a data point is in respect to the mean (i.e. how many standard deviations, sd, above or below the mean; thus, the hotspots receive >2.58 sd of fishing intensity in respect to other locations; the coldspots were the least fished < –2.58 sd)

Fig. 9-12 Tuna fishing trends around the Galápagos Marine Reserve from 2011 to 2015: spatial hotspot analysis (data source: Inter-American Tropical Tuna Commission). Layout: GPC; redrawn and modified after Boerder et al. (2017).

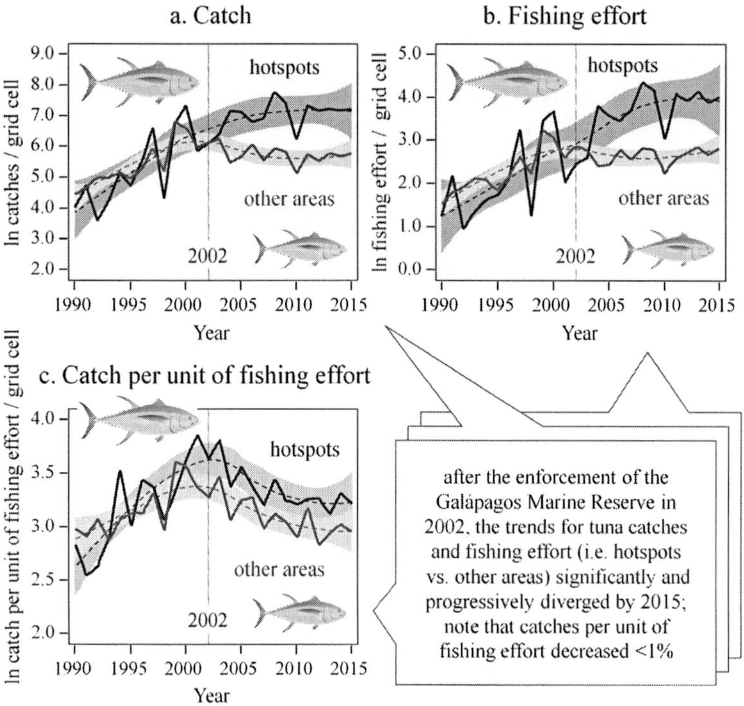

* natural logarithm (ln) of temporal (1990 to 2015) tuna catches, fishing effort, and catch per unit of fishing effort (as per 1° x 1° latitude-longitude grid cell, see Figs. 9.11 and 9.12); shaded areas and dashed lines (mean) correspond to polynomial regression models for hotspots (darker gray) or other areas (lighter gray) with 95% confidence intervals

Fig. 9-13 Tuna fishing trends* within the hotspots versus other areas around the Galápagos Marine Reserve (1990-2015). Data source: Inter-American Tropical Tuna Commission. Layout: GPC; redrawn and modified after Boerder et al. (2017).

to 2015, catches per unit of fishing effort decreased at both hotspots and other areas (but see below). Why? Researchers hypothesized that the creation of the reserve around the archipelago might have enhanced favorable conditions for tuna spawning and breeding within the protected area (e.g. for yellow-fin tuna *Thunnus albacares*, bigeye tuna *T. obesus*, and skipjack tuna *Katsuwonus pelamis*), as well as benefited the fisheries. Thus, GMR was acting as a "well-protected and well-placed reserve" (Boerder et al. 2017)! After the reinforcement of GMR in 2002, but not before, tuna catches increased 52% and fishing effort 75%, while catches per unit of fishing effort decreased only <1% (Fig. 9.13). It is conceivable, however, that tuna fisheries might be benefiting temporarily from a "spill-over effect", i.e., the export of fish biomass from the protected area (GMR) to the adjacent fishing grounds (the current fishery "hotspots" in Figs. 9.11-9.13; for review on ecological or fishery spillovers, see Di Lorenzo et al. 2020). This issue requires long-term documentation.

6 The Decline of Sea Lions and Fur Seals

In our book, *A Synthesis of the Galápagos*, we introduce readers to sea-lion (*Zalophus wollebaeki*) population's status in the Galápagos Archipelago (i.e. sea lions as the most conspicuous, large vertebrates on the islands' shores). Here, we expand on the topic and also include fur seals (*Arctocephalus galapagoensis*). Note that the counts and estimates in this section range from 2014 to 2018. *Zalophus* and *Arctocephalus* are carnivore marine mammals (Pinnipeds) in the family Otariidae (see Figs. 8.19-8.21 in Chapter Eight of this volume).

In Fig. 9.14, we plot the across-islands location of breeding rookeries of sea lions (triangles) and fur seals (squares) based on individual counts from 2014 to 2018 (after Páez-Rosas et al. 2021). Sea lion rookeries were common in the Central and Southeast bioregions of the archipelago, while fur seals' were frequent in the West and North, suggesting habitat partitioning between the species. [The North, Central, West, and Southeast bioregions were consistent with those proposed and characterized by Edgar et al. (2004), based on differential species composition (fish and macro-invertebrates) in the subtidal communities on the shallow reefs of the Galápagos Marine Reserve (see Box 1.3 in Chapter One)]. Smaller-size sea-lion rookeries ranged from 22 to 200 individuals; larger-size rookeries reached 500 to 658 individuals. Fur seals' smaller-size rookeries ranged from 11 to 400 individuals; larger-size rookeries reached 1,200 to 1,577 individuals (Fig. 9.14, top).

Examine Fig. 9.15a. It summarizes actual counts of both the total population (solid trend line) and pup abundance (dashed trend line) of Galápagos sea lions (left plot) and fur seals (right) during 2014-2018 (after Páez-Rosas et al. 2021). Sea lions' data corresponds to annual counts at 32 breeding rookeries on 13 islands, while fur seals' at 11 rookeries on 7 islands (as per map in Fig. 9.14). Note the slight trend of increase from 2014 to 2018 in *total population* or *pups* for both sea lions and fur seals. However, the drop in numbers is evident in 2015, a severe [strong] El Niño year, particularly for sea lions' pups, for which the impact continued during 2016, with a slow, gradual recovery by 2018. By contrast, the fur seals' pups recovered vigorously and much faster.

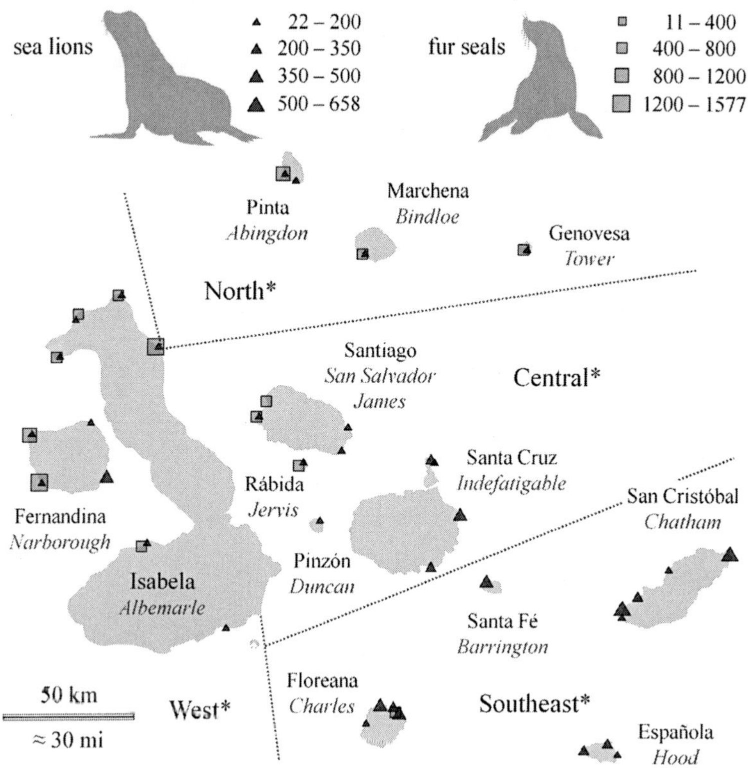

* North, Central, West and Southeast correspond to bioregions consistent with those proposed and characterized by Edgar et al. (2004) based on differential species composition (fish and macro-invertebrates) in the subtidal communities on the shallow reefs of the Galápagos Marine Reserve (see Box 1.3 in Chapter One); note how the sea lion rookeries are common in the Central and Southeast bioregions, while the fur seals' are frequent in the West and North, suggesting habitat partitioning between the species

Fig. 9-14 Breeding rookeries of Galápagos sea lions (*Zalophus wollebaeki*) and fur seals (*Arctocephalus galapagoensis*): individual counts 2014-2018. Layout: GPC; redrawn and modified after Páez-Rosas et al. (2021).

Bring your attention to Fig. 9.15b, in which we plot pup abundance (sea lions left plot; fur seals right) in relation to El Niño intensity (weak, moderate, or strong) and the location of breeding rookeries as per bioregions in the archipelago (map in Fig. 9.14): North (N), Central (C), West

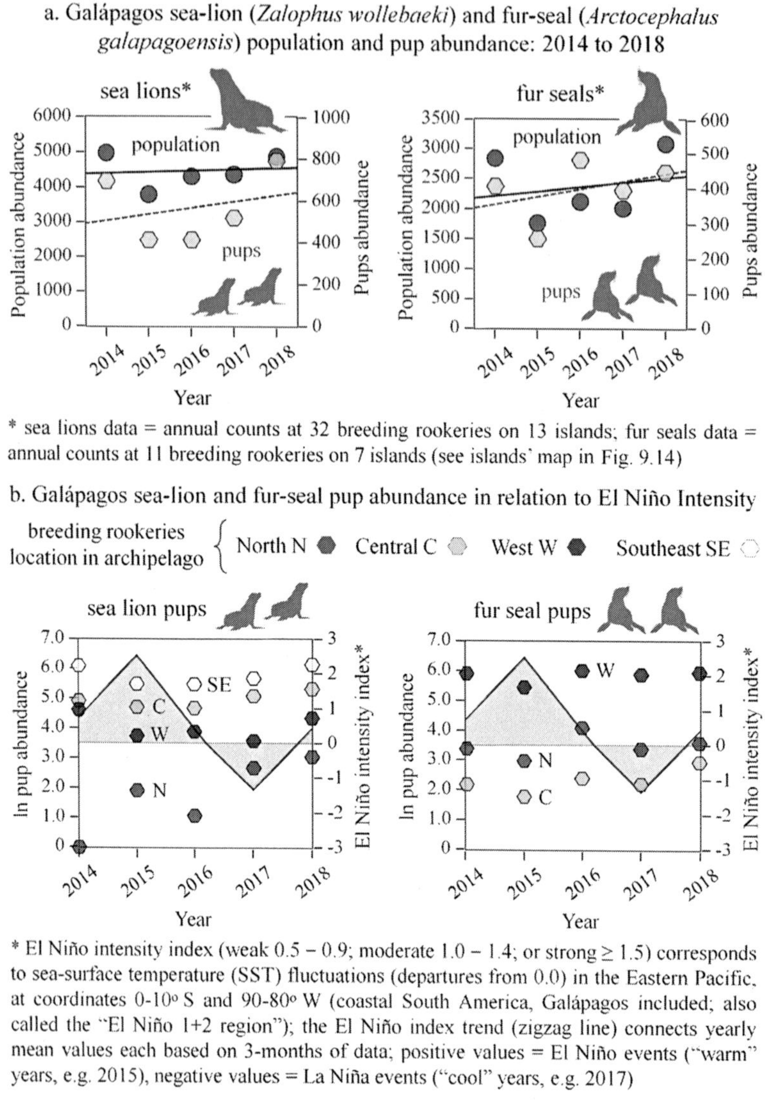

Fig. 9-15 Population and pup abundance of Galápagos sea lions (*Zalophus wollebaeki*) and fur seals (*Arctocephalus galapagoensis*) during 2014-2018 (a); and in relation to El Niño intensity (b). Layout: GPC; redrawn and modified after Páez-Rosas et al. (2021).

(W), or Southeast (SE). Spot the zigzag line on each plot; it corresponds to strong El Niño intensity in 2015 (a warm year) or weak intensity in 2017 (a cool year=La Niña). Note how in 2015, sea-lion pup abundance was particularly affected in the North (N) bioregion, followed by West (W), Central (C), and Southeast (SE) [+60% of the sea lion population is located in the Southeast islands of the Galápagos]. In the case of fur seals, pups from the Central (C) and North (N) bioregions were the most affected by the 2015 El Niño, while pups from the West (W) bioregion were the least affected [76% of the fur seal population is located in Fernandina's/Isabela's shores] (Fig. 9.15b; Páez-Rosas et al. 2021).

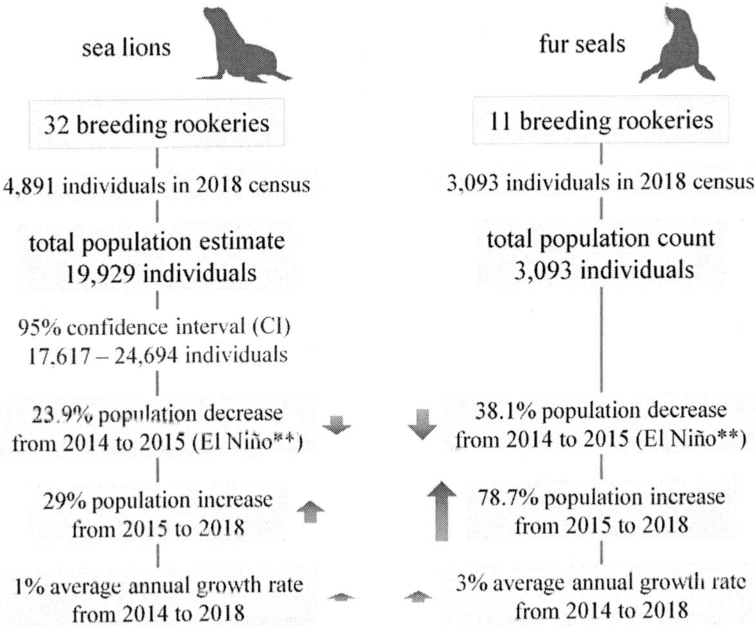

* data comes from annual census of populations at breeding rookeries (i.e. each October, the peak of pup births); ** El Niño corresponds to 2015 with an intensity index >2.0 (strong)

Fig. 9-16 Galápagos sea-lion (*Zalophus wollebaeki*) and fur-seal (*Arctocephalus galapagoensis*) summary demographics*: 2014 to 2018. Layout GPC; data extracted from Páez-Rosas et al. (2021).

In Fig. 9.16, we summarize the demographic trends 2014-2018 of both sea lions and fur seals. Realize that by 2018, the actual count of sea lions (at 32 breeding rookeries) was 4,891 individuals. From this count, researchers estimated a total population of 19,929 individuals for the entire archipelago.

[At 95% confidence interval (CI), such estimated total population ranged from 17,617 to 24,694 individuals]. By contrast, the fur seal total count (at 11 breeding rookeries) was 3,093 individuals for the entire archipelago (Páez-Rosas et al. 2021). El Niño impact of 2015 caused a 23.9% population decrease (in respect to the previous year) among sea lions versus 38.1% among fur seals. After that, and from 2015 to 2018, the sea lion population increased by 29%, while the fur seal population increased by 78.7%. The relevant fact about both sea lions and fur seals is that, from 2014 to 2018, their populations increased at an average annual growth rate of 1% and 3%, respectively. Despite this growth, scientists alert that because severe El Niño events can disrupt sea-lion and fur-seal populations—as climate change continues to intensify the strength and frequency of El Niño years in Galápagos—both species could become critically endangered or even extinct (Páez-Rosas et al. 2021). [*Zalophus wollebaeki* and *Arctocephalus galapagoensis* are currently listed as endangered by IUCN]. Indeed, extreme weather fluctuations and higher air/sea surface temperatures will disrupt nutrient cycles and fish-food availability for marine-mammal predators like sea lions and fur seals, and sea level rise will cover breeding areas—for colonies—ashore (Friess et al. 2019; see also Riofrío-Lazo et al. 2017; discussion in Paz-y-Miño-C and Espinosa 2024b). Thus, protection and management of crucial areas within the Galápagos Marine Reserve (GMR) will benefit Pinnipeds. As shown in Fig. 9.17, researchers have determined that in areas in which the spatial distributions of sea lions and fur seals overlap substantially (i.e. Western part of the archipelago), sea lions often forage proximate to the continental shelf, while fur seals do it offshore, in deep waters, and with "more heterogeneous space usage over time" (i.e. cool versus warm seasons or versus El Niño years; Ventura et al. 2019). Based on these observations, as well as characterizations of habitat use and preference by *Zalophus wollebaeki* and *Arctocephalus galapagoensis*, three priority areas for conservation were identified: (1) deep waters Northwest and shallow waters West of Fernandina Island; (2) deep waters Southwest of Isabela Island; and (3) shallow waters over the continental shelf to the East of Fernandina Island (Fig. 9.17). Although these areas were located within GMR (i.e. the 133,000-km^2/51,352-sq-mi of protected ocean in number 1, Fig. 7.3), only 8% of them overlapped with the recently (as per 2016) proposed "conservation zones" (no commercial fishing, no tourism) in the archipelago. Thus, neither GMR nor its proposed "conservation zonation" might be optimal for Galápagos' Pinniped long-term preservation, at least in its West bioregion (Ventura et al. 2019). This topic needs further research to identify, protect, and manage areas of high conservation priority [and value] for Pinnipeds in the entire archipelago.

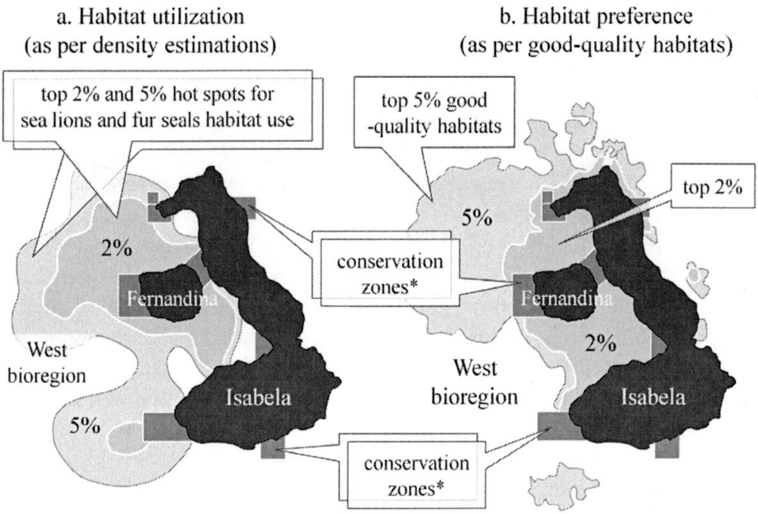

a. Density estimations of sea lions and fur seals based on GPS-tracked tagged individuals. b. Habitat preference determined by modeling environmental variables (physiographic, biological, oceanographic). *Conservation zones corresponded to proposed (as per 2016 management plans by the Galápagos National Park Directorate) "no commercial fishing", "no tourism" areas, although artisanal fishing and navigation of tourism vessels, local/cargo boats was allowed.

Fig. 9-17 Areas of high conservation priority as per (a) habitat utilization or (b) habitat preference by Galápagos sea lions (*Zalophus wollebaeki*) and fur seals (*Arctocephalus galapagoensis*) on the Western bioregion of the archipelago (for location of bioregions, see Fig. 9.14). Layout: GPC; redrawn and modified after Ventura et al. (2019).

7 Wildlife Trade: Tortoises and Iguanas for Sale

Wildlife trade is a multibillion-dollar business, one of 10 most lucrative criminal activities worldwide (Sun-Wyler and Sheikh 2013). From a dataset of 31,500 terrestrial vertebrate species globally distributed, investigators identified 5,579 (18%) as "traded in the global markets". The majority were birds *N*=2,345 (23% of the 10,190 bird species in the dataset), followed by mammals *N*=1,441 (27% of 5,330 mammals in the dataset), reptiles *N*=1,184 (27% of 9,540 reptiles in the dataset), and amphibians *N*=609 (27% of 6,470 amphibians in the dataset; Fig. 9.18 top; after Scheffers et al. 2019). Note in the histograms of Fig. 9.18-bottom-left that >25% of all vertebrates traded (*N*=5,579 species) were listed by IUCN (International Union for Conservation of Nature) as either vulnerable (VU), near threatened (NT), endangered (EN), or critically endangered

(CR). Likewise, as shown in the semi-circular plot in Fig. 9.18-bottom-right, >25% of all the bird species traded (*N*=2,345) were listed as CR, EN, VU, or NT. In the case of mammals, >50% of all the species traded (*N*=1,441) were listed as CR, EN, or VU, while >25% were listed as NT. Among reptiles, 25% of all the species traded (*N*=1,184) were listed as CR, EN, VU, or NT. And among amphibians, >10% of all the species traded (*N*=609) were listed as VU, NT, or least concern (LC).

Researchers established that supply hotspots in the global wildlife trade were South and Central America, Southeast Africa, the Himalayas, Southeast Asia, and Australia. The networks of illegal trade interconnected hotspots with local, regional, and global markets, particularly in Asia (Scheffers et al. 2019; Hughes 2021; Brookes et al. 2022).

Despite its internationally recognized conservation tradition (see Chapter Eight), Galápagos is not immune to the global wildlife trade. Examine Fig. 9.19 as follows: (1) Ecuadorian authorities have uncovered that baby giant tortoises [and presumably other reptiles, below] are illegally—and recurrently—extracted from the wild or captive breeding centers. (2-3) The tortoises are then smuggled by boat or plane cargo and transported to Guayaquil (mainland Ecuador) and later to trade intermediaries in Lima (Peru). (4) Wildlife markets in Asia seem to be the main destination of the baby tortoises and possibly other reptiles (land and marine iguanas). (5) In 2017, 29 baby tortoises were confiscated while being transported (via bus) to Piura, Northern Peru. Newspaper El Universo (Ecuador) reported that smugglers could make $300 USD per tortoise; the estimated cost of global trade could reach $50,000 USD per tortoise. Other reptiles in demand, as reported by El Universo (2019), included land iguanas ($200 USD each) and marine iguanas ($100 USD each). (6) In 2018, Ecuador's Ministry of the Environment confirmed that 123 baby tortoises (*Chelonoidis vicina* and *C. guntheri*) were stolen from the Captive Breeding Center "Arnaldo Tupiza" in Isabela (*Albemarle*) Island. (7) By 2021, 185 additional baby tortoises were found concealed in ordinary baggage during a routine x-ray inspection at Baltra Airport (Galápagos); the tortoises were repatriated. Thus, from 2017 to 2021, 337 baby giant tortoises, with a total trade price ranging from $101,000 USD to $16,850,000 USD, were seized by authorities. Some traffickers ended up jailed (AP 2021; Galápagos Conservancy 2023).

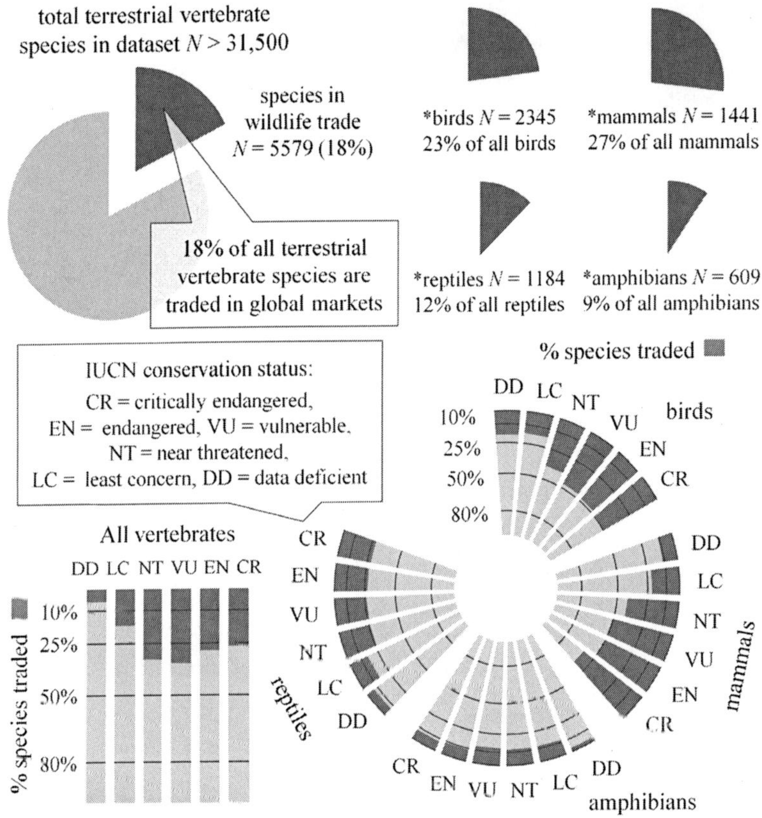

* species in dataset: birds ≈10,190; mammals ≈5330; reptiles ≈9540; amphibians ≈6470

Fig. 9-18 Share of the wildlife trade among terrestrial vertebrates. Layout: GPC; redrawn and modified after Scheffers et al. (2019).

The instances above suggest that an organized network of wildlife traffickers, with local (Galápagos), in-country (Ecuador), and international intermediaries (Perú and other nations), operates to target-smuggle from UNESCO-World-Heritage Galápagos the most iconic endemic reptiles: giant tortoises, whose species are listed by IUCN as either critically endangered (CR: *Chelonoidis darwini*, *C. donfaustoi*, *C. guntheri*, *C. hoodensis*, *C. phantasticus*, *C. porteri*), endangered (EN: *C. chathamensis*, *C. microphyes*, *C. vicina*), or vulnerable (VU: *C. duncanensis*, *C. becki*, *C. vandenburghi*), with two lineages extinct in the recent past (EX: *C. abingdonii*, *C. niger*; IUCN 2025)! [We discussed Galápagos giant tortoises in

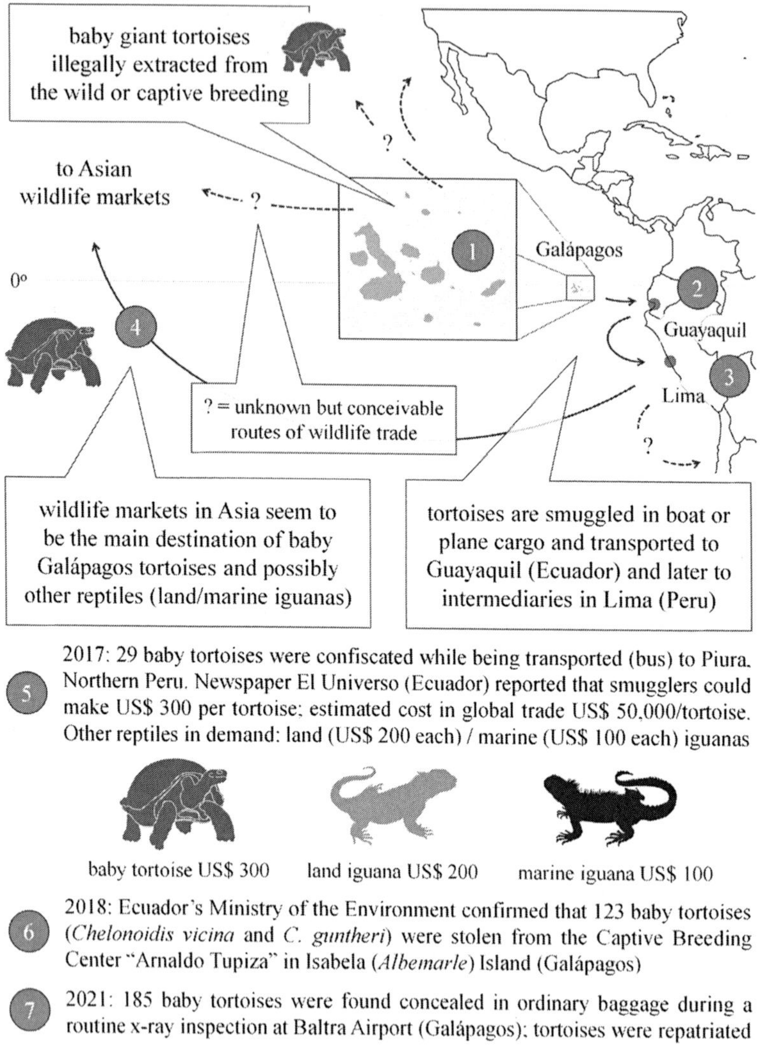

Fig. 9-19 Suspected routes of illegal trade of baby Galápagos giant tortoises and possibly land and marine iguanas. Layout: GPC; redrawn after El Universo (2019).

Chapter Eight, see Figs. 8.5-8.8, 8.39d,e,f, and 8.55-8.58; for natural history discussions, see Paz-y-Miño-C and Espinosa (2024,a,b)].

Conservationists estimate that the global illegal wildlife trade (IWT) of animal and plant species reaches $8 billion to $21 billion USD annually

(Sun-Wyler and Sheikh 2013; Scheffers et al. 2019; but see 't Sas-Rolfes et al. 2019). [For an assessment of wildlife-trade revenues, see Hughes 2021]. IWT imposes additional pressure on already-threatened species and frustrates and/or corrupts conservation efforts. [Note that wildlife trade has been cause-and-effect linked to global species extinctions and even to the imminent "Sixth Mass Extinction" (Ceballos et al. 2020; Hinsley et al. 2023)]. The problem is intricate since it involves wildlife harvesters, trade intermediaries, and consumers of wildlife, all interconnected at local, regional, and global scales (Phelps et al. 2016; Hughes 2021). In the Tropical Eastern Pacific, where the Galápagos Marine Reserve is located, an additional concern is intensifying: drug smuggling and its concealed networks of delivery with which wildlife shipments could merge. [We discussed this situation in section *2019: US Military Planes Back in Galápagos*; Chapter Five, Figs. 5.20-5.22]. Thus, improving wildlife-trade control interventions (policing) and conservation policy responses (management) will continue to be a challenge.

[Note that the networks of trade of live Galápagos iguanas (i.e. land iguanas, *Conolophus*, and marine iguanas, *Amblyrhynchus*) have been identified. Part of the information is available in the CITES Trade Database (=Convention on International Trade in Endangered Species of Wild Fauna and Flora). Importer countries (N=18) include Canada, China, Ecuador, Japan, India, Indonesia, Italy, Malaysia, North Korea, South Korea, Sudan, Sweden, Switzerland, Taiwan, Thailand, Togo, Uganda, Ukraine, and United States of America. Exporter countries (N=15) include Australia, Canada, Ecuador, Guyana, Indonesia, Japan, Mali, Mexico, Peru, Solomon Islands, Togo, Uganda, Sudan, Switzerland, and United States of America (Auliya et al. 2025)].

8 Wildlife Roadkills: *Homo* the Driver in Galápagos

Where modern humans go, motor vehicles follow; it is a global phenomenon (=anthropogenic disturbance via colonization/technology). Transportation is still incipient in Galápagos (≈1,500 vehicles total), but *Homo the Driver* is having an impact on local wildlife, particularly on birds. Field biologists assessed bird roadkills caused by motor vehicles along two road stretches in Santa Cruz (*Indefatigable*) Island: from Puerto Ayora to Bellavista (road stretch 1: 6.84 km/4.25 mi) and from Bellavista to El Cascajo (road stretch 2: 6.96 km/4.32 mi; Fig. 9.20a; after García-Carrasco et al. 2020). [For maps of Santa Cruz Island, its towns, and roads, see Fig. 1.28 in Chapter One, and Figs. 6.3-6.4 in Chapter Six].

Examine Fig. 9.20b; it summarizes bird-roadkill counts on census-road stretches 1 and 2, during July 2017 (14 alternate day counts). A total of 125 bird carcasses were counted [and removed] along road stretch 1 and 43 bird carcasses along road stretch 2 (grand total 168 birds for one month). The most common species encountered was *Setophaga petechia* (yellow warbler), with 66.7% (*N*=112; 92 on road stretch 1; 20 on road stretch 2) of all carcasses; followed by *Geospiza fuliginosa* (small ground-finch), with 9.5% (*N*=16; 6 on road stretch 1; 10 on road stretch 2) of all carcasses; and unknown finch with 9.5% (*N*=16; 10 on road stretch 1; 6 on road stretch 2) of all carcasses. The rest of the bird species (each <5% of all carcasses) were *Crotophaga ani* (smooth-billed ani), *Mimus parvulus* (Galápagos' mockingbird), *Geospiza fortis* (medium ground-finch), *Camarhynchus parvulus* (small tree-finch), and *Platyspiza crassirostris* (vegetarian finch). The share of bird roadkills per kilometer (or mile) of road stretch varied widely, as follows (Fig. 9.20c): (1) Road stretch 1 accumulated 3-times more bird roadkills (*N*=125) than road-stretch 2 (*N*=43) during the one-month census period. (2) Road stretch 1 had more vehicle traffic at higher speed than road stretch 2. (3) When speed limit was higher (e.g. 50-km/30-mi per hour), like between km 2-3/mi 1-2 of road stretch 1, bird mortality was much higher than at other sections of the road where speed limit was either lower (e.g. 30-km/20-mi per hour) or the road stretch had close-to-one-another variable speed limits (e.g. 30-50-30-km/20-30-20-mi per hour), like at km 4-to-6/mi 2.5-to-3.7 of the road. (4) Researchers extrapolated the census data to 220 bird carcasses per km per year on road stretch 1, and 74 bird carcasses per km per year on road stretch 2. (5) This implied that ≈1,500 birds could potentially be killed per year in entire road stretch 1, and ≈500 in entire road stretch 2 (total ≈2,000 birds in both road stretches per year)!

Although the seven species listed in Fig. 9.20 are considered "least conservation concern" (LC) by IUCN (i.e. their populations are robust in the archipelago, plus one taxon, *Crotophaga ani*, is an introduced invasive omnivore and predator of other birds), it is obvious that thousands of birds are being killed by motor vehicles on Santa Cruz Island. It might be the case that local birds lack rapid, adaptive response to avoid collisions with vehicles, a common behavior linked to the overall tameness of Galápagos animals (i.e. birds' inability to respond quickly enough to human-induced environmental change, often called "maladaptive response"; Paz-y-Miño-C 2014). The roads may be acting as "sinking" areas where birds "produced in nearby healthy habitats" (away from the roads) end up replacing those living close to the roads but later dying on them. This growing problem in Galápagos needs long-term monitoring.

a. Roadkill of Birds Caused by Motor Vehicles in Santa Cruz (*Indefatigable*) Island, Galápagos: a month-census suggests high yearly mortality in respect to world islands

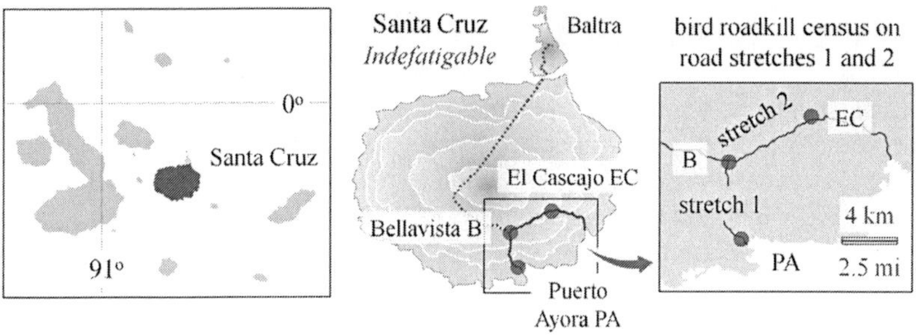

b. Bird roadkill counts on census road stretches 1 and 2 (July 2017: 14 alternate day counts)

bird species		census on road: stretch 1	census on road: stretch 2	total (%)
yellow warbler	*Setophaga petechia*	92	20	112 (66.7)
small ground-finch	*Geospiza fuliginosa*	6	10	16 (9.5)
unknown finch	–	10	6	16 (9.5)
smooth-billed ani	*Crotophaga ani*	3	5	8 (4.8)
Galápagos mockingbird	*Mimus parvulus*	8	0	8 (4.8)
medium ground-finch	*Geospiza fortis*	4	0	4 (2.4)
small tree-finch	*Camarhynchus parvulus*	1	2	3 (1.8)
vegetarian finch	*Platyspiza crassirostris*	1	0	1 (0.6)
total		125	43	168

c. Share of bird roadkills per km in road stretches 1 and 2 (estimated carcasses/km/year *)

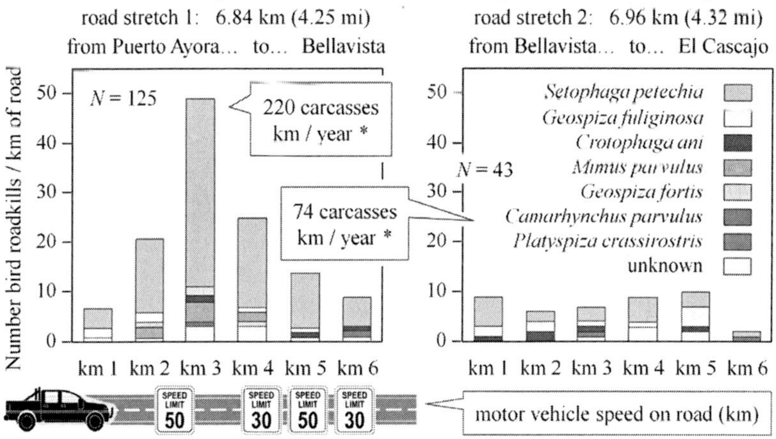

Fig. 9-20 Roadkill of birds on Santa Cruz (*Indefatigable*) Island (the most populated island), Galápagos. Layout: GPC; redrawn and modified after García-Carrasco et al. (2020).

9 Oceans' Microplastics Reach Galápagos

Adrift plastics—discarded human trash—circulate in the oceans. Exposed to seawater, sunlight, and temperature fluctuations, large pieces of plastic (macroplastics) break into smaller ones and continue to disintegrate; they eventually become microplastics (size <5 mm/<0.2 in). Marine currents and waves take plastics on predictable paths: they arrive at beaches, pollute landscapes and habitats, and are ingested by wildlife.

Scientists computer-simulated the trajectory of virtual particles [making them behave as microplastics in the simulations] that could arrive in Galápagos from diverse oceans (Fig. 9.21; after van Sebille et al. 2019). Researchers tracked the movement patterns of 62,000 adrift particles, which varied in concentration (from lower to higher; see gray-color range in Fig. 9.21 top) and were influenced by either ocean currents alone (Fig. 9.21a) or by both ocean currents and waves (Fig. 9.21b). Note in Fig. 9.21a-b how most particle trajectories were confined to the Tropical Eas tern Pacific Ocean, the coasts of South America, and the Antarctic Circumpolar Current. When trajectory simulations were based on ocean currents only (Fig. 9.21a), scientists noticed that some particles arriving in Galápagos could originate in Southern California, although most particles came from coastal Central America and Northwestern South America. Interestingly, when trajectory simulations were based on both ocean currents and waves (Fig. 9.21b), some virtual particles arriving in Galápagos originated as far as the Indian Ocean (van Sebille et al. 2019)! [Keep in mind that we characterized ocean currents around/within the Galápagos Archipelago in Chapter One, see Figs. 1.32, 1.34].

Zooming in. Examine Fig. 9.22. In a separate computer simulation, scientists assessed the amount (% fraction) of adrift virtual ocean particles (*N*=120,000) that once leaving the coasts of Central and South America passed through Galápagos (van Sebille et al. 2019). Again, the simulations included ocean currents only (dark-gray-solid lines connecting data points in Fig. 9.22) as well as both ocean currents and waves (light-gray-dotted lines connecting data points). Most particles that did pass through Galápagos came from coastal Northern Peru (+20%) and Southern Ecuador (≈10%). Some particles originated in coastal Colombia (5%).

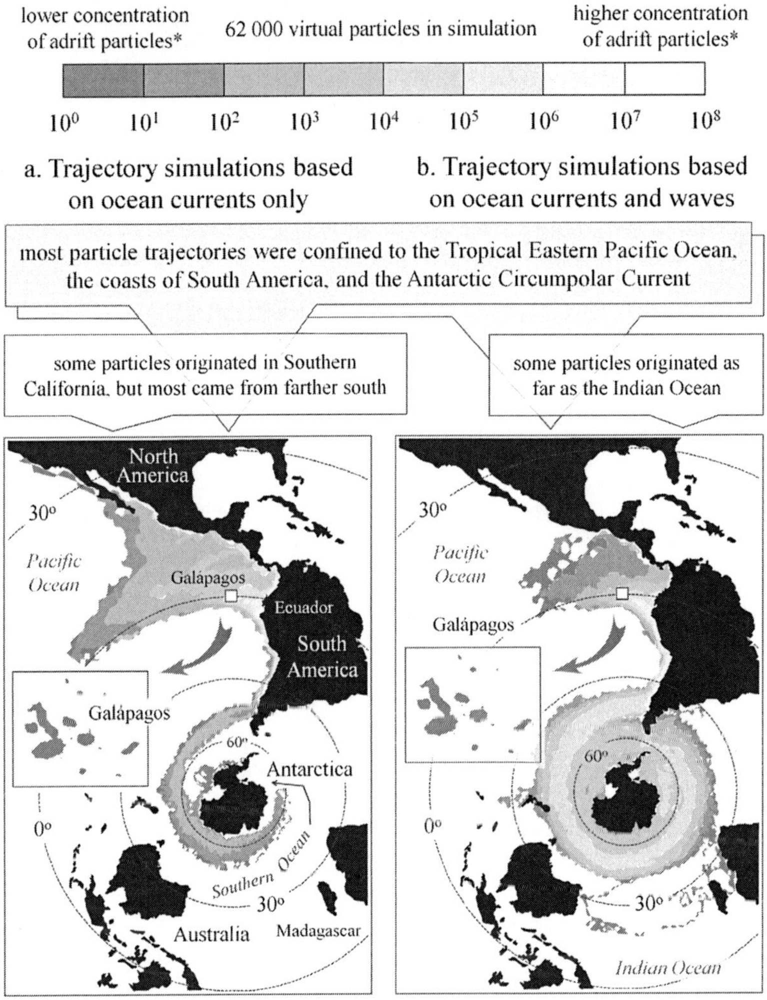

* the gray-colored scale (logarithmic) corresponds to the number of adrift virtual particle crossings per each 1 x 1 degree cells on a background map (computer simulation)

Fig. 9-21 Density of adrift ocean particle trajectories that end up in the Galápagos Marine Reserve (computer simulations of virtual particles). Layout GPC; redrawn and modified after van Sebille et al. (2019).

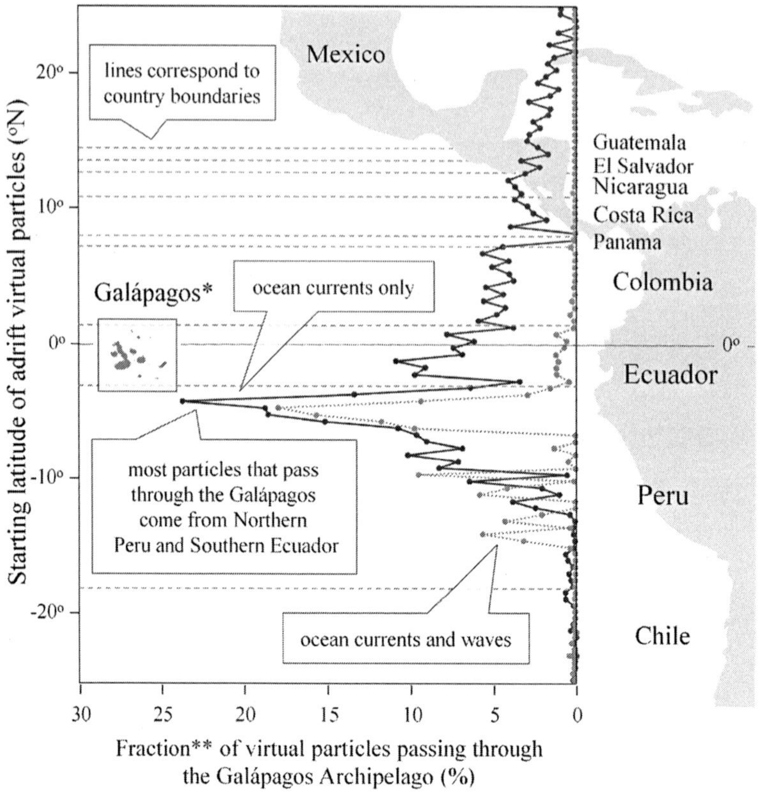

* position and size of the Galápagos not to scale in respect to Central and South America;
** dependent variable plotted on the x-axis for didactic reasons

Fig. 9-22 Fraction (%) of adrift ocean particles that once leaving the coasts of Central and South America pass through the Galápagos (computer simulations included 120,000 virtual particles). Layout: GPC; redrawn and modified after van Sebille et al. (2019).

Could intense La Niña or El Niño years impact the flow of adrift virtual ocean particles that once leaving the coasts of Central and South America pass through the Galápagos Archipelago? Researchers did model such scenarios (Fig. 9.23; after van Sebille et al. 2019). Look at Fig. 9.23a; it corresponds to simulations based on ocean currents only. Notice how the amount (% fraction) of virtual particles that originated on the coasts of Peru, and later passed through the Galápagos Archipelago, remained steady (range 4 to 8%) from 2000 to 2010, and across both La Niña (LN) or El Niño (EN) years (vertical gray-color bands in Fig. 9.23). In the case of virtual particles originating in coastal Ecuador, the amount passing through Galápagos ranged from 10 to 14% between 2000 and 2002-2003, then dropped by mid-2003 and 2004 (to about 8%), and continued to decline to 4-6% between 2006 and 2010. Such fluctuations entailed seasonal effects of La Niña and/or El Niño.

The strong El Niño years of 2003 and 2005 did have a noticeable effect on the amount of virtual particles that originated on the coasts of Colombia and later passed through Galápagos [consistent with ocean currents' seasonal fluctuations]. The flow of particles increased from ≈3% in 2000 to 12% by 2005 and later decreased to <2% by late 2007 to 2010.

Examine Fig. 9.23b. It corresponds to simulations based on both ocean currents and waves. Note how the fraction of particles passing through Galápagos remained at low, steady levels during the entire decade (2000-2010), regardless of their place of origin, either Colombia, Ecuador, or Peru. Taken together, the simulations in Fig. 9.23ab suggested that ocean currents alone, when interacting with fluctuations between La Niña and El Niño off the shores of Colombia, Ecuador, or Peru, did have a differential effect on the flow direction and amount of particles that eventually passed through Galápagos (details in van Sebille et al. 2019).

What would be the fate of microplastic virtual particles if they originated in actual locations of fisheries off the coasts of South America? Researchers computer-simulated the trajectory of 520,000 virtual particles "released" from 3,885 locations in actual fisheries (Fig. 9.24). [The fisheries corresponded to on-record positions of vessels' "fishing efforts" monitored by Global Fishing Watch (GFW 2023)]. Interestingly, <1% of all adrift particles originating in the fishing vessels' positions reached Galápagos. Of those particles that did reach the archipelago, +5% to +10% came from the North, East, and Southeastern regions of the Tropical Eastern Pacific, and +15% to +30% of particles came from fisheries located off the coasts of Southwestern Ecuador and Northwestern Peru (Fig. 9.24a). The simulations adjusted the release of virtual particles, as per vessel/fishing effort, to 5 years. Spot the light-gray color dots in Fig. 9.24; they correspond to virtual particles with zero probability of reaching Galápagos within 5 years. In other words, a great proportion of particles were not expected to pass through the archipelago either under trajectory simulations based on ocean currents only (Fig. 9.24a) or based on both ocean currents and waves (Fig. 9.24b).

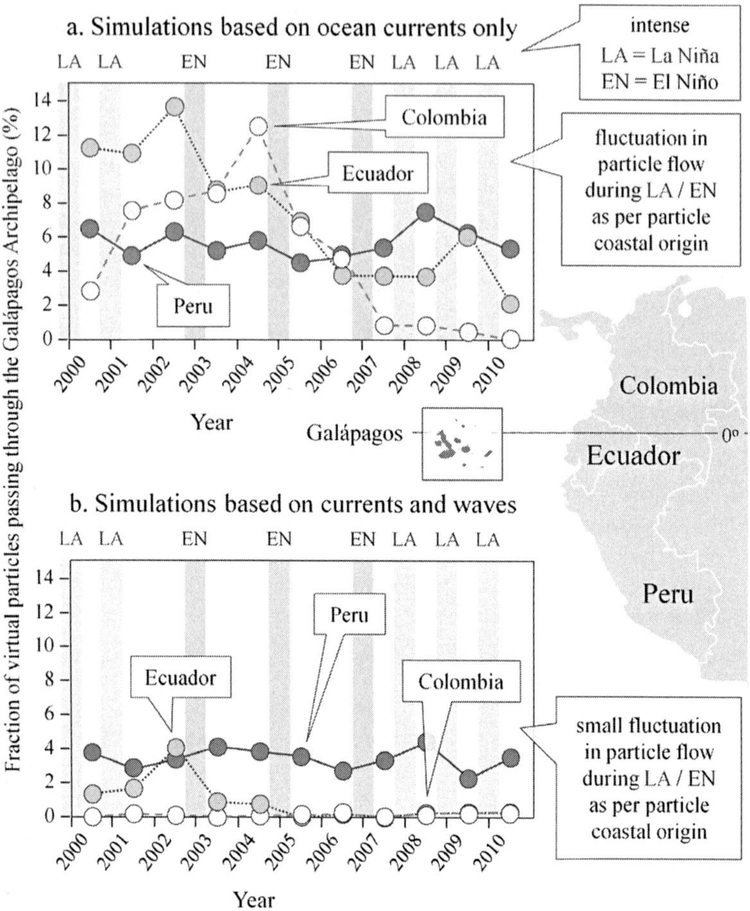

Fig. 9-23 Fraction (%) of adrift virtual ocean particles that once leaving the coasts of Colombia, Ecuador, or Peru pass through the Galápagos: emphasis on intense La Niña- or El Niño-year effects. Layout: GPC; redrawn/modified after van Sebille et al. (2019).

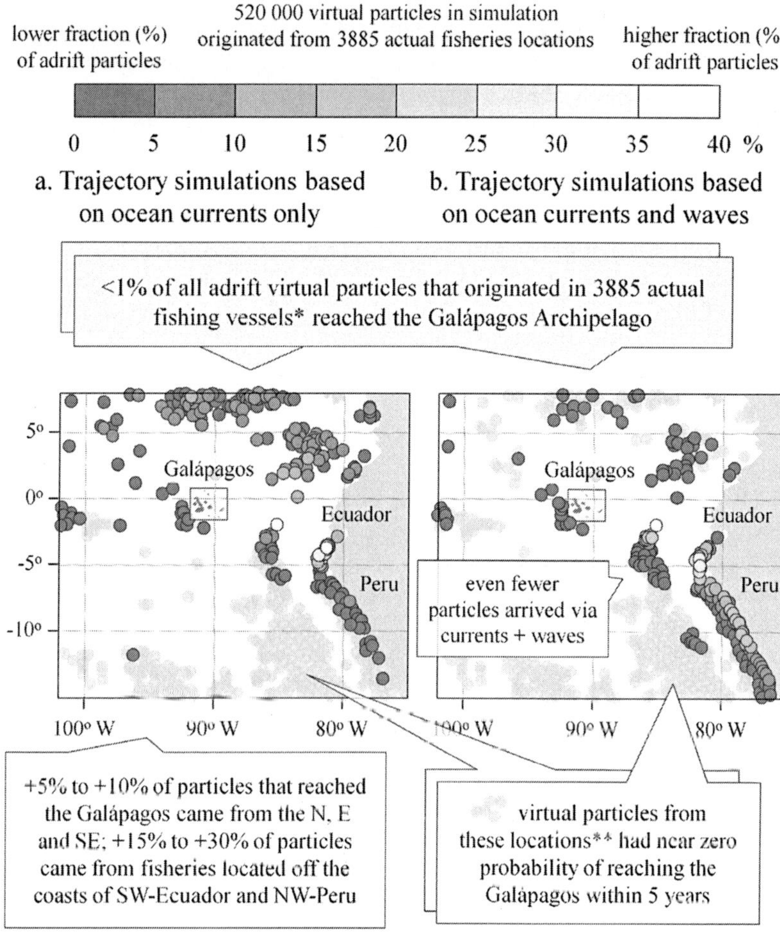

* fishing vessels corresponded to actual "fishing effort" recorded by Global Fishing Watch up to 2018, i.e. locations with 24-hour fishing activity from Jan 2012 to Dec 2016; the simulations adjusted the release of virtual particles as per vessel / fishing effort for up to 5 years; ** locations with near zero probability of releasing particles that reach the Galápagos are depicted as pale circles with no circumference delineation

Fig. 9-24 Fraction (%) of adrift virtual ocean particles that, once leaving actual fisheries' locations, reach the Galápagos. Particles in simulation (*N*=520,000) originated from 3,885 actual fisheries. Layout: GPC; redrawn/modified after van Sebille et al. (2019).

The take-home message of Figs. 9.21-9.24 is: (1) Adrift ocean-plastics, particularly microplastics, can arrive in Galápagos primarily from the proximate ocean (i.e. the Tropical Eastern Pacific), although computer simulations suggest globally circulating microplastics [i.e. virtual particles mimicking them] can come from as far as Southern California (Northern hemisphere) or even the Southern and Indian Oceans (via the South Pole). (2) Most particles that do pass through Galápagos come from coastal Northern Peru and Southern Ecuador. (3) La Niña and/or El Niño events can influence, seasonally and differentially, ocean-particle-flow off the coasts of Colombia, Ecuador, and Peru. (4) If virtual particles originate at locations of actual fisheries in the Tropical Eastern Pacific, only a small fraction (<1%) of them would eventually pass through Galápagos; among those particles reaching Galápagos, +15% to +30% of them can come from fisheries located off the coasts of Ecuador and Peru.

10 Microplastics in the Guts and Tissues of Marine Life: Coastal Ecuador and Galápagos

To estimate microplastics' particle concentrations in surface ocean waters off coastal Ecuador and around the Galápagos Archipelago, marine biologists sampled 40 locations within an area of 453,000 km^2/175,000 sq mi (Fig. 9.25a). Samples were obtained along meridians 92°, 89°, 86°, and 82° West; and latitudes 2° North and 3° South (after Alfaro-Núñez et al. 2021). Thus, sampling covered Western and Eastern Galápagos (92°W and 89°W, respectively), international waters (86°W), and Ecuador's continental waters (82°W). [Ecuador has exclusive economic zones (EEZ) in both its continental coasts and around Galápagos (EEZs expand 370 km off the coastline; EEZs are called "the 200 nautical miles" of a country's territorial sea]. The highest microplastics' particle concentration (0.51 μp/m^3) was found in international waters; this region also had the highest mean value of 0.36 μp/m^3 (as per samples along meridian 86°W); followed by Ecuador's continental waters (mean 0.26 μp/m^3), Eastern Galápagos (mean 0.24 μp/m^3), and Western Galápagos (mean 0.22 μp/m^3). With this data, researchers inferred the spatial distribution of microplastics' concentration along the surface-water contour lines (i.e. currents) for the entire area (453,000 km^2/175,000 sq mi; Fig. 9.25b). The highest concentration (range 0.42-0.51 μp/m^3) occurred in international waters, followed by coastal Ecuador, Eastern Galápagos, and Western Galápagos. Microplastics were everywhere!

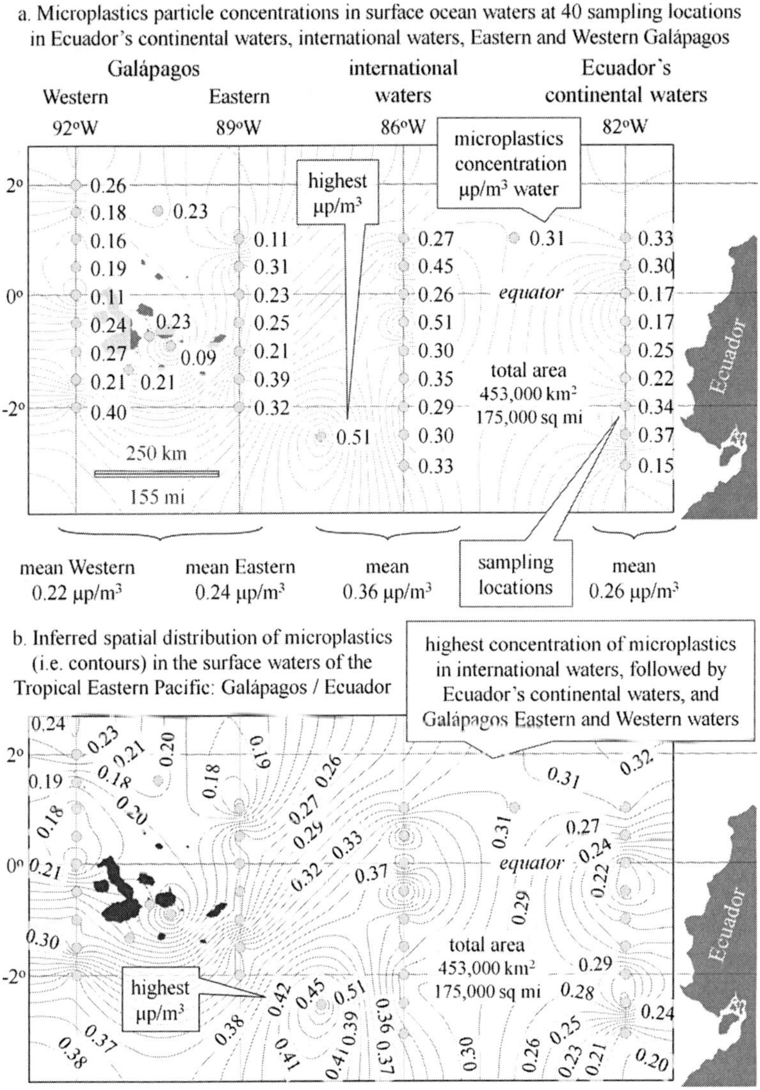

Fig. 9-25 Microplastics' pollution in surface waters proximate to Galápagos (total area in maps 453,000 km²/175,000 sq mi). a. Sampling locations (N=40). b. Inferred microplastics' particle concentration as per surface-water contour lines (currents); highest-value range 0.42-0.51 µp/m³. Layout: GPC; redrawn and modified after Alfaro-Núñez et al. (2021).

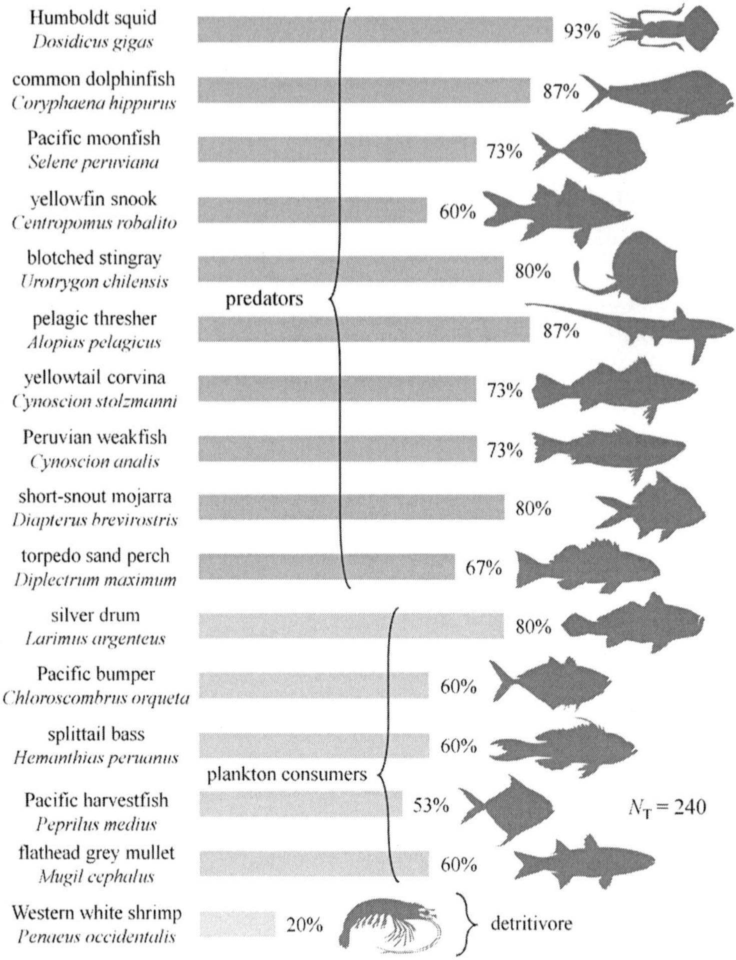

* fish markets where located along the Pacific coast of Ecuador, in the provinces of Esmeraldas, Manabí, Santa Elena and El Oro (just 900 km or 600 mi from the Galápagos)

Fig. 9-26 Prevalence of microplastic particles in the gut/muscles of exemplar marine species of economic value in the fish markets of coastal* Ecuador (N=15 per species; N_T=240). Layout: GPC; redrawn and modified after Alfaro-Núñez et al. (2021).

If microplastics were ubiquitous in the ocean, could they be present inside life forms? Scientists assessed the prevalence of microplastic particles in the guts and muscles of marine species of economic value in the fish markets of coastal Ecuador (Fig. 9.26; Alfaro-Núñez et al. 2021). The fish markets were located along the Western, Pacific shores of the country, in the provinces of Esmeraldas, Manabí, Santa Elena, and El Oro, located just 900 km/600 mi from the Galápagos Archipelago. Researchers evaluated 16 species (N=15 samples per species; N_T=240), including predators (10 species), plankton consumers (5 species), and one detritivore. Note in Fig. 9.26 how all the species in the list carried microplastics in their guts and/or muscles! Among predators, prevalence of microplastics ranged from 60% in yellowfin snook (*Centropomus robalito*) to 93% in Humboldt squid (*Dosidicus gigas*). In the case of plankton consumers, prevalence ranged from 53% in harvestfish (*Peprilus medius*) to 80% in silver drum (*Larimus argenteus*). And in the only case of a detritivore, Western white shrimp (*Penaeus occidentalis*), prevalence reached 20% (complete data in Alfaro-Núñez et al. 2021).

What about Galápagos? Another team of researchers assessed presence of macro and microplastics' contamination in shore- (sandy beaches) and off-shore marine environments around San Cristóbal (*Chatham*) Island (Fig. 9.27; after Jones et al. 2021). Biologists focused their sampling on 17 sites, either visited by tourists (light-gray-color squares, N=7; Fig. 9.27) or not visited by tourists (dark-gray-color circles, N=10). Sites 1-7 were located in the North and West parts of the island; sites 8-11 in the South; and sites 12-17 in the East. Note that the latter was directly exposed to the Humboldt current, which arrived from coastal South America [see map of currents around the Galápagos Archipelago in Figs. 1.32, 1.34, Chapter One]. North and West sides were not directly exposed to Humboldt. And the South side was a mix.

In Fig. 9.28a, we summarize the amount of macroplastics found on the beaches of San Cristóbal (after Jones et al. 2021). 88% of them were weathered hard fragments (49%) or "unsourced" items (39%), likely originating outside the Galápagos Archipelago; they included caps/lids, drink-bottles, plastic-bags fragments, bottle rings, films, polystyrene fragments, toothbrushes, pens, or food containers. 10% were maritime items, including ropes, strings, cords, or eel traps. And 2% were tourist-sites' items, including cutlery, straws, or cups. In Fig. 9.28b, we show the partition of macroplastics found as per location on the island's beaches: North and West, East, or South. The histograms correspond to number of macroplastics per distance surveyed along the beach; e.g., most items were found on the East side of San Cristóbal, about 4,000 items along 400-m/1,300-ft of beach surveyed, with a density of fragments = 0.27 per m^2 (0.27 per 10.7 sq-ft). Although macroplastics were also found along the beaches on the North/West and South sides of the island, the amount and density of items were a fraction in comparison to the East side.

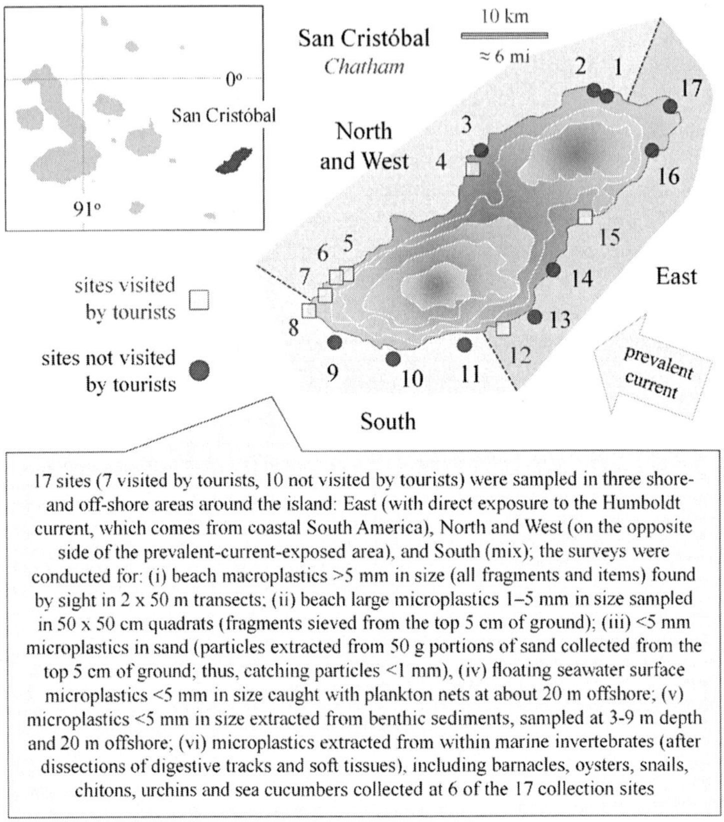

Fig. 9-27 Documentation of macro- and micro-plastic contamination in shore- (sandy beaches) and off-shore marine environments, as well as in invertebrates sampled at 17 sites around San Cristóbal (*Chatham*) Island, Galápagos. Layout: GPC; redrawn and modified after Jones et al. (2021).

Bring your attention to Fig. 9.29. It reports the detection of microplastics in the soft tissues of exemplar invertebrates sampled around San Cristóbal Island. Microplastics corresponded to minute fragments or fibers (polypropylene- or synthetic-cellulosic fragments/fibers) extracted from within the digestive track or other soft tissues (via dissections) of barnacles, oysters, snails, chitons, urchins, and sea cucumbers collected at 6 sites on the North/West and East coasts of San Cristóbal (as per map in Fig. 9.27). Polymer composition of these fragments of fibers included: polyethylene, polypropylene, polystyrene, cellulosic, polyester, polyvinyl chloride, and polyacrylamide (Jones et al. 2021). On average, 52% of all

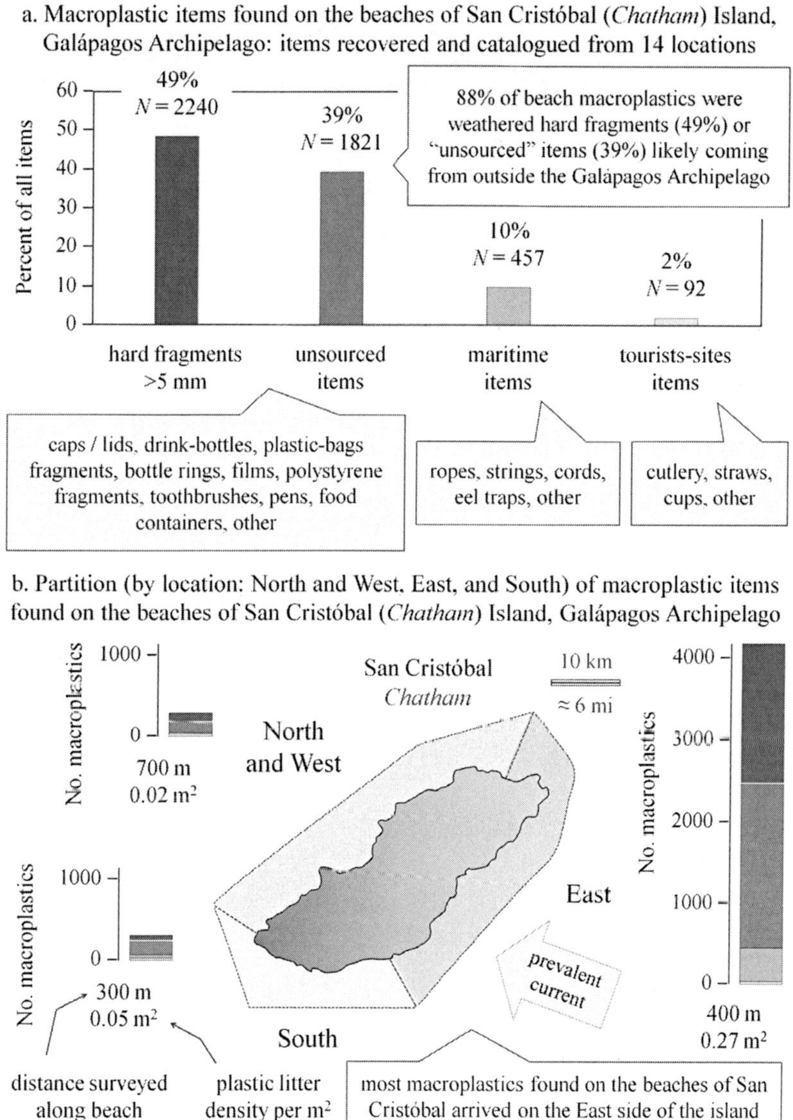

Fig. 9-28 Macroplastics found on the beaches of San Cristóbal (*Chatham*) Island, Galápagos. Layout: GPC; redrawn and modified after Jones et al. (2021).

invertebrate specimens (from 7 species) had microplastics in their guts or soft tissues, with a grand total (N_T) of 123 specimens examined. The species included filter feeders (N_T=25; goose barnacles, giant barnacles, and palmate oysters); grazers (N_T=49; rough-ribbed nerite snails, sculptured chiton, and slate pencil urchins); and a single deposit feeder (N_T=49; sea cucumber; complete data in Jones et al. 2021).

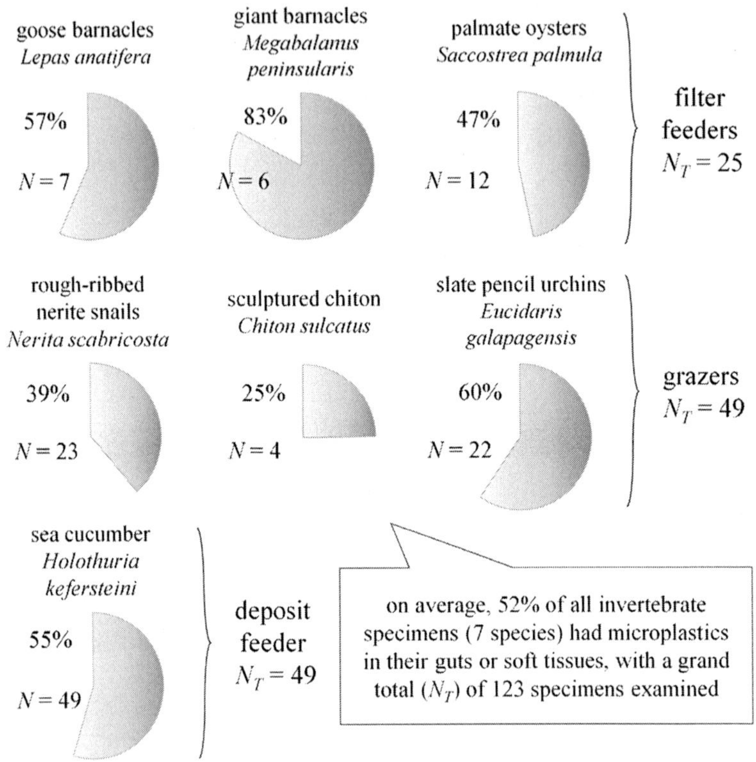

* correspond to microscopic fragments or fibers (polypropylene- or synthetic-cellulosic fragments/fibers) extracted from within the digestive track or other soft tissues (via dissections) of marine invertebrates collected at 6 sites on the North-and-West and East coasts of San Cristóbal (see map in Figure 1.27, Chapter One); polymer composition of these fragments of fibers included: polyethylene, polypropylene, polystyrene, cellulosic, polyester, polyvinyl chloride, and polyacrylamide (Jones et al. 2021)

Fig. 9-29 Detection of microplastics in the soft tissues* of exemplar invertebrates sampled around San Cristóbal (*Chatham*) Island, Galápagos. Layout: GPC; re-drawn and modified after Jones et al. (2021).

Together, Figs. 9.25-9.29 suggest that both ocean macro and microplastics are ubiquitously present from coastal Ecuador (continental waters) to international waters and to the Galápagos Archipelago. Plastics drift in predictable directions (as per ocean currents). Microplastics can be found in the gut and/or other tissues of marine vertebrates (including species of economic value in Ecuador's fish markets) and invertebrates. These observations imply that plastics are being ingested and taken up by all trophic levels in the ecosystem.

11 Macroplastics Present in All Galápagos Islands

Researchers and "citizen scientists" (=local communities' members involved in Galápagos' conservation) assessed plastic pollution along shores and beaches of 11 islands in the archipelago (Fig. 9.30; after Muñoz-Pérez et al. 2023). Teams of volunteers removed macroplastics wherever possible, while researchers coordinated the technical inventory of plastic pollution. As shown in Fig. 9.30, all the sampled islands had macroplastics contaminating their shores/beaches. The highest concentration of items was found on sites exposed directly to prevalent winds year-round (Northwest-West, NWW), which directed ocean water—carrying adrift plastics in it—toward the coast. The highest density of macroplastics was found on Santiago Island (2.870 items per m^2). Relying on a subset of macroplastics (*N*=466) of those collected across islands, researchers inferred (based on labels still readable on items) that they likely originated in Perú, Ecuador, or China, among 11 other countries [the Chinese fragments possibly came from fisheries active outside or proximate to the Galápagos Marine Reserve rather than from Asia].

Based on field observations and incidents' reports involving wildlife being affected by plastics' pollution in Galápagos, scientists alerted that entanglement with fishnets, fishing lines, or ropes (floating adrift in the ocean), as well as ingestion of plastics, were immediate threats to local fauna (Muñoz-Pérez et al. 2023). The most threatened marine/terrestrial taxa included 52 species (20 endemic, 24 native, 5 migratory, 3 unknown), among them: green sea turtles (*Chelonia mydas*), flightless cormorants (*Phalacrocorax harrisi*), marine iguanas (*Amblyrhynchus cristatus*), lava gulls (*Larus fuliginosus*), sea lions (*Zalophus wollebaeki*), whale sharks (*Rhincodon typus*), scalloped hammerhead sharks (*Sphyrna lewini*), spine-tail mobulas (*Mobula japanica*), black-striped salemas (*Xenocys jessiae*), white-spotted sand bass (*Paralabrax albomaculatus*), sea urchins (*Lytechinus semituberculatus*), hermit crabs (*Coenobita compressus*), sally lightfood crabs (*Grapsus grapsus*), Santa Cruz's giant tortoises (*Chelonoidis porteri*), and medium-ground finches (*Geospiza fortis*).

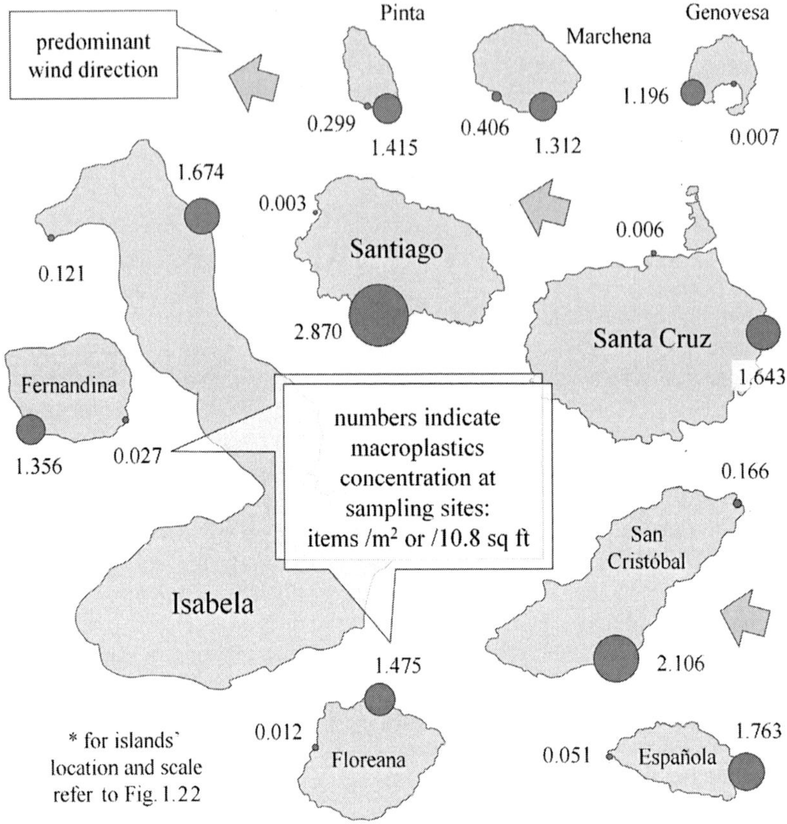

Fig. 9-30 Macroplastics concentration (i.e. items per m^2 or per 10.8 sq ft) at diverse shore/beach sampling sites in the Galápagos Archipelago. Numbers correspond to concentration of macroplastics' items (pieces >5-mm/>0.2-in size) per m^2. Arrows point in the predominant wind direction year-round (NWW). Density of plastics' items was higher on shores/beaches exposed to predominant winds (highest density in Santiago Island, 2.870/m^2). Based on analysis of a subset of plastics (N=466) of those collected across islands, researchers determined that they likely originated in Perú, Ecuador, or China, among 11 other countries (the Chinese fragments possibly came from fisheries active outside or proximate to the Galápagos Marine Reserve rather than from Asia). *For islands' location and scale, refer to satellite image of the archipelago in Fig. 1.22, Chapter One. Layout: GPC; redrawn and modified after Muñoz-Pérez et al. (2023).

The distinction between macro and microplastics is—although practical and didactic—arguably arbitrary. The relevant issue is that plastics, from macroscopic to microscopic (=continuum), are ubiquitous on Earth and entering all trophic levels. Scientists have found microplastic fibers in 37% of Galápagos sea-lions' scats (N_T=180) sampled from rookeries on four islands: San Cristóbal (El Malecón, Punta Pitt), Española (Punta Suárez), Floreana (Post Office), and Santa Fé (Bahía Santa Fé; Moreira-Mendieta et al. 2023). The results were not surprising when compared to other Pinnipeds (sea lions, fur seals) distributed along the Pacific coasts of North, Central, and South America, among which microplastics' prevalence in scats ranged from 21% in *Arctocephalus australis* (coastal Peru and Chile) to 100% in *Callorhinus ursinus* (North America).

12 Plastics Ingested and Uptaken by All Trophic Levels

Why do organisms ingest plastics? Scientists hypothesize that the abrupt appearance of plastics in the environment has triggered maladaptive feeding responses in animals (H_i=plastic ingestion trap). This is because plastics resemble, visually and/or chemically, food items (Fig. 9.31a; Santos et al. 2021). ["Some behaviors become maladaptive when selective pressures change, usually because of human intervention, and animals are no longer able to assess the change or respond appropriately to it" (Paz-y-Miño-C 2014)]. As shown in Fig. 9.31b, the risk [low to high] of plastic ingestion by organisms may be a function of: (1) availability of plastics in the environment; (2) nutritional state of an organism; (3) resemblance between plastics and natural prey/food items; and (4) prey/food item selection by individuals [hypothetical individuals=gray-color dots in Fig. 9.31b]. The highest risk of plastic ingestion, as pointed out by the dialog box in Fig. 9.31b, corresponds to the center overlap of these variables. Researchers reason that consumption of plastics by animals might result from the "mistaken identity" of items that ultimately resemble food (Santos et al. 2021). The maladaptive response emerges in the context of how abundant (frequency) the foreign items are in the environment; how hungry (motivation) individual animals are to consume plastics as "food"; the immediate and/or longer-term consequences of eating the "wrong thing" (if no immediate negative outcome occurs, plastics will be more frequently ingested); how other conspecifics respond to the ingestion of plastics (social facilitation or deterrence in the case of gregarious animals); or how perception, learning, memory, or decision making, particularly among taxa with com-

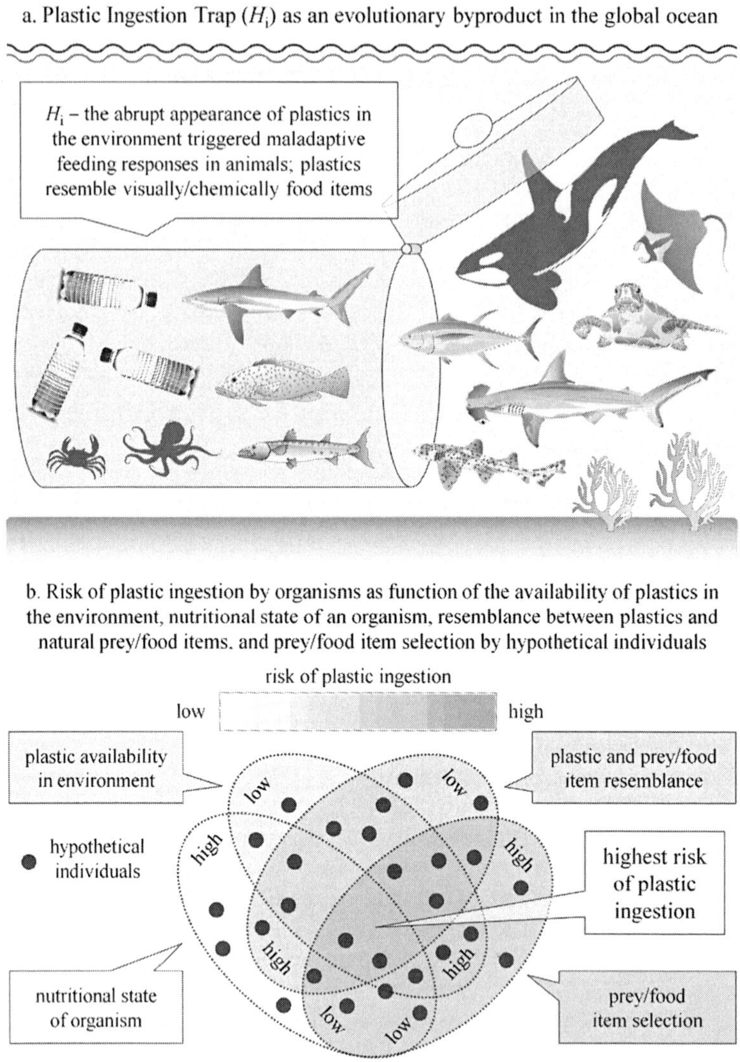

Fig. 9-31 Why do organisms ingest plastics? a. The maladaptive response hypothesis (i.e. when exposed to sudden, human-induced environmental change, animals cannot assess the change or respond appropriately to it). b. Risk of plastic ingestion (low to high) by organisms as per four variables (plastic availability in the environment; nutritional state of organisms; plastic and prey/food item resemblance; and prey/food item selection by organisms). Layout: GPC; conceptualized, redrawn, and modified after Santos et al. (2021).

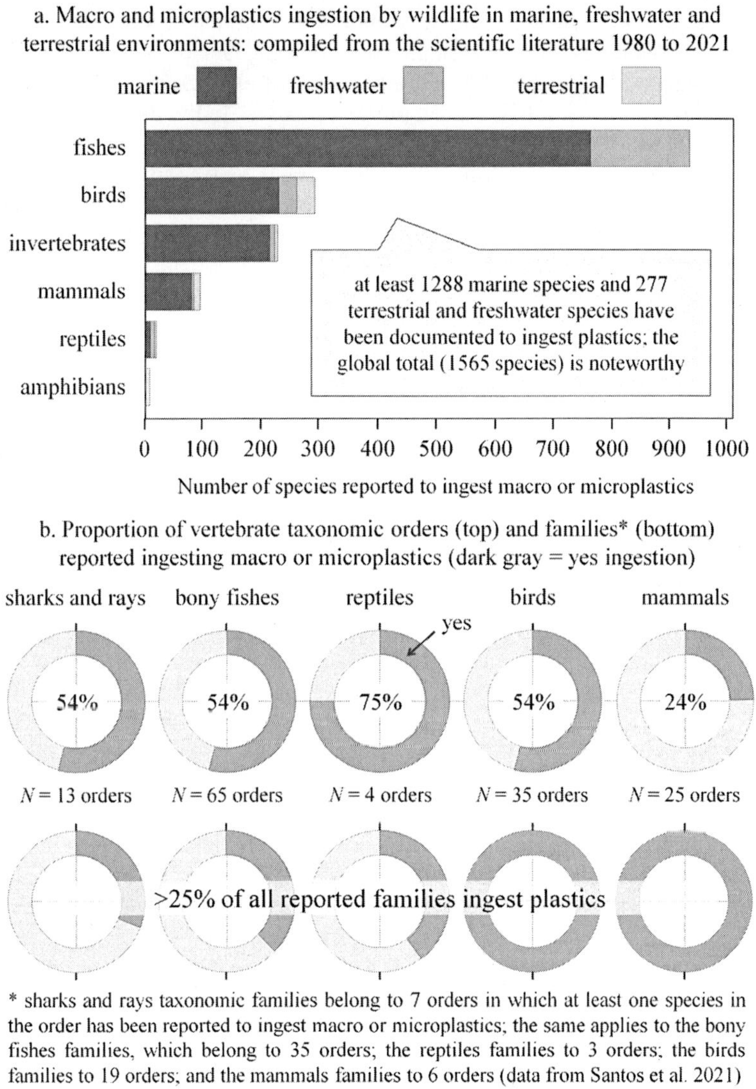

Fig. 9-32 Ingestion of macro and microplastics by wildlife. a. Marine, freshwater, and terrestrial environments. b. Vertebrates reported ingesting macro and micro-plastics. Layout: GPC; redrawn and modified after Santos et al. (2021).

plex cognition, influence the way individuals obtain information about their surroundings, retain such information, and act on it (Shettleworth 2001; Paz-y-Miño-C 2014). But the ingestion of plastics, as readers should infer, goes beyond solely maladaptive responses. Individual animals can consume the right prey/food item and still ingest plastics through that meal. Global plastics' pollution is systemic, and, as we will highlight later in this section, it has permeated all trophic levels.

Examine Fig. 9.32a, in which we summarize the ingestion of macro and microplastics by wildlife in marine, freshwater, and terrestrial environments (data compiled by Santos et al. 2021). Notice how at least 1,288 marine species and 277 terrestrial and freshwater species have been documented to ingest plastics; the global total (1,565 species) is noteworthy. As sketched in Fig. 9.32b, five groups of vertebrates (sharks and rays, bony fishes, reptiles, birds, and mammals) have been reported to ingest macro or microplastics. The proportion of taxonomic orders within these groups—that ingest plastics—ranges from 24% (*N*=25 orders) in mammals to 75% (*N*=4 orders) in reptiles. And >25% of all reported taxonomic families within the five groups of vertebrates ingest plastics! As indicated in Fig. 9.32, sharks and rays' taxonomic families belong to 7 orders in which at least one species in the order has been reported to ingest macro or microplastics; the same applies to bony fishes' families, which belong to 35 orders; reptiles' families to 3 orders; birds' families to 19 orders; and mammals' families to 6 orders (data from Santos et al. 2021). [We discussed the relevance of sharks and rays to the Galápagos Marine Reserve in Chapter Eight, see Figs. 8.64-8.67; see also *Thirty-Five Species of Sharks and Counting*, Figs. 8.68-8.77 and Table 8.9].

Based on scientific-literature information about macro- and microplastics' consumption by marine organisms worldwide (1,288 species), scientists have constructed conceivable "food webs" (=sketches representing hierarchical connections among trophic levels) for oceanic tropical and subtropical ecological communities (Fig. 9.33a), as well as for coral and rocky reefs (Fig. 9.33b; data from Santos et al. 2021). It is obvious that macro- and microplastics are being ingested by most, or all, organisms within communities and transferred from one trophic level to another (as arrows indicate in Fig. 9.33): from "microscopic primary producers and detritus" (Fig. 9.33a bottom) or "suspended/deposited organic matter" (Fig. 9.33b bottom) to "primary consumers and detritivores" to "secondary consumers" to "predators" and "top predators". Readers will infer that the same or comparable species to those listed in Fig. 9.33 (ecological equivalents) exist in the Galápagos Marine Reserve (its oceanic, and coral and rocky-reef food webs) with the potential to ingest and/or uptake plastics systemically.

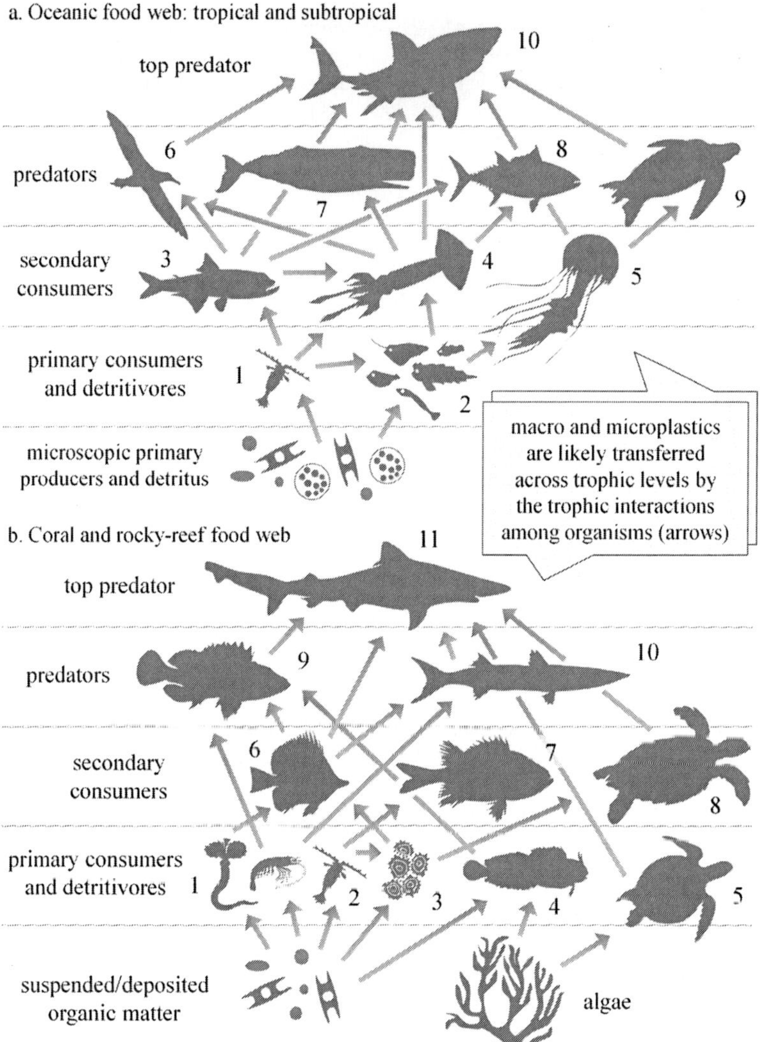

Fig. 9-33 Macro and microplastics are ingested and uptaken by all trophic levels. a. Oceanic food web (tropical and subtropical); numbers correspond to species in which ingestion of macro- or microplastics has been documented (data from Santos et al. 2021): (1) copepods; (2) fish larvae; (3) Benoit's lanternfish (*Hygophum benoit*); (4) neon flying squid (*Ommastrephes bartrami*); (5) mauve stinger (*Pelagia noctiluca*); (6) Laysan albatross (*Phoebastria immutabilis*); (7) sperm whale or cachalot (*Physeter macrocephalus*); (8) Atlantic bluefin tuna (*Thunnus*

thynnus); (9) leatherback sea turtle (*Dermochelys coriacea*); and (10) great white shark (*Carcharodon carcharias*). b. Coral and rocky-reef food web; numbers correspond to species in which ingestion of macro- or microplastics has been documented (data from Santos et al. 2021): (1) tube/fanworm (*Serpula vermicularis*); (2) copepods; (3) zoanthid cnidarians; (4) combtooth blenny (*Scartichthys viridis*); (5) green sea turtle (*Chelonia mydas*); (6) oblique butterflyfish (*Prognathodes obliquus*); (7) soldierfishes (*Myripristis* spp.); (8) hawksbill sea turtle (*Eretmochelys imbricata*); (9) honeycomb grouper (*Epinephelus merra*); (10) blackfin barracuda (*Sphyraena qenie*); and (11) Brazilian sharpnose shark (*Rhizoprionodon lalandii*). Readers would infer that the same or comparable species (ecological equivalents) exist in the Galápagos Marine Reserve (its oceanic, and coral and rocky-reef food webs) with the potential to ingest and/or uptake plastics. Layout: GPC; redrawn and modified after Santos et al. (2021).

Researchers think that, long-term, the "plastic ingestion trap" (*Hi* in Fig. 9.31a) could affect wildlife populations' viability (Santos et al. 2021). Realize that individuals, populations, or species will differentially respond [behaviorally, physiologically] to plastics' pollution (or to any inducer of maladaptive responses; Paz-y-Miño-C 2014). It is a challenge to forecast the magnitude of impact resulting from the invasion of plastics at all trophic levels or entire ecological communities, or to reverse such impact effectively at local, regional, or global scales.

Conclusions

We omit citations in this section to abridge the text. Each conclusion point summarizes topics substantiated by citations in this chapter.

Galápagos has multiple conservation-related problems. In this chapter, we discussed 12 case studies regarding iconic conservation problems in the archipelago and its marine reserve. Some were local problems, others were regional or global, and affected Galápagos directly or indirectly.

(1) Introduced Species: 1535 Onwards. Since the islands were discovered +490 years ago, about 1,579 alien species have been introduced, of which 1,476 are now established, with 868 of the latter (59%) naturalized (i.e. species that now spread and reproduce without active human involvement).

(2) Invasive Species at World Heritage Sites. Environmental scientists characterized the "threat level from invasive species" at 241 of UNESCO's World Heritage Sites, including 7 exemplar case studies; Galápagos was one of them. ≈300 species of invasive alien taxa were found to be a threat to about 50 (21%) of the 241 WHS. The Galápagos Islands were threatened by 60 invasive species; the threat level was considered "very high" and the management efforts to reduce the threat

"moderately effective (some concern)". Galápagos' most threatening alien "invasives" were: *Cedrela odorata* (tree), *Felis catus* (cat), *Rattus norvegicus* (brown rat), *Rattus rattus* (black rat), *Philornis downsi* (fly), *Polistes versicolor* (wasp), *Psidium guajava* (shrub/tree), *Solenopsis geminata* (ant), *Rubus niveus* (shrub), and *Wasmannia auropunctata* (ant).

(3) The Global Decline of Shark Populations. The waters around Darwin (*Culpepper*) and Wolf (*Wenman*), the Northwestern islands in the Galápagos Marine Reserve, have the largest shark biomass in the world. Why is this relevant? Because global shark populations are declining. Sharks are indicators of the health and wellbeing of reef communities. Thus, sharks and their habitats (e.g. Galápagos) are central to marine environments' conservation. The global presence of 18 shark and ray species has declined by 71% since 1970. Twelve of the 18 species occur in the Galápagos Marine Reserve. Sharks' habitats have declined in both tropical and temperate regions; sharks and rays of all sizes (particularly large and medium ones) have become less abundant; and sharks and rays of all generation cohorts have become less abundant. What caused this global decline in sharks and rays? An 18-fold increase in relative fishing pressure. Global abundance of sharks has decreased as relative fishing pressure on sharks has increased. The current shark populations are likely below biomass or abundance levels to produce sustainable yield.

(4) The *Fu Yuan Yu Leng 999* and the Shark-Fin Market. In August 2017, Ecuadorian authorities detained the Chinese Reefer *Fu Yuan Yu Leng 999* within the Galápagos Marine Reserve. The vessel carried—illegally—572 tons of fish and fish products, yet only +1 of its 6 freezers was full. The reefer contained 7,639 sharks; 2,114 bony fishes; and 537 sacks with shark fins (cuts). From a sample of 929 juvenile/adult sharks, researchers identified morphologically and genetically-barcoded 12 species. Scientists alerted that the *Fu Yuan Yu Leng 999* "exemplif[ied] how industrial fisheries [were] likely depleting shark populations in the Eastern Tropical Pacific Ocean". In fact, sharks—and particularly their fins—were, still are, the target of fishing efforts.

(5) Tuna Fisheries Around the Galápagos Marine Reserve. Researchers gathered historical records on tuna catches, fishing effort, and catches per unit of fishing effort and plotted the data on a 1x1 degree latitude-longitude grid covering the Galápagos Marine Reserve (GMR is a non-commercial fishing zone) and within the Galápagos Exclusive Economic Zone (GEEZ). The grid expanded from 6°N to 6°S and from 85°W to 97°W. Scientists determined that catches, fishing effort, and catches per unit of fishing effort were all concentrated to the West of GMR (=fishing "hotspots"). Upon the enforcement of GMR in 2002, trends for tuna catch-

es and fishing effort (hotspots versus other areas outside GMR) significantly and progressively diverged by 2015, with more catches and fishing effort at hotspots (but not at other areas). The creation of the reserve around the archipelago apparently enhanced favorable conditions for tuna spawning and breeding within the protected area (GMR), as well as benefited the fisheries. Tuna fisheries were likely benefiting temporarily from a “spillover effect”, i.e., the export of fish biomass from the protected area (GMR) to the adjacent fishing grounds (the current fishery hotspots). Thus, GMR was acting as a “well-protected and well-placed reserve”!

(6) The Decline of Sea Lions and Fur Seals. Scientists have alerted that because severe El Niño events can disrupt Galápagos’ sea-lion (*Zalophus wollebaeki*) and fur-seal (*Arctocephalus galapagoensis*) populations—as climate change continues to intensify El Niño impacts—both species could become critically endangered or even extinct. [*Zalophus wollebaeki* and *Arctocephalus galapagoensis* are currently listed as endangered by IUCN]. Extreme weather fluctuations and higher air/sea surface temperatures will disrupt nutrient cycles and fish-food availability for marine-mammal predators, and sea level rise will cover breeding areas—for colonies—ashore. Thus, protection and management of crucial areas within GMR will benefit Pinnipeds. Researchers have determined that in areas in which the spatial distributions of sea lions and fur seals overlap substantially (Western part of the archipelago), sea lions often forage proximate to the continental shelf, while fur seals do it offshore, in deep waters, and with “more heterogeneous space usage over time”. Based on these observations, as well as characterizations of habitat use and preference by sea lions and fur seals, three priority areas for conservation were identified: (i) deep waters Northwest and shallow waters West of Fernandina Island; (ii) deep waters Southwest of Isabela Island; and (iii) shallow waters over the continental shelf to the East of Fernandina Island.

(7) Wildlife Trade: Tortoises and Iguanas for Sale. Wildlife trade is a multibillion-dollar business and one of the 10 most lucrative criminal activities worldwide. Networks of trade connect “hotspots” of wildlife providers (in South and Central America, Southeast Africa, the Himalayas, Southeast Asia, and Australia) with local, regional, and global markets. Galápagos is not immune to the global wildlife trade. Ecuadorian authorities have uncovered that baby giant tortoises are illegally—and recurrently—extracted from the wild or captive breeding centers. The tortoises are then smuggled by boat or plane cargo and transported to Guayaquil (mainland Ecuador) and later to trade intermediaries in Lima (Peru). Wildlife markets in Asia seem to be the main destination for baby tortoises and possibly other reptiles (land and marine iguanas). From 2017 to 2021, 337

baby giant tortoises, with a trade price ranging from $101,000 USD to $16,850,000 USD, were seized by authorities. Some traffickers were jailed. These instances suggest that an organized network of wildlife traffickers, with local (Galápagos), in-country (Ecuador), and international intermediaries (Perú and other nations), operates to target-smuggle from UNESCO-World-Heritage Galápagos the most iconic endemic reptiles: giant tortoises!

(8) Wildlife Roadkills: *Homo* the Driver in Galápagos. Where modern humans go, motor vehicles follow; it is a global phenomenon (=anthropogenic disturbance via colonization/technology). Transportation is still incipient in Galápagos (≈1,500 vehicles total), but *Homo the Driver* is having an impact on local wildlife, particularly on birds. During July 2017, biologists assessed bird roadkills caused by motor vehicles along two road stretches in Santa Cruz (*Indefatigable*) Island: from Puerto Ayora to Bellavista (road stretch 1: 6.84 km/4.25 mi) and from Bellavista to El Cascajo (road stretch 2: 6.96 km/4.32 mi). The share of bird roadkills per kilometer (or mile) of road stretch varied, as follows: (i) Road stretch 1 accumulated 3-times more bird roadkills (*N*=125) than road-stretch 2 (*N*=43) during the one-month census period. (ii) Road stretch 1 had more vehicle traffic at a higher speed than road stretch 2. (iii) When speed limit was higher (e.g. 50-km/30-mi per hour), bird mortality was much higher than at other sections of the road where speed limit was either lower (e.g. 30-km/20-mi per hour) or the road stretch had close-to-one-another variable speed limits (e.g. 30-50 30-km/20-30-20-mi per hour). (iv) Researchers extrapolated the census data to 220 bird carcasses per km per year on road stretch 1, and 74 bird carcasses per km per year on road stretch 2. (v) This implied that ≈1,500 birds could potentially be killed per year in entire road stretch 1, and ≈500 in entire road stretch 2 (total ≈2,000 birds in both road stretches per year)!

(9) Oceans' Microplastics Reach Galápagos. Scientists computer-simulated the trajectory of virtual particles [making them behave as microplastics in the simulations] that could arrive in Galápagos from diverse oceans or locations. (i) Adrift ocean plastics, particularly microplastics, could arrive in Galápagos primarily from the proximate ocean (i.e. the Tropical Eastern Pacific), although simulations also suggested that globally circulating microplastics could arrive in Galápagos from as far as Southern California or even the Southern and Indian Oceans. (ii) Most particles that did pass through Galápagos came from coastal Northern Peru and Southern Ecuador. (iii) La Niña and/or El Niño events did influence, seasonally and differentially, ocean particle flow off the coasts of Colombia, Ecuador, and Peru. (iv) If virtual particles originated at locations of

actual fisheries in the Tropical Eastern Pacific, only a small fraction (<1%) of them eventually passed through Galápagos; among those particles reaching Galápagos, +15% to +30% could come from fisheries located off the coasts of Ecuador and Peru.

(10) Microplastics in the Guts and Tissues of Marine Life: Coastal Ecuador and Galápagos. To estimate microplastics' particle concentrations in surface ocean waters off coastal Ecuador and the Galápagos Archipelago, biologists sampled 40 locations within an area of 453,000 km^2/175,000 sq mi. Samples were obtained along meridians 92°, 89°, 86°, and 82° West; and latitudes 2° North and 3° South. The highest microplastics' particle concentration (0.51 $\mu p/m^3$) was found in international waters; this region also had the highest mean value of 0.36 $\mu p/m^3$ (as per samples along meridian 86°W), followed by Ecuador's continental waters (mean 0.26 $\mu p/m^3$ at 82°W), Eastern Galápagos (mean 0.24 $\mu p/m^3$ at 89°W), and Western Galápagos (mean 0.22 $\mu p/m^3$ at 92°W). Microplastics were everywhere! Scientists assessed the prevalence of microplastic particles in the guts and muscles of marine species of economic value in the fish markets of coastal Ecuador, located just 900 km/600 mi from the Galápagos Archipelago. Researchers evaluated 16 species (N_T=240 samples), including predators (10 species), plankton consumers (5 species), and one detritivore. All the species assessed carried microplastics in their guts and/or muscles!

(11) Macroplastics Present in All Galápagos Islands. Researchers and "citizen scientists" (=local communities' members involved in Galápagos' conservation) assessed plastic pollution along shores and beaches of 11 islands in the Galápagos archipelago. The highest concentration of items was found on islands' sites exposed to prevalent winds (Northwest-West, NWW), which directed ocean water—carrying adrift plastics in it—toward the coast. Relying on a subset of macroplastics (N=466) of those collected across islands, researchers inferred (based on labels still readable on items) that they likely originated in Perú, Ecuador, or China, among 11 other countries [the Chinese fragments possibly came from fisheries active outside or proximate to the Galápagos Marine Reserve rather than from Asia].

(12) Plastics Ingested and Uptaken by All Trophic Levels. Scientists have hypothesized that the abrupt appearance of plastics in the environment has triggered maladaptive feeding responses in animals (H_i=plastic ingestion trap). This is because plastics resemble, visually and/or chemically, food items. Some behaviors become maladaptive when selective pressures change, usually because of human intervention, and animals are no longer able to assess the change or respond appropriately to it. The risk of plastic ingestion by organisms may be function of: (i) availability of plastics in the environment; (ii) nutritional state of an organism; (iii)

resemblance between plastics and natural prey/food items; and (iv) prey/food item selection by individuals. Global plastics' pollution is systemic, and it has permeated all trophic levels: 1,288 marine species and 277 terrestrial and freshwater species have been documented to ingest plastics. Five groups of vertebrates (sharks and rays, bony fishes, reptiles, birds, and mammals) have been reported to ingest macro or microplastics. The proportion of taxonomic orders within these groups—that ingest plastics—ranges from 24% (*N*=25 orders) in mammals to 75% (*N*=4 orders) in reptiles. And >25% of all reported taxonomic families within the five groups of vertebrates ingest plastics. It has become obvious that macro- and microplastics are being ingested by most, or all, organisms within communities and transferred from one trophic level to another: from "microscopic primary producers and detritus" or "suspended/deposited organic matter" to "primary consumers and detritivores" to "secondary consumers" to "predators" and "top predators"!

Questions

The following questions (Q1–3) merit generous written answers. You might need to do additional reading or library work to enrich your answers (we suggest exploring Paz-y-Miño-C and Espinosa 2024a,b).

Question 1: When examining Fig. 9.2, we indicated that the Galápagos Islands were threatened by numerous invasive species, the threat level of such invasions was "very high", and the ongoing management efforts to reduce the threat were considered—by scientists—to be "moderately effective (some concern)". We explained that "moderately effective (some concern)" referred to "*management… taking place but* [was] *not effectively reducing the threat and impact of IAS* [invasive alien species]—*this could affect the values and integrity of the* [World Heritage Site] *in the long term—adaptive management could potentially improve the situation. There* [was] *funding* (for management programs), *but it might not be adequate or sustained in the long run*". (a) What exactly is being done in Galápagos to mitigate the threats and impacts of invasive alien species? Short-term or long-term? (b) Will these efforts be enough to secure—long-term—Galápagos' status as a World Heritage Site? Why?

Question 2: In this chapter, we briefly discussed the "spillover effect", i.e., the export of fish biomass from the protected area (i.e. the Galápagos Marine Reserve) to the adjacent [tuna] fishing grounds (i.e. the current fishery hotspots proximate to GMR; Figs. 9.11-9.13). We also indicated that GMR was acting as a "well-protected and well-placed reserve"! It might cause you anxiety to realize that the fishes [tuna, sharks, and others of economic value] that find refuge within GMR can easily be hunted outside the reserve and in international waters. (a) What efforts are being made in the Tropical Eastern Pacific—where GMR is located—to ensure a large-scale protection of the fishes of economic value, their habitats, and the ecological and evolutionary processes essential to their persistence? (b) How do these efforts lead to the persistence of the overall marine biodiversity in the Tropical Eastern Pacific?

Question 3: In our book, A Synthesis of the Galápagos (Paz-y-Miño-C and Espinosa 2024b), we explain that "scientists have warned us that, as plastic debris continues to build up in Earth's environments (particularly in the oceans), microorganisms might 'hitchhike' on them to proliferate. Plastics in the 'plastisphere' [the global human-made plastic habitats] will offer additional substrates for both non-pathogenic and pathogenic unicellular aggregations to thrive. Heavy metals present in the environment and mutagens sorbed onto plastics would alter the cells' genetic material and enhance the emergence of antimicrobial-resistance genes. On a warming planet, which favors pathogen proliferation, all trophic levels would be affected, including humans as consumers of marine life" (for review, see Bowley et al. 2021). (a) How exactly will such interactions between microorganisms and microplastics alter the chemical and nutrient cycles in the ocean? (b) Will the "plastisphere" become the "norm biosphere" in which microplastics [and all plastics] are a ubiquitous and ordinary component of the environment? (c) Are microplastics a new "environmental unknown"? Why? [Note that Norman Myers formulated the concept of "environmental unknowns" in 1995: "Among the environmental problems ahead, the most important ones could be those that are still unknown to us". Since Myers (1995), much conceptual progress has been made with respect to this topic].

How to Answer the Questions Above

The questions above have various possible answers (A1–3). Here we suggest perspectives from which to address the queries.

Answer 1: There are numerous and ongoing conservation programs in Galápagos to eradicate invasive alien species. Some programs are successful; others confront challenges. You have to identify them by doing library work. The Galápagos National Park, together with non-governmental organizations (local and international), dedicates resources and personnel to address the problems caused by invasive alien species. Your answer to Q1 a–b should be comprehensive and documented. Note that, in Chapter

Eight, we already discussed several topics regarding the Galápagos designation and status as a World Heritage Site.

Answer 2: (a–b) Keep in mind the following three points to address Q2. First, yes, the data presented in Figs. 9.11-9.13 suggests that the Galápagos Marine Reserve (GMR) is a "well-protected and well-placed reserve" (i.e. from the perspective of the nearby tuna fishing grounds). Second, the "spillover effect" seems real, and the data in Figs. 9.11-9.13 supports it (again, from the perspective of the nearby tuna fishing grounds). And third, you need to find out information about the 2021 pledge by Ecuador, Costa Rica, Panama, and Colombia to join efforts to protect the Tropical Eastern Pacific. Various formal agreements were signed after the 2021 commitment. Your answer should be comprehensive.

Answer 3: To answer Q3 a–c, you need to explore the literature about the "plastisphere" (e.g. Bowley et al. 2021) and link it to the concept of "environmental unknowns" (Myers 1995). Now, realize that both concepts have developed since they were first proposed (look for additional references). Your answer should be analytical rather than descriptive.

Discussion Topics

The discussion topics below (D1–D2) should be addressed with your peers. We encourage college instructors to facilitate the dialogue.

Discussion 1: As noted in this chapter [and in Paz-y-Miño-C and Espinosa (2024b)], an organized network of wildlife traffickers, with local (Galápagos), in-country (Ecuador), and international intermediaries (Perú and other nations), operates to target-smuggle from UNESCO-World-Heritage Galápagos the most iconic endemic reptiles: giant tortoises (Fig. 9.19), whose species are listed by IUCN as either critically endangered, endangered, or vulnerable, with two lineages extinct in the recent past (IUCN 2025). The problem is intricate since it involves wildlife harvesters, trade intermediaries, and consumers of wildlife, all interconnected at local, regional, and global scales (Paz-y-Miño-C and Espinosa 2024a,b; see also Phelps et al. 2016; Hughes 2021). International wildlife trade imposes additional pressure on already-threatened species and frustrates and/or corrupts conservation efforts. As we pointed out in this chapter, as well as in Chapter Five [and in Paz-y-Miño-C and Espinosa (2024a,b)], parallel to wildlife trafficking, an additional concern is intensifying in the Tropical Eastern Pacific: drug smuggling and its concealed networks of delivery with which wildlife shipments could merge. [Ecuador is primarily a transit country in regional and global drug smuggling (UNODC 2021)]. Your DISCUSSION should be based on the following questions: (a) What are the *ultimate factors* that cause the emergence of wildlife trafficking in Galápagos? (b) How exactly does wildlife trade *impose additional pressure on already threatened species* like giant tortoises or land/marine iguanas? (c) How exactly does wildlife trade *frustrate and/or corrupt conservation efforts*? (d) What should be done locally (Galápagos), in-country (Ecuador), and internationally to stop the illegal trade of Galápagos wildlife? (e) What can be done to dismantle the networks of wildlife trade and prevent their merging with drug trafficking in the Tropical Eastern Pacific?

Discussion 2: In this chapter, we introduced readers to the concept of "maladaptive response", which is often used by conservation behaviorists and cognitive ethologists (Shettleworth 2001; Paz-y-Miño-C 2014). We explained that, when exposed to sudden, human-induced environmental change, [some] animals cannot assess the change or respond appropriately to it. We discussed two examples: birds' collisions with vehicles (Fig. 9.20) and ingestion of plastics by [animals] at all trophic levels (Figs. 9.31-9.33). The scientific literature provides hundreds of case studies regarding maladaptive responses. Your DISCUSSION should be based on the following questions: (a) Can you provide five additional examples of maladaptive responses by iconic Galápagos animals (terrestrial or marine)? (b) Which of your exemplar species will be most vulnerable to the adverse consequences of maladaptive responses? Why? (c) Can such maladaptive responses by your exemplar Galápagos species lead to their population decline? To their vulnerability or even extinction? How? (d) If maladaptive responses are behavioral or cognitive phenomena, can scientists condition or train animals to no longer show maladaptive responses? This discussion needs previous library work.

Suggested Observation Projects

The suggested observation projects below (P1–P2) take into consideration that you will travel to Ecuador and the Galápagos legally; that you will visit the islands as a tourist [holding a Transit Control Card, a mandatory document issued by Ecuador and specifically by the Governing Council of Galápagos]; and that you will respect the regulations that protect the Galápagos National Park and the Marine Reserve (see complete list of rules in Appendix B). Only accredited scientists, academics, and/or naturalists who have obtained official permits can do research in Galápagos. You can take pictures, but flash photography is not permitted when taking photos of wildlife. Professional photography and videos recorded for commercial purposes must be authorized by the Galápagos National Park

Directorate (GNPD). The ideas below are for enthusiastic college explorers interested in learning about the archipelago.

Project 1: Each of the human-inhabited islands in the Galápagos has an interpretation center or an office run by the Galápagos National Park Directorate (GNPD). GNPD's headquarters and the central offices of the Charles Darwin Research Station are located on Santa Cruz Island. Each island also has world-famous giant tortoise breeding facilities: Fausto Llerena (Santa Cruz), Galapaguera de Cerro Colorado or David Rodríguez/former Jacinto Gordillo (San Cristóbal), Arnaldo Tupiza (Isabela), and Asilo de la Paz, part of Floreana's Restoration Project (Floreana; Paz-y-Miño-C and Espinosa 2024b). [The latter aims at recovering the ecological function of the tortoises—as top herbivores—since the local endemic species (*Chelonoidis elephantopus*) went extinct in the mid-1800s (but see Miller et al. 2017). Tortoises from other islands were translocated to Floreana and are currently kept in fenced areas at Asilo de la Paz (as discussed in Paz-y-Miño-C and Espinosa 2024a,b)]. During your trip to Galápagos, you will likely explore various interpretation centers and the giant tortoise breeding facilities. Here is your OBSERVATION PROJECT: Put together a 20-minute-long amateur documentary [video] about the conservation threats affecting the iconic and emblematic Galápagos giant tortoises. Use the settings of both the interpretation centers and the breeding facilities [hopefully of all islands] to structure your documentary. Combine video with still images; mix clips from the interpretation centers with those from the breeding facilities and scenes captured in the wild (if you hike in the islands, you will certainly see giant tortoises in their turf). Be the host/narrator of your documentary. Keep your amateur documentary rigorous yet fun and didactic. [As stated above, only accredited scientists, academics, and/or naturalists who have obtained official permits can do research in Galápagos. Professional photography and videos recorded for commercial purposes must be authorized by GNPD].

Project 2: Do you want to help clean up the Galápagos beaches and remove their macroplastics? You might want to contact in advance the Galápagos National Park Directorate (GNPD) and offer to participate—as a volunteer—in their "citizen scientists" programs, also called "*ciencia ciudadana*" (=local communities' members involved in Galápagos' conservation *https://galapagos.gob.ec/voluntarios/*). Note that non-governmental organizations (NGOs and universities), such as the Charles Darwin Foundation, University San Francisco of Quito and its Galápagos Science Center, Galápagos Conservancy, or Galápagos Conservation Trust (among others), also offer opportunities for volunteers. As discussed in Fig. 9.30, researchers and "citizen scientists" periodically assess plastic pollution along shores and beaches in the archipelago. Teams of locals and international helpers remove macroplastics wherever possible, while researchers coordinate the technical inventory of plastic pollution. You can document your team's work and outcomes in a notebook; take pictures to illustrate the methods followed. Organize the data in an Excel file and rely on descriptive statistics to summarize the results. [As stated above, only accredited scientists, academics, and/or naturalists who have obtained official permits can do research in Galápagos]. Put together a slideshow presentation (for a 20-minute talk) about "Cleaning Up Galápagos' Macroplastics: My 'Citizen Science Work' as a 'Citizen-of-the-World Volunteer'" to be shared with a general audience. Avoid slides with "text only" [they are often terrible]. Keep it rigorous yet fun and didactic.

References

Alfaro-Núñez, Alonzo, Diana Astorga, Lenin Cáceres-Farías, Lisandra Bastidas, Cynthia Soto Villegas, Kewrin Macay, and Jan H. Christensen. 2021. "Microplastic Pollution in Seawater and Marine Organisms Across the Tropical Eastern Pacific and Galapagos". *Scientific Reports*, No. 11: 6424.

AP (Associated Press). 2021. "Ecuador: Prisión a Policía por Tráfico de Tortugas Galápagos". Accessed May 1, 2025. *https://apnews.com/article/noticias-f3c273bcefe917210e28016e0b7fd7cc*.

Auliya, Mark, Vincent Nijman, Sandra Altherr, Washington T. Aguilera, Daniel Ariano-Sanchéz, Juan Carlos Cantu, Giuliano Colosimo, Gabriele Gentile, Glenn Gerber, Tandora Grant, Evert Henningheim, Alice Hughes, Charles R. Knapp, Susan Lieberman, Catherine Malone, Stesha A. Pasachnik, Gohar Petrossian, Christian Sevilla, Monique Sosnowski, and Bruce Weissgold. 2025. "Trafficking of Galápagos Iguanas as an Example of a Global Problem: CITES Permits, Laundering and the Role of Transit Countries in Europe and Africa". *Biological Conservation*, No. 305: 111104.

Boerder, Kristina, Andrea Bryndum-Buchholz, and Boris Worm. 2017. "Interactions of Tuna Fisheries with the Galápagos Marine Reserve". *Marine Ecology Progress Series*, No. 585: 1-15.

Bonaccorso, Elisa, Nicté Ordóñez-Garza, Diana A. Pazmiño, Alex Hearn, Diego Páez-Rosas, Sebastián Cruz, Juan Pablo Muñoz-Pérez, Eduardo Espinoza, Jenifer Suárez, Lauren D. Muñoz-Rosado, Andrea Vizuete, Jaime A. Chaves, Maria de Lourde Torres, Walter Bustos, Danny Rueda, Maximilian Hirschfeld, and Juan M. Guayasamin. 2021. "International Fisheries Threaten Globally Endangered Sharks in the Eastern Tropical Pacific Ocean: the Case of the Fu Yuan Yu Leng 999 Reefer Vessel Seized within the Galápagos Marine Reserve". *Scientific Reports*, No. 11: 14959.

Bowley, Jake, Craig Baker-Austin, Adam Porter, Rachel Hartnell, and Ceri Lewis. Oceanic Hitchhikers–Assessing Pathogen Risks from Marine Microplastic". *Trends in Microbiology*, No. 29: 107-116.

Brookes, Victoria J., Okta Wismandanu, Etih Sudarnika, Justin A. Roby, Lynne Hayes, Michael P. Ward, Chaerul Basri, Hendra Wibawa, Joan Davis, Dikky Indrawan, Jennifer Manyweathers, Widagdo S. Nugroho, Sarasati Windria, and Marta Hernandez-Jover. 2022. "A Scoping Review of Live Wildlife Trade in Markets Worldwide". *Science of the Total Environment*, No. 819: 153043.

Carlton, James T., Inti Keith, and Gregory M. Ruiz. 2019. "Assessing Marine Bioinvasions in the Galápagos Islands: Implications for Conservation Biology and Marine Protected Areas". Aquatic Invasions, No. 14: 1-20.

Ceballos, Gerardo, Paul R. Ehrlich, and Peter H. Raven. 2020. "Vertebrates on the Brink as Indicators of Biological Annihilation and the Sixth Mass Extinction". *Proceedings of the National Academy of Sciences of the United States of America*, No. 117: 13596-13602.

Di Lorenzo, Manfredi, Paolo Guidetti, Antonio Di Franco, Antonio Calò, and Joachim Claudet. 2020. "Assessing Spillover from Marine Protected Areas and Its Drivers: A Meta-Analytical Approach". *Fish and Fisheries*, No. 21: 906-915.

Edgar, Graham J., Samuel Banks, José M. Farina, Mónica Calvopiña, and Camilo Martínez. 2004. "Regional Biogeography of Shallow Reef Fish and Macro-Invertebrate Communities in the Galapagos Archipelago". *Journal of Biogeography*, No. 31: 1107-1124.

El Universo. 2019. "Tortugas Bebés Galápagos, Sacadas de Islas por Mar a Perú". Accessed August 16, 2023. *https://www.eluniverso.com/ noticias/2019/09/15/nota/7518766/tortugas-bebes-sacadas-islas-mar-piura/.*

Fields, Andrew T., Gunter A. Fischer, Stan K. H. Shea, Huarong Zhang, Kevin A. Feldheim, and Demian D. Chapman. 2020. "DNA Zipcoding: Identifying the Source Populations Supplying the International Trade of a Critically Endangered Coastal Shark". *Animal Conservation*, No. 23: 670-678.

Friess, Daniel A., Kerrylee Rogers, Catherine E. Lovelock, Ken W. Krauss, Stuart E. Hamilton, Shing Yip Lee, Richard Lucas, Jurgenne Primavera, Anusha Rajkaran, and Suhua Shi. 2019. "The State of the World's Mangrove Forests: Past, Present, and Future". *Annual Review of Environment and Resources*, No. 44: 89-115.

Galápagos Conservancy. 2023. "Galápagos Wildlife Trafficker Sentenced to Three Years in Prison". Accessed September 3, 2023. *https://www.galapagos.org/newsroom/galapagos-wildlife-trafficker/*.

García-Carrasco, José-María, Washington Tapia, and Antonio-Román Muñoz. 2020. "Roadkill of Birds in Galapagos Islands: A Growing Need for Solutions". *Avian Conservation and Ecology*, No. 15: 19.

GFW (Global Fishing Watch). 2023. "Revolutionizing Ocean Monitoring and Analysis". Accessed September 6, 2023. *https://globalfishingwatch.org/*.

Hinsley, Amy, Jasmin Willis, Abigail R. Dent, Rodrigo Oyanedel, Takahiro Kubo, and Daniel W. S. Challender. 2023. "Trading Species to Extinction: Evidence of Extinction Linked to the Wildlife Trade". *Cambridge Prisms: Extinction*, No.1: e10.

Hughes, Alice. 2021. "Wildlife Trade". *Current Biology*, No. 31: R1141-R1224.

IUCN (International Union for Conservation of Nature). 2025. "The IUCN Red List of Threatened Species". Accessed May 7, 2025. *https://www.iucnredlist.org/*.

Jones, Jen S., Adam Porter, Juan Pablo Muñoz-Pérez, Daniela Alarcón-Ruales, Tamara S. Galloway, Brendan J. Godley, David Santillo, and Jessica Vagg, Ceri Lewis. 2021. "Plastic Contamination of a Galapagos Island (Ecuador) and the Relative Risks to Native Marine Species". *Science of the Total Environment*, No. 789: 147704.

Lenzner, Bernd, Guillaume Latombe, César Capinha, Céline Bellard, Franck Courchamp, Christophe Diagne, Stefan Dullinger, Marina Golivets, Severin D. H. Irl, Ingolf Kühn, Brian Leung, Chunlong Liu, Dietmar Moser, Núria Roura-Pascual, Hanno Seebens, Anna Turbelin, Patrick Weigelt, and Franz Essl. 2020. "What Will the Future Bring for Biological Invasions on Islands? An Expert-Based Assessment". *Frontiers in Ecology and Evolution*, No. 8: 280.

Miller, Joshua M., Maud C. Quinzin, Nikos Poulakakis, James P. Gibbs, Luciano B. Beheregaray, Ryan C. Garrick, Michael A. Russello, Claudio Ciofi, Danielle L. Edwards, Elizabeth A. Hunter, Washington Tapia, Danny Rueda, Jorge Carrión, Andrés A. Valdivieso, and Adalgisa Caccone. 2017. "Identification of Genetically Important Individuals of the Rediscovered Floreana Galápagos Giant Tortoise (*Chelonoidis elephantopus*) Provides Founders for Species Restoration Program". *Scientific Reports*, No. 7: 11471.

Moreira-Mendieta, Andrés, Odei Garcia-Garin, Juan Pablo Muñoz-Pérez, Diego O. Urquía, Massimiliano Drago, Asunción Borrell, and Diego Páez-Rosas. 2023. "Detection and Quantification of Microplastic Pollution in the Endangered Galapagos Sea Lion". *Science of The Total Environment*, No. 00: 166223.

Muñoz-Pérez, Juan Pablo, Gregory A. Lewbart, Daniela Alarcón-Ruales, Alice Skehel, Esteban Cobos, Roberth Rivera, Alexis Jaramillo, Henry Vivanco, Leo Zurita-Arthos, Bryan Wallace, Carlos A. Valle, and Kathy A. Townsend. 2023. "Galápagos and the Plastic Problem". *Frontiers in Sustainability*, No. 4: 1091516.

Myers, Norman. 1995. "Environmental Unknowns". *Science*, No. 269: 358-360.

Pacoureau, Nathan, Cassandra L. Rigby, Peter M. Kyne, Richard B. Sherley, Henning Winker, John K. Carlson, Sonja V. Fordham, Rodrigo Barreto, Daniel Fernando, Malcolm P. Francis, Rima W. Jabado, Katelyn B. Herman, Kwang-Ming Liu, Andrea D. Marshall, Riley A. Pollom, Evgeny V. Romanov, Colin A. Simpfendorfer, Jamie S. Yin, Holly K. Kindsvater, and Nicholas K. Dulvy. 2021. "Half a Century of Global Decline in Oceanic Sharks and Rays". *Nature*, No. 589: 567-571.

Páez-Rosas, Diego, Jorge Torres, Eduardo Espinoza, Adrian Marchetti, Harvey Seim, and Marjorie Riofrío-Lazo. 2021. "Declines and Recovery in Endangered Galapagos Pinnipeds During the El Niño Event". *Scientific Reports*, No. 11: 8785.

Paz-y-Miño-C, Guillermo. 2014. "Cognitive Repertoire, Cognitive Legacies, and Collective Reasoning: The Next Frontiers in Conservation Behavior". In *Animal Behavior: How and Why Animals Do the Things They Do,* edited by Ken Yasukawa, Volume 3, 289-310. Santa Barbara: Praeger, ABC-CLIO, LLC.

—. and Avelina Espinosa. 2024a. *Arrivals of Life to the Galápagos: From Berlanga to Darwin, Melville, and World War II.* Newcastle upon Tyne: Cambridge Scholars Publishing.

—. 2024b. *A Synthesis of the Galápagos: Natural History, Human Conflict, and the Destiny of an Archipelago*. Newcastle upon Tyne: Cambridge Scholars Publishing.

Phelps, Jacob, Duan Biggs, and Edward L. Webb. 2016. "Tools and Terms for Understanding Illegal Wildlife Trade". *Frontiers in Ecology and the Environment*, No. 14: 479-489.

Riofrío-Lazo, Marjorie, Francisco Arreguín-Sánchez, and Diego Páez-Rosas. 2017. "Population Abundance of the Endangered Galapagos Sea Lion *Zalophus wollebaeki* in the Southeastern Galapagos Archipelago". *PLoS ONE*, No. 12: e0168829.

Salinas-de-León, Pelayo, Sonia Andrade, Camila Arnés-Urgellés, Jorge R. Bermudez, Santiago Bucaram, Sarah Buglass, Florencia Cerutti, William Cheung, Carmen De la Hoz, Vicky Hickey, Gustavo Jiménez-

Uzcátegui, Inti Keith, José R. Marín Jarrín, Patricia Martí-Puig, Mónica Medina, Andrés Moya, Daniel Pauly, Daniel Orellana, Regina Ostergaard-Klem, Charles Stock, Jon Witman, and Boris Worm. 2020. "Evolution of the Galapagos in the Anthropocene". *Nature Climate Change*, No. 10: 380-384.

—. David Acuña-Marrero, Etienne Rastoin, Alan M. Friedlander, Mary K. Donovan, and Enric Sala. 2016. "Largest Global Shark Biomass Found in the Northern Galápagos Islands of Darwin and Wolf". *PeerJ*, No. 4: e1911.

Santos, Robson G., Gabriel E. Machovsky-Capuska, and Ryan Andrades. 2021. "Plastic Ingestion as an Evolutionary Trap: Toward a Holistic Understanding". Science, No. 373: 56-60.

Scheffers, Brett R., Brunno F. Oliveira, Ieuan Lamb, and David P. Edwards. 2019. "Global Wildlife Trade Across the Tree of Life". Science, No. 366, 71-76.

Schiller, Laurenne, Juan José Alava, Jack Grove, Günther Reck, and Daniel Pauly. 2015. "The Demise of Darwin's Fishes: Evidence of Fishing Down and Illegal Shark Finning in the Galápagos Islands". *Aquatic Conservation: Marine and Freshwater Ecosystems*, No. 25: 431-446.

Shackleton, Ross T., Bastian Bertzky, Louisa E. Wood, Nancy Bunbury, Heinke Jäger, Remco van Merm, Christian Sevilla, Kevin Smith, John R. U. Wilson, Arne B. R. Witt, and David M. Richardson. 2020. "Biological Invasions in World Heritage Sites: Current Status and a Proposed Monitoring and Reporting Framework". Biodiversity and Conservation, No. 29: 3327-3347.

Shettleworth, Sara J. 2001. Animal Cognition and Animal Behaviour. *Animal Behaviour*, No. 61: 277-286.

Sun-Wyler, Liana, and Pervaze A. Sheikh. 2013. *International Illegal Trade in Wildlife: Threats and U.S. Policy*. Washington DC: Congressional Research Service.

Toral-Granda, M. Verónica, Charlotte E. Causton, Heinke Jäger, Mandy Trueman, Juan Carlos Izurieta, Eddy Araujo, Marilyn Cruz, Kerstin K. Zander, Arturo Izurieta, and Stephen T. Garnett. 2017. "Alien Species Pathways to the Galapagos Islands, Ecuador". *PLoS ONE*, No. 12: e0184379.

't Sas-Rolfes, Michael, Daniel W. S. Challender, Amy Hinsley, Diogo Veríssimo, and E. J. Milner-Gulland. 2019. "Illegal Wildlife Trade: Scale, Processes, and Governance". *Annual Review of Environment and Resources*, No. 44: 201-228.

UNODC (United Nations Office on Drugs and Crime). 2021. *World Drug Report–Drug Market Trends: Cocaine Amphetamine-Type Stimulants*. Vienna: United Nations.

van Sebille, Erik, Philippe Delandmeter, John Schofield, Britta Denise Hardesty, Jen Jones, and Andy Donnell. 2019. "Basin-Scale Sources and Pathways of Microplastic that Ends Up in the Galápagos Archipelago". Ocean Science, No. 15: 1341-1349.

Ventura, Francesco, Jason Matthiopoulos, and Jana W. E. Jeglinski. 2019. "Minimal Overlap between Areas of High Conservation Priority for Endangered Galapagos Pinnipeds and the Conservation Zone of the Galapagos Marine Reserve". *Aquatic Conservation: Marine and Freshwater Ecosystems*, No. 29: 115-126.

CHAPTER TEN

HOW WILL CLIMATE CHANGE AFFECT GALÁPAGOS?

This chapter is short, yet crucial. Earlier in this book, we discussed some direct or indirect impacts of climate change on Galápagos natural history. For example, in section *The Ocean and Its Turbulent Waters*, Chapter One (see Fig. 1.32), we described ocean currents in the Tropical Eastern Pacific as they were influenced by yearly variation in regional and global weather patterns. Also in Chapter One, section *The Upwelling*, we examined the hypotheses (*Hi* and *Hii*) about the drivers of the Galápagos' oceanic upwelling on the West side of the archipelago (see Figs. 1.33-1.34) and indicated that the intensity in the upwelling depended either on *Hi* remote/stochastic events (currents' interactions; seasonality El Niño/La Niña) or on *Hii* local atmosphere-and-oceanic interactions (wind). And in section *The "Hot" and "Cool" Seasons* (Figs. 1.43-1.44, Chapter One), we reintroduced readers to El Niño and La Niña events and how they characterized Galápagos' seasonality and were subject, too, to variation in regional and global weather. [For ample analysis of these topics, see Paz-y-Miño-C and Espinosa (2024a,b)]. In Chapter Nine, section *The Decline of Sea Lions and Fur Seals* (see Figs. 9.14-9.16), we highlighted that scientists have alerted that because severe El Niño events can disrupt Galápagos' sea lion (*Zalophus wollebaeki*) and fur seal (*Arctocephalus galapagoensis*) populations—as climate change continues to intensify El Niño impacts—both species could become critically endangered or even extinct. [*Zalophus wollebaeki* and *Arctocephalus galapagoensis* are currently listed as endangered by IUCN (2025)]. We explained that extreme weather fluctuations and higher air/sea surface temperatures will disrupt nutrient cycles and fish-food availability for marine-mammal predators like sea lions and fur seals, and sea level rise will cover breeding areas—for colonies—ashore (for a list of references, see Chapter Nine). Lastly, we also referred to "climate change" at the end of several of the chapters of this book, i.e., in the sections on *questions*, *answers*, *discussion topics*, and *suggested observation projects*.

Readers may ask, can we summarize—in a single paragraph—probable climate-change impacts on the Galápagos? Here we do it in 468 words! From 1981 to 2017, the average land-surface temperature increased by 0.6ºC/1.08ºF in the lowlands of the Galápagos Islands. Scientists expect an increase between 1.4ºC/2.52ºF and 1.9ºC/3.42ºF by 2050, which will correlate with an estimated rise in average precipitation of 30% to 45% (Paltán et al. 2021). Lots of torrential rain is coming! Extreme weather (severe hot-rainy and cool-dry seasons) and sea-level rise (globally expected to reach 33.3 cm/1.1 ft in 100 years; NASA Sea Level Change 2025) will bring stress to both natural habitats (e.g. mangroves, seashores, sandy beaches, deciduous shrublands/forests, and misty forests) and human settlements (i.e. cities, towns, rural villages, and farmers' houses and storage facilities). Frequent rainstorms will affect buildings, harbors, homes, and infrastructure connectivity (runways, roads, and bridges); sewage pipes will break and lead to further contamination of the drinking water (already a problem in Galápagos, as discussed in our volume, *A Synthesis of the Galápagos*; Paz-y-Miño-C and Espinosa 2024b). Nature hiking or biking trails in the areas open to visitors will have to be repaired or rebuilt, as will viewing platforms, observation decks, and interpretive signs (additional costs to the Galápagos National Park). Economic losses will impact tour/expedition providers, hotels, restaurants, groceries and fish markets, shops (e.g. hiking, biking, scuba, or snorkeling gear, and souvenirs), and cultivated areas (including the increasingly popular eco-farms and agro-tourism sites). [Remember that Galápagos' economy is powered by tourism, as discussed in Chapter Seven, *Why Is Nature Tourism Essential to Galápagos?* Figs. 7.10-7.13]. Food supply to the islands, either coming from the continent (e.g. cargo ships) or being distributed within the archipelago (i.e. small boats), will be affected by extreme weather (e.g. more turbulent seas will lead to longer journeys, extra use of fuel, and higher market prices for goods and supplies). Depending on the severity of climate change impacts on the human population (e.g. if tourism drops significantly or collapses due to natural disasters, combined with large-scale loss of farmlands to urbanization as well as inflated costs of living), future sizeable emigration back to the continent could result (reviews in Hauer et al. 2020; Zickgraf 2020). Climate change will accentuate differential financial stress among dissimilar income groups. Socio-economic inequality and unemployment could grow, as could social tensions and the lack of safety and security for the public and tourists (review in Nguyen et al. 2023). Keep in mind that Galápagos continues to be one of the safest international destinations (with almost nonexistent violent offenses), but, in recent years, civil unrest, crime, and kidnapping have increased in continental Ecuador (see more about this topic in Chapters Five and Nine; GOV.UK 2025; OECO 2025; US Department of State 2025). Will Galápagos remain safe? It is hard to be sure. The point here is that climate change will impact almost all aspects of human life and people's interactions with nature!

Below, we further elaborate on two topics regarding climate-change impacts on both terrestrial and marine ecosystems in the Galápagos: *Extreme Weather and Its Sequels* (Figs. 10.1-10.2) and *Impacts of Climate Change on Coral Reefs* (Fig. 10.3 and Table 10.1). [Remember that we discussed the conservation value of coral reefs in Chapter Eight: sections *Darwin and Wolf: Source of Coral Larvae* (Figs. 8.60-8.62; Table 8.8) and *Interactions between Coral Reefs and Sharks* (Figs. 8.63-8.67)].

Extreme Weather and Its Sequels

Climate change is expected to generate extreme weather (globally, regionally, and locally), from very hot to very cold and from very wet to very dry (Schleussner et al. 2018). The Galápagos Islands are warming up and—for now—becoming rather dry. Scientists have monitored temperature and rainfall trends from 1981 to 2017 at meteorological stations in the highlands and lowlands of Santa Cruz (*Indefatigable*) and San Cristóbal (*Chatham*) Islands (Fig. 10.1; Paltán et al. 2021). During the observation period (35 years), mean annual land surface temperature (both islands) ranged from 22ºC/72ºF (cool-dry season) to 26ºC/79ºF (hot-wet season). In the highlands of both islands (>250-m/>820-ft elevation), mean land surface temperature increased by 0.21ºC/0.38ºF; the main increase occurred during the cool-dry season (Jun-Nov). In the lowlands of both islands (<250-m/<820-ft elevation), mean land surface temperature increased by 0.6ºC/1.08ºF; the main increase occurred during the hot-wet season (Dec-May). [Remember that we already discussed Galápagos' seasonality in Chapter One; see Figs. 1.43-1.44]. Precipitation was highly variable, particularly during El Niño years. In the highlands of both islands, mean annual rainfall ranged from 1,050 mm/41 in (cool-dry season) to 1,670 mm/66 in (hot-wet season), with no accentuated differences between seasons. By contrast, in the lowlands of both islands, mean annual rainfall reached ≈500 mm/≈20 in, concentrated mainly in the hot-wet season; during the cool-dry season, mean annual rainfall reached ≈130 mm/≈5 in, with accentuated difference between seasons (Fig. 10.1). Note that during strong El Niño years (1982-1983 or 1997-1998), mean annual precipitation reached ≈3,000 mm/≈118 in versus non-El Niño years ≈1,100 mm/≈43 in (Paltán et al. 2021)!

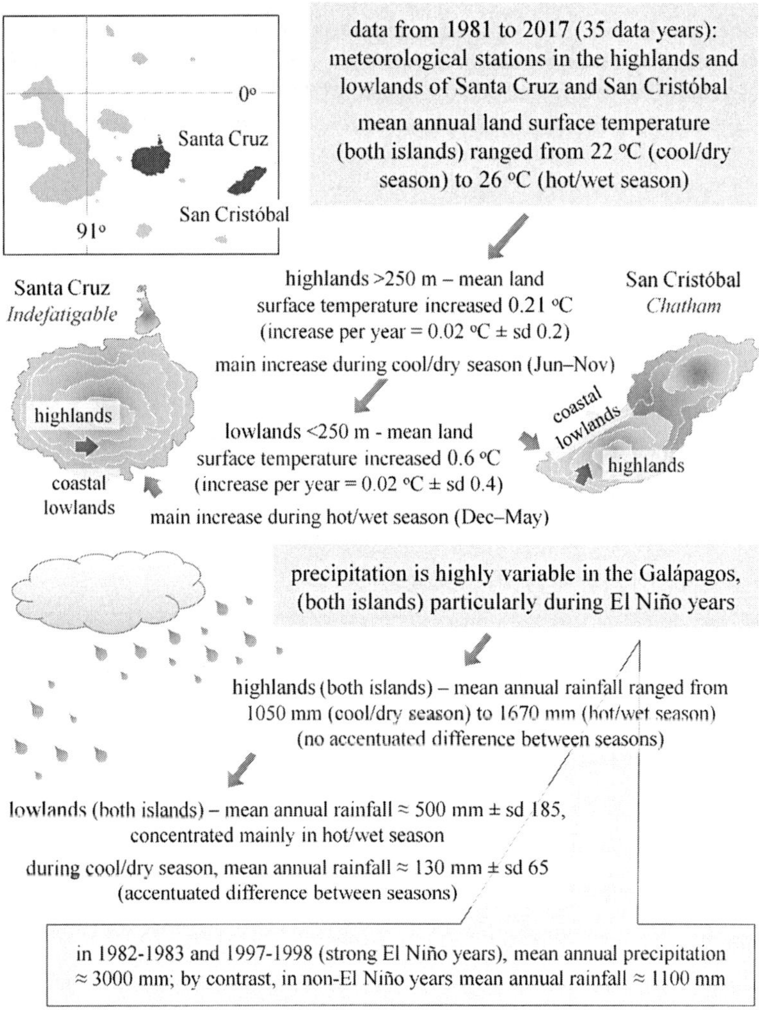

Fig. 10-1 Temperature and rainfall trends in the highlands and lowlands of Santa Cruz (*Indefatigable*) and San Cristóbal (*Chatham*) Islands, Galápagos: 1981 to 2017. Layout: GPC; data extracted from Paltán et al. (2021).

Examine Fig. 10.2. In it, we show cumulative annual precipitation as a function of time (from 0 to 365 days of the year) in the highlands (Fig. 10.2a) and lowlands (Fig. 10.2b) of Santa Cruz Island (after Paltán et al. 2021). The data (1981-2019) can be used to infer the “onset of the [hot-wet] rainy season", i.e., the day since December 1st when the cumulative 10% of the total annual rainfall has been achieved. Note how during the 38-year observation period, the onset of the rainy [wet] season in the highlands had a 7-day delay, while in the lowlands, it had a ≈20-day delay.

Galápagos appears to be destined to have (1) less rain during the cool-dry season (drier archipelago for now, but see below), combined with (2) a delay in the onset of the hot-wet season (rains will start later), plus a mix of (3) severe hot-wet seasons during more frequent El Niño years and (4) intense cool-dry seasons during La Niña years. Climate change shall bring extreme variation in seasonality to Galápagos!

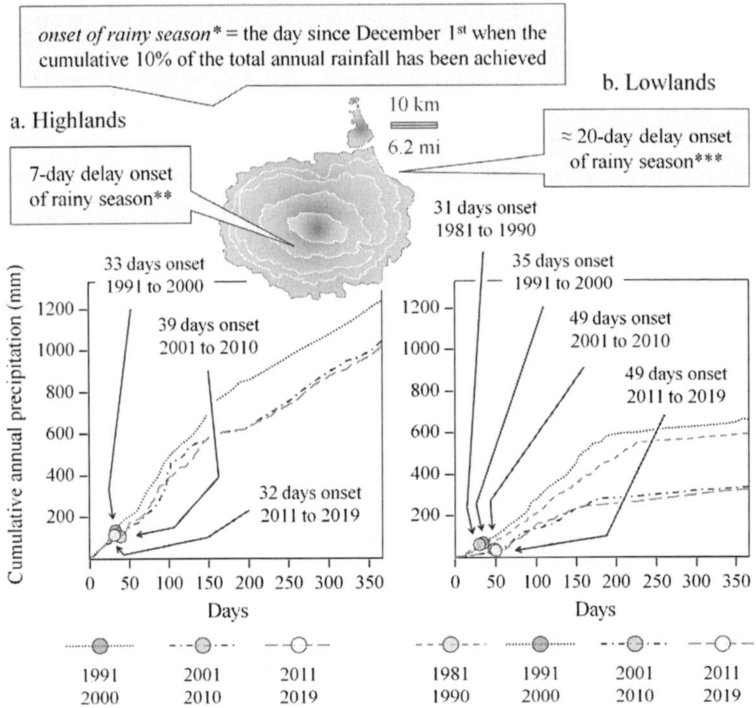

* the beginning of the rainy season is conventionally set to December 1st; ** the 7-day delay in the onset of the rainy season for highlands comes from the difference between the 33/32 days versus 39 days corresponding to 1991 to 2000/2001 to 2010 versus 2011 to 2019; *** the ≈ 20-day delay in the onset of the rainy season for lowlands corresponds to the difference between 1981 to 1990 versus 2011 to 2019

Fig. 10-2 Onset of the rainy [wet] season in the highlands and lowlands of Santa Cruz (*Indefatigable*) Island, Galápagos: data from 1981 to 2019. Layout: GPC; redrawn and modified after Paltán et al. (2021).

How is climate change affecting the Galápagos marine environments? In Fig. 10.3, we summarize sea surface temperature (SST) trends in the Galápagos Marine Reserve (GMR), as well as warming or cooling trends in SST detected at five bioregions within GMR (years 2002-2018). [We discussed marine bioregions in Chapter One; see Box 1.3, *Proposed Bioregions within the Galápagos Marine Reserve*, after Edgar et al. (2004)]. Note how mean SST has steadily increased by 1.2°C/2.16°F in GMR during the 16-year observation period (Fig. 10.3a). The warming or cooling trends in SST detected at the five marine bioregions within GMR are described in Fig. 10.3b (data from Paltán et al. 2021) as follows: (1) researchers observed high spatio-temporal heterogeneity in SST from 2002 to 2018 in the entire GMR (=all bioregions); (2) the differential exposure to warm or cool ocean currents during El Niño or La Niña years influenced the spatio-temporal SST heterogeneity (as in number 1); (3) frequency of hot years over time increased, e.g., 2002-2003, 2007-2008, 2010-2011, 2013-2014, and 2016-2017; (4) the Far North, North, and Southeast bioregions showed significant SST warming trends from 2002 to 2018; (5) after El Niño of 2015 (hot-rainy year) and La Niña of 2016 (cool-dry year), the Southeast bioregion had a warming trend until 2018; and (6) the West and Elizabeth Bay bioregions had no significant warming trends, but cooling trends were detected from 2002 to 2018. Notably, at Elizabeth Bay, researchers found instances of both unusually cool (–7.8°C/–14.0°F) or warm (+2.3°C/+4.14°F) SST, suggesting wide variations did occur and/or could become frequent in the future. In the longer term, these trends (1-6) shall alter marine habitats (particularly coral reefs), accelerate invasion of alien species in GMR (as documented in Chapter Nine, section *Introduced Species: 1535 Onwards*, Fig. 9.1b), and impact the local human population, which depends on a healthy ocean for subsistence, e.g., artisanal fishing and tourism (Carlton et al. 2019; Paltán et al. 2021; see also Paz-y-Miño-C and Espinosa 2024b).

Impacts of Climate Change on Coral Reefs

Extreme changes in seawater temperature can cause “bleaching” in corals. Under heat stress, photosynthetic microbes—symbionts of coral polyps—die and coral colonies turn white. Coral bleaching is a global phenomenon (van Woesik et al. 2022). [We also discuss coral reefs in our Galápagos books: Paz-y-Miño-C and Espinosa (2024a), chapter *Galápagos versus Other Archipelagos*, as well as Paz-y-Miño-C and Espinosa (2024b), chapters *Crowded Shores, Murky Waters, and the Open Sea*, and *Microbes and Viruses, Always Present*].

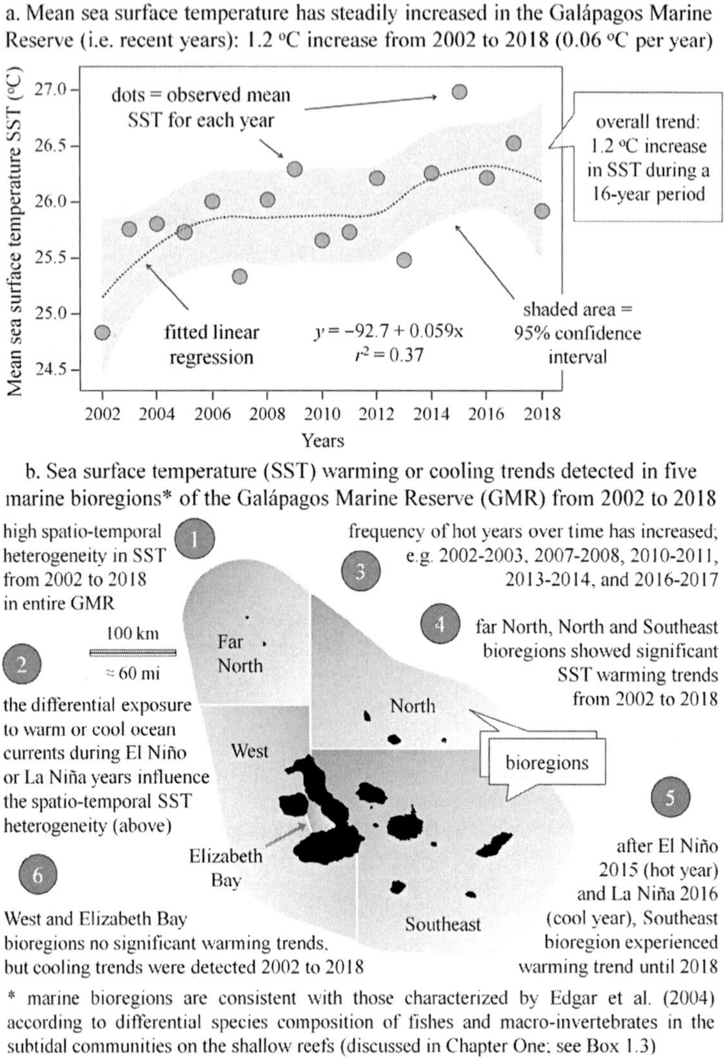

Fig. 10-3 Sea surface temperature (SST) trends in the Galápagos Marine Reserve (GMR) as well as warming or cooling trends in SST detected at five bioregions within GMR (2002-2018). Note in a the overall trend of 1.2°C/2.16°F increase in SST during a 16-year period, and in b the variable warming or cooling trends in SST detected at the five marine bioregions within GMR. At Elizabeth Bay, researchers found instances of both unusually cool or warm SST. Layout: GPC; redrawn and modified after Paltán et al. (2021).

Table 10.1 Beginning of severe coral-reef bleaching events at World Heritage Sites under three scenarios: current trends in global CO_2 emissions (projected 4.6 °C global mean temperature increase by 2100), half the current trends by 2050 (projected 2.4 °C increase by 2100), and significant reduction in global CO_2 emissions (i.e. The Paris Agreement) to limit the global mean temperature increase to 1.5 °C by 2100 (after Heron et al. 2018)

beginning of severe coral-reef bleaching events (year range): twice/decade or once/year	≤ 2025	> 2025 to ≤ 2040	> 2040 to ≤ 2055	> 2055 to ≤ 2099	not projected to occur by 2100
					–

World Heritage Sites with reefs	current trends 2/decade	current trends 1/year	half current trends 2/decade	half current trends 1/year	sig. reduction 2/decade	sig. reduction 1/year
Great Barrier Reef	2035	2044	2041	2051	–	–
Low Howe Island Group	2034	2043	2036	2055	–	–
Ningaloo Coast	2041	2049	2052	2074	–	–
Shark Bay, Western Australia	2038	2047	2045	2074	–	–
Belize Barrier Reef Reserve System	2028	2036	2036	2044	–	–
Brazilian Atlantic Islands	2028	2039	2035	2049	–	–
Malpelo Fauna and Flora Sanctuary	2038	2050	2056	2077	–	–
Cocos Island National Park	2019	2032	2028	2036	2062	–
Guanacaste Conservation Areas	2030	2043	2040	2055	–	–
Galápagos Islands	**2017**	**2036**	**2027**	**2042**	**2070**	**–**
Lagoons of New Caledonia	2031	2040	2039	2050	–	–
Komodo National Park	2017	2025	2021	2032	–	–
Ujungu Kulon National Park	2032	2043	2042	2053	–	–
Ogasawara Islands	2030	2038	2041	2049	–	–
Phoenix Islands Protected Area	2020	2035	2028	2040	2038	–
Gulf of California	2044	2052	–	–	–	–
Revillagigedo Archipelago	2031	2042	2043	2052	–	–
Sian Ka'an	2025	2033	2033	2041	–	–
Rock Islands Southern Lagoon	2028	2036	2032	2044	–	–
Coiba National Park	2030	2043	2040	2053	2053	–
Tubbataha Reefs Natural Park	2030	2039	2037	2048	–	–
Aldabra Toll	2038	2036	2034	2042	–	
East Rennell	2025	2033	2030	2044	–	–
iSimangaliso Wetland Park	2031	2040	2036	2048	–	–
Sanganeb and Dungonab Bay Mukkawar Island	2037	2046	2055	2069	–	–
Everglades National Park	2036	2044	2056	2071	–	–
Papahanaumokuakea	2029	2041	2044	2052	–	–
Ha Long Bay	2077	2086	–	–	–	–
Socotra Archipelago	2040	2048	2061	2077	–	–

Researchers modeled the onset of severe coral reef bleaching events at 30 World Heritage Sites (including the Galápagos Islands) resulting from conceivable global warming trends caused by variable rates of carbon dioxide (CO_2) emissions (Table 10.1; data from Heron et al. 2018). The models included three scenarios: (1) current trends in global CO_2 emissions (which projected a 4.6ºC/8.3ºF global mean temperature increase by the year 2100); (2) half the current trends by 2050 (projected 2.4ºC/4.3ºF increase by 2100); and (3) significant reduction in global CO_2 emissions (as per The Paris Agreement, TPA) to limit the global mean temperature increase to 1.5ºC/2.7ºF by 2100. [TPA=international treaty on climate change adopted at the United Nations Climate Change Conference in Paris, France, on December 12, 2015 (UNCC 2025)].

Examine the gray-color codes in Table 10.1. They highlight the projected beginning of coral reef bleaching events [at the 30 World Heritage Sites, WHS] from most severe bleach (dark gray) to least severe bleach (light gray) and "not projected to occur [bleach] by year 2100" (white). Each of the three scenarios in the models (i.e. current trends in global CO_2 emissions; half the current trends by the year 2050; and significant reduction in CO_2 emissions by 2100) covers a specific year range (below) during which coral reef bleaching events shall occur either twice per decade (coded 2/decade) or once per year (coded 1/year). The year range corresponds to ≤2025 or >2025 to ≤2040 (for "current trends in global CO_2 emissions"); >2040 to ≤2055 or >2055 to ≤2099 (for "half the current trends by year 2050"); and 2100 (for "significant reduction in CO_2 emissions by 2100"; rationale in Heron et al. 2018).

Locate the Galápagos Islands in Table 10.1. Note how, under current trends in global CO_2 emissions, most severe coral reef bleaching events, twice per decade, are expected to begin by the year 2017; or once per year by 2036. If global current CO_2 emissions were cut by half by the year 2050, the onset of severe coral reef bleaching events, twice per decade, would begin by the year 2027; or once per year by 2042. And if global CO_2 emissions were significantly reduced as per The Paris Agreement (above), mild coral reef bleaching events, twice per decade, would begin by the year 2070, and no once-per-year bleaching events would occur by 2100!

Compare the Galápagos Islands with Komodo National Park (Lesser Sunda Islands, Indonesia), Cocos Island National Park (Costa Rica), and Phoenix Islands Protected Area (Central Pacific Ocean, Republic of Kiribati). These WHS shall face comparable coral reef bleaching events as per the modeling scenarios of global CO_2 emissions above. Readers should realize that coral reefs at these sites are particularly vulnerable to global increases in seawater temperature! Still, if nothing is done to reduce global CO_2 emissions, 21 other WHS listed in Table 10.1 would be destined, too, to impactful bleaching events, at least twice per decade, during the 2020s to 2030s. And if CO_2 emissions were cut in half by 2050, 18 other WHS listed in Table 10.1 shall experience bleaching during the 2030s to 2040s!

Readers might understand why world environmental scientists insist that only a significant reduction in global CO_2 emissions, as per The Paris Agreement, will limit the global mean temperature increase to 1.5°C/2.7°F by the year 2100 and, thus, lessen the sequels of climate change (long-term shifts in temperatures and weather patterns) on coral reef communities.

Conclusions

We omit citations in this section to abridge the text. Each conclusion point summarizes topics substantiated by citations in this chapter.

(1) In Galápagos, climate change will impact almost all aspects of human life and people's interactions with nature. Remember that Galápagos' economy is powered—almost entirely—by nature tourism.

(2) Extreme weather (severe hot-rainy and cool-dry seasons) and sea-level rise will bring stress to both natural habitats and human settlements.

(3) Frequent rainstorms will affect buildings, harbors, homes, and infrastructure connectivity; sewage pipes will break and lead to further contamination of the drinking water (already a problem in Galápagos).

(4) Nature hiking or biking trails in the areas open to visitors will have to be repaired or rebuilt, as will viewing platforms, observation decks, and interpretive signs (additional costs to the Galápagos National Park).

(5) Economic losses will impact tour/expedition providers, hotels, restaurants, groceries and fish markets, shops, and cultivated areas.

(6) Food supply to the islands, either coming from the continent or being distributed within the archipelago, will be affected by extreme weather (e.g. more turbulent seas will lead to longer journeys, extra use of fuel, and higher market prices for goods and supplies).

(7) Depending on the severity of climate change impacts on the human population (e.g. if tourism drops significantly or collapses due to natural disasters, combined with large-scale loss of farmlands to urbanization as well as inflated costs of living), future sizeable emigration back to the continent could result—the eco-exodus or environmental refugees.

(8) Climate change will accentuate differential financial stress among dissimilar income groups. Socio-economic inequality and unemployment could grow, as could social tensions and the lack of safety and security for the public and tourists. [Keep in mind that Galápagos continues to be one of the safest international destinations, but, in recent years, civil unrest, crime, and kidnapping have increased in continental Ecuador].

(9) In this chapter, we elaborated on two topics regarding climate-change impacts on both terrestrial and marine ecosystems in the Galápagos: (i) *Extreme Weather and Its Sequels* and (ii) *Impacts of Climate*

Change on Coral Reefs. Points (10) to (12) summarize our conclusions with respect to these topics.

(10) The Galápagos Islands are warming up and—for now—becoming rather dry [but "extreme wet" will prevail long-term].

(11) Galápagos appears to be destined to: (i) less rain during the cool-dry season (drier archipelago for now); combined with (ii) a delay in the onset of the hot-wet season (rains will start later); plus, a mix of (iii) severe hot-wet seasons during more frequent El Niño years; and (iv) intense cool-dry seasons during La Niña years. Climate change will bring extreme variation in seasonality to Galápagos and affect both wildlife and humans!

(12) Extreme changes in sea-water temperature can cause "bleaching" in corals. Under heat stress, photosynthetic microbes—symbionts of coral polyps—die and coral colonies turn white. Coral bleaching is a global phenomenon. Researchers modeled the onset of severe coral reef bleaching at 30 World Heritage Sites—including Galápagos—resulting from conceivable global warming trends caused by variable rates of carbon dioxide (CO_2) emissions. The models explored three scenarios: (i) current trends in global CO_2 emissions (which projected 4.6°C/8.3°F global mean temperature increase by year 2100); (ii) half the current trends by 2050 (projected 2.4°C/4.3°F increase by 2100); and (iii) significant reduction in global CO_2 emissions (as per The Paris Agreement) to limit the global mean temperature increase to 1.5°C/2.7°F by 2100. In the case of Galápagos, under the current trends in global CO_2 emissions, most severe coral reef bleaching events, twice per decade, were expected to begin by year 2017; or once per year by 2036. If global current CO_2 emissions were cut by half by year 2050, the onset of severe coral reef bleaching events, twice per decade, would begin by year 2027; or once per year by 2042. And if global CO_2 emissions were significantly reduced as per The Paris Agreement, mild coral reef bleaching events, twice per decade, would begin by year 2070; no once-per-year bleaching events would occur by 2100!

Questions

The following questions (Q1–2) merit generous written answers. You might need to do additional reading or library work to enrich your answers (we suggest exploring Paz-y-Miño-C and Espinosa 2024a,b).

Question 1: Imagine severe climate change impacts on the Galápagos Archipelago and its Marine Reserve within the next 100 years [this chapter should help you foresee that]. (a) What sequence of events will lead to an inevitable "eco-exodus" and the emergence of "environmental refugees"?

(b) Considering that Galápagos is a UNESCO World Heritage Site [something discussed in Chapter Eight, *What Is the Conservation Value of the Galápagos?*], how exactly should the local administrative entities (*in situ*), the Ecuadorian government (which has jurisdiction over the archipelago), and world organizations respond to an "*inevitable eco-exodus* [in Galápagos] *and the emergence of environmental refugees*"? (c) Are Galápagos (*in situ*), the country of Ecuador (i.e. the sovereign nation), and the world prepared to manage an "*inevitable eco-exodus* [in Galápagos] *and the emergence of environmental refugees*"? How?

Question 2: You know that the Galápagos' economy is powered almost entirely by nature tourism [as discussed in Chapter Seven]. You are also aware that climate change is expected to impact all Galápagos environments, from the bottom of the ocean to the top of the volcanoes. (a) How exactly will nature tourism be affected in Galápagos due to severe climate change impacts? (b) If the iconic Galápagos flora and fauna (e.g. prickly-pear trees, *Opuntia*; Darwin's giant daisies, *Scalesia*; snapdragon shrub, *Galvezia*; sea lions, fur seals, whale sharks, hammerhead sharks, killer whales, spinner dolphins, marine iguanas, land iguanas, giant tortoises, or the Galápagos/Darwin finches) become rare or—in some cases—extinct due to severe climate change impacts, what type of tourism will Galápagos be left to offer? (c) Will severe climate change lessen Galápagos' "outstanding universal value" as a World Heritage Site? How? Will the Galápagos nature-tourism industry survive the sequels of severe climate change? [Remember that we discussed the "outstanding universal value" of a World Heritage Site in Chapter Eight, *What Is the Conservation Value of the Galápagos?*].

How to Answer the Questions Above

The questions above have various possible answers (A1–2). Here we suggest perspectives from which to address the queries.

Answer 1: Q1 is a complex question. We are not asking for your opinion but rather an educated assessment of probable scenarios leading to an eco-exodus and the emergence of environmental refugees in the Galápagos as a consequence of severe climate change impacts. (a–c) Several chapters in this book will help you find arguments to document your answers, i.e., Chapter Six, *Who Lives in Today's Galápagos Islands?* Chapter Seven, *Why Is Nature Tourism Essential to Galápagos?* Chapter Eight, *What Is the Conservation Value of the Galápagos?* And Chapter Nine, *What Are the Main Conservation Problems in Galápagos?* Our Galápagos books shall also give you hints to address Q1 (Paz-y-Miño-C and Espinosa 2024a,b).

Answer 2: Three concepts should be clear to you before answering Q2: nature tourism, ecotourism, and conventional tourism [we discussed them in Chapter Seven, *Why Is Nature Tourism Essential to Galápagos?*]. You should also be familiar with UNESCO's criteria to designate a World Heritage Site [we discussed them in Chapter Eight, *What Is the Conservation Value of the Galápagos?*]. (a) You need to be specific about how nature tourism will be affected in Galápagos due to severe climate change impacts. Knowing the diverse modalities of tourism offered in Galápagos and the types of environments in which tourism takes place should also help you address Q2a. Now, for Q2b–c, keep in mind that Galápagos' uniqueness is central to having been designated a World Heritage Site. Today's nature-tourism industry is intertwined with the very fact that Galápagos is a World Heritage Site. Plus, there are attributes that a World Heritage Site like Galápagos has [intrinsically] and on which the nature-tourism industry rides (e.g. the spectacular underwater wildlife and seascape features of the archipelago; its plate tectonics, volcanism, and ongoing geomorphological processes; its ecological, evolutionary, and biogeographic characteristics of terrestrial and marine environments; and its local biodiversity and endemism and the conservation status of the flora and fauna; for discussion, see Chapter Eight).

Discussion Topics

The discussion topics below (D1–D2) should be addressed with your peers. We encourage college instructors to facilitate the dialogue.

Discussion 1: In Chapter One, section The Upwelling, we indicated that "seasonal upward surges of cold, nutrient-rich deep-sea waters—upwelling—fuel the biological cycles of microorganisms (primary production), which, in turn, sustain all other life forms in the islands" (see Figs. 1.33-1.34, Chapter One). In this chapter (Chapter Ten), we discussed the Impacts of Climate Change on Coral Reefs (Fig. 10.3). And in Box 1.3, also in Chapter One, we described the Proposed bioregions within the Galápagos Marine Reserve (as per Edgar et al. 2004). To DISCUSS the question below, you will need to integrate three topics: The Upwelling with the Impacts of Climate Change on Coral Reefs and with the Proposed bioregions within the Galápagos Marine Reserve. How exactly will extreme weather (severe hot-rainy and cool-dry seasons) and sea-level rise impact [mechanistically] the interactions among the upwelling, the coral reef communities, and the marine biodiversity in the bioregions within the Galápagos Marine Reserve?

Discussion 2: In this chapter, we stated that "…climate change will accentuate differential financial stress among dissimilar income groups. Socio-economic inequality and unemployment could grow, as could social tensions and the lack of safety and security for the public and tourists". We also highlighted that although "…Galápagos continues to be one of the safest international destinations (with almost nonexistent violent offenses), …in recent years, civil unrest, crime, and kidnapping have increased in continental Ecuador". Here is your DISCUSSION: Hypothesize how climate change will intensify [mechanistically] social unrest in the Galápagos. Note that we are asking you to formulate university-level alternative hypotheses rather than to speculate based on private opinions. Some references cited in this chapter will help you address this point of discussion (e.g. Hauer et al. 2020; Zickgraf 2020; Nguyen et al. 2023) as well as the topics discussed in Chapter Five, i.e., 2019: US Military Planes Back in Galápagos. Remember that in the Tropical Eastern Pacific, where the Galápagos Marine Reserve is located, drug smuggling is intensifying, which will likely affect the Galápagos in the coming years (see Figs. 5.20-5.22 in Chapter Five). You will need to do additional library work to formulate and discuss your hypotheses.

Suggested Observation Projects

The suggested observation projects below (P1–P2) take into consideration that you will travel to Ecuador and the Galápagos legally; that you will visit the islands as a tourist [holding a Transit Control Card, a mandatory document issued by Ecuador and specifically by the Governing Council of Galápagos]; and that you will respect the regulations that protect the Galápagos National Park and the Marine Reserve (see complete list of rules in Appendix B). Only accredited scientists, academics, and/or naturalists who have obtained official permits can do research in Galápagos. You can take pictures, but flash photography is not permitted when taking photos of wildlife. Professional photography and videos recorded for commercial purposes must be authorized by the Galápagos National Park Directorate (GNPD). The ideas below are for enthusiastic college explorers interested in learning about the archipelago.

Project 1: Prepare a short survey with questions to be asked to as many people as you run into while visiting the Galápagos (i.e. locals, other tourists, some naturalist guides, diverse people of all ages [adults], and anyone you think could answer your friendly questions). For example, you could ask, Do you know how climate change will impact Galápagos? The islands? The marine reserve? Subsistence fishing? Nature tourism? The local residents? You could ask locals; Do you think that climate change will impact you personally or your family as Galápagos residents? If climate change is severe and affects you personally or your family, will you stay in Galápagos or move somewhere else? Where to? If climate change disrupts Galápagos tourism, do you think that the residents will be able to withstand a significant reduction in yearly visitors? Why? [Keep in mind that the Covid-19 pandemic already halted Galápagos tourism (see Figs. 7.3 and 7.9 in Chapter Seven); locals will remember that]. You can write your own questions, but keep in mind that surveying should be friendly, short, and for personal interest only. Organize your data in an Excel file and rely on descriptive statistics to summarize the outcomes. [As stated above, only accredited scientists, academics, and/or naturalists who have obtained official permits can do research in Galápagos]. Put together a slideshow presentation (for a 20-minute talk) about "Peoples' Level of Awareness about Climate Change Impacts on Galápagos" to be shared with a general audience. Avoid slides with "text only" [they are often terrible]. Keep it rigorous yet fun and didactic.

Project 2: This project combines studio/library work with *in situ* observations. Compare El Niño vs. La Niña impacts on Galápagos biodiversity. Keep in mind that we discussed Galápagos seasonality in Chapter One (see Figs. 1.43-1.44) and have referred to it in several chapters of this book. Climate change is expected to cause extreme weather fluctuations in the Galápagos, i.e., severe El Niño hot-rainy years and severe La Niña cool-dry years; both will bring stress to natural habitats. Here is your project: Create a database [Excel file or an alternative to it] in which you characterize how El Niño vs. La Niña will impact Galápagos biodiversity, particularly its iconic or emblematic species (marine and land iguanas, sharks, whales and dolphins, penguins, flightless cormorants, sea lions, fur seals, flamingos, giant tortoises, hawks and owls, or finches) in aquatic and terrestrial environments. You can rely on Dueñas et al. (2021) as a guiding reference to organize your data, but you will likely need other studies. Document on your Excel file how exactly severe hot-rainy years vs. severe cool-dry years will impact your iconic or emblematic species. While visiting the Galápagos, verify *in situ* the population status of these species [i.e. a snapshot observation] by simply registering where and when you find them (plot this on a map as per island and specific locations) and the overall abundance of individuals you encounter (e.g. you can assign *ad hoc* categories as *very common*, *common*, or *rare* and quantify the overall observations by using descriptive statistics). Register the time of the year when you visited Galápagos; was it during a particularly hot-rainy or cool-dry season? Ask your naturalist guide whether the iconic or emblematic species that you observed and counted were as abundant as usual or if they were somewhat atypical. Put together a slideshow presentation (for a 20-minute talk) about "Expected El Niño vs. La Niña Impacts on Iconic and Emblematic Galápagos Species" to be shared with a general audience. Avoid slides with "text only" [they are often terrible]. Keep it rigorous yet fun and didactic.

References

Carlton, James T., Inti Keith, and Gregory M. Ruiz. 2019. "Assessing Marine Bioinvasions in the Galápagos Islands: Implications for Conservation Biology and Marine Protected Areas". *Aquatic Invasions*, No. 14: 1–20.

Dueñas, Alejandra, Gustavo Jiménez-Uzcátegui, and Thijs Bosker. 2021. "The Effects of Climate Change on Wildlife Biodiversity of the Galapagos Islands". *Climate Change Ecology*, No. 2: 100026.

Edgar, Graham J., Samuel Banks, José M. Farina, Mónica Calvopiña, and Camilo Martínez. 2004. "Regional Biogeography of Shallow Reef Fish and Macro-Invertebrate Communities in the Galapagos Archipelago". *Journal of Biogeography*, No. 31: 1107-1124.

GOV.UK (Government United Kingdom). 2025. "Foreign Travel Advice Ecuador". Accessed May 13, 2025. *https://www.gov.uk/foreign-travel-advice/ecuador/safety-and-security*.

Hauer, Mathew E., Elizabeth Fussell, Valerie Mueller, Maxine Burkett, Maia Call, Kali Abel, Robert McLeman, and David Wrathall. 2020. "Sea-Level Rise and Human Migration". *Nature Reviews Earth & Environment*, No. 1: 28-39.

Heron, Scott F., Ruben van Hooidonk, Jeffrey Maynard, Kristen Anderson, Jon C.Day, Erick Geiger, Ove Hoegh-Guldberg, Terry Hughes, Paul Marshall, David Obura, and C. Mark Eakin. 2018. *Impacts of Climate Change on World Heritage Coral Reefs: Update to the First Global Scientific Assessment*. Paris: UNESCO World Heritage Centre.

IUCN (International Union for Conservation of Nature). 2025. "The IUCN Red List of Threatened Species". Accessed May 7, 2025. *https://www.iucnredlist.org/*.

NASA Sea Level Change. 2025. "Understanding Sea Level". Accessed May 12, 2025. *https://sealevel.nasa.gov/understanding-sea-level/key-indicators/global-mean-sea-level*.

Nguyen, Trung Thanh, Ulrike Grote, Frank Neubacher, Dil B. Rahut, Manh Hung Do, and Gokul P. Paudel. 2023. "Security Risks from Climate Change and Environmental Degradation: Implications for Sustainable Land Use Transformation in the Global South". *Current Opinion in Environmental Sustainability*, No. 63: 101322.

OECO (Observatorio Ecuatoriano de Crimen Organizado). 2025. "Visualizador de Datos: Número de Homicidios". Accessed May 12, 2025. *https://oeco.padf.org/visualizador-de-datos-numero-de-homicidios/.*

Paltán, Homero A., Fátima L. Benitez, Paulina Rosero, Daniel Escobar‑Camacho, Francisco Cuesta, and Carlos F. Mena. 2021. "Climate and Sea Surface Trends in the Galapagos Islands". *Scientific Reports*, No. 11: 14465.

Paz-y-Miño-C, Guillermo, and Avelina Espinosa. 2024a. *Arrivals of Life to the Galápagos: From Berlanga to Darwin, Melville, and World War II*. Newcastle upon Tyne: Cambridge Scholars Publishing.

—. 2024b. *A Synthesis of the Galápagos: Natural History, Human Conflict, and the Destiny of an Archipelago*. Newcastle upon Tyne: Cambridge Scholars Publishing.

Schleussner, Carl-Friedrich, Delphine Deryng, Sarah D'haen, William Hare, Tabea Lissner, Mouhamed Ly, Alexander Nauels, Melinda Noblet, Peter Pfleiderer, Patrick Pringle, Martin Rokitzki, Fahad Saeed, Michiel Schaeffer, Olivia Serdeczny, and Adelle Thomas. 2018. "1.5°C Hotspots: Climate Hazards, Vulnerabilities, and Impacts". Annual Review of Environment and Resources, No. 43: 135-163.

UNCC (United Nations Climate Change). 2025. "The Paris Agreement". Accessed May 12, 2025. *https://unfccc.int/process-and-meetings/the-paris-agreement.*

US Department of State (United States Department of State Bureau of Consular Affairs). 2025. "Ecuador Travel Advisory". Accessed May 12, 2025. *https://travel.state.gov/content/travel/en/traveladvisories/traveladvisories/ecuador-travel-advisory.html.*

van Woesik, Robert, Tom Shlesinger, Andréa G. Grottoli, Rob J. Toonen, Rebecca Vega Thurber, Mark E. Warner, Ann Marie Hulver, Leila Chapron, Rowan H. McLachlan, Rebecca Albright, Eric Crandall, Thomas M. DeCarlo, Mary K. Donovan, Jose Eirin-Lopez, Hugo B. Harrison, Scott F. Heron, Danwei Huang, Adriana Humanes, Thomas Krueger, Joshua S. Madin, Derek Manzello, Lisa C. McManus, Mikhail Matz, Erinn M. Muller, Mauricio Rodriguez-Lanetty, Maria Vega-Rodriguez, Christian R. Voolstra, and Jesse Zaneveld. 2022. "Coral-Bleaching Responses to Climate Change Across Biological Scales". *Global Change Biology*, No. 28: 4229-4250.

Zickgraf, Caroline. 2020. "Climate Change, Slow Onset Events and Human Mobility: Reviewing the Evidence". *Current Opinion in Environmental Sustainability*, No. 50: 21-30.

CHAPTER ELEVEN

WHAT CAN YOU DO TO HELP GALÁPAGOS?

The principle "think globally, act locally" applies here. Now, to be most effective helping the Galápagos, you will need to think and act globally as well as think and act regionally and locally. The global context will be, of course, planet Earth. The regional setting will be the Tropical Eastern Pacific (TEP), where Galápagos is located; TEP will also include the proximate countries, those with jurisdiction over the TEP maritime zones (i.e. territorial waters/sea and exclusive economic zones), like Ecuador, Colombia, Panama, and Costa Rica. And the local scale will correspond to both Galápagos itself, whenever you visit it and explore it as a citizen of the world, and your own community wherever you reside.

Think and Act Globally: Planet Earth

As a global citizen, you have rights and responsibilities. Your rights are stated in the Universal Declaration of Human Rights, as per the United Nations General Assembly of 1948 and later modifications (see its current 30 articles in United Nations 2025). In addition, the United Nations Environment Programme acknowledges your environmental rights, including the right to a healthy, clean, safe, favorable, wholesome, or ecologically balanced environment (UNEP 2025). Some of your responsibilities as a citizen of the world include to understand: (i) "*the different perspectives* [that almost every global issue has, i.e., cultural, social, political, and economic] *and promote problem-solving consensus among the different perspectives and the building of common ground solutions*"; (ii) "*the many ways in which* [your life is] *inter-connected with people and countries in different parts of the world*"; (iii) "*how* […] *income inequalities across the world affect the quality of* [people's] *lives* [and the environmental implications of such inequalities]"; (iv) "*the ways in which the global environment affects* [you] *where* [you] *live, and how the environmental lifestyle* [you] *choose affect*[s] *the environment in other parts of the world*"; and (v) "[how] *to advocate for having* [your and other] *countries ratify and implement the global* [environmental] *agreements, conventions, and treaties that* [your and other countries] *have signed*" (adapted from The Global Citizens' Initiative; TGCI 2025). Thus, with respect to the environment, it is your right and responsibility to think and act globally and locally and care about Planet Earth. Helping to preserve World Heritage Galápagos can be one of your goals. How? Become an *ad hoc* Galápagos world citizen guardian; organize your own team of college-educated protectors of the Galápagos Archipelago and its Marine Reserve; inform yourselves about the past and present of the islands and their international relevance as a UNESCO World Heritage Site; learn about the global organizations working on behalf of the Galápagos [see abridged list in Appendix A] and in coordination with their regional (the TEP countries) and *in situ* partners (continental Ecuador and the Galápagos Islands); volunteer quality labor for such organizations at the same time that you assess their success on Galápagos' conservation and sustainable development; donate yearly to their fieldwork and local communities' programs; and ask for periodical reports about how exactly the donors' monies were spent. Do not be shy and offer positive criticism; most international non-governmental organizations (NGOs) will appreciate your involvement and vigilance. Keep in mind that we live in an interconnected twenty-first-century world and that, as a World Heritage Site, Galápagos belongs to all of us (see Chapter Eight, *What Is the Conservation Value of the Galápagos?*)!

In Chapter Nine, *What Are the Main Conservation Problems in Galápagos?* we listed eight topics of global, regional, and local relevance. In today's world, all these problems have global implications: (i) The impacts of climate change—a threat to the planet—on the Galápagos' habitats, the people living on the islands, and their means of subsistence (details in Chapter Ten). (ii) Large-scale predatory fishing in the vicinity of the Galápagos Marine Reserve—mostly by international fisheries. (iii) Macro- and microplastics in the ocean—which arrive in the Galápagos from all over the world. (iv) Unmanageable local population growth—associated with global and local tourism expansion (details in Chapter Six). (v) Increasing eco-unfriendly tourism—particularly, entertainment tourism (details in Chapter Seven). (vi) Wildlife trade (e.g. baby giant tortoises for the international market). (vii) Illegal drug trafficking in the Tropical Eastern Pacific that has targeted the archipelago as a steppingstone on its way to international consumers (details in Chapter Five). And (viii) geopolitical instability in the [TEP] region—partly derived from drug trafficking and linked to Galápagos' historical role in the world's geopolitics (details in Chapter Five). These problems pose unprecedented threats to the Galápagos, and all entail human conflict with nature (Paz-y-Miño-C and Espinosa 2024a,b)! What can you do to help solve—or minimize—these complex challenges? Global problems require large-scale solutions. The impacts of climate change—in the Galápagos or your hometown—ought to be acknowledged as a reality by all nations. Global citizens—with their voice, vote, and constructive environmental activism—can lead their communities and governments to be sincere participants in "*the global* [environmental] *agreements, conventions, and treaties that* [their countries] *have signed*" (above). For example, 196 parties pledged to adhere to the 2015 Paris Agreement, or Paris Climate Accords (United Nations Climate Change 2025), and 193 member states are part of the United Nations Environmental Programme (UNEP 2025). Chances are that your country is part of both. If that is not the case, advocate for it! Be aware that the International Environmental Agreements Database Project lists 3,950 entries (The IEA Database 2025). Thus, basic legislation already exists to support your global pro-healthy-Earth mission, and the topic "climate change" is at the center of it.

As noted earlier, another problem with "global roots" is the large-scale predatory fishing around the Galápagos Marine Reserve (GMR). Fishing efforts themselves are regional (the Tropical Eastern Pacific) and local (proximate to GMR), but the international fisheries are global! In Chapter Nine, we discussed three topics associated with large-scale predatory fishing: Global Decline of Shark Populations, The *Fu Yuan Yu Leng 999* and the Shark-Fin Market, and Tuna Fisheries Around the Galápagos Marine Reserve. Yes, populations of sharks—and rays—are declining worldwide, likely due to habitat alteration and overfishing! The *Fu Yuan Yu Leng 999* tale illustrates how a single vessel containing 572 tons of fish and fish products can boldly enter Ecuador's maritime zone and GMR. The reefer contained 7,639 sharks, 2,114 bony fishes, and 537 sacks with shark fins (see Fig. 9.8 in Chapter Nine). Only +1 of its 6 freezers was full! World citizens, environmental NGOs, and international government officials were outraged over the incident and supported Ecuador's response; the fish catch, and the Chinese vessel were confiscated by local authorities. Researchers alerted that the *Fu Yuan Yu Leng 999* "exemplif[ied] how industrial fisheries [were] likely depleting shark populations [and other fishes] in the Eastern Tropical Pacific Ocean", where GMR is located (Bonaccorso et al. 2021). And researchers have also called to our attention that global fisheries operating nearby GMR might be benefitting—although temporarily—from a phenomenon called "spillover effect", i.e., the export of fish biomass from a protected area [in our case GMR] to the adjacent fishing grounds (=international waters in TEP). Here is the scenario: Since its creation, GMR has functioned as a "well-protected and well-placed reserve" (Boerder et al. 2017), enhancing favorable conditions for tuna spawning and breeding within GMR (e.g. for yellow-fin tuna *Thunnus albacares*, bigeye tuna *T. obesus*, and skipjack tuna *Katsuwonus pelamis*). Today's global fisheries concentrate their fishing effort just outside GMR (see Figs. 9.11-9.13 in Chapter Nine). Interestingly, after the reinforcement of GMR in 2002, but not before, tuna catches have increased 52% and fishing effort 75%, while catches per unit of fishing effort have decreased only <1% (see Fig. 9.13 in Chapter Nine; for a review on ecological or fishery spillovers, see Di Lorenzo et al. 2020). As readers will infer, large-scale predatory fishing has all the incentives to stay and thrive proximate to GMR. But the current bonanza shall be short-lived if global, regional, and local habitats continue to be altered and resources depleted. What can you do, as a global citizen, to help minimize large-scale predatory fishing around GMR? A few things. Monitor Galápagos news (i.e. follow on social media the Galápagos National Park Directorate), environmental NGOs' work on behalf of the Galápagos (e.g. Charles Darwin Foundation, Galapagos Conservancy, Galapagos Conservation Trust, and many others), and Ecuador's frequent requests of international support and action to protect its maritime zone and GMR. Again, think and act globally, regionally, and locally. Globally, remember that you have a voice from wherever you reside: Galápagos is a UNESCO World Heritage Site, sponsored by the World Heritage Fund (UNESCO 2025a), and multiple international environmental organizations that invest in the broad conservation of the Tropical Eastern Pacific, where Ecuador, Colombia, Panama, and Costa Rica have maritime jurisdiction and protected areas, including GMR (below and Box 11.1). Remember that with your voice, vote, and constructive environmental activism, you can lead your community and government to be sincere participants in "*the global* [environmental] *agreements, conventions, and treaties that* [your country has] *signed*" (above). And you can advocate for the environmental rights of all people (i.e. in your own community or those living in the TEP countries), including the right to a healthy, clean, safe, favorable, wholesome, or ecologically balanced environment (above)!

Think and Act Regionally: TEP and Proximate Countries

Ecuador, Colombia, Panama, and Costa Rica have protected areas in the Tropical Eastern Pacific (TEP; Box 11.1; Fig. 11.1). Readers will infer that GMR is one of them (for a review on GMR, see Reck 2014). Other areas include Cocos Island National Park (Costa Rica), Coiba Island National Park (Panama), and Malpelo Island Sanctuary (Colombia). What can you do as a global citizen to help the TEP countries safeguard these areas? Advocate with your voice, vote, and constructive environmental activism for these nations' sovereignty over their territorial sea [often violated by fisheries], which includes their exclusive economic zones, protected areas, and regional treaties to protect them (e.g. Eastern Tropical Pacific Marine Corridor, CMAR; Box 11.1). Monitor that Ecuador, Colombia, Panama, and Costa Rica remain sincerely committed to the long-term sustainable management of TEP. Their people, governments, and NGOs ought to pursue legislation and prompt action for the safeguarding of TEP. But keep in mind that it is your responsibility to understand "*the different perspectives* [that almost every global issue has, i.e., cultural, social, political, and economic] *and promote problem-solving consensus among the different perspectives and the building of common ground solutions*"; "*the many ways in which* [your life is] *inter-connected with people and countries in different parts of the world*"; "*how* […] *income inequalities across the world affect the quality of* [people's] *lives* [and the environmental implications of such inequalities]"; as well as "[how] *to advocate for having* [your and other] *countries ratify and implement the global* [environmental] *agreements, conventions, and treaties that* [your and other countries] *have signed*" (as per TGCI 2025).

As explained in Box 11.1, via CMAR, Ecuador, Colombia, Panama, and Costa Rica will protect 1,374,496 km² (530,695 sq mi) of an area rich in marine life, with its *on-going ecological and biological processes*, its *significant natural habitats for in-situ conservation of biological diversity*, and its *threatened species of outstanding universal value from the point of view of science or conservation* (as per UNESCO 2025b,c). The Galápagos Marine Reserve, including a 2021 proposed expansion (see Fig. 11.1), will be the Westernmost part of CMAR. Scientists have documented unique biodiversity and abundance of life in CMAR's region, particularly large, charismatic vertebrates like sea turtles, giant manta rays, whale sharks, humpbacks, and blue whales; their movement patterns in the region are discussed and sketched in Box 11.1 (Figs. 11.2-11.3). In Table 11.1, we list 28 documented and/or suspected migratory marine vertebrates in the Galápagos Archipelago, which are common to CMAR. Each species' IUCN conservation status between 1998-2000 and 2020 is provided as either worsening or staying the same or improving over time. The taxa include 14 sharks, 5 birds, 4 turtles, 2 whales, 1 giant manta ray, 1 sea lion, and 1 fur seal (after Hearn et al. 2021). By 2020, 8 species were catalogued as endangered (EN), 7 as critically endangered (CR), and 6 as vulnerable (V). Thus, the protection of TEP (i.e. its connectivity as a region) by the regional countries is a priority!

Box 11-1 The Galápagos Marine Reserve (GMR) in the context of the Tropical Eastern Pacific (TEP).

In 2021, at the United Nations Climate Change Conference, Ecuador announced its plan to protect 60,000 km^2 (23,166 sq mi) adjacent to the current Galápagos Marine Reserve (GMR). And in 2022, the country created the Hermandad Marine Reserve as part of its "30 by 30" Global Alliance pledge to protect 30 percent of its marine environments by year 2030 (US-DFC 2023).

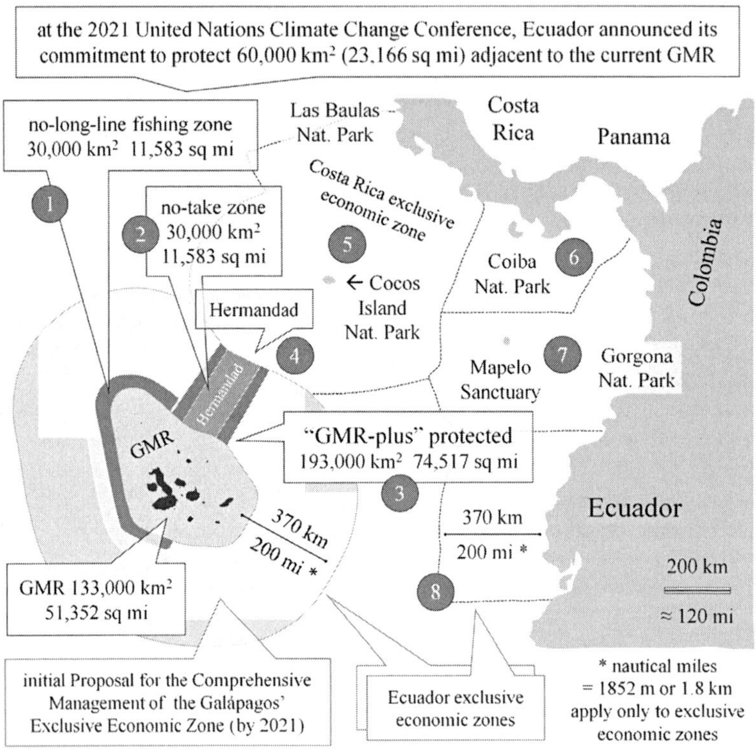

Ecuador made the announcement in the context of the 26th Global Climate Summit –Conference of the Parties (COP26), Glasgow, Scotland; part of the plan included a debt-for-nature swap to renegotiate a fraction of Ecuador's national debt (see Paz-y-Miño-C and Espinosa 2024b), which will be destined to long-term conservation/management programs for GMR, Galápagos National Park, and with participation of the local communities

Fig. 11-1 Official agreement of Ecuador's government to add 60,000 km^2 (23,166 sq mi) of protected areas to the Galápagos Marine Reserve (GMR). Layout: G. Paz-y-Miño-C.

As shown in Fig. 11.1, Ecuador's 2021 commitment to protect additional 60,000 km^2 (23,166 sq mi) adjacent to the current GMR (133,000 km^2/51,352 sq mi)

consisted of: (1) a no-long-line fishing zone (30,000 km^2/11,583 sq mi) and (2) a no-take zone (30,000 km^2/11,583 sq mi). (3) The "GMR-plus" protected area totaled 193,000 km^2 /74,517 sq mi around the archipelago. (4) The Hermandad ("brotherhood") Marine Reserve (HMR), announced in 2022, corresponded to a "corridor" within number 2 (MAATEE 2023). Note the connectivity created by the HMR with (5) Cocos Island National Park and the overall Costa Rican exclusive economic zone (including Las Baulas National Park). Note also how (6) Coiba National Park in Panama and (7) Malpelo Sanctuary and Gorgona National Park in Colombia became part of a multi-nation pledge to protect a large portion of the Tropical Eastern Pacific (i.e. the exclusive economic zones of these countries). (8) Realize that Ecuador has exclusive economic zones (EEZ) in both its continental coasts and around the Galápagos Archipelago (EEZs expand 370 km off the coastline; EEZs are called "the 200 nautical miles" of a country's territorial sea).

Together with Ecuador and in the context of the 2021 United Nations Climate Change Conference in Glasgow, three other countries—Colombia, Panama, and Costa Rica—announced their adherence to protecting the connectivity of marine environments under their jurisdiction (the multi-nation protected area, above). Thus, the Eastern Tropical Pacific Marine Corridor, or CMAR (Spanish for "Corredor Marino" del Pacífico Este Tropical), was formally conceived. CMAR will protect 1,374,496 km^2 (530,695 sq mi) of an area rich in marine life, with its *on-going ecological and biological processes*, its *significant natural habitats for in-situ conservation of biological diversity*, and its *threatened species of outstanding universal value from the point of view of science or conservation* (as per UNESCO 2025b,c). The Galápagos Marine Reserve, including its 2021 proposed expansion (Fig. 11.1), will be the Westernmost part of CMAR. Scientists have documented unique biodiversity and abundance of life in CMAR's region, particularly large, charismatic vertebrates like sea turtles, giant manta rays, whale sharks, humpbacks, and blue whales (as discussed in Chapter Four, Figs. 4.11-4.12, and Chapter Eight, Figs. 8.57-8.59, 8.63-8.77, Table 8.9).

In Figs. 11.2 and 11.3, we sketch the movement patterns of exemplar marine vertebrates in the Tropical Eastern Pacific and highlight the connectivity of GMR with CMAR. Note in Fig. 11.3: (A) green sea turtle *Chelonia mydas*; (B) leatherback turtle *Dermochelys coriacea*; (C) giant manta *Manta birostris*; (D) scalloped hammerhead shark *Sphyrna lewini*; (E) Galápagos shark *Carcharhinus galapagensis*; (F) silky shark *Carcharhinus falciformis*; (G) smalltooth sand tiger shark *Odontaspis ferox*; (H) bigeye thresher shark *Alopias superciliosus*; (I) tiger shark *Galeocerdo cuvier*; (J) whale shark *Rhincodon typus*; (K) humpback whale *Megaptera novaeangliae*; and (L) blue whale *Balaenoptera musculus* (data Mission Blue 2020).

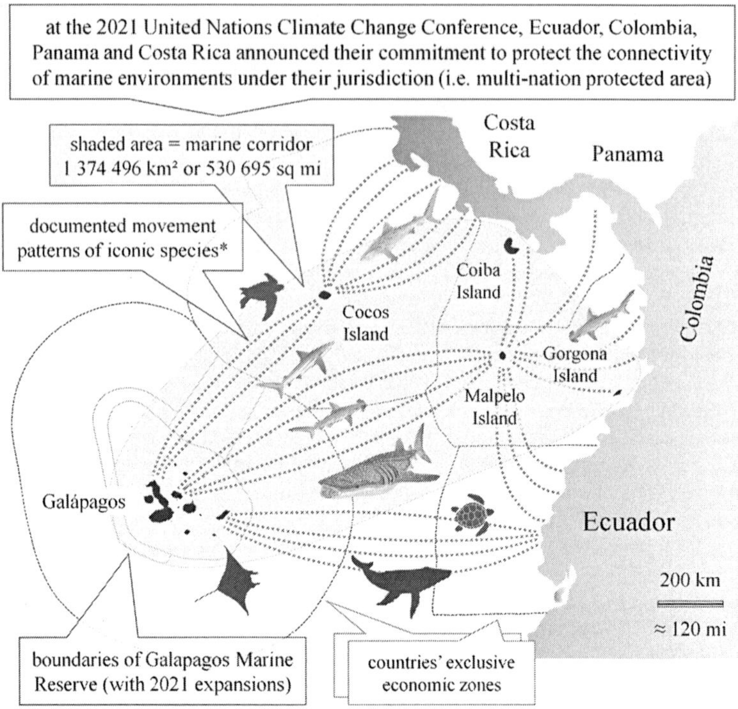

* for movement patterns of exemplar marine vertebrates in the tropical Eastern Pacific, see Fig. 11.3; for list of documented and/or suspected migratory marine vertebrates in the Galápagos Archipelago, as well as their IUCN conservation status, see Table 11.1

Fig. 11-2 The Eastern Tropical Pacific Marine Corridor. Movement of exemplar vertebrates determined by satellite tracking of tagged individuals. Layout: GPC; redrawn and modified after Enright et al. (2021) with input from MAATEE (2021a,b) and CMAR (2023).

[Note that in Chapter Eight we discussed various topics related to exemplar marine vertebrates in TEP, including: *Tiger Sharks Come to Hunt, Green Sea Turtles to Nest* (Figs. 8.57-8.59), *Interactions between Coral Reefs and Sharks* (Figs. 8.66-8.67), and *Thirty-Five Species of Sharks and Counting* (Figs. 8.68-8.77 and Table 8.9); whales were discussed in the historical context of whaling in Chapter Four, topic *Galápagos: Whaling Ground 38* (see Figs. 4.11-4.12 for most commonly observed whales and dolphins in the Galápagos and proximate areas)].

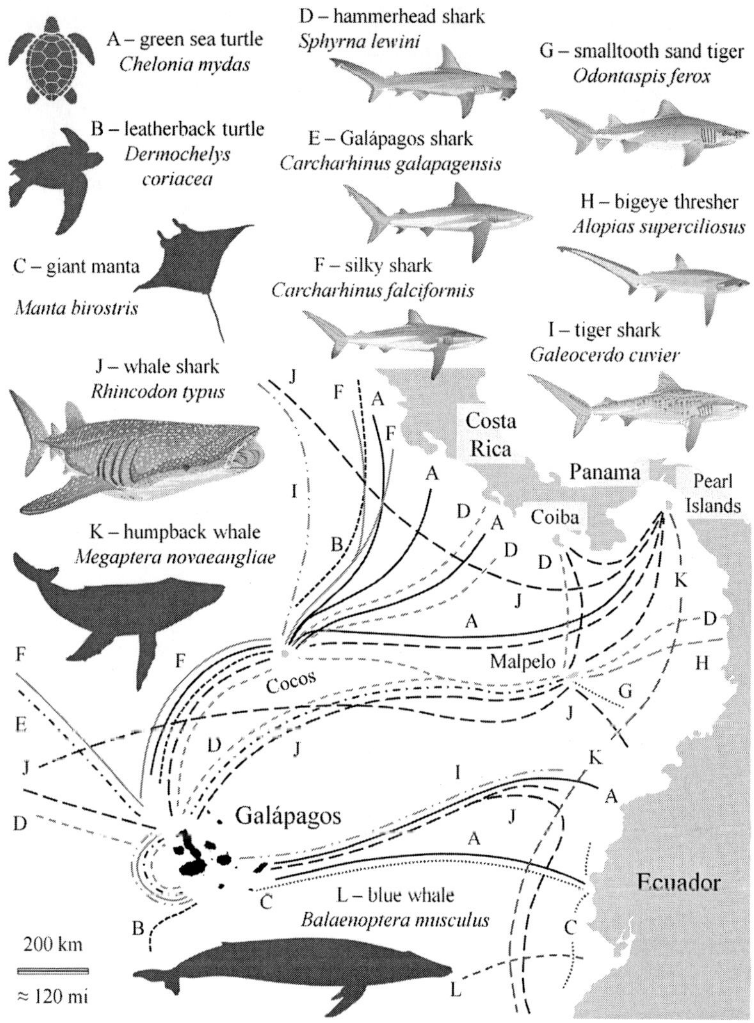

Fig. 11-3 Movement patterns of exemplar marine vertebrates in the Tropical Eastern Pacific: connectivity of the Galápagos Marine Reserve (GMR). Layout: GPC; redrawn and modified after Mission Blue (2020).

In Table 11.1, we list 28 documented and/or suspected migratory marine vertebrates in the Galápagos Archipelago [common to CMAR]. Each species' IUCN conservation status between 1998-2000 and 2020 is provided as either worsening or staying the same or improving over time. Taxa include 14 sharks, 5 birds, 4 turtles, 2 whales, 1 giant manta ray, 1 sea lion, and 1 fur seal (after Hearn et al. 2021). By 2020, 8 species were endangered (EN), 7 critically endangered (CR), and 6 vulnerable (V).

Table 11-1 Documented and/or suspected migratory marine vertebrates (exemplars) in the Galápagos Archipelago; each species' IUCN conservation status* between 1998-2000 and 2020 is provided (i.e. worsening, staying the same, or improving status over time is highlighted in different tones of gray)

species		IUCN conservation status* 1998 - 2000	2020
scalloped hammerhead	*Sphyrna lewini*	LC / NT	CR
oceanic whitetip shark	*Carcharhinus longimanus*	LC / NT	CR
great hammerhead	*Sphyrna mokarran*	DD	CR
leatherback turtle	*Dermochelys coriacea*	EN	CR
Galápagos albatross	*Phoebastria irrorata*	VU	CR
whale shark	*Rhincodon typus*	VU	EN
pelagic thresher	*Alopias pelagicus*	VU (2009)	EN
shortfin mako shark	*Isurus oxyrinchus*	VU	EN
longfin mako shark	*Isurus paucus*	VU	EN
Galápagos sea lion	*Zalophus wollebaeki*	VU	EN
Galápagos fur seal	*Arctocephalus galapagoensis*	VU	EN
silk shark	*Carcharhinus falciformis*	LC / NT	VU
smooth hammerhead	*Sphyrna zygaena*	LC / NT	VU
hawksbill sea turtle	*Eretmochelys imbricata*	CR	CR
Galápagos petrel	*Pterodroma phaeopygia*	CR	CR
green sea turtle	*Chelonia mydas*	EN	EN
blue whale	*Balaenoptera musculus*	EN	EN
bigeye thresher	*Alopias superciliosus*	VU (2009)	VU
giant manta	*Manta birostris*	VU (2011)	VU
sperm whale	*Physeter macrocephalus*	VU	VU
olive ridley sea turtle	*Lepidochelys olivacea*	EN	VU
blacktip shark	*Carcharhinus limbatus*	–	NT
tiger shark	*Galeocerdo cuvier*	–	NT
blue shark	*Prionace glauca*	–	NT
great frigatebird	*Fregata minor*	–	LC
blue-footed booby	*Sula nebouxii*	–	LC
swallow-tailed gull	*Creagrus furcatus*	–	LC
Galápagos shark	*Carcharhinus galapagensis*	–	LC

* IUCN = International Union for Conservation of Nature; the IUCN Red List of Threatened Species uses quantitative parameters to evaluate a species extinction risk (from lowest to highest risk): DD = data deficient; LC = least concern; NT = near threatened; VU = vulnerable; EN = endangered; CR = critically endangered (from nine total risk categories); data comes from Hearn et al. (2021)

What are the main environmental problems in the Tropical Eastern Pacific? Plenty. From the sequels of climate change (i.e. extreme weather fluctuations, severe El Niño and La Niña, and anticipated sea-level rise) to large-scale predatory fishing (e.g. tuna and sharks), macro- and micro-plastic pollution (items discarded *in situ* by fisheries or brought to TEP by ocean currents; see Fig. 9.21 in Chapter Nine), introduction of invasive species (particularly to nature reserves), illegal drug trafficking (e.g. 74% of trafficking in cocaine takes place via the Eastern Pacific Vector; see Fig. 5.22 in Chapter Five), wildlife trade (e.g. baby Galápagos giant tortoises, land and marine iguanas; see Figs. 9.18-9.19 in Chapter Nine), and geopolitical instability in the region (part of it derived from drug trafficking). We examined these topics in Chapters Five, Nine, and Ten (see also Paz-y-Miño-C and Espinosa 2024a,b). Thus, the challenges are huge. Still, with your voice, vote, and constructive environmental activism, you can make a difference from wherever you reside. For example, go online and contribute to social media discussions in which scientists—not pop influencers—lead the dialogue about the conservation and management of natural resources in any [or all] of the TEP countries, the conflicts between conservation efforts and predatory fishing (often by powerful fisheries operating worldwide), the policies that the TEP countries ought to adopt or reinforce to secure the long-term preservation of the local biodiversity (i.e. its high endemism), as well as the success and disappointment of regional NGOs working on sustainable development with the local communities. Yes, hold everyone accountable for their much-admired work on behalf of safeguarding TEP. Now, remember that one of your responsibilities as a global citizen includes understanding "*the ways in which the global environment affects* [you] *where* [you] *live, and how the environmental lifestyle* [you] *choose affect*[s] *the environment in other parts of the world*" (as per TGCI 2025). Hold yourself accountable for how impactful your constructive environmental activism is. Plus, stay informed about the news in Ecuador, Colombia, Panama, and Costa Rica by following well-established conservation and environmental science news organizations, including MONGABAY, EcoAméricas, The Nature Conservancy, among many others. Daily, weekly, or periodically, these outlets report the progress and setbacks in terms of environmental protection in TEP. Share their coverage on social media and invite your network of friends to become part of the conversation. Start your own podcast and rely on facts to motivate discussion. And send a loud message to the entire TEP region that you and hundreds of thousands of university students across the world—perhaps millions—care about the future of the Tropical Eastern Pacific!

Think and Act Locally: Galápagos and Your Own Community Wherever You Reside

How can you be effective *in situ* and help the Galápagos? You might want to start by visiting the archipelago and its Marine Reserve. If you have already done it, then turn this UNESCO World Heritage Site into a long-term commitment: return periodically to *The Encantadas* [as per Melville (1856)] to pursue saving them forever. Become a connoisseur of the islands and the ocean around them, the local towns, streets, and roads; administrative buildings; offices of the Galápagos National Park (GNP) and its museums and interpretation or captive-breeding centers (giant tortoises and land iguanas); monuments; university campuses; schools; restaurants and cafes; souvenir shops; harbors; airports; and all possible nature-visiting sites on each island, which are stunning (see Figs. 6.1-6.10 in Chapter Six and 7.8 and 7.14 in Chapter Seven). Volunteer for GNP and local NGOs, like the Charles Darwin Foundation, Galapagos Conservancy, Galapagos Conservation Trust, and others (i.e. find information online; you might need to apply to become a volunteer, this will require time and paperwork); there are activities year-round in which volunteers participate, e.g., removing plastics piled up on the rocky shores and sandy beaches (i.e. macro plastics pollution), working with local communities on environmental education and community involvement, motivating elementary-school and high-school students to learn about Galápagos conservation (public outreach and workshops for the youth), recycling campaigns, "responsible pet owners" campaigns (to encourage proper care of dogs and cats; feral animals are a serious problem in the islands), or assisting local NGOs' staff, scientists and veterinarians in their daily work (fieldwork, lab work, or both). Establish a long-lasting network of key contacts and friends, including naturalist guides, park rangers, and personnel at museums or interpretation centers (GNP and NGOs). Stay connected with them on social media and track their work on behalf of the islands. Follow homegrown organizations [social media] and monitor their activities and frequent pleas for support to their campaigns and programs, e.g., Governing Council of the Special Regime of Galápagos (CGREG or *Consejo de Gobierno del Régimen Especial de Galápagos*), GNP (*Parque Nacional Galápagos*), Galápagos Biosecurity Agency (*Agencia de Bioseguridad para Galápagos*), and the local municipality offices in Santa Cruz, San Cristóbal (note that Floreana is, administratively, part of San Cristóbal), and Isabela Islands. Learn about the Galápagos governmental structure (CGREG is at the top), private-sector involvement in tourism and its development (tourism drives the Galápagos economy; see Chapter Seven), conservation investments, as well as citizens' organizations. [As listed above, there are two conservation-related issues central to today's Galápagos governing reality, and although both have global and regional connections, they manifest *in situ*: unmanageable local population growth—associated with global and local tourism expansion (discussed in Chapter Six); and increasing eco-unfriendly tourism—particularly, entertainment tourism (discussed in Chapter Seven)]. Empower your network of *in situ* contacts by magnifying [again, social media] the message that Galápagos belongs to all of us. Remember that effective management of any nature sanctuary is possible when people participate with their voice, vote, and constructive environmental activism (above). And—in today's globalized world—you can do all of this from wherever you reside. Put your cellphone, tablet, or computer to good use!

Conclusion

With your voice, vote, and constructive environmental activism, you can do much to help preserve—forever—the Galápagos Archipelago and its Marine Reserve. The principle "think globally, act locally" applies here. To be most effective, you will need to think and act globally as well as think and act regionally and locally. The global context will be planet Earth. The regional setting will be the Tropical Eastern Pacific (TEP), where Galápagos is located; TEP will also include the proximate countries, those with jurisdiction over the TEP maritime zones, like Ecuador, Colombia, Panama, and Costa Rica. And the local scale will correspond to both Galápagos itself, whenever you visit it and explore it as a citizen of the world, and your own community wherever you reside.

Question, Answer, Discussion, and Project

Here is your last, yet comprehensive, assignment of this book: an activity in which you will integrate a question (formulated by us, the authors) with a possible answer (in fact, a network of answers generated by you and your team of students) that requires discussion (facilitated by a college instructor) and a final product (a concept map to be built by you and your team for a poster presentation, i.e., the project itself). Read the text below and follow the instructions afterwards.

According to human-cognition researchers Joseph Novak and Alberto Cañas (2008), "Concept Maps are graphical tools for organizing and representing knowledge. They include concepts, usually enclosed in circles or boxes of some type, and relationships between concepts indicated by a

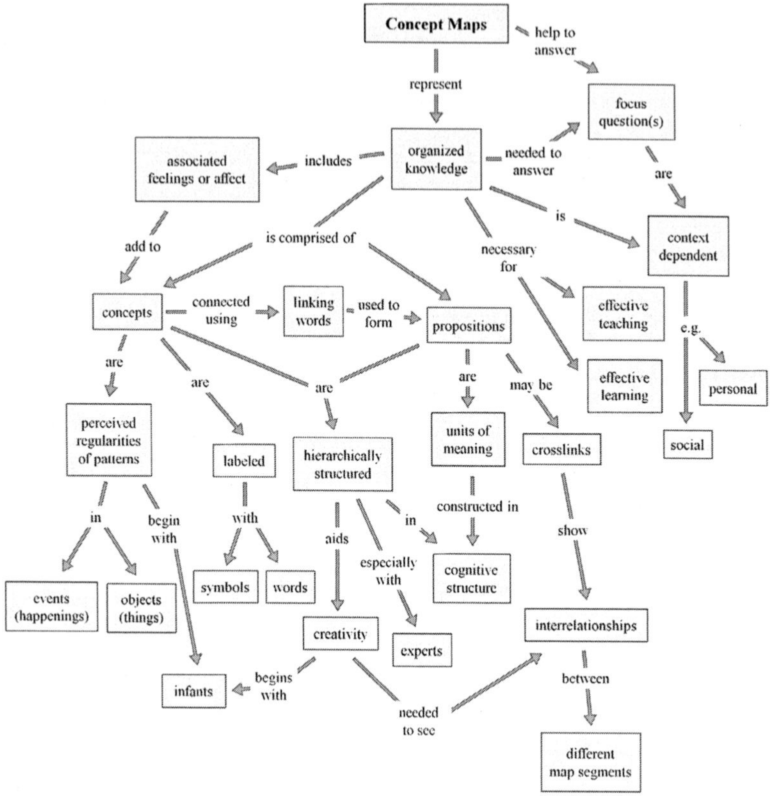

Fig. 11-4 Concept map depicting what "concept maps" are (redrawn and modified after Novak and Cañas 2008). Novak and Cañas (2008) explain that "*concept maps tend to be read progressing from the top downward*" and advise that "[it] *is best to construct concept maps with reference to some particular question we seek to answer... called a focus question*" [for example, How can you help the Galápagos at a global, regional, and local level?].

connecting line linking two [or more] *concepts* [see Figure 11.4]. *Words on the line, referred to as linking words or linking phrases, specify the relationship between the two concepts. Another characteristic of concept maps is that the concepts are represented in a hierarchical fashion with the most inclusive, most general concepts at the top of the map and the more specific, less general concepts arranged hierarchically below...* [It] *is best to construct concept maps with reference to some particular question we seek to answer... called a focus question* [for example, How can you help the Galápagos at a global, regional, and local level?]... *Another important characteristic of concept maps is the inclusion of cross-links. These are relationships or links between concepts in different segments or domains of the concept map. Cross-links help us see how a concept in one domain of knowledge represented on the map is* [associated with] *a concept in another domain shown on the map...*"

[Be alert that concept maps differ from flowcharts. As shown in Fig. 11.4, concept maps depict associations among concepts (i.e. conceptual components on the map), whereas flowcharts depict connections in a sequence of steps, e.g., the typical "step-by-step process" or "component-to-component connection" or "organization chart of a company" illustrating hierarchical positions, i.e., president, vice president, and others].

Here is your task: Put together a computer-generated poster presentation (vertical/portrait format: width 90 cm/36 in x height 122 cm/48 in) in which you depict a concept map addressing the focus question, *How can you help the Galápagos at a global, regional, and local level?* Keep it rigorous yet fun and didactic (e.g. include some attractive images). Share your poster presentation with the world (again, social media)!

References

Boerder, Kristina, Andrea Bryndum-Buchholz, and Boris Worm. 2017. "Interactions of Tuna Fisheries with the Galápagos Marine Reserve". *Marine Ecology Progress Series*, No. 585: 1-15.

Bonaccorso, Elisa, Nicté Ordóñez-Garza, Diana A. Pazmiño, Alex Hearn, Diego Páez-Rosas, Sebastián Cruz, Juan Pablo Muñoz-Pérez, Eduardo Espinoza, Jenifer Suárez, Lauren D. Muñoz-Rosado, Andrea Vizuete, Jaime A. Chaves, Maria de Lourde Torres, Walter Bustos, Danny Rueda, Maximilian Hirschfeld, and Juan M. Guayasamin. 2021. "International Fisheries Threaten Globally Endangered Sharks in the Eastern Tropical Pacific Ocean: the Case of the Fu Yuan Yu Leng 999 Reefer Vessel Seized within the Galápagos Marine Reserve". *Scientific Reports*, No. 11: 14959.

CMAR (Corredor Marino del Pacífico Este Tropical). 2023. "Sobre Pacífico Este Tropical". Accessed June 18, 2023. *https://www.cmarpacifico.org/.*

Di Lorenzo, Manfredi, Paolo Guidetti, Antonio Di Franco, Antonio Calò, and Joachim Claudet. 2020. "Assessing Spillover from Marine Protected Areas and Its Drivers: A Meta-Analytical Approach". *Fish and Fisheries*, No. 21: 906-915.

Enright, Sarah R., Ricardo Meneses-Orellana, and Inti Keith. 2021. "The Eastern Tropical Pacific Marine Corridor (CMAR): The Emergence of a Voluntary Regional Cooperation Mechanism for the Conservation and Sustainable Use of Marine Biodiversity Within a Fragmented Regional Ocean Governance Landscape". Frontiers in Marine Science, No. 8: 674825.

Hearn, Alex, Susana Cárdenas, Harriet Allen, Mauricio Castrejón, Sebastián Cruz, Eduardo Espinoza, Alex Forryan, María-Virginia Gabela, Dan Kelley, Alberto Naveira-Garabato, Bethan O'Leary, Diana Pazmiño, Cesar Peñaherrera-Palma, Josué Picho, Gunther Reck, Harry Reyes, Franz Smith, Sandy Tudhope, Andrea Vera, Diana Vinueza, Meriwether Wilson, and Leo Zurita. 2021. *Propuesta de Ordenamiento Espacial de la Zona Económica Exclusiva Insular de Ecuador*. Puerto Ayora: Más Galápagos.

MAATEE (Ministerio del Ambiente, Agua y Transición Ecológica del Ecuador). 2021a. "Corredor Marino del Pacífico Este Tropical". Accessed June 18, 2023. *https://twitter.com/Ambiente_Ec/status/1455624816300601344*.

—. 2021b. "Un millón de dólares para conservar el Corredor Marino del Pacífico". Accessed June 18, 2023. *https://www.ambiente.gob.ec/un-millon-de-dolares-para-conservar-el-corredor-marino-del-pacifico/*.

—. 2023. *Plan de Manejo de la Reserva Marina Hermandad*. Quito: Ministerio del Ambiente, Agua y Transición Ecológica del Ecuador; Dirección del Parque Nacional Galápagos; Subsecretaría de Patrimonio Natural; Fundación de Conservación Jocotoco; and Biogennia Cía. Ltda.

Melville, Herman. 1856. *The Piazza Tales*. New York: Dix Edwards and Company.

Mission Blue. 2020. "World's First Marine Migratory Species Hope Spot Declared between Cocos and the Galápagos". Accessed June 16, 2023. *https://mission-blue.org/2020/05/worlds-first-bi-national-marine-protected-area-gains-momentum-as-a-hope-spot/*.

Novak, Joseph D., and Alberto J. Cañas. 2008. "The Theory Underlying Concept Maps and How to Construct and Use Them". Technical Report IHMC CmapTools 2006-01 Rev 01-2008, Institute for Human and Machine Cognition. Accessed August 17, 2025. *https://cmap.ihmc.us/docs/pdf/TheoryUnderlyingConceptMaps.pdf*.

Paz-y-Miño-C, Guillermo, and Avelina Espinosa. 2024a. *Arrivals of Life to the Galápagos: From Berlanga to Darwin, Melville, and World War II*. Newcastle upon Tyne: Cambridge Scholars Publishing.

—. 2024b. *A Synthesis of the Galápagos: Natural History, Human Conflict, and the Destiny of an Archipelago*. Newcastle upon Tyne: Cambridge Scholars Publishing.

Reck, Günther. 2014. "Development of the Galápagos Marine Reserve". In *The Galápagos Marine Reserve: A Dynamic Social-Ecological System*, edited by Denkinger, Judith, and Luis Vinueza, 139-158. New York: Springer.

TGCI (The Global Citizens' Initiative). 2025. "The Rights & Responsibilities of Global Citizenship". Accessed July 23, 2025. *https://www.theglobalcitizensinitiative.org/the-rights-and-responsibilities-of-global-citizenship/.*

The IEA Database. 2025. "International Environmental Agreements Database Project". Accessed July 29, 2025. *https://www.iea.ulaval.ca/en.*

UNEP (United Nations Environment Programme). 2025. "What Are Environmental Rights?" Accessed July 23, 2025. *https://www.unep.org/explore-topics/environmental-rights-and-governance/what-we-do/advancing-environmental-rights/what.*

UNESCO (United Nations Educational, Scientific and Cultural Organization). 2025a. "World Heritage Fund". Accessed August 8, 2025. *https://whc.unesco.org/en/world-heritage-fund.*

UNESCO World Heritage Convention. 2025b. "Galápagos Islands". Accessed May 28, 2025. *https://whc.unesco.org/en/list/1/.*

—. 2025c. "The Criteria For Selection". Accessed May 28, 2025. *https://whc.unesco.org/en/criteria/.*

United Nations. 2025. "Universal Declaration of Human Rights". Accessed July 7, 2025. *https://www.un.org/en/about-us/universal-declaration-of-human-rights.*

United Nations Climate Change. 2025. "The Paris Agreement". Accessed July 29, 2025. *https://unfccc.int/process-and-meetings/the-paris-agreement.*

US-DFC (United States International Development Finance Corporation). 2023. "Financial Close Reached in Largest Debt Conversion for Marine Conservation to Protect the Galápagos". Accessed June 26, 2023. *https://www.dfc.gov/media/press-releases/financial-close-reached-largest-debt-conversion-marine-conservation-protect.*

Chapter Twelve

Can the "Little World Within Itself" Be Saved Forever?

> "...I was always amused, when overtaking one of these great monsters [a giant tortoise] as it was quietly pacing along, to see how suddenly, the instant I passed, it would draw in its head and legs, and uttering a deep hiss fall to the ground with a heavy sound, as if struck dead… The natural history of this archipelago is very remarkable: it seems to be a little world within itself; the greater number of its inhabitants, both vegetable and animal, being found nowhere else..." (Charles Darwin 1839).

"Galápagos, the volcanic archipelago, will persist for millions of years into the future. As long as the hotspot beneath the islands remains active, it will continue to fuel them with magma. New volcanoes will rise while the present ones gradually drift to the East on top of the Nazca Plate, become dormant, and sink to the bottom of the ocean as seamounts. If left alone and undisturbed [i.e. the islands], ecological and evolutionary processes will go on in terrestrial and aquatic environments. *Opuntia* and candelabra cactuses will grow as imposing trees. Their fruits will be eaten by birds and reptiles, which will disperse the seeds all over. The finches will adaptively radiate into newer lineages; some will hybridize with seemingly distant phylogenetic relatives. The giant tortoises will vigorously migrate to the highlands of the tall islands during the hot-rainy season, where they will find plenty of browsing grounds. The land iguanas will reach the calderas of the volcanoes and compete for nesting sites, while their offshore counterparts, the marine iguanas, will dive even deeper to feast on the algae growing on the submerged lava rocks. Tiger sharks will come to hunt while green sea turtles, their favorite prey, swim along the sandy beaches looking for suitable places to lay eggs. Whale sharks will gently cruise along the water surface with their mouths open, attempting to filter minute organisms. Killer whale orcas will keep on ambushing other whales or catching distracted sea lions and fur seals. And entire deep-sea hydrothermal vent communities will manage to thrive in the dark, powered by hot water and chemosynthesis rather than sunlight. Their microbes and viruses; their giant

Riftia worms, clams, and mussels; and their crabs, octopuses, and blind fishes will endure and rule in the blackest depths. But the immediate, 21st-century destiny of the Encantadas will be in the hands of *Homo*, the maker of the Anthropocene. Climate change; large-scale predatory fishing in the vicinity of the Galápagos Marine Reserve; macro- and microplastics in the ocean, permeating all trophic levels; unmanageable local population growth in the islands of Santa Cruz, San Cristóbal, and Isabela; increasing eco-unfriendly tourism; wildlife trade of baby giant tortoises, which are smuggled for the global market; illegal drug trafficking in the Tropical Eastern Pacific that has targeted the archipelago as to steppingstone on its way to international consumers; and geopolitical instability in the region pose unprecedented threats to the Galápagos. Will this World Heritage Site be the one that forever-survives? Will Galápagos be different? Will its caretakers and policymakers not mimic the mistakes that nature custodians have made in other volcanic archipelagos, where pristine ecosystems and endemic wildlife have substantially disappeared while being replaced by large towns and cities, conventional tourism resorts, highways, and introduced species? Will the Galápagos' people, Ecuador, and the world commit to safeguarding and managing the archipelago and its marine reserve so that sound sustainable development practices limit human population expansion on the islands, sponsor solely eco-friendly tourism, and secure revenue fairness for the residents?" (Paz-y-Miño-C and Espinosa 2024a,b). In this book, we have addressed these and multiple other questions. We have challenged readers to answer them, to discuss them, and to envision observation projects—inspired by these queries—while visiting the Galápagos Archipelago and its Marine Reserve.

The Destiny of a World Heritage Site: Galápagos

"Forty years ago, as high school graduates, we visited Galápagos for the first time. We were among ≈20,000 yearly tourists at that time. In retrospect, exploring and sightseeing on the islands turned us into biologists. Few places on our planet can have such an impact. Later in our lives, we have tried to pass on our passion for Galápagos to students" (Chapter Seven, *Why Is Nature Tourism Essential to Galápagos?* see also Paz-y-Miño-C and Espinosa 2024a,b). We were—of course—presided by Tomás de Berlanga and his crew, who accidentally arrived in unknown islands in 1535 and discovered the later to be called "*y^{a}s de galapagos*" or "*isles of galapagos*" (Chapters One, *What Are the Galápagos Islands?*, and Two, *How Were the "y^{a}s de galapagos" Discovered?*); also by Captain Robert FitzRoy and Charles Darwin, who surveyed the archipelago on board

HMS Beagle in 1835 (Chapter Three, *Why Was Charles Darwin's Visit to Galápagos Significant?*); and by Herman Melville, who arrived as a whaler in 1841 and later described *The Encantadas, or Enchanted Isles*, as a desolate place in which "*No voice, no low, no howl is heard; the chief sound of life here is a hiss*" (Melville 1856; also Chapter Four, *What Happened in Galápagos Before and After Darwin?*). [Melville authored *Moby-Dick; or, The Whale* in 1851; comprehensive analysis in Paz-y-Miño-C and Espinosa (2024a)].

From 1934 until our first trip to the Galápagos (around the early 1980s), about 54,000 tourists [cumulative number] had arrived in the archipelago. From 1934 to 2023, about 5,000,000 tourists [cumulative number] visited the islands; that is, a two-orders-of-magnitude increase from the 1980s to the 2020s (data extracted from DPNG and OTG 2018; DPNG 2019, 2020, 2021, 2022; Discovering Galapagos 2023). By 1982, the resident population in the islands was 6,119; by 2022, it had reached 28,583 (i.e. 4.67 times larger; INEC 1982, 2015, 2023, 2024). Thus, the chronological data suggests that unmanageable local population growth combined with unsustainable tourism practices (both on the rise) are a serious threat to the long-term preservation of World Heritage Galápagos (Chapters Six, *Who Lives in Today's Galápagos Islands?* and Seven, *Why Is Nature Tourism Essential to Galápagos?*). Moreover, as listed in Chapter Nine, *What Are the Main Conservation Problems in Galápagos?*, 21st-century Galápagos is destined to be affected by: (i) the impacts of climate change (=a human-induced phenomenon) on the local habitats, the people living on the islands, and their means of subsistence (agriculture, artisanal fishing, and tourism; also discussed in Chapter Ten, *How Will Climate Change Affect Galápagos?*). (ii) Large-scale predatory fishing in the vicinity of the Galápagos Marine Reserve (particularly by international fisheries; Chapter Nine). (iii) Macro- and microplastics in the ocean, permeating all trophic levels (globally and in Galápagos; Chapter Nine). (iv) Unmanageable local population growth in the islands of Santa Cruz, San Cristóbal, and Isabela (also discussed in Chapter Six). (v) Increasing eco-unfriendly tourism (also discussed in Chapter Seven). [Note that iv and v correlate with introductions of invasive species as discussed in Chapter Nine]. (vi) Wildlife trade of iconic endemic species (Chapter Nine). (vii) Illegal drug trafficking in the Tropical Eastern Pacific (also discussed in Chapter Five, *Why Did the United States Establish a Military Base in Galápagos During World War II?*). And (viii) geopolitical instability in the region (part of it derived from drug trafficking; also discussed in Chapter Five).

"Going back to the questions formulated above, will this World Heritage Site be the one that forever survives? Will Galápagos be different? The answers lay before our eyes. We still have the Galápagos Islands that we do not want to lose—the ones with half a century of conservation and management efforts, many of them successful (Chapter Eight, *What is the Conservation Value of the Galápagos?*). Hence, we either keep Galápagos unique and extraordinary, as per UNESCO's World Heritage standards, or we turn it into another leisure destination. The choice—or the responsibility—is between a matchless, well-preserved, and environmentally managed Galápagos versus an altered and unrecognizable Galápagos. All will depend on the type of future adopted and the speed at which irreparable change is imposed" (Paz-y-Miño-C and Espinosa 2024b).

"Will Galápagos' caretakers and policymakers not mimic the mistakes that nature custodians have made in other volcanic archipelagos, where pristine ecosystems and endemic wildlife have substantially disappeared while being replaced by large towns and cities, conventional tourism resorts, highways, and introduced species? We addressed these points in our book, *Arrivals of Life to the Galápagos*, particularly in the chapter, *Galápagos versus Hawai'i: Lessons to Never Forget* (Paz-y-Miño-C and Espinosa 2024a). But we still need to ask: can we save the Galápagos we have? Likely yes. If most of its ecosystems remain legally protected and restoration and reintroduction programs continue, the essence of Galápagos has a chance to make it into the future. If the state of Ecuador continues to own and administer the land under low-impact development ideals, the exceptional Galápagos will be exposed to manageable habitat alteration. If human presence in the islands remains at low numbers and density, Galápagos could persist long-term. If tourism is entirely eco-friendly, small-scale, and respectful of both the fragile terrestrial/marine habitats as well as the local people, Galápagos could offer for centuries the possibility of being explored by those who love and value nature (Paz-y-Miño-C and Espinosa 2024a,b). But if Galápagos chooses the path of fast, unsustainable change, urban crowding leading to occupation of additional land ["overflow effect"], and eco-unfriendly tourism (amusement, entertainment), or if it is persuaded by local or international investors to mimic or resemble Hawai'i's socio-economic trajectory or that of other oceanic, volcanic archipelagos (e.g. Canaries, Mascarenes, Madeira & Selvagens, Azores, Cape Verde, or Seychelles; as discussed in Paz-y-Miño-C and Espinosa 2024a), its destiny will be comparable. It will be a matter of time".

"Will the Galápagos' people, Ecuador, and the world commit to safeguarding and managing the archipelago and its marine reserve so that sound sustainable development practices limit human population expansion on the islands, sponsor solely eco-friendly tourism, and secure revenue fairness for the residents? (Paz-y-Miño-C and Espinosa 2024a). The Galápagos' people, the nation of Ecuador and its government, and the world have the responsibility of preserving, restoring, and managing the fragile marine and terrestrial habitats of the Galápagos Islands (Chapter Eight, *What is the Conservation Value of the Galápagos?*). Yet, as a UNESCO World Heritage Site, Galápagos belongs to all of us. Our collective duty is to protect: '(i) its *superlative natural phenomena and exceptional natural beauty;* (ii) its *record of life, on-going geological processes, geomorphic or physiographic features;* (iii) its *on-going ecological and biological processes in the* [context of] *evolution and development of terrestrial, fresh water, coastal and marine ecosystems and communities of plants and animals;* and (iv) its *significant natural habitats for in-situ conservation of biological diversity, including* [their] *threatened species of outstanding universal value from the point of view of science or conservation'"* (UNESCO 2025a,b).

From the geologic point of view, as we indicated above and discussed in Chapter One, *What Are the Galápagos Islands?*, the Galápagos "archipelagos of the past" are long gone; the existing archipelago [≈4 million years old] will continue to drift East on top of the Nazca Plate; its volcanoes will sink to the bottom of the ocean as seamounts; and future archipelagos, with different spatial configurations, will emerge. That is the geological fate of the Galápagos. [As long as the Galápagos hotspot remains active and pierces the lithosphere above it, new islands will continue to emerge, that is, if there is an ocean above the spot; Paz-y-Miño-C and Espinosa (2024a)].

Conclusion

"Regardless of humans surviving the Anthropocene (e.g. the consequences of climate change, mass extinctions, or warfare) or global catastrophic events (e.g. Earth's collisions with asteroids, which will probabilistically happen long-term), ecological and evolutionary processes will continue in future Galápagos archipelagos, of course on a planet already scarred by human presence. But these realities can inspire us to pledge our existence to the health of the Earth" (Paz-y-Miño-C and Espinosa 2024b). As Charles Darwin reflected in the *Journal of Researchers* (1839), just four years after exploring the Galápagos in 1835, *[t]he natural history of this archipelago is very remarkable: it seems to be a little world within itself; the greater number of its inhabitants, both vegetable and animal, being found nowhere else* (Darwin 1839; Browne 2018). Indeed, a little world, the only one we have, is worth saving forever".

References

Browne, Janet. 2018. *The Quotable Darwin*. Oxford: Princeton University Press.

Darwin, Charles. 1839. 1845. 1890. *Journal of Researches into the Natural History and Geology of the Countries Visited During the Voyage of H.M.S. Beagle Round the World, under the Command of Capt. Fitz Roy, R.N.* London: John Murray.

Discovering Galapagos. 2023. "Galapagos: History of Tourism". Galapagos Conservation Trust and Royal Geographical Society. Accessed November 8, 2023. *https://www.discoveringgalapagos.org.uk/discover/sustainable-development/sustainable-tourism/history-of-tourism/*.

DPNG, and OTG (Dirección del Parque Nacional Galápagos, and Observatorio de Turismo de Galápagos). 2018. *Informe Anual de Visitantes a las Áreas Protegidas de Galápagos del Año 2018*. Puerto Ayora: Dirección del Parque Nacional Galápagos and Observatorio de Turismo de Galápagos.

DPNG (Dirección del Parque Nacional Galápagos). 2019. *Informe Anual de Visitantes a las Áreas Protegidas de Galápagos del Año 2019*. Puerto Ayora: Dirección del Parque Nacional Galápagos.

—. 2020. *Informe Anual de Visitantes a las Áreas Protegidas de Galápagos del Año 2020*. Puerto Ayora: Dirección del Parque Nacional Galápagos.

—. 2021. *Informe Anual de Visitantes a las Áreas Protegidas de Galápagos del Año 2021*. Puerto Ayora: Dirección del Parque Nacional Galápagos.

—. 2022. *Informe Anual de Ingreso de Visitantes a las Áreas Protegidas de Galápagos del Año 2022*. Puerto Ayora: Dirección del Parque Nacional Galápagos.

INEC (Instituto Nacional de Estadísticas y Censos). 1982. *IV Censo de Población 1982: Galápagos*. Quito: INEC.

—. 2015. *Análisis de Resultados Definitivos Censo de Población y Vivienda Galápagos 2015*. Quito: INEC.

—. 2023. "Censo Ecuador: Galápagos 2022". Accessed November 6, 2023. *https://censoecuador.ecudatanalytics.com/*.

—. 2024. *Resultados Principales Galápagos*. Quito: INEC.

Melville, Herman. 1851. *Moby-Dick; or, The Whale*. New York: Harper & Brothers, Publishers.

—. 1856. *The Piazza Tales*. New York: Dix Edwards and Company.

Paz-y-Miño-C, Guillermo, and Avelina Espinosa. 2024a. *Arrivals of Life to the Galápagos: From Berlanga to Darwin, Melville, and World War II*. Newcastle upon Tyne: Cambridge Scholars Publishing.
—. 2024b. *A Synthesis of the Galápagos: Natural History, Human Conflict, and the Destiny of an Archipelago*. Newcastle upon Tyne: Cambridge Scholars Publishing.
UNESCO (United Nations Educational, Scientific and Cultural Organization) World Heritage Convention. 2025a. "Galápagos Islands". Accessed May 16, 2025. *https://whc.unesco.org/en/ list/1/.*
—. 2025b. "The Criteria For Selection". Accessed May 16, 2025. *https://whc.unesco.org/en/criteria/.*

APPENDIX A

FIGURES' NOTES AND SOURCES

All figures in this book were built on Power Point (background slide = 150 x 100 cm), processed in Adobe Photoshop, and saved as TIFF files (horizontal and vertical resolution = 600 dpi); the illustrations included 280 labeled figures, the majority with multiple sub-figures (total sub-figures = 733). The images' setting/layout was sketched, drawn, or entirely redrawn based on scientific references as indicated in each caption:

(1) Depending on the figure, we clarified in the caption that some components were obtained from public domain sources (details in this Appendix), for example, in Fig. 1.1, Chapter One, we stated, "Location of the Galápagos Islands with respect to the Tropical Eastern Pacific and the proximate mainland (South and Central America)". In this Appendix, we included the following supplementary information: "Fig. 1.1 Background map https://www.freeusandworldmaps.com/world-projections-blank-printable-clip-art-maps/. Curved arrow https://openclipart.org/detail/161695/curved-arrow". Here is another example: In Fig. 3.1, Chapter Three, we stated, "Redrawn by GPC (as per Paz-y-Miño-C and Espinosa 2024a), after Journal of Researchers (1890)." And, as noted in this Appendix, Fig. 3.1 included various Public Domain silhouettes: casks/barrels, jugs, wine bottles, ladder, chair, stool, rope, etc. WWW links to such free downloads are provided below. Realize, however, that we digitally sketched the majority of the silhouette-looking components on the figures.

(2) When two distinctive components of a figure (i.e. image and data) were inspired by more than one source, as, for example, Fig. 8.5 (a,b), Chapter Eight, we indicated, "Layout GPC [G. Paz-y-Miño-C]. a. Data from Poulakakis et al. (2008, 2012, 2015); Gaughran et al. (2017); Miller et al. (2017, 2018). b. 50% majority rule consensus tree (16S rRNA, cyt *b* mtDNA) for Maximum Parsimony (Santa Fé excluded); modified after Caccone et al. (1999)."

(3) When the design of a single figure was inspired by more than one source, for example, in Fig. 4.17, Chapter Four, we stated: "Data extracted from Townsend (1935), Woolmer (2013) and Halpin et al. (2009). Number of whale-oil-barrels comes either from whaleships logbooks or from esti-

mates based on the literature, as follows: The 30 barrels of oil per whale corresponds to the Galápagos, after Whitehead and Hope (1991). Data on a whaleship carrying 500 barrels on a complete voyage comes from Smith et al. (2008)". Similarly, in figures in which we generated the digital sketches based on multiple sources, e.g. all shark drawings (35 species) in Figs. 8.70-8.77, Chapter Eight, we stated, "Layout, computer-generated sketches: G. Paz-y-Miño-C; redrawn and modified after Grove and Lavenberg (1997); Jiménez-Prado and Béarez (2004); Ebert (2016); Dando (2020); Cantú and Levi-Méndez (2022); Shark Research Institute (2022)".

(4) When figures came directly from a source (e.g. Open Access Journal like *PLoS ONE*) but were modified and/or adjusted for this book (which happened solely in Figs. 4.27-4.29, Chapter Four), we indicated, "Maps extracted from open-access Smith et al. (2012); link in Appendix A". And in References, we listed "Smith, Tim D., Randall R. Reeves, Elizabeth A. Josephson, and Judith N. Lund. 2012. "Spatial and Seasonal Distribution of American Whaling and Whales in the Age of Sail". *PLoS ONE*, 7: e34905". Note that for this particular article, PLoS ONE's disclaimer states, "*This is an open-access article distributed under the terms of the Creative Commons Attribution License, which permits unrestricted use, distribution, and reproduction in any medium, provided the original author and source are credited*".

(5) We aimed at simplifying all figures as much as we could; in some instances, we omitted error scale-bars due to the complexity of the figure; or omitted complex statistical analyses while encouraging the reader to find details in the original sources; all these adjustments were made for simplicity. We provided comprehensive lists of references at the end of each chapter; the reader should search for them.

(6) Finally, to abridge content in both main text and figure captions, particularly in longer chapters, we alerted readers that "We omit[ed] citations [or allusions to specific figures] in this [particular] section [e.g. Conclusions] to abridge the text." And we clarified that "Each conclusion [or figure caption] point summarize[d] topics [already] substantiated by citations in this chapter".

We thank the hundreds of scientists, conservation biologists, and wildlife managers who authored the peer-reviewed articles cited in this book (496 references), which we used as sources to design the figures (in all cases, as indicated in captions, "redrawn and modified"). Below we provide a list of public-domain images included in some illustrations.

Chapter One: What Are the Galápagos Islands?

Fig. 1-1 Background map https://www.freeusandworldmaps.com/world-projections-blank-printable-clip-art-maps/. Curved arrow https://openclipart. org/detail/161695/curved-arrow.

Fig. 1-3 Nazca Plate and neighboring plates sketched from compilation of various sources: Map of South American Plate https://en.wikipedia.or g/wiki/South_American_Plate#/media/File:SouthAmericanPlate.png; Map of Nazca Plate https://en.wikipedia.org/wiki/Nazca_Plate#/ media/File: Nazca Plate.png; and Nazca Plate https://www.britannica. com/science/ Nazca-Plate.

Fig. 1-5 Map sketched after GEBCO, NOAA, NCEI and Charles Darwin Foundation https://fcdgps.maps.arcgis.com/apps/MapSeries/index.html? appid=d202f27c344142f49e18f2d19cdd3988; https://geodata-fcdgps. opendat a.arcgis.com/.

Fig. 1-13 Galapagos Islands Flyby. NASA/Goddard Space Flight Center, the SeaWiFS Project and GeoEye, Scientific Visualization Studio; July 17, 2009. https://svs.gsfc.nasa.gov/3628.

Fig. 1-21 Galapagos Islands Image. NASA ID: PIA10121. Mercury-bound MESSENGER spacecraft; August 2 2005. https://images.nasa.gov/ details-PIA10121.html; Secondary Creator Credit: NASA/Johns Hopkins University Applied Physics Laboratory/Carnegie Institution of Washington. Curved arrow clipart: https://openclipart.org/detail/161695/ curved-arrow.

Fig. 1-22a Galápagos. NASA's Moderate-resolution Imaging Spectroradiometer (MODIS), Terra satellite, March 12, 2002. https:// eoimages.gsfc.nasa.gov/images/imagerecords/57000/57939/Galapagos.A2002 071.1625.250m.jpg. Inset magnification of Santiago/San Salvador/ James: photograph taken from the International Space Station (ISS), Expedition crew 33, October 18, 2012. https://earthobservatory.nasa. gov/images/79530/isla-santiago-galapagos-islands.

Fig. 1-23b North of Isabela/Albemarle (volcanoes Wolf, Darwin and Alcedo) and Fernandina/Narborough (volcano La Cumbre) as capture by NASA's Landsat 7 satellite, October 16, 2001. https://eoimages. gsfc.nasa.gov/images/imagerecords/37000/37003/wolfvolcano_l7_200 1299_lrg.jpg.

Fig. 1-24a Eruption of Wolf Volcano, Galapagos Islands. Advanced Spaceborne Thermal Emission and Reflection Radiometer (ASTER) on board of the Terra Satellite; NASA, June 11, 2015. https://www. nasa.gov/image-feature/eruption-of-wolf-volcano-galapagos-islands.

Fig. 1-24b Interferogram; Sentinel 1A satellite, May 3 and June 23, 2015 (processed by EFFIGIS GeoSolutions; NASA, US Geological Survey and European Space Agency; Copernicus program). https://effigis.com/en/volcanic-eruption-galapagos-islands-observed-28-centimetre-soil-displacement/.

Fig. 1-24c View of Galapagos Islands from International Space Station (ISS). November 11, 2013. NASA-ISS Expedition 38 crew. https://www. nasa.gov/content/view-of-galapagos-islands-from-space-station.

Fig. 1-25 Galápagos Islands. Copernicus Sentinel-2 of the European Space Agency (ESA), September 23, 2020. https://www.esa.int/ESA_Multimedia/Images/2021/03/Galapagos_Islands.

Fig. 1-26 Proba-Vegetation (Proba-V) view of Galápagos, February 27, 2017. European Space Agency/Belspo, Produced by VITO Remote Sensing Belgium. https://www.esa.int/ESA_Multimedia/Images/2019/03/Proba-V_view_of_Galapagos.

Figs. 1-27 to **1-31** Silhouettes of the Galápagos Islands. Sketches by G. Paz-y-Miño-C with impute from multiple sources: Galapagos Islands topographic map-de.svg, Wikimedia Commons https://commons.wikimedia.org/wiki/File:Galapagos_Islands_topographic_map-de.svg. McBirney, A. R., and Howel Williams. 1969. *Geology and Petrology of the Galápagos Islands*. Boulder: The Geological Society of America Inc. Peñaherrera-Palma, César, Karen Harpp, and Stuart Banks. 2016. "Rapid Seafloor Mapping of the Northern Galápagos Islands, Darwin and Wolf". *Galapagos Research*, 68: 2-9. Silhouette of two engine jet: https://www.freepik.com/premium-vector/airplane-silhouette-passenger-plane-back-front-bottom-views-aircraft-jet-silhouettes-illustration-icons-set-jet-monochrome-plane-airplane-commercial-passenger-flight_9602097.htm. Silhouette of small aircraft: https://fr.wikipedia.org/wiki/Fichier:Airplane_GA_Black.svg. Silhouette of helicopter: https://flyclipart.com/collection-of-helicopter-silhouette-clip-art-download-them-blackhawk-clipart-558707.

Fig. 1-32 Curved arrow https://openclipart. org/detail/161695/curved-arrow.

Fig. 1-33 Silhouettes of crab https://www.pinterest.com/pin/56013589092970415/; and octopus https://www.seekpng.com/ipng/u2q8q8e6e6q8q8 t4_octopus-silhouette-octopus-png/.

Fig. 1-34 Silhouettes from free downloads: flat arrow https://www.nicepng.com/ourpic/u2q8q8i1w7u2r5w7_stock-arrow-up-png-3d-arrow-png/; curved arrow https://openclipart.org/ detail/161695/ curved-arrow.

Fig. 1-35 Curved arrows https://openclipart. org/detail/161695/curved-arrow.

Fig. 1-36 is similar to **Fig. 1-33**.

Fig. 1-45 to **1-47** Maps on these figures were redrawn and simplified for didactic purpose; sketched in black-and-white (distinctive gray-color tones and textures) and grouped them as per Northwestern and Northern

Islands (Fig. 1.45); Western Islands (Fig. 1.46); and Central, East, and Southeastern Islands (Fig. 1.47). Source: Rivas-Torres, Gonzalo, Fátima L. Benítez, Danny Rueda, Christian R. Sevilla, and Carlos F. Mena. 2018. "A Methodology for Mapping Native and Invasive Vegetation Coverage in Archipelagos: An Example from the Galápagos Islands". *Progress in Physical Geography*, 42: 83-111. Moity, Nicolas, Byron Delgado, and Pelayo Salinas-de-León. 2019. "Mangroves in the Galapagos Islands: Distribution and Dynamics". *PLoS ONE*, 14: e0209313. Important links: https://journals.sagepub.com/doi/abs/10.1177/0309133317752278; www. institutodegeografia.org/vega; https://journals.plos.org/plosone/article?id=10.1371/journal.pone.0209313.

Fig. 1-49 and Fig. **1-50** Curved arrows https://openclipart. org/detail/161695/curved-arrow.

Chapter Two: How Were the *y^a s de galapagos* Discovered?

Fig. 2-1 Photo of Statue of Fray Tomás de Berlanga at Berlanga de Duero, Spain. Courtesy: Francisco Dousdebés (with permission).

Fig. 2-2 Fray Tomás de Berlanga source Galapagos Conservation Trust https://galapagosconservation.org.uk/wp-content/uploads/2014/03/tomc3a 1s-de-berlanga.jpg; ship silhouette, modified from free clip art https://www.pinterest.com/pin/300122762662072827/?d=t&mt=login.

Figs. 2-2a,b and **2-3** Ship silhouette, modified from free clip art https://www.pinterest. com/pin/300122762662072827/?d=t&mt=login.

Fig. 2-4 Portolan Chart of the Pacific Coast from Guatemala to Northern Peru 1565. Library of Congress of the United States of America. https://www.loc.gov/item/2006627298/.

Fig. 2-5 Galápagos Islands, Surveyed by Captain FitzRoy, R. N. and the Officers of H. M. S. Beagle, 1836, with Additions and Corrections to 1926. Wikimedia Commons https://commons.wikimedia.org/wiki/File:Galapagos_1836.jpg.

Chapter Three: Why Was Charles Darwin's Visit to Galápagos Significant?

Fig. 3-1 HMS Beagle redrawn after Darwin, Charles. *Journal of Researchers into the Natural History and Geology of various Countries Visited by HMS Beagle*. London: John Murray, 1890, http://darwin-online.org.uk/content/frameset?pageseq=1&itemID=F59&viewtype=side; Captain FitzRoy portrait https://en.wikipedia.org/wiki/Robert_FitzRoy#/media/File:Robert_Fitzroy.jpg; Darwin portrait https://en.

wikipedia.org/wiki/Charles_Darwin#/media/File: Charles_Darwin_by_G._Richmond.png; HMS Beagle includes free-downloads clipart from multiple sources: casks/barrels https://imgbin.com/ png/pUW8HyCc/beer-wine-barrel-keg-drawing-png; jugs silhouettes https://www.freevector.com/jug-silhouette-21896; wine bottles https://www.1001freedownloads.com/free-clipart/wine-bottle; ladder, chair, stool, rope, chain, anchor, sac, and sailor.

Fig. 3-2 Sketch after General Chart showing the Principal Tracks of H.M.S. Beagle 1831-6. http://darwin-online.org.uk/content/frameset?item ID=F10.2a&viewtype=image&pageseq=393; In FitzRoy, R. 1839. *Narrative of the Surveying Voyages of His Majesty's Ships Adventure and Beagle between the Years 1826 and 1836, Describing their Examination of the Southern Shores of South America, and the Beagle's Circumnavigation of the Globe*. Appendix to Volume II. London: Henry Colburn. http://darwin-online.org.uk/content/frameset?pageseq=393&itemID=F10.2 a&viewtype=side; and Rookmaaker, Kees. 2009. *Darwin's Itinerary on the Voyage of the Beagle*. http://darwin-online.org.uk/content/frameset? view type=text& itemID=A575&pageseq=1.

Fig. 3-3 After *Galápagos Islands*, Surveyed by Captain FitzRoy, R. N. and the Officers of H. M. S. Beagle, 1836, with Additions and Corrections to 1926. https://commons.wikimedia.org/wiki/File:Galapagos_1836.jpg.

Fig. 3-4 Modified after Estes, Gregory, K. Thalia Grant, and Peter R. Grant. "Darwin in Galápagos: His Footsteps through the Archipelago." *Notes and Records of the Royal Society of London* 54 (2000): 343-368; FitzRoy and Darwin portraits same is in Fig. 3.1; sailing ship silhouette modified after https://www.clipartkey.com/view/iiRxTib_sailing-ship-ship -model-clip-art-sailing-ship/.

Fig. 3-5 Data from Grant and Estes (2009), as cited in Chapter; sailing ship silhouette as in **Fig. 3-4**.

Fig. 3-6 Data from Rookmaaker (2009), as cited in Chapter; sailing ship silhouette as in **Fig. 3-4**.

Fig. 3-7 Charles Frederick Holder's book (1891) electronic copy https://archive.org/details/charlesdarwinhis00holduoft/page/n11/mode/2up?view=theater.

Fig. 3-8 Modified after photo (taken by GPC) of map at Charles Darwin Research Station, Puerto Ayora, Santa Cruz Island, Galápagos; Jan 2022.

Fig. 3-9 Earth sphere https://www.kissclipart.com/global-women-chamber-of-commerce-clipart-balcolm-c-kx3356/; curved arrow https://openclipart.org/detail/161695/curved-arrow; sailing ship silhouette as in **Fig. 3-4**.

Fig. 3-11 HMS Beagle navigating the Straits of Magellan. Illustration by Robert Taylor Pritchett. Wikipedia, public domain, and open access. https://en.wikipedia.org/wiki/HMS_Beagle#/media/File:PSM_V57_D097_Hms_beagle_in_the_straits_of_magellan.png.

Fig. 3-12 Image of the "I think" tree and cover of Darwin's Notebook B: Transmutation (images with permission of the Syndics of Cambridge University Library): Files MS-DAR-00121-000-00001: MS DAR 121, cover, www link https://cudl.lib.cam.ac.uk/view/MS-DAR-00121/1 and MS-DAR-00121-000-00038: MS DAR 121, p. 36, www link https://cudl.lib.cam.ac.uk/view/MS-DAR-00121/38.

Fig. 3-15 Darwin statue photo (taken by GPC), Natural History Museum, London, 2010.

Chapter Four: What Happened in Galápagos Before and After Darwin?

Fig. 4-1 Sketch of Whaling Bark Alice Knowles of New Bedford was based on https://mysite.du.edu/~ttyler/ploughboy/whalingbarkbot.htm; sperm whale silhouette after http://mysite.du.edu/~ttyler/ploughboy/starbuck.htm; smoking chimney https://www.kindpng.com/imgv/JhTTmm_cartoon-smoke-png-smoke-clipart-png-transparent-png/; sailing ship silhouette https://www.clipart key.com/view/iiRxTib_sailing-ship-ship-model-clip-art-sailing-ship/.

Fig. 4-2 to **4-9** Photos taken by G. Paz-y-Miño-C at the New Bedford Whaling Museum, Massachusetts, United States. See *acknowledgments* to the New Bedford Whaling Museum and particularly to Bob Rocha.

Fig. 4-4 (top) Harpoons: Wikimedia Commons https://commons.wikimedia.org/wiki/File:Whaling_Harpoons_1887.jpg.

Fig. 4-10 Open access digitized maps by Whale History (2017; Mystic Seaport Museum and New Bedford Whaling Museum). Whale History. 2017. "Townsend's Whaling Charts". Accessed February 27, 2025. https://whalinghistory.org/townsends-whaling-charts/.

Fig. 4-14 same as **Fig. 4-10**.

Figs. 4-15, 4-16, 4-18 to **4-24** Sperm whale silhouette: public domain https://www.freepik.com/free-icon/sperm-whale-silhouette_722818.htm.

Figs. 4-16, 4-18 and **4-21** ship silhouette, modified from free clip art https://www.pinterest.com/ pin/300122762662072827/?d=t&mt=login.

Figs. 4-17, 4-22 and **4-23** casks/barrels, public domain: https://imgbin.com/png/pUW8HyCc/beer-wine-barrel-keg-dr awing-png.

Fig. 4.25 Earth Globe Map sphere, public domain: https://www.clipartmax.com/middle/m2i8G6N4b1i8b1b1_earth-globe-world-map-earth-globe-world-map/.

Figs. 4-26 to **4-29** Maps come from open access material: Smith, Tim D., Randall R. Reeves, Elizabeth A. Josephson, and Judith N. Lund. 2012. "Spatial and Seasonal Distribution of American Whaling and Whales in the Age of Sail". *PLoS ONE*, 7: e34905. https://journals.plos.org/plosone/article?id=10.1371/journal.pone.0034905; https://journals.plos.org/plosone/article/figure?id=10.1371/journal.pone.0034905.g001 ; https://doi.org/10.1371/journal.pone.0034905.g002.

Fig. 4-30 Silhouette of sword fish comes from clipart free downloads https://creazilla.com/nodes/414-swordfish-silhouette; same applies to squid https://www.pinclipart.com/pindetail/ihmoiTh_colossal-squid-misc-art-giant-squid-clipart/; humpback whale skeleton https://www.3dhorse.com/products/humpback-whale-skeleton.

Fig. 4-31 Photo of blue whale (*Balaenoptera musculus*); courtesy of Ian Wiese (with permission), Geographe Bay, Western Australia.

Chapter Five: Why Did the United States Establish a Military Base in Galápagos During World War II?

Several of the historic photographs in Chapter Five are broadly available online at multiple depositories, as we provide links below. We highlight that book author and historian John Woram compiled a great initial collection of these images on his Galápagos website (www.galapagos.to), unfortunately no longer available.

Box 5-1 Eleanor Roosevelt interviews President Galo Plaza of Ecuador. Public Domain image from the Harry S. Truman Library and Museum https://www.trumanlibrary.gov/photograph-records/66-8162.

Fig. 5-1 Images come from Life magazine 1942 https://www.oldlifemagazines.com/march-02-1942-life-magazine.html; as well as Open Access PDF of Life's 1942 entire article https://imgur.com/a/UeDra#sEVbDz0.

Fig. 5-2 "Special Map, Galapagos Islands, Seymour Island" by the Corps of Engineers United States Army, Panama Canal Department, 1943. Image comes from book author and Galápagos historian John Woram's collection www.galapagos.to (link no longer available on www).

Fig. 5-3 Seymour Island "The Rock". Aerial photo by the United States Air Force 1959; available at multiple sources: 970th Airborne Air Control Squadron https://military-history.fandom.com/wiki/970th_Airborne_

Air_Control_Squadron; Revisionismo Histórico del Ecuador https://guerrade1941.blogspot.com/ p/base-beta-galapagos.html.

Fig. 5-5 Eleanor Roosevelt visiting the United States troops at "Base Beta" in Seymour Island or "The Rock" Galápagos (March 1944). Top: source "Day by Day" archives from the Franklin D. Roosevelt Presidential Library http://www.fdrlibrary.marist.edu/daybyday/resource/march-1944-4/. Bottom: source National Archives and Records Administration via WikiMedia Commons https://commons.wikimedia.org/wiki/File:Eleanor_ Roosevelt_in_Galapagos_Island_-_NARA_-_196623.jpg.

Figs. 5-6 to **5-10** and **5-12** Images come from multiple sources, e.g.: Revisionismo Histórico del Ecuador https://guerrade1941.blogspot.com /p/base-beta-galapagos.html; 970th Airborne Air Control Squadron https://military-history.fandom.com/wiki/970th_Airborne_Air_Control_ Squadron; Base Aérea Galápagos https://www.youtube.com/ watch?v= RS3O8UCKUMQ; THE ROCK - Galápagos en La II Guerra Mundial VIDEO https://www.youtube. com/watch?v=nYPibVzbPwA.

Fig. 5-11 Background silhouette map of the Americas comes from WikiMedia Commons https://en.wikipedia.org/wiki/Americas#/media/ File: Americas_(orthographic_projection).svg; photo of USS Houston (CA-30) comes from https://www.tapatalk.com/groups/flattopshistorywarpolitics/ambush-in-the-sunda-strait-t2060.html. Information about the *Logs of the President's Trips: Inspection Cruise and Fishing Expedition on Board USS Houston, July 16 - August 9, 1938*, comes from http://www. fdrlibrary.marist.edu/_resources/images/tully/7_03.pdf.

Fig. 5-13 Post World-War-Two Military Air Base Requirements [Western Hemisphere] by the United States, approved November 23, 1943, comes from https://www.rand.org/pubs/monographs/MG1244.html.

Fig. 5-15 Ecuador's President Carlos Arroyo del Río with US President Franklin D. Roosevelt (FDR) Bettmann Collection, via Getty Images, with permission.

Fig. 5-18 National parks and exemplar protected areas in continental and insular Ecuador after http://www.forosecuador.ec/forum/ecuador/ educaci%C3%B3n-y-ciencia/164669-51-%C3%A1reas-protegidas-del-ecuador-mapa-lista-de-zonas-naturales-protegidas.

Fig. 5-19 Clipart arrow https://openclipart.org/detail/161695/curved-arrow.

Fig. 5-20 Background map http://www.freeusandworldmaps.com/images /WorldPrintable/WorldRobinCountryNoLinesPrint.jpg; fishing boats silhouettes https://www.pngfind.com/mpng/ixxxbRo_fmc-dockyard-fishing-silhouette-of-a-fishing-boat/; P3 Orion Plane https://www.flickr.com/ photos/cbpphotos/5764837880; clipart arrow https://openclipart.org/detail/ 161695/curved-arrow; with input from El Universo. 2019. Aviones de Vigilancia de los Estados Unidos Operarán en Galápagos https://www. eluniverso.com/noticias/2019/05/30/ nota/7352517/ministro-defensa-anuncia-planes-incrementar-vigilancia/.

Fig. 5-21 Ecuador's Minister of Defense, Oswaldo Jarrín (left), and US Admiral of the Southern Command, Craig S. Faller. Photo Rodrigo Buendía/AFP, via Getty Images, with permission.

Fig. 5-22 Redrawn and modified after UNODC (United Nations Office on Drugs and Crime). 2021. World Drug Report–Drug Market Trends: Cocaine Amphetamine-Type Stimulants. Vienna: United Nations, with input from Primicias. 2020. El 74% de la Cocaína que Va Hacia Estados Unidos Es Enviada desde Colombia y Ecuador https://www.primicias.ec/noticias/en-exclusiva/cocaina-eeuu-sale-ecuador-colombia/.

Chapter Six: Who Lives in Today's Galápagos Islands?

Fig. 6-2 Two-engine jet silhouette https://www.pngegg.com/en/png-nsurr. Small aircraft https://fr.wikipedia.org/wiki/Fichier:Airplane_GA_ Black. svg. Tortoises and helicopter sketched by G. Paz-y-Miño-C.

Fig. 6-3 Photo of young Charles Darwin statue at the Charles Darwin Research Station in Puerto Ayora, Galápagos, by G. Paz-y-Miño-C.

Fig. 6-3-6-11 Curved arrows https://openclipart.org/detail/ 161695/curved-arrow. Drawings or silhouettes of yacht, bus, gas-station machine, zodiac boat, swimmer, and graduation hat come from multiple free download www sites. Tourism vessel, water taxi, tortoises' silhouettes and photos, as well as helicopter sketches by G. Paz-y-Miño-C.

Fig. 6-16 Curved arrow free download clipart from https://openclipart.org/ detail/161695/curved-arrow.

Chapter Seven: Why is Nature Tourism Essential to Galápagos?

Fig. 7-1 Silhouettes of small aircraft https://fr.wikipedia.org/wiki/Fichier: Airplane_GA_Black.svg; large aircraft https://www.freepik.com/ premium-vector/airplane-silhouette-passenger-plane-back-front-bottom-views-aircraft-jet-silhouettes-illustration-icons-set-jet-monochrome-plane-airplane-commercial-passenger-flight_9602097.htm.

Fig. 7-4 and **7-11** Silhouettes of palm tree https://openclipart.org/detail/ 261353/palm-tree; cactus https://www.freepik.com/free-photos-vectors/

cactus-silhouette; houses https://publicdomainvectors.org/en/free-clipart/Vector-graphics-of-silhouette-of-a-simple-hou se/29863.html.

Fig. 7-6 World sphere clip art https://www.clipartmax.com/middle/m2K9A0m2d3b1i8A0_black-and-white-globe-clipart-black-and-white-globe-clipart/.

Fig. 7-8 Boat https://pngtree.com/; inflatable boat (zodiac) https://www.boatcoversdirect.com/products/boat_covers/styled_to_fit/style/67/; scuba diver silhouette https://gallery.yopriceville.com/Free-Clipart-Pictures/Silhouettes-PNG/Scuba_Diver_Silhouette_PNG_Clip_Art_Ima ge#.XIkWfCJKj3g.

Fig. 7-10 Arrows https://openclipart. org/detail/161695/curved-arrow; map section of Northwestern South America and Central America http://www.freeusandworldmaps.com/images/WorldPrintable/WorldRobinCountryNoLinesPrint.jpg.

Fig. 7-11 Wine bottles, wine glass, and utensils https://www.vecteezy.com/vector-art/5154747-wine-and-glass-vector; https://creazilla.com/nod es/3161385-utensils-clipart.

Chapter Eight: What Is the Conservation Value of the Galápagos?

Fig. 8-1 Digital, computer generated sketches based on collaged www sources: redrawn and substantially modified by us.

Fig. 8-2 Similar to **Fig. 8-1** and also **Fig. 1-36** in Chapter One.

Figs. 8-36, 8-47 Clipart arrow; https://openclipart.org/detail/161695/curved-arrow.

Fig. 8-63 Clipart silhouettes from free download sources: crab https://www.pinterest.com/pin/560135890929704151/; and octopus https://www.seekpng.com/ipng/u2q8q8e6e6q8q8t4_octopus-silhouette-octopus-png/.

Fig. 8-73 Scuba diver silhouette clip art, free download https://gallery.yopriceville.com/Free-Clipart-Pictures/Silhouettes-PNG/Scuba_Diver_Silhouette_PNG_Clip_Art_Image#.XIkWfCJKj3g.

Chapter Nine: What Are the Main Conservation Problems in Galápagos?

Figs. 9-9 World map with countries outlined. Public domain image from J. Bruce Jones; free download for educational projects. http://www.freeusandworldmaps.com/images/WorldPrintable/WorldRobinCountryNoLinesPrint.jpg.

Figs. 9-21 curved arrow https://openclipart. org/detail/161695/curved-arrow.

Figs. 9-31, 9-33 Crab silhouette https://www.pinterest.com/pin/56013 5890 92970415/; octopus https://www.seekpng.com/ipng/u2q8q8e6e6q 8q8t4_ octopus-silhouette-octopus-png/; salmon https://www.netclipart .com/isee/mmJwJJ_salmon-png-salmon/; plastic containers https://www.vecteezy. com/vector-art/638254-plastic-waste-cartoon-vector-illustration; shrimp https:// pixabay.com/ illustrations/shrimp-lobster-prawn-logo-7987323/.

Chapter Ten: How Will Climate Change Affect Galápagos?

Figs. 10-1 to **10-3** Computer generated sketches by G. Paz-y-Miño-C; data extracted from Paltán, Homero A., Fátima L. Benitez, Paulina Rosero, Daniel Escobar‐Camacho, Francisco Cuesta, and Carlos F. Mena. 2021. "Climate and Sea Surface Trends in the Galapagos Islands". *Scientific Reports*, No. 11: 14465.

Chapter Eleven: What Can You Do to Help Galápagos?

Figs. 11-1 to **11-3** Layout and sketches GPC, after references therein.
Fig. 11-4 Redrawn and modified after Novak, Joseph D., and Alberto J. Cañas. 2008. "The Theory Underlying Concept Maps and How to Construct and Use Them". Technical Report IHMC CmapTools 2006-01 Rev 01-2008, Institute for Human and Machine Cognition. Accessed August 17, 2025. https://cmap.ihmc.us/docs/pdf/Theory UnderlyingConceptMaps.pdf.
Global organizations working on behalf of the Galápagos (keep in mind that hundreds of NGOs and dozens of world-class universities work in the Galápagos; the list below is abridged):
Charles Darwin Foundation https://www.darwinfoundation.org/en/
Durrell Wildlife Conservation Trust https://www.durrell.org/conservation/ places/gala-pagos/
Galapagos Conservancy https://www.galapagos.org/
Galapagos Conservation Trust https://galapagosconservation.org.uk/
Galápagos Life Fund https://galapagoslifefund.org.ec/
Re:wild Galápagos https://www.rewild.org/get-to-know/the-galapagos
University San Francisco of Quito https://www.usfq.edu.ec/en/galapagos
University of North Carolina https://galapagos.unc.edu/

APPENDIX B

GALÁPAGOS NATIONAL PARK RULES*

If you travel to Ecuador and the Galápagos, you must do it legally (i.e. with identification documents, a passport, and a visa, the latter if it applies to your country). Most likely, you will visit the islands as a tourist and hold a Transit Control Card (a mandatory document issued by Ecuador and specifically by the Governing Council of Galápagos). You will respect the regulations that protect the Galápagos National Park and the Marine Reserve. Only accredited scientists, academics, and/or naturalists who have obtained official permits can do research in Galápagos.

(1) ***Only authorized naturalist guides***. Visitors to any protected areas within the Galápagos National Park must be accompanied by a naturalist guide authorized by the Galápagos National Park Directorate (GNPD).

(2) ***Only authorized tour operators***. Travel only with tour operators and/or boats authorized to work in the protected areas of Galápagos.

(3) ***Respect trails and signs***. Remain on marked trails at visitor sites and always respect signs for the protection of wildlife and for your safety.

(4) ***Don't touch wildlife***. Keep a six-foot (two-meter) distance from wildlife to avoid disturbing the animals, even if they approach you.

(5) ***Don't feed wildlife***. Never feed wildlife, as this can cause health problems or trigger unnatural behaviors.

(6) ***No drones or flash photography***. Flash photography is not permitted when taking photos of wildlife. Professional photography and videos recorded for commercial purposes must be authorized by the GNPD.

---

* Modified from Galápagos Conservancy https://www.galapagos.org/travel/park-rules/

(7) ***Only authorized camping***. Camping is only allowed in authorized areas on the islands. Request authorization—for camping—at the Galápagos National Park's offices at least 48 hours in advance.

(8) ***Don't introduce food, animals, or plants***. It is your responsibility not to introduce food, animals, or plants into the archipelago. Cooperate fully with all environmental inspection and quarantine officials during your visit. Authorities could inspect your luggage multiple times.

(9) ***No illegal trading***. Do not take or buy any products or souvenirs made from banned substances, including black coral, shells, lava rock, animal parts, or any native wood or vegetation. This is illegal and must be reported.

(10) ***Don't alter the environment***. Practice "leave no trace" (e.g. graffiti or rock piles) principles to maintain the beauty of the environment.

(11) ***No littering***. Pack out all trash and dispose of or recycle it in the populated areas or on your tour boat. Plastic trash is a serious concern!

(12) ***Never smoke in the wilderness; no campfires***. Smoking and/or campfires are strictly prohibited within the Galápagos National Park, as fires pose a serious risk to the local flora and fauna.

(13) ***Only authorized fishing***. Fishing is only permitted on specific sites or recreational tour boats authorized by the GNPD.

(14) ***No motorized aquatic entertainment***. Motorized aquatic sports, mini-subs, and aerial tourism activities are not permitted in the Galápagos National Park or Marine Reserve.

ABOUT THE AUTHORS

Paz-y-Miño-C and **Espinosa** visited Galápagos–independently–for the first time ≈40 years ago. Nowadays, they explore the archipelago yearly.

Guillermo Paz-y-Miño-C, PhD Ecology, Evolution, and Behavior: His research programs include the link between different levels of social organization and the cognitive abilities of animals, as well as the origin and evolution of signals for aggregation behavior, including taxa, clone, and kin discrimination in unicellular and multicellular eukaryotes. He also conducts hypothesis-based research on the incompatibility between scientific rationalism/empiricism and the belief in supernatural causation. Paz-y-Miño-C is the recipient of a Citation for Outstanding Performance by the Commonwealth of Massachusetts and author of 150 publications, including peer-reviewed articles, book chapters, editorials, and the books *Arrivals of Life to the Galápagos* (2024); *A Synthesis of the Galápagos* (2024); *Kin Recognition in Protists and Other Microbes: Genetics, Evolution, Behavior and Health* (2018); *Measuring the Evolution Controversy: A Numerical Analysis of Acceptance of Evolution at America's Colleges and Universities* (2016); and *Evolution Stands Faith Up: Reflections on Evolution's Wars* (2013).

Avelina Espinosa, PhD Molecular Microbiology: Professor of Biology and faculty lead of Biotechnology Programs (graduate and undergraduate) at Roger Williams University. Her extramurally funded research includes the evolution of anaerobic dehydrogenases (*Entamoeba* systems); behavioral characterizations to resolve taxa crypticity in Amoebozoans; the mechanisms of taxa, clone, and kin discrimination in unicellular eukaryotes (protists); and scientific analyses of diverse audiences' views on evolution, science, and pseudoscience. Espinosa works with diverse audiences to advance scientific literacy in society. She is the recipient of a Young Investigator Award by the American Society of Tropical Medicine and Hygiene and author of 50 publications, including peer-reviewed papers (two cover-journal articles), book chapters, popular articles, and the books (coauthored with GPC) *Arrivals of Life to the Galápagos*; *A Synthesis of the Galápagos*; *Kin Recognition in Protists and Other Microbes*; and *Measuring the Evolution Controversy*.